A Passage to Modern Analysis

Pure and Applied
UNDERGRADUATE TEXTS · 41

The Sally SERIES

A Passage to Modern Analysis

William J. Terrell

AMERICAN MATHEMATICAL SOCIETY
Providence, Rhode Island USA

2010 *Mathematics Subject Classification.* Primary 26-01, 54C30, 28-01.

For additional information and updates on this book, visit
www.ams.org/bookpages/amstext-41

Library of Congress Cataloging-in-Publication Data

Cataloging-in-Publication Data has been applied for by the AMS.
See http://www.loc.gov/publish/cip/.

∞ The paper used in this book is acid-free and falls within the guidelines
established to ensure permanence and durability.
Visit the AMS home page at `https://www.ams.org/`

10 9 8 7 6 5 4 3 2 1 24 23 22 21 20 19

To my wife Maria Holperin Terrell
and
To the memory of my father and mother
Edgar Allen Terrell, Jr.
and
Emmie Jennings Terrell

Contents

List of Figures

Preface

Thanks for turning to the Preface.

This introductory text is written for students at the advanced undergraduate level and beyond in mathematics and its applications and for those in the sciences and engineering who desire a rigorous introduction to mathematical analysis. The major part of the book provides a motivated introduction to analysis in Euclidean space, beginning with the single variable case and properties of the real number field. Later chapters include topics that are helpful in the study of more advanced areas such as ordinary and partial differential equations, Fourier series, Lebesgue measure and integration, and Hilbert space. These later chapters are intended as a springboard for such studies, and the applications in the book are there to spark interest rather than delve deeply into specific application areas. My purpose has been to write a book that students will find interesting and useful.

The genesis of the book was a collection of written supplements I used in teaching advanced courses in ordinary and partial differential equations and applied analysis for more than twenty years. Most of the students in these courses were majoring in mathematics or applied mathematics, with possibly a quarter of the class majoring in one of the sciences or engineering or having come to mathematics from another undergraduate major. Most of the students had no exposure to basic analysis in Euclidean space or normed spaces in general. In an effort to provide an appropriate language for making and understanding mathematical statements about differential equations and dynamical systems, I supplied written handouts on the basic analysis background required. I found the process of filling in the gaps in those supplements to be long but enjoyable, and it led me to complete this book.

This book addresses three major goals of analysis instruction in the undergraduate curriculum. The first goal is to present a careful, rigorous study of real valued functions of a real variable and vector valued functions of a vector variable, starting from the properties of the real number system. The second goal is to help students develop the mathematical maturity and critical thinking skills necessary for success throughout the upper division of an undergraduate program and beyond. A third goal is to provide a passage, a transit point, to a few developments of modern analysis that have had an important influence in both its theoretical and applied aspects. With these goals in view, the core of the book is grounded solidly in the world of Euclidean space, but at appropriate places, the book introduces and applies inner product, normed, and metric spaces. Thus, readers are made aware that there are interesting and useful developments beyond Euclidean space in which the basic concepts of analysis play important roles and may be studied further.

The prerequisite for beginning the book is two semesters of the standard university undergraduate curriculum in elementary single variable calculus and an introductory course in proof technique often having titles such as Transition to Advanced Mathematics or Introduction to Mathematical Reasoning. This should suffice for Chapters 1–7. However, undergraduate introductions to multivariable calculus and linear algebra are prerequisites for the material from Chapter 8 onward, where the focus is on n-dimensional Euclidean space $\mathbf{R}^n$ and a few function spaces. At most American universities, students will have taken a third semester calculus course in introductory multivariable calculus before entry into a course such as the present book. Readers with more substantial undergraduate mathematics backgrounds than this will probably make more rapid progress in the book. In particular, introductory courses in elementary differential equations or numerical methods may provide some readers with additional motivation for some of the topics considered.

This text provides a bridge from one-dimensional analysis to more general spaces, building on the core topics of differentiation and integration and a few well chosen application areas such as solving equations, inverting functions, measuring the volume of sets, and understanding basic properties of differential equations, including some basic Fourier analysis for application to partial differential equations. The text culminates with two chapters on the Lebesgue theory and a chapter on inner product spaces and Fourier expansion in Hilbert space.

The book is suitable for both classroom instruction and self-study and provides students with a solid background to build on if they wish to move on to more advanced studies or applications in their areas of interest.

There are several unique features of this text.

(1) The book combines expansive coverage of analysis on the real line and Euclidean space and detailed coverage of the Lebesgue theory suitable for motivated and advanced undergraduates or first-year graduate students. It has topics chapters on fundamental aspects of ordinary differential equations, Fourier series, and basic problems in partial differential equations.

    The book offers three successive bridges in this passage to modern analysis:

    Lower Bridge: Chapters 2–7 provide rigorous coverage of real valued functions of a real variable. There is enough material here for a comprehensive semester course.

    Middle Bridge: With Chapters 2–7 providing background in basic concepts, Chapters 8–13 cover analysis in $\mathbf{R}^n$, including differentiation and integration of vector functions and the extension of the Riemann integral to functions on bounded subsets of $\mathbf{R}^n$. The inverse function theorem and implicit function theorem apply the derivative and linearization ideas to the local solution of systems of equations. There is enough coverage of general metric space and normed space ideas to allow for discussions of some basic applications, including matrix norms in Chapter 9 and the contraction mapping theorem in complete metric spaces in Chapter 11. The discussion of matrix norms is important for work in numerical analysis, which we have indicated in some exercises. Matrix norms also play an important role in some estimates in Chapter 13. The contraction mapping theorem is applied to the existence and uniqueness theorem for ordinary differential equations in Chapter 14. Chapters 8–13 provide an appropriate springboard for more advanced studies of analysis, including the background for the differential equations and Fourier series in Chapters 14 and 15.

Upper Bridge: Chapters 14 and 15 can be used as topics chapters that draw especially on material from Chapters 7–11. Chapter 15 can also be viewed as an introduction to ideas that can be pursued in more detail in the final three chapters of the book. Chapters 16 and 17 cover Lebesgue measure and the Lebesgue integral, respectively. These topics are motivated and covered in enough generality so that the spaces of most interest, the integrable functions (and square integrable functions) on measurable subsets of $\mathbf{R}$ or $\mathbf{R}^n$, can be discussed. Applications of these ideas appear in an introduction to Hilbert space in Chapter 18, which also explores and clarifies, in a more general setting, some issues arising in Chapter 15 concerning Fourier series.

(2) The concepts of geometric series and contraction mappings are introduced early in the book. Differentiation of vector functions, the multivariable mean value theorem, and the inverse and implicit function theorem are all given full consideration due to their importance in applications and more advanced studies. Applications of these ideas appear throughout the text. Many introductory analysis texts do not place enough emphasis on these ideas.

(3) The final five chapters cover topics beyond the standard undergraduate coverage: aspects of ordinary differential equations, Fourier series and partial differential equations, Lebesgue measure, the Lebesgue integral and its comparison with the Riemann integral, and the study of the Hilbert space $L^2[-\pi, \pi]$ and its isometric isomorphism with the sequence space l^2 established with the help of the Lebesgue theory.

The features of this book make it useful as a text in several ways. The book can be useful for students who wish to cross only the first, or the first two, or all three, of the bridges described earlier.

a. First, it can be a comprehensive text for a semester course in undergraduate analysis based on Chapters 1–7.

b. Second, a follow-up semester in analysis, emphasizing Euclidean space, can be based on the material in Chapters 8–13. For a semester-length honors course, this second course might include some differential equations or Fourier analysis with selections from Chapters 14 and 15 for interested students.

c. Third, an undergraduate honors or topics section could be formed by selections from Chapters 14–18, which go beyond the standard undergraduate coverage of analysis in $\mathbf{R}^n$. Such a course could benefit undergraduates interested in proceeding to graduate school.

d. First-year graduate students in mathematics or applied mathematics, the sciences, or engineering may want a refresher course, either guided or self-study, in analysis. Such a course could be based on selected portions of Chapters 8–18, with reference to Chapters 2–7 as needed. Basic analysis is important for many students in these areas.

The book has enough material for three academic semesters of coursework, including topics for individual reading or an honors course. It presents material at a level appropriate for advanced undergraduates and some first-year graduate students, depending on instructor choices and student backgrounds and interests. An introduction that covers $\mathbf{R}^n$ in a comprehensive manner and discusses metric spaces that are important in applications will serve these students well.

There are more than 570 exercises. Almost every section of the book includes an Exercises section at the end. Some of the longer sections include exercises at the end of subsections. Many of the exercises are supplied with a hint or presented as guided exercises with multiple parts. The exercises reinforce the reading of the text and provide opportunities to develop skills in mathematical reasoning, analysis, and writing. The index has more than 1000 entries.

Remark on Item Numbering: The numbered items within any section are Definitions, Lemmas, Propositions, Theorems, Corollaries, and Examples; these items are numbered consecutively as they appear by **chapnum.secnum.itemnum** to indicate chapter, section, and item number. Since Exercises appear in blocks at the end of sections (and a few subsections) they are numbered separately within a similar scheme, **chapnum.secnum.exernum**. Numbered equations and figures are numbered using **chapnum.itemnum**.

Descriptive chapter summaries follow:

Chapter 1 Sets and Functions. This chapter will be a review for many readers, but it includes basic notation and essential results on sets and cardinality that everyone will need.

Chapters 2–7 provide sufficient material for a semester course in the analysis of real valued functions of a real variable. Their main purpose is to impart a solid working knowledge of the concepts of analysis so that the student can proceed to the study of Euclidean metric space $\mathbf{R}^n$.

Chapter 2 The Complete Ordered Field R. This chapter presents the axioms for the real numbers, including the completeness axiom in the form of the least upper bound property. (References are given for the *construction* of a field having this completeness property.) The focus is on motivating and proving the main properties of the field of real numbers: the Archimedean property, the nested interval property and decimal representations, the Bolzano-Weierstrass theorem, and the convergence of Cauchy sequences. The chapter ends with demonstrations that some results familiar from elementary calculus cannot hold using only the field of rational numbers; these are included to emphasize the importance of the least upper bound property and to motivate interested readers to read about the construction of the real numbers from the rationals.

Chapter 3 Introduction to Series. After basic definitions, this chapter introduces the geometric series and applies it in the discussion of the Cantor set. The Euler number e and alternating series are followed by the simplest versions of the ratio and root tests, which are based on the geometric series. General versions of the ratio and root tests appear in Section 3.11 after covering limit inferior and limit superior in Section 3.10. Other convergence tests include Abel's test and Dirichlet's test. (The integral test appears in Chapter 6.) Absolute versus conditional convergence is introduced in Section 3.7 after alternating series, while a more complete discussion of the contrast between absolute and conditional convergence appears in Section 3.12, which includes Riemann's rearrangement theorem.

Chapter 4 Basic Topology, Limits, and Continuity. The coverage of basic topology of the real line includes open sets, closed sets, compact sets, and connected sets. We define the limit of a function and continuity at a point and then discuss continuity on an interval, uniform continuity, and the continuous image of compact sets. The chapter ends with a classification of discontinuities of functions.

Chapter 5 Differentiation. After basic definitions, we establish the mean value theorem and the scalar inverse function theorem, as well as Darboux's theorem on the intermediate value property of derivatives. We point out the role of the mean value theorem in helping to establish certain mappings as contraction mappings, and thus show its usefulness for the solution of equations as well as for basic estimates of differences in function values. Other important results include Cauchy's mean value theorem with applications to l'Hôpital's rules for indeterminate forms, the single variable Taylor's theorem, and the extreme value theorem.

Chapter 6 The Riemann Integral. After defining partitions and Riemann-Darboux sums, we discuss the integral of a bounded function, the integrability of continuous functions and monotone functions, Lebesgue measure zero, and the criterion for Riemann integrability. The coverage then proceeds with integral properties and mean value theorems, the fundamental theorem of calculus, Taylor's theorem with integral remainder, and improper integrals.

Chapter 7 Sequences and Series of Functions. The major topics are pointwise convergence and its importance and limitations, uniform convergence and its advantages; integration and differentiation of series, and the Weierstrass test for uniform convergence; the existence of a continuous nowhere differentiable function; power series and Taylor series; series definitions for the elementary transcendental functions and proofs of some of their properties; and, in the final section, the Weierstrass approximation theorem.

Chapters 8 and 10–13 are primarily on $\mathbf{R}^n$*, though Chapter 8 shows a variety of vector spaces of interest in analysis. Chapter 9 is the metric space preparation for Chapters 11 and 14–18.*

Chapter 8 The Metric Space $\mathbf{R}^n$. After a vector space review in Section 8.1, the coverage proceeds with the inner product and norm structure of n-dimensional Euclidean space, and Fourier expansion with respect to an orthogonal basis, in Sections 8.2–8.4. Section 8.5 presents the spectral theorem for real symmetric matrices in complete detail, which provides a good application and extension of the discussion in Section 8.4. Section 8.6 discusses the metric distance on $\mathbf{R}^n$. Section 8.7 establishes the completeness of $\mathbf{R}^n$, defined as the convergence of all Cauchy sequences. Basic topological definitions are collected in Section 8.8, which includes the relative topology of a subset of $\mathbf{R}^n$. The nested intervals property and the Bolzano-Weierstrass theorem appear in Section 8.9. Mappings of Euclidean spaces are covered in Section 8.10 in six short subsections, beginning with limits and continuity and moving on to topological properties of continuous mappings, including continuous images of compact sets and connected sets. A result on differentiation under the integral is included here since it uses facts about continuous mappings of compact sets.

Chapter 9 Metric Spaces and Completeness. Metric spaces are essential; bounded subsets of normed spaces are metric spaces but not normed vector spaces. With the experience of $\mathbf{R}^n$ well in hand, the basic topology in Section 9.1 is a direct generalization of what is now familiar for n-dimensional space. There is only a slight increase in abstraction here, with some notation required for general metric space. Section 9.2 presents the contraction mapping theorem for complete metric spaces in a direct generalization of the scalar version established in Chapter 5. Section 9.3 proves the completeness of the function space $C[a, b]$ with the maximum norm (uniform norm) and the completeness of the l^2 sequence space, both introduced in Section 8.3. Other topics include the l^p sequence spaces, matrix norms, and the completeness of the space of $n \times n$ real matrices. These function spaces are involved in the study of ordinary differential equations and Fourier series in Chapters 14 and 15.

Chapter 10 Differentiation in $\mathbf{R}^n$. In Sections 10.1–10.5 we define and discuss partial derivatives, the derivative as a linear mapping, the matrix representation of the derivative for given bases in domain and range, sufficient conditions for the existence of the derivative, and the chain rule. We prove the mean value theorem first for real valued functions in Section 10.6 and apply it in Section 10.7 with other single variable calculus ideas to prove the two-dimensional case of the implicit function theorem. The mean value theorem for vector valued functions appears in Section 10.8. The chapter winds up with a presentation of Taylor's theorem in Section 10.9 and relative extrema without constraints in Section 10.10.

Chapter 11 Inverse and Implicit Function Theorems. In Section 11.1, the scalar geometric series motivates a convergence result for matrix geometric series. We apply this result to prove that matrix inversion is a continuous mapping of the set of invertible $n \times n$ real matrices. Section 11.2 proves the inverse function theorem in $\mathbf{R}^n$ as an application of the contraction mapping theorem, and the continuity of matrix inversion is used in the proof of continuous differentiability of the local inverse function. Section 11.3 uses the inverse function theorem to prove the implicit function theorem. The problem of constrained extrema and the Lagrange multiplier theorem appear in Section 11.4. Section 11.5 presents the Morse lemma, another application of the inverse function theorem.

Chapters 12 and 13 cover the Riemann integral; the theory of Jordan measurable sets in $\mathbf{R^n}$, the bounded sets that have a well-defined volume; and the C^1 change of variables formula.

Chapter 12 The Riemann Integral in $\mathbf{R}^n$. Sections 12.1 and 12.2 describe the extension of the Riemann integral to closed intervals and certain bounded subsets of $\mathbf{R}^n$. This extension allows a theory of measurable sets (called Jordan measurable sets or sets with volume), introduced in Section 12.3. This may be the reader's first exposure to a type of measure theory. Sections 12.4 and 12.5 develop the criterion that a function is Riemann integrable if and only if it is continuous except possibly on a set of Lebesgue measure zero. (The proof includes the case of real functions of a real variable, which is stated without proof in Section 6.4.) The Riemann integral and Jordan measure are limited from the larger point of view required by modern analysis, as they are restricted to certain bounded sets and bounded functions with limited amounts of discontinuity. Consequently, the integral does not behave well under pointwise limits, as seen already in the single variable case. However, the Riemann integral is adequate for many purposes, including areas that lie beyond this book, such as the study of finite-dimensional smooth manifolds and the integration of smooth functions on them. Thus, the Riemann integral in $\mathbf{R}^n$ and Jordan measure deserve a place in an introductory text. (Later in the text, the Lebesgue integral is seen to be a significant extension of the Riemann integral and one that behaves well under limit processes without the strong assumption of uniform convergence. Moreover, the Lebesgue theory does not require a separate theory of improper integrals to handle unbounded functions and unbounded domains, as the Riemann theory does.)

Chapter 13 Transformation of Integrals. Space-filling curves are intrinsically interesting, and we include an example of one in Section 13.1 to motivate interest in appropriate conditions for coordinate transformations (variable substitutions) in the integral. Section 13.2 considers the transformation of integrable functions and sets with volume under C^1 transformations. We develop the change of variables formula in Sections 13.3 and 13.4. The rather involved proof is as geometric as we can make it, with the argument building in a fairly natural way, starting from the case of linear mappings. In Section 13.5, we show that the surface integrals familiar from introductory multivariable calculus are well defined by virtue of the change of variables formula. In the same section, we recall the divergence theorem and then establish a coordinate-free interpretation for the divergence of a vector field $\mathbf{F}$; if the vector field is a gradient field, $\mathbf{F} = \nabla f$, then we obtain a coordinate-free interpretation of the Laplacian of f. This is helpful in understanding the physical significance of Laplace's equation in Chapter 15.

Chapter 14 Ordinary Differential Equations. Section 14.1 presents the existence and uniqueness theorem for initial value problems for scalar differential equations as an application of the contraction mapping theorem. A similar mathematical setup for systems appears in Section 14.2, including the equivalent integral equation, the completeness of a space of solution candidates, and the local Lipschitz condition, followed by the proof of existence and uniqueness for initial value problems. We cover the extension of solutions to a maximal interval of existence in Section 14.3 and the continuous dependence of solutions on initial conditions and parameters in Section 14.4. Section 14.5 covers the special case of linear autonomous systems and the matrix exponential solution.

Chapter 15 The Dirichlet Problem and Fourier Series. In terms of background, this chapter depends primarily on the reader's knowledge of uniform convergence and the ideas of orthogonal Fourier expansion from Section 8.4. Section 15.1 provides motivation for studying Laplace's partial differential equation and introduces the Dirichlet problem for the unit disk. Section 15.2 introduces the standard trigonometric set of functions on $[-\pi, \pi]$ and establishes their orthogonality. Section 15.3 constructs a solution of the Dirichlet problem using separation of variables and Fourier series. This section includes Poisson's theorem, which shows that the solution in the interior of the disk, constructed by the separation of variables method, matches up continuously with the given boundary data. Uniqueness of the solution is also established. Section 15.4 explores in guided exercises the application of the separation of variables method to some Sturm-Liouville boundary value problems for the heat equation and the wave equation. Section 15.5 establishes the best mean square approximation property of the Fourier coefficients. A trigonometric version of the Weierstrass approximation theorem follows easily, and we use it to prove Parseval's equality for continuous functions of period 2π. Sections 15.6 and 15.7 discuss pointwise convergence of the Fourier series for piecewise smooth functions and Fejér's theorem on the uniform convergence of the Cesàro means for continuous functions of period 2π.

Chapter 16 Measure Theory and Lebesgue Measure. The introduction provides some motivation for the study of measure theory and the form it takes. Sections 16.1–16.3 motivate and discuss σ-algebras, arithmetic in the extended real numbers, and basic properties of measures. We attempt to indicate that probability problems are a natural motivator for σ-algebras, measure theory, and measurable functions. Section 16.4 describes the construction of a measure from an outer measure. Section 16.5 applies this construction to define Lebesgue measure on $\mathbf{R}$ and on $\mathbf{R}^n$, considering the cases where $n > 1$ separately for those who wish to focus only on Lebesgue measure on the real line. We prove that all Borel sets (hence all open sets and all closed sets) are Lebesgue measurable. Vitali's example of a nonmeasurable set appears at the end of the chapter.

Chapter 17 The Lebesgue Integral. The introduction describes Lebesgue's approach to the integral and contrasts it with that of Riemann in order to motivate the definition of measurable function. Section 17.1 gives the definition and basic properties of real valued measurable functions on a measurable space; the section is mostly set-theoretic and no measure is yet involved. Section 17.2 defines the simple functions and their integrals on a measure space. Section 17.3 continues with the definition of the integral for nonnegative measurable functions and then for general measurable functions, and defines the class of integrable functions. Section 17.4 establishes the fundamental limit theorems: the monotone convergence theorem (which implies the linearity of the integral), Fatou's lemma, and the dominated convergence theorem. Section 17.5 shows that the Lebesgue integral is a true extension of the Riemann integral for bounded functions on $[a, b]$. Section 17.6 shows that the space of integrable functions on a given measure space is a complete normed space, a Banach space.

Chapter 18 Inner Product Spaces and Fourier Series. This chapter places the concrete setting of Chapter 15 into the proper abstract framework of infinite-dimensional inner product spaces. Sections 18.1 and 18.2 include examples of orthonormal sets and orthonormal expansions that generalize the simpler setting of Fourier expansion in $\mathbf{R}^3$ from Section 8.4. In particular, Section 18.2 characterizes complete orthonormal sets in a complete inner product space (a Hilbert space). Section 18.3 discusses mean square convergence (convergence in the L^2 norm) for the Fourier series of continuous functions and Riemann integrable functions on $[a, b]$. Finally, in Section 18.4, we define the Lebesgue space $L^2[-\pi, \pi]$ of square integrable functions, prove that it is a Hilbert space, and show that the standard trigonometric set is a complete orthonormal set in $L^2[-\pi, \pi]$. The Riesz-Fischer theorem is also included, showing that the Hilbert spaces $L^2[-\pi, \pi]$ and l^2 are isometrically isomorphic.

Appendix A The Schroeder-Bernstein Theorem. A formal proof of the useful Schroeder-Bernstein theorem appears here. The theorem appears in the text in Section 1.4 with a plausibility argument at that point.

Appendix B Symbols and Notations. This appendix provides a quick reference for some symbols and notations, including the Greek alphabet.

Acknowledgments. Many people deserve thanks and recognition for helping to create this book.

Thanks are due to my students in classes of all types over many years at Virginia Commonwealth University for reminding me to express important ideas in a clear manner.

The bibliography includes all the resources that were helpful in some specific way in the writing of this book. Notes and References at the end of many of the chapters record either a general influence or more specific acknowledgments or suggestions for further reading. I am grateful to all of these authors and publishers for the availability of their work; this book could not have been written without them.

Many thanks to my editor, Stephen F. Kennedy, for his interest, time, effort, and guidance in this project. My thanks also to several anonymous reviewers and the review board of the textbook program of the American Mathematical Society (AMS) for their time and effort in consideration of the book for inclusion in the Sally Series on Pure and Applied Undergraduate Texts. My thanks to the entire staff at AMS for their expert handling and guidance of the book through the publication process.

My thanks to Robert E. Terrell for supplying graphics files for this book and for his support during the project.

Much appreciation and many thanks to Stephen L. Campbell of North Carolina State University for his support over many years.

Many thanks to my wife, Maria Holperin Terrell, for supporting this and all my efforts. Thanks also to friends Lucy, Sheba, and Wyatt for careful attention.

William J. Terrell
Richmond, Virginia
January 27, 2019

Chapter 1

Sets and Functions

Readers of this book will have some previous experience with the language of sets and familiarity with basic logic and proof techniques. The exercises at the end of each section of this preliminary chapter provide opportunities for practice in these areas. Some references for set theory and proof technique are given at the end of the chapter.

1.1. Set Notation and Operations

Our purpose in this section is to introduce some terminology and notation that is helpful in making clear and concise statements in mathematical analysis. In order to provide examples and to make the discussion of some interest, we rely on the reader's previous experience to provide some intuition about real numbers, integers, rational numbers, and functions defined on subsets of real numbers.

By the term **set** we mean a collection of objects of our conception or imagination, including a means to distinguish unambiguously the members (or elements) of a set from objects that are not in the set. This meaning of the term set can lead to paradoxes if carried too far; see Exercise 1.1.1. In this book we will not encounter such paradoxes beyond Exercise 1.1.1.

Definition 1.1.1. *The* **empty set**, *denoted* $\emptyset$, *is the set which contains no elements.*

We imagine the empty set every day: we have no emails from home today (the collection of emails from home today is the empty set); we have no credit cards (the set of our credit cards is empty); the set of integer solutions of the equation $x^2 = 2$ is the empty set.

A useful concept in any discussion is the idea of the universal set. The universal set is the set of all objects being considered in a specific discussion. For example, when thinking about the properties of integers, the universal set is the set $\mathbf{Z}$ of all integers.[1] Another specific discussion may involve all the functions that are defined and continuous on the real number interval $0 \leq x \leq 1$ and which take values in the range interval $0 \leq y \leq 1$.

We generally denote sets by uppercase letters. These could be Roman letters $A, B, C, \ldots, X, Y, Z$ or script letters $\mathcal{A}, \mathcal{B}, \mathcal{C}, \ldots, \mathcal{X}, \mathcal{Y}, \mathcal{Z}$. We generally denote elements of sets by lower case letters $a, b, c, \ldots, x, y, z$. In later parts of the book we denote certain vector quantities by boldface letters, $\mathbf{a}, \mathbf{b}, \mathbf{c}, \ldots, \mathbf{x}, \mathbf{y}, \mathbf{z}$. The symbolic statement

$$a \in A$$

means that a is an element of (belongs to, is contained in) the set A. The statement

$$a \notin A$$

means that a is not an element of the set A. Sets are often defined or identified by a notation of the form

$$\{x \in X : \text{x has property P}\}.$$

For example, the set of even integers can be described by

$$\mathbf{E} = \{x \in \mathbf{Z} : x = 2n, \ n \in \mathbf{Z}\},$$

since we already understand $\mathbf{Z}$ is the set of all integers.

Definition 1.1.2. *If X and Y are sets, then the set*

$$X - Y = \{x \in X : x \notin Y\}$$

is called the **complement of Y relative to X** *or the* **complement of Y in X**. *If the universal set U is understood in a specific discussion, then we may write*

$$Y^c = U - Y$$

and call it the **complement** *of Y.*

Definition 1.1.3. *If X and Y are sets, the statement X is a* **subset** *of Y, written $X \subseteq Y$, means that if $x \in X$, then $x \in Y$.*

This definition of $X \subseteq Y$ allows the possibility that $X = Y$. If it is important in a specific discussion that Y contain an element not in X, it is usually clear from the context, or can be mentioned explicitly, or the notation $X \subset Y$ can be used.

The empty set is a subset of every set, as follows from the definition of subset. (See Exercise 1.1.2.)

The statement that sets X and Y are equal, written $X = Y$, means that $X \subseteq Y$ and $Y \subseteq X$. This statement of set equality involves two implications: $x \in X$ implies $x \in Y$, and $x \in Y$ implies $x \in X$. To establish set equality, both implications must be established.

Definition 1.1.4. *Let A and B be sets. The* **union** *of A and B, denoted $A \cup B$, is the set*

$$A \cup B = \{x : x \in A \text{ or } x \in B\}.$$

[1] The set of integers is denoted $\mathbf{Z}$ with a nod to the German word *Zahlen* for *numbers*.

The **intersection** *of A and B, denoted $A \cap B$, is the set*

$$A \cap B = \{x : x \in A \text{ and } x \in B\}$$

We say that sets A and B are **disjoint** *if $A \cap B$ is empty.*

For example, let us write $\mathbf{O}$ for the set of odd integers. With $\mathbf{E}$ denoting the set of even integers, we have

$$\mathbf{Z} = \mathbf{E} \cup \mathbf{O} \quad \text{and} \quad \mathbf{E} \cap \mathbf{O} = \emptyset.$$

We may also say that $\mathbf{Z}$ is the disjoint union of $\mathbf{E}$ and $\mathbf{O}$.

We will denote the set of real numbers by $\mathbf{R}$. The standard notations for intervals of real numbers are probably already familiar. We use these real interval notations for bounded intervals:

$$[a, b] = \{x \in \mathbf{R} : a \leq x \leq b\},$$
$$(a, b) = \{x \in \mathbf{R} : a < x < b\},$$
$$(a, b] = \{x \in \mathbf{R} : a < x \leq b\},$$
$$[a, b) = \{x \in \mathbf{R} : a \leq x < b\}.$$

For unbounded intervals, we use these interval notations:

$$(a, \infty) = \{x \in \mathbf{R} : a < x\} \quad \text{and} \quad [a, \infty) = \{x \in \mathbf{R} : a \leq x\},$$
$$(-\infty, b) = \{x \in \mathbf{R} : x < b\} \quad \text{and} \quad (-\infty, b] = \{x \in \mathbf{R} : x \leq b\}.$$

Thus the notation (a, ∞) specifies the set of all real numbers x such that $x > a$. The interval (a, ∞) may also be indicated by writing the inequality $a < x < \infty$ to emphasize that no upper bound is intended for x. The interval $(-\infty, b)$ may also be indicated by writing the inequality $-\infty < x < b$ to emphasize that no lower bound is intended for x. We may occasionally write $\mathbf{R} = (-\infty, \infty)$.[2]

There are useful generalizations of the union and intersection operations on pairs of sets. We employ the idea of an arbitrary index set $\mathcal{I}$, which may be either finite or infinite.

Definition 1.1.5. *Let $\mathcal{I}$ be an index set (finite or infinite), and suppose that for each $i \in \mathcal{I}$ there is associated a set A_i. The* **union** *of the sets A_i is the set*

$$\bigcup_{i \in \mathcal{I}} A_i = \{x : x \in A_i \textit{ for } \text{some } i \in \mathcal{I}\}.$$

The **intersection** *of the sets A_i is the set*

$$\bigcap_{i \in \mathcal{I}} A_i = \{x : x \in A_i \textit{ for } \text{all } i \in \mathcal{I}\}.$$

De Morgan's laws relate the operations of complementation and general unions and intersections.

[2] These uses of the symbol ∞ are intended as convenient reminders (for emphasis) of the unboundedness of an interval, in one direction or the other, when writing inequality notations. We do not consider them as number elements until Section 16.2, where we admit the two symbols $\pm\infty$ as elements of the extended real number system for use in measure theory, where they enter into arithmetic laws with the real numbers. At that point, the new elements $\pm\infty$ are also defined to obey the ordering requirement that $-\infty < x < \infty$ for every real number x.

Theorem 1.1.6 (De Morgan's laws). *Let $\mathcal{I}$ be an index set. Given the sets A_i, for $i \in \mathcal{I}$, we have*

$$\left(\bigcap_{i\in\mathcal{I}} A_i\right)^c = \bigcup_{i\in\mathcal{I}} A_i^c \tag{1.1}$$

and

$$\left(\bigcup_{i\in\mathcal{I}} A_i\right)^c = \bigcap_{i\in\mathcal{I}} A_i^c. \tag{1.2}$$

In other words, the complement of the intersection equals the union of the complements, and the complement of the union equals the intersection of the complements.

Proof. We will prove (1.1) and leave (1.2) as Exercise 1.1.6.

The complement of the intersection is contained in the union of the complements: If $x \in (\bigcap_{i\in\mathcal{I}} A_i)^c$, then $x \notin \bigcap_{i\in\mathcal{I}} A_i$, so there is an i_0 such that $x \notin A_{i_0}$, hence $x \in A_{i_0}^c$. Therefore $x \in \bigcup_{i\in\mathcal{I}} A_i^c$.

The union of the complements is contained in the complement of the intersection: If $x \in \bigcup_{i\in\mathcal{I}} A_i^c$, then for some i_0, $x \in A_{i_o}^c$, hence $x \notin A_{i_0}$. Therefore $x \notin \bigcap_{i\in\mathcal{I}} A_i$, so $x \in (\bigcap_{i\in\mathcal{I}} A_i)^c$. This completes the proof of (1.1). □

Exercises.

Exercise 1.1.1. *The Russell Paradox*
A situation in which an object of our imagination or conception has a certain property if and only if it does not have that property is called a **paradox**. Paradoxes are logically unacceptable, of course. The example here will show that we cannot say just anything at all in order to define a set, and that some care is required. Let us say that a set is **respectable** if it does not contain itself as an element. Now let $\mathcal{B}$ be the set of all respectable sets. Try to answer this question: Is $\mathcal{B}$ a respectable set? The attempt to answer it yields a paradox, as follows:
1. Show that if $\mathcal{B}$ is respectable, then $\mathcal{B}$ is not respectable.
2. Show that if $\mathcal{B}$ is not respectable, then $\mathcal{B}$ is respectable.

Exercise 1.1.2. Show that Definition 1.1.3 implies that the empty set is a subset of every set. *Hint*: The implication, *If A then B*, is false only when A is true and B is false.

Exercise 1.1.3. Prove that $A = B$, if $B = \{(1,0),(0,1)\}$ and $A = \{(x,y) : x \in \mathbf{R},\ y \in \mathbf{R} \text{ and } x^2+y^2 = 1,\ x+y = 1\}$. *Hint*: Proving $B \subseteq A$ means *verifying* that $(1,0)$ and $(0,1)$ are solutions of the two equations in the definition of A. Proving $A \subseteq B$ means finding (constructing) all solutions of the equations defining A and showing they are in the list given by B.

Exercise 1.1.4. Prove the following statements about sets A and B:

1. $A - B$ is empty if and only if $A \subseteq B$.
2. If $A \cap B$ is empty, then $A - B = A$ and $B - A = B$.
3. If $A \subseteq B$, then $B^c \subseteq A^c$.

Exercise 1.1.5. Prove the following statements about sets A, B, C:

1. $A - B = A \cap B^c$.
2. $A \cap (B \cup C) = (A \cap B) \cup (A \cap C)$ and $A \cup (B \cap C) = (A \cup B) \cap (A \cup C)$.
3. $(A \cup B) - C = (A - C) \cup (B - C)$ and $(A \cap B) - C = (A - C) \cap (B - C)$.

Exercise 1.1.6. Prove De Morgan's law (1.2).

1.2. Functions

An important construction with sets is the Cartesian product.

Definition 1.2.1. *If X and Y are sets, then the* **Cartesian product** *of X and Y is the set*

$$X \times Y = \{(x, y) : x \in X \text{ and } y \in Y\},$$

which consists of all **ordered pairs** *(x, y), with $x \in X$ and $y \in Y$.*

Now think about the collection of all the functions that are defined and continuous on the real number interval $0 \leq x \leq 1$ and take values in the range interval $0 \leq y \leq 1$. You may think about the **graph** of one of these functions f as a subset of the Cartesian product $[0, 1] \times [0, 1]$; the graph is

$$\operatorname{graph} f = \{(x, y) \in [0, 1] \times [0, 1] : y = f(x)\}.$$

In fact, the function itself may be defined as this particular subset of the Cartesian product.

Let X and Y be sets. We define a **function** f from X into Y to be a subset of the Cartesian product $X \times Y$ such that for each $x \in X$, there is associated a *unique* $y \in Y$ such that $(x, y) \in f$; this y is denoted $f(x)$. Informally we can think of a function as mapping elements $x \in X$ to elements $y = f(x) \in Y$; more precisely, each $x \in X$ is mapped to a *unique* element $f(x) \in Y$. Consequently we shall typically write $f : X \to Y$ to indicate that f is a function mapping X into Y. Our visual image of a function graph provides a geometric image to illustrate the rule associating the value $f(x) = y \in Y$ with each $x \in X$.

Definition 1.2.2. *Let X and Y be sets and $f : X \to Y$ a function. Let A be a subset of X and B be a subset of Y. The* **direct image** *of A under f (or simply, the* **image** *of A) is the set*

$$f(A) = \{f(x) \in Y : x \in A\}.$$

The **inverse image** *of B under f is the set*

$$f^{-1}(B) = \{x \in X : f(x) \in B\}.$$

We record the basic properties of inverse images and direct images in Theorem 1.2.3 and Theorem 1.2.4, respectively. These properties are left as exercises for the reader. In particular, the behavior of inverse images under f is easy to describe.

Theorem 1.2.3. *Let $f : X \to Y$ be a function. The following properties hold:*

1. *For every $B \subseteq Y$, $f(f^{-1}(B)) \subseteq B$.*
2. *If $B_1 \subseteq B_2 \subseteq Y$, then $f^{-1}(B_1) \subseteq f^{-1}(B_2)$.*

3. *If* $B_1 \subseteq Y$ *and* $B_2 \subseteq Y$*, then* $f^{-1}(B_1 \cup B_2) = f^{-1}(B_1) \cup f^{-1}(B_2)$.
4. *If* $B_1 \subseteq Y$ *and* $B_2 \subseteq Y$*, then* $f^{-1}(B_1 \cap B_2) = f^{-1}(B_1) \cap f^{-1}(B_2)$.
5. *For every* $B \subseteq Y$, $f^{-1}(B^c) = [f^{-1}(B)]^c$*, where* $B^c = Y - B$ *is the complement of* B *in* Y.

The behavior of direct images requires some care in the case of intersections and complements.

Theorem 1.2.4. *Let* $f : X \to Y$ *be a function. The following properties hold:*

1. *For every* $A \subseteq X$, $A \subseteq f^{-1}(f(A))$.
2. *If* $A_1 \subseteq A_2 \subseteq X$*, then* $f(A_1) \subseteq f(A_2)$.
3. *If* $A_1 \subseteq X$ *and* $A_2 \subseteq X$*, then* $f(A_1 \cup A_2) = f(A_1) \cup f(A_2)$.
4. *If* $A_1 \subseteq X$ *and* $A_2 \subseteq X$*, then* $f(A_1 \cap A_2) \subseteq f(A_1) \cap f(A_2)$.

Consider property 4 on the image of an intersection. For example, if there are points $a \neq b$ in X such that $f(a) = f(b)$, then with $A_1 = \{a\}$ and $A_2 = \{b\}$, the intersection $A_1 \cap A_2$ is empty, and hence $f(A_1 \cap A_2)$ is the empty set, but $f(A_1) \cap f(A_2)$ has one element.

Definition 1.2.5 (One-to-One and Onto). *A function* $f : X \to Y$ *is* **one-to-one** *(or* **injective***) if for any* $x_1, x_2 \in X$ *with* $x_1 \neq x_2$*, we have* $f(x_1) \neq f(x_2)$. *Equivalently, for all* $x_1, x_2 \in X$, $f(x_1) = f(x_2)$ *implies* $x_1 = x_2$. *The function* f *is* **onto** *(or* **surjective***) if* $f(X) = Y$. *The function* f *is a* **bijection** *if it is one-to-one and onto.*

If $f : X \to Y$ is *bijective*, then the direct image preserves complements, that is, if $A \subseteq X$, then

$$f(X - A) = f(A^c) = [f(A)]^c = Y - f(A).$$

However, $f : \mathbf{R} \to [-1, 1] = Y$ defined by $f(x) = \sin x$ is onto Y but not one-to-one, and if $A = [0, \pi]$, then $f(A^c) = [-1, 1]$ and $[f(A)]^c = [-1, 1] - f(A) = [-1, 1] - [0, 1] = [-1, 0)$. The function $g : \mathbf{N} \to \mathbf{N}$ defined by $g(n) = n^2$ is one-to-one but not onto, and if $A = \{2\}$, then $g(A^c) = \{1, 3^2, 4^2, \ldots\}$; however, $[g(A)]^c = \mathbf{N} - \{4\}$.

Definition 1.2.6 (Inverse Function). *A function* $f : A \subseteq X \to Y$ *is* **invertible** *if it is one-to-one on the set* A. *If* $y \in f(A)$*, then there is a unique* $x \in A$ *such that* $f(x) = y$. *We write* $x = f^{-1}(y)$*, and this correspondence defines a function* $f^{-1} : f(A) \to A$ *called the inverse of* f*, or the inverse of* f *restricted to* A. *The domain of this inverse is* $f(A)$ *and the range is* A.

Exercises.

Exercise 1.2.1. Prove Theorem 1.2.3. In addition, show that f is onto Y if and only if for every $B \subset Y$, $f(f^{-1}(B)) = B$.

Exercise 1.2.2. Prove Theorem 1.2.4. In addition, show that f is one-to-one if and only if for every $A \subset X$, $A = f^{-1}(f(A))$.

1.3. The Natural Numbers and Induction

The natural numbers are the numbers used for counting objects. The set of natural numbers is denoted by

$$\mathbf{N} := \{1, 2, 3, \ldots\}.$$

The set $\mathbf{N}$ is ordered by the *less than or equal to* relation (denoted by the symbol $\leq$), which has the following properties:

For every k, n and m in $\mathbf{N}$,

1. $k \leq k$; (reflexive)
2. if $k \leq n$ and $n \leq k$, then $k = n$; (antisymmetric)
3. if $k \leq n$ and $n \leq m$, then $k \leq m$. (transitive)

These three properties define what is called a *partial ordering*, but the ordering on $\mathbf{N}$ has an additional property that should be remembered as the most important property for our purposes in this book.

The most important property of the set of natural numbers is that $\mathbf{N}$ is *well ordered* by the *less than or equal to* relation. A partially ordered set is called **well ordered** if every nonempty subset of it has a least element. (A nonempty subset S has a least element if it contains an element a such that $a \leq y$ for every element y in S. By the antisymmetry property of a partial ordering, if a least element of S exists, there cannot be more than one.)

An important consequence of $\mathbf{N}$ being well ordered is that the natural numbers are **totally ordered**, which means that for any natural numbers n and m, either $n \leq m$ or $m \leq n$. The reason is that the nonempty set $S = \{n, m\}$ has a least element; if it is n, then $n \leq m$, and if it is m, then $m \leq n$.

Proofs by mathematical induction are based on the following result, which is a consequence of well ordering.

Theorem 1.3.1. *Suppose S is a subset of the set $\mathbf{N}$ of natural numbers such that*

(1) *$1 \in S$;*

(2) *for each $n \in \mathbf{N}$ with $n \geq 1$, $n \in S$ implies $n + 1 \in S$.*

Then $S = \mathbf{N}$.

Proof. The proof is by contradiction. Suppose (1) and (2) hold and that $S \neq \mathbf{N}$. Let $F = \mathbf{N} - S$. Then F is nonempty, and by well ordering, F has a least element $t \in F$. Since $1 \in S$, $t \neq 1$, hence $t > 1$. Let $s = t - 1$. Then $s \in \mathbf{N}$ and $s < t$. Moreover, s cannot be an element of F since t is the least element of F. Therefore $s \in S$. But then $s + 1 = (t - 1) + 1 = t \in S$ by (2). This is the desired contradiction. □

Many arithmetical statements can be proved with the help of Theorem 1.3.1.

Example 1.3.2. Suppose we wish to prove that the following formula, denoted $A(n)$, holds for each natural number n:

$$A(n) : \sum_{k=1}^{n} k = \frac{n(n+1)}{2}.$$

We can proceed this way: Let $S = \{n \in \mathbf{N} : A(n) \text{ is true}\}$. We want to show that $S = \mathbf{N}$. Thus we want to show that properties (1) and (2) (the hypotheses of Theorem 1.3.1) hold. Note that $1 \in S$ because statement $A(1)$ is the statement that $\sum_{k=1}^{1} k = \frac{1(1+1)}{2}$, and this is certainly true. Now suppose that $A(n)$ is true, that is, $n \in S$, so that

$$\sum_{k=1}^{n} k = \frac{n(n+1)}{2}.$$

Then compute that

$$\sum_{k=1}^{n+1} k = \Big(\sum_{k=1}^{n} k\Big) + (n+1) = \frac{n(n+1)}{2} + (n+1),$$

by the hypothesis that $A(n)$ is true. Now rearrange the right-hand side of this result to obtain

$$\sum_{k=1}^{n+1} k = \frac{(n+2)(n+1)}{2},$$

which is precisely the statement $A(n+1)$. Hence the truth of $A(n)$ (that is, $n \in S$) implies the truth of $A(n+1)$ (that is, $n+1 \in S$). Thus $A(n)$ is true for all $n \in \mathbf{N}$. △

Most applications of Theorem 1.3.1 follow the general pattern of this example. Assuming that step (1) holds, the work consists in showing the **induction step** in (2): $n \in S$ *implies* $n+1 \in S$.

The principle of mathematical induction can be restated in the following useful form.

Theorem 1.3.3 (Mathematical Induction). *Suppose that for each positive integer n we are given a statement $A(n)$. Suppose further that we can prove the following two properties:*

(1) *$A(1)$ is true;*

(2) *for each positive integer n, the truth of $A(n)$ implies the truth of $A(n+1)$.*

Then for all positive integers n, statement $A(n)$ is true.

Proof. The proof is by contradiction, and has essentially the same structure as the proof of Theorem 1.3.1, allowing only for the change in the language of the properties (1) and (2) given here. Let $S = \{n \in \mathbf{N} : A(n) \text{ is true}\}$. Thus, we assume (1) and (2) hold, and that the set $F = \mathbf{N} - S$, consisting of the positive integers n for which the statement $A(n)$ is false, is nonempty. By well ordering of the positive integers, there is a least element m in F, and statement $A(m)$ is false. By (1), $m \neq 1$, so $m > 1$. Since m is the least element of F, $m - 1 \notin S$; that is, statement $A(m-1)$ is true. But then, by (2), statement $A(m)$ is true since $m = (m-1) + 1$. The desired contradiction is that statement $A(m)$ is both true and false. Therefore hypotheses (1) and (2) imply that for all positive integers n, statement $A(n)$ is true. □

The following statement of the induction principle has a modified property (2) which appears to be stronger than (2) in Theorem 1.3.3. In fact, Theorem 1.3.3 is equivalent to Theorem 1.3.4.

Theorem 1.3.4 (Mathematical Induction II). *Suppose that for each positive integer n we are given a statement $A(n)$. Suppose further that we can prove the following two properties:*

(1) *The statement $A(1)$ is true.*

(2) *For each positive integer n, the truth of $A(k)$ for all k with $1 \leq k < n$ implies the truth of $A(n)$.*

Then the statement $A(n)$ is true for all positive integers n.

Proof. Assume (1) and (2) hold. Let F be the set of integers $n \geq 0$ for which the statement $A(n)$ is false, and assume that F is nonempty, having least element m. Then $A(m)$ is false. By hypothesis (1), $m \neq 1$, so $m > 1$. Since m is the least element of F, for every $k < m$ statement $A(k)$ is true. Then by hypothesis (2), statement $A(m)$ is true. The deduction that $A(m)$ is both true and false is the contradiction which completes the proof. □

Theorem 1.3.4 is helpful, for example, in the proof that all real symmetric matrices are diagonalizable by a real orthogonal matrix; see Theorem 8.5.7.

Sometimes it is convenient to start the indexing of a list of statements with the index 0 rather than 1. In the proof of the induction principle, we could have replaced the initial number 1 by the number 0, and the argument would have proceeded just as well. (The nonnegative integers are well ordered.)

Example 1.3.5. With practice, we find that induction is lurking behind many simple statements we want to make. Suppose $\{A_j : j \in \mathbf{N}\}$ is a collection of sets (of numbers or other objects) and we wish to talk about the union of all these sets, denoted $\bigcup_{j=1}^{\infty} A_j$. Some, or many, of the sets A_j may overlap (have nonempty intersection), or not, as the case may be, but in some situations, any overlap is a mere inconvenience. For example, at several places later in the book we find it is useful to express the same union as a union of sets that are pairwise disjoint. In other words, we want to write

$$\bigcup_{j=1}^{\infty} A_j = \bigcup_{j=1}^{\infty} B_j$$

where the sets B_j are *pairwise disjoint*, meaning that $B_i \cap B_j$ is empty for $i \neq j$. We may also say that the collection $\{B_j : j \in \mathbf{N}\}$ is a disjoint collection or a disjoint sequence (see Definition 1.3.6 below). We also say that the union of the sets B_j is a disjoint union. A simple definition that yields such sets B_j is to define $B_1 = A_1$, and then to say we define

$$B_n = A_n - \bigcup_{j=1}^{n-1} A_j, \quad \text{for } n \geq 2,$$

or equivalent wording. It is the principle of induction that assures us that the sets B_n have truly been defined for every positive integer n. △

The concept of a *sequence* is important throughout this book.

Definition 1.3.6. *A* **sequence** *in a set X is a function $f : \mathbf{N} \to X$.*

Defining $f_n = f(n)$ (by induction), we often denote sequences by enclosing the elements in parentheses, $(f_1, f_2, f_3, \ldots)$, though this is not a strict rule. It is legitimate to simply refer to *a sequence* f_k, for example. The notation, $(f_1, f_2, f_3, \ldots)$, carries with it the natural ordering of the images inherited from the ordering of $\mathbf{N}$. A notation such as $(f_k)_{k=1}^{\infty}$ can be used if it is important to indicate explicitly the starting value of the index. For uniformity of notation when discussing general properties of sequences, it is assumed that $k = 1$ is the starting value unless specified otherwise. This is the reason for using $\mathbf{N}$ as the domain in Definition 1.3.6. But it would be counterproductive to insist on always starting with index $k = 1$. The index set $\{0, 1, 2, 3, \ldots\} = \{0\} \cup \mathbf{N}$ is frequently used in the study of infinite series. For a sequence f, it is not only the range of f, denoted $\{f(k) : k \in \mathbf{N}\}$, that matters, but often the ordered listing of elements is of interest as well. As shown in the section on Equivalence of Sets and Cardinality, in principle any countably infinite set can serve as the domain (or index set) of a sequence. Finite sequences in X can also be useful. They normally have domain $\{1, \ldots, n\}$ for some $n \in \mathbf{N}$.

The set $\mathbf{Z}$ of integers, the numbers that can represent the results of transactions involving whole currency units, is denoted by

$$\mathbf{Z} := \{\ldots, -3, -2, -1, 0, 1, 2, 3, \ldots\}.$$

The set $\mathbf{Z}$ is totally ordered by $\leq$, but not well ordered.

The algebraic operations of addition and multiplication of integers operate on the product set

$$\mathbf{Z} \times \mathbf{Z} = \{(a, b) : a \in \mathbf{Z} \text{ and } b \in \mathbf{Z}\},$$

and are viewed as functions from $\mathbf{Z} \times \mathbf{Z}$ to $\mathbf{Z}$. Any two integers a, b can be added to give another integer, $a + b$, or multiplied to give an integer ab. Each of these binary algebraic operations is *commutative*,

$$a + b = b + a \quad \text{and} \quad ab = ba, \quad \text{for all } a, b \in \mathbf{Z},$$

and *associative*,

$$a + (b + c) = (a + b) + c \quad \text{and} \quad a(bc) = (ab)c, \quad \text{for all} \quad a, b, c \in \mathbf{Z}.$$

Multiplication is *distributive* over addition:

$$a(b + c) = ab + ac, \quad \text{for all} \quad a, b, c \in \mathbf{Z}.$$

The number 0 is the unique additive identity for the set $\mathbf{Z}$: $a + 0 = a$ for any $a \in \mathbf{Z}$, and 0 is the only number with that property. For each $a \in \mathbf{Z}$, there is a unique additive inverse, denoted $-a$, such that $a + (-a) = 0$. The number 1 is the unique multiplicative identity for $\mathbf{Z}$: $1a = a$ for any $a \in \mathbf{Z}$. The set $\mathbf{Z}$ does not include multiplicative inverses for every nonzero integer. Only the numbers 1 and -1 have multiplicative inverses in $\mathbf{Z}$.

Exercises.

Exercise 1.3.1. Prove by induction: For each natural number n,

$$\sum_{k=1}^{n} k^2 = \frac{n(n+1)(2n+1)}{6}.$$

Exercise 1.3.2. Prove by induction: For each natural number n,

$$\sum_{k=1}^{n} k^3 = \frac{n^2(n+1)^2}{4} = \Big(\sum_{k=1}^{n} k\Big)^2.$$

Exercise 1.3.3. Recall that a positive integer is a **prime number** if its only positive integer factors are 1 and itself. Prove: Every positive integer is either a prime number or the product of prime numbers. *Hint*: Use Theorem 1.3.4.

Exercise 1.3.4. Prove the finite geometric sum formula: If $r \neq 1$, then for any positive integer n,

$$\sum_{k=0}^{n} r^k = \frac{1-r^{n+1}}{1-r}.$$

Exercise 1.3.5. Let $h > 0$. Use induction to prove *Bernoulli's inequality*: For all positive integers n, $(1+h)^n \geq 1 + nh$.

Exercise 1.3.6. Let k and n be nonnegative integers with $0 \leq k \leq n$. The **binomial coefficients** are defined by

$$\binom{n}{k} = \frac{n!}{k!(n-k)!}$$

where $0! := 1$, and for $n \geq 1$, $n!$ is the product of the first n positive integers: $n! = n(n-1)(n-2)\cdots(3)(2)(1)$.

1. Prove that $\binom{n}{k-1} + \binom{n}{k} = \binom{n+1}{k}$ for $1 \leq k \leq n$.
2. Prove the *special binomial theorem*: For real y and any positive integer n,
$$(1+y)^n = \sum_{k=0}^{n} \binom{n}{k} y^k.$$
Hint: Use part (1) for the induction step. Observe that for a general finite sum such as $\sum_{k=0}^{n} a_k$, equivalent expressions are $a_0 + \sum_{k=1}^{n} a_k$ and $a_n + \sum_{k=1}^{n} a_{k-1}$.
3. Show that Bernoulli's inequality, $(1+h)^n \geq 1 + nh$ for positive real h and positive integer n, follows from the special binomial theorem.
4. Prove the *general binomial theorem*: For real $x \neq 0$ and y, and any positive integer n,
$$(x+y)^n = \sum_{k=0}^{n} \binom{n}{k} x^{n-k} y^k.$$

Exercise 1.3.7. For the index set $\mathbf{N}$ of natural numbers, define the real number intervals $J_n = (1, 1 + 1/n)$ for $n \in \mathbf{N}$. Find $\bigcup_{n\in\mathbf{N}} J_n$ and $\bigcap_{n\in\mathbf{N}} J_n$. Then find $(\bigcup_{n\in\mathbf{N}} J_n)^c$ and $(\bigcap_{n\in\mathbf{N}} J_n)^c$.

Exercise 1.3.8. Repeat the previous exercise with the real number intervals $J_n = [1, 1 + 1/n]$, $n \in \mathbf{N}$.

1.4. Equivalence of Sets and Cardinality

Let $J_n \subset \mathbf{N}$ be the set of the first n positive integers, $J_n = \{1, 2, 3, \ldots, n\}$. A set X is **finite**, by definition, if and only if there exists an $n \in \mathbf{N}$ such that there is a bijection $f : J_n \to X$. If a set X is not finite, then X is **infinite**. For finite sets, we can refer to the number of elements in the set; the number of elements is some positive integer n. For infinite sets, we cannot speak so easily about the number of elements. For example, it might seem that the set E of even positive integers has only half as many elements as the set $\mathbf{N}$ of positive integers. It might seem that there must be more numbers on the real number line than there are numbers in the real interval $[0, 1]$. If we are thinking of the inclusion mapping $x \mapsto x$ from $[0, 1]$ to $(-\infty, \infty)$ or from E to $\mathbf{N}$, then of course there are numbers not in the image of this injective mapping. But this does not rule out the possibility of a bijection between $[0, 1]$ and $(-\infty, \infty)$ or between E and $\mathbf{N}$.

The way to approach these questions is by means of bijections and the concept of two sets X and Y having the same cardinality.

Definition 1.4.1. *Let U be a set. Subsets X and Y of U have the* **same cardinality** *if there exists a bijection $h : X \to Y$, in which case we write* $\text{card}(X) = \text{card}(Y)$.

It is useful to have the concept of an *equivalence relation* on a set M, which gives a classification of the elements which are alike in a specific way. Let M be a set or collection. An **equivalence relation** on M is a relation, denoted $\sim$, between selected pairs of elements of M such that (i) $x \sim x$ for all x; (ii) if $x \sim y$, then $y \sim x$; and (iii) if $x \sim y$ and $y \sim z$, then $x \sim z$. These properties are known as the (i) reflexive, (ii) symmetric, and (iii) transitive, properties of the relation. A more formal definition of an equivalence relation $\sim$ on M is to define it as that subset R of the Cartesian product $M \times M$ consisting of all pairs (x, y) such that $x \sim y$. Then (i) $(x, x) \in R$ for all x; (ii) if $(x, y) \in R$, then $(y, x) \in R$; and (iii) if $(x, y) \in R$ and $(y, z) \in R$, then $(x, z) \in R$.

If U is a set and M is the collection of all subsets of U, then the relation defined by $X \sim Y$ if and only if $\text{card}(X) = \text{card}(Y)$ is an equivalence relation on M. Reflexivity and symmetry are clear from the definition. The relation is also transitive, because if there are bijections $h : X \to Y$ and $g : Y \to W$, so that X and Y have the same cardinality, and Y and W have the same cardinality, then $g \circ h : X \to W$ is also a bijection, and thus X and W have the same cardinality.

Example 1.4.2. Let E be the set of even positive integers, and let $f : \mathbf{N} \to E$ be $f(k) = 2k$. Then f is a bijection, with inverse $f^{-1} : E \to \mathbf{N}$ given by $f^{-1}(k) = k/2$. So E and $\mathbf{N}$ have the same cardinality. Define $g : \mathbf{Z} \to \mathbf{N}$ by setting $g(n) = 2n$ if $n > 0$, and $g(n) = -2n+1$ if $n \leq 0$. We can visualize the one-to-one correspondence given by g and its inverse g^{-1}:

$$(0, 1, -1, 2, -2, 3, -3, \ldots) \quad \xrightarrow{g} \quad (1, 2, 3, 4, 5, 6, 7, \ldots),$$

$$(1, 2, 3, 4, 5, 6, 7, \ldots) \quad \xrightarrow{g^{-1}} \quad (0, 1, -1, 2, -2, 3, -3, \ldots).$$

Thus, $\mathbf{Z}$ and $\mathbf{N}$ have the same cardinality, and by transitivity of the relation, $\mathbf{Z}$ and E also have the same cardinality. △

Let X and Y be sets. If there exists an injective (one-to-one) function $f : X \to Y$, then, in the absence of further knowledge, there is the possibility that there are elements in Y that are not in the image $f(X)$. Without further knowledge, the question remains whether there exists a bijection between X and Y. It can often happen that injective mappings $f : X \to Y$ and $g : Y \to X$ are available, but neither f nor g is onto. In such a case, the knowledge gap associated with specific injective mappings can always be closed. This is the assertion of the Schroeder-Bernstein theorem.

Theorem 1.4.3 (Schroeder-Bernstein). *Let X and Y be sets. If there exists a one-to-one mapping $f : X \to Y$ and a one-to-one mapping $g : Y \to X$, then X and Y have the same cardinality.*

Plausibility argument. The rigorous proof of this theorem might be a hard sell, and therefore it is deferred to an appendix for those who might be interested. Instead, consider the following plausibility argument:

To help in thinking about the situation, we can think of the elements of X as cats and the elements of Y as dogs. We say that each cat $x \in X$ picks a dog $f(x) \in Y$ to chase, and different cats chase different dogs. Similarly, each dog $y \in Y$ picks a cat $g(y) \in X$ to chase, and different dogs chase different cats. With each dog and cat chasing a unique cat and dog, respectively, we note four possible types of patterns, or chasing chains:

- ⋆ Chasing chains that form a finite loop consisting of an even number of animals; within such a loop, we match each cat with the dog it chases.
- ⋆ Chasing chains that are doubly infinite, with no start and no end; within such a chain, we match each cat with the dog it chases.
- ⋆ Chasing chains that start with a cat, but have no end; within such chains, we match each cat with the dog it chases.
- ⋆ Chasing chains that start with a dog, but have no end; in such chains, we match each cat with the dog chasing it.

Given these four possibilities, we may convince ourselves that the elements of X and Y may be put into one-to-one correspondence. But the construction of such a correspondence is not obvious. For example, the number of chasing chains of each type is not known. As noted, this is best considered a plausibility argument. If you are satisfied, it is perfectly fine, and the examples after the theorem are recommended. □

The Schroeder-Bernstein theorem is an important tool in the study of cardinality. We consider some examples.

Example 1.4.4. The real intervals $[0, 1]$ and $[0, 1)$ have the same cardinality. Let $f : [0, 1) \to [0, 1]$ be the inclusion mapping, $f(x) = x$, and let $g : [0, 1] \to [0, 1)$ be $g(x) = x/2$. Then f are g are both injective, hence $[0, 1]$ and $[0, 1)$ have the same cardinality by Theorem 1.4.3. △

Theorem 1.4.3 and elementary functions can be used to show that any two of the intervals (a, b), $[a, b]$, $(a, b]$, $[a, b)$, (a, ∞), $[a, \infty)$, $(-\infty, b)$, $(-\infty, b]$, and $(-\infty, \infty)$

have the same cardinality. Stated more briefly, any two nontrivial real intervals are in one-to-one correspondence.

Example 1.4.5. We show that $(0,1)$ and $(-\infty,\infty)$ have the same cardinality. The function $f:(-\pi/2,\pi/2)\to(-\infty,\infty)$ given by $f(x)=\tan x$ is strictly increasing, hence one-to-one, and f is onto $(-\infty,\infty)$. The function $g:(0,1)\to(-\pi/2,\pi/2)$ defined by $g(x)=-\pi/2+\pi x$ is a bijection. Hence, $(0,1)$ and $(-\infty,\infty)$ have the same cardinality since $f\circ g$ is a bijection. △

Definition 1.4.6. *A set X is* **denumerable** *if it has the same cardinality as the set $\mathbf{N}$ of natural numbers, that is, there exists a bijection $h:\mathbf{N}\to X$.*

A set X is denumerable if and only if there is a bijection $h:\mathbf{N}\to X$. The bijection h provides an *enumeration* or listing, of the set X, given by the sequence $(h_1,h_2,h_3,\ldots)$, where $h_k:=h(k)$ for each $k\in\mathbf{N}$. We say that X is *enumerated* by the given sequence.

Example 1.4.7. Consider the bijection $g:\mathbf{Z}\to\mathbf{N}$ given earlier where $g(n)=2n$ if $n>0$, and $g(n)=-2n+1$ if $n\le 0$. Thus,

$$(0,1,-1,2,-2,3,-3,\ldots)\quad\xrightarrow{g}\quad(1,2,3,4,5,6,7,\ldots).$$

In fact, g^{-1} provides the enumeration of $\mathbf{Z}$ by mapping the sequence on the right, which is $\mathbf{N}$, one-to-one and onto $\mathbf{Z}$, the sequence on the left. △

The next result says that denumerable sets frequently appear.

Proposition 1.4.8. *Every infinite set S has a denumerable subset.*

Proof. We define a denumerable subset D of S and give its enumeration. From each nonempty subset of S we can select a specific element of that subset. For S itself, let x_1 be the chosen element of S. For the subset $S-\{x_1\}=\{x_1\}^c$, let x_2 be the chosen element. Suppose that we have selected elements $x_1,x_2,x_3,\ldots,x_{n-1}$ such that for each $k=2,\ldots,n-1$, element $x_k\in\{x_1,\ldots,x_{k-1}\}^c=S-\{x_1,\ldots,x_{k-1}\}$. Then from the subset $S-\{x_1,\ldots,x_{n-1}\}=\{x_1,\ldots,x_{n-1}\}^c$, we select an element x_n. By induction we have defined a set $\{x_1,x_2,x_3,\ldots\}$ of distinct elements of S indexed by the set of positive integers. Then $D=\{x_1,x_2,x_3,\ldots\}$ is a denumerable subset of S. □

Proposition 1.4.9. *If S is an infinite subset of $\mathbf{N}$, then S is denumerable and there is a unique enumeration of S, $(k_1,k_2,k_3,\ldots,k_n,k_{n+1},\ldots)$ such that $k_1<k_2<k_3<\cdots<k_n<k_{n+1}<\cdots$.*

Proof. There is a smallest element of S, which we denote by k_1. Suppose we have defined $k_1,k_2,\ldots,k_n$ such that $k_1<k_2<\cdots<k_n$ and such that if $k\in S$ but $k\notin\{k_1,k_2,\ldots,k_n\}$, then $k>k_n$. Then define k_{n+1} to be the smallest element of S which is greater than k_n. Then the mapping $h(n)=k_n$, $n\in\mathbf{N}$, gives the desired enumeration of S. □

The next corollary explains why denumerable sets may be considered the smallest infinite sets.

Corollary 1.4.10. *If D is a denumerable set and S is an infinite subset of D, then S is denumerable.*

Proof. Given any enumeration of D, its infinite subset S corresponds to an infinite subset of $\mathbf{N}$ via the given enumeration. Then by the previous proposition, S is denumerable. □

Corollary 1.4.11. *If D is a denumerable set and $f : D \to S$ is onto S, then S is either denumerable or finite.*

Proof. Since f is onto S, for each $y \in S$ there is some element $x_y \in D$ such that $f(x_y) = y$. Define $g : S \to D$ by $g(y) = x_y$. Then g is one-to-one, because if $y, z \in S$ and $g(y) = g(z)$, then

$$y = f(x_y) = f(g(y)) = f(g(z)) = f(x_z) = z.$$

Now $g(S) \subset D$ and since g is one-to-one, g provides a bijection of S and $g(S)$. If $g(S)$ is infinite, then $g(S)$, and hence S, is denumerable by the previous corollary. Otherwise, $g(S)$, and hence S, is finite. □

Proposition 1.4.12. *The Cartesian product $\mathbf{N} \times \mathbf{N}$ is denumerable.*

Proof. Let $h : \mathbf{N} \times \mathbf{N} \to \mathbf{N}$ be the mapping

$$h(n, m) = 2^n 3^m.$$

It can be shown that h is one-to-one (Exercise 1.4.1). Hence, by Proposition 1.4.9, $\mathbf{N} \times \mathbf{N}$ is denumerable. □

Corollary 1.4.13. *If D is a denumerable set, then the Cartesian product $D \times D$ is denumerable.*

Proof. If $h : \mathbf{N} \to D$ is a bijection, then so is the mapping $H : \mathbf{N} \times \mathbf{N} \to D \times D$ defined by $H(n, m) = (h(n), h(m))$. □

The proof of the next proposition is left to Exercise 1.4.2.

Proposition 1.4.14. *If for each $k \in \mathbf{N}$, D_k is a denumerable set, then the union $D = \bigcup_{k=1}^{\infty} D_k$ is denumerable.*

Finally, we say that a set is **countable** if it is either finite or denumerable. In either case, the set may be put into a one-to-one correspondence with a subset of the counting numbers $\mathbf{N}$. The context should make the meaning clear as to whether a finite or infinite set is intended.

Exercises.

Exercise 1.4.1. Complete the proof of Proposition 1.4.12 by showing that the mapping $h : \mathbf{N} \times \mathbf{N} \to \mathbf{N}$ given by $h(n, m) = 2^n 3^m$ is one-to-one. *Hint*: Use the Fundamental Theorem of Arithmetic (the Unique Factorization Theorem for positive integers).

Exercise 1.4.2. Prove Proposition 1.4.14. *Hint*: Enumerate each of the sets D_k as follows: $D_1 = (x_{11}, x_{12}, x_{13}, \ldots)$, $D_2 = (x_{21}, x_{22}, x_{23}, \ldots)$, and so on. Define $f : \mathbf{N} \times \mathbf{N} \to D$ by $f(i, j) = x_{ij}$ and show that f is onto $D = \bigcup_{k=1}^{\infty} D_k$. Then apply Corollary 1.4.11.

Exercise 1.4.3. Suppose that for each $k \in \mathbf{N}$, D_k is nonempty and is either a finite or a denumerable set. Show that the union $D = \bigcup_{k=1}^{\infty} D_k$ is denumerable.

Exercise 1.4.4. Let $\sim$ be an equivalence relation on a set M. Show that $\sim$ determines a partition of M into disjoint subsets, called the equivalence classes relative to $\sim$, such that the elements within each class are all equivalent under the relation, and M is the union of the equivalence classes.

1.5. Notes and References

The description of the Russell paradox in Exercise 1.1.1 is from Sagan [**54**]. The Unique Factorization Theorem for positive integers referenced in Exercise 1.4.1 is in Birkhoff and Mac Lane [**4**] or any modern algebra text.

Halmos [**26**] is an excellent and readable presentation of the essentials of set theory. For more on basic logic and methods of proof, see Krantz [**39**].

Chapter 2

The Complete Ordered Field of Real Numbers

The algebraic structure known as a *field* is an efficient way of organizing and expressing the properties of the rational numbers, the real numbers, and the complex numbers, among other important number systems such as the finite fields that play an important role in modern cryptography. The main goal of this chapter is to explain the algebraic structure of fields and, in particular, the special structure of the field of real numbers. The special structure of the real numbers is described in the statement that the real number system is a *complete ordered field*. It will take some space, time, and effort to explain the terms involved in that statement, but the result will be a deeper understanding of the real numbers as the foundation for all of modern analysis.

It might seem appropriate to *define* the real numbers before discussing their properties. The real number field can be *defined* by means of constructions based on the rational numbers. We give references later for the standard constructions. Those constructions are important because they establish that the real number system, endowed with its special properties needed for the success of calculus, actually exists as a consistent number system. The rigorous construction of a complete ordered field ensures that the definition we give of the real number field is more than an empty exercise.

However, for an understanding of analysis, what is needed is an understanding of the *properties* of the real numbers, and for that purpose we do not detour to study the rigorous proof of existence and uniqueness of a complete ordered field. In short, you can proceed to study the properties of the real numbers without worry that the effort is an empty exercise. After all, you have previous experience with the problem-solving power of calculus, so there must be something substantial at the foundations. That something is the existence of a complete ordered field. The properties of the field of real numbers are all presented in the axioms for a complete ordered field, and these axioms appear in the first two sections of the chapter.

Sections 2.3-2.7 explore the properties of the real numbers that are of fundamental importance for analysis. For interested readers, Section 2.8 includes references for the construction of a complete ordered field and provides encouragement for the study of such a construction.

2.1. Algebra in Ordered Fields

Readers of this book have years of experience in dealing with the field operations and properties of the rational numbers and the real numbers. A **field** is a set F together with two binary operations, defined by functions $A : F \times F \to F$ and $M : F \times F \to F$ called addition and multiplication, respectively, which satisfy the axioms set out in items A1 - A4, M1 - M4, and D1 in the next subsection. We shall denote the addition and multiplication operations on elements of F in the familiar way indicated by $A(x, y) = x + y$ and $M(x, y) = xy$.

2.1.1. The Field Axioms.

The axioms for addition in F are as follows:

A1. For all $x, y \in F$, $x + y = y + x$. (**commutativity**)

A2. For all $x, y, z \in F$, $(x + y) + z = x + (y + z)$. (**associativity**)

A3. There exists an element $0 \in F$ (an **additive identity**) such that $x + 0 = x$ for all $x \in F$.

A4. For every $x \in F$ there is an element $y \in F$ such that $x + y = 0$.

The additive identity 0 in **A3** is uniquely determined. Also, an additive inverse y for x in **A4** is uniquely determined by x, and we write it as $-x$, so that $x + (-x) = 0$.

We can form finite sums for elements $x_1, x_2, \ldots, x_n \in F$ by defining inductively

$$x_1 + x_2 + \cdots + x_n = (x_1 + x_2 + \cdots + x_{n-1}) + x_n.$$

Finite sums do not depend on the ordering of the terms; one can give a proof by induction of this fact, but we shall not do so. The sum $x_1 + x_2 + \cdots + x_n$ can be written using the concise and unambiguous summation notation

$$\sum_{k=1}^{n} x_k.$$

The axioms for multiplication in F are as follows.

M1. For all $x, y \in F$, $xy = yx$. (**commutativity**)

M2. For all $x, y, z \in F$, $(xy)z = x(yz)$. (**associativity**)

M3. There exists an element $1 \in F$ (a **multiplicative identity**) different from the additive identity 0, such that $x1 = 1x = x$ for all $x \in F$.

M4. If $x \in F$ and $x \neq 0$, there exists an element $v \in F$ such that $xv = vx = 1$. (**multiplicative inverse**)

The multiplicative identity 1 is unique. Given $x \in F$ with $x \neq 0$, the multiplicative inverse v of x in **M4** is uniquely determined by x, and we write it x^{-1} or $1/x$. In a field, we can never divide by the additive identity (Exercise 2.1.1).

We can form the product of finitely many elements $x_1, x_2, \ldots, x_n \in F$ by defining inductively

$$x_1 x_2 \cdots x_{n-1} x_n = (x_1 x_2 \cdots x_{n-1}) x_n.$$

The product does not depend on the ordering of the terms; again, one can give a proof by induction of this fact. The product $x_1 x_2 \cdots x_n$ can be written using the concise and unambiguous product notation

$$\prod_{k=1}^{n} x_k.$$

If n is a positive integer and $x \in F$, then we define $x^n = xx \cdots x = \prod_{k=1}^{n} x$. If $a \neq 0 \in F$, then we define $a^0 = 1$, the multiplicative identity. Then for any integers $m, n \geq 0$ and $a \in F$,

$$a^{m+n} = a^m a^n.$$

Also we define $a^{-m} = (a^{-1})^m$; with this definition, one can show that the law of exponents $a^{m+n} = a^m a^n$ holds for *all* integers m, n.

The axiom of distributivity relates addition and multiplication, asserting that multiplication distributes over addition:

D1. For all $x, y, z \in F$, $x(y + z) = xy + xz$. (**distributivity**)

We observe that by the commutativity of multiplication, we also have

$$(y + z)x = x(y + z) = xy + xz = yx + zx.$$

Example 2.1.1. The set $\mathbf{Q}$ of rational numbers, with the usual operations of addition and multiplication, satisfies the field axioms and is therefore a field. △

Example 2.1.2. The set $\mathbf{R}$ of real numbers, which you may think of as the set of decimal expansions if you like, with the usual operations of addition and multiplication you have always used, is assumed to be a field. (Later in the chapter, we define decimal representations for the real numbers.) If $q \in \mathbf{Q}$, then q is a real number, hence $\mathbf{R}$ contains $\mathbf{Q}$ as a subset. (We show later in the chapter how this containment follows from the field axioms and the order axiom discussed below.) In fact, the addition and multiplication operations on $\mathbf{R}$, when restricted to $\mathbf{Q}$, are exactly the operations of $\mathbf{Q}$, so $\mathbf{Q}$ is a subfield of $\mathbf{R}$. In the remainder of this chapter, we shall identify the property or properties of $\mathbf{R}$ that distinguish it from $\mathbf{Q}$ and allow us to do analysis. △

There are only two elements whose existence is specifically demanded by the field axioms: the additive identity and the multiplicative identity. These two elements alone describe a particular field, as follows.

Example 2.1.3. The set $\mathbf{Z}_2 = \{0, 1\}$ is made into a field by defining addition by $0 + 0 = 0$, $0 + 1 = 1$, $1 + 0 = 1$, and $1 + 1 = 0$, and defining multiplication by $(0)(0) = 0$, $(0)(1) = 0$, $(1)(0) = 0$, and $(1)(1) = 1$. One can verify directly that the field axioms hold. △

Here are some basic properties that follow from the field axioms. In any field, $(-1)x = -x$, so the additive inverse of x equals the product of x and the additive inverse of the multiplicative unit. Also, $-(ab) = (-a)b = a(-b)$ follows easily from the axioms and the uniqueness of additive inverses.

Example 2.1.4. The field $\mathbf{C}$ of complex numbers is the set of ordered pairs (a, b), where $a, b \in \mathbf{R}$, with the operations of addition and multiplication for $(a, b), (c, d) \in \mathbf{C}$ defined by

$$(a, b) + (c, d) = (a + c, b + d)$$

and

$$(a, b)(c, d) = (ac - bd, bc + ad),$$

respectively. Although it may not be tremendously exciting to do so, one can check directly that both of these operations are commutative and associative, and multiplication distributes over addition:

$$[(a, b) + (c, d)](e, f) = (a, b)(e, f) + (c, d)(e, f).$$

The additive identity is $(0, 0)$. The additive inverse of (a, b) is $(-a, -b)$. The multiplicative identity is $(1, 0)$, and the multiplicative inverse of $(a, b) \neq (0, 0)$ is $(a/(a^2 + b^2), -b/(a^2 + b^2))$. These facts can all be verified directly from the stated definitions. Instead of doing that, we wish to show the reasonableness of the definitions. Let us write $a = (a, 0)$. If we set $i = (0, 1)$, then by the definition of multiplication, $i^2 = (0, 1)(0, 1) = (-1, 0) = -1$. Again by the definition of multiplication and our agreed notations, $(0, b) = (b, 0)(0, 1) = bi$. Thus, for any $(a, b) \in \mathbf{C}$, the definition of addition and our agreed notations allow us to write

$$(a, b) = (a, 0) + (0, b) = a + bi.$$

One can now verify that the definition of multiplication is exactly the one needed to ensure that if $i^2 = -1$ and multiplication and addition are indeed commutative, then $(a + bi)(c + di)$ can be expanded by the usual rules to yield the product $(ac - bd) + (bc + ad)i = (ac - bd, bc + ad)$. Finally, if $z = a + 0i = a$, then z is a real number, hence $\mathbf{C}$ contains $\mathbf{R}$ as a subset. In fact, the addition and multiplication operations on $\mathbf{C}$, when restricted to $\mathbf{R}$, are exactly the operations of $\mathbf{R}$, by definition, so $\mathbf{R}$ is a subfield of $\mathbf{C}$. $\triangle$

2.1.2. The Order Axiom and Ordered Fields. Let F be a field and $P \subset F$ a subset that satisfies the following conditions:

O1. If $x, y \in P$, then $x + y \in P$ and $xy \in P$.

O2. For each $x \in F$, exactly one of the following is true:

$$x \in P, \quad \text{or} \quad x = 0, \quad \text{or} \quad -x \in P.$$

Then P is called a **positive set**.

An **ordered field** is a field F that contains a positive set P. The idea of a positive set is that an order comparison between two elements of F only requires a determination of whether an element is in the positive set, because in a field we have additive inverses and therefore can subtract one element from another element. (Is it true that $b - a > 0$, that is, is it true that $b - a \in P$?) We begin by learning a little bit about the positive set P. First, if F is an ordered field with positive set P, and $a \in F$ with $a \neq 0$, then $a^2 \in P$. Here is the proof: If $a \neq 0$, then either $a \in P$ or $-a \in P$. If $a \in P$, then $a^2 \in P$ by the definition of a positive set. If $-a \in P$, then $(-a)^2 \in P$, and $(-a)^2 = (-a)(-a) = (-1)(-1)a^2 = -(-1)a^2 = a^2$, so again $a^2 \in P$. Since $1 \neq 0$ in F and $1^2 = 1$, we know that $1 \in P$.

Other important properties of ordered fields include the following ones.

The product of a positive element and a negative element of F must be negative: If $a \in P$ ($a > 0$) and $-b \in P$ ($b < 0$), then $a(-b) = (-a)b = -(ab) \in P$, and hence $ab < 0$.

In an ordered field, $x \in P$ implies $x^{-1} \in P$. Here is the proof: If $x \in P$, then $x \neq 0$, so x^{-1} exists and $x^{-1} \neq 0$. If $-x^{-1} \in P$, then P contains

$$x(-x^{-1}) = -(xx^{-1}) = -1,$$

but this contradicts the fact that $1 \in P$. Hence, $x^{-1} \in P$.

In order to derive further properties involving order, it is best to define the usual order symbols $<$, $>$, $\leq$ and $\geq$ and work with them. In an ordered field, we write $a < b$ if and only if $b + (-a) \in P$. We also write $b > a$ to mean $a < b$. In particular, $x > 0$ if and only if $x \in P$. We write $a \leq b$ if and only if $a = b$ (meaning $b - a = 0$) or $a < b$.

Condition O2 implies that the positive set of an ordered field induces a total ordering by $\leq$. Any ordered field is totally ordered.

Example 2.1.5. The set of positive rational numbers

$$\mathbf{Q}^+ = \{p : p \in \mathbf{Q}, p > 0\}$$

is the set of positive elements of the rational number field $\mathbf{Q}$. $\mathbf{Q}$ is totally ordered by $\leq$, but not well ordered. △

Example 2.1.6. The set of positive real numbers

$$\mathbf{R}^+ = \{r : r \in \mathbf{R}, r > 0\}$$

is the set of positive elements of the real number field $\mathbf{R}$. $\mathbf{R}$ is totally ordered by $\leq$, but not well ordered. △

In an ordered field, we have seen that -1 is not an element of the positive set P. This fact allows us to see that there is no possible positive set P for the field of complex numbers. Thus, the complex field $\mathbf{C}$ is not an ordered field.

Proposition 2.1.7. *The field* $\mathbf{C}$ *of complex numbers is not an ordered field. (No positive set exists in* $\mathbf{C}$*.)*

Proof. Suppose $\mathbf{C}$ has a positive set. Since $i^2 = -1$ and i^2 is the square of a nonzero element, -1 is an element of the positive set, which is a contradiction of our earlier deduction that -1 is not in the positive set of an ordered field. Therefore a positive set for $\mathbf{C}$ does not exist. □

Theorem 2.1.8. *Let x, y and z be elements of an ordered field F with positive set P. The following properties hold:*

1. *If $x < y$, then $x + z < y + z$.*
2. *If $x < y$ and $z \in P$, then $xz < yz$.*
3. *If $x < y$ and $-z \in P$, then $yz < xz$.*
4. *If $x < y$, then $x < (x + y)/2 < y$.*
5. *If $x < y$ and $y < z$, then $x < z$.*

Proof. 1. Since $x < y$, $y - x \in P$. We want to show that $(y+z) - (x+z) \in P$. But $(y+z) - (x+z) = y - x \in P$, hence $x + z < y + z$.

2. Since $x < y$, $y - x \in P$. Hence $yz - xz = (y-x)z \in P$ since $y - x$ and z are positive elements. Part 3 is left as an exercise for the reader.

4. Using part 1 twice, we have $2x = x + x < x + y < y + y = 2y$. Hence, on multiplication by $2^{-1} \in P$, two applications of part 2 imply that $x < (x+y)/2 < y$.

5. This follows from the transitivity of $\leq$. Or we can say that by hypothesis, $y - x \in P$ and $z - y \in P$, hence $(z-y) + (y-x) = z - x \in P$. □

We know that **Q** and **R** are ordered fields but **C** is not. Our next goal is to see in what ways **R** and **Q** are different.

The earliest geometers knew that the hypotenuse of a right triangle having two unit-length sides is incommensurate with the side, that is, the side cannot be subdivided into a whole number of standard lengths, such that the ratio of the unit-length side to the hypotenuse is an exact ratio of whole numbers. In modern language, the real number $\sqrt{2}$ is not a rational number. We prove this now, but before beginning the proof, we note that a positive integer is even if its square is even (Exercise 2.1.5).

Lemma 2.1.9. *There is no rational number x such that $x^2 = 2$.*

Proof. Suppose there exists a rational number x with $x^2 = 2$. Write $x = p/q$ with $p, q \in \mathbf{Z}$ and suppose that p and q have no common integer factor (other than 1). Then

$$2 = x^2 = p^2/q^2 \Longrightarrow 2q^2 = p^2.$$

From this we conclude that p^2 is an even integer, so we may write $p^2 = 2k$ for a positive integer k. But then $q^2 = p^2/2 = 4k^2/2 = 2k^2$ must also be even. So both p^2 and q^2 are even. If the square of a positive integer is even, then the integer itself is even. (The equivalent contrapositive statement is: If an integer is odd, then its square is odd.) Thus both p and q are even, and this contradicts the assumption that p and q have no common factor other than 1. □

Lemma 2.1.9 points out a specific deficiency of the rational number field that we will return to later on. (There are many other gaps; for example, $\sqrt[3]{2}$ is not rational, and neither are π and the Euler number e.) For now we continue the general discussion of properties of fields and, in particular, ordered fields.

Every *ordered* field F contains a copy of the natural numbers, under the identification

$$n \leftrightarrow n \cdot 1 := \underbrace{1 + \cdots + 1}_{n \text{ terms}}, \quad n \in \mathbf{N},$$

where 1 is the multiplicative identity of F. In order for this statement to make useful sense, we must show that all these elements are actually distinct, in an ordered field. (Note that these elements in $\mathbf{Z}_2$ simply give the two elements $0, 1 \in \mathbf{Z}_2$, since $1 \cdot 1 = 1$, $2 \cdot 1 = 1 + 1 = 0$, $3 \cdot 1 = 1$, and so on. But $\mathbf{Z}_2$ is not an ordered field.)

Theorem 2.1.10. *In an ordered field, the elements $n \cdot 1$, $n \in \mathbf{N}$, that is, the elements $1, 1+1, 1+1+1, \ldots$ are all positive and distinct.*

Proof. Let $A(n)$ be the statement $A(n) : \left(\sum_{j=1}^{n} 1\right) \in P$, where P is the positive set of the ordered field. In the summation, the index runs through the positive integers from 1 to n, and the term being summed is the multiplicative unit of the field. Then $A(1)$ is true since $\sum_{j=1}^{1} 1 = 1 \in P$. Assume that $A(n) : \left(\sum_{j=1}^{n} 1\right) \in P$ is true. Then

$$\sum_{j=1}^{n+1} 1 = \left(\sum_{j=1}^{n} 1\right) + 1 \in P,$$

since P is closed under addition. By the induction principle, for each positive integer n, statement $A(n)$ is true, that is, for each n, $\left(\sum_{j=1}^{n} 1\right) \in P$.

In order to see that these elements are all distinct, consider the difference between any two of them. The difference is either (i) one of these sums and hence an element of P, or (ii) the additive inverse of one of these sums. In either case, an element of P or the additive inverse of an element of P cannot be the zero element of F. □

An immediate corollary of this theorem is that *any ordered field is infinite.*

We are now justified in making the identification

$$n \leftrightarrow n \cdot 1 := \underbrace{1 + \cdots + 1}_{n \text{ terms}}, \quad n \in \mathbf{N},$$

where 1 is the multiplicative identity of F, and asserting that every ordered field contains a copy of the natural numbers. With this understanding we write $\mathbf{N} \subset F$. Consequently, every ordered field F also contains a copy of the integers $\mathbf{Z}$ and the rational numbers $\mathbf{Q}$, since for any positive integers m and n in F, the elements 0, $-m$, and $m/n = mn^{-1}$ are in F. It is with this understanding that we write $\mathbf{N} \subset \mathbf{Z} \subset \mathbf{Q} \subset F$, where F is any ordered field. In particular, the ordered field of most interest to us is the field $\mathbf{R}$ of real numbers, and $\mathbf{N} \subset \mathbf{Z} \subset \mathbf{Q} \subset \mathbf{R}$.

We are assuming that both $\mathbf{R}$ and $\mathbf{Q}$ satisfy the field axioms. The order relation on $\mathbf{Q}$ is embedded in the order relation on $\mathbf{R}$, and the positive set of $\mathbf{Q}$ is contained in the positive set of $\mathbf{R}$. It will be essential for us to isolate the way in which $\mathbf{R}$ differs from $\mathbf{Q}$, and that is the subject of the next section.

Exercises.

Exercise 2.1.1. *No division by zero.*
Prove: In a field, the additive identity 0 has no multiplicative inverse. *Hint*: Proof by contradiction.

Exercise 2.1.2. Show that $\mathbf{Z}_2$ is not an ordered field.

Exercise 2.1.3. Prove property 3 of Theorem 2.1.8. Then prove the following properties in an ordered field F as an extension of Theorem 2.1.8:

6. If $x < 0$, then $1/x < 0$.
7. If $xy > 0$, then either ($x > 0$ and $y > 0$) or ($x < 0$ and $y < 0$).
8. If $x^2 + y^2 = 0$, then $x = 0$ and $y = 0$.
9. If $x < y$, then there are infinitely many elements z with $x < z < y$.

Exercise 2.1.4. Let F be an ordered field and $a \in F$ with $a > 0$. Prove: If $a < b$, then for each positive integer n, $a^n < b^n$.

Exercise 2.1.5. Show that a positive integer is even if its square is even. *Hint*: The contrapositive statement: If a positive integer is odd, then its square is odd.

2.2. The Complete Ordered Field of Real Numbers

Many people identify the set $\mathbf{R}$ of real numbers with the set of all points on a number line. We may also think of the real numbers as the collection of all decimal numbers.[1] This is a valid point of view, as we will see later. However, our main goal in this section is to describe the property that distinguishes the rational number field from the real number field, because it is this distinguishing property that allows us to fill the gaps in the *rational* number line such as $\sqrt{2}$. To describe this property, we need some additional concepts.

Definition 2.2.1. *Let S be a subset of an ordered field F.*

1. *We say that S is* **bounded above** *if there is an element $b \in F$ such that $s \le b$ for all $s \in S$, and b is then called an* **upper bound** *for S.*
2. *We say that S is* **bounded below** *if there is an element $a \in F$ such that $a \le s$ for all $s \in S$, and a is then called a* **lower bound** *for S.*
3. *The set S is* **bounded** *if it is bounded above and bounded below.*

The subset of the rational numbers defined by

$$S = \{s \in \mathbf{Q} : \text{either } s < 0, \text{ or } s \ge 0 \text{ and } s^2 < 2\}$$

is nonempty, since $1 \in S$. S is not bounded below, but it is bounded above. For example, $b = 2$ is an upper bound for S: The proof is by contradiction, for if $s \in S$ and $s > 2 = b$, then $s^2 > 2^2 = 4$, which contradicts the assumption that $s \in S$. So we clearly have $s \le b = 2$ for all $s \in S$. In a similar way, one can show that $b = 3/2$ is also a rational upper bound for S, and so is $142/100 = 71/50$. In fact, for any rational number b which is an upper bound for S, there is another rational number β which is an upper bound for S and $\beta < b$. In fact there is *no least upper bound* for this set S of rational numbers.

Definition 2.2.2 (Supremum and Infimum). *Let F be an ordered field and $S \subset F$.*

1. *If S is bounded above, then an element $b \in F$ is the* **least upper bound** *or* **supremum** *of S if b is an upper bound for S and $b \le u$ for all upper bounds u for S.*
2. *If S is bounded below, then an element $m \in F$ is the* **greatest lower bound** *or* **infimum** *for S if m is a lower bound for S and $l \le m$ for all lower bounds l for S.*

If S is a nonempty set which has no upper bound, we may write $\sup S = \infty$, and if S is nonempty and has no lower bound, we may write $\inf S = -\infty$. This is a notational convenience.

[1] There is actually a certain restriction which we shall note later.

Using the concept of least upper bound, it is possible to show that if F is an ordered field and we define the set S as above, that is,

$$S = \{s \in F : \text{either } s < 0, \text{ or } s \geq 0 \text{ and } s^2 < 2\},$$

and *if* $\sup S$ *exists in* F, then $(\sup S)^2 = 2$. (This argument will be carried out in the proof of Theorem 2.3.11 below.) However, since we have shown that there is no rational number x such that $x^2 = 2$, it follows that the set S, defined as a subset of the field $F = \mathbf{Q}$, does not have a least upper bound in $\mathbf{Q}$.

If a least upper bound for S exists, then it must be unique. Indeed, if b_1 and b_2 are least upper bounds for S, then we have $b_1 \leq b_2$, since b_2 is also an upper bound for S, and $b_2 \leq b_1$, since b_1 is also an upper bound for S; hence, $b_1 = b_2$. If a greatest lower bound for S exists, then it must be unique (Exercise 2.2.1).

The intervals $(0, \pi]$ and $(0, \pi)$ both have least upper bound π; it is the maximum of $(0, \pi]$, whereas $(0, \pi)$ has no maximum. Standard terminology for the least upper bound of S is the **supremum** of S, written $\sup S$.

The intervals $[-\pi, 0)$ and $(-\pi, 0)$ both have greatest lower bound $-\pi$; it is the minimum of $[-\pi, 0)$, whereas $(-\pi, 0)$ has no minimum. Standard terminology for the greatest lower bound of S is the **infimum** of S, written $\inf S$.

The one remaining axiom of the real number field is the *least upper bound property* of $\mathbf{R}$.

LUB. (**Least Upper Bound**) Every nonempty subset $S \subset \mathbf{R}$ that is bounded above has a least upper bound in $\mathbf{R}$; that is, there is a real number b such that $b = \sup S$.

The least upper bound property is often called the Completeness Axiom for $\mathbf{R}$, since it is this property of the real numbers that guarantees there are no gaps in the real number line.

Definition 2.2.3. *An ordered field F is called* **complete** *if it has the least upper bound property.*

The fields $\mathbf{Q}$ and $\mathbf{R}$ both satisfy the axioms for an ordered field, and $\mathbf{R}$ also satisfies the least upper bound property. However, we have seen that $\mathbf{Q}$ is not a complete ordered field, since the set

$$S = \{s \in \mathbf{Q} : \text{either } s < 0, \text{ or } s \geq 0 \text{ and } s^2 < 2\}$$

is bounded above but does not have a least upper bound in $\mathbf{Q}$.

We are assuming that the field of real numbers is complete, since we are assuming the least upper bound property for $\mathbf{R}$. In the last analysis, this is perfectly legitimate, but it may be somewhat unsatisfying. If you are one of the readers for whom this assumption is unsatisfying, then congratulations to you and rest assured that it is legitimate, because it is possible to show that a complete ordered field exists. In other words, the deductions we make from the least upper bound property are not an empty exercise. A complete ordered field can be constructed, starting from the rational field $\mathbf{Q}$. Moreover, the constructed complete ordered field can be represented by the more familiar decimal representation of numbers on the (real) number line. We can establish the decimal representation based (ultimately) on the least upper bound property. That is, the decimal representation depends on the least upper bound property and some of its consequences. We do not pause here to construct **R** from **Q** and *prove* that **R** has the least upper bound property. For some guidance and references for this construction, see the last section of this chapter.

We will eventually see that there are several properties of **R** equivalent to the least upper bound property. One of them is the *greatest lower bound property* of **R**.

GLB. (**Greatest Lower Bound**) Every nonempty subset $S \subset \mathbf{R}$ that is bounded below has a greatest lower bound in **R**; that is, there is a real number m such that $m = \inf S$.

Indeed, it is not difficult to show that S is bounded below if and only if the set

$$-S := \{-s \,:\, s \in S\}$$

is bounded above, and in this case we have

$$\inf S = -\sup(-S). \tag{2.1}$$

We conclude that if there exists a supremum for any set that is bounded above, then there exists an infimum for any set that is bounded below. For example, the interval $S = (\pi, 4]$ has the number π as greatest lower bound, even though S has no minimum number. For this set S, note that

$$\inf S = \inf(\pi, 4] = -\sup[-4, -\pi) = -\sup(-S).$$

On the other hand, if S is bounded above then $-S$ is bounded below, and $-\inf(-S) = \sup(S)$. Thus, if there exists an infimum for any set that is bounded below, then there exists a supremum for any set that is bounded above. See Exercise 2.2.4. It is only necessary to assume one of these two properties, and then the other property can be deduced as a theorem, as we have just argued.

We will continue to refer to the least upper bound property of **R** as the Completeness Axiom, the one additional axiom that makes **R** very different from **Q**. The deepest and most interesting results about real valued functions of a real variable depend on the least upper bound property. A study of the consequences of this property will lead us to several more properties of **R** that are equivalent to the least upper bound property, and we should view these equivalent properties as different aspects or different expressions of the completeness of **R**.

The following result is a fundamental characterization of the least upper bound and greatest lower bound for bounded subsets of an ordered field.

Theorem 2.2.4. *Let S be a subset of the ordered field F.*

1. *An upper bound M of S is the least upper bound for S if and only if for every positive element ϵ there is an element $x \in S$ such that $M - \epsilon < x \le M$.*
2. *A lower bound m of S is the greatest lower bound for S if and only if for every positive element ϵ there is an element $y \in S$ such that $m \le y < m + \epsilon$.*

Proof. We prove statement 1 here and leave statement 2 to Exercise 2.2.3.

Suppose $M = \sup S$ and suppose that there exists an $\epsilon_0 \in F$ with $\epsilon_0 > 0$ such that $x \le M - \epsilon_0$ for every $x \in S$. Then $M - \epsilon_0$ is an upper bound for S which is strictly less than M, a contradiction of the definition of M. Thus $M = \sup S$

implies that for every $\epsilon \in F$ with $\epsilon > 0$ there is an element $x \in S$ such that $M - \epsilon < x \leq M$.

Now let M be an upper bound for S such that for every $\epsilon \in F$ with $\epsilon > 0$ there is an element $x \in S$ such that $M - \epsilon < x \leq M$. If M is not the least upper bound for S, then there is an $M_0 < M$ such that M_0 is an upper bound for S, so that $x \leq M_0$ for all $x \in S$. Given $\epsilon = M - M_0$, there is an $x \in S$ such that $M - \epsilon = M - (M - M_0) < x \leq M$, which says that $M_0 < x$, contrary to the assumption that M_0 was an upper bound for S. This completes the proof of statement 1. □

We end the section with some basic results on supremum and infimum.

Theorem 2.2.5. *The following properties hold for the supremum and infimum of subsets of real numbers:*

1. *Let B be a bounded set of real numbers. If $\alpha > 0$, then*

$$\sup\{\alpha b : b \in B\} = \alpha \sup B, \qquad \inf\{\alpha b : b \in B\} = \alpha \inf B. \tag{2.2}$$

2. *If Γ is a given index set and sets $A = \{a_\gamma : \gamma \in \Gamma\}$, $B = \{b_\gamma : \gamma \in \Gamma\}$ are bounded, then*

$$\sup\{a_\gamma + b_\gamma : \gamma \in \Gamma\} \leq \sup A + \sup B, \tag{2.3}$$
$$\inf\{a_\gamma + b_\gamma : \gamma \in \Gamma\} \geq \inf A + \inf B. \tag{2.4}$$

3. *If Γ, Δ are given index sets and $A = \{a_\gamma : \gamma \in \Gamma\}$, $B = \{b_\delta : \delta \in \Delta\}$ are bounded, then*

$$\sup\{a_\gamma + b_\delta : \gamma \in \Gamma, \delta \in \Delta\} = \sup A + \sup B, \tag{2.5}$$
$$\inf\{a_\gamma + b_\delta : \gamma \in \Gamma, \delta \in \Delta\} = \inf A + \inf B. \tag{2.6}$$

Proof. We prove only (2.4) and (2.6) and leave the remaining properties for Exercise 2.2.7.

Proof of (2.4): From the definition of infimum, for every $\gamma \in \Gamma$ we have $a_\gamma \geq \inf A$ and $b_\gamma \geq \inf B$, hence $a_\gamma + b_\gamma \geq \inf A + \inf B$. Now (2.4) follows immediately.

Proof of (2.6): By the definition of $\inf A$ and $\inf B$, the right-hand side of (2.6) must be a lower bound for the set $\{a_\gamma + b_\delta : \gamma \in \Gamma, \delta \in \Delta\}$. Given any $\epsilon > 0$, there is an $a_{\gamma_0} \in A$ such that

$$a_{\gamma_0} < \inf A + \epsilon/2$$

and a $b_{\delta_0} \in B$ such that

$$b_{\delta_0} < \inf B + \epsilon/2.$$

Then $a_{\gamma_0} + b_{\delta_0}$ is an element of $\{a_\gamma + b_\delta : \gamma \in \Gamma, \delta \in \Delta\}$ such that

$$a_{\gamma_0} + b_{\delta_0} < \inf A + \inf B + \epsilon.$$

By Theorem 2.2.4, the greatest lower bound of $\{a_\gamma + b_\delta : \gamma \in \Gamma, \delta \in \Delta\}$ must be $\inf A + \inf B$. □

Exercises.

Exercise 2.2.1. Let S be a subset of an ordered field with S bounded below. Show that if a greatest lower bound for S exists, then it must be unique.

Exercise 2.2.2. Show: If a set contains one of its upper bounds, then that bound must be the supremum of the set. If a set contains one of its lower bounds, then that bound must be the infimum of the set. A finite set contains its supremum and infimum.

Exercise 2.2.3. Prove statement 2 of Theorem 2.2.4.

Exercise 2.2.4. Prove: If S is a set which is bounded above and bounded below, then $\inf(-S) = -\sup S$ and $\sup(-S) = -\inf S$.

Exercise 2.2.5. Prove: If S is nonempty and bounded, then $\inf S \le \sup S$. What must be true of S if S is nonempty and $\inf S = \sup S$?

Exercise 2.2.6. Let $F = \{a + b\sqrt{2} : a, b \in \mathbf{Q}\}$, considered as a subset of $\mathbf{R}$. Show that F is a field which contains an element x such that $x^2 = 2$. (This example should dispel any idea that the real numbers are introduced merely to have a field in which 2 has a positive square root.)

Exercise 2.2.7. Prove the parts of Theorem 2.2.5 not proved in the text.

2.3. The Archimedean Property and Consequences

We have seen that $\mathbf{N} \subset \mathbf{Z} \subset \mathbf{Q} \subset \mathbf{R}$, where $\mathbf{R}$ is the complete ordered field of real numbers. We wish to understand the distribution of the integers and the rationals within the field of real numbers. Understanding the distribution of the integers $\mathbf{Z}$ within any ordered field (and hence within $\mathbf{Q}$ and $\mathbf{R}$) requires no new concepts, so we deal with that issue first.

Lemma 2.3.1. *Let F be an ordered field. For any integer n, there are no integers in the open interval $(n, n+1) \subset F$.*

Proof. For the case $n = 0$, note that every positive integer $k \in F$ satisfies $k \ge 1$. (This can be established by induction.) So the interval $(0,1)$ contains no positive integer, and since it clearly contains no negative integer, the interval $(0,1)$ contains no integer at all. Now for the case of general n. Suppose there exists an integer k in the interval $(n, n+1)$. Then

$$n < k < n+1 \quad \Longrightarrow \quad 0 < k - n < 1.$$

Thus $k - n$ is an integer in the interval $(0,1)$, which contradicts the conclusion we just reached above. The contradiction is due to the assumption that there existed an integer k in the interval $(n, n+1)$. So the interval $(n, n+1)$ contains no integers. □

In $\mathbf{R}$, any bounded subset S consisting only of integers must contain both $\sup S$ and $\inf S$, as the next lemma shows.

Lemma 2.3.2. *If $S \subset \mathbf{Z} \cap \mathbf{R}$ and S is bounded above, then S contains a maximum element; that is, $\sup S \in S$. If $S \subset \mathbf{Z} \cap \mathbf{R}$ is bounded below, then S contains a minimum element; that is, $\inf S \in S$.*

Proof. If $S \subset \mathbf{Z}$ is bounded above, then $m = \sup S \in \mathbf{R}$ exists by the completeness axiom. By definition of supremum, $m-1$ is not an upper bound for S, so there is an integer n in S such that $m-1 < n \leq m$. Then $m < n+1$, and since m is an upper bound for S, we must have $S \subset (-\infty, m]$. By Lemma 2.3.1, there is no integer in the interval $(n, n+1)$, so there is no element of S in $(n, n+1)$. Since $n \leq m < n+1$, n itself is an upper bound for S, and therefore we have $n = m = \sup S \in S$.

If $S \subset \mathbf{Z}$ is bounded below, then $-S = \{-s : s \in S\}$ is bounded above, hence $\sup(-S) = -s_0 \in -S$ for some $s_0 \in S$, by the argument above. But then

$$\inf S = -\sup(-S) = -(-s_0) = s_0,$$

which shows that S contains its minimum element. $\square$

In general, the supremum of a set that is bounded above need not be an element of that set, since the set need not contain a maximum element. The infimum of a set that is bounded below need not be an element of that set, since the set need not contain a minimum element. Remember that Lemma 2.3.2 is about bounded sets of *integers*.

There is a property of $\mathbf{Q}$ that is also possessed by $\mathbf{R}$. This is the *Archimedean property*, which holds for some, but not all, ordered fields. The Archimedean property is the key to understanding the distribution of $\mathbf{Q}$ within $\mathbf{R}$.

Definition 2.3.3. *An ordered field F is called* **Archimedean** *if for every $x \in F$ there is an $n \in \mathbf{N}$ such that $n > x$.*

The Archimedean property asserts, for those fields F that have it, that the set $\mathbf{N}$ within F is unbounded. There is no upper bound for $\mathbf{N}$ in an Archimedean field.

Proposition 2.3.4. *The rational field $\mathbf{Q}$ is Archimedean.*

Proof. It is sufficient to consider positive $x = p/q \in \mathbf{Q}$ with both $p > 0$ and $q > 0$. Since $q \geq 1$, we have $x = p/q \leq p$, and hence $x = p/q < p+1 \in \mathbf{N}$. $\square$

Example 2.3.5. Let $S = \{\frac{1}{2}, \frac{1}{3}, \frac{1}{4}, \ldots, \frac{1}{n}, \ldots\}$ be considered as a subset of the rational field $\mathbf{Q}$. Then $\inf S$ exists and $\inf S = 0$. In order to see this, we use the Archimedean property of $\mathbf{Q}$ and apply statement 1 of Theorem 2.2.4. For every $\epsilon \in \mathbf{Q}$ with $\epsilon > 0$, there is an element $1/n \in S$ such that $0 < 1/n < 0 + \epsilon$; this is true since for any given $\epsilon \in \mathbf{Q}$ with $\epsilon > 0$ there is an $n \in \mathbf{N}$ such that $n > 1/\epsilon$. Note that $0 = \inf S \notin S$. $\triangle$

The Archimedean property of $\mathbf{Q}$ is an easy consequence of the order properties in $\mathbf{Q}$. We want to show that $\mathbf{R}$ is also Archimedean; in fact, it will be essential for us to know this. The Archimedean property of $\mathbf{R}$ follows from the least upper bound property of $\mathbf{R}$ and Lemma 2.3.2.

Theorem 2.3.6. *The field $\mathbf{R}$ of real numbers is Archimedean.*

Proof. The proof is by contradiction. Suppose that $\mathbf{R}$ is not Archimedean. Then there is an element $x \in \mathbf{R}$ such that $n \leq x$ for all $n \in \mathbf{N}$. Then $\mathbf{N}$ is a bounded subset of $\mathbf{R}$ and x is an upper bound for $\mathbf{N}$. By Lemma 2.3.2,

$$n_0 := \sup \mathbf{N} \in \mathbf{N},$$

and therefore $n \le n_0$ for every $n \in \mathbf{N}$. But $n_0 + 1 \in \mathbf{N}$, and $n_0 + 1 > n_0$, which is a contradiction of the fact that $n_0 = \sup \mathbf{N}$. Thus $\mathbf{R}$ is Archimedean. $\square$

It is the Archimedean property of $\mathbf{R}$ that allows us to make the following statement: Given any real $\epsilon > 0$ there exists a positive integer n such that $1/n < \epsilon$. (Given $\epsilon > 0$, there is an n such that $1/\epsilon < n$ by the Archimedean property, and hence $1/n < \epsilon$.)

It is clear from Lemma 2.3.1 that if n is an *integer*, then the real interval $[n, n+1)$ contains exactly one integer, namely n. It is also true that for *any real number* α, the real interval $[\alpha, \alpha + 1)$ contains exactly one integer. The proof calls on the Archimedean property of the real numbers.

Theorem 2.3.7. *For any real number α, the interval $[\alpha, \alpha + 1)$ contains exactly one integer.*

Proof. Let
$$S = \{n \in \mathbf{Z} : n < \alpha + 1\}.$$
Since $-(\alpha + 1)$ is a real number, the Archimedean property implies that there is a positive integer n such that $n > -(\alpha + 1)$, and thus $-n < \alpha + 1$. Since $-n$ is an integer, S is nonempty. Since S is bounded above by $\alpha + 1$, Lemma 2.3.2 implies there is a maximum member m of S. If $m < \alpha$, then $m + 1 < \alpha + 1$, so $m + 1 \in S$, which contradicts the fact that m is the largest member of S. Thus, $m \ge \alpha$ and the integer m satisfies $\alpha \le m < \alpha + 1$.

Now suppose there are integers m_1 and m_2 with $\alpha \le m_1 < m_2 < \alpha + 1$. Then $m_2 - m_1 > 0$. Since $m_1 \ge \alpha$ and $m_2 < \alpha + 1$,
$$0 < m_2 - m_1 < (\alpha + 1) - \alpha = 1,$$
which says that $m_2 - m_1$ is an integer in the interval $(0, 1)$, a contradiction of Lemma 2.3.1. So there is exactly one integer in the interval $[\alpha, \alpha + 1)$. $\square$

We can now describe the distribution of the rational numbers and the irrational numbers along the real number line.

Definition 2.3.8. *A subset S of the real numbers is* **dense** *in* $\mathbf{R}$ *if for any two real numbers $a < b$, there is an $s \in S$ such that $a < s < b$.*

A statement equivalent to Definition 2.3.8 is that a subset $S \subseteq \mathbf{R}$ is dense in $\mathbf{R}$ if every nonempty open interval $(a, b) \subseteq \mathbf{R}$ intersects S.

We will prove the existence of the real number square roots $\sqrt{2}$ and $-\sqrt{2}$ later in the section; we know that they are irrational numbers. The next theorem shows that the set $\mathbf{Q}$ of rational numbers is dense in $\mathbf{R}$, and that as a consequence, the set $\mathbf{I}$ of irrational numbers is also dense in $\mathbf{R}$.

Theorem 2.3.9. *The complete ordered field of real numbers has the following properties:*

1. *(The Density of Rationals): For any two real numbers $a < b$, there is a rational number q such that $a < q < b$.*
2. *(The Density of Irrationals): For any two real numbers $a < b$, there is an irrational number x such that $a < x < b$.*

Proof. (The Density of Rationals): Given real numbers a and b with $a < b$, let $\delta = b - a > 0$. By the Archimedean property, there is a natural number n such that $n(b-a) > 1$. By Theorem 2.3.7 there is an integer m in the interval $[nb-1, nb)$. Since $n \neq 0$,

$$nb - 1 \leq m < nb \quad \Longrightarrow \quad b - 1/n \leq m/n < b.$$

Since $1/n < b - a$, $-1/n > a - b$, so that

$$a = b + (a - b) < b - 1/n \leq m/n < b,$$

which says that the rational number m/n is in the interval (a, b).

(The Density of Irrationals): The density of the irrationals follows from the density of the rationals together with the fact that irrational numbers exist. To see this, let a, b satisfy $a < b$. Now $\sqrt{2} > 0$, hence $1/\sqrt{2} > 0$, so $a/\sqrt{2} < b/\sqrt{2}$. By the density of the rationals, there is a rational number q such that $a/\sqrt{2} < q < b/\sqrt{2}$. Hence, $a < q\sqrt{2} < b$. The number $q\sqrt{2}$ must be irrational, since it is the product of a rational and an irrational number. □

The proof of the density of the rationals and the irrationals should make us pause and ponder the result. It is impossible for us to minimize the importance of the Archimedean property or the existence of even a single irrational number.

A few final thoughts on the Archimedean property are in order before we move on to the absolute value function and its basic properties.

An Archimedean ordered field need not be complete. The rational field $\mathbf{Q}$ is an example.

There exist ordered fields that are not Archimedean. They play no role in this book. However, the interested reader can see Exercise 2.3.2 for an example.

We now discuss the absolute value function on an ordered field F. The *absolute value* of $a \in F$, denoted $|a|$, is defined by

$$|a| := \begin{cases} a & \text{if } a \geq 0, \\ -a & \text{if } a < 0. \end{cases}$$

By considering the cases $a < 0$ and $a \geq 0$, it should be clear that we may write $\pm a \leq |a|$ for any $a \in F$. The next theorem lists the main properties of the absolute value function.

Theorem 2.3.10. *The absolute value in an ordered field F has the following properties for $a, b \in F$:*

1. $|a| \geq 0$, *and* $|a| = 0$ *if and only if* $a = 0$;
2. $|ab| = |a||b|$;
3. $|a + b| \leq |a| + |b|$ *(the triangle inequality);*
4. $\big||a| - |b|\big| \leq |a + b|$ *(the reverse triangle inequality).*

Proof. 1. If $a = 0$, then $|a| = 0$, by definition. If $a \neq 0$, then $|a| > 0$, by the two cases in the definition. Now given that $|a| \geq 0$, the contrapositive of the last statement amounts to saying that if $|a| = 0$, then $a = 0$.

2. Consider cases. If $a > 0$ and $b > 0$, then $ab > 0$, hence $|ab| = ab = |a||b|$. If either $a = 0$ or $b = 0$, then $ab = 0$, so $|ab| = 0 = |a||b|$. If $a < 0$ and $b > 0$, then $ab < 0$, so $|ab| = -ab = |a||b|$. A similar argument applies if $a > 0$ and $b < 0$.

3. If $a + b > 0$, then $|a + b| = a + b \leq |a| + |b|$ since $a \leq |a|$ and $b \leq |b|$. If $a + b < 0$, then $|a + b| = -(a + b) = -a - b \leq |a| + |b|$ since $-a \leq |a|$ and $-b \leq |b|$. If $a + b = 0$, then $|a + b| = 0 \leq |a| + |b|$ since $|a| \geq 0$ and $|b| \geq 0$.

4. We have $a = a + b - b$, so $|a| \leq |a + b| + |-b| = |a + b| + |b|$, using statement 3. Consequently, $|a| - |b| \leq |a + b|$. Now reverse the roles of a and b, and write $b = b + a - a$. Then $|b| \leq |a + b| + |a|$, and hence $|b| - |a| \leq |a + b|$. The two results together yield $||a| - |b|| \leq |a + b|$. □

Theorem 2.3.10 applies to $\mathbf{R}$ since $\mathbf{R}$ is ordered.

We have seen that there is no *rational* number x such that $x^2 = 2$. The completeness of the real number field guarantees that there is a real number solution to this equation. The proof of this result uses the algebraic properties of a field and the Archimedean property of the real numbers.

Theorem 2.3.11. *There exists a positive real number x such that $x^2 = 2$.*

Proof. Define the set

$$S = \{s \in \mathbf{R} : s \geq 0 \text{ and } s^2 \leq 2\}.$$

S is nonempty, since $1 \in S$. S is bounded above by 3, for if $s \in S$ and $s > 3$, then $s^2 > 3^2 = 9$, a contradiction of $s^2 \leq 2$. Therefore $s \in S$ implies $s \leq 3$. We want to show that $\sup S$ is the desired positive square root of 2. Now let $r = \sup S$, which exists by the least upper bound property. The three possibilities are that $r^2 < 2$, $r^2 > 2$, or $r^2 = 2$. We show that the first two options cannot occur.

Suppose that $r^2 < 2$, and let $\delta = 2 - r^2 > 0$. In order to reach a contradiction, we want to show that there is a positive integer m such that $(r + \frac{1}{m})^2 < 2$. By the binomial theorem, for any positive integer m we have

$$\left(r + \frac{1}{m}\right)^2 = r^2 + 2r\frac{1}{m} + \frac{1}{m^2}.$$

By the Archimedean property of $\mathbf{R}$, there exists a positive integer m such that

$$\frac{2r}{\delta/2} < m \quad \text{and} \quad \frac{1}{\delta/2} < m < m^2.$$

Hence,

$$2r\frac{1}{m} + \frac{1}{m^2} < \frac{\delta}{2} + \frac{\delta}{2} = \delta, \tag{2.7}$$

and we have

$$\left(r + \frac{1}{m}\right)^2 = r^2 + 2r\frac{1}{m} + \frac{1}{m^2} < r^2 + \delta = 2.$$

But this says that $r + \frac{1}{m} \in S$ and thus contradicts $r = \sup S$. We may conclude that $r^2 \geq 2$.

If $r^2 > 2$, then we let $\delta = r^2 - 2 > 0$. One can then show that there is a positive integer m such that $(r - \frac{1}{m})^2 > 2$: To see this, note that for any positive integer m,

$$\left(r - \frac{1}{m}\right)^2 = r^2 - 2r\frac{1}{m} + \frac{1}{m^2} > r^2 - \left(2r\frac{1}{m} + \frac{1}{m^2}\right),$$

and, as above, we may choose m such that (2.7) holds, and hence

$$\left(r - \frac{1}{m}\right)^2 \geq r^2 - \left(2r\frac{1}{m} + \frac{1}{m^2}\right) > r^2 - \delta = 2.$$

But this implies that $r - \frac{1}{m}$ is an upper bound for S (for if not, then there is an $s \in S$ such that $r - \frac{1}{m} < s$ and then $(r - \frac{1}{m})^2 < s^2 \leq 2$). However, $r - \frac{1}{m} < r$, and r is the least upper bound for S. This contradiction rules out the option $r^2 > 2$.

The only remaining possibility is that $r^2 = 2$. This proves the existence of a positive square root of 2. Uniqueness follows easily from the fact that $0 < x < y$ implies $x^2 < y^2$. $\square$

The *positive* real number x such that $x^2 = 2$ is not rational. Note that $-x$ also satisfies $(-x)^2 = 2$, since $(-x)(-x) = x^2 = 2$. Of course we write $x = \sqrt{2}$ and $-x = -\sqrt{2}$. (The use of the radical sign as a root notation is defined more generally by a Remark after Theorem 2.3.12.) Theorem 2.3.11 confirms the existence of real numbers that are not rational, as $\sqrt{2}$ and $-\sqrt{2}$ are not rational. If x is a real number and x is not rational, then x is called an **irrational** number. (We anticipated this proof of existence of irrationals in Theorem 2.3.9 when we proved the density of the rationals and the irrationals.) We denote the set of irrational real numbers by $\mathbf{I}$. Thus, $\mathbf{I} = \mathbf{R} - \mathbf{Q}$ and $\mathbf{R} = \mathbf{Q} \cup \mathbf{I}$.

The argument given to prove Theorem 2.3.11 actually shows that if F is an ordered field, and

$$S = \{s \in F : \text{either } s < 0, \text{ or } s \geq 0 \text{ and } s^2 \leq 2\},$$

and *if* $\sup S$ *exists in* F, then $(\sup S)^2 = 2$. However, since we have shown that there is no rational number x such that $x^2 = 2$, it follows that the bounded set $S \subset \mathbf{Q} = F$ does not have a least upper bound in $\mathbf{Q}$. Thus $\mathbf{Q}$ is not complete; $\mathbf{Q}$ does not have the least upper bound property.

We now apply the least upper bound property of the field of real numbers to prove the existence of n-th roots.

Theorem 2.3.12. *If a is a real number such that $a \geq 0$ and n is a positive integer, then there exists a unique real number $r > 0$ such that $r^n = a$.*

Proof. The unique n-th root of $a = 0$ is, of course, $r = 0$. Now assume $a > 0$. Define the set

$$S = \{s \in \mathbf{R} : s \geq 0 \text{ and } s^n \leq a\}.$$

We want to show that S is bounded above and $\sup S$ is the desired n-th root of a. The number $a + 1$ is an upper bound for S. For if not, then there is an $s \in S$ such that $s > a + 1$, which implies $s^n > (a + 1)^n \geq 1 + na > a$ by Bernoulli's inequality, and this contradicts the definition of S. Now let $r = \sup S$. The three possibilities are that $r^n < a$, $r^n > a$, or $r^n = a$. We show that the first two options cannot occur.

Suppose that $r^n < a$, and let $\delta = a - r^n > 0$. In order to reach a contradiction, we want to show that there is a positive integer m such that $(r + \frac{1}{m})^n < a$. By the binomial theorem, for any positive integer m we have

$$\left(r + \frac{1}{m}\right)^n = \sum_{k=0}^{n} \binom{n}{k} r^k \frac{1}{m^{n-k}} = r^n + \sum_{k=0}^{n-1} \binom{n}{k} r^k \frac{1}{m^{n-k}}.$$

By the Archimedean property of $\mathbf{R}$, for each integer k in the range $0 \le k \le n-1$ there exists a positive integer m_k such that

$$\frac{\binom{n}{k} r^k}{\delta/n} < m_k < m_k^{n-k}.$$

Hence, for $k = 0, 1, \ldots, n-1$,

$$\binom{n}{k} r^k \frac{1}{m_k^{n-k}} < \frac{\delta}{n}.$$

By choosing $m = \max\{m_0, m_1, \ldots, m_{n-1}\}$, we have

$$\left(r + \frac{1}{m}\right)^n = r^n + \sum_{k=0}^{n-1} \binom{n}{k} r^k \frac{1}{m^{n-k}} \le r^n + n\frac{\delta}{n} = r^n + \delta = a.$$

But this says that $r + \frac{1}{m} \in S$, and thus contradicts $r = \sup S$. We may conclude that $r^n \ge a$.

If $r^n > a$, then we let $\delta = r^n - a > 0$. One can then show that there is a positive integer m such that $(r - \frac{1}{m})^n > a$. (See Exercise 2.3.6.) But this implies that $r - \frac{1}{m}$ is an upper bound for S (for if not, then there is an $s \in S$ such that $r - \frac{1}{m} < s$ and then $(r - \frac{1}{m})^n < s^n \le a$). However, $r - \frac{1}{m} < r$, and r is the least upper bound for S. This contradiction rules out the option $r^n > a$.

The only remaining possibility is that $r^n = a$. This proves the existence of a positive n-th root of a. Uniqueness follows easily from the fact that $0 < x < y$ implies $x^n < y^n$. □

Remark (Radical Sign Notation for n-th Roots). *The unique n-th root of a number $a \ge 0$ may be written using the common radical sign notation as $\sqrt[n]{a}$.*

Corollary 2.3.13. *If $a < 0$ and n is an odd positive integer, then there exists a unique real number $b < 0$ such that $b^n = a$.*

Proof. The positive number $|a| = -a$ has a unique n-th root r, with $r^n = |a| = -a$ and $r > 0$. Let $b = -r$. Then $b^n = (-1)^n r^n = -r^n = -(-a) = a$. The proof of uniqueness is left to the reader. □

We summarize the discussion so far with the following definition.

Definition 2.3.14 (Roots). *Let n be a positive integer.*

1. *If $x \ge 0$, then we define $x^{1/n}$ to be the unique nonnegative real number y such that $y^n = x$.*
2. *If x is real and n is odd, then $x^{1/n}$ is the unique real number y such that $y^n = x$.*

We can now define power expressions x^r for rational r. If x is real and n is a positive integer, then, provided $x^{1/n}$ is defined and $x^{1/n} \neq 0$, we write

$$x^{-1/n} = \frac{1}{x^{1/n}}.$$

If x is a real number and r is rational, say $r = p/q$ where $p, q \in \mathbf{Z}$ and p and q have no common factors other than 1 (so r is in *lowest terms*), we define

$$x^r = (x^{1/q})^p,$$

provided $x^{1/q}$ is defined. With these definitions, the usual laws of rational exponents hold. Instead of listing these laws here, the reader can view them in Section 7.5, where the definition of b^r, for $b > 0$, is extended to arbitrary real exponents r.

We turn to the complex field $\mathbf{C}$ for a moment. Even though $\mathbf{C}$ is not an ordered field, there is a very useful function on $\mathbf{C}$ that has the same basic properties as the absolute value of real numbers. This absolute value function for $\mathbf{C}$ must be defined without reference to an order relation. If $z = a + bi \in \mathbf{C}$, then define

$$|z| = |a + bi| = \sqrt{a^2 + b^2}.$$

The real number $|z|$ is called the *absolute value* of z (or the *modulus* of z). If we think of $z = a + bi$ as the point (a, b) in the coordinate plane, then $|z|$ is the Euclidean distance from (a, b) to the origin. If z is real, that is, $z = a + 0i$, then one can verify that $|z| = \sqrt{a^2} = |a|$.

The *conjugate* of the complex number $z = a + bi$ is denoted $\bar{z}$ and is defined by $\bar{z} = a - bi$. Geometrically, $\bar{z}$ is the reflection of z in the x-axis of the coordinate plane. If $z = a + bi$, then

$$z\bar{z} = (a + bi)(a - bi) = a^2 + b^2 + 0i = a^2 + b^2.$$

Consequently, if $z = a + bi \neq 0$, then

$$z^{-1} = \frac{1}{z} = \frac{\bar{z}}{z\bar{z}} = \frac{a - bi}{a^2 + b^2} = \frac{a}{a^2 + b^2} - \frac{bi}{a^2 + b^2},$$

which confirms the formula stated in Example 2.1.4 for the multiplicative inverse of a nonzero complex number.

Before proceeding, we prove explicitly a simple result that will be used repeatedly.

Lemma 2.3.15. *Let L be a real or complex number. If $|L| \leq \epsilon$ for every $\epsilon > 0$, then $L = 0$.*

Proof. Clearly $|L| \geq 0$ by the definition of absolute value. If $|L| > 0$, then take $\epsilon = |L|/2$. By hypothesis, $|L| \leq |L|/2$, hence $2|L| \leq |L|$, but $|L| > 0$ implies that $2 \leq 1$, which is a contradiction. Hence $|L| = 0$, so $L = 0$. □

Exercises.

Exercise 2.3.1. Prove: Given any real number $\epsilon > 0$ there exists a positive integer n such that $1/2^n < \epsilon$.

Exercise 2.3.2. Let $\mathbf{Q}[x]$ denote the collection of rational expressions of the form $f(x)/g(x)$, where

$$f(x) = a_n x^n + a_{n-1}x^{n-1} + \cdots + a_1 x + a_0, \quad \text{where } n \in \mathbf{N},$$
$$g(x) = b_m x^m + b_{m-1}x^{m-1} + \cdots + b_1 x + b_0, \quad \text{where } m \in \mathbf{N}, \quad b_m \neq 0,$$

$a_i, b_j \in \mathbf{Q}$, and $f(x), g(x)$ have no linear factor in common.

(a) Show that $\mathbf{Q}[x]$ is a field.

(b) Show that $\mathbf{Q}[x]$ is ordered by the positive set P consisting only of those elements $f(x)/g(x)$ for which the product of the lead coefficients of $f(x)$ and $g(x)$ is positive, that is, $a_n b_m > 0$.

(c) Show that $\mathbf{Q}[x]$ does not have the Archimedean property.

Exercise 2.3.3. Show that for all a, b in an ordered field, $|a - b| \leq |a| + |b|$ and $-|a| \leq a \leq |a|$. Given that $-|a| \leq a \leq |a|$ and $-|b| \leq b \leq |b|$, add the corresponding parts of these inequalities to obtain

$$-|a| - |b| \leq a + b \leq |a| + |b|,$$

and show that this gives another proof of the triangle inequality.

Exercise 2.3.4. Show that properties 1-4 of Theorem 2.3.10 hold for the absolute value of *complex* numbers.

Exercise 2.3.5. Prove by induction: For any positive integer n and any real numbers $a_1, a_2, \ldots, a_n$, $|a_1 + a_2 + \cdots + a_n| \leq |a_1| + |a_2| + \cdots + |a_n|$.

Exercise 2.3.6. Show that if $r^n > a$, then there is a positive integer m such that $(r - \frac{1}{m})^n > a$. *Hint*: For any positive integer m,

$$\left(r - \frac{1}{m}\right)^n = \sum_{k=0}^{n} \binom{n}{k} r^k \frac{(-1)^{n-k}}{m^{n-k}} \geq r^n - \sum_{k=0}^{n-1} \binom{n}{k} r^k \frac{1}{m^{n-k}}.$$

Exercise 2.3.7. Prove: If $x \in \mathbf{Q}$ and $y \in \mathbf{I}$, then $x + y \in \mathbf{I}$. If $x \in \mathbf{Q}$, $x \neq 0$, and $y \in \mathbf{I}$, then $xy \in \mathbf{I}$.

2.4. Sequences

Many of the basic concepts of analysis involve processes of approximation. A fundamental tool for discussing approximation is the concept of a sequence, and a solid understanding of the convergence of sequences will enable us to understand other limit processes that occur later.

Recall the definition of a sequence in a set X (Definition 1.3.6). A **sequence** in a field F is a function $a : \mathbf{N} \to F$. A sequence will generally be indicated by writing (a_k). (Recall that we distinguish the sequence (a_k) from the range of the sequence, $\{a(k) : k \in \mathbf{N}\}$.)

We present the common basic properties of sequences in any of the fields of essential interest to us: $\mathbf{Q}$, $\mathbf{R}$ and $\mathbf{C}$. Most of the properties we present for sequences also hold for sequences in $\mathbf{C}$, since they depend on an absolute value function to measure distance between elements. We just need to remember that $\mathbf{C}$ is not ordered.

The values of a sequence inherit a natural ordering from the ordering of the natural numbers. The notation a_k indicates the particular value of the k-th element of a sequence with respect to this ordering, whereas the full ordered sequence is indicated by writing (a_k); however, we may occasionally write "the sequence a_k" and then the meaning is clear. More complete notations, such as (a_k), $(a_k)_{k=1}^{\infty}$ or $(a_k)_1^{\infty}$, may be used for sequences if it is essential to indicate the starting index.

Definition 2.4.1. *A sequence (a_k) in $\mathbf{R}$ has* **limit** *$L \in \mathbf{R}$ if for every real $\epsilon > 0$ there exists a natural number $M = M(\epsilon)$ such that if $k \geq M$, then $|a_k - L| < \epsilon$. We indicate the limit by writing $\lim_{k\to\infty} a_k = L$, and we also say that the sequence* **converges** *to L or is* **convergent** *(to L).*

Although the complex field $\mathbf{C}$ is not an ordered field, the absolute value function allows a similar definition for complex number sequences: A sequence (a_k) in $\mathbf{C}$ has limit $L \in \mathbf{C}$ if for every real $\epsilon > 0$ there exists a natural number $M = M(\epsilon)$ such that if $k \geq M$, then $|a_k - L| < \epsilon$.

It is essential to know that if a real or complex sequence has a limit, then the limit is uniquely determined.

Theorem 2.4.2. *A sequence of real or complex numbers that converges has a unique limit.*

Proof. Let $(a_k)_{k=1}^{\infty}$ be a real or complex sequence that converges to a limit L_1, that is, $\lim_{k\to\infty} a_k = L_1$. Suppose there is another number L_2 such that $\lim_{k\to\infty} a_k = L_2$. We want to show that $L_1 = L_2$.

Choose any $\epsilon > 0$. Since $\lim_{k\to\infty} a_k = L_1$, there is an $N_1(\epsilon/2) > 0$ such that

$$|a_k - L_1| < \epsilon/2 \quad \text{for all } k \geq N_1(\epsilon/2).$$

Since $\lim_{k\to\infty} a_k = L_2$, there is an $N_2(\epsilon/2) > 0$ such that

$$|a_k - L_2| < \epsilon/2 \quad \text{for all } k \geq N_2(\epsilon/2).$$

Then for all $k \geq N(\epsilon) := \max\{N_1(\epsilon/2), N_2(\epsilon/2)\}$,

$$|L_1 - L_2| = |L_1 - a_k + a_k - L_2| \leq |L_1 - a_k| + |a_k - L_2| < \epsilon/2 + \epsilon/2 = \epsilon.$$

Since $\epsilon > 0$ was arbitrary, we conclude that $|L_1 - L_2| = 0$, by Lemma 2.3.15. □

Instead of writing $\lim_{k\to\infty} a_k = L$, we may sometimes indicate this limit by writing

$$a_k \to L \quad \text{as} \quad k \to \infty,$$

or sometimes simply by $a_k \to L$, when it is understood that $k \to \infty$. If a sequence has no limit, then we say that the sequence **diverges**, or is a **divergent** sequence.

Here is a simple, but very important, example of a convergent sequence.

Example 2.4.3. The real number sequence $(1/n)$, $n \in \mathbf{N}$ converges with limit 0. We must use the Archimedean property of $\mathbf{R}$. Given $\epsilon > 0$, there is an $M \in \mathbf{N}$ such that $1/M < \epsilon$, because $1/\epsilon > 0$, and thus there is an $M \in \mathbf{N}$ such that $M > 1/\epsilon$ by the Archimedean property. Then for $k \geq M$ we have $|1/k - 0| = 1/k < 1/M < \epsilon$. Since $\epsilon > 0$ was arbitrary, $\lim_{n\to\infty} 1/n = 0$. △

We say that a set S of real numbers is **bounded** if there is a real number M such that $|x| \leq M$ for all $x \in S$. A sequence (a_k) in $\mathbf{R}$ or $\mathbf{C}$ is **bounded** if the range of the sequence is a bounded set. Thus (a_k) is a bounded sequence if and only if there is a real number M such that $|a_k| \leq M$ for each k.

Theorem 2.4.4. *A sequence of real or complex numbers that converges is bounded.*

Proof. Suppose (a_k) is a sequence that converges to the limit L. Consider a specific value of $\epsilon > 0$, say $\epsilon = 1$. There is an $N(\epsilon) = N(1)$ such that

$$|a_k - L| < 1 \quad \text{for all } k \geq N(1).$$

Since $|a_k| - |L| \leq |a_k - L|$ we have

$$|a_k| \leq |L| + 1 \quad \text{for all } k \geq N(1).$$

Now let

$$M := \max\left\{|L| + 1, |a_1|, |a_2|, \ldots, |a_{N(1)}|\right\}.$$

Then for each $k \in \mathbf{N}$, $|a_k| \leq M$, so the sequence (a_k) is bounded. □

Limit calculations routinely use the following results on constant multiples, sums, products, and quotients of convergent sequences.

Theorem 2.4.5. *Let $(a_k)_1^\infty$ and $(b_k)_1^\infty$ be convergent sequences (in $\mathbf{R}$ or $\mathbf{C}$), with*

$$\lim_{k\to\infty} a_k = a \quad \text{and} \quad \lim_{k\to\infty} b_k = b.$$

Then the following statements are true:

1. *If c is any real or complex constant, then the sequence $c(a_k) := (ca_k)$ is convergent, and*

$$\lim_{k\to\infty} ca_k = ca = c \lim_{k\to\infty} a_k.$$

2. *The sequence $(a_k + b_k)_1^\infty$ is convergent, and*

$$\lim_{k\to\infty} (a_k + b_k) = a + b = \lim_{k\to\infty} a_k + \lim_{k\to\infty} b_k.$$

3. *The sequence $(a_k b_k)_1^\infty$ is convergent, and*

$$\lim_{k\to\infty} (a_k b_k) = ab = [\lim_{k\to\infty} a_k][\lim_{k\to\infty} b_k].$$

4. *If $b \neq 0$, then the sequence $(a_k/b_k)_1^\infty$ is convergent, and*

$$\lim_{k\to\infty} \frac{a_k}{b_k} = \frac{a}{b} = \frac{\lim_{k\to\infty} a_k}{\lim_{k\to\infty} b_k}.$$

Proof. We prove statements 2 and 3 here, and leave statements 1 and 4 to Exercise 2.4.1.

2. Let $\epsilon > 0$. By hypothesis, there is an $N_1 = N_1(\epsilon)$ such that if $k \geq N_1$, then $|a_k - a| < \epsilon/2$, and there is an $N_2 = N_2(\epsilon)$ such that if $k \geq N_2$, then $|b_k - b| < \epsilon/2$. Thus, if $k \geq \max\{N_1, N_2\}$, then

$$\begin{aligned} |a_k + b_k - (a + b)| &= |a_k - a + b_k - b| \\ &\leq |a_k - a| + |b_k - b| < \frac{\epsilon}{2} + \frac{\epsilon}{2} = \epsilon. \end{aligned}$$

Since $\epsilon > 0$ is arbitrary, this shows that $\lim_{k\to\infty}(a_k + b_k) = a + b$.

3. By Theorem 2.4.4, there are bounds $M_1 > 0$, $M_2 > 0$ such that $|a_k| \leq M_1$ and $|b_k| \leq M_2$ for all k. We estimate as follows:

$$\begin{aligned} |a_k b_k - ab| &= |a_k b_k - ab_k + ab_k - ab| \\ &\leq |a_k - a||b_k| + |a||b_k - b| \\ &\leq |a_k - a| M_2 + M_1 |b_k - b|. \end{aligned}$$

Given $\epsilon > 0$, there is an $N_1 = N_1(\epsilon)$ such that $k \geq N_1$ implies $|a_k - a| < \epsilon/(2M_2)$, and there is an $N_2 = N_2(\epsilon)$ such that $k \geq N_2$ implies $|b_k - b| < \epsilon/(2M_1)$. Then $k \geq \max\{N_1, N_2\}$ implies

$$|a_k b_k - ab| < \frac{\epsilon}{2M_2} M_2 + M_1 \frac{\epsilon}{2M_1} = \epsilon.$$

Therefore $\lim_{k\to\infty}(a_k b_k) = ab = [\lim_{k\to\infty} a_k][\lim_{k\to\infty} b_k]$. □

Limit computations often require some preliminary estimates. A comparison result such as the following one is often useful.

Theorem 2.4.6 (The Squeeze Theorem). *Let (a_k), (b_k) and (c_k) be sequences of real numbers such that for each k,*

$$a_k \leq b_k \leq c_k.$$

If $\lim_{k\to\infty} a_k = \lim_{k\to\infty} c_k = L$ for some real number L, then

$$\lim_{k\to\infty} b_k = L.$$

Proof. Write $b_k - L = b_k - a_k + a_k - L$. For each k, $a_k \leq b_k \leq c_k$, hence

$$|b_k - L| \leq |b_k - a_k| + |a_k - L| \leq |c_k - a_k| + |a_k - L|.$$

Since $c_k - a_k = c_k - L + L - a_k$, we may also write

$$|b_k - L| \leq |c_k - L| + 2|a_k - L|.$$

Given $\epsilon > 0$, there exists M_1 such that $k \geq M_1$ implies $|c_k - L| < \epsilon/2$, and there exists M_2 such that $k \geq M_2$ implies $|a_k - L| < \epsilon/4$. If $k \geq \max\{M_1, M_2\}$, then

$$|b_k - L| \leq |c_k - L| + 2|a_k - L| < \epsilon/2 + 2\epsilon/4 = \epsilon.$$

Consequently, $\lim_{k\to\infty} |b_k - L| = 0$, and hence $\lim_{k\to\infty} b_k = L$. □

The squeeze theorem for real sequences is quite useful.

Example 2.4.7. For each $k \in \mathbf{N}$, $0 < 1/(1+k) < 1/k$, and since $1/k \to 0$ as $k \to \infty$, the squeeze theorem implies that $\lim_{k\to\infty} 1/(1+k) = 0$. △

Example 2.4.8. Suppose we wish to determine the limit indicated here:

$$\lim_{k\to\infty} \frac{k + 3\sin k}{2 + k^2}.$$

Note that for each $k \in \mathbf{N}$,

$$0 \leq \left| \frac{k + 3\sin k}{2 + k^2} \right| \leq \frac{k + 3|\sin k|}{2 + k^2} < \frac{k+3}{k^2} = \frac{1}{k} + \frac{3}{k^2}.$$

Since $\lim_{k\to\infty} 1/k = 0 = \lim_{k\to\infty} 3/k^2$, the squeeze theorem implies that the indicated limit is 0. △

The ordering of the real numbers plays an important role in the squeeze theorem for limits of real sequences. Since $\mathbf{C}$ is not an ordered field, in general it is only the absolute values of complex numbers that obey order relations inherited from $\mathbf{R}$. Is there a squeeze theorem for complex sequences? In other words, if (a_k), (b_k), and (c_k) are complex sequences with

$$|a_k| \le |b_k| \le |c_k| \quad \text{for each } k,$$

and there exists a complex number L such that $\lim_{k\to\infty} a_k = L = \lim_{k\to\infty} c_k$, is it true that $\lim_{k\to\infty} b_k = L$? The answer is negative; Exercise 2.4.2 requests a counterexample.

A *subsequence* of a given sequence (a_k) is a sequence obtained from (a_k) by deleting some terms from (a_k) and then reindexing the remaining terms such that the k-th term that remains had index n_k in the original sequence. Thus if the subsequence is (b_k), we have

$$b_1 = a_{n_1}, \quad b_2 = a_{n_2}, \quad \dots \quad , b_k = a_{n_k}, \quad \dots,$$

where $n_1 < n_2 < n_3 < \cdots$, that is, $n_k < n_{k+1}$ for every k. We may then denote the subsequence (b_k) also by (a_{n_k}) by considering the composition indicated here:

$$k \mapsto n(k) =: n_k \mapsto a(n(k)) =: a_{n_k}.$$

In other words, the sequence $(b_k) = (a_{n_k})$ is a subsequence of (a_k) if (n_k) is a strictly increasing sequence of positive integers. This gives the formal definition of subsequence.

Definition 2.4.9. *A* **subsequence** *of a sequence $a : \mathbf{N} \to \mathbf{R}$ is a composition of a with a strictly increasing function from $\mathbf{N}$ into $\mathbf{N}$. If the elements of the sequence are denoted a_k (so $k \mapsto a(k) = a_k$), then the elements of a subsequence may be denoted a_{n_k} (so $k \mapsto n_k \mapsto a(n_k) = a_{n_k}$).*

A useful fact to remember about the notation for any subsequence (a_{n_k}) of (a_k) is that $n_k \ge k$ for every $k \in \mathbf{N}$; this may be proved by induction (Exercise 2.4.3).

Example 2.4.10. The odd positive integers form a subsequence of the positive integers: If $a_k = k$ for $k \in \mathbf{N}$, and $n_k = 2k - 1$ for $k \in \mathbf{N}$, then $a_{n_k} = 2k - 1$ yields the sequence of odd positive integers. △

Example 2.4.11. The sequence $(a_k) = (1, 1/2, 1/3, \dots)$ converges with limit 0. In fact, any subsequence of this sequence also converges with limit 0. △

The statement in the last example is generalized in the next theorem.

Theorem 2.4.12. *If the sequence (a_k) converges to L, then every subsequence of (a_k) also converges to L.*

Proof. If $a_k \to L$ as $k \to \infty$, then for every $\epsilon > 0$ there is an $N(\epsilon)$ such that

$$|a_k - L| < \epsilon \quad \text{for all } k > N(\epsilon).$$

For any subsequence (a_{n_k}), we have $n_k \ge k$, so $k > N(\epsilon)$ implies $n_k > N(\epsilon)$, and therefore

$$|a_{n_k} - L| < \epsilon \quad \text{for all } k > N(\epsilon).$$

Since $\epsilon > 0$ is arbitrary, we conclude that $\lim_{k\to\infty} a_{n_k} = L$. □

There is a simple corollary of this theorem: If the sequence (a_k) converges with limit L, then for each fixed natural number m, $\lim_{k\to\infty} a_{k+m} = L$, since the sequence (a_{k+m}) is a subsequence of (a_k). Thus, sequential convergence does not depend on the first few terms of the sequence; in fact, sequential convergence does not depend on any finite number of leading terms of a sequence.

In the next example we determine some candidates for a sequential limit.

Example 2.4.13. Define $a_{k+1} = \sqrt{2 + a_k}$. The sequence is completely determined if we specify a value for a_1, say $a_1 = \sqrt{2}$. Note that if $\lim_{k\to\infty} a_k = L$ exists, then $\lim_{k\to\infty} a_{k+1} = L$ as well, and hence, by writing the equation as $a_{k+1}^2 = 2 + a_k$, the limit laws imply that $L^2 = 2 + L$. Thus the limit L, if it exists, satisfies $L^2 = 2 + L$, so either $L = -1$ or $L = 2$. By the definition of the sequence, we cannot have L negative. But we should not jump to the conclusion that $L = 2$ is the limit, because the existence question really matters and we have not yet settled it. Recall that we stated that *if the limit L exists* then L must satisfy $L^2 = 2 + L$. So our conclusion thus far is this: *If a real sequence is determined by the equation $a_{k+1} = \sqrt{2 + a_k}$ and the specification of $a_1 > -2$, and if the limit $L = \lim_{k\to\infty} a_k$ exists, then $L = 2$.* △

The concept of a *monotone sequence* is key to the understanding of more general sequences.

Definition 2.4.14. *A sequence (a_k) of real numbers is* **monotone increasing** *if $a_{k+1} \geq a_k$ for all k, and* **monotone decreasing** *if $a_{k+1} \leq a_k$ for all k. A sequence is* **monotone** *if it is either monotone increasing or monotone decreasing.*

Given a monotone sequence $(a_k)_1^\infty$, we may indicate that it is monotone increasing by writing $a_1 \leq a_2 \leq \ldots$, or that it is monotone decreasing by writing $a_1 \geq a_2 \geq \ldots$. If a sequence is specified (or known) to be either increasing or decreasing, then the term *monotone* is not strictly necessary.

Theorem 2.4.15 (Monotone Sequence Theorem). *Let (b_k) be a sequence of real numbers. If the sequence (b_k) is monotone increasing and bounded, then it converges and*

$$\lim_{k\to\infty} b_k = \sup_k \{b_k\}.$$

If (b_k) is monotone decreasing and bounded, then it converges and

$$\lim_{k\to\infty} b_k = \inf_k \{b_k\}.$$

Proof. Suppose (b_k) is monotone increasing and bounded, and let $B = \sup_k\{b_k\}$; by assumption, $B < \infty$. Given any $\epsilon > 0$, by the definition of supremum there is an $N = N(\epsilon)$ such that $B - \epsilon < b_N$, and hence $|b_N - B| < \epsilon$. Since the sequence is monotone increasing, for $k \geq N$ we have $b_N \leq b_k \leq B$. Therefore $|b_k - B| < \epsilon$ if $k \geq N$. Thus, $\lim_{k\to\infty} b_k = B$.

The statement on bounded decreasing sequences is proved by a similar argument, which is left as Exercise 2.4.4. □

Theorem 2.4.15 asserts that any bounded monotone sequence is convergent. It allows us to complete the discussion of convergence from Example 2.4.13.

Example 2.4.16. The sequence in Example 2.4.13 such that $a_{k+1} = \sqrt{2 + a_k}$ and $a_1 = \sqrt{2}$ can be shown to be monotone increasing and bounded above (Exercise 2.4.5). It converges with the limit $L = 2 = \sup\{a_k\}$. △

We have seen that every subsequence of a convergent sequence with limit L must converge to the same limit L. However, the convergence of a subsequence does not generally imply the convergence of the original sequence. The next result is an important exception.

Theorem 2.4.17. *If (b_k) is an increasing sequence and if some subsequence (b_{n_k}) of (b_k) converges and $\lim_{k\to\infty} b_{n_k} = b$, then (b_k) itself converges to the same limit, $\lim_{k\to\infty} b_k = b$.*

Proof. Since the subsequence (b_{n_k}) is increasing, the convergence assumption implies that

$$b = \lim_{k\to\infty} b_{n_k} = \sup_{k\in\mathbf{N}}\{b_{n_k}\}.$$

Thus, for every $\epsilon > 0$, there is an $N = N(\epsilon)$ such that

$$b - \epsilon < b_{n_{N(\epsilon)}} \le b_{n_k} \le b$$

for all $k > N(\epsilon)$. Since $n_k \ge k$ for all $k \in \mathbf{N}$, we have $b_{n_k} \ge b_k$ for all $k \in \mathbf{N}$. Hence, $b_k \le b$ for all k, and $b_{n_{N(\epsilon)}} \le b_k \le b_{n_k} \le b$ for all $k > n_{N(\epsilon)}$. Therefore

$$|b_k - b| < \epsilon \quad \text{for all } k > n_{N(\epsilon)}.$$

Since ϵ is arbitrary, this shows that $\lim_{k\to\infty} b_k = b$. □

Exercises.

Exercise 2.4.1. Prove parts 1 and 4 of Theorem 2.4.5.

Exercise 2.4.2. Give an example of complex number sequences (a_k), (b_k) and (c_k) such that for each k, $|a_k| \le |b_k| \le |c_k|$, and $\lim_{k\to\infty} a_k = \lim_{k\to\infty} c_k = L$ for some complex number L, but $\lim_{k\to\infty} b_k \ne L$. *Hint*: Consider an example where both (a_k) and (c_k) are constant sequences.

Exercise 2.4.3. Show by induction that if (a_{n_k}) is a subsequence of (a_k), then for each $k \in \mathbf{N}$, $n_k \ge k$.

Exercise 2.4.4. Show that if the sequence (b_k) is monotone decreasing and bounded, then it converges, and $\lim_{k\to\infty} b_k = \inf_k\{b_k\}$.

Exercise 2.4.5. Show that the sequence defined by $a_{k+1} = \sqrt{2 + a_k}$, $a_1 = \sqrt{2}$, is monotone increasing and bounded above. (*Hint*: Use induction to establish both properties.) Then conclude that $\lim_{k\to\infty} a_k = 2$.
What if $a_1 = 1$ or $a_1 = 2$? What if $a_1 > -2$? What happens if $a_1 = -2$?

Exercise 2.4.6. Prove: If $a_n > 0$ and $a_n \to L > 0$, then $\sqrt{a_n} \to \sqrt{L}$.

Exercise 2.4.7. Show that the sequence (a_n), defined by

$$a_1 = 1, \quad \text{and} \quad a_{n+1} = -1 + \sqrt{8 + a_n} \quad (n \ge 1),$$

is increasing and bounded. Find $\lim_{n\to\infty} a_n$.

2.5. Nested Intervals and Decimal Representations

The least upper bound property implies the monotone sequence theorem, and the monotone sequence theorem implies the next result called *the nested interval theorem.*

Theorem 2.5.1 (Nested Interval Theorem). *For each positive integer n let a_n and b_n be real numbers such that $a_n < b_n$. Let $I_n := [a_n, b_n]$ and suppose that for all n,*

$$I_{n+1} \subseteq I_n.$$

Then $\bigcap_{n=1}^{\infty} I_n$ is nonempty. If, in addition,

$$\lim_{n\to\infty} (b_n - a_n) = 0,$$

then there is exactly one point x that belongs to I_n for every n, and the sequences (a_n) and (b_n) both converge to x.

Proof. The hypothesis that the intervals are nested, that is, $I_{n+1} \subseteq I_n$ for each n, means that for every n,

$$a_n \leq a_{n+1} < b_{n+1} \leq b_n.$$

The sequence (a_n) is increasing and bounded above by b_1, and the sequence (b_n) is decreasing and bounded below by a_1. By Theorem 2.4.15, there are numbers a and b such that, for all n,

$$a_n \leq a \quad \text{and} \quad b \leq b_n, \tag{2.8}$$

and $\lim_{n\to\infty} a_n = a$, $\lim_{n\to\infty} b_n = b$. By (2.8), $a, b \in \bigcap_{n=1}^{\infty} I_n$. By hypothesis, $(b_n - a_n) \to 0$, and the difference law for limits implies that

$$0 = \lim_{n\to\infty} (b_n - a_n) = b - a.$$

Thus $a = b$, and if we set $x = a = b$, then by (2.8), $x \in I_n$ for every n. Finally, the existence of two distinct points that belong to every interval I_n would contradict the hypothesis that $(b_n - a_n) \to 0$. □

The remaining goal of this section is to define a decimal representation for each real number and to show that for each decimal representation of a certain type, there is a unique real number associated with it. The argument for Theorem 2.5.2 below shows that the elements of a complete ordered field have unique representations as nonterminating decimal expansions, and this helps to confirm one of the long-held intuitions about the real numbers that most of us develop. But first we must clarify some terminology.

A **decimal expansion** is an expression of the form $d_0.d_1d_2d_3\ldots$, where for each positive integer n, the digit $d_n \in \{0, 1, 2, \ldots, 9\}$, and d_0 can be any nonnegative integer. A **terminating decimal expansion** is one for which there is an $m \in \mathbf{N}$ such that for each $k \geq m$, $d_k = 0$. A decimal expansion is **nonterminating** if it is not terminating. It will be convenient to define and work with nonterminating expansions. For the moment, we need not give any arithmetic meaning to a decimal expansion in its entirety (this is best done later, using the basic concepts of infinite series). Right now, it is only important to note that Theorem 2.5.2 uses concepts and facts about $\mathbf{R}$ that are familiar thus far in the book.

Given a positive real number $x > 0$, there is a unique integer $n(x)$ such that $n(x) < x \le n(x) + 1$. The integer $n(x)$ is called the **integer part** of $x > 0$, for the purpose of decimal representation using nonterminating expansions. Note that $n(x)$ is the largest integer less than x, and $x - n(x) \in (0, 1]$. For example, $x = .1$ implies $n(x) = 0$, and $x = 2$ implies $n(x) = 1$, which allows the nonterminating expansion $2 = 1.99\overline{9}$. (A bar over a digit or group of digits indicates that the pattern repeats thereafter.) With this definition of the integer part of a positive real number in hand, we can focus on the decimal representation of real numbers in the interval $(0, 1]$.

Theorem 2.5.2. *There is a one-to-one correspondence between the interval* $(0, 1]$ *of real numbers and the set of nonterminating decimal expansions of the form* $0.d_1d_2d_3\ldots$, *with each* $d_k \in \{0, 1, 2, \ldots, 9\}$.

Proof. The first part of the proof is the definition of the mapping from $(0, 1]$ to the set of decimal representations and the proof that this mapping is one-to-one. Let $x \in (0, 1]$. Then $n(x) = 0$. The selection of the digit d_1 is as follows. Express the interval $(0, 1]$ as the union of 10 disjoint intervals of the form $(a, b]$, each having length equal to $1/10$. These are the intervals $(0, 1/10]$, $(1/10, 2/10]$, and so on, up to $(9/10, 1]$. Then x is in exactly one of these intervals, and if it is the k-th interval counting from the left, denoted $I_1 := ((k-1)/10, k/10]$, then set $d_1 = k - 1$. (This is equivalent to choosing d_1 equal to $n(10x)$.)

In order to select the digit d_2 in the representation of x, express the previous interval $I_1 = ((k-1)/10, k/10]$ as the union of 10 disjoint intervals of the form $(a, b]$, each having length equal to $1/10^2 = 1/100$. Then x is in exactly one of these intervals, and if it is the j-th one, then set $d_2 = j - 1$. (This j-th interval of I_1 is the interval we denote $I_2 := ((k-1)/10 + (j-1)/100, (k-1)/10 + j/100]$.)

Continue in the way indicated above. More specifically, after selecting the m-th digit d_m in the interval I_m having length $1/10^m$, divide I_m into 10 subintervals of the form $(a, b]$ having length $1/10^{m+1}$; if x is in the i-th subinterval, then set $d_{m+1} = i - 1$.

This procedure associates with $x \in (0, 1]$ a decimal representation of the form $0.d_1d_2d_3\ldots$.

Let $\overline{I}_m$ be the union of I_m and its left-hand endpoint, so that $\overline{I}_m$ is a closed interval. The association $x \mapsto 0.d_1d_2d_3\ldots$ is one-to-one because by construction of the nested sequence of closed intervals $(\overline{I}_m)$, with $x \in \overline{I}_m$ for each m, the length of $\overline{I}_m$ is $1/10^m$, and $\lim_{m\to\infty} 1/10^m = 0$. (In this limit statement we have used the Archimedean property of $\mathbf{R}$.) Thus there cannot be distinct points in the intersection of the intervals I_m.

The next part of the proof establishes that the mapping $x \mapsto 0.d_1d_2d_3\ldots$ defined above is onto the set of decimal representations of the indicated form. Given a decimal expansion $0.d_1d_2d_3\ldots$, we must 'decode it' to show there is an $x \in (0, 1]$ which is associated with it via the mapping. The given digits determine the appropriate intervals within which x should be found. Starting with d_1 and proceeding left to right with the given digits, we may select the appropriate closed interval $\overline{I}_m$ of length $1/10^m$ based on the digit d_m. The resulting sequence of closed intervals $\overline{I}_m$ is a nested sequence, and the limit of the lengths is $\lim_{m\to\infty} 1/10^m = 0$, by construction. Hence, by the nested interval theorem, $\bigcap_m \overline{I}_m = \{x\}$ for some unique $x \in \mathbf{R}$. By the selection of these intervals, the element x is associated, via the mapping defined in the first part of the proof, with the representation $0.d_1d_2d_3\ldots$, with which we started. □

Based on the digit selection procedure of Theorem 2.5.2, the decimal representation of the number 1 is given by $0.99\overline{9}$. The decimal representation of $1/2$ is $0.499\overline{9}$.

This digit selection procedure produces nonterminating expansions. Consequently, rational numbers are represented by nonterminating expansions that repeat, rather than by terminating expansions. (The one exception to this rule is the expansion for 0, which is $0.00\overline{0}$, but this is the only terminating expansion we use.) Conversely, each repeating expansion represents a rational number, a fact that is perhaps easiest to realize after the basic concepts of infinite series are introduced.

What about positive x not in $(0, 1]$? If $x > 0$, then we define the decimal representation of x to be $n(x).d_1d_2d_3\ldots$, where $0.d_1d_2d_3\ldots$ is the representation of $x - n(x) \in (0, 1]$.

For $x < 0$, we take the integer part of x to be the least integer greater than x, instead of the greatest integer less than x. Then we can use the decimal representation of the positive number $-x$ and introduce a negative sign in front. For example, we want $-1/2$ to be represented by $-.499\overline{9}$. Thus, the expansion of -1 is $-0.99\overline{9}\ldots$.

With a good understanding of the proof of Theorem 2.5.2, it is clear that there is nothing in the structure of the argument that requires the use of the digits in the set $\{0, 1, 2, \ldots, 9\}$. It is really no more difficult to consider expansions using the digits from the set $\{0, 1, 2, \ldots, b-1\}$, where $b \in \mathbf{N}$ and $b \geq 2$. This means that at each step, we subdivide into b subintervals rather than 10 subintervals. Then the appropriate nested intervals I_m in the modified argument will have length $1/b^m$, and again the crucial fact that $\lim_{m\to\infty} 1/b^m = 0$ holds true, by the Archimedean property. With $b = 2$, we obtain **binary expansions** $0.d_1d_2d_3\ldots$ for $x \in (0, 1]$ with each $d_n \in \{0, 1\}$. With $b = 3$ we have **tertiary expansions** $0.d_1d_2d_3\ldots$ for $x \in (0, 1]$ with each $d_n \in \{0, 1, 2\}$. Proceeding as indicated above for positive and negative elements, the binary expansion of 1 is $0.11\overline{1}$, and the binary expansion of -1 is $-0.11\overline{1}$; the tertiary expansion of 1 is $0.22\overline{2}$, and the tertiary expansion of -1 is $-0.22\overline{2}$. (For simplicity in what follows, we simply refer to the d_0 digit as an integer.)

If $b \in \mathbf{N}$ and $b \geq 2$, we call the unique nonterminating expansion of a nonzero real number x using only the digits in the set $\{0, 1, 2, \ldots, b-1\}$ **the expansion of x in the number base b**. We can state the following result.

Theorem 2.5.3. *Let $b \in \mathbf{N}$ with $b \geq 2$. There is a one-to-one correspondence between $\mathbf{R} - \{0\}$ and the set of nonterminating base b expansions of the form $d_0.d_1d_2d_3\ldots$ where $d_k \in \{0, 1, 2, \ldots, b-1\}$ for $k \in \mathbf{N}$, and d_0 is an integer.*

We now show that $\mathbf{R}$ and $\mathbf{Q}$ do not have the same cardinality.

Theorem 2.5.4. *The complete ordered field $\mathbf{R}$ is uncountably infinite.*

Proof. We choose to work with the base two nonterminating binary expansions.

Suppose to the contrary that $\mathbf{R}$ is countably infinite, and let an enumeration of $\mathbf{R}$ be given by the following listing of base 2 expansions from Theorem 2.5.3 (plus the expansion for 0, listed first here):

$$\begin{aligned} r_0 &= a_{00}.a_{01}a_{02}a_{03}\ldots = 0.000\ldots, \\ r_1 &= a_{10}.a_{11}a_{12}a_{13}\ldots, \\ r_2 &= a_{20}.a_{21}a_{22}a_{23}\ldots, \\ \cdots &= \cdots, \end{aligned}$$

where, for $j \geq 1$, a_{ij} is either 0 or 1, and for each i, a_{i0} is an integer. Now define a binary expansion $r^* = d_0.d_1d_2d_3\ldots$ as follows:
Step 0. Set $d_0 = 1$.
Step 1. Set $d_1 = 1$ if $a_{11} = 0$; set $d_1 = 0$ if $a_{11} = 1$.
Step 2. Set $d_2 = 1$ if $a_{22} = 0$; set $d_2 = 0$ if $a_{22} = 1$, and so on.

The digits d_k of r^* are defined for all k, and r^* differs from r_k in the k-th digit, since for each $k \geq 0$, $d_k \neq a_{kk}$. Therefore r^* is not in the assumed listing of all elements of $\mathbf{R}$. The listing is assumed to consist of the nonterminating expansions guaranteed by Theorem 2.5.3. Can r^* be a terminating expansion? If that were the case, then all elements in the above listing past some index j have their diagonal digit equal to 1, since $d_k = 0$ for $k \geq j$ implies $a_{kk} = 1$ for $k \geq j$. However, in that case the listing does not include all the expansions guaranteed by Theorem 2.5.3, since there would be only finitely many r_k, specifically those for $0 \leq k < j$, which could have a zero diagonal entry a_{kk}. However, this is absurd, as we may easily specify infinitely many binary nonterminating expansions which have zero diagonal entry a_{kk}. This contradiction implies that r^* is nonterminating, and thus r^* is one of the expansions guaranteed by Theorem 2.5.3. However, r^* is not in the listing, as we have seen, and this contradiction proves the theorem. □

It is immediate from Theorem 2.5.4 that the set $\mathbf{I}$ of irrational numbers is uncountably infinite: We have $\mathbf{R} = \mathbf{Q} \cup \mathbf{I}$, and if $\mathbf{I}$ is countable, then so is $\mathbf{R}$, a contradiction. We have deduced the fact that there are uncountably many irrationals from the the least upper bound property (by way of the Archimedean property and the nested interval theorem). Let us see how the nested interval theorem fails for $\mathbf{Q}$. We know there is a $\sqrt{2}$ gap in the rational number line. The number $x = \sqrt{2}$ has a decimal expansion $1.d_1d_2d_3\ldots$. For each n, let r_n be the truncation of this expansion after the digit d_n, so

$$r_n = 1.d_1d_2d_3\ldots d_n = 1.d_1d_2d_3\ldots d_n.$$

It is clear from the digit selection procedure in Theorem 2.5.2 that this truncated expansion represents the number

$$r_n = 1 + \frac{d_1}{10} + \frac{d_2}{10^2} + \cdots + \frac{d_n}{10^n}.$$

For each n, we have $\sqrt{2} \in I_n$, where I_n is the n-th interval chosen in the digit selection procedure for $\sqrt{2}$. If $J_n = I_n \cap \mathbf{Q}$, then J_n is an interval in the ordered field $\mathbf{Q}$ and the endpoints of J_n are in $\mathbf{Q}$. However,

$$\bigcap_{n=1}^{\infty} J_n = \Big(\bigcap_{n=1}^{\infty} I_n\Big) \cap \mathbf{Q} = \{\sqrt{2}\} \cap \mathbf{Q},$$

which is empty. The nested interval theorem fails in $\mathbf{Q}$.

An **algebraic number** is a real or complex number that satisfies a polynomial equation with integer coefficients. (An equivalent definition is that an algebraic number is a real or complex number that satisfies a polynomial equation with rational coefficients.) For example, $\sqrt{2}$ and $\sqrt[3]{2}$ are real algebraic numbers, and $i = \sqrt{-1}$ is a complex algebraic number. Every rational number is a real algebraic number, and $\sqrt{2}$ and $\sqrt[3]{2}$ are algebraic irrationals. The **algebraic irrationals** can be described as the set of real numbers x_0 that satisfy an equation of the form

$$a_0 x_0^n + a_1 x_0^{n-1} + \cdots + a_{n-1} x_0 + a_n = 0,$$

for integers a_j and some $n \geq 2$, but no linear equation of the form

$$b_0 x + b_1 = 0,$$

with integer b_0 and b_1. Denote the set of algebraic irrationals by $\mathbf{I}_a$. Let $\mathbf{I}_t = \mathbf{I} - \mathbf{I}_a$; the elements of $\mathbf{I}_t$ are called **transcendental numbers**. Then $\mathbf{R} = \mathbf{Q} \cup \mathbf{I}_a \cup \mathbf{I}_t$, a disjoint union.

Exercises.

Exercise 2.5.1. Give an example to show that a nested sequence of open intervals (or half-open intervals) can have empty intersection.

Exercise 2.5.2. Show that the argument of Theorem 2.5.2 does not assign any number $x \in (0, 1]$ a terminating expansion. Thus, according to the discussion following the theorem, no nonzero real number is assigned a terminating expansion.

Exercise 2.5.3. A nonterminating decimal expansion r is **repeating** if for some nonnegative integers $n \leq m$, $r = d_0.d_1 \ldots d_{n-1}\overline{d_n \ldots d_m}$. Otherwise, it is **nonrepeating**.

1. Show that each repeating decimal expansion represents a rational number, and each rational number is represented by a repeating decimal expansion.
2. Show that there are countably many nonterminating, repeating decimal expansions.
3. Show that there are uncountably many nonterminating, nonrepeating decimal expansions.

Exercise 2.5.4. Let $x > 0$ and suppose that the portion $0.d_1 d_2 d_3 \ldots d_n$ of the expansion for $x - n(x) \in (0, 1]$ has been selected as in the proof of Theorem 2.5.2, where $d_0 = n(x)$. Show that

$$\frac{d_1}{10} + \frac{d_2}{10^2} + \cdots + \frac{d_n}{10^n} < x \leq \frac{d_1}{10} + \frac{d_2}{10^2} + \cdots + \frac{d_n + 1}{10^n}.$$

Exercise 2.5.5. Let $p > 0$ be a prime number. Show that $\sqrt{p}$ is an algebraic irrational number. *Hint*: If p divides a product $q_1 q_2$ of integers q_1 and q_2, then either p divides q_1 or p divides q_2.

Exercise 2.5.6. Show that the set $\mathbf{I}_a$ of algebraic irrationals is countably infinite. Show that the set of real algebraic numbers $\mathbf{I}_a \cup \mathbf{Q}$ is countably infinite. Conclude that the set $\mathbf{I}_t$ of transcendental numbers is uncountable. (Thus transcendental numbers exist, but it is difficult to prove that specific real numbers are transcendental. In particular, e and π are known to be transcendental.)

2.6. The Bolzano-Weierstrass Theorem

We have seen that any bounded, infinite monotone sequence in $\mathbf{R}$ has a limit. On the other hand, bounded nonmonotonic sequences need not converge.

Example 2.6.1. Consider the bounded sequence given by[2]

$$a_k = (-1)^k \frac{e^k}{1+e^k},$$

which does not converge: The subsequence $a_{2k} = e^{2k}/(1+e^{2k})$ has limit 1, and the subsequence $a_{2k-1} = -e^{2k-1}/(1+e^{2k-1})$ has limit -1, as $k \to \infty$. △

Perhaps intuition suggests that a bounded infinite subset of the real numbers should have the property that its points tend to cluster or accumulate about *some* point. We will see in this section that such an intuition is sound. It is not difficult to see that the sequence in Example 2.6.1 has convergent subsequences. But the sequence $(\sin k)_{k=1}^{\infty}$ is more difficult to think about. It is bounded. Its range has infinitely many elements, otherwise the sequence would be eventually constant, and that is not possible, as the sine function has least period 2π and therefore cannot eventually repeat itself over a unit interval. It is not clear if $\sin k$ can get arbitrarily close to ± 1, for example, unless we know how closely positive integers can approximate the numbers $n\pi/2$, $n \in \mathbf{N}$. We cannot seem to pin down any other specific point where the elements of this sequence cluster. Are there any?

In order to clarify the issue, we need the next definition.

Definition 2.6.2. *Let S be a subset of an ordered field F. A point $p \in F$ is a* **cluster point** *(or* **accumulation point***) of S if for each $\epsilon \in F$ with $\epsilon > 0$ the interval $(p-\epsilon, p+\epsilon)$ in F contains infinitely many points of S distinct from p.*

We note that there is a distinction between a bounded sequence with infinite range (it is an ordered infinite set with infinitely many elements) and general bounded infinite sets. We will soon clarify that the distinction makes no difference as far as the question of existence of a cluster point.

We can now state the Bolzano-Weierstrass theorem for bounded infinite sets in $\mathbf{R}$. It says that an infinite search is guaranteed to find a cluster point for the set $\{\sin k : k \in \mathbf{N}\}$.

Theorem 2.6.3 (Bolzano-Weierstrass I). *Every bounded infinite set of real numbers has a cluster point, which need not be an element of the set.*

---

[2]Basic properties of the exponential function e^x and the other elementary functions are established in Section 7.5.

Proof. Let S be a bounded infinite subset of $\mathbf{R}$. Then there is a finite interval $[a, b]$, $a < b$, such that $S \subseteq [a, b]$. Bisect the interval $[a, b]$ into its left and right halves, $[a, (a + b)/2]$ and $[(a + b)/2, b]$, each of which is a closed interval. At least one of these closed intervals must contain infinitely many elements of S. (If both halves contain infinitely many, then for definiteness choose the left half.) Denote the chosen half by I_1. Now bisect I_1. Either the left half or the right half of I_1 contains infinitely many elements of S. Call that half I_2. Assuming I_n has been chosen, we choose the half (either left or right) of I_n which contains infinitely many elements of S and call that half I_{n+1}. By induction, this defines a sequence of closed intervals I_n, $n \in \mathbf{N}$, such that for each n, $I_{n+1} \subset I_n$, and each I_n contains infinitely many elements of S. For each n, the length of I_{n+1} is half the length of I_n. In fact, for each n, the length of I_n is $(b - a)/2^n$. By the nested interval theorem (Theorem 2.5.1), there is exactly one point x_0 that is an element of each I_n. We want to show that x_0 is a cluster point of S. Given any $\epsilon > 0$, there is an n such that the length of I_n is $(b - a)/2^n < \epsilon$. Since $x_0 \in I_n$, if $y \in I_n$, then $|y - x_0| < \epsilon$, and hence I_n is contained in the interval $(x_0 - \epsilon, x_0 + \epsilon)$. Since I_n contains infinitely many elements of S, this proves that x_0 is a cluster point of S. □

There is also a useful version of the Bolzano-Weierstrass theorem for infinite sequences. Before we state and prove it, note that the sequence $(b_k) = ((-1)^k) = (-1, 1, -1, 1, \ldots)$ has convergent subsequences, but the range is the finite set $\{1, -1\}$, which clearly has no cluster points. Any sequence with finite range has a convergent subsequence, whether or not the sequence is eventually constant. Thus, for the sequential Bolzano-Weierstrass theorem we only need to consider bounded infinite sequences (that is, bounded sequences with infinite range). An example is the sequence $(\sin(1/k))_{k=1}^{\infty}$, for which the range set $\{\sin(1/k) : k \in \mathbf{N}\}$ is infinite. Observe that the range has the cluster point 0, since $\lim_{k\to\infty} \sin(1/k) = 0$. In fact, we have the following basic consequence of the definition of sequential convergence (Definition 2.4.1).

Proposition 2.6.4. *Suppose that $(a_k)_{k=1}^{\infty}$ is an infinite sequence in $\mathbf{R}$, that is, the range $\{a_k : k \in \mathbf{N}\}$ of the sequence is an infinite subset of $\mathbf{R}$. If $\lim_{k\to\infty} a_k = L \in \mathbf{R}$, then L is a cluster point of the infinite set $\{a_k : k \in \mathbf{N}\}$.*

Proof. If L is the limit of the sequence, then given $\epsilon_k = 1/k > 0$, there is an $N(\epsilon_k)$ such that if $k \geq N(\epsilon_k)$, then $|a_k - L| < 1/k$. Since the sequence has infinitely many elements, there must be infinitely many of them within a distance $1/k$ of L that are distinct from L. This is true for each positive integer k, hence L is a cluster point of the range of the sequence. □

Here is the sequential Bolzano-Weierstrass theorem.

Theorem 2.6.5 (Bolzano-Weierstrass II). *Every bounded infinite sequence of real numbers (that is, every bounded sequence with infinite range) has a convergent subsequence.*

Proof. If (a_k) is a bounded infinite sequence, then the range $\{a_k\}$ has a cluster point a, by Theorem 2.6.3. Thus, given $\epsilon_k = 1/k$, there is an element a_{n_k} such that $|a_{n_k} - a| < 1/k$. Moreover, for each k we may choose $n_{k+1} \geq n_k$, since each interval about a contains infinitely many elements of the range of the sequence. This construction yields a subsequence (a_{n_k}) that converges to a. □

These two results, Bolzano-Weierstrass I (for bounded infinite sets) and Bolzano-Weierstrass II (for bounded infinite sequences), are actually equivalent. As an exercise, one can show that Theorem 2.6.3 follows from Theorem 2.6.5.

There is no Bolzano-Weierstrass theorem for $\mathbf{Q}$. For example, one can define a monotone increasing sequence (s_k) in the set

$$S = \{s \in \mathbf{Q} : \text{either } s < 0, \text{ or } s \geq 0 \text{ and } s^2 < 2\}$$

having no limit in $\mathbf{Q}$, since $\sup S$ does not exist in $\mathbf{Q}$. Thus, we also see that there is no analogue of the monotone sequence theorem (Theorem 2.4.15) for $\mathbf{Q}$.

Exercises.

Exercise 2.6.1. 1. Does the set $\{1/n : n \in \mathbf{N}\}$ have a cluster point?
2. Does the sequence

$$\tan^{-1}\left((-1)^k \frac{e^k}{1+e^k}\right)$$

have a convergent subsequence? How many cluster points does the range of this sequence have?

Exercise 2.6.2. Show that if every bounded infinite sequence of real numbers has a convergent subsequence, then every bounded infinite set of real numbers has a cluster point.

Exercise 2.6.3. Show that the sequential Bolzano-Weierstrass Theorem 2.6.5 implies the nested interval Theorem 2.5.1.

2.7. Convergence of Cauchy Sequences

If an infinite sequence (a_k) of real numbers converges, then the elements of the sequence must get arbitrarily close to each other as $k \to \infty$, since the limit must be the only cluster point of the range of the sequence. We want to investigate whether the converse is true: If the elements of an infinite sequence (a_k) in $\mathbf{R}$ get arbitrarily close to each other as $k \to \infty$, does the sequence converge to a limit in $\mathbf{R}$? First, we need the definition of what it means for the elements of a sequence in $\mathbf{R}$ to get arbitrarily close to each other as $k \to \infty$.

The next definition applies to both real and complex number sequences.

Definition 2.7.1 (Cauchy Sequence). *A sequence (a_k) in $\mathbf{R}$ or $\mathbf{C}$ is called a* **Cauchy sequence** *if for each $\epsilon > 0$, there is a natural number $N = N(\epsilon)$ such that for all $m, n \geq N$,*

$$|a_m - a_n| < \epsilon.$$

In fact, the concept of Cauchy sequence makes sense in any ordered field F, since the absolute value function is defined in F. (We see later in the book that the concept of Cauchy sequence makes sense in any normed vector space; for example, $\mathbf{C}$ is a complex vector space normed by the absolute value function.) As we noted above, *a convergent sequence must be a Cauchy sequence.* The proof is left as an exercise.

There are Cauchy sequences of rational numbers that do not have a rational number limit. In other words, not every Cauchy sequence of rationals converges to an element of $\mathbf{Q}$. Think of the sequence of finite decimal approximations of $\sqrt{2}$, which is a Cauchy sequence of rational numbers that does not converge to an element of $\mathbf{Q}$.

Theorem 2.7.2. *Every Cauchy sequence in* $\mathbf{R}$ *converges to a limit in* $\mathbf{R}$.

Proof. We may assume that the sequence has infinite range. Suppose that (a_k) is a Cauchy sequence. Since the sequence is Cauchy, it is bounded: Let $\epsilon = 1$ in the definition of Cauchy sequence; then there is an integer N such that if $n, m > N$, then $|a_n - a_m| < 1$, hence $|a_n| - |a_m| < 1$. Fix $m = N + 1$. Then for $n > N$,

$$|a_n| < |a_{N+1}| + 1.$$

If we let $M = \max\{|a_j| : 1 \le j \le N\}$, then for all positive integers k,

$$|a_k| \le \max\{M, |a_{N+1}| + 1\}.$$

By the Bolzano-Weierstrass theorem (Theorem 2.6.5), the sequence (a_k) has a subsequence (a_{n_k}) that converges to a limit L. So we want to show that a Cauchy sequence that has a subsequence with limit L must itself converge to L. Let $\epsilon > 0$. Since (a_{n_k}) converges, there is an integer N_1 such that

$$|a_{n_k} - L| < \frac{\epsilon}{2} \quad \text{if } k \ge N_1.$$

Since (a_k) is Cauchy, there is an integer N_2 such that

$$|a_n - a_m| < \frac{\epsilon}{2} \quad \text{if } n, m \ge N_2.$$

Thus, if we take $k \ge \max\{N_1, N_2\}$, then, since $n_k \ge k$ for each k,

$$|a_k - a_{n_k}| < \frac{\epsilon}{2} \quad \text{and} \quad |a_{n_k} - L| < \frac{\epsilon}{2}.$$

Consequently, for $k \ge \max\{N_1, N_2\}$,

$$|a_k - L| \le |a_k - a_{n_k}| + |a_{n_k} - L| < \frac{\epsilon}{2} + \frac{\epsilon}{2} = \epsilon,$$

which proves that (a_k) converges with limit L. □

It can be shown that the convergence of Cauchy sequences in $\mathbf{R}$ (Theorem 2.7.2) implies the least upper bound property, hence these two properties of $\mathbf{R}$ are actually equivalent.

Corollary 2.7.3. *Every Cauchy sequence in* $\mathbf{C}$ *converges to a limit in* $\mathbf{C}$.

Proof. If $z_k = a_k + ib_k$ defines a Cauchy sequence in $\mathbf{C}$, then (a_k) and (b_k) are real Cauchy sequences, since

$$|a_m - a_n| \le |z_m - z_n| \qquad \text{and} \qquad |b_m - b_n| \le |z_m - z_n|.$$

Hence, $a_k \to a \in \mathbf{R}$ and $b_k \to b \in \mathbf{R}$, by Theorem 2.7.2. Since

$$|z_k - (a + ib)| = |(a_k - a) + i(b_k - b)| \le |a_k - a| + |i(b_k - b)| = |a_k - a| + |b_k - b|,$$

the squeeze theorem for real sequences implies that $\lim_{k\to\infty} |z_k - (a+ib)| = 0$ (Exercise 2.7.2). Hence, $z_k \to (a+ib) \in \mathbf{C}$. □

The result of Corollary 2.7.3 is what is meant by the **completeness** of the complex field $\mathbf{C}$: Every Cauchy sequence in $\mathbf{C}$ converges to an element of $\mathbf{C}$.

We have not proved a Bolzano-Weierstrass theorem for the complex field $\mathbf{C}$. However, there is such a theorem, and the reader is invited to prove it in Exercise 2.7.3.

Exercises.

Exercise 2.7.1. A convergent sequence must be a Cauchy sequence.

Exercise 2.7.2. In the proof of Corollary 2.7.3 show that the squeeze theorem for real sequences implies that $\lim_{k\to\infty} |z_k - (a+ib)| = 0$.

Exercise 2.7.3 (Bolzano-Weierstrass for $\mathbf{C}$).

1. Show that every bounded infinite subset S of the complex field $\mathbf{C}$ has a cluster point which need not be contained in S. Thus, every bounded infinite sequence in $\mathbf{C}$ has a convergent subsequence with limit in $\mathbf{C}$. *Hint*: S is contained in some square $[a,b]\times[a,b]$; subdivide it repeatedly to obtain two real Cauchy sequences, and then use Theorem 2.7.2.
2. Give a proof of the completeness of $\mathbf{C}$ based on part 1 of this exercise.

Exercise 2.7.4. Let $(a_k)_{k=1}^{\infty}$ be a Cauchy sequence. Show that there is a subsequence $(a_{n_k})_{k=1}^{\infty}$ such that $|a_{n_{k+1}} - a_{n_k}| < 1/2^k$ for all k.

2.8. Summary: A Complete Ordered Field

Here we summarize some thoughts and facts about the real numbers that may serve as a guide for further reading.

2.8.1. Properties that Characterize Completeness. Several important facts about the complete ordered field $\mathbf{R}$ have been presented in this chapter. The completeness axiom is the least upper bound property in (a) below. The main results established thus far are listed in (b)-(e).

(a) $\mathbf{R}$ has the least upper bound property;
(b) the monotone sequence theorem;
(c) the nested interval theorem;
(d) the Bolzano-Weierstrass theorem for bounded infinite sets;
(e) the convergence of Cauchy sequences.

We proved the following sequence of implications:

$$(a) \Longrightarrow (b) \Longrightarrow (c) \Longrightarrow (d) \Longrightarrow (e)$$

Recall that we proved the important Archimedean property from the least upper bound axiom (a). It is possible to show that (e) implies (d).

Theorem 2.8.1. *Property (e) implies (d): If every Cauchy sequence in an ordered field F converges to an element of F, then every bounded infinite subset of F has a cluster point (and hence every bounded infinite sequence in F has a convergent subsequence).*

A proof of this result appears in Schramm [**57**] (Theorem 10.17, page 176).

It is also possible to show that (d) together with the Archimedean property implies (a).

Theorem 2.8.2. *If the ordered field F has the Archimedean property and if every bounded infinite subset of F has a cluster point, then F has the least upper bound property.*

For a proof of this result, see Schramm [**57**] (Theorem 7.6, page 110).

The somewhat mysterious reappearance of the Archimedean property as a hypothesis in closing the loop back to (a) suggests a longer story. The following theorem gives something more of the full story; its proof forms a major theme throughout the text by Schramm [**57**].

Theorem 2.8.3. *Let F be an ordered field. Then the following are equivalent:*

1. *F has the least upper bound property.*
2. *F has the Archimedean property and the nested interval theorem holds in F.*
3. *F has the Archimedean property and the Bolzano-Weierstrass theorem on bounded infinite sets holds in F.*
4. *F has the Archimedean property and every Cauchy sequence in F converges to an element of F.*

Each of these properties serves to characterize the complete ordered field $\mathbf{R}$, and hence each item describes a property of $\mathbf{R}$ which does not hold in $\mathbf{Q}$. There is a longer list of equivalences on offer in Schramm [**57**] than stated here.

2.8.2. Why Calculus Does Not Work in Q. Several important results in introductory calculus courses are consequences of the least upper bound property. The emphasis here is to show that some familiar facts of introductory calculus do depend on the least upper bound property of the reals, by showing that the same statements fail to be true when working with the rational number field $\mathbf{Q}$. The message to take away is that it is the least upper bound property of $\mathbf{R}$, the completeness of $\mathbf{R}$, that makes calculus and modern analysis possible. The theorems mentioned by name below should prompt some memory of introductory calculus, and they are proved (for the real field!) later in the text.

Let us consider why calculus does not work in $\mathbf{Q}$.[3] Let $a, b \in \mathbf{Q}$ with $a < b$, and let $[a, b]$ denote the interval of rational numbers x such that $a \leq x \leq b$. We define continuity and differentiability for functions $f : [a, b] \to \mathbf{Q}$ as follows:

Let $x \in [a, b] \subset \mathbf{Q}$. The function $f : [a, b] \to \mathbf{Q}$ is continuous at $x \in \mathbf{Q}$ if and only if for every sequence (x_k) in $\mathbf{Q}$ such that $\lim_{k\to\infty} x_k = x$, it is true that $\lim_{k\to\infty} f(x_k) = f(x)$, that is, for every rational $\epsilon > 0$ there is a positive integer

[3] This discussion is motivated in part by examples in Chapter 1 of Körner [**38**].

$N = N(\epsilon)$ such that if $k \geq N$, then $|f(x_k) - f(x)| < \epsilon$. If f is continuous at each $x \in [a, b] \subset \mathbf{Q}$, then we say f is continuous on $[a, b] \subset \mathbf{Q}$.

If $f : (a, b) \to \mathbf{Q}$ is defined on an open interval $(a, b) \subset \mathbf{Q}$ and $x \in (a, b) \subset \mathbf{Q}$, then the derivative of f at x is defined by the limit

$$f'(x) = \lim_{h \to 0} \frac{f(x + h) - f(x)}{h}, \qquad h \text{ rational},$$

provided this limit exists in $\mathbf{Q}$. Then it is true that differentiability of f at x implies continuity of f at x, because one can argue, correctly, that

$$\lim_{h \to 0} [f(x + h) - f(x)] = \lim_{h \to 0} \frac{f(x + h) - f(x)}{h} \lim_{h \to 0} h = f'(x) \cdot 0 = 0.$$

Also, it is still true that the derivative of a sum of two differentiable functions equals the sum of their derivatives. Good so far.

Example 2.8.4. Let $f : \mathbf{Q} \to \mathbf{Q}$ be defined by

$$f(x) = \begin{cases} 1 & \text{if } x^2 < 2, \\ -1 & \text{if } x^2 > 2. \end{cases}$$

Then f is continuous at each $x \in \mathbf{Q}$. Indeed, f is differentiable at each $x \in \mathbf{Q}$, and $f'(x) = 0$ for all $x \in \mathbf{Q}$. Despite the fact that f is continuous, we have $f(1) = 1$ and $f(2) = -1$, and there is no $x \in (1, 2) \subset \mathbf{Q}$ with $f(x) = 0$. Thus *the intermediate value theorem for continuous functions* fails for $f : \mathbf{Q} \to \mathbf{Q}$. △

Exercise 2.8.1 shows that *the mean value theorem* also fails for f on $[1, 2]$. (It is also worth recalling that the part of the fundamental theorem of calculus dealing with the evaluation of definite integrals depends on the mean value theorem.)

Let I be an interval in $\mathbf{Q}$. A function $h : I \to \mathbf{Q}$ is increasing if $x, y \in \mathbf{Q}$, $x < y$ implies $h(x) < h(y)$. However, there is a function $h : \mathbf{Q} \to \mathbf{Q}$ with $h'(x) > 0$ for all $x \in \mathbf{Q}$, and yet h is not an increasing function!

Example 2.8.5. Consider the function $h : \mathbf{Q} \to \mathbf{Q}$ defined by

$$h(x) = x + f(x),$$

where f is the function in Example 2.8.4. Then $h'(x)$ exists for each $x \in \mathbf{Q}$, and $h'(x) = 1 + f'(x) = 1 > 0$. But consider that $4/3 < 3/2$, while $h(4/3) = 4/3 + 1 = 7/3$, and $h(3/2) = 3/2 - 1 = 1/2$. So $h(4/3) > h(3/2)$, and therefore h is not an increasing function. △

2.8.3. The Existence of a Complete Ordered Field. The existence of a complete ordered field is essential so that calculus is not an empty exercise; of course, it is not. There are two standard ways of showing that a complete ordered field $\mathbf{R}$ exists. Both methods construct $\mathbf{R}$ from the rational number field $\mathbf{Q}$. It is probably fair to say that serious students of mathematics should have some familiarity with both constructions.

1. The complete ordered field $\mathbf{R}$ can be constructed as the set of Dedekind cuts of the rational numbers. Schramm [**57**] (Chapter 22) offers a nice discussion with exercises for the reader.

2. The complete ordered field $\mathbf{R}$ can be constructed as the set of equivalence classes of Cauchy sequences in $\mathbf{Q}$. Strichartz [**64**] has a nice exposition of this approach.

The end result of either of these approaches is the following theorem.

Theorem 2.8.6. *There exists an ordered field* $\mathbf{R}$ *which contains the rational field* $\mathbf{Q}$ *and is complete, that is,* $\mathbf{R}$ *has the property that any nonempty subset of* $\mathbf{R}$ *which is bounded above has a least upper bound in* $\mathbf{R}$.

Familiarity with the construction of $\mathbf{R}$ from $\mathbf{Q}$ can deepen the understanding and appreciation of the real numbers, but it will not, in and of itself, help the reader to understand analysis. Understanding basic analysis depends largely on working with the properties of the real number field that follow from completeness.

2.8.4. The Uniqueness of a Complete Ordered Field. Two ordered fields F and F' are order-isomorphic if there is a one-to-one mapping $h : F \to F'$ onto F' such that $x, y \in F$ implies

$$h(xy) = h(x)h(y) \quad \text{and} \quad h(x + y) = h(x) + h(y)$$

and moreover, $x, y \in F$ and $x < y$ implies $h(x) < h(y)$. This means that F and F' are simply relabeled versions of each other and the field operations and order relations are preserved under the relabeling.

Theorem 2.8.7. *Any two complete ordered fields* F *and* F' *are order-isomorphic.*

A proof of this theorem is given in McShane and Botts [**46**] (Theorem 6.1, pages 22-24).

The success of mathematicians in establishing the uniqueness under order isomorphism of a complete ordered field is a very satisfying result, as there is essentially only one complete ordered field, and that is the complete ordered field of real numbers.

Exercise.

Exercise 2.8.1. Show that the mean value theorem fails for f on $[1, 2]$ in Example 2.8.4: Even though f is continuous on $[1, 2] \subset \mathbf{Q}$ and differentiable on $(1, 2) \subset \mathbf{Q}$, there is no point $\xi \in (1, 2)$ such that $f(2) - f(1) = f'(\xi)(2 - 1)$.

Chapter 3

Basic Theory of Series

This chapter contains the basic theory of numerical series. A central role is played by geometric series.

3.1. Some Special Sequences

Before introducing infinite series, there are some sequences whose limits are useful enough to single out for attention in this section. We establish these limits in detail and in the process illustrate several useful techniques.

The first result provides a key to understanding geometric series later in Theorem 3.3.1.

Lemma 3.1.1. *If a is a real or complex number and $|a| < 1$, then*

$$\lim_{n\to\infty} a^n = 0.$$

Proof. If $a = 0$, then (a^n) is the constant sequence consisting of zeros only, so the statement is true. Otherwise, we have $0 < |a| < 1$. For each $n \in \mathbf{N}$, $|a^{n+1}| = |a||a^n| < |a^n|$, so the sequence $(|a^n|)$ is monotone decreasing and bounded below by 0. Thus $L = \lim_{n\to\infty} |a^n|$ exists and, in fact, $L = \inf\{|a^n| : n \in \mathbf{N}\}$. Since $|a| > 0$, $|a^n| = |a|^n > 0$ for each n, and therefore $L \geq 0$. We want to show that $L = 0$. Suppose to the contrary that $L > 0$. For each positive integer n, we have

$$|a^n| = \frac{|a^{n+1}|}{|a|} \geq \frac{L}{|a|},$$

by the definition of L. Therefore $L/|a|$ is also a lower bound for $\{|a^n| : n \in \mathbf{N}\}$. But $0 < |a| < 1$ implies that $L/|a| > L$, and this contradicts the fact that L is the greatest lower bound. Hence, $L = 0 = \lim_{n\to\infty} |a^n|$, and this is equivalent to $\lim_{n\to\infty} a^n = 0$. □

Other proofs of Lemma 3.1.1 may be of interest in showing different techniques. After the limit L is established, one can argue as follows. Since $\lim_{n\to\infty} |a^n| = L$, we also have $\lim_{n\to\infty}|a^{n+1}| = L$, since $(|a^{n+1}|)$ is a subsequence of $(|a^n|)$. Hence,

$$L = \lim_{n\to\infty}|a^{n+1}| = |a|\lim_{n\to\infty}|a^n| = |a|L,$$

and therefore $(1-|a|)L = 0$. By hypothesis, $|a| < 1$, so $L = 0$.

Still another proof of Lemma 3.1.1 can be based on the use of Bernoulli's inequality. Take $0 < |a| < 1$ and write $|a| = 1/(1+h)$ where $h > 0$, that is, choose $h = (1-|a|)/|a|$. By Bernoulli's inequality, for each positive integer n,

$$\frac{1}{|a^n|} = (1+h)^n \geq 1 + nh,$$

so we have

$$0 < |a^n| \leq \frac{1}{1+nh} < \frac{1}{nh}.$$

Given $\epsilon > 0$, there is an n_0 such that $n_0 > 1/(\epsilon h)$ (by the Archimedean property). Then for $n \geq n_0$,

$$|a^n| < \frac{1}{nh} \leq \frac{1}{n_0 h} < \epsilon.$$

Since this is true for every $\epsilon > 0$, $\lim_{n\to\infty}|a^n| = 0$ if $0 < |a| < 1$.

It is useful to define what is meant by a sequence approaching $\pm\infty$. This is different from saying that the sequence is unbounded (in either the positive or negative direction).

Definition 3.1.2. *A sequence (a_k)* **diverges to** $+\infty$, *written conveniently (with a slight abuse of limit notation) as*

$$\lim_{k\to\infty} a_k = +\infty,$$

if for every $M \in \mathbf{N}$ there is an $N = N(M)$ such that if $k \geq N$, then $a_k > M$. Similarly, (a_k) **diverges to** $-\infty$, *written conveniently as*

$$\lim_{k\to\infty} a_k = -\infty,$$

if for every $M \in \mathbf{N}$ there is an $N = N(M)$ such that if $k \geq N$, then $a_k < -M$.

As we mentioned, divergence to $\pm\infty$ is different from saying that a sequence is unbounded in either the positive or the negative direction. For example, the sequence $a_k = (-1)^k 2^k$ is unbounded in both the positive and negative directions, in the sense that for every $M \in \mathbf{N}$ we can find indices k, l such that $a_k > M$ and $a_l < -M$, but this sequence does not diverge to either $+\infty$ or to $-\infty$. However, we *can* say that $|a_k|$ diverges to $+\infty$.

Another example to consider is the sequence

$$(a_k)_{k=1}^{\infty} = \left(\left|k \sin\frac{k\pi}{2}\right|\right) = (1, 0, 3, 0, 5, 0, 7, 0, \ldots),$$

which is unbounded in the positive direction, but does not diverge to $+\infty$ since $a_k = 0$ for all even k.

Lemma 3.1.3. *If $a > 0$, then $\lim_{n\to\infty} a^{1/n} = 1$.*

Proof. If $a = 1$, the statement is clearly true. If $a > 1$, then for each positive integer n, $a^{1/n} > 1$. (Otherwise, with $a^{1/n} \leq 1$, we have $a = (a^{1/n})^n \leq 1^n = 1$.) Let $b_n = a^{1/n} - 1$. Then $b_n > 0$, and $(1 + b_n)^n = a$. By the binomial theorem,

$$a = (1 + b_n)^n \geq 1 + nb_n,$$

and hence

$$0 < b_n \leq \frac{a-1}{n}.$$

This shows that $\lim_{n\to\infty} b_n = 0$, which is equivalent to $\lim_{n\to\infty} a^{1/n} = 1$.

If $0 < a < 1$, then $1/a > 1$, and therefore by the case already considered,

$$\lim_{n\to\infty} (1/a)^{1/n} = 1.$$

Then, by the quotient limit law, $\lim_{n\to\infty} a^{1/n} = \lim_{n\to\infty} \frac{1}{(1/a)^{1/n}} = 1$. □

The next lemma is used to help define the Euler number e in Theorem 3.1.5 below.

Lemma 3.1.4. *Let a and b be real numbers such that $0 \leq a < b$. Then*

$$\frac{b^{n+1} - a^{n+1}}{b - a} < (n+1)b^n.$$

Proof. One verifies directly that for any a and b,

$$b^{n+1} - a^{n+1} = (b-a)(b^n + ab^{n-1} + a^2b^{n-2} + \cdots + a^{n-1}b + a^n).$$

If $0 \leq a < b$, then

$$\begin{aligned} \frac{b^{n+1} - a^{n+1}}{b - a} &= b^n + ab^{n-1} + a^2b^{n-2} + \cdots + a^{n-1}b + a^n \\ &< b^n + bb^{n-1} + b^2b^{n-2} + \cdots + b^{n-1}b + b^n \\ &= (n+1)b^n, \end{aligned}$$

as desired. □

Theorem 3.1.5. *The sequence $(1 + 1/n)^n$ is increasing and convergent. The limit is called the* **Euler number** *and is denoted by e. Thus,*

$$e = \lim_{n\to\infty} \left(1 + \frac{1}{n}\right)^n.$$

Proof. By Lemma 3.1.4, if $0 \leq a < b$, then

$$b^{n+1} - a^{n+1} < (n+1)b^n(b-a),$$

which is equivalent to

$$b^n[b - (n+1)(b-a)] < a^{n+1}.$$

If $a = 1 + \frac{1}{n+1}$ and $b = 1 + \frac{1}{n}$, then $0 \leq a < b$ and $b - (n+1)(b-a) = 1$, so we have

$$b^n < a^{n+1}, \quad \text{which is} \quad \left(1 + \frac{1}{n}\right)^n < \left(1 + \frac{1}{n+1}\right)^{n+1}.$$

This shows that the sequence $(1 + 1/n)^n$ is increasing.

If we let $a = 1$ and $b = 1 + \frac{1}{2n}$, then $b - (n+1)(b-a) = \frac{1}{2}$, so

$$\frac{1}{2}b^n < a^{n+1}, \quad \text{which is} \quad \frac{1}{2}\left(1 + \frac{1}{2n}\right)^n < 1^{n+1} = 1.$$

Multiply the latter inequality by 2 and then square the result to obtain

$$\left(1 + \frac{1}{2n}\right)^{2n} < 4.$$

Since the sequence $(1 + \frac{1}{n})^n$ is increasing, for each positive integer n we have

$$\left(1 + \frac{1}{n}\right)^n < \left(1 + \frac{1}{2n}\right)^{2n} < 4.$$

So $(1 + \frac{1}{n})^n$ is bounded above and hence convergent. □

Theorem 3.1.6. *The sequence* $(n^{1/n})$, *for* $n = 3, 4, \ldots$, *is decreasing and convergent, and* $\lim_{n\to\infty} n^{1/n} = 1$.

Proof. We have seen, at the end of the proof of Theorem 3.1.5, that for each positive integer n,

$$\left(1 + \frac{1}{n}\right)^n < 4.$$

So we certainly have

$$\frac{(n+1)^n}{n^n} \le n$$

for $n \ge 4$. Hence, for $n \ge 4$, we have

$$(n+1)^n \le n^{n+1} \implies (n+1)^{n/(n+1)} \le n \implies (n+1)^{1/(n+1)} \le n^{1/n}.$$

This shows that the sequence $(n^{1/n})$ is decreasing for $n \ge 4$, and since it is bounded below by 1, $\lim_{n\to\infty} n^{1/n} = L$ exists and $L \ge 1$. Then the sequence of squares $(n^{2/n})$ has limit L^2. Now,

$$\lim_{n\to\infty} \left(\frac{n}{2}\right)^{2/n} = \lim_{n\to\infty} n^{2/n}\left(\frac{1}{2}\right)^{2/n} = L^2,$$

by Lemma 3.1.1 and the product limit law. The subsequence $(2n/2)^{2/2n}$ of $(n/2)^{2/n}$ also converges to L^2. Hence,

$$\left(\frac{2n}{2}\right)^{2/2n} = n^{1/n} \to L^2,$$

from which we conclude that $L^2 = L$. Since $L \neq 0$, we have $L = 1$. □

Exercises.

Exercise 3.1.1. Find $\lim_{n\to\infty}(\frac{1}{2} - i\frac{1}{3})^n$.

Exercise 3.1.2. *Prove*: If $a > 1$, then $\lim_{k\to\infty} a^k = +\infty$ (Definition 3.1.2). *Hint*: Let $a = 1 + \delta$, where $\delta > 0$, and use the estimate $(1+\delta)^k > 1 + k\delta$ which follows from the binomial expansion of $(1+\delta)^k$.

Exercise 3.1.3. If q is a real number such that $|q| < 1$, then $\lim_{n\to\infty} nq^n = 0$. *Hint*: If $q > 0$, write $q = 1/(1+h)$ where $h > 0$. *Hint*: Use the previous exercise.

Exercise 3.1.4. Find the indicated limits:

1. $\lim_{n\to\infty}(1+\frac{1}{n^2})^{n^2}$;
2. $\lim_{n\to\infty}(1+\frac{1}{n^2})^{n}$;
3. $\lim_{n\to\infty}(1+\frac{1}{3n})^{3n}$;
4. $\lim_{n\to\infty}(1+\frac{1}{3n})^{n}$.

Exercise 3.1.5. Show that $\lim_{n\to\infty}(n^2)^{1/n}=1$.

Exercise 3.1.6. Show that for any $k\in\mathbf{N}$, $\lim_{n\to\infty}(n^k)^{1/n}=1$.

3.2. Introduction to Series

A finite series of real or complex numbers is a sum indicated using standard summation notation, as in the expression

$$\sum_{k=1}^{N} a_k = a_1 + a_2 + \cdots + a_N.$$

The elements of the finite sequence (a_k) can be real or complex numbers. The numerical value of a *finite* sum is always well-defined. We must define what is meant by an *infinite series* of real or complex numbers a_k, denoted $\sum_{k=1}^{\infty} a_k$, where $(a_k)_{k=1}^{\infty}$ is an infinite sequence.

Definition 3.2.1. *An infinite series of real or complex numbers, $\sum_{k=1}^{\infty} a_k$, is the* **sequence of partial sums** $(s_n)_{n=1}^{\infty}$ *where*

$$s_n := \sum_{k=1}^{n} a_k = a_1 + a_2 + \cdots + a_{n-1} + a_n.$$

The numbers a_k are called the **terms** *of the series.*

Our concept of summation of an infinite series is essentially the idea of adding the terms one by one, accumulating a running total for the sum. The appropriate concept for a running total is the *partial sum*, which respects the given ordering of the terms as indexed by $k\in\mathbf{N}$.

Definition 3.2.2. *An infinite series $\sum_{k=1}^{\infty} a_k$* **converges** *if $\lim_{n\to\infty} s_n$ exists, and we define the* **sum** *of the series to be $s=\lim_{n\to\infty} s_n$. If (s_n) does not converge, then we say the series* **diverges**.

Example 3.2.3. If the real number x has decimal representation $a_0.a_1a_2\ldots$, then the series $\sum_{k=0}^{\infty} a_k 10^{-k}$ converges with sum x. △

Example 3.2.4. Consider the infinite series

$$\sum_{k=0}^{\infty}\frac{1}{2^k} = 1 + \frac{1}{2} + \frac{1}{2^2} + \frac{1}{2^3} + \cdots.$$

Since the summation begins at $k=0$, it is convenient to write s_n for the sum of the terms through the nth power term:

$$s_n = \sum_{k=0}^{n}\frac{1}{2^k} = 1 + \frac{1}{2} + \frac{1}{2^2} + \cdots + \frac{1}{2^n}.$$

By Exercise 1.3.4 on finite geometric sums, we have

$$s_n = \sum_{k=0}^{n} \left(\frac{1}{2}\right)^k = \frac{1-(1/2)^{n+1}}{1-(1/2)} = 2[1-(1/2)^{n+1}].$$

Letting $n \to \infty$, we have $(1/2)^{n+1} \to 0$ by Lemma 3.1.1, and therefore

$$\lim_{n\to\infty} s_n = 2.$$

The sequence of partial sums converges with limit 2. Therefore the series $\sum_{k=1}^{\infty} 1/2^k$ converges, with sum 2. We write $\sum_{k=1}^{\infty} 1/2^k = 2$. △

There is a *Cauchy criterion* for convergence of series.

Theorem 3.2.5. *An infinite series $\sum_{k=1}^{\infty} a_k$ of real or complex numbers converges if and only if for every $\epsilon > 0$ there is an integer $N = N(\epsilon)$ such that if $m > n \geq N$, then*

$$|s_m - s_n| = |a_{n+1} + a_{n+2} + \cdots + a_m| < \epsilon.$$

Proof. The series converges if and only if the sequence of partial sums converges, if and only if (s_n) is a Cauchy sequence in **R** or **C**. □

Example 3.2.6 (The Harmonic Series)**.** The infinite series

$$\sum_{k=1}^{\infty} \frac{1}{k} = 1 + \frac{1}{2} + \frac{1}{3} + \frac{1}{4} + \cdots$$

is known as the *harmonic series*. We will show that the harmonic series diverges by showing that the sequence of partial sums is not a Cauchy sequence. Let

$$s_n = \sum_{k=1}^{n} \frac{1}{k} = 1 + \frac{1}{2} + \frac{1}{3} + \cdots + \frac{1}{n-1} + \frac{1}{n}.$$

Consider, in particular, the partial sums s_n, s_m where $n = 2^k$ and $m = 2^{k+1} = 2^k + 2^k$. Then the difference $s_m - s_n$ consists of 2^k nonzero terms,

$$|s_m - s_n| = s_m - s_n = \frac{1}{2^k+1} + \frac{1}{2^k+2} + \cdots + \frac{1}{2^k+2^k}.$$

Each of these terms is greater than or equal to $1/2^{k+1}$, so we have

$$s_m - s_n > 2^k \frac{1}{2^{k+1}} = \frac{1}{2}.$$

This is true for any choice of k in defining the numbers $n = 2^k$ and $m = 2^{k+1}$. Hence (s_n) is not a Cauchy sequence, and in fact the partial sums are unbounded. Therefore the harmonic series diverges. △

In each of the last two examples, the general summand a_k of the series satisfied the condition that $\lim_{k\to\infty} a_k = 0$. Notice that the sequence $1/2^k$ converges to zero faster than $1/k$. The more rapid convergence of $1/2^k$ to zero enabled the convergence of the partial sums in Example 3.2.4, while the relatively slower convergence of $1/k$ allows the partial sums of the harmonic series to become unbounded. But there is no sharp dividing line between convergent series of positive terms and divergent series of positive terms with regard to the rate of convergence of a_k to zero.

(See Exercise 3.2.1.) We have the following important *necessary* condition for the convergence of an infinite series of real or complex numbers a_k.

Theorem 3.2.7. *If the series $\sum_{k=1}^{\infty} a_k$ converges, then $\lim_{k\to\infty} a_k = 0$.*

Proof. If the series converges, then the sequence $(s_n)_{n=1}^{\infty}$ of partial sums converges. Note that $a_n = s_n - s_{n-1}$ and

$$\lim_{n\to\infty} a_n = \lim_{n\to\infty} (s_n - s_{n-1}) = \lim_{n\to\infty} s_n - \lim_{n\to\infty} s_{n-1} = 0,$$

which completes the proof. □

The condition $\lim_{k\to\infty} a_k = 0$ is *not sufficient* for convergence of $\sum_{k=1}^{\infty} a_k$. This failure of the converse of Theorem 3.2.7 is illustrated clearly by the harmonic series in Example 3.2.6. We can never conclude solely on the basis of $a_k \to 0$ that the series $\sum_{k=1}^{\infty} a_k$ converges.

A statement equivalent to Theorem 3.2.7 is its contrapositive, which provides *a direct test for divergence of infinite series*: If $\lim_{k\to\infty} a_k$ does not exist or, if existing, $\lim_{k\to\infty} a_k \neq 0$, then the series $\sum_{k=1}^{\infty} a_k$ diverges.

Example 3.2.8. The series $\sum_{k=1}^{\infty}(k-3)/2k$ diverges, because

$$\lim_{k\to\infty} a_k = \lim_{k\to\infty} \frac{k-3}{2k} = \frac{1}{2} \neq 0.$$

The series $\sum_{k=1}^{\infty} k\cos(1/k)$ diverges, because $\lim_{k\to\infty} k\cos(1/k)$ does not exist. △

It is important to have some basic facts about sums, differences and constant multiples of series.

Theorem 3.2.9 (Sums and Constant Multiples of Series). *Let (a_k) and (b_k) be sequences of real or complex numbers, and let c be a real or complex constant. If $\sum_{k=1}^{\infty} a_k$ and $\sum_{k=1}^{\infty} b_k$ converge, then their sum, difference, and constant multiple by c also converge; moreover, $\sum_{k=1}^{\infty}(a_k \pm b_k) = \sum_{k=1}^{\infty} a_k \pm \sum_{k=1}^{\infty} b_k$ and $\sum_{k=1}^{\infty} ca_k = c\sum_{k=1}^{\infty} a_k$.*

Proof. Each result follows from the corresponding parts of Theorem 2.4.5 on sums, differences and constant multiples of convergent sequences. The details are left to Exercise 3.2.2. □

Theorem 3.2.10. *Suppose $a_k \geq 0$ for all $k = 1, 2, \ldots$. Then $\sum_{k=1}^{\infty} a_k$ converges if and only if the sequence of partial sums s_n is bounded above.*

Proof. The partial sums s_n form a monotone increasing sequence since $a_k \geq 0$ for all k. If (s_n) converges, then the partial sums are bounded above, by Theorem 2.4.4. If the partial sums are bounded above, then (s_n) converges, by Theorem 2.4.15. □

Theorem 3.2.11. *Let numbers b_k be given, for $k = 1, 2, \ldots$. The telescoping series $\sum_{k=1}^{\infty}(b_{k+1} - b_k)$ converges if and only if $\lim_{n\to\infty} b_n$ exists, and when convergent, the series sum is*

$$\sum_{k=1}^{\infty}(b_{k+1} - b_k) = \lim_{n\to\infty} b_n - b_1.$$

Proof. The partial sum s_n is $s_n = \sum_{k=1}^{n}(b_{k+1} - b_k) = b_{n+1} - b_1$. Therefore the telescoping series converges if and only if $\lim_{n\to\infty} b_{n+1} = \lim_{n\to\infty} b_n$ exists. □

Example 3.2.12. We may use partial fractions to write

$$\sum_{k=1}^{\infty} \frac{1}{k(k+1)} = \sum_{k=1}^{\infty} \left(\frac{1}{k} - \frac{1}{k+1}\right) = -\sum_{k=1}^{\infty} \left(\frac{1}{k+1} - \frac{1}{k}\right).$$

With $b_k = 1/k$, we have $\lim_{k\to\infty} b_k = 0$, so $\sum_{k=1}^{\infty} 1/[k(k+1)] = -(-b_1) = 1$. △

Exercises.

Exercise 3.2.1. Let $a_k > 0$ for all $k \in \mathbf{N}$.

1. Show: If $\sum_{k=1}^{\infty} a_k$ converges, then there are numbers c_k with $0 < a_k < c_k$ such that $\sum_{k=1}^{\infty} c_k$ converges. (There is no "largest" convergent series.)
2. Show: If $\sum_{k=1}^{\infty} a_k$ diverges, then there are numbers d_k with $0 < d_k < a_k$ such that $\sum_{k=1}^{\infty} d_k$ diverges. (There is no "smallest" divergent series.)

Exercise 3.2.2. Provide the details in the proof of Theorem 3.2.9.

Exercise 3.2.3. Determine whether these series converge, and find the sum if the series does converge:

1. $(1/\sqrt{2} - 1) + (1/\sqrt{3} - 1/\sqrt{2}) + (1/\sqrt{4} - 1/\sqrt{3}) + \cdots$.
2. $\frac{1}{1\cdot 3} + \frac{1}{2\cdot 4} + \frac{1}{3\cdot 5} + \cdots$. *Hint*: $1/(k+2) - 1/k = 1/(k+2) - 1/(k+1) + 1/(k+1) - 1/k$.

Exercise 3.2.4. Set $m = n + 1$ in the Cauchy criterion for series to deduce the necessary condition, $a_n \to 0$, for convergence of the series.

Exercise 3.2.5. Consider the series

$$\sum_{k=1}^{\infty} \log\left(1 + \frac{1}{k}\right) = \log 2 + \log\frac{3}{2} + \log\frac{4}{3} + \cdots .$$

1. Show that the sequence of general terms $a_k = \log(1 + \frac{1}{k})$ converges to zero.
2. Find an expression for the sum of the first n terms of the series.
3. Show that the series diverges.

3.3. The Geometric Series

Let q be a real or complex number. The infinite series $\sum_{k=0}^{\infty} q^k$ is called a **geometric series**. It is convenient to define

$$s_n = \sum_{k=0}^{n} q^k = 1 + q + q^2 + q^3 + \cdots + q^n,$$

so that s_n is the sum of the first $n + 1$ terms of the geometric series. There is an interesting and useful fact about these partial sums of a geometric series. For any $n \in \mathbf{N}$, the partial sum s_n of the first $n + 1$ terms in a geometric series is given by

$$s_n = \frac{1 - q^{n+1}}{1 - q}. \tag{3.1}$$

In order to see this, note that the product $(1-q)s_n$ yields a telescoping or collapsing sum, so that

$$(1-q)s_n = 1 - q^{n+1},$$

from which (3.1) follows, if $q \neq 1$. We now recall (Lemma 3.1.1) that if q is a number such that $|q| < 1$, then $\lim_{n\to\infty} q^{n+1} = 0$. The next result completely describes the convergence properties of a geometric series.

Theorem 3.3.1 (Geometric Series). *The geometric series $\sum_{k=0}^{\infty} q^k$ converges if $|q| < 1$ and diverges if $|q| \geq 1$. If $|q| < 1$, then*

$$\sum_{k=0}^{\infty} q^k = \frac{1}{1-q}.$$

Proof. The geometric series converges if and only if its sequence (s_n) of partial sums converges. The partial sum $s_n = \sum_{k=0}^{n} q^n$, for $q \neq 1$, is given by the formula in (3.1). If $|q| < 1$, then $\lim_{n\to\infty} q^{n+1} = 0$, and by the quotient limit law,

$$\lim_{n\to\infty} s_n = \lim_{n\to\infty} \frac{1-q^{n+1}}{1-q} = \frac{1}{1-q}.$$

Therefore the series converges, with sum equal to $1/(1-q)$. On the other hand, if $|q| \geq 1$, then $|q^{n+1}| = |q||q^n| \geq |q^n| \geq 1$ for each $n \in \mathbf{N}$, so $\lim_{n\to\infty} q^n \neq 0$ and the series diverges. □

Example 3.3.2. We notice that

$$\sum_{k=1}^{\infty} (9)10^{-k} = 9\sum_{k=0}^{\infty} 10^{-k} - 9 = 9\frac{1}{1-(1/10)} - 9 = 9\frac{10}{9} - 9 = 1.$$

Therefore the repeating base 10 decimal $0.99\overline{9}$ represents the number 1. △

Exercises.

Exercise 3.3.1. Determine whether these geometric series converge or diverge, and find the sum if the series converges:

$$\text{(a)} \sum_{k=0}^{\infty} \left(\frac{\sqrt{11}}{\sqrt{2}+\sqrt{3}}\right)^k \qquad \text{(b)} \sum_{k=0}^{\infty} \frac{1}{(1-i)^k} \qquad \text{(c)} \sum_{k=0}^{\infty} \left(\frac{\sqrt{2}}{1+i}\right)^k.$$

Exercise 3.3.2. Use Theorem 3.3.1 to determine whether these series converge or diverge, and find the sum if the series converges:

$$\text{(a)} \sum_{k=1}^{\infty} \frac{1}{2^k} \qquad \text{(b)} \sum_{k=2}^{\infty} \frac{1}{(1-i)^k} \qquad \text{(c)} \sum_{k=2}^{\infty} \left(\frac{\sqrt{2}}{3}\right)^k.$$

Exercise 3.3.3. Suppose the number $.0202\overline{02}$ is a ternary (base-3) representation. What number is represented?

3.4. The Cantor Set

The Cantor set is a remarkable set of real numbers. Its study helps us to begin to see the richness of the real number line. The Cantor set C is a subset of the interval $[0,1]$ constructed by a process of *extracting middle thirds*. We now describe exactly what this means. We begin by considering the following sequence of sets, each of which is a finite union of open intervals:

$$\begin{aligned} D_0 &= \emptyset, \\ D_1 &= \left(\frac{1}{3},\frac{2}{3}\right), \\ D_2 &= \left(\frac{1}{9},\frac{2}{9}\right)\cup\left(\frac{7}{9},\frac{8}{9}\right), \\ D_3 &= \left(\frac{1}{27},\frac{2}{27}\right)\cup\left(\frac{7}{27},\frac{8}{27}\right)\cup\left(\frac{19}{27},\frac{20}{27}\right)\cup\left(\frac{25}{27},\frac{26}{27}\right), \\ &\cdots. \end{aligned}$$

Each of the sets D_n is extracted, or carved away, from $[0,1]$. For each $n \geq 1$, D_n is a union of 2^{n-1} open intervals, each of which is the middle third of an interval that remains after the points in D_{n-1} have been removed.

Definition 3.4.1. *The* **Cantor set** C *is the complement in* $[0,1]$ *of the union of the sets* D_n*:*

$$C = \left\{x \in [0,1] : x \notin \bigcup_{k=0}^{\infty} D_k\right\} = [0,1] - \bigcup_{k=0}^{\infty} D_k.$$

Remark. *The Cantor set* C *is also called the* **Cantor middle thirds set** *or the* **Cantor ternary set**.

We may write

$$C = \left(\bigcup_{k=0}^{\infty} D_k\right)^c$$

with the understanding that the complement is taken with respect to $[0,1]$. By DeMorgan's law,

$$C = \left(\bigcup_{k=0}^{\infty} D_k\right)^c = \bigcap_{k=0}^{\infty} D_k^c.$$

Since we form the Cantor set C by this intersection, let us write $C_0 = [0,1]$ and define $C_k := C_{k-1} - D_k$, for $k \geq 1$. Then it is immediate that

$$C = \bigcap_{k=0}^{\infty} C_k. \tag{3.2}$$

The first few sets C_k are given by

$$\begin{aligned} C_0 &= \left[0,1\right], \\ C_1 &= \left[0,\frac{1}{3}\right]\cup\left[\frac{2}{3},1\right], \\ C_2 &= \left[0,\frac{1}{9}\right]\cup\left[\frac{2}{9},\frac{1}{3}\right]\cup\left[\frac{2}{3},\frac{7}{9}\right]\cup\left[\frac{8}{9},1\right], \end{aligned}$$

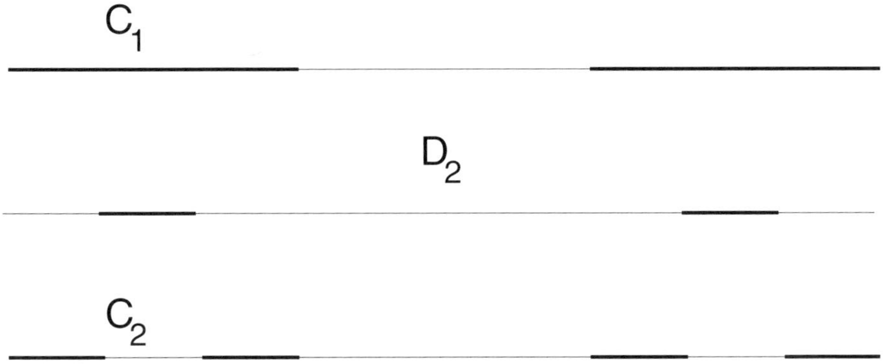

Figure 3.1. The first few sets in the construction of the Cantor set. The indicated sets are shown with bold line segments within the unit interval. The set D_2 consists of the middle thirds of the segments comprising C_1. Then $C_2 = C_1 - D_2$.

and

$$C_3 = \left[0, \frac{1}{27}\right] \cup \left[\frac{2}{27}, \frac{1}{9}\right] \cup \left[\frac{2}{9}, \frac{7}{27}\right] \cup \left[\frac{8}{27}, \frac{1}{3}\right] \cup \left[\frac{2}{3}, \frac{19}{27}\right] \cup \left[\frac{20}{27}, \frac{7}{9}\right] \cup \left[\frac{8}{9}, \frac{25}{27}\right] \cup \left[\frac{26}{27}, 1\right].$$

(See Figure 3.1 for the first few sets in this sequence.) The **length**, denoted by $\lambda(I)$, of a bounded interval $I = [a, b]$ (or $I = (a, b]$ or $I = [a, b)$) is $\lambda(I) = b - a$. One can prove by induction that each closed set C_k is the union of 2^k disjoint closed intervals I_j, each with length $\lambda(I_j) = 1/3^k$. For example, C_{20} is the union of $1,048,576$ disjoint closed intervals of length $1/3^{20} \approx 2.868(10)^{-10}$. Since $C \subset C_k$, C is covered by 2^k disjoint closed intervals whose lengths sum to $2^k/3^k$. But this is true for each positive integer k, so it seems that C takes up no space whatever in $[0, 1]$. The reader is invited to examine this in Exercise 3.4.1.

If the Cantor set takes up no space in $[0, 1]$, then which points, and how many, can it contain? It is clear that the endpoints of the open intervals removed in forming C are in C, so C is an infinite set. For example, the numbers $1/3^k$, for $k \in \mathbf{N}$, are elements of C, as are the numbers $1 - 1/3^k$. What *else* is in C? (See Exercise 3.4.2 for a few more numbers in C.)

It may be surprising that C is not countable. But it is possible to show that C is an uncountable set. To see why this is so, it is helpful to understand the Cantor set from a different point of view.

Every real number x in $[0, 1]$ has a base-3 decimal expansion, which we write as

$$x = \sum_{j=1}^{\infty} b_j 3^{-j}, \quad \text{where } b_j \in \{0, 1, 2\}.$$

This base-3 representation is unique unless x is of the form $q3^{-k}$ for some integers q and k, in which case there can be only two such expansions; one expansion ends in an infinite sequence of 2s and the other expansion ends in an infinite sequence of 0s. (This situation also occurs with decimal expansions and repeated nines versus repeated zeros.) Two examples with $k = 3$ are given by the numbers

$$x = (5)3^{-3} = .011\overline{222}\ldots = .012\overline{000}\ldots$$

and

$$y = (4)3^{-3} = .010\overline{222}\ldots = .011\overline{000}\ldots.$$

Note that one of the expansions for each number has $b_k = b_3 = 1$ and the other has either $b_3 = 0$ or $b_3 = 2$. For such points, we agree to use the representation where $b_k = 0$ or $b_k = 2$, *not* the one where $b_k = 1$. With this agreement, the base-3 expansion of any x in $[0, 1]$ is unique. For example, the expansion for $1/3$ is then $.0\overline{222}\ldots$, for $2/3$ it is $.2\overline{000}\ldots$, for $1/9$, $.00\overline{222}\ldots$, and for $7/9$, $.20\overline{222}\ldots$. The chosen expansion for $(5)3^{-3}$ is $.012\overline{000}\ldots$, and for $(4)3^{-3}$, $.010\overline{222}\ldots$.

Now let $x \in [0, 1]$ and let $\sum_{j=1}^{\infty} b_j 3^{-j}$, where $b_j \in \{0, 1, 2\}$, be the unique base-3 expansion for x, determined by our convention when necessary. Referring to the excluded sets D_n in the construction of the Cantor set, we see that

1. $b_1 = 1$ if and only if $x \in D_1$;
2. $b_1 \neq 1$ and $b_2 = 1$ if and only if $x \in D_2$;
3. $b_1 \neq 1$, $b_2 \neq 1$ and $b_3 = 1$ if and only if $x \in D_3$;

and, in general, $b_j \neq 1$ for $1 \leq j \leq k - 1$ and $b_k = 1$, if and only if $x \in D_k$. We conclude that a number x in $[0, 1]$ is in the Cantor set C if and only if the base-3 expansion for x has $b_j \neq 1$ for all j. This proves the following theorem.

Theorem 3.4.2. *The Cantor set C is the set of all points in $[0, 1]$ that have a base-3 ternary expansion $\sum_{j=1}^{\infty} b_j 3^{-j}$ with $b_j \neq 1$ for all j. Hence,*

$$C = \Big\{x \in [0, 1] : x = \sum_{j=1}^{\infty} \frac{b_j}{3^j}, \text{ where } b_j \in \{0, 2\}\Big\}.$$

We can now show that the Cantor set is uncountable.

Theorem 3.4.3. *The Cantor set C is uncountable.*

Proof. The proof is by a diagonal argument like that in the proof that $\mathbf{R}$ is uncountable. Every $x \in C$ has a unique base-3 ternary expansion $\sum_{j=1}^{\infty} b_j 3^{-j}$ where for each j, either $b_j = 0$ or $b_j = 2$. We know that C is an infinite set. If C is assumed to be countably infinite, then one can derive a contradiction. The details are left as Exercise 3.4.4. □

Exercises.

Exercise 3.4.1. Consider the representation $C = \bigcap_{k=0}^{\infty} C_k$ of the Cantor set.

1. Prove by induction: Each set C_k is the union of 2^k disjoint closed intervals I_j, with length $\lambda(I_j) = 1/3^k$ for $j = 1, \ldots, 2^k$.
2. Prove: For every $\epsilon > 0$ there is a finite collection of closed intervals $\{I_j\}_{j=1}^m$ such that $C \subset \bigcup_{j=1}^m I_j$ and $\sum_{j=1}^m \mu(I_j) < \epsilon$. *Hint*: Use Lemma 3.1.1.

Exercise 3.4.2. Show that the numbers $1/3^k$, $1-1/3^k$, $k \in \mathbf{N}$, are in the Cantor set C. Show also that $1/4$ and $3/4$ are in C. Then show that $1/(3^k 4)$ and $1 - 1/(3^k 4)$, $k \in \mathbf{N}$, are in C. *Hint*: C is symmetric with respect to the midpoint $1/2$ of $[0, 1]$.

Exercise 3.4.3. Write the base-3 expansion of $\frac{1}{9}$, $\frac{2}{9}$, $\frac{7}{9}$, and $\frac{8}{9}$.

Exercise 3.4.4. Prove that the Cantor set C is uncountable. *Hint*: Use a diagonal argument.

Exercise 3.4.5. Show that the sets D_n, removed from the interval $[0, 1]$ in the construction of the Cantor set, have lengths that sum to 1. Compare this with Exercise 3.4.1.

Exercise 3.4.6. *A Cantor set with positive total length*
Let $C_0 = [0, 1]$ and let $0 < \alpha < 1$. Starting with C_0 at step $n = 0$, we remove the open middle portion of C_0 of length $\alpha/3$. For each integer $n \geq 1$, at step n we remove 2^n open intervals, each with length $\alpha 3^{-(n+1)}$, each one the open middle portion of one of the remaining 2^n closed intervals of C_0. (The Cantor set construction had $\alpha = 1$.) Show that the sum of the lengths of all the open intervals removed in forming F is α. Therefore F can be said to have total length $1 - \alpha$. (Later in the text, we will understand this statement to mean that F has Lebesgue measure $1 - \alpha$.)

3.5. A Series for the Euler Number

We give another representation of the important Euler number e, which has been defined by

$$e = \lim_{n\to\infty} \left(1 + \frac{1}{n}\right)^n .$$

(See Theorem 3.1.5.) First, we note that for each positive integer $k \geq 4$, we have $k! \geq 2^k$ and hence $1/k! \leq 1/2^k$. Thus for each $n \geq 4$,

$$s_n = \sum_{k=4}^{n} \frac{1}{k!} \leq \sum_{k=4}^{n} \frac{1}{2^k},$$

where the sums on the right-hand side are the partial sums of a convergent geometric series. Therefore $(s_n)_{n=4}^{\infty}$ is a bounded increasing sequence of positive numbers. We conclude that the series $\sum_{k=0}^{n} 1/k!$ converges.

The following notation is standard for finite products of numbers a_k:

$$\prod_{k=1}^{n} a_k = a_1 a_2 \cdots a_n .$$

Theorem 3.5.1. *The number $e := \lim_{n\to\infty} \left(1 + \frac{1}{n}\right)^n$ is the sum of the following series:*

$$e = \sum_{k=0}^{\infty} \frac{1}{k!} = 1 + 1 + \frac{1}{2!} + \frac{1}{3!} + \cdots + \frac{1}{k!} + \cdots .$$

Proof. For each integer $n \geq 0$, let $s_n = \sum_{k=0}^{n} 1/k!$, and for $n \geq 1$, let $b_n = (1 + \frac{1}{n})^n$. We want to show that $\lim_{n\to\infty} s_n = \lim_{n\to\infty} b_n$. Starting from the binomial theorem, we reduce each term in b_n by the common factors in the quotient

$n!/(n-k)!$, and then divide each of the remaining factors by n, as follows:

$$\begin{aligned}
b_n &= \left(1+\frac{1}{n}\right)^n = \sum_{k=0}^{n} \frac{n!}{k!(n-k)!}\left(\frac{1}{n}\right)^k \\
&= 1+1+\frac{1}{2!}\frac{n(n-1)}{n^2}+\frac{1}{3!}\frac{n(n-1)(n-2)}{n^3}+\cdots+\frac{1}{n!}\frac{n!}{n^n} \\
&= 1+1+\frac{1}{2!}\left(1-\frac{1}{n}\right)+\frac{1}{3!}\left(1-\frac{1}{n}\right)\left(1-\frac{2}{n}\right)+\cdots+\frac{1}{n!}\prod_{k=1}^{n-1}\left(1-\frac{k}{n}\right) \\
&\leq 1+1+\frac{1}{2!}+\frac{1}{3!}+\cdots+\frac{1}{n!} = s_n.
\end{aligned}$$

Hence, $e = \lim_{n\to\infty} b_n \leq \lim_{n\to\infty} s_n$. Now we need the inequality in the other direction. Fix n and a positive integer $m \leq n$. In the expansion above for b_n, by retaining only the terms through the $1/m!$ term, we have

$$b_n \geq 1+1+\frac{1}{2!}\left(1-\frac{1}{n}\right)+\cdots+\frac{1}{m!}\prod_{k=1}^{m-1}\left(1-\frac{k}{n}\right) =: c_{nm}.$$

For each fixed m, we have

$$e = \lim_{n\to\infty} b_n \geq \lim_{n\to\infty} c_{nm} = 1+1+\frac{1}{2!}+\cdots+\frac{1}{m!} = s_m.$$

Thus $e \geq \lim_{m\to\infty} s_m$. We conclude that $e = \sum_{k=0}^{\infty} 1/k!$. □

By using either the partial sums s_n of the series $\sum_{k=0}^{\infty} 1/k!$ or the sequence $b_n = (1+\frac{1}{n})^n$, one can verify that $e \approx 2.718$ to three decimal places.

In general, it can be quite difficult to show that specific numbers are irrational. However, we now have the tools available to show that the Euler number e is irrational. We need the series expression for e, the geometric series, and the fact that the interval $(0,1)$ contains no integer (Lemma 2.3.1).

Theorem 3.5.2. *The Euler number e is irrational.*

Proof. Let n be a positive integer and let $s_n = \sum_{k=0}^{n} 1/k!$. Then $e - s_n > 0$, since all the terms of the series for e are positive. We estimate $e - s_n$ as follows:

$$\begin{aligned}
e - s_n &= \frac{1}{(n+1)!}+\frac{1}{(n+2)!}+\frac{1}{(n+3)!}+\cdots \\
&= \frac{1}{(n+1)!}\left[1+\frac{1}{n+2}+\frac{1}{(n+2)(n+3)}+\cdots\right] \\
&< \frac{1}{(n+1)!}\left[1+\frac{1}{n+1}+\frac{1}{(n+1)^2}+\frac{1}{(n+1)^3}+\cdots\right] \\
&= \frac{1}{(n+1)!}\,\frac{1}{1-(1/(n+1))} \\
&= \frac{1}{(n+1)!}\,\frac{n+1}{n} = \frac{1}{n!\,n}.
\end{aligned}$$

Thus, $0 < e - s_n < 1/(n!\,n)$ for every positive integer n.

Now suppose that e is rational. Then $e = m/n$ for some positive integers m and n. By the estimate just proven, we have

$$0 < e - s_n < \frac{1}{n!\,n},$$

hence

$$0 < n!\,(e - s_n) < \frac{1}{n}.$$

Since

$$e - s_n = \frac{m}{n} - \Big(1 + 1 + \frac{1}{2!} + \frac{1}{3!} + \cdots + \frac{1}{n!}\Big),$$

it follows that $n!\,(e - s_n)$ is a sum of integers, and hence must be an integer. But since $0 < n!\,(e - s_n) < 1$, this contradicts Lemma 2.3.1 which asserts that the interval $(0, 1)$ contains no integer. Therefore e is irrational. $\square$

3.6. Alternating Series

We have already seen an example of a series whose terms have alternating signs, for example, the convergent geometric series

$$\sum_{k=0}^{\infty} (-1/2)^k = 1 - \frac{1}{2} + \frac{1}{4} - \cdots .$$

Definition 3.6.1. *Infinite series of the form*

$$\sum_{k=1}^{\infty} (-1)^{k+1} a_k = a_1 - a_2 + a_3 - \cdots \quad \textit{or} \quad \sum_{k=1}^{\infty} (-1)^k a_k = -a_1 + a_2 - a_3 + \cdots ,$$

where $a_k > 0$ for all positive integers k, are called **alternating series**.

Theorem 3.6.2. *If (a_k) is a decreasing sequence of positive numbers such that $\lim_{k\to\infty} a_k = 0$, then the alternating series $\sum_{k=1}^{\infty} (-1)^{k+1} a_k$ and $\sum_{k=1}^{\infty} (-1)^k a_k$ converge. For either of these convergent series, if s is the sum and s_n is the partial sum of the first n terms, then*

$$|s_n - s| < |a_{n+1}|.$$

Proof. We shall work with the series $\sum_{k=1}^{\infty} (-1)^{k+1} a_k = a_1 - a_2 + a_3 - \cdots$. The argument for the other series is similar. Consider the partial sums s_{2n+1} with an odd number of terms and s_{2n} with an even number of terms. Since (a_k) is decreasing, the subsequence (s_{2n}) of the partial sums is increasing, and the subsequence (s_{2n+1}) is decreasing. Also notice that we have $s_2 \le s_3$ (for $n = 1$), $s_4 \le s_5$ $(n = 2)$, and, in general, $s_{2n} \le s_{2n+1}$. Thus, we have a nested sequence of closed intervals, $[s_{2n}, s_{2n+1}]$, which are nonempty since $a_k \ne 0$ for each k. Since $|s_{2n+1} - s_{2n}| = |a_{2n+1}| \to 0$ as $n \to \infty$, the intersection of these closed intervals is a number s such that $\lim_{n\to\infty} s_{2n} = s = \lim_{n\to\infty} s_{2n+1}$. It follows that $\lim_{n\to\infty} s_n = s$ and the series converges with sum s. For the error estimate in using the partial sum s_n, notice that if $n = 2m$, then $s_{2m} \le s \le s_{2m+1}$, so

$$|s_n - s| = |s_{2m} - s| < |s_{2m+1} - s_{2m}| = |a_{2m+1}|,$$

while if $n = 2m + 1$, then $s_{2m+2} \le s \le s_{2m+1}$, so

$$|s_n - s| = |s_{2m+1} - s| < |s_{2m+2} - s_{2m+1}| = |-a_{2m+2}|.$$

This competes the proof. $\square$

Notice that with $q < 0$ and $|q| < 1$, convergent geometric series $\sum_{k=0}^{\infty} q^k = \sum_{k=0}^{\infty}(-1)^k|q|^k$ are covered by Theorem 3.6.2. In the next example we consider the harmonic series with alternating signs.

Example 3.6.3. The **alternating harmonic series** is

$$\sum_{k=1}^{\infty}(-1)^{k+1}\frac{1}{k} = 1 - \frac{1}{2} + \frac{1}{3} - \frac{1}{4} - \cdots .$$

Theorem 3.6.2 applies, and this alternating series converges. △

Example 3.6.4. Consider the series

$$\sum_{k=1}^{\infty}(-1)^k\frac{1}{\sqrt[3]{k}} \quad \text{and} \quad \sum_{k=1}^{\infty}(-1)^{k+1}\frac{1}{\sqrt{k^2+k}}.$$

Theorem 3.6.2 applies to each series, and we conclude that each converges. △

Exercises.

Exercise 3.6.1. Let $a_k = k^3$ for $1 \le k \le 10^{10}$, and $a_k = (-1)^k/k^{4/3}$ for $k > 10^{10}$. Show that $\sum_{k=1}^{\infty} a_k$ converges.

Exercise 3.6.2. Give an example of convergent series $\sum a_k$ and $\sum b_k$ such that $\sum a_k b_k$ does not converge.

Exercise 3.6.3. Show that the series $\sum_{k=1}^{\infty}(-1)^{k+1}x^k/k$ converges for $0 < x < 1$. (The sum is $\log(1+x)$ for these values of x, where log denotes the natural logarithm function.) Show that the series converges when $x = 1$. (The sum when $x = 1$ can be shown to equal $\log 2$.) What about $x = -1$?

3.7. Absolute Convergence and Conditional Convergence

A useful approach to the question of series convergence involves an examination of the series of absolute values of the original terms.

Definition 3.7.1. *Let a_k be real or complex numbers. The series $\sum_{k=1}^{\infty} a_k$* **converges absolutely** *(is* **absolutely convergent***) if the series with nonnegative terms $\sum_{k=1}^{\infty} |a_k|$ converges.*

The following result is of fundamental importance.

Theorem 3.7.2. *If $\sum_{k=1}^{\infty} a_k$ converges absolutely, then it converges.*

Proof. Assume $\sum_{k=1}^{\infty} a_k$ converges absolutely, that is, $\sum_{k=1}^{\infty} |a_k|$ converges. Denoting the partial sums of the latter series by

$$S_n = \sum_{k=1}^{n} |a_k|,$$

we have that (S_n) is a Cauchy sequence of real numbers. We wish to show that the partial sums

$$s_n = \sum_{k=1}^{n} a_k$$

for the original series form a Cauchy sequence, from which we may conclude that (s_n) converges. Given $\epsilon > 0$, there is an N such that if $m > n \geq N$, then

$$|S_m - S_n| = S_m - S_n < \epsilon.$$

Since $s_m - s_n = a_{n+1} + \cdots + a_m$, we have the estimate

$$|s_m - s_n| = |a_{n+1} + \cdots + a_m| \leq |a_{n+1}| + \cdots + |a_m| = S_m - S_n.$$

Thus, if $m > n \geq N$, then

$$|s_m - s_n| \leq S_m - S_n < \epsilon.$$

Since $\epsilon > 0$ is arbitrary, (s_n) is Cauchy and hence $\lim_{n\to\infty} s_n$ exists. □

The potential for wide applicability of Theorem 3.7.2 depends on having a good supply of sufficient conditions for the convergence of series with nonnegative terms. In the next section we consider several useful convergence tests for such series.

Let us consider some standard terminology for the convergent series not covered by Theorem 3.7.2.

Definition 3.7.3. *The series $\sum_{k=0}^{\infty} a_k$ is said to be* **conditionally convergent** *if it converges, but $\sum_{k=0}^{\infty} |a_k|$ does not converge. Thus a series is conditionally convergent if it converges, but does not converge absolutely.*

An example of a conditionally convergent series is the alternating harmonic series given by $\sum_{k=1}^{\infty}(-1)^k\frac{1}{k}$. This series converges, but not absolutely, since $\sum_{k=1}^{\infty}|(-1)^k\frac{1}{k}| = \sum_{k=1}^{\infty}\frac{1}{k}$ is the harmonic series, which diverges.

The term *conditional* convergence is appropriate, due to the fact that the convergence is conditioned on the stated ordering of the terms a_k in the series. As we will see, if a series of real numbers converges conditionally, then the terms of the series may be reordered in such a way as to create a new series that converges to any pre-specified real number. On the other hand, a series that converges absolutely will still converge, and to the same sum, if the terms of the series are rearranged in any order whatever. These matters are discussed in greater detail in the final section of this chapter.

Exercise.

Exercise 3.7.1. Prove the following statements:

1. If $b_k \geq 0$ for all k, $\sum_{k=1}^{\infty} b_k$ converges with sum b, and $|a_k| \leq b_k$ for all k, then $\sum_{k=1}^{\infty} a_k$ converges.
2. If the a_k are real numbers and $a = \sum_{k=1}^{\infty} a_k$, then $-b \leq a \leq b$, where $b = \sum_{k=1}^{\infty} b_k$ from part 1.
3. If the a_k are complex numbers and the sum $a = \sum_{k=1}^{\infty} a_k$ is not a real number, then $|a| \leq b$. *Hint*: Denote the n-th partial sums for the two series by $s_n = \sum_{k=1}^{n} a_k$ and $S_n = \sum_{k=1}^{n} b_k$, and use the triangle inequality to compare them. In particular, show that $|s_n| \leq S_n$ and that for $n > m$, $|s_n - s_m| \leq |S_n - S_m|$.

3.8. Convergence Tests for Series with Positive Terms

We begin with simple direct comparisons for series with nonnegative terms, giving a basic convergence test and a divergence test.

Theorem 3.8.1. *Let $a_k \geq 0$, $c_k \geq 0$ and $d_k \geq 0$ for each k. The following statements are true:*

1. *If there is an N such that $a_k \leq c_k$ for all $k \geq N$, and $\sum_{k=1}^{\infty} c_k$ converges, then $\sum_{k=1}^{\infty} a_k$ converges.*
2. *If there is an N such that $a_k \geq d_k$ for all $k \geq N$, and $\sum_{k=1}^{\infty} d_k$ diverges, then $\sum_{k=1}^{\infty} a_k$ diverges.*

Proof. It suffices to consider the series as beginning with index value $k = N$, since the convergence or divergence of a series is independent of the first N terms. Series with nonnegative terms have monotone increasing sequences of partial sums, and these partial sums converge if and only if they are bounded. If $a_k \leq c_k$ for $k \geq N$, and $\sum_{k=N}^{\infty} c_k$ converges, then there is a number B such that

$$a_N + \cdots + a_{N+n} \leq c_N + \cdots + c_{N+n} \leq B \quad \text{for all } n \in \mathbf{N}.$$

Therefore $\sum_{k=N}^{\infty} a_k$ converges since its partial sums are bounded. On the other hand, if $a_k \geq d_k$ for $k \geq N$, and $\sum_{k=N}^{\infty} d_k$ diverges, then

$$a_N + \cdots + a_{N+n} \geq d_N + \cdots + d_{N+n} \to \infty \quad \text{as } n \to \infty.$$

Therefore $\sum_{k=N}^{\infty} a_k$ diverges since its partial sums are unbounded. □

Example 3.8.2. Consider the series $\sum_{k=1}^{\infty} 5/(2^k + 3)$. We have

$$\frac{5}{2^k + 3} \leq \frac{5}{2^k}.$$

Since $\sum_{k=1}^{\infty} 1/2^k$ converges, we conclude that $\sum_{k=1}^{\infty} 5/2^k = 5 \sum_{k=1}^{\infty} 1/2^k$ converges. △

Example 3.8.3. For the series $\sum_{k=1}^{\infty} k/(k^2 + k + 3)$, we have

$$\frac{k}{k^2 + k + 3} \leq \frac{k}{k^2} = \frac{1}{k},$$

but $\sum 1/k$ *diverges* and is termwise larger than our series; this comparison is inconclusive. It might appear difficult to get a series that is termwise smaller than $k/(k^2 + k + 3)$ and known to diverge. But notice that

$$\frac{k}{k^2 + k^2 + k^2} < \frac{k}{k^2 + k + 3} \quad \text{for } k \geq 2,$$

and since $\sum k/3k^2 = \frac{1}{3} \sum \frac{1}{k}$ diverges, our original series must diverge. △

When a direct comparison is difficult, the limit comparison test of Exercise 3.8.2 is sometimes helpful.

Exercises.

Exercise 3.8.1. Prove: If a_k, c_k are positive for each k and there are constants $\alpha > 0$ and $N > 0$ such that $a_k < \alpha c_k$ for all $k \geq N$, then the convergence of $\sum_{k=1}^{\infty} c_k$ implies the convergence of $\sum_{k=1}^{\infty} a_k$.

Exercise 3.8.2. *A limit comparison test*
Prove: If $a_k > 0$ and $b_k > 0$ for each k, and if

$$\lim_{k\to\infty} \frac{a_k}{b_k} = L > 0,$$

then $\sum_{k=1}^{\infty} a_k$ converges if and only if $\sum_{k=1}^{\infty} b_k$ converges. *Hint*: By the limit hypothesis, there exists N such that if $k \geq N$, then $L/2 < a_k/b_k < 3L/2$. Use the result of Exercise 3.8.1.

Exercise 3.8.3. *More limit comparisons*

1. Prove: If $a_k > 0$ and $b_k > 0$ for each k, and if $\lim_{k\to\infty} a_k/b_k = 0$ and $\sum b_k$ converges, then $\sum a_k$ converges.
2. Prove: If $a_k > 0$ and $b_k > 0$ for each k, and if $\lim_{k\to\infty} a_k/b_k = \infty$ and $\sum b_k$ diverges, then $\sum a_k$ diverges.

Exercise 3.8.4. Which series converge and which diverge?

$$\text{(a)} \sum_{k=1}^{\infty} \frac{k+2}{k\sqrt{k}} \qquad \text{(b)} \sum_{k=1}^{\infty} \frac{1+1/k}{5^k} \qquad \text{(c)} \sum_{k=1}^{\infty} \frac{1}{|\sin(k)|}.$$

3.9. Geometric Comparisons: The Ratio and Root Tests

Since geometric series are completely understood they are a good choice as the dominating series for comparison purposes. We obtain some useful general tests for absolute convergence in this way. The following special case of Theorem 3.8.1 is important.

Theorem 3.9.1. *If there is a $B > 0$ and $q \in (0, 1)$ such that $|a_k| \leq Bq^k$ for all $k \in \mathbf{N}$, then the series $\sum_{k=1}^{\infty} a_k$ converges absolutely.*

Proof. The geometric series $\sum_{k=1}^{\infty} Bq^k$ converges (to $Bq/(1-q)$), hence $\sum_{k=1}^{\infty} a_k$ converges absolutely by Theorem 3.8.1. □

There are useful variations of this result. For example, the hypothesis that $|a_k| \leq Bq^k$ for all $k \geq M$ for some positive integer M also implies that $\sum_{k=1}^{\infty} a_k$ converges absolutely. From the geometric series comparison in Theorem 3.9.1 we obtain two important sufficient conditions for absolute convergence. These are the basic versions of the ratio test and the root test usually discussed in introductory calculus.

Theorem 3.9.2. *Let $\sum_{k=1}^{\infty} a_k$ be a series of nonzero real numbers.*

1. *If $\lim_{k\to\infty} |a_{k+1}|/|a_k| < 1$, then the series converges absolutely.*
2. *If $\lim_{k\to\infty} |a_{k+1}|/|a_k| > 1$, then the series diverges.*
3. *If $\lim_{k\to\infty} |a_{k+1}|/|a_k| = 1$, then the test is inconclusive.*

Proof. To prove statement 1, let $\alpha = \lim_{k\to\infty} |a_{k+1}|/|a_k| < 1$ and choose q such that $\alpha < q < 1$. There exists $N = N(\epsilon)$ such that if $k \geq N$, then $|a_{k+1}|/|a_k| < q < 1$. Hence, $|a_k| \leq q^{k-N}|a_N|$ for $k \geq N$. Since $q = q^{k+1}/q^k$, we have

$$\frac{|a_{k+1}|}{q^{k+1}} < \frac{|a_k|}{q^k} \leq q^{-N}|a_N| \quad \text{for } k \geq N,$$

and therefore $|a_k| \leq Bq^k$ if $k > N$, where $B = q^{-N}|a_N|$. Thus the series converges absolutely by Theorem 3.9.1. The reader is invited to prove part 2 in Exercise 3.9.1. See also Exercise 3.9.3 for an illustration of part 3. □

Theorem 3.9.3. *Let* $\sum_{k=1}^{\infty} a_k$ *be a series of real numbers.*

1. *If* $\lim_{k\to\infty} |a_k|^{1/k} < 1$, *then the series converges absolutely.*
2. *If* $\lim_{k\to\infty} |a_k|^{1/k} > 1$, *then the series diverges.*
3. *If* $\lim_{k\to\infty} |a_k|^{1/k} = 1$, *then the test is inconclusive.*

Proof. We prove part 1 here. If $\lim |a_k|^{1/k} = \alpha < 1$, choose q such that $\alpha < q < 1$. There exists N such that $|a_k|^{1/k} < q < 1$ for $k \geq N$, hence $|a_k| \leq q^k$ for $k \geq N$. Since $\sum_{k=N}^{\infty} q^k$ converges, $\sum_{k=1}^{\infty} a_k$ converges absolutely. Exercise 3.9.2 asks for a proof of part 2, and Exercise 3.9.3 provides an illustration of part 3. □

More general versions of the ratio test and root test allow the maximum applicability from conditions similar to those in Theorem 3.9.2 and Theorem 3.9.3. These general versions of the ratio test and root test appear in Section 3.11. They use the concepts of *limit superior* and *limit inferior* of a sequence of real numbers, discussed in Section 3.10.

Exercises.

Exercise 3.9.1. Prove part 2 of Theorem 3.9.2.

Exercise 3.9.2. Prove part (2) of Theorem 3.9.3.

Exercise 3.9.3. Consider the divergent series $\sum_{k=1}^{\infty} 1/k$ and the convergent series $\sum_{k=1}^{\infty} 1/k^2$. Show that $\lim_{k\to\infty} |a_{k+1}/a_k| = 1$ and $\lim_{k\to\infty} |a_k|^{1/k} = 1$ for each of these series. This establishes part 3 for both Theorem 3.9.2 and Theorem 3.9.3.

Exercise 3.9.4. Show that the series $\sum_{k=0}^{\infty} x^k/k!$ converges absolutely for all real x. (The sum is, of course, e^x, established later in the text.) What about $\sum_{k=0}^{\infty} z^k/k!$ for $z \in \mathbf{C}$?

Exercise 3.9.5. Consider the following series:

$$\text{(i)} \ \sum_{k=1}^{\infty} x^k \qquad \text{(ii)} \ \sum_{k=1}^{\infty} \frac{x^k}{k} \qquad \text{(iii)} \ \sum_{k=1}^{\infty} \frac{x^k}{k^2}.$$

Show that each of these series converges when $|x| < 1$ and diverges when $|x| > 1$. Then consider $|x| = 1$: Show that series (i) does not converge if $|x| = 1$; series (ii) converges for one value of x where $|x| = 1$ and diverges for the other; and series (iii) converges when $|x| = 1$.

3.10. Limit Superior and Limit Inferior

This section covers the concepts of limit superior and limit inferior of a sequence. Even a sequence that does not converge may have subsequences that converge, and it is useful to describe the behavior of these subsequences.

Example 3.10.1. Consider the sequence

$$\left(2, -1, \frac{1}{2}, -\frac{1}{2}, \frac{2}{3}, -\frac{1}{3}, \frac{3}{4}, -\frac{1}{4}, \frac{4}{5}, \dots\right).$$

It is not difficult to see that the largest subsequential limit of (a_k) is 1 and the least subsequential limit is 0. This conclusion follows from the observation that any convergent subsequence must have its limit enclosed by the intervals $[-1/m, 1]$ for each $m \in \mathbf{N}$, together with the observation that there are indeed specific subsequences which converge to 1 and 0, respectively. Notice also that these limit values are different from the supremum and infimum of the range of the sequence, given by 2 and -1, respectively. △

Before proceeding, let us first observe that a sequence that is unbounded below has a subsequence that diverges to $-\infty$, and a sequence that is unbounded above has a subsequence that diverges to ∞. The reader should verify this observation as an exercise.

The concepts of *limit superior* and *limit inferior* of a bounded sequence (a_k) are designed for the purpose of describing subsequential limits of (a_k) in a systematic way. We now set up some notation to help with the introduction of these concepts.

Let (a_k), $k \in \mathbf{N}$, be a bounded sequence of real numbers, so that $|a_k| \leq M$ for some M and all $k \in \mathbf{N}$; thus, the sequence is bounded above and below. Letting $l_1 = \inf\{a_k : k \in \mathbf{N}\}$ and $u_1 = \sup\{a_k : k \in \mathbf{N}\}$, we observe that the interval $[l_1, u_1]$ contains every term of (a_k). For each m, we consider the infimum and supremum of the tail end of (a_k) for $k \geq m$. Thus, define

$$\begin{aligned} l_m &= \inf\{a_k : k \geq m\}, \\ u_m &= \sup\{a_k : k \geq m\}. \end{aligned}$$

Then (l_m) is monotone increasing, (u_m) is monotone decreasing, and therefore we obtain a nested sequence of intervals $[l_m, u_m]$, $m \in \mathbf{N}$, such that the mth interval contains the entire tail of (a_k) for $k \geq m$. Now let

$$\alpha = \sup_m l_m = \sup_m \inf\{a_k : k \geq m\}, \tag{3.3}$$

$$\omega = \inf_m u_m = \inf_m \sup\{a_k : k \geq m\}. \tag{3.4}$$

The numbers α and ω both exist by the monotone sequence theorem. We make the following definitions.

Definition 3.10.2. *Let (a_k) be a bounded sequence of real numbers.*

1. *The* **limit superior** *of (a_k) is defined by*

$$\limsup a_k := \inf_m \sup\{a_k : k \geq m\}.$$

2. *The* **limit inferior** *of (a_k) is defined by*

$$\liminf a_k := \sup_m \inf\{a_k : k \geq m\}.$$

We emphasize that every *bounded* sequence of real numbers has a limit infimum and a limit supremum and each of these numbers is finite. If (a_k) is not bounded below, then we may indicate this by writing $\liminf a_k = -\infty$. If (a_k) is not bounded above, then we may write $\limsup a_k = \infty$.

It is possible to have a sequence bounded above for which the number ω in (3.4) is $-\infty$; for example, consider the sequence $(-1, -2, -3, \ldots)$. Of course, then the number α in (3.3) is $-\infty$ as well. And it is possible to have a sequence bounded below for which the number α in (3.3) is ∞; for example, consider the sequence $(1, 2, 3, \ldots)$. Of course, then the number ω in (3.4) is ∞ as well. Notice that neither of these examples has any convergent subsequences.

We think of $\limsup a_k$ and $\liminf a_k$ as the largest and smallest subsequential limits of (a_k), respectively, and this thought is justified by Theorem 3.10.3 and Theorem 3.10.4 below. These theorems give several equivalent characterizations of the limit superior and limit inferior, respectively.

First we make a few more observations about subsequential limits. If the number b is the limit of a subsequence of (a_k), then by (3.3) and (3.4), b belongs to the interval $[\alpha, \omega]$. The sequence (a_k) converges if and only if $\alpha = \omega$, in which case there is exactly one subsequential limit for (a_k). Thus, if S is the set of subsequential limits of (a_k), then $S \subset [\alpha, \omega]$. (The theorems below show that, in fact, α and ω are elements of S, as we indicated above.)

A helpful observation in the description of subsequential limits of (a_k) is as follows. If β is a number such that $a_k > \beta$ for at most finitely many terms of (a_k), then no subsequence of (a_k) can converge to a limit greater than β, since such a limit would require that infinitely many terms of (a_k) be greater than β. Stated in a different but logically equivalent form, we observe that if β has the property that there exists an N_β such that $a_k \le \beta$ for all $k \ge N_\beta$, then no number greater than β can be a subsequential limit of (a_k).

We may characterize $\limsup a_k$ as follows.

Theorem 3.10.3. *Let (a_k) be a bounded sequence of real numbers. Then the following statements for a real number x^* are equivalent:*

1. *If $u_m = \sup\{a_k : k \ge m\}$, then $x^* = \inf_m u_m = \lim_m u_m$, that is, $x^* = \limsup a_k$ (Definition 3.10.2).*
2. *If S is the set of all subsequential limits of (a_k), then $x^* = \sup S$.*
3. *x^* is the infimum of the set B consisting of all real numbers β such that $a_k > \beta$ for at most finitely many terms of (a_k).*
4. *For any $\epsilon > 0$, there are at most finitely many terms of (a_k) such that $a_k > x^* + \epsilon$, but infinitely many terms of (a_k) such that $a_k > x^* - \epsilon$.*

Proof. 1 implies 2: Suppose (a_{n_k}) is a convergent subsequence of (a_k) with limit b. Then $b \le u_m$ for each m, hence $b \le \lim_m u_m$. On the other hand, by definition of u_1, there exists $n_1 \ge 1$ such that $u_1 - 1 < a_{n_1} \le u_1$. By induction, we may choose $n_{k+1} > n_k$ such that for all $k \in \mathbf{N}$,

$$u_k - \frac{1}{k} < a_{n_k} \le u_k.$$

Since $\lim_k u_k = x^*$, we conclude that $\lim_k a_{n_k} = x^*$, and hence $x^* \in S$. Therefore $x^* = \sup S$. (This proves that $\limsup a_k$ as defined by Definition 3.10.2 really is the largest subsequential limit of (a_k).)

2 implies 3: Let $x^* = \sup S$. Then given any $\epsilon > 0$, there are at most finitely many terms of (a_k) with $a_k > x^* + \epsilon$. Thus, $x^* + \epsilon$ belongs to the set B of 3. Since $x^* = \sup S$, there is a subsequence of (a_k) that converges to a number greater than $x^* - \epsilon$, and therefore $x^* - \epsilon$ does not belong to B. Since $\epsilon > 0$ is arbitrary, we conclude that $x^* = \inf B$. Thus, 3 holds.

3 implies 4: Let $\epsilon > 0$. Since $x^* = \inf B$, there exists $\beta \in B$ such that $x^* \leq \beta < x^* + \epsilon$. Then by definition of B, $x^* + \epsilon$ also belongs to B. So there are at most finitely many terms of (a_k) such that $a_k > x^* + \epsilon$. But $x^* - \epsilon$ does not belong to B, so there are infinitely many terms of (a_k) such that $a_k > x^* - \epsilon$. Thus, 4 holds.

4 implies 1: If 4 holds, then for any $\epsilon > 0$, we have $u_m < x^* + \epsilon$ for all sufficiently large m. Hence, $\inf\{u_m : m \in \mathbf{N}\} \leq x^* + \epsilon$. Also by 4, there are infinitely many terms of (a_k) such that $a_k > x^* - \epsilon$, so $x^* - \epsilon < u_m$ for all $m \in \mathbf{N}$. Hence, $x^* - \epsilon \leq \inf\{u_m : m \in \mathbf{N}\}$. Since $\epsilon > 0$ is arbitrary, we conclude that $x^* = \inf\{u_m : m \in \mathbf{N}\}$. And since (u_m) is decreasing, we also have $x^* = \lim_m u_m$. Therefore 1 holds. $\square$

The next theorem gives a characterization of $\liminf a_k$. The proof is left to the interested reader as Exercise 3.10.5.

Theorem 3.10.4. *Let (a_k) be a bounded sequence of real numbers. Then the following statements for a real number x^* are equivalent:*

1. *If $l_m = \inf\{a_k : k \geq m\}$, then $x^* = \sup_m l_m = \lim_m l_m$, that is, $x^* = \liminf a_k$ (Definition 3.10.2).*
2. *If S is the set of all subsequential limits of (a_k), then $x^* = \inf S$.*
3. *x^* is the supremum of the set B consisting of all real numbers β such that $a_k < \beta$ for at most finitely many terms of (a_k).*
4. *For any $\epsilon > 0$, there are at most finitely many terms of (a_k) such that $a_k < x^* - \epsilon$, but infinitely many terms of (a_k) such that $a_k < x^* + \epsilon$.*

Exercises.

Exercise 3.10.1. Find $\liminf a_k$ and $\limsup a_k$ for these sequences:

1. $a_k = (1 - \frac{1}{k}) \cos(k\pi)$;
2. $a_k = (1 + \frac{1}{k}) \sin(k\frac{\pi}{3})$;
3. $a_k = 2\sin(k\frac{\pi}{2}) \cos(k\pi)$.

Exercise 3.10.2. Suppose (a_k) and (b_k) are sequences bounded above. Prove that $\limsup(a_k + b_k) \leq \limsup a_k + \limsup b_k$.

Exercise 3.10.3. Suppose (a_k) and (b_k) are sequences bounded below. Prove that $\liminf(a_k + b_k) \geq \liminf a_k + \liminf b_k$.

Exercise 3.10.4. Suppose $a_k \leq b_k$ for all $k \geq N$ for some positive integer N. Show that $\limsup a_k \leq \limsup b_k$ and $\liminf a_k \leq \liminf b_k$.

Exercise 3.10.5. Prove Theorem 3.10.4.

3.11. Additional Convergence Tests

This section provides some additional convergence tests for numerical series.

For absolute convergence, the coverage includes the general versions of the root test and ratio test, and geometric series continue to play an important role here. (The integral test can be found in Section 6.3.)

For conditional convergence, we consider Abel's test and Dirichlet's test.

3.11.1. Absolute Convergence: The Root and Ratio Tests. Earlier in Theorem 3.9.2 and Theorem 3.9.3 we saw applications of a comparison test with geometric series, which gave us the basic ratio test and root test usually seen in calculus courses. In this section we present more general versions of the root test and ratio test by employing the concepts of limit superior and limit inferior, as well as geometric series.

Theorem 3.11.1 (Root Test). *Let $\sum a_k$ be a series of real numbers.*

1. *If* $\limsup |a_k|^{1/k} < 1$, *then the series converges absolutely.*
2. *If* $\limsup |a_k|^{1/k} > 1$, *then the series diverges.*

Proof. 1. Assume $L = \limsup |a_k|^{1/k} < 1$. By hypothesis, we can choose an $\epsilon > 0$ such that $L + \epsilon < 1$. Then there exists an $N(\epsilon)$ such that

$$|a_k|^{1/k} < L + \epsilon \quad \text{for all } k > N(\epsilon).$$

Thus, $|a_k| < (L + \epsilon)^k$ for all $k > N(\epsilon)$. Since $L + \epsilon < 1$, the terms on the right are the terms of a convergent geometric series. By the comparison test, the series $\sum_N^\infty |a_k|$ converges, so $\sum a_k$ converges absolutely.

2. If $L = \limsup |a_k|^{1/k} > 1$, then we can choose an $\epsilon > 0$ such that $L - \epsilon > 1$. There exists a subsequence (a_{n_k}) such that

$$|a_{n_k}|^{1/n_k} > L - \epsilon > 1,$$

and hence $|a_{n_k}| > (L - \epsilon)^{n_k} > 1$. Consequently, we cannot have $a_k \to 0$ as $k \to \infty$, and therefore $\sum a_k$ diverges. □

Theorem 3.11.2 (Ratio Test). *Let $\sum a_k$ be a series of real numbers.*

1. *If* $\limsup |a_{k+1}/a_k| < 1$, *then the series converges absolutely.*
2. *If* $\liminf |a_{k+1}/a_k| > 1$, *then the series diverges. (An example is given below to show that the condition* $\limsup |a_{k+1}/a_k| > 1$ *does not imply divergence.)*

Proof. 1. Assume $L = \limsup |a_{k+1}/a_k| < 1$. By hypothesis, we can choose an $\epsilon > 0$ such that $L + \epsilon < 1$. Then there exists an $N = N(\epsilon)$ such that

$$\left|\frac{a_{k+1}}{a_k}\right| < L + \epsilon \quad \text{for all } k \geq N(\epsilon).$$

Thus, for each positive integer j, we have

$$|a_{N+j}| < |a_N|\,(L + \epsilon)^j.$$

Since $L + \epsilon < 1$, the terms on the right are the terms of a convergent geometric series. By the comparison test, the series $\sum_{j=1}^{\infty} |a_{N+j}|$ converges, so $\sum a_k$ converges absolutely.

2. Suppose $L = \liminf |a_{k+1}/a_k| > 1$. Then we may choose an $\epsilon > 0$ such that $L - \epsilon > 1$. Then there exists an $N = N(\epsilon)$ such that

$$\left|\frac{a_{k+1}}{a_k}\right| > L - \epsilon \quad \text{for all } k \geq N(\epsilon).$$

Thus, for each positive integer j, we have

$$|a_{N+j}| > |a_N| \, (L - \epsilon)^j.$$

Since $L - \epsilon > 1$, the terms on the right are the terms of a divergent geometric series. Consequently, $\lim_{j\to\infty} |a_{N+j}| \neq 0$ if the limit exists at all, and so $\sum a_k$ diverges. □

There is an interesting relation between the ratio test and root test. It is possible to establish the inequalities

$$\liminf \left|\frac{a_{k+1}}{a_k}\right| \leq \liminf |a_k|^{1/k}$$

and

$$\limsup |a_k|^{1/k} \leq \limsup \left|\frac{a_{k+1}}{a_k}\right|.$$

These inequalities imply that if either part of the ratio test applies, then the corresponding part of the root test applies. Thus the root test is the more general test in that it will apply in some cases where the ratio test does not apply. However, there are many series for which the ratio test may be easier to apply. These inequalities also help to explain the need for the limit inferior in the divergence test portion of the ratio test. Also consider the next example.

Example 3.11.3. We know that the alternating harmonic series

$$\sum_{k=1}^{\infty} (-1)^{k+1}\frac{1}{k} = 1 - \frac{1}{2} + \frac{1}{3} - \frac{1}{4} + \cdots$$

converges. (The sum equals $\log 2$, where log indicates the natural logarithm function.) By Theorem 3.2.9 on the sum of convergent series, we may write

$$\begin{aligned}
\frac{1}{2}\sum_{k=1}^{\infty}(-1)^{k+1}\frac{1}{k} &= \sum_{k=1}^{\infty} 0 + \frac{1}{2}\sum_{k=1}^{\infty}(-1)^{k+1}\frac{1}{k} \\
&= 0 + \frac{1}{2} + 0 - \frac{1}{4} + 0 + \frac{1}{6} + 0 - \frac{1}{8} + \cdots = \frac{1}{2}\log 2.
\end{aligned}$$

Since $\frac{3}{2}\log 2 = \log 2 + \frac{1}{2}\log 2$, we may also conclude from Theorem 3.2.9 that

$$\begin{aligned}
\frac{3}{2}\log 2 &= \left(1 - \frac{1}{2} + \frac{1}{3} - \frac{1}{4} + \frac{1}{5} - \frac{1}{6} + \frac{1}{7} - \frac{1}{8} + \frac{1}{9} - \frac{1}{10} + \frac{1}{11} - \frac{1}{12} + \cdots\right) \\
&\quad + \left(0 + \frac{1}{2} + 0 - \frac{1}{4} + 0 + \frac{1}{6} + 0 - \frac{1}{8} + 0 + \frac{1}{10} + 0 - \frac{1}{12} + \cdots\right) \\
&= 1 + 0 + \frac{1}{3} - \frac{1}{2} + \frac{1}{5} + 0 + \frac{1}{7} - \frac{1}{4} + \frac{1}{9} + 0 + \frac{1}{11} - \frac{1}{6} + \cdots.
\end{aligned}$$

The reader may verify that the pattern continues as follows:

$$\frac{3}{2}\log 2 = \left(1+\frac{1}{3}\right) - \frac{1}{2} + \left(\frac{1}{5}+\frac{1}{7}\right) - \frac{1}{4} + \left(\frac{1}{9}+\frac{1}{11}\right) - \frac{1}{6} + \left(\frac{1}{13}+\frac{1}{15}\right) + \cdots,$$

giving a series having sum $\frac{3}{2}\log 2$. For the latter series, which we label (a_k), the subsequence of ratios given by

$$\left(\frac{|a_{3k}|}{|a_{3k-1}|}\right)_{k=1}^{\infty} = (3/2, 7/4, 11/6, \ldots, (4k-1)/2k, \ldots)$$

shows that $\limsup |a_{k+1}|/|a_k| \geq 2$. From this we conclude that the condition $\limsup |a_{k+1}|/|a_k| > 1$ does not imply divergence. △

In the next two examples, the limits in the ratio test and root test do not exist, but the limit superior does exist.

Example 3.11.4. For the series

$$\sum_{k=1}^{\infty} a_k = 1 + \frac{2}{3} + \frac{1}{3} + \frac{2}{3^2} + \frac{1}{3^2} + \frac{2}{3^3} + \frac{1}{3^3} + \frac{2}{3^4} + \cdots,$$

we have $|a_{k+1}|/|a_k| = 2/3$ if k is odd, and $|a_{k+1}|/|a_k| = 1/2$ if k is even. Therefore $\lim_{k\to\infty} |a_{k+1}|/|a_k|$ does not exist. But $\limsup |a_{k+1}|/|a_k| = 2/3 < 1$, so the series converges by part 1 of the ratio test. △

Example 3.11.5. For the series

$$\sum_{k=1}^{\infty} a_k = \frac{1}{10} + \frac{1}{2^2} + \frac{1}{10^3} + \frac{1}{2^4} + \frac{1}{10^5} + \frac{1}{2^6} + \cdots,$$

we have $\sqrt[k]{a_k} = 1/10$ if k is odd, and $\sqrt[k]{a_k} = 1/2$ if k is even. Therefore $\lim_{k\to\infty} \sqrt[k]{a_k}$ does not exist, but $\limsup \sqrt[k]{a_k} = 1/2 < 1$, so the series converges by part 1 of the root test. △

Example 3.11.6. The series $\sum_{k=0}^{\infty} 1/k!$ converges, and by definition its sum is the Euler number e. Let us show that this series converges by the root test. If m is a positive integer and $k > m$, then

$$k! = k(k-1)\cdots(m+1)(m)\cdots(2)(1) > m^{k-m}m = m^{k-m+1}.$$

By taking k sufficiently large, we have

$$(k!)^{1/k} > (m^{k-m+1})^{1/k} = m\, m^{(1-m)/k} > \frac{m}{2}.$$

Then

$$\left(\frac{1}{k!}\right)^{1/k} < \frac{2}{m}.$$

Since m was arbitrary, this shows that $\lim_{k\to\infty}(1/k!)^{1/k}$ exists and equals zero. So $\sum_{k=0}^{\infty} 1/k!$ converges by the root test. Note that the ratio test also applies to this series, since $k!/(k+1)! = 1/(k+1) \to 0$ as $k \to \infty$. △

3.11.2. Conditional Convergence: Abel's and Dirichlet's Tests. In this section we present the Abel summation formula and show that it leads to two additional tests that help with conditionally convergent series: Abel's Test and Dirichlet's Test.

The Abel summation formula is an elementary result; it is also called the *summation by parts* formula.

Theorem 3.11.7 (Abel's Summation by Parts). *Let $(a_k)_0^\infty$ and $(b_k)_0^\infty$ be sequences of real or complex numbers. For any positive integer n, let*

$$\sigma_n(a) := \sum_{j=0}^{n} a_j.$$

Then

$$\sum_{k=0}^{n} a_k b_k = b_{n+1}\sigma_n(a) - \sum_{k=0}^{n}(b_{k+1} - b_k)\sigma_k(a).$$

Proof. For any k we have $a_k = \sigma_k(a) - \sigma_{k-1}(a)$, where we set $\sigma_{-1}(a) = 0$. Then we have

$$\begin{aligned}
\sum_{k=0}^{n} a_k b_k &= \sum_{k=0}^{n}(\sigma_k(a) - \sigma_{k-1}(a))b_k \\
&= \sum_{k=0}^{n}\sigma_k(a)b_k - \sum_{k=0}^{n}\sigma_{k-1}(a)b_k \\
&= \sum_{k=0}^{n}\sigma_k(a)b_k - \sum_{k=1}^{n}\sigma_{k-1}(a)b_k \quad (\text{since } \sigma_{-1}(a) = 0) \\
&= \sum_{k=0}^{n}\sigma_k(a)b_k - \sum_{k=0}^{n}\sigma_k(a)b_{k+1} + b_{n+1}\sigma_n(a) \quad (\text{by a shift of index}).
\end{aligned}$$

A simple rearrangement yields the stated result. □

The summation by parts formula leads to Abel's test for convergence.

Theorem 3.11.8 (Abel's Test). *If the series $\sum_{k=0}^\infty a_k$ is convergent and $(b_k)_{k=0}^\infty$ is a monotonic convergent sequence, then the series $\sum_{k=0}^\infty a_k b_k$ is convergent.*

Proof. We wish to apply Abel's summation by parts formula. Let us write $\sigma_n(a) := \sum_{k=0}^n a_k$. Since $(\sigma_n(a))$ and (b_n) are convergent sequences, the sequence $(b_{n+1}\sigma_n(b))$ is also convergent. From the summation by parts formula, it remains to show that the sequence of sums

$$\sum_{k=0}^{n}(b_{k+1} - b_k)\sigma_k(a), \tag{3.5}$$

indexed by n, converges, where, as before, $\sigma_k(a) := \sum_{j=0}^k a_j$. Since the sequence $(\sigma_k(a))$ is bounded, we may write

$$|\sigma_k(a)| \le M \quad \text{for some } M.$$

Now we estimate

$$\sum_{k=0}^{n} |b_{k+1} - b_k||\sigma_k(a)| \leq M \sum_{k=0}^{n} |b_{k+1} - b_k|,$$

and the summation on the right must be bounded since (b_k) is bounded. In fact, since (b_k) is monotonic,

$$\sum_{k=0}^{n} |b_{k+1} - b_k||\sigma_k(a)| \leq M|b_{n+1} - b_0|.$$

Since (b_n) is bounded, the sums on the left side here form a bounded, monotone increasing sequence, which converges. Therefore the sequence (3.5) converges, and this completes the proof. □

In the proof of Abel's test, the convergence of the series $\sum_{k=0}^{\infty} a_k$ and of the sequence (b_k) was needed only to ensure convergence of the sequence $b_{n+1}\sigma_n(a)$ appearing in the summation by parts formula. *Dirichlet's test* relaxes the assumption of convergence of $\sum_{k=0}^{\infty} a_k$ to boundedness of the partial sums, but requires that the monotone sequence (b_k) converge to zero.

Theorem 3.11.9 (Dirichlet's Test). *The series $\sum_{k=0}^{\infty} a_k b_k$ converges if the following conditions hold:*
1. the partial sums $\sigma_n(a) = \sum_{k=0}^{n} a_k$ form a bounded sequence;
2. $b_0 \geq b_1 \geq b_2 \geq \cdots$;
3. $\lim_{k\to\infty} b_k = 0$.

Proof. The partial sums of the series are given by the summation by parts formula, and the convergence of the sequence of sums

$$\sum_{k=0}^{n} (b_{k+1} - b_k)\sigma_k(a)$$

follows in exactly the same way as in the proof of Abel's test. It remains to show that the sequence $b_{n+1}\sigma_n(a)$ converges. If we have the bound $|\sigma_n(a)| \leq M$ for all n, then

$$|b_{n+1}\sigma_n(a)| \leq M|b_{n+1}| \to 0 \quad \text{as } n \to \infty.$$

Hence, $b_{n+1}\sigma_n(a) \to 0$ as $n \to \infty$, and from the summation by parts formula we conclude that the series $\sum_{k=0}^{\infty} a_k b_k$ converges. □

We have already considered alternating series in Theorem 3.6.2, but it is interesting to note that the earlier theorem follows from Dirichlet's test. For convenience we recall that Theorem 3.6.2 states: If the sequence (a_k) is monotone decreasing with limit zero, then the series with alternating signs

$$\sum_{k=1}^{\infty} (-1)^k a_k = -a_1 + a_2 - a_3 + \cdots$$

converges. The partial sums of $\sum_{k=1}^{\infty}(-1)^k$ take only the values -1 and 0, hence they are bounded, while the a_k are decreasing with limit zero. Therefore the convergence of the alternating series follows directly from Dirichlet's test.

The next example can be helpful with some trigonometric series. To discuss it, we call on some facts about the sine and cosine functions and the real exponential function. These functions are discussed more completely in Section 7.5, but are likely to be familiar from introductory calculus courses.

Example 3.11.10. If x is real, we define

$$e^{ix} = \cos x + i \sin x.$$

(See Exercise 3.11.2 for a motivation of this formula using the complex exponential series, which has the same form as the real exponential series discussed in detail in Section 7.5; see also Definition 7.5.1 and the comments following it.) Recall that $\cos(-x) = \cos x$ and $\sin(-x) = -\sin x$. Then, from the definition above for e^{ix}, it follows by induction, together with the trigonometric identities for the sine and cosine of the sum of two angles, that for all integers k, we have

$$(e^{ix})^k = e^{ikx} = \cos kx + i \sin kx.$$

(See Exercise 3.11.2.) Notice that for any real x, $e^{-ix} = \cos x - i \sin x = \overline{e^{ix}}$, the conjugate of e^{ix}, and thus

$$\cos x = \frac{e^{ix} + e^{-ix}}{2} \qquad \text{and} \qquad \sin x = \frac{e^{ix} - e^{-ix}}{2i}.$$

By the formula for the sum of a finite geometric series, we have

$$\begin{aligned}
\sum_{k=1}^{n} e^{ikx} &= \frac{1 - e^{i(n+1)x}}{1 - e^{ix}} - 1 \\
&= \frac{1 - e^{i(n+1)x}}{1 - e^{ix}} - \frac{1 - e^{ix}}{1 - e^{ix}} \\
&= e^{ix}\frac{1 - e^{inx}}{1 - e^{ix}} \\
&= e^{ix}\frac{e^{inx/2}(e^{-inx/2} - e^{inx/2})}{e^{ix/2}(e^{-ix/2} - e^{ix/2})} \\
&= e^{i(n+1)x/2}\frac{e^{-inx/2} - e^{inx/2}}{e^{-ix/2} - e^{ix/2}} \\
&= \left[\cos\frac{1}{2}(n+1)x + i \sin\frac{1}{2}(n+1)x\right]\left(\frac{-2i\sin(nx/2)}{-2i\sin(x/2)}\right) \\
&= \left[\cos\frac{1}{2}(n+1)x + i \sin\frac{1}{2}(n+1)x\right]\frac{\sin\frac{1}{2}nx}{\sin\frac{1}{2}x}.
\end{aligned}$$

This computation is valid as long as $e^{ix} \neq 1$, that is, $x \neq 2n\pi$, $n \in \mathbf{Z}$. By extracting the real and imaginary parts from each side, we have

$$\sum_{k=1}^{n} \cos kx = \frac{\cos\frac{1}{2}(n+1)x \, \sin\frac{1}{2}nx}{\sin\frac{1}{2}x}$$

and

$$\sum_{k=1}^{n} \sin kx = \frac{\sin\frac{1}{2}(n+1)x \, \sin\frac{1}{2}nx}{\sin\frac{1}{2}x}.$$

Consequently, these sums have the following bounds:

$$\left|\sum_{k=1}^{n} \cos kx\right| \leq \frac{1}{|\sin \frac{1}{2}x|} = \left|\csc \frac{1}{2}x\right|$$

and

$$\left|\sum_{k=1}^{n} \sin kx\right| \leq \frac{1}{|\sin \frac{1}{2}x|} = \left|\csc \frac{1}{2}x\right|,$$

with both inequalities holding for $x \neq 2n\pi$, $n \in \mathbf{Z}$. △

Theorem 3.11.11. *If the sequence (b_k) decreases with $\lim_{k\to\infty} b_k = 0$, then the following statements are true:*

1. *The series $\sum_{k=1}^{\infty} b_k \sin kx$ converges for all real x.*
2. *The series $\sum_{k=1}^{\infty} b_k \cos kx$ converges for all real x except possibly for $x = 2n\pi$, $n \in \mathbf{Z}$.*

Proof. By the result of Example 3.11.10, the hypotheses of Dirichlet's test are satisfied, since for real x the partial sums $\sum_{k=1}^{n} \sin kx$ and $\sum_{k=1}^{n} \cos kx$ are bounded by $|\csc \frac{1}{2}x|$, with the only possible exceptions being those stated for the cosine series. (Observe that $\sin kx = 0$ for $x = 2n\pi$, $n \in \mathbf{Z}$.) □

Exercises.

Exercise 3.11.1. Show that neither the ratio test nor the root test applies to the series $\sum_{k=1}^{\infty} 1/\sqrt{k}$.

Exercise 3.11.2. Let x be a real number.

1. Use the complex exponential series defined by $\exp(z) = \sum_{k=0}^{\infty} z^k/k!$, convergent for all $z \in \mathbf{C}$, to define the expression e^{ix} as $\exp(ix)$, and show that $e^{ix} = \cos x + i \sin x$. (This is called *Euler's identity.*)
2. Show that $e^{i\pi} = -1$, relating four of the most important numbers in mathematics.
3. Show that $(e^{ix})^k = e^{ikx} = \cos kx + i \sin kx$ for each positive integer k. Conclude that this identity also holds for negative integers k.

Exercise 3.11.3. Show that the series $\sum_{k=1}^{\infty} \frac{1}{k} \sin k$ converges.

Exercise 3.11.4. Show that the series $\sum_{k=1}^{\infty} \frac{1}{k} \sin^2 k$ diverges. *Hint*: $\sin^2 x = (1 - \cos 2x)/2$.

3.12. Rearrangements and Riemann's Theorem

It is implicit in the definition of an infinite series $\sum a_k$ that the terms in the series are ordered. If we rearrange the terms of $\sum a_k$, for example by interchanging the numbers in each successive pair from (a_k), then we obtain the new series

$$a_2 + a_1 + a_4 + a_3 + a_6 + a_5 + \cdots,$$

which is a series different from $\sum a_k$ because the sequence of its partial sums differs from that of $\sum a_k$.

Definition 3.12.1. *Let $\sum_{k=1}^{\infty} a_k$ be a given series. Let (p_k) be a sequence in which every positive integer occurs exactly once, that is, $p : \mathbf{N} \to \mathbf{N}$ is 1-1 and onto. Then (p_k) is called a* **permutation** *of* $\mathbf{N}$ *and the series $\sum_{k=1}^{\infty} a_{p_k}$ is called a* **rearrangement** *of $\sum_{k=1}^{\infty} a_k$.*

Any rearrangement of an absolutely convergent series must converge to the same sum.

Theorem 3.12.2. *If $\sum_{k=1}^{\infty} a_k$ converges absolutely and $S = \sum_{k=1}^{\infty} a_k$, and if $\sum_{k=1}^{\infty} a_{p_k}$ is any rearrangement of $\sum_{k=1}^{\infty} a_k$, then we also have $S = \sum_{k=1}^{\infty} a_{p_k}$.*

Proof. Let (p_k) be any permutation of $\mathbf{N}$, and $\sum_{k=1}^{\infty} a_{p_k}$ the corresponding rearrangement of $\sum_{k=1}^{\infty} a_k$. We use the Cauchy criterion for the convergence of $\sum_{k=1}^{\infty} |a_k|$ to show that $S = \sum_{k=1}^{\infty} a_{p_k}$. Let $\epsilon > 0$. We may choose an integer $M > 0$ such that if $n, m \in \mathbf{N}$ and $n \geq m > M$, then

$$\sum_{k=m}^{n} |a_k| < \epsilon. \tag{3.6}$$

We can choose a positive integer $K > M$ such that all the integers $1, 2, \ldots, M$ are contained in the list $p_1, p_2, \ldots, p_K$. If we then choose a positive integer $N > K$, then the partial sum

$$\sum_{k=1}^{N} a_k$$

will include each of the terms $a_1, a_2, \ldots, a_M$. The partial sum

$$\sum_{k=1}^{N} a_{p_k}$$

will also include each of the terms $a_1, a_2, \ldots, a_M$. Thus the difference of these partial sums,

$$\sum_{k=1}^{N} a_k - \sum_{k=1}^{N} a_{p_k},$$

includes the original terms a_k only for $k > M$. Thus, let $m = M + 1$ in (3.6) and take any $n \geq N > K \geq M + 1$, such that n is larger than all indices appearing in the two summations up to N. Then we have

$$\left| \sum_{k=1}^{N} a_k - \sum_{k=1}^{N} a_{p_k} \right| \leq 2 \sum_{k=M+1}^{n} |a_k| < 2\epsilon.$$

Letting $n \to \infty$, we have

$$\left| \sum_{k=1}^{N} a_k - \sum_{k=1}^{N} a_{p_k} \right| \leq 2 \sum_{k=M+1}^{\infty} |a_k| \leq 2\epsilon.$$

We can achieve this inequality (that is, such an N exists) for every $\epsilon > 0$. Therefore $\lim_{N\to\infty} \left(\sum_{k=1}^{N} a_k - \sum_{k=1}^{N} a_{p_k} \right)$ exists and equals zero. By hypothesis,

$$\lim_{N\to\infty} \sum_{k=1}^{N} a_k = S$$

exists, so that

$$\lim_{N\to\infty} \sum_{k=1}^{N} a_{p_k} = S$$

as well. □

We wish to prove Riemann's rearrangement theorem, which illustrates the stark contrast between absolute convergence and conditional convergence. Riemann's Theorem 3.12.4, below, follows from Theorem 3.12.3.

As usual we write the series as $\sum_{k=1}^{\infty} a_k$. Let

$$a_k^+ = \max\{a_k, 0\} \quad \text{and} \quad a_k^- = \max\{-a_k, 0\}.$$

Then $a_k^+ = a_k$ if a_k is positive and $a_k^+ = 0$ otherwise, and $a_k^- = |a_k|$ if a_k is negative and $a_k^- = 0$ otherwise. Thus the nonzero a_k^+ are the positive terms of the series, and the nonzero a_k^- are the absolute values of the negative terms. It is useful to have the placeholder zeros, however, as we observe that

$$a_k = a_k^+ - a_k^- \quad \text{and} \quad |a_k| = a_k^+ + a_k^-.$$

Theorem 3.12.3. *If $\sum_{k=1}^{\infty} a_k$ is absolutely convergent, then the series $\sum_{k=1}^{\infty} a_k^+$ and $\sum_{k=1}^{\infty} a_k^-$ are both convergent. If $\sum_{k=1}^{\infty} a_k$ is conditionally convergent, then the series $\sum_{k=1}^{\infty} a_k^+$ and $\sum_{k=1}^{\infty} a_k^-$ are both divergent.*

Proof. Suppose $\sum_{k=1}^{\infty} a_k$ is absolutely convergent. By definition of a_k^+ and a_k^-, we have

$$0 \le a_k^+ \le |a_k| \quad \text{and} \quad 0 \le a_k^- \le |a_k|,$$

hence the series $\sum_{k=1}^{\infty} a_k^+$ and $\sum_{k=1}^{\infty} a_k^-$ are both convergent by a direct comparison.

Now suppose that $\sum_{k=1}^{\infty} a_k$ is conditionally convergent, and thus the series $\sum_{k=1}^{\infty} |a_k|$ diverges. We want to show that the series $\sum_{k=1}^{\infty} a_k^+$ and $\sum_{k=1}^{\infty} a_k^-$ are both divergent. Since $|a_k| = a_k^+ + a_k^-$, the series

$$\sum_{k=1}^{\infty} (a_k^+ + a_k^-)$$

diverges. We cannot have both $\sum_{k=1}^{\infty} a_k^+$ and $\sum_{k=1}^{\infty} a_k^-$ convergent, since that would imply that their sum $\sum_{k=1}^{\infty} |a_k|$ is convergent. Therefore *at least one* of $\sum_{k=1}^{\infty} a_k^+$ and $\sum_{k=1}^{\infty} a_k^-$ diverges. We seek a contradiction by assuming that one of these series converges and the other diverges. Without loss of generality, let us assume that $\sum_{k=1}^{\infty} a_k^+ = S < \infty$ and $\sum_{k=1}^{\infty} a_k^- = \infty$. Write s_n and $s_n^{\pm}$ for the n-th partial sum of the series $\sum_{k=1}^{\infty} a_k$ and $\sum_{k=1}^{\infty} a_k^{\pm}$, respectively; thus, we have $s_n = s_n^+ - s_n^-$ for each n. By our hypothesis, for any positive M, no matter how large, there is an $N = N(M)$ such that for $n \ge N$, we have $s_n^- > M + S$, and $s_n^+ \le S$ (since s_n^+ increases with limit S). It follows that $s_n = s_n^+ - s_n^- < S-(M+S) = -M$, and hence that $s_n \to -\infty$, so that $\sum_{k=1}^{\infty} a_k$ diverges. This contradicts the original hypothesis that $\sum_{k=1}^{\infty} a_k$ is conditionally convergent. Therefore the second statement of the theorem is true. □

Theorem 3.12.3 indicates how different conditional convergence is from absolute convergence. If a series converges absolutely, then the series of its positive terms and the series of its negative terms both converge. A conditionally convergent series can converge only due to a strong dependence on the cancellation between its positive and negative terms.

Theorem 3.12.3 also tells us that a conditionally convergent series tends to converge slowly; the divergence of $\sum_{k=1}^{\infty} |a_k|$ implies that a_k does not approach zero very rapidly as $k \to \infty$, and thus it might take a very large value of n for the partial sum s_n to approximate the series sum to a specified accuracy.

We now prove the rearrangement theorem due to B. Riemann.

Theorem 3.12.4. *Suppose $\sum_{k=1}^{\infty} a_k$ is conditionally convergent. Given any real number S, there is a rearrangement $\sum_{k=1}^{\infty} a_{p_k}$ that converges to S.*

Proof. By Theorem 3.12.3, the series $\sum_{k=1}^{\infty} a_k^+$ and $\sum_{k=1}^{\infty} a_k^-$ both diverge. Since $\sum_{k=1}^{\infty} a_k$ converges, $\lim_{k\to\infty} a_k = 0$.

Suppose first that $S \geq 0$. We rearrange $\sum_{k=1}^{\infty} a_k$ to converge to S as follows:

1. Add the positive terms from the series $\sum_{k=1}^{\infty} a_k$, in their original order, up to and including the first positive term such that the sum exceeds S. This step is possible since $\sum_{k=1}^{\infty} a_k^+$ diverges.

2. Then add the negative terms from $\sum_{k=1}^{\infty} a_k$, in their original order, up to and including the first negative term such that the sum is less than S. This step is possible since $\sum_{k=1}^{\infty} a_k^-$ diverges.

3. Repeat steps 1 and 2. This process never terminates since $\sum_{k=1}^{\infty} a_k^+$ and $\sum_{k=1}^{\infty} a_k^-$ are both divergent.

Any given term a_k of the original series is eventually captured in a repetition of either step 1 or step 2 and included in the new series. Therefore this algorithm results in a rearrangement $\sum_{k=1}^{\infty} a_{p_k}$ of the original series.

We now show that the rearranged series converges to S. Given $\epsilon > 0$, there exists an $N = N(\epsilon)$ such that if $k \geq N$, then $|a_k| < \epsilon$. Choose $K = K(\epsilon)$ sufficiently large that all the terms $a_1, \ldots, a_N$ are included among the list $a_{p_1}, \ldots, a_{p_K}$. Then for $k \geq K$, $|a_{p_k}| < \epsilon$. Then for $n \geq K$, the switching specified in steps 1 and 2 implies that the partial sums of the rearranged series satisfy

$$\left| S - \sum_{k=1}^{n} a_{p_k} \right| < \epsilon.$$

To see this, note that if we had $|S - \sum_{k=1}^{n} a_{p_k}| \geq \epsilon$ for some $n \geq K$, then we added in too many terms of the same sign, in contradiction to the algorithm specifications. This proves that $S = \lim_{n\to\infty} \sum_{k=1}^{n} a_{p_k}$.

If the given real number S is negative, we interchange steps 1 and 2 so that we begin the algorithm with the step that adds negative terms until we first undershoot S, and proceed from there. □

Here is one example of a rearranged conditionally convergent series converging to a value different from the sum of the original series.

Example 3.12.5. The convergent series of Example 3.11.3,

$$\left(1+\frac{1}{3}\right)-\frac{1}{2}+\left(\frac{1}{5}+\frac{1}{7}\right)-\frac{1}{4}+\left(\frac{1}{9}+\frac{1}{11}\right)-\frac{1}{6}+\left(\frac{1}{13}+\frac{1}{15}\right)-\frac{1}{8}+\left(\frac{1}{17}+\frac{1}{19}\right)+\cdots,$$

which has sum $\frac{3}{2}\log 2$, is a rearrangement of the conditionally convergent series

$$\sum_{k=1}^{\infty}(-1)^{k+1}\frac{1}{k}=1-\frac{1}{2}+\frac{1}{3}-\frac{1}{4}+\frac{1}{5}-\frac{1}{6}+\frac{1}{7}-\frac{1}{8}+\cdots,$$

which has sum $\log 2$. The parentheses in the rearranged series indicates the grouping specified by the algorithm in the proof of Theorem 3.12.4 in order to achieve the sum $S=\frac{3}{2}\log 2$. △

Exercise.

Exercise 3.12.1. Show that if $\sum_{k=1}^{\infty} a_k$ is conditionally convergent, there are rearrangements of the series whose partial sums diverge to $+\infty$ or $-\infty$.

3.13. Notes and References

The material of this chapter was influenced by many sources, but especially by Folland [**16**], Rudin [**52**], and Sagan [**54**]. In particular, the presentation of Theorem 3.12.3 and Theorem 3.12.4 in Section 3.12 follows Folland [**16**]. One of the topics we did not discuss here is the product (Cauchy product) of series, which can be found in Folland [**16**] or Rudin [**52**]. Many additional topics and interesting exercises can be found in these books.

Chapter 4

Basic Topology, Limits, and Continuity

Topological concepts are based on the concept of an *open set*. The collection of open subsets of $\mathbf{R}$ provides us with a language for working with local properties of real valued functions defined on $\mathbf{R}$. Examples of local properties are continuity at a point and differentiability at a point. The language of open sets can also help us to describe global properties of sets and functions. Mathematicians define the *topology* of $\mathbf{R}$ as the collection of all open sets in $\mathbf{R}$.

The first three sections of the chapter define and discuss the basic properties of open sets and closed sets, compact sets, and connected sets. The remainder of the chapter presents basics on limits, continuity, uniform continuity, and the classification of discontinuities of a function.

4.1. Open Sets and Closed Sets

The following terminology concerning bounded intervals of real numbers is probably familiar from elementary calculus. If a and b are real numbers with $a < b$, then the interval (a, b) is an *open interval*, and $[a, b]$ is a *closed interval*. (Intervals of the form $[a, b)$ or $(a, b]$ are sometimes called *half-open* or *half-closed*, but this terminology will not be very helpful to us.) The characteristic property of an open interval that distinguishes it from the other three types of interval is this: Each point $x_0 \in (a, b)$ can be surrounded by another open interval $(c, d) \subset (a, b)$ such that $x_0 \in (c, d)$. In other words, for each point x_0 of (a, b) there is some open interval $(c, d) \subset (a, b)$ containing x_0 and consisting only of points from the set (a, b) itself. That is, x_0 is surrounded only by other points of (a, b). By considering the included endpoint(s) of the other three types of interval above, we can see that this property does not hold for those intervals.

Definition 4.1.1. *Let S be a set of real numbers.*

1. *A point $x_0 \in S$ is an* **interior point** *of S if there exists an $\epsilon > 0$ such that the open interval $(x_0 - \epsilon, x_0 + \epsilon)$ is contained in S. The set of all interior points of S, sometimes denoted* Int S, *is called the* **interior** *of S.*
2. *A point $x_0 \in \mathbf{R}$ is a* **boundary point** *of S if for every $\epsilon > 0$ the interval $(x_0 - \epsilon, x_0 + \epsilon)$ has nonempty intersection with both S and its complement S^c. The set of all boundary points of S, denoted ∂S, is called the* **boundary** *of S.*

Note that a boundary point of S need not be an element of S.

The property that distinguishes an *open* interval of real numbers from other types of intervals is that every point in the interval is an interior point of it. The interval $[2, 3)$ is *not* open since the endpoint 2 is not an interior point of $[2, 3)$. There is no $\epsilon > 0$ such that the interval $(2 - \epsilon, 2 + \epsilon)$ is contained in $[2, 3)$.

There are sets other than open intervals having the property that all of their points are interior points. We want to call these sets *open sets* as well.

Definition 4.1.2. *A set $O \subseteq \mathbf{R}$ is* **open** *if every point in O is an interior point of O.*

Thus a set $O \subseteq \mathbf{R}$ is open if, and only if, for every $x \in O$ there is an $\epsilon > 0$ such that the interval $(x - \epsilon, x + \epsilon)$ is contained in O. Equivalently, S is open if and only if S equals its interior.

Let us verify that an open interval (a, b) is indeed an open set. This may seem obvious since the definition of and open set was patterned on the distinguishing property of open intervals as opposed to other types of intervals. However, given any $x \in (a, b)$, we should be able to find an ϵ which shows that x is an interior point of (a, b). For example, let

$$\epsilon = \frac{1}{2}\min\{|x - a|, |x - b|\} = \frac{1}{2}\min\{x - a, b - x\} > 0.$$

(Sketch the interval (a, b), the point $x \in (a, b)$ and ϵ.) This definition of ϵ implies that

$$a \leq x - \epsilon < x + \epsilon \leq b$$

and hence

$$(x - \epsilon, x + \epsilon) \subset (a, b).$$

This construction of $\epsilon > 0$ (which is dependent on x) works for any $x \in (a, b)$, so (a, b) is an open set.

Example 4.1.3. The finite union of open intervals given by

$$(1, 2) \cup (3, 5) \cup (5, 7)$$

is an open set. The countable union of open intervals given by

$$\bigcup_{n=1}^{\infty} \left(n, n + \frac{1}{n}\right) = \left(1, 2\right) \cup \left(2, \frac{5}{2}\right) \cup \left(3, \frac{10}{3}\right) \cup \cdots$$

is an open set. Each of these statements is true, since any point in the union must be a point of one of the open intervals in the collection, and hence an interior point of the union since it is an interior point of that interval. △

Theorem 4.1.4. *The union of any collection (finite, countable, or uncountable) of open sets in* $\mathbf{R}$ *is an open set in* $\mathbf{R}$.

Proof. Let A be any index set and let $O := \bigcup_{\alpha \in A} O_\alpha$ where, for each $\alpha \in A$, O_α is an open set of real numbers. If $x \in O = \bigcup_{\alpha \in A} O_\alpha$, then $x \in O_{\alpha_0}$ for some $\alpha_0 \in A$. Since O_{α_0} is open, there is an $\epsilon > 0$ such that $(x - \epsilon, x + \epsilon) \subset O_{\alpha_0} \subset O$. Therefore every point of the union is an interior point and hence O is open. □

The set $\mathbf{R} = (-\infty, \infty)$ is an open set. This follows directly from the definition of an open set. We should also address the other extreme case of a subset of the reals: The empty set is an open set. (Why? Explain why the definition of an open set implies that the empty set is open.)

Are intersections of open sets always open? Note that $(0, 1) \cap (1, 2) = \emptyset$, which is one reason we needed to decide whether or not the empty set was open.

Observe that we can write a set consisting of a single point, such as $\{1\}$, as the intersection of infinitely many open intervals, for example

$$\{1\} = \bigcap_{j=1}^{\infty} (1 - 1/j, 1 + 1/j),$$

and hence the intersection of infinitely many open sets need not be open. However, note that the finite intersection

$$\bigcap_{j=1}^{10} (1 - 1/j, 1 + 1/j) = (1 - 1/10, 1 + 1/10) = (9/10, 11/10)$$

is an open set.

Consider open sets that are not nested as in the last paragraph. The nonempty intersection

$$(0, 1) \cap (.5, 2) \cap (.4, .75)$$

is easily seen to be open, as a simple sketch will suggest. The intersection is $(.5, .75)$. If x is in this intersection $(.5, .75)$, then x has a neighborhood (open interval) $(x - \epsilon_1, x + \epsilon_1)$ surrounding it and contained in $(0, 1)$, and another neighborhood $(x - \epsilon_2, x + \epsilon_2)$ surrounding it and contained in $(.5, 2)$, and still another neighborhood $(x - \epsilon_3, x + \epsilon_3)$ surrounding it and contained in $(.4, .75)$. Taking $\epsilon := \min\{\epsilon_i : i = 1, 2, 3\}$, we get an ϵ-interval about x contained entirely within the intersection $(.5, .75)$. This type of argument works for the intersection of any finite number of open sets.

Theorem 4.1.5. *The intersection of any finite collection of open sets in* $\mathbf{R}$ *is an open set in* $\mathbf{R}$.

Proof. Let $n \in \mathbf{N}$ and let $O_1, \ldots, O_n$ be open sets of real numbers. We want to show that the intersection

$$V := \bigcap_{j=1}^{n} O_j$$

is also an open set. Let $x \in V$. Since $x \in O_j$ for all $j = 1, \ldots, n$, there are numbers $\epsilon_j > 0$ such that $(x - \epsilon_j, x + \epsilon_j) \subset O_j$ for $j = 1, \ldots, n$. If we let $\epsilon = \min\{\epsilon_1, \ldots, \epsilon_n\}$, then $(x-\epsilon, x+\epsilon) \subset O_j$ for $j = 1, \ldots, n$. Thus $(x-\epsilon, x+\epsilon) \subset V$ and x is an interior point of V. Hence V is open. □

Since arbitrary unions of open sets are open, it is easy to construct a variety of open sets by taking unions of pairwise disjoint collections of open intervals. For example, both

$$(0,1) \cup (2,3) \cup (4,6)$$

and

$$\bigcup_{j=2}^{\infty} (j + 1/j^2, j + 1/j) = (9/4, 5/2) \cup (28/9, 10/3) \cup \cdots$$

are unions of pairwise disjoint collections of open sets, and hence they are open. The union

$$\bigcup_{j=2}^{\infty} (j + 1/j, j + 3/j) = (5/2, 7/2) \cup (10/3, 4) \cup (17/4, 19/4) \cup (26/5, 28/5) \cup \cdots$$

is also open since it is a union of open sets. However, these open sets are not pairwise disjoint. Note that there is some overlap in the first two intervals shown, but for $j \geq 4$ the intervals are pairwise disjoint. Therefore this open set can be expressed as the disjoint union

$$(5/2, 4) \cup (17/4, 19/4) \cup (26/5, 28/5) \cdots .$$

If you have difficulty providing an example of an open set which cannot be expressed as a countable union of disjoint open intervals, you might be motivated to try to prove the following theorem.

Theorem 4.1.6 (Structure of Open Sets). *Let O be an open subset of real numbers. Then there are countably many pairwise disjoint open intervals I_k such that*

$$O = \bigcup_k I_k.$$

Proof. The result is true in a trivial way when $O = \emptyset$. Therefore we assume that O is nonempty and open. The intersection of O with the set of rational numbers is countable. Let $r_1, r_2, \ldots$ be an enumeration of $O \cap \mathbf{Q}$. For each k there exist real numbers a and b such that $r_k \in (a,b) \subset O$, since O is open. Using the least upper bound and greatest lower bound properties of the reals, we may define

$$a_k = \inf\{a : r_k \in (a,b) \subset O \text{ for some } b > a\}$$

and

$$b_k = \sup\{b : r_k \in (a,b) \subset O \text{ for some } a < b\}.$$

It is possible to have $a_k = -\infty$ or $b_k = \infty$. Let $I_k = (a_k, b_k)$. Then, by the definition of a_k and b_k, $I_k \subset O$, hence $\bigcup_k I_k \subset O$. Now $r_k \in I_k$ and $r_1, r_2, \ldots$ is an enumeration of $O \cap \mathbf{Q}$, so $\bigcup_k I_k$ contains every rational number in O. If s is irrational and $s \in O$, then there exist a and b such that $s \in (a,b) \subset O$, since O is open. Since Q is dense in $\mathbf{R}$, there is a j such that $r_j \in (a,b)$, and thus $s \in I_j \subset \bigcup_k I_k$. This proves that $\bigcup_k I_k = O$. □

Any proof of the structure theorem for open sets in **R** must use the completeness of the reals in some way. For an alternative proof based on an equivalence relation with the I_k being the equivalence classes, see Exercise 4.1.3.

Definition 4.1.7. *A set $F \subseteq \mathbf{R}$ is* **closed** *if its complement $F^c \subseteq \mathbf{R}$ is open.*

Since **R** is open, its complement $\emptyset$ is closed, and since $\emptyset$ is open, its complement **R** is closed. So **R** and $\emptyset$ are both open and closed. We will see immediately after Theorem 4.1.13 below that there are no nonempty proper subsets of **R** that are both open and closed.

Any closed interval $[a, b]$ with a and b finite is a closed set, since the complement, $(-\infty, a) \cup (b, \infty)$, is open. For the same reason, intervals of the form $(-\infty, b]$ and $[a, \infty)$ are closed sets.

It is left as an exercise to show that every finite set of real numbers is closed. The next example is more interesting.

Example 4.1.8. The set

$$F = \{0\} \cup \{1/j : j \in \mathbf{N}\} = \{0, 1, 1/2, 1/3, \ldots\}$$

is closed. Note that the complement of this set is

$$(-\infty, 0) \cup \left[\bigcup_{j=1}^{\infty} \left(\frac{1}{j+1}, \frac{1}{j}\right)\right] \cup (1, \infty)$$

which is open since it is a union of open intervals. △

Theorem 4.1.9. *The intersection of any collection (finite, countable, or uncountable) of closed sets in* **R** *is a closed set in* **R**.

Proof. Let F_α be a closed set of real numbers for each α in some index set A, and let

$$F = \bigcap_{\alpha \in A} F_\alpha.$$

We show that the complement of F is open. By one of DeMorgan's laws,

$$F^c = \left[\bigcap_{\alpha \in A} F_\alpha\right]^c = \bigcup_{\alpha \in A} F_\alpha^c.$$

For each α, F_α is closed, hence F_α^c is open. Thus F^c is open, since a union of open sets is open. Hence F is closed. □

Are arbitrary unions of closed sets always closed? No. Think of an example like the following countable union of closed intervals,

$$\bigcup_{j=2}^{\infty} [1/j, 1 - 1/j] = (0, 1),$$

which is not closed.

Theorem 4.1.10. *The union of any finite collection of closed sets in* **R** *is a closed set in* **R**.

Proof. Let $n \in \mathbf{N}$, let $F_1, \ldots, F_n$ be closed sets of real numbers, and let $F = \bigcup_{j=1}^{n} F_j$. Then $F^c = \left(\bigcup_{j=1}^{n} F_j\right)^c = \bigcap_{j=1}^{n} F_j^c$ is an intersection of a finite collection of open sets in $\mathbf{R}$, and by Theorem 4.1.5, F^c is open. $\square$

Recall that we defined the concept of *cluster point* (or accumulation point) in an ordered field in Definition 2.6.2 in the discussion of the Bolzano-Weierstrass theorem. For convenience we repeat that definition here for the ordered field $\mathbf{R}$.

Definition 4.1.11. *Let S be a set of real numbers.*

1. *A point $x_0 \in \mathbf{R}$ is a* **cluster point** *(or,* **accumulation point***) of S if for every $\epsilon > 0$ the interval $(x_0 - \epsilon, x_0 + \epsilon)$ contains infinitely many points of S distinct from x_0.*
2. *If $x_0 \in S$ and x_0 is not a cluster point of S, then it is an* **isolated point** *of S.*

We usually opt for two syllables rather than five, and use the term *cluster point.* Note that a cluster point of S need not be an element of S. A point $x_0 \in S$ is an isolated point of S if and only if there exists some $\epsilon > 0$ such that the interval $(x_0 - \epsilon, x_0 + \epsilon)$ contains no point of S other than x_0 (Exercise 4.1.4). The reader should also verify the following statements: Every interior point of S is a cluster point of S. Every isolated point of S is a boundary point of S. Every nonisolated boundary point of S is a cluster point of S.

Theorem 4.1.12. *A point $x_0 \in S$ is a cluster point of S if and only if there is a nonconstant sequence (x_n) of points of S distinct from x_0 such that $x_n \to x_0$ as $n \to \infty$.*

Proof. Suppose x_0 is a cluster point of S. Let $\epsilon_n = 1/n$. Then for each n we may choose a point x_n of S such that $x_n \neq x_0$ and $|x_n - x_0| < 1/n$. By the definition of a cluster point, we may even arrange that for each n, $x_n \neq x_k$ for $1 \leq k < n$. This gives an infinite, nonconstant sequence (x_n) of points distinct from x_0 such that $x_n \to x_0$ as $n \to \infty$.

For the converse, if there is a nonconstant sequence (x_n) of points of S distinct from x_0 such that $x_n \to x_0$ as $n \to \infty$, then x_0 satisfies the definition of a cluster point of S. $\square$

Note that any sequence in S that converges to an isolated point of S must eventually be constant.

Theorem 4.1.13. *A subset of the real numbers is closed if and only if it contains all its cluster points.*

Proof. Suppose S is a closed subset of $\mathbf{R}$. Then S^c is open. If a is a cluster point of S and $a \in S^c$, then there is a $\delta > 0$ such that $(a - \delta, a + \delta) \subset S^c$, but this contradicts a being a cluster point of S. Hence every cluster point of a closed set S is in S.

Suppose now that S contains all its cluster points. Let $b \in S^c$. Then b is not a cluster point of S, so there exists a $\delta > 0$ such that $(b - \delta, b + \delta)$ contains no point of

S, and hence $(b-\delta, b+\delta) \subset S^c$. (See Exercise 4.1.4.) This is true for each $b \in S^c$, so S^c is an open set. Therefore S is closed. □

Theorem 4.1.13 easily explains why the set F in Example 4.1.8 is closed, since the only cluster point of F is 0 and $0 \in F$.

From the structure Theorem 4.1.6 and Theorem 4.1.13, we can see that there is no nonempty *proper* subset of the reals that is both open and closed, as follows: If O is an open, nonempty proper subset of $\mathbf{R}$, then by Theorem 4.1.6 there is at least one endpoint b of a nonempty open interval $I_k \subset O$ such that $b \notin O$. The point b is a cluster point of O since it is a cluster point of I_k; however, $b \notin I_k$, and b cannot be an element of any other open interval in the disjoint union of the structure theorem. Therefore O does not contain all its cluster points. Hence O is not closed, by Theorem 4.1.13.

Definition 4.1.14. *The* **closure** *of a set $S \subseteq \mathbf{R}$, denoted $\overline{S}$, is the union of S and its set of cluster points.*

Theorem 4.1.15. *A subset S of the real numbers is closed if and only if it equals its closure, that is, $\overline{S} = S$.*

Proof. S is closed if and only if it contains all its cluster points, by Theorem 4.1.13, so by definition of the closure of S, S is closed if and only if $\overline{S} = S$. □

Recall that we defined a subset S of $\mathbf{R}$ to be dense in $\mathbf{R}$ provided that between any two real numbers there is an element of S (Definition 2.3.8). Thus, a subset S is dense in $\mathbf{R}$ if every nonempty open interval (a, b) intersects S. An equivalent statement is that S is dense in $\mathbf{R}$ if the closure of S equals $\mathbf{R}$, $\overline{S} = \mathbf{R}$: Let $x \in \mathbf{R}$; then $x \in \overline{S}$ if and only if there is a sequence of points $s_k \in S$ such that $s_k \to x$ as $k \to \infty$, and this occurs if and only if every nonempty open interval about x intersects S. Thus, $\overline{S} = \mathbf{R}$ if and only if every nonempty open interval in $\mathbf{R}$ intersects S. We have seen that the set of rational numbers is dense in the real line, $\overline{\mathbf{Q}} = \mathbf{R}$, and the set of irrational numbers is dense as well, $\overline{\mathbf{I}} = \mathbf{R}$.

More generally, we say that a set S is **dense in an open set** U if $U \subset \overline{S}$. For example, the rational numbers are dense in $(0, 1)$ since $(0, 1) \subset \overline{\mathbf{Q}}$. The irrational numbers are also dense in $(0, 1)$.

The following definition is also useful.

Definition 4.1.16. *A set $S \subset \mathbf{R}$ is* **nowhere dense** *if its closure $\overline{S}$ has no interior point.*

It follows from the definition that a set $S \subset \mathbf{R}$ is nowhere dense if and only if its closure $\overline{S}$ contains no open interval of positive length.

There is another interesting property of the Cantor set C we can now consider.

Theorem 4.1.17. *The Cantor set C is closed and nowhere dense.*

Proof. By its definition, C is the intersection $\bigcap_{k=1}^{\infty} C_k$ of a countable collection of closed sets, hence C is a closed set.

Since C is closed, it equals its closure, so we must show that C has no interior points, that is, C contains no open interval of positive length. Suppose to the contrary that C does have an interior point and thus does contain an open interval J of positive length, $\lambda(J)$. Let k be a positive integer such that $2^{-k} \le \lambda(J)$. Since C is contained in the set C_k which is a union of 2^k disjoint intervals, each of length 3^{-k}, J must be contained in one of these intervals. Thus, $\lambda(J) \le 3^{-k}$. But then $2^{-k} \le \lambda(J) \le 3^{-k}$, which is the contradiction we seek. Therefore C is nowhere dense. □

Recall that C has total length zero. This is a good opportunity to look again, or for the first time, at Exercise 3.4.6, which describes a Cantor set F in $[0, 1]$ with positive total length. The set F is a countable intersection of closed sets C_k, hence F is closed. By the construction in Exercise 3.4.6, each closed set C_k consists of 2^k disjoint closed intervals of length $\alpha/3^{k+1}$, where $0 < \alpha < 1$. If F has an interior point, it contains an interval J of positive length, say $\lambda(J)$. Let k be a positive integer such that $2^{-k} \le \lambda(J)$. Then J must be contained in one of these closed intervals of length $\alpha/3^{k+1}$, and hence

$$2^{-k} \le \lambda(J) \le \alpha/3^{k+1} < 3^{-(k+1)},$$

a contradiction. Therefore F contains no open interval of positive length. Hence F is a nowhere dense set which has positive total length.

It may be useful to briefly summarize the four types of points we have defined to describe basic topological notions concerning sets of real numbers:

- ⋆ interior points,
- ⋆ isolated points,
- ⋆ cluster points,
- ⋆ boundary points.

Given a set S, every point of S is either an interior point or a boundary point. Interior points and isolated points belong to S, by definition. Boundary points need not belong to S. Cluster points are either interior points or boundary points, but never isolated points. Isolated points are boundary points.

Exercises.

Exercise 4.1.1. Show: $S \subset \mathbf{R}$ is an open set if and only if $\partial S = \emptyset$.

Exercise 4.1.2. Show: If a is a real number and $\delta > 0$, then $\{x : |x - a| > \delta\}$ is an open set.

Exercise 4.1.3. This exercise outlines another proof of the structure theorem for open sets in $\mathbf{R}$. Let $O \subseteq \mathbf{R}$ be an open set.

1. Define a relation on O as follows: $x \sim y$ if all real numbers strictly between x and y are elements of O. Show that this defines an equivalence relation on O.
2. Let $\{O_\alpha\}_{\alpha \in A}$ be the collection of equivalence classes for this equivalence relation. Show that each O_α is a nonempty open interval.
3. Show that there are countably many equivalence classes.

Exercise 4.1.4. Prove: A point $x_0 \in S$ is an isolated point of S (Definition 4.1.11) if and only if there exists $\epsilon > 0$ such that the interval $(x_0 - \epsilon, x_0 + \epsilon)$ contains no point of S other than x_0.

Exercise 4.1.5. A subset $F \subset \mathbf{R}$ is closed if and only if every Cauchy sequence of elements of F has its limit in F.

Exercise 4.1.6. Prove: For any set S of real numbers, $\partial S = \partial(\mathbf{R} - S)$.

Exercise 4.1.7. Prove: For any set S of real numbers, $\partial S = \overline{S} - \operatorname{Int} S$.

Exercise 4.1.8. Prove: If $S = \partial S$, then S cannot be open unless S is empty. Give an example of such a set S that is closed, and an example that is neither closed nor open.

Exercise 4.1.9. Give an example of an unbounded set S for which ∂S is bounded.

Exercise 4.1.10. Prove: S is closed if and only if S contains all its boundary points.

4.2. Compact Sets

One of the most useful concepts in analysis is that of a compact set. The definition of compact set uses the concept of an open cover of a set.

Definition 4.2.1. *Let S be a subset of the real numbers and let O_γ be an open set for each γ in some index set Γ. If*

$$S \subset \bigcup_{\gamma \in \Gamma} O_\gamma,$$

then the collection $\{O_\gamma\}_{\gamma\in\Gamma}$ is called an **open cover** *of S. If $\{O_\gamma\}_{\gamma\in\Gamma}$ is an open cover of S, $\Gamma_0 \subset \Gamma$, and*

$$S \subset \bigcup_{\gamma \in \Gamma_0} O_\gamma,$$

then the collection $\{O_\gamma\}_{\gamma\in\Gamma}$ contains the **subcover** *$\{O_\gamma\}_{\gamma\in\Gamma_0}$ of S. If a subcover of S has finitely many elements, it is a* **finite subcover** *of S.*

Definition 4.2.2. *A set K of real numbers is* **compact** *if every open cover of K contains a finite subcover of K.*

We first observe that a compact set must be bounded. For suppose $K \subset \mathbf{R}$ is compact, and consider the union of all open intervals of the form $(-k, k)$ where $k \in \mathbf{N}$. This union certainly contains K, so there is a finite subcover, say $\{(-k_j, k_j) : j = 1, \ldots, m\}$. Since

$$K \subset \bigcup_{j=1}^{m} (-k_j, k_j) = (-k_0, k_0)$$

where $k_0 = \max\{k_1, k_2, \ldots, k_m\}$, it follows that K is bounded. Since compact sets must be bounded, $\mathbf{R}$ is not compact. Also, no *unbounded* interval can be compact.

Example 4.2.3. The collection of open intervals

$$\Omega = \{(1/k, 1 + 1/k) : k \in \mathbf{N}\}$$

is an open cover of $S = (0, 1]$, but this collection contains no finite subcover of S. For any finite subcollection $\hat{\Omega} \subseteq \Omega$ there is a maximum value of k, say k_0, such that the interval $(1/k_0, 1 + 1/k_0) \in \hat{\Omega}$. But then $1/(2k_0) \in S$ and $1/(2k_0)$ is not contained in any interval of the collection $\hat{\Omega}$, so $\hat{\Omega}$ cannot cover S. Hence $S = (0, 1]$ is not compact. △

Example 4.2.4. Let S be the set of reals defined by

$$S = \left\{\frac{1}{2^n} : n \in \mathbf{N}\right\} = \left\{\frac{1}{2}, \frac{1}{4}, \frac{1}{8}, \frac{1}{16}, \ldots\right\}.$$

Let $\Omega = \{O_n : n \in \mathbf{N}\}$ be the collection of open intervals

$$O_n = \left(\frac{1}{2^n} - \frac{1}{2}\frac{1}{2^{n+1}}, \frac{1}{2^n} + \frac{1}{2}\frac{1}{2^{n+1}}\right) = \left(\frac{1}{2^n} - \frac{1}{2^{n+2}}, \frac{1}{2^n} + \frac{1}{2^{n+2}}\right),$$

that is,

$$O_1 = \left(\frac{3}{8}, \frac{5}{8}\right), \quad O_2 = \left(\frac{3}{16}, \frac{5}{16}\right), \quad \ldots \quad , O_n = \left(\frac{3}{2^{n+2}}, \frac{5}{2^{n+2}}\right), \quad \ldots .$$

Then for each $n \in \mathbf{N}$, O_n contains $\frac{1}{2^n}$ and no other element of S. The collection $\{O_n\}_{n \in \mathbf{N}}$ is an open cover of S. Since the sets in the cover are pairwise-disjoint, no subcover from the collection $\{O_n\}_{n \in \mathbf{N}}$ can cover S. Hence S is not compact. The only cluster point of S is 0, so the closure of S is $\bar{S} = S \cup \{0\}$. Consider the open cover Ω_1 of $\bar{S}$ obtained by augmenting the collection $\{O_n\}_{n \in \mathbf{N}}$ with the interval $(-\frac{1}{10}, \frac{1}{10})$, that is, $\Omega_1 = \Omega \cup \{(-\frac{1}{10}, \frac{1}{10})\}$. Note that $\bar{S}$ can be covered by the finite subcover of Ω_1 given by

$$\left\{\left(-\frac{1}{10}, \frac{1}{10}\right), O_1, O_2, O_3\right\}.$$

that is, $\bar{S} \subset (-\frac{1}{10}, \frac{1}{10}) \cup O_1 \cup O_2 \cup O_3$. Of course, the interval $(-\frac{1}{10}, \frac{1}{10})$ in this discussion could be replaced by any open interval $(-\delta, \delta)$ with $\delta > 0$, and still the open cover $\Omega \cup \{(-\delta, \delta)\}$ of $\bar{S}$ contains a finite subcover. △

The last two examples suggest a relation between the property of compactness and the properties of being closed and bounded. We clarify this relation with the following result.

Theorem 4.2.5. *If $K \subset \mathbf{R}$ is compact, then K is closed and bounded.*

Proof. Suppose $K \subset \mathbf{R}$ and K is compact. Then K is bounded, as we saw in the comments immediately following Definition 4.2.2. We will show K is closed by showing that K^c is open. Let $a \in K^c$. We want to show that a is an interior point of K^c. Consider the collection of open sets O_k, $k \in \mathbf{N}$, defined by

$$O_k = \{x \in \mathbf{R} : |x - a| > 1/k\}.$$

Then $\bigcup_{k=1}^{\infty} O_k = \mathbf{R} - \{a\}$, so $K \subset \bigcup_{k=1}^{\infty} O_k$. Since K is compact, there is a finite subcover, say

$$K \subset \bigcup_{j=1}^{m} O_{k_j}.$$

Let $N = \max\{k_1, \ldots, k_m\}$. Then

$$K \subset \bigcup_{j=1}^{m} O_{k_j} = \{x \in \mathbf{R} : |x - a| > 1/N\}.$$

By taking complements, we have

$$\{x \in \mathbf{R} : |x - a| < 1/N\} \subset \{x \in \mathbf{R} : |x - a| \le 1/N\} \subset K^c,$$

which shows that a is an interior point of K^c. Since every point of K^c is an interior point, K^c is open, and hence K is closed. □

We show next that any closed subset of a compact set is compact.

Theorem 4.2.6. *If F is closed, K is compact, and $F \subset K \subset \mathbf{R}$, then F is compact.*

Proof. Let $\{O_\alpha\}$ be an open cover of F. Then

$$\{F^c\} \cup \{O_\alpha\}$$

is an open cover of K. Since K is compact, there is a finite subcover of K. If the finite subcover of K includes F^c, then we may omit F^c from the subcover and still have a finite subcover of F from the open cover $\{O_\alpha\}$. □

Theorem 4.2.7. *Every closed interval $[a, b]$ is compact.*

Proof. The proof is by contradiction. We suppose that $\{O_\alpha\}$ is an open cover of $I_0 = [a, b]$ for which there is no finite subcover of $[a, b]$. Then one of the subintervals $[a, (a+b)/2]$ or $[(a+b)/2, b]$ (or both) cannot be covered by a finite subcollection from $\{O_\alpha\}$; choose one and label it I_1. By our assumption on I_1, either the left-half or the right-half closed subinterval of I_1 cannot be covered by a finite subcollection from $\{O_\alpha\}$. Choose one and call it I_2. If this process cannot be continued indefinitely by some choice of subintervals at each stage, there is a contradiction of the assumption that $\{O_\alpha\}$ has no finite subcover. Thus, under our assumption, there is a sequence of closed and nonempty intervals I_n such that $I_{n+1} \subset I_n$ for each $n = 0, 1, 2, \ldots$, with length $\lambda(I_n) = (a + b)/2^n$. Since $\lim_{n\to\infty} \lambda(I_n) = 0$, Theorem 2.5.1 implies there is a unique point $x \in \bigcap_{n=1}^{\infty} I_n \subset [a, b]$. Since $x \in [a, b]$, $x \in O_{\alpha_0}$ for some α_0. Since O_{α_0} is open, there is a $\delta > 0$ such that $(x - \delta, x + \delta) \subset O_{\alpha_0}$. But there is a positive integer N such that $\lambda(I_N) = (a + b)/2^N < \delta/2$. (This follows from the Archimedean property of $\mathbf{R}$.) But then $x \in I_N \subset O_{\alpha_0}$, which contradicts the fact that I_N cannot be covered by a finite subcollection from $\{O_\alpha\}$. This contradiction shows that $[a, b]$ is compact. □

Theorem 4.2.8. *If $K \subset \mathbf{R}$ is closed and bounded, then K is compact.*

Proof. Suppose K is closed and bounded. Then $K \subseteq [-b, b]$ for some $b > 0$. Since $[-b, b]$ is compact by Theorem 4.2.7, its closed subset K is compact by Theorem 4.2.6. □

Theorems 4.2.5 and 4.2.8 together prove the following result called the *Heine-Borel theorem.*

Theorem 4.2.9. *A set $K \subset \mathbf{R}$ is compact if and only if it is closed and bounded.*

Compact sets in $\mathbf{R}$ may also be characterized as those sets K such that every infinite subset of K contains a nonconstant convergent sequence with limit in K; in other words (by Theorem 4.1.12) every infinite subset of K contains a cluster point in K.

Theorem 4.2.10. *A subset $K \subset \mathbf{R}$ is compact if and only if every infinite subset of K contains a nonconstant convergent sequence with limit in K.*

Proof. If K is compact, then K is closed and bounded. Let S be an infinite subset of K. By the Bolzano-Weierstrass theorem, the bounded infinite set S has a cluster point p. By Theorem 4.1.12, there is a nonconstant sequence (a_k) of points of S, with $a_k \neq p$, such that $\lim_{k\to\infty} a_k = p$. Since $S \subset K$ and K is closed, $p \in K$.

Suppose that every infinite subset of K contains a nonconstant convergent sequence with limit in K. If K is not bounded, then for each positive integer n, there is an element $a_n \in K$ such that $|a_n| > n$. The resulting sequence (a_n) in K is unbounded and hence diverges to ∞. More to the point, every subsequence of a sequence that diverges to ∞ must also diverge to ∞. But this is a contradiction of our hypothesis. Thus K must be bounded. In order to show that K is closed, we want to show that K contains all its cluster points. Let $p \in \mathbf{R}$ be a cluster point of K. Then, by Theorem 4.1.12, there is a nonconstant sequence (x_k) of points of K such that $x_k \neq p$, and $\lim_{k\to\infty} x_k = p$. By hypothesis, the infinite set $\{x_k\}$ contains a nonconstant convergent sequence (a subsequence of (x_k)) that converges to a limit in K; however, that limit must equal p, so $p \in K$. Therefore K is closed. By Theorem 4.2.8, K is compact. □

We will see in Chapter 8 that the characterizations of compactness in Theorem 4.2.9 and Theorem 4.2.10 extend directly to n-dimensional space $\mathbf{R}^n$. However, in the more general context of metric spaces considered in Chapter 9, we will see that compactness is not equivalent to being closed and bounded.

Exercises.

Exercise 4.2.1. Show that if $A \subset \mathbf{R}$ is bounded, then its closure $\overline{A}$ is compact.

Exercise 4.2.2. Show that if $S \subset \mathbf{R}$ and S is compact, then $\sup S \in S$ and $\inf S \in S$.

Exercise 4.2.3. Suppose F and K are subsets of $\mathbf{R}$ with F closed and K compact. Show that $F \cap K$ is compact.

Exercise 4.2.4. A nested sequence $K_{n+1} \subset K_n$ of nonempty, compact subsets $K_n \subset \mathbf{R}$ has a nonempty, compact intersection. *Hint*: $K_n \subseteq I_n = [\inf K_n, \sup K_n]$.

Exercise 4.2.5. Show that $K \subset \mathbf{R}$ is compact if and only if every infinite sequence in K has a convergent subsequence with limit in K.

4.3. Connected Sets

The property of *connectedness* is an important one in topology because, as we show later, it is a property preserved by continuous mappings. In order to introduce the concept of connectedness, it is easiest to state what is meant by a set being *disconnected*.

Definition 4.3.1. *A subset S of* $\mathbf{R}$ *is* **disconnected** *if there exist open sets A and B such that*

$$A \cap S \neq \phi, \quad B \cap S \neq \phi, \quad (A \cap S) \cap (B \cap S) = \emptyset, \quad (A \cap S) \cup (B \cap S) = S.$$

In this case, we say that A and B provide a **disconnection** *of S, or that A and B* **disconnect** *S. A set $S \subset \mathbf{R}$ is* **connected** *if it is not disconnected.*

The connected subsets of $\mathbf{R}$ are fairly easy to characterize.

Theorem 4.3.2. *A subset of* $\mathbf{R}$ *is connected if and only if it is an interval. In particular,* $\mathbf{R}$ *is connected.*

Proof. Each of the required implications is proved by proving the contrapositive statement.

S connected implies S is an interval. The contrapositive is: If $S \subset \mathbf{R}$ is not an interval, then S is disconnected. If S is not an interval, then there are distinct points x, y in S with $x < y$ and a point $z \notin S$ with $x < z < y$. Let $A = \{w \in \mathbf{R} : w < z\} = (-\infty, z)$ and $B = \{w \in \mathbf{R} : z < w\} = (z, \infty)$. Then A and B are open sets with $A \cap B = \emptyset$, $A \cap S \neq \emptyset$ since $x \in A$, $B \cap S \neq \emptyset$ since $y \in B$, and $(A \cap S) \cup (B \cap S) = S$. Hence A and B disconnect S.

S is an interval implies S is connected. (This ensures that $\mathbf{R}$ is connected, since $\mathbf{R}$ is an interval.) The contrapositive is: If $S \subseteq \mathbf{R}$ is disconnected, then S is not an interval. Thus, we suppose the open sets A and B disconnect S. By the structure theorem for open subsets of $\mathbf{R}$, we may express A and B as disjoint countable unions of open intervals, say $A = \bigcup_k A_k$ and $B = \bigcup_j B_j$ where A_k and B_j are open intervals for each k, j. Since $A \cap S \neq \emptyset$, $B \cap S \neq \emptyset$, and $(A \cap S) \cap (B \cap S) = \emptyset$, we may assume, by relabeling A and B if necessary, that for some k_0 and j_0, A_{k_0} lies to the left of B_{j_0}. Clearly, $(A_{k_0} \cap S) \cap (B \cap S) = \emptyset$ and $(A \cap S) \cap (B_{j_0} \cap S) = \emptyset$. Since $S \subset A \cup B$, but $\sup A_{k_0} \notin S$ and $\inf B_{j_0} \notin S$, S is not an interval. $\square$

The proof of Theorem 4.3.2 used the least upper bound property of $\mathbf{R}$ implicitly when we used the structure theorem for open sets. As a corollary of Theorem 4.3.2, the reader can deduce a result we noted earlier in the chapter; see Exercise 4.3.1.

Exercise.

Exercise 4.3.1. Deduce from Theorem 4.3.2 that the only subsets of $\mathbf{R}$ that are both open and closed are the empty set and $\mathbf{R}$.

4.4. Limit of a Function

In this section we define the limit of a function at a cluster point of its domain D. Recall that a cluster point of D need not be an element of D.

Definition 4.4.1 (Limit at a Point). *Let $D \subset \mathbf{R}$ and $f : D \to \mathbf{R}$. Let a be a cluster point of D, and let $L \in \mathbf{R}$. We say that f has the limit L as x approaches a, and we write*

$$\lim_{x \to a} f(x) = L$$

if for every $\epsilon > 0$ there is a $\delta = \delta(\epsilon, a) > 0$ such that

$$x \in D \text{ and } 0 < |x - a| < \delta \quad \Longrightarrow \quad |f(x) - L| < \epsilon.$$

Notice that the definition of limit allows for one-sided limits at an endpoint of a domain like $[a, b]$. In Section 4.9 we give a general definition of one-sided limits that is helpful in discussing discontinuities of a function.

The uniqueness of a limiting value as defined by Definition 4.4.1 is an important issue considered below in Theorem 4.4.6.

That the δ in the definition generally must depend on both ϵ and a will be clear from the next example.

Example 4.4.2. Consider $f(x) = 1/x$ on $D = (0, 1]$. Suppose we want to show that $\lim_{x \to 1/2} f(x) = 2$ (which it surely is) from the definition. We have to show that for every $\epsilon > 0$ there is a $\delta = \delta(\epsilon, a = 1/2) > 0$ such that

$$x \in D \text{ and } 0 < |x - 1/2| < \delta \quad \Longrightarrow \quad |f(x) - 2| < \epsilon.$$

We begin by writing

$$|f(x) - 2| = \left|\frac{1}{x} - 2\right| = \left|\frac{1 - 2x}{x}\right| = \frac{2|1/2 - x|}{|x|},$$

where we have isolated the factor $|x - 1/2|$ that must be restricted by a suitable δ. In the spirit intended by the question of a limiting value, we can agree to consider only those x with, say, $1/4 \leq x \leq 3/4$, which are in D. This is useful because it provides us with a bound for the factor $1/|x| = 1/x$. Consequently, for these values of x, we have

$$|f(x) - 2| = \frac{2|1/2 - x|}{|x|} \leq 8|x - 1/2|.$$

The upper bound helps because it isolates the crucial factor $|x - 1/2|$ in an upper bound for $|f(x) - 2|$, and because we may now achieve

$$8|x - 1/2| < \epsilon$$

and hence $|f(x) - 2| < \epsilon$, by choosing

$$|x - 1/2| < \delta$$

where δ is any positive number less than or equal to $\min\{1/4, \epsilon/8\}$. On the other hand, consider the verification that

$$\lim_{x \to 1/4} f(x) = 4.$$

Similar reasoning and calculations show that we will have

$$|f(x) - 4| < \epsilon$$

provided we choose $1/8 < x < 3/8$ and

$$|x - 1/4| < \delta$$

where δ is any positive number less than or equal to $\min\{1/8, \epsilon/32\}$. (As an exercise, check this statement.) Clearly, the values of a function may change more rapidly as we vary x in some regions of its domain than in other regions. From this standpoint, it is no surprise that the choice of δ for a given ϵ in the definition of $\lim_{x \to a} f(x) = L$ generally depends on both ϵ and a. △

Let us emphasize another important feature of the definition of limit. The essential issue about a limit is the behavior of the function for values of x *near to, but not equal to,* the point a, hence the consideration of those x with $0 < |x - a| < \delta$ in the definition. A function f may have a limiting value at a whether or not $f(a)$ is defined, and the value $f(a)$ (if defined) is immaterial in determining the existence of the limit. (See Exercise 4.4.1.) As we will see later, the concept of continuity of a function at a point a simplifies the question of the limiting value of the function at a.

The computation of some limits using the definition provides good practice with both the definition of limit and some elementary estimates. We present one example here and leave others for the exercises.

Example 4.4.3. We show that $\lim_{x\to 1} x^2 = 1$ from the definition. We have

$$|x^2 - 1| = |(x+1)(x-1)| = |(x+1)|\,|(x-1)| \leq 3|x-1|$$

for all x in the interval defined by $|x-1| \leq 1$. Let $\epsilon > 0$. If $\epsilon < 3$, then we may take $\delta = \epsilon/3$ so that $|x-1| < \delta$ implies $|x-1| < 1$, hence $|x^2-1| < \epsilon$. (Of course a similar statement can be made for any $\epsilon > 0$, with an appropriate δ being $\delta = \min\{\epsilon/3, 1\}$.) Hence, $\lim_{x\to 1} x^2 = 1$. △

As this simple example shows, the key in applying Definition 4.4.1 directly is to relate the difference $|f(x) - f(a)|$ to the difference $|x - a|$ by an inequality, by supplying appropriate constant bounds on certain terms that arise. A sketch of the graph of f for x near a can often be helpful in determining an appropriate candidate for $\lim_{x\to a} f(x)$. See Exercise 4.4.2.

We may also define limits at infinity.

Definition 4.4.4 (Limit at Infinity). *Let $D \subseteq \mathbf{R}$ and $f : D \to \mathbf{R}$. Let $L \in \mathbf{R}$. We write*

$$\lim_{x\to\infty} f(x) = L \quad (\lim_{x\to -\infty} f(x) = L)$$

if for every $\epsilon > 0$ there is an $M > 0$ such that

$$x \in D \text{ and } x \geq M \ (x \leq -M) \quad \Longrightarrow \quad |f(x) - L| < \epsilon.$$

Example 4.4.5. Let $f : [1,\infty) \to \mathbf{R}$ be the function $f(x) = 1/x$. Then $\lim_{x\to\infty} f(x) = 0 = L$, since for every $\epsilon > 0$ we may choose $N = N(\epsilon) = 1/\epsilon$, and then for all $x > N$, $|f(x)| = 1/x < \epsilon$. Similarly, for each positive integer n, $f(x) = 1/x^n$ has limit 0 as $x \to \infty$. △

Before doing anything else, we should show that limits are unique.

Theorem 4.4.6 (Uniqueness of Limits). *Let $f : D \to \mathbf{R}$ and let a be a cluster point of D. If $\lim_{x\to a} f(x) = L_1$ and $\lim_{x\to a} f(x) = L_2$ according to Definition 4.4.1, then $L_1 = L_2$.*

Proof. By the triangle inequality,

$$|L_1 - L_2| \leq |L_1 - f(x)| + |f(x) - L_2|.$$

By hypothesis, given any $\epsilon > 0$ there is a $\delta_1 > 0$ such that if $0 < |x-a| < \delta_1$, then $|L_1 - f(x)| < \epsilon/2$; and, there is a $\delta_2 > 0$ such that if $0 < |x-a| < \delta_2$, then $|f(x) - L_2| < \epsilon/2$. Thus, if $0 < |x-a| < \delta := \min\{\delta_1, \delta_2\}$, then $|L_1 - L_2| < \epsilon/2 + \epsilon/2 = \epsilon$. Since $\epsilon > 0$ was arbitrary, the result follows. □

The proof of uniqueness of limits at $a = \pm\infty$ is similar and is left as an exercise.

We are now ready to establish some basic properties of limits that are used as a matter of routine. The proofs of some of these facts will be left as exercises for the reader.

Theorem 4.4.7. *Let a be a cluster point of the domain of f, or $\pm\infty$. Let $L \in \mathbf{R}$. The following properties hold:*

(a) *If $\lim_{x\to a} f(x) = L$ exists, then for any $c \in \mathbf{R}$, $\lim_{x\to a} cf(x) = cL$.*

(b) *$\lim_{x\to a} f(x) = L$ exists if and only if $\lim_{x\to a} |f(x) - L| = 0$.*

(c) *If $g(x) \leq f(x) \leq h(x)$ for x in a common domain having a as a cluster point, and $\lim_{x\to a} g(x) = \lim_{x\to a} h(x) = L$ exists, then $\lim_{x\to a} f(x) = L$.*

Proof. We will prove (a) and (c) when a is a cluster point and leave the remaining parts to Exercise 4.4.5.

(a) The result is clear if $c = 0$, so we assume that $c \neq 0$. By hypothesis, for every $\epsilon > 0$ there is a $\delta > 0$ such that if $0 < |x - a| < \delta$, then $|f(x) - L| < \epsilon/|c|$. Thus if $0 < |x - a| < \delta$, then

$$|cf(x) - cL| = |c|\,|f(x) - L| < |c|\,\frac{\epsilon}{|c|} = \epsilon.$$

Since $\epsilon > 0$ was arbitrary, this proves (a).

(c) We may write

$$\begin{aligned} |f(x) - L| &= |f(x) - g(x) + g(x) - L| \\ &\leq |f(x) - g(x)| + |g(x) - L| \\ &\leq |h(x) - g(x)| + |g(x) - L| \\ &\leq |h(x) - L| + |L - g(x)| + |g(x) - L|, \end{aligned}$$

where we used $g(x) \leq f(x) \leq h(x)$ in the third line. By hypothesis, given $\epsilon > 0$, there is a $\delta_1 > 0$ such that if $0 < |x - a| < \delta_1$, then $|g(x) - L| < \epsilon/3$; and, there is a $\delta_2 > 0$ such that if $0 < |x - a| < \delta_2$, then $|h(x) - L| < \epsilon/3$. Thus, if $0 < |x - a| < \delta := \min\{\delta_1, \delta_2\}$, then

$$|f(x) - L| < \epsilon/3 + \epsilon/3 + \epsilon/3 = \epsilon.$$

The proofs of (a) and (c) when $a = \pm\infty$, and the proof of (b) when a is either a cluster point or $\pm\infty$, are left to Exercise 4.4.5. □

The result in Theorem 4.4.7(c) is often called the *squeeze theorem* and it is quite useful. An example will be considered below.

As a matter of routine, we need to find the limits of sums, differences, products, and quotients of functions. Suppose that the functions f and g are defined on a common domain D. Then the sum $f + g$ and difference $f - g$ are defined by

$$(f \pm g)(x) := f(x) \pm g(x), \quad x \in D.$$

The product fg is defined on D by

$$(fg)(x) := f(x)g(x), \quad x \in D.$$

Finally, the quotient f/g is defined by

$$(f/g)(x) := f(x)/g(x), \quad \{x \in D : g(x) \neq 0\}.$$

Establishing the limits for these combinations directly from the limit definition requires a significant amount of work. The definition of limit of a function f at a point a makes no reference to sequences as the method of approach to the limit point and limiting function value. However, the next result characterizes the limit of a function in terms of sequential convergence and is often a labor-saving device.

Theorem 4.4.8. *Let $D \subset \mathbf{R}$ and $f : D \to \mathbf{R}$. Let a be a cluster point of D. Then $\lim_{x\to a} f(x) = L$ if and only if for every sequence $(x_n)_{n=1}^{\infty}$ satisfying $\lim_{n\to\infty} x_n = a$, we have $\lim_{n\to\infty} f(x_n) = L$.*

Proof. (*only if part*) Suppose $\lim_{x\to a} f(x) = L$, and let (x_n) be any sequence that converges to a. By hypothesis, for every $\epsilon > 0$ there is a $\delta = \delta(\epsilon) > 0$ such that if $0 < |x - a| < \delta$, then $|f(x) - L| < \epsilon$. Given $\delta > 0$ there is an $N = N(\delta(\epsilon))$ such that if $n \geq N$, then $0 < |x_n - a| < \delta$. Thus if $n \geq N$, then $|f(x_n) - L| < \epsilon$. This shows that $\lim_{n\to\infty} f(x_n) = L$.

In order to establish the *if part*, we will prove its contrapositive. Thus, assume that it is not true that $\lim_{x\to a} f(x) = L$. Then there exists an $\epsilon > 0$ such that for every $\delta_n = 1/n$, there is some point x_n such that $0 < |x_n - a| < \delta_n = 1/n$ and $|f(x_n) - L| \geq \epsilon$. Then the sequence (x_n) satisfies $\lim_{n\to\infty} x_n = a$, and it is not the case that $\lim_{n\to\infty} f(x_n) = L$. □

We use the sequential characterization of limit in the next result, where we employ the limit laws for sequences in Theorem 2.4.5.

Theorem 4.4.9. *Let f and g be functions defined on a common domain D, and let a be a cluster point of D. Suppose*

$$\lim_{x\to a} f(x) = L_1 \quad \text{and} \quad \lim_{x\to a} g(x) = L_2.$$

Then

1. $\lim_{x\to a}(f \pm g)(x) = L_1 \pm L_2$;
2. $\lim_{x\to a}(fg)(x) = L_1 L_2$;
3. $\lim_{x\to a}(f/g)(x) = L_1/L_2$, *if* $L_2 \neq 0$.

Proof. Since

$$\lim_{x\to a} f(x) = L_1 \quad \text{and} \quad \lim_{x\to a} g(x) = L_2,$$

by invoking Theorem 4.4.8 we have that if $a_k \to a$, then $f(a_k) \to L_1$ and $g(a_k) \to L_2$. Then the limit laws of parts 2 and 3 Theorem 2.4.5 imply, respectively, that

$$\lim_{a_k\to a} (f \pm g)(a_k) = L_1 \pm L_2$$

and

$$\lim_{a_k\to a} (fg)(a_k) = L_1 L_2.$$

Thus, by Theorem 4.4.8, statements 1 and 2 of the present theorem hold.

In addition, if $L_2 \neq 0$, then part 4 of Theorem 2.4.5 implies that

$$\lim_{a_k\to a} (f/g)(a_k) = L_1/L_2.$$

Thus, by Theorem 4.4.8, statement 3 of the present theorem holds. □

We consider some examples.

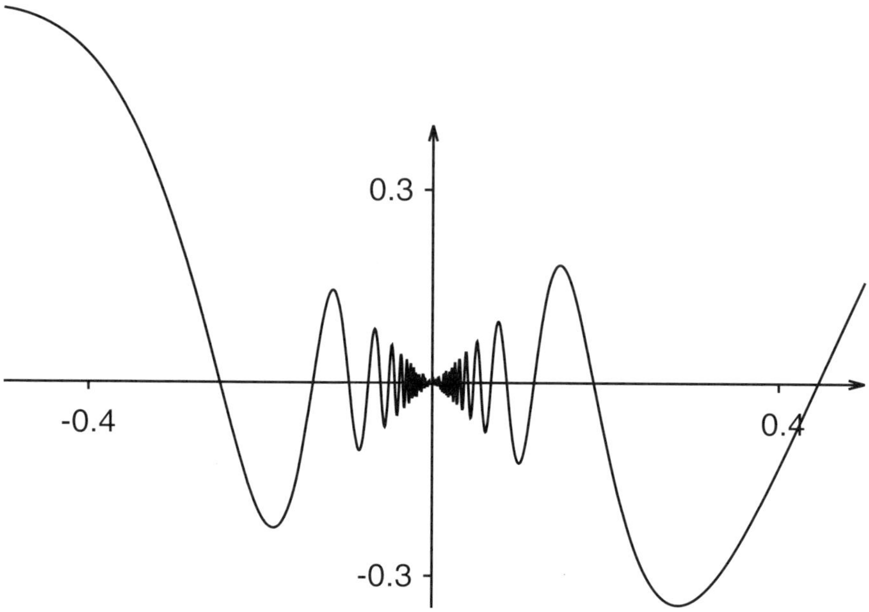

Figure 4.1. The function f in Example 4.4.11 exhibits increasingly rapid oscillation and limit zero as x approaches 0.

Example 4.4.10. By Theorem 4.4.7(b) with $f(x) = x$, we clearly have $\lim_{x\to a} x = a$ for any number a. It then follows from an induction argument using the product limit law that any power function x^n with positive integer exponent n satisfies $\lim_{x\to a} x^n = a^n$ for any a. We can then conclude from the constant multiple and the sum limit laws that any *polynomial* function

$$p(x) = a_n x^n + a_{n-1}x^{n-1} + \cdots + a_1 x + a_0$$

satisfies $\lim_{x\to a} p(x) = p(a)$ for any a. Any quotient p/q of polynomial functions p and q is called a *rational function*. By the quotient limit law, any rational function p/q satisfies $\lim_{x\to a} p(x)/q(x) = p(a)/q(a)$ at every point a where it is defined. △

Example 4.4.11. The graph of the function

$$f(x) = x\cos(1/x) + \frac{x^3 + 3x}{4 + x^5}\sin(1/x)$$

oscillates in a complicated manner as $x \to 0$. (See Figure 4.1.) However, it is not difficult to see from the triangle inequality that

$$0 \le |f(x)| \le |x| + \left|\frac{x^3 + 3x}{4 + x^5}\right|,$$

and both terms in the bound on the right side approach 0 as $x \to 0$. Thus, we have $\lim_{x\to 0} f(x) = 0$ by the squeeze Theorem 4.4.7(c). △

In view of the previous definitions of limits, it would seem paradoxical to speak of infinite limits; however, the following conventional notation is often useful and worth setting out as a definition.

Definition 4.4.12. *Let $D \subset \mathbf{R}$ and $f : D \to \mathbf{R}$. Let a be a cluster point of D. We say that $\lim_{x \to a} f(x) = \infty$ if for every $N > 0$ there is a $\delta = \delta(N) > 0$ such that if $|x - a| < \delta$, then $f(x) > N$. Similarly, we say that $\lim_{x \to a} f(x) = -\infty$ if for every $N > 0$ there is a $\delta = \delta(N) > 0$ such that if $|x - a| < \delta$, then $f(x) < -N$.*

The notational convention of this definition can be extended to the case of $a = \pm\infty$ as a "limit point". For example, $\lim_{x \to \infty} f(x) = \infty$ if for every $N > 0$ there is an $M = M(N) > 0$ such that if $x > M$, then $f(x) > N$.

Exercises.

Exercise 4.4.1. Consider the following functions on the domains indicated:

1. $f(x) = x^2$ for $x \in (-2, 2)$;
2. $g(x) = (x^2 - x)/(x - 1)$ for $x \in (-2, 2)$ and $x \neq 1$, and $g(1) = 4$;
3. $h(x) = x^2$ for $x \in (-2, 2)$ and $x \neq 1$.

Graph each function and show that

$$\lim_{x \to 1} f(x) = \lim_{x \to 1} g(x) = \lim_{x \to 1} h(x) = 1.$$

Exercise 4.4.2. Suppose $f(x) = 3x^2 \sin(1/x)$ for $x \neq 0$. Graph f for x near 0 and then verify your candidate for $\lim_{x \to 0} f(x)$ using Definition 4.4.1.

Exercise 4.4.3. Show that $\lim_{x \to 0}(x^3 + 3x)/(4 + x^5) = 0$. (See Example 4.4.11.)

Exercise 4.4.4. Find $\lim_{x \to 0}(\sqrt{1 + x^2} - 1)/x$.

Exercise 4.4.5. Refer to Theorem 4.4.7. Prove parts (a) and (c) when $a = \pm\infty$, and prove (b) when a is either a cluster point or $\pm\infty$.

4.5. Continuity at a Point

Intuitively, the idea of continuity of a function f at a point a is that as x approaches a, the function values $f(x)$ should approach a well-defined limit value and that limit value should be the value of the function at a. Thus continuity at a rules out behavior like that in Exercise 4.4.1, where the limit of a function exists but is different from the function value, or the limit exists but the function is not defined at the limit point. Continuity of f at a should also rule out unbounded asymptotic behavior or oscillating behaviors as $x \to a$ such that a well-defined function limit does not exist.

Definition 4.5.1. *Let $D \subset \mathbf{R}$ and $f : D \to \mathbf{R}$. Let $a \in D$. Then f is* **continuous** *at the point a if $\lim_{x \to a} f(x) = f(a)$.*

By the definition of limit, then, f is continuous at $a \in D$ if and only if for every $\epsilon > 0$ there is a $\delta = \delta(\epsilon, a) > 0$ such that

$$x \in D \text{ and } 0 < |x - a| < \delta \quad \Longrightarrow \quad |f(x) - f(a)| < \epsilon.$$

If the domain D has any isolated point(s) a, this definition implies that f is continuous at a.

Example 4.5.2. It follows from the discussion in Example 4.4.10 that polynomial functions

$$p(x) = a_n x^n + a_{n-1}x^{n-1} + \cdots + a_1 x + a_0$$

and rational functions

$$r(x) = \frac{a_n x^n + a_{n-1}x^{n-1} + \cdots + a_1 x + a_0}{b_m x^m + b_{m-1}x^{m-1} + \cdots + b_1 x + b_0}$$

are continuous at every point at which they are defined. Thus $p(x)$ is continuous on $\mathbf{R}$, and $r(x)$ is continuous at every point where its denominator is not zero. (Exercise 4.5.1.) △

Example 4.5.3. Let $f(x) = x^2$ for $0 \le x < 1$ and $f(x) = 1 + 1/n$ for $x = 1 + 1/n$, where n is any positive integer. The domain of this function is the set $D = [0,1) \cup \{1 + 1/n\}_{n=1}^{\infty}$ and f is continuous at every point of this domain. Each point of the set $\{1 + 1/n\}_{n=1}^{\infty}$ is an isolated point of D. Note that f has no chance of being continuous at $x = 1$ since f is undefined at $x = 1$. △

The next result is an immediate consequence of the theorem on limits of sums, products, and quotients of functions.

Theorem 4.5.4. *Suppose f and g are functions defined in an open set D containing the point a. If f and g are both continuous at a, then the following conditions hold:*

1. *$f \pm g$ is continuous at a.*
2. *fg is continuous at a.*
3. *f/g is continuous at a if $g(a) \neq 0$.*

Proof. Apply the theorem on the limit of a sum, product, and quotient of functions. □

We also have a sequential characterization of continuity at a point.

Theorem 4.5.5. *Let $f : D \to \mathbf{R}$ and $a \in D$. The function f is continuous at a if and only if for every sequence (x_n) in D such that $\lim_{n\to\infty} x_n = a$,*

$$\lim_{n\to\infty} f(x_n) = f(a).$$

Proof. Apply the sequential characterization of limit in Theorem 4.4.8. □

Recall that if $g : U \to \mathbf{R}$ and $f : V \to \mathbf{R}$, then the composition $f \circ g : U \to \mathbf{R}$ is the function defined by $(f \circ g)(x) = f(g(x))$ for $x \in g^{-1}(V) \cap U$. Under what conditions is a composite function continuous at a point?

Example 4.5.6. It is possible to have $\lim_{x\to x_0} g(x_0) = L$ and $\lim_{z\to L} f(z) = M$, but $\lim_{x\to x_0} f(g(x)) \neq M$. As a simple example of this behavior, consider $g(x) \equiv 1$ and

$$f(x) = \begin{cases} 2 & \text{if } x \neq 1, \\ -1 & \text{if } x = 1. \end{cases}$$

Then $g(x) \to 1 = L$ as $x \to 1 = x_0$, and $f(z) \to 2$ as $z \to 1 = L$, but $f \circ g(x) = f(g(x)) = -1$ as $x \to 1$. △

In the example, f is not continuous at $g(1) = 1 = L$.

Theorem 4.5.7. *Suppose $g : U \to \mathbf{R}$ and $f : V \to \mathbf{R}$. If g is continuous at $a \in U$ and f is continuous at $g(a) \in V$, then $f \circ g$ is continuous at a.*

Proof. We use the sequential characterization of continuity. Let (x_n) be any sequence such that $x_n \to a$ as $n \to \infty$. Then, by continuity of g at a, $g(x_n) \to g(a)$, and hence by continuity of f at $g(a)$, $f(g(x_n)) \to f(g(a))$. □

Under the hypotheses of Theorem 4.5.7, we have

$$f(g(a)) = f(\lim_{n\to\infty} g(x_n)) = \lim_{n\to\infty} f(g(x_n))$$

where the first equality is due to continuity of g at a and the second equality is due to continuity of f at $g(a)$.

There is another characterization of continuity at a point that is useful later on in describing the class of Riemann integrable functions. This characterization is just as easily described in the more general setting of functions defined on subsets of $\mathbf{R}^n$ for any fixed $n \geq 1$, so it is presented later, in Exercise 8.10.7. (For interested readers there is no harm in looking ahead and working that exercise for $f : D \subset \mathbf{R} \to \mathbf{R}$.)

Exercises.

Exercise 4.5.1. Verify the continuity statements made in Example 4.5.2.

Exercise 4.5.2. In Example 4.5.3, if we include the point $x = 1$ in the domain of f, how should $f(1)$ be defined so as to make f continuous at 1?

Exercise 4.5.3. Prove: If $f : D \to \mathbf{R}$ is continuous on D, then $|f| : D \to \mathbf{R}$, defined by $|f|(x) = |f(x)|$, is also continuous on D.

Exercise 4.5.4. Let $f : D \to \mathbf{R}$ and suppose $a \in D$. Show that f is discontinuous at a if and only if there exists an $\epsilon > 0$ and a sequence $x_k \to a$ as $k \to \infty$ such that for every k, $|f(x_k) - f(a)| \geq \epsilon$.

Exercise 4.5.5. Let $f(x) = x \sin(1/x)$ if $x \neq 0$, and let $f(0) = 0$. Show that f is continuous on $[0, 1]$.

4.6. Continuous Functions on an Interval

We say that a function $f : D \to \mathbf{R}$ is **continuous on** D if f is continuous at each $x \in D$. With this definition and Theorem 4.5.4, it is easy to talk about the continuity on $[a, b]$ of the sum, difference, and product of functions that are continuous on $[a, b]$. If f and g are continuous on $[a, b]$ and $g(x) \neq 0$ for any $x \in [a, b]$, then f/g is continuous on $[a, b]$. If the composition $f \circ g$ is defined on $[a, b]$, g is continuous on $[a, b]$ and f is continuous on the image $g([a, b])$, then, by Theorem 4.5.7, $f \circ g$ is continuous on $[a, b]$.

Proposition 4.6.1. *Let $f : D \to \mathbf{R}$ and let $a \in D$. If f is continuous at a and $f(a) \neq 0$, then there is an interval $(a - \delta, a + \delta)$, where $\delta > 0$, such that $f(x) \neq 0$ and $f(x)$ has the same sign as $f(a)$, for all $x \in (a - \delta, a + \delta)$.*

Proof. Let $\epsilon < |f(a)|/2$. By continuity of f at a, there is a $\delta > 0$ such that if $x \in D$ and $|x - a| < \delta$, then $|f(x) - f(a)| < \epsilon$, and therefore

$$-\frac{|f(a)|}{2} < -\epsilon < f(x) - f(a) < \epsilon < \frac{|f(a)|}{2}.$$

Adding $f(a)$ to both sides gives

$$f(a) - \frac{|f(a)|}{2} < f(x) < f(a) + \frac{|f(a)|}{2}.$$

If $f(a) > 0$, then for $x \in D$ and $|x - a| < \delta$, we have $0 < \frac{f(a)}{2} < f(x)$. On the other hand, if $f(a) < 0$, then for $x \in D$ and $|x - a| < \delta$, we have $f(x) < \frac{f(a)}{2} < 0$. □

The next result is a special case of the intermediate value theorem, and it is of special interest in root finding problems.

Proposition 4.6.2. *If f is continuous on $[a, b]$ and $f(a)$ and $f(b)$ are nonzero and have opposite sign, then there is some point $z \in (a, b)$ such that $f(z) = 0$.*

Proof. We will assume that $f(a) < 0$ and $f(b) > 0$. (The argument is similar if $f(b) < 0$ and $f(a) > 0$.) Let $A = \{x \in [a, b] : f(x) < 0\}$. Then A is nonempty since $a \in A$, and A is bounded above by b. Hence, $\sup A$ exists, and we let $z = \sup A$. Then $z > a$ by Proposition 4.6.1, and $z < b$, also by Proposition 4.6.1, since $f(b) > 0$. Thus $z \in (a, b)$. Since $z = \sup A$, for any positive integer n, there is an $x_n \in A$ such that $z - 1/n < x_n \leq z$. Then $x_n \to z$ as $n \to \infty$, so $f(x_n) \to f(z)$, by continuity of f. Since $x_n \in A$, $f(x_n) < 0$, and therefore $f(z) = \lim_{n\to\infty} f(x_n) \leq 0$. If $f(z) < 0$, then Proposition 4.6.1 implies that f takes negative values throughout some open interval about z, and this contradicts the fact that $z = \sup A$. Hence, $f(z) = 0$. □

The next result is the *intermediate value theorem*.

Theorem 4.6.3 (Intermediate Value Theorem). *If $f : [a, b] \to \mathbf{R}$ is a continuous function and c is any real number between $f(a)$ and $f(b)$, then there exists a point $z \in (a, b)$ such that $f(z) = c$.*

Proof. We may assume that $f(a) < f(b)$. Given c between $f(a)$ and $f(b)$, the function $F(x) = f(x) - c$ is continuous on $[a, b]$ and satisfies the hypothesis of Proposition 4.6.2, since $F(a) = f(a) - c < 0$ and $F(b) = f(b) - c > 0$. Thus there is some point $z \in (a, b)$ such that $F(z) = 0$, that is, $f(z) = c$. □

In the next section we continue to study continuity as a global property of a function.

Exercises.

Exercise 4.6.1. Apply Proposition 4.6.2 to show that $f(x) = x^3 - x - 1$ has a root between $a = 0$ and $b = 2$.

Exercise 4.6.2. Assume 0 is in the range of $f : [a, b] \to \mathbf{R}$. Prove: If f is continuous, then the set $f^{-1}(0) = \{x \in [a, b] : f(x) = 0\}$ is compact. What if $f^{-1}(0) = \emptyset$?

Exercise 4.6.3. Suppose that $f : [a, b] \to \mathbf{R}$ is continuous. Show that the image $f([a, b])$ must be an interval.

Exercise 4.6.4. *Continuity and inverse images of open sets*
This exercise provides a characterization of continuity of a real function on an open interval. This property generalizes to functions of several variables. Suppose $f : (a, b) \to \mathbf{R}$, where we may have $a = -\infty$ or $b = \infty$.

1. Prove: If f is continuous on (a, b), then the inverse image $f^{-1}(O)$ is open for any open set O.
2. Prove: If the inverse image $f^{-1}(O)$ is open for any open set O, then f is continuous on (a, b).

Exercise 4.6.5. *Continuity of an inverse function*
Suppose f is continuous on (a, b) and f is either strictly increasing or strictly decreasing on (a, b). Show that f is invertible and the inverse function $f^{-1} : f((a, b)) \to (a, b)$ is continuous. *Hint*: If $(a, b) \neq (-\infty, \infty)$, then it is straightforward to extend f to all of $\mathbf{R}$ so that f is strictly monotone on $\mathbf{R}$, hence f is invertible on $\mathbf{R}$. Argue that f maps open intervals to open intervals using the intermediate value theorem. Then draw a conclusion for the inverse restricted to $f((a, b))$.

4.7. Uniform Continuity

We know that $f : D \to \mathbf{R}$ is continuous *at a point* x_0 if and only if for every $\epsilon > 0$ there is a $\delta > 0$ such that

$$x \in D \text{ and } |x - x_0| < \delta \quad \Longrightarrow \quad |f(x) - f(x_0)| < \epsilon.$$

If this statement holds for *all* $x_0 \in D$, then f is continuous on D. As we have seen, the δ generally depends on ϵ *and* the point x_0. Suppose that for every $\epsilon > 0$, a $\delta = \delta(\epsilon) > 0$ can be found, dependent only on ϵ, such that

$$x_1, x_2 \in D \text{ and } |x_1 - x_2| < \delta \quad \Longrightarrow \quad |f(x_1) - f(x_2)| < \epsilon.$$

Then we have a stronger form of continuity of f on the domain D called *uniform continuity* of f on D.

Definition 4.7.1. *Let $D \subset \mathbf{R}$. A function $f : D \to \mathbf{R}$ is* **uniformly continuous on** *D if for every $\epsilon > 0$ there is a $\delta = \delta(\epsilon) > 0$ such that if $x, y \in D$ and $|x - y| < \delta(\epsilon)$, then $|f(x) - f(y)| < \epsilon$.*

It is clear from this definition that if f is uniformly continuous on D, then f is continuous at every point in D.

Example 4.7.2. Let us show that $f(x) = x^2$ is uniformly continuous on $(0, 2)$. Given $\epsilon > 0$, we can ensure that, for $x, y \in (0, 2)$,

$$|f(x) - f(y)| = |x^2 - y^2| = |(x + y)(x - y)| = |x + y||x - y| \le 4|x - y| < \epsilon$$

provided we choose

$$|x - y| < \delta = \frac{\epsilon}{4}.$$

Since ϵ was arbitrary, $f(x) = x^2$ is uniformly continuous on $(0, 2)$. △

Example 4.7.3. The function $f(x) = 1/x$ is not uniformly continuous on $(0, 1]$, because we can satisfy the negation of the definition statement, as follows. Let $\epsilon = 1/2$. Let $x_n = 1/n$ and $y_n = 2/n$ for $n \in \mathbf{N}$. Then, for all n,

$$|f(x_n) - f(y_n)| = \left|\frac{1}{x_n} - \frac{1}{y_n}\right| = \left|\frac{y_n - x_n}{x_n y_n}\right| = \frac{n}{2} \geq 1/2 = \epsilon.$$

This shows that $f(x) = 1/x$ is not uniformly continuous on $(0, 1]$. △

Theorem 4.7.4. *Let $f : K \to \mathbf{R}$ be continuous on a compact set $K \subset \mathbf{R}$. Then f is uniformly continuous on K.*

Proof. Let $\epsilon > 0$. For each $y \in K$ there is a $\delta = \delta(\epsilon, y) > 0$ such that

$$x \in D \text{ and } |x - y| < \delta(\epsilon, y) \quad \Longrightarrow \quad |f(x) - f(y)| < \frac{\epsilon}{2}.$$

Let $J_y = \{x \in K : |x - y| < \frac{1}{2}\delta(\epsilon, y)\}$. Then the collection $\{J_y : y \in K\}$ is an open cover of K. Since K is compact, there is a finite collection of points $y_1, y_2, \ldots, y_m \in K$ such that

$$K \subset J_{y_1} \cup \cdots \cup J_{y_m}.$$

Now let

$$\delta := \frac{1}{2} \min\{\delta(\epsilon, y_1), \delta(\epsilon, y_2), \ldots, \delta(\epsilon, y_m)\}.$$

Then $\delta > 0$ since it is the minimum of a finite set of positive numbers. Suppose now that $x_1, x_2 \in K$ with $|x_1 - x_2| < \delta$. Since $x_1 \in K$, $x_1 \in J_{y_j}$ for some j with $1 \leq j \leq m$, hence

$$|x_1 - y_j| < \frac{1}{2}\delta(\epsilon, y_j).$$

By the definition of δ, we also have

$$|x_2 - y_j| \leq |x_2 - x_1| + |x_1 - y_j| < \delta + \frac{1}{2}\delta(\epsilon, y_j) < \delta(\epsilon, y_j).$$

Then it follows from the definition of $\delta(\epsilon, y_j)$ that

$$|f(x_1) - f(x_2)| \leq |f(x_1) - f(y_j)| + |f(y_j) - f(x_2)| < \frac{\epsilon}{2} + \frac{\epsilon}{2} = \epsilon.$$

Since $\epsilon > 0$ is arbitrary, we have shown that for every $\epsilon > 0$ there is a $\delta > 0$ such that if $x_1, x_2 \in K$ and $|x_1 - x_2| < \delta$, then $|f(x_1) - f(x_2)| < \epsilon$. Therefore f is uniformly continuous on K. □

An immediate corollary of Theorem 4.7.4 is that every real valued function that is continuous on a closed and bounded interval $[a, b]$ is uniformly continuous on $[a, b]$.

Exercise 4.7.3 offers a proof of Theorem 4.7.4 by contradiction.

Exercises.

Exercise 4.7.1. Suppose $g : D \to E$ is uniformly continuous on D and $f : E \to \mathbf{R}$ is uniformly continuous on E. Is $f \circ g : D \to \mathbf{R}$ uniformly continuous on D?

Exercise 4.7.2. A function $f : D \to \mathbf{R}$ satisfies a **Lipschitz condition** on D if there is a number L such that $|f(x) - f(y)| \leq L|x - y|$ for all $x, y \in D$. Then L is called a **Lipschitz constant** for f on D. (*Note*: Using the mean value theorem, a function with a bounded derivative on an interval J can be shown to satisfy a Lipschitz condition on J; see Exercise 5.2.1.)
Show that a function that satisfies a Lipschitz condition on D is uniformly continuous on D.

Exercise 4.7.3. By the negation of Definition 4.7.1 of uniform continuity, the function $f : D \to \mathbf{R}$ is not uniformly continuous on D if and only if there is an $\epsilon > 0$ and sequences x_n, z_n in D such that $\lim_{n\to\infty} |x_n - z_n| = 0$ but $|f(x_n) - f(z_n)| \geq \epsilon$. Give a proof of Theorem 4.7.4 by contradiction using this statement together with Theorem 4.2.10.

Exercise 4.7.4. Suppose $f : \mathbf{R} \to \mathbf{R}$ is continuous on $\mathbf{R}$, and there is a number p such that $f(x + p) = f(x)$ for all x. Show that f is uniformly continuous on $\mathbf{R}$.

4.8. Continuous Image of a Compact Set

The *extreme value theorem*, Theorem 4.8.2 of this section, is one of the most frequently applied results in introductory calculus.

We show first that the continuous image of a compact set is compact. Recall that if $f : K \to \mathbf{R}$, $K \subset \mathbf{R}$, then the image of K under f is $f(K) = \{y : y = f(x), x \in K\}$.

Theorem 4.8.1. *If $f : K \to \mathbf{R}$ is continuous on K and K is compact, then the image $f(K)$ is compact.*

Proof. Assume that $f : K \to \mathbf{R}$ is continuous on K and K is compact. We wish to show that $f(K)$ is closed and bounded.

Suppose that $f(K)$ is not bounded. Without loss in generality we assume that $f(K)$ is not bounded above. Then there is a sequence of points $f(x_n)$, $x_n \in K$, such that for each n, $f(x_n) > n$. Since K is compact, by Theorem 4.2.10 there is a subsequence x_{n_k} such that $\lim_{k\to\infty} x_{n_k} = a \in K$. Since f is continuous at a, $f(x_{n_k}) \to f(a)$. But we have $f(x_{n_k}) > n_k$ for all k, by our assumption that $f(K)$ is not bounded above. This contradiction shows that $f(K)$ must be bounded above. By a similar argument by contradiction, one can show that $f(K)$ must be bounded below. Hence, $f(K)$ is bounded.

If $f(K)$ consists of isolated points only, then $f(K)$ is closed and we are done. Otherwise, let b be a cluster point of $f(K)$. Then there is a sequence x_n in K such that $\lim_{n\to\infty} f(x_n) = b$. Since K is compact, Theorem 4.2.10 implies that the sequence x_n in K has a convergent subsequence x_{n_k} such that $\lim_{k\to\infty} x_{n_k} = a \in K$. Since $(f(x_{n_k}))$ is a subsequence of $(f(x_n))$, we have $\lim_{k\to\infty} f(x_{n_k}) = b$. Since f is continuous at a, we have $f(a) = b$, which shows that $b \in f(K)$. Since b is an arbitrary cluster point of $f(K)$, $f(K)$ is closed. $\square$

A more general version of Theorem 4.8.1 for functions of several variables is considered in Theorem 8.10.19.

If $f : D \to \mathbf{R}$, $D \subset \mathbf{R}$, and there is an $x_M \in D$ such that $f(x) \leq f(x_M)$ for all $x \in D$, then $f(x_M)$ is called the **maximum** value of f on D. If there is an $x_m \in D$ such that $f(x) \geq f(x_m)$ for all $x \in D$, then $f(x_m)$ is called the **minimum** value of f on D.

The following result is called the *extreme value theorem.*

Theorem 4.8.2 (Extreme Value Theorem)**.** *If K is compact and $f : K \to \mathbf{R}$ is continuous on K, then f assumes its maximum and its minimum values on K.*

Proof. If K is compact and f is continuous on K, then $f(K)$ is compact, by Theorem 4.8.1. Since $f(K)$ is compact, $\sup f(K) \in f(K)$ and $\inf f(K) \in f(K)$ (Exercise 4.2.2). Thus there exist $x_M, x_m \in K$ such that $f(x_M) = \sup f(K)$ and $f(x_m) = \inf f(K)$. Consequently, for every $x \in K$, $f(x_m) \leq f(x) \leq f(x_M)$, so f assumes its maximum and minimum values on K. □

An immediate corollary of Theorem 4.8.2 is that every function continuous on a closed and bounded interval $[a, b]$ achieves its maximum and minimum values on $[a, b]$.

Continuous functions on noncompact sets need not achieve a maximum or a minimum value (Exercise 4.8.2). Functions that are defined on a compact set K, if not continuous on K, need not achieve a maximum or a minimum value there (Exercise 4.8.3).

Exercises.

Exercise 4.8.1. Supply the proof by contradiction in Theorem 4.8.1 that $f(K)$ must be bounded below.

Exercise 4.8.2. 1. Give an example of a function $f : [0, 1) \to \mathbf{R}$ which is continuous but does not achieve a maximum value on $[0, 1)$.
2. Give an example of a function on $[0, \infty)$ (or on another noncompact set) which is continuous but does not achieve a minimum value on that set.

Exercise 4.8.3. Let $f : [-1, 1] \to \mathbf{R}$ be defined by $f(x) = x^2$ for $x \in (-1, 0) \cup (0, 1)$, and $f(-1) = 2$, $f(0) = 2$, $f(1) = 2$. Show that f has no minimum value on $[-1, 1]$, but that f does have a maximum value there.

Exercise 4.8.4. Give an example of a function $f : \mathbf{R} \to \mathbf{R}$ that has a bounded range and achieves a minimum value, but has no maximum value.

Exercise 4.8.5. Give a different (and possibly shorter) proof of Theorem 4.8.1 by starting with an open cover of the image $f(K)$ and using the idea of Exercise 4.6.4 in this slightly modified form: A function $f : K \to \mathbf{R}$ is continuous on K if and only if for any open set $O \subseteq \mathbf{R}$, we have $f^{-1}(O) = K \cap U$ for some open set U of real numbers. (A set of the form $K \cap U$, where U is open, is said to be *open relative to K*.)

4.9. Classification of Discontinuities

A general definition of one-sided limits helps in discussing discontinuities of a function. We extend the earlier definition of limit (Definition 4.4.1) as follows. Let $f : (a, b) \to \mathbf{R}$. We say that f has a **right-hand limit** at a, denoted $f(a+)$, if for every $\epsilon > 0$ there is a $\delta > 0$ such that

$$0 < x - a < \delta \quad \Longrightarrow \quad |f(x) - f(a+)| < \epsilon.$$

We write $f(a+) = \lim_{x \to a+} f(x)$ when this limit exists. We say that f has a **left-hand limit** at b, denoted $f(b-)$, if for every $\epsilon > 0$ there is a $\delta > 0$ such that

$$0 < b - x < \delta \quad \Longrightarrow \quad |f(x) - f(b-)| < \epsilon.$$

We write $f(b-) = \lim_{x \to b-} f(x)$ when this limit exists.

It should be clear that if x_0 is an interior point of the domain of f, then $\lim_{x \to x_0} f(x)$ exists if and only if f has both a left-hand limit and a right-hand limit at x_0 and $\lim_{x \to x_0+} f(x) = \lim_{x \to x_0-} f(x)$. Then $\lim_{x \to x_0} f(x)$ equals the common value of these one-sided limits.

Definition 4.9.1. *A function f is said to have a* **discontinuity of the first kind**, *or* **jump discontinuity**, *at the point a if the one-sided limits $f(a+)$ and $f(a-)$ both exist but are unequal: $f(a+) \neq f(a-)$.*

Example 4.9.2. The sawtooth function defined by

$$f(x) = \begin{cases} x + 1 & \text{if } -1 \leq x < 0, \\ x - 1 & \text{if } 0 \leq x < 1, \end{cases}$$

has a jump discontinuity at $a = 0$. Note that $f(0+) = \lim_{x \to 0+} f(x) = -1$, and $f(0-) = \lim_{x \to 0-} f(x) = 1$. △

If f is not defined at a, but the one-sided limits $f(a+)$ and $f(a-)$ exist and are equal, then a is a *removable discontinuity* of f, in the sense that if we define $f(a)$ to be this common limiting value, then f is made continuous at a. (Of course the discontinuity is also removable if $f(a)$ is already defined but is not equal to the common limit.) See Exercise 4.4.1.

Definition 4.9.3. *A function f is said to have a* **discontinuity of the second kind** *at the point a if either of the one-sided limits $f(a+)$ or $f(a-)$, or both, fail to exist.*

Example 4.9.4. The function defined by $f(x) = 1/x$ for $x > 0$ and $f(x) = e^x$ for $x \leq 0$. Then $\lim_{x \to 0+} f(x) = \infty$. That is, the right-hand limit at 0 does not exist. Therefore f has a discontinuity of the second kind at 0. △

We can also say that a function f has an **infinite discontinuity** at a if $\lim_{x \to a+} f(x) = \pm\infty$ or $\lim_{x \to a-} f(x) = \pm\infty$, or both.

Example 4.9.5. For a different kind of function behavior which exhibits a discontinuity of the second kind, consider the function $y = f(x) = \sin(1/x)$ for $|x| \neq 0$. The graph of f oscillates more and more rapidly between its bounds $y = \pm 1$ as $x \to 0$ from either direction, and $\lim_{x \to 0} f(x)$ does not exist. Consequently, there is no way to define a value for $f(0)$ so as to make f continuous at 0. △

We consider one more example. Let us write $B_\delta(a) := \{x : |x-a| < \delta\}$.

Example 4.9.6. Define $f : \mathbf{R} \to \mathbf{R}$ by

$$f(x) = \begin{cases} 0 & \text{if } x \in \mathbf{I}, \\ 1/n & \text{if } x = m/n,\ m, n \in \mathbf{Z},\ (m,n) = 1. \end{cases}$$

Here (m, n) means the GCD (greatest common divisor) of m and n. Thus if x is rational and $x = m/n$ in "lowest terms", then $f(x) = 1/n$. We will show that f is continuous at every irrational and discontinuous at every rational. Let $a \in \mathbf{I}$. Then $a \in (k, k+1)$ for some integer k. Let $\epsilon > 0$ and choose $n_0 \in \mathbf{N}$ such that $1/n_0 < \epsilon$. There are only finitely many rational numbers m/n in $(k, k+1)$ that have $n < n_0$, so one of them is closest to a, say at distance $\delta = \delta(\epsilon) > 0$. We have

$$|f(x) - f(a)| = \begin{cases} |0-0| = 0, & \text{if } x \in \mathbf{I}, \\ |1/n - 0| = 1/n, & \text{if } x = m/n. \end{cases}$$

By our definition of δ, $1/n < 1/n_0 < \epsilon$ for all $m/n \in B_\delta(a) \cap \mathbf{Q}$. Therefore $|f(x) - f(a)| < \epsilon$ for all $x \in B_\delta(a)$. Therefore f is continuous at a. Since a was arbitrary in $\mathbf{I}$, f is continuous at every irrational. On the other hand, if $a = m/n$ is rational, with $(m, n) = 1$, then there is a sequence of irrationals y_k such that $y_k \to m/n$, while $|f(y_k) - f(m/n)| = |0 - 1/n| = 1/n$. Thus f is not continuous at $a = m/n$. $\triangle$

The discontinuities of monotone functions are relatively easy to describe and we turn to them now.

Definition 4.9.7. *Let $f : I \to \mathbf{R}$ where I is an interval. Then*

1. *f is* **monotone increasing** *on I if $x_1 \le x_2$ implies $f(x_1) \le f(x_2)$;*
2. *f is* **monotone decreasing** *on I if $x_1 \le x_2$ implies $f(x_1) \ge f(x_2)$.*

A more strict definition is needed when we discuss invertible real functions. We say that a function $f : I \to \mathbf{R}$ is **strictly increasing** if $x_1 < x_2$ implies $f(x_1) < f(x_2)$, and **strictly decreasing** if $x_1 < x_2$ implies $f(x_1) > f(x_2)$.

Theorem 4.9.8. *Monotone functions on an open interval have discontinuities only of the first kind, that is, jump discontinuities.*

Proof. To show that the discontinuities of f are all of the first kind, it suffices to show that at each point of I the left-hand and right-hand limits of f exist.

Let f be a monotone decreasing function on an open interval I. Let $p \in I$. If $p < x$, then $f(p) \ge f(x)$. Therefore the set

$$R = \{f(x) : p < x\}$$

is bounded above. Let $M = \sup R$. If $\epsilon > 0$, then there is a point s such that $p < s$ and

$$|f(s) - M| < \epsilon.$$

If $p < x < s$, then $M \ge f(x) \ge f(s)$, so $|f(x) - M| < \epsilon$. Therefore $\lim_{x \to p+} f(x)$ exists and equals M. Now let

$$L = \{f(x) : x < p\},$$

which is bounded below with $m = \inf L$. An argument similar to the previous one using the definition of infimum and the fact that f is decreasing shows that $\lim_{x \to p-} f(x)$ exists and equals m.

The arguments required for the case of an increasing function are similar to those given for the case of decreasing f (Exercise 4.9.1). $\square$

It can also be shown that monotone functions on an interval have at most countably many discontinuities; see Exercise 4.9.2.

Exercises.

Exercise 4.9.1. Complete the proof of Theorem 4.9.8 for the case of a monotone increasing function.

Exercise 4.9.2. Show that a monotone function on an interval has at most countably many discontinuities. *Hint*: All discontinuities must be jump discontinuities, by Theorem 4.9.8. For each discontinuity, choose a rational number in the "jump interval" not contained in the range. Show that this choice yields a one-to-one mapping of the set of discontinuities into $\mathbf{Q}$.

Chapter 5

The Derivative

This chapter presents the basic properties of the derivative of a real valued function of a real variable. We assume the reader has experience from introductory calculus courses with the geometric idea of the derivative at an interior point of the domain as the slope of the tangent line to the graph of a function. We also assume familiarity with the elementary functions (polynomials, rational functions, trigonometric functions, exponential functions, logarithmic functions, and their inverses) and we shall use facts about derivatives of elementary functions in examples. For reference, we note that the natural logarithm function, denoted $\log(x)$ in this book, is defined in Theorem 6.7.9. The exponential function $\exp(x) = e^x$ (the inverse of the natural logarithm function), as well as exponential and logarithm functions for other bases $b > 0$, and the sine and cosine functions, are defined and discussed in detail in Section 7.5.

5.1. The Derivative: Definition and Properties

Geometrically, the slope of a function graph at a point $(a, f(a))$ on the graph indicates the rate of change of the function with respect to the independent variable as that variable approaches the point a. This rate of change is the limiting value of the slopes

$$\frac{f(x) - f(a)}{x - a}$$

of chords joining the points $(x, f(x))$ and $(a, f(a))$, as x approaches a, when this limiting value exists. This limit process makes sense whenever the point a is an interior point of the domain.

Definition 5.1.1. *Let D be an interval of real numbers, let $f : D \to \mathbf{R}$, and suppose $a \in D$ is an interior point. If the limit*

$$\lim_{x \to a} \frac{f(x) - f(a)}{x - a}$$

exists, then f is said to be **differentiable** *at a, and the limiting value is denoted by $f'(a)$ and called the* **derivative** *of f at a. If D is an open interval, and if f is differentiable at every $a \in D$, then we say f is* **differentiable on** *D.*

In this limit process we always have x approaching a through values $x \in D$. If $f'(a)$ exists, then we have

$$f'(a) = \lim_{x \to a} \frac{f(x) - f(a)}{x - a} = \lim_{h \to 0} \frac{f(a+h) - f(a)}{h}.$$

This is easily seen by setting $x = a + h$ and noting that $x \to a$ if and only if $h \to 0$. The derivative of f at a is uniquely determined when it exists, since the function (of x or of h, as indicated) defined by the difference quotient has a unique limit when the limit exists.

If the domain of f is a closed interval $D = [a, b]$, then we can discuss the possibility of derivatives at the endpoints, and this is sometimes useful. These are called *one-sided derivatives* when they exist. Specifically, at the left-hand endpoint a, the derivative of f, if it exists, is a *right-sided derivative* of f at a,

$$f'(a) = f'_R(a) = \lim_{x \to a+} \frac{f(x) - f(a)}{x - a},$$

the limit notation indicating that x approaches a from the *right* within the domain D. At the right-hand endpoint b, the derivative of f, if it exists, is a *left-sided derivative* of f at b,

$$f'(b) = f'_L(b) = \lim_{x \to b-} \frac{f(x) - f(b)}{x - b},$$

the limit notation indicating that x approaches b from the *left* within the domain D. One-sided derivatives are defined at interior points of the domain in the same way, provided the appropriate limits exist. It is an exercise to show that at an *interior* point $d \in D$, $f'(d)$ exists if and only if $f'_L(d)$ and $f'_R(d)$ both exist and are equal. (See Exercise 5.1.1.)

Example 5.1.2. The function $f[-1, 1] \to \mathbf{R}$ given by $f(x) = |x|$ has a left-sided derivative and a right-sided derivative at $a = 0$. In particular,

$$f'_L(0) = \lim_{x \to 0-} \frac{|x| - 0}{x - 0} = \lim_{x \to 0-} \frac{-x}{x} = -1,$$

which is also clear from the graph of f. By a similar calculation, $f'_R(0) = 1$. But f is not differentiable at 0, since $f'_L(0) \neq f'_R(0)$. △

Suppose we need to consider the continuity of a derivative function at an endpoint. For example, suppose $f : [a, b] \to \mathbf{R}$ is differentiable on $[a, b]$. The right-sided limit of f' at a is denoted

$$f'(a+) = \lim_{x \to a+} f'(x),$$

and the left-sided limit of f' at b is denoted

$$f'(b-) = \lim_{x \to b-} f'(x).$$

Then f' is continuous at a if $f'_R(a) = f'(a+)$, and f' is continuous at b if $f'_L(b) = f'(b-)$.

If a function has a well-defined slope at a given point $(a, f(a))$ of its graph, then the function must be continuous at the point a.

Theorem 5.1.3. *Let $f : D \to \mathbf{R}$ and let $a \in D$ be an interior point of D. If $f'(a)$ exists, then f is continuous at a.*

Proof. For $x \neq a$, we can write

$$f(x) - f(a) = (x - a)\frac{f(x) - f(a)}{x - a}.$$

Now let $x \to a$ on both sides and use the product law for limits to deduce that

$$\lim_{x\to a}[f(x) - f(a)] = \lim_{x\to a}\left[(x - a)\frac{f(x) - f(a)}{x - a}\right] = 0 \cdot f'(a) = 0.$$

Thus $\lim_{x\to a} f(x) = f(a)$ and f is continuous at a. □

Our definitions of continuity and differentiability at endpoints allow for one-sided limits at endpoints of an interval domain. By an argument similar to that in Theorem 5.1.3, we can say that if f is differentiable on $[a, b]$ (or on $[a, b)$ or $(a, b]$), then f is also continuous on $[a, b]$ (or $[a, b)$ or $(a, b]$, respectively).

Differentiability at a point is strictly stronger than continuity at a point; that is, continuity at a point does not imply differentiability at that point. For an example where continuity does not imply differentiability, think of a function graph having a sharp corner point where there is no well-defined slope, although continuity holds. In fact, Example 5.1.2 will do, since $|x|$ is continuous at 0 (since $|x| \to 0$ as $x \to 0$) but $|x|$ is not differentiable at 0.

To emphasize just how different the concepts of continuity and differentiability really are, we note here that a real valued continuous function on $\mathbf{R}$ may be nowhere differentiable. A specific example is given in Section 7.3.

The continuity property assured by Theorem 5.1.3 is useful in the next example.

Example 5.1.4. Suppose g is differentiable at x and $g(x) \neq 0$. Then by continuity of g at x (Theorem 5.1.3), $g(x + h) \neq 0$ for sufficiently small $|h|$. We compute the derivative of the reciprocal function $1/g(x)$ at the point x directly from the definition and the facts deduced thus far. The difference quotient can be expressed as

$$\frac{1}{h}\left[\frac{1}{g(x + h)} - \frac{1}{g(x)}\right] = \frac{1}{g(x + h)g(x)}\frac{g(x) - g(x + h)}{h}.$$

We take the limit as $h \to 0$ and use appropriate limit laws to find the derivative:

$$\begin{aligned}
\frac{d}{dx}\left[\frac{1}{g(x)}\right] &= \lim_{h\to 0}\frac{1}{g(x + h)g(x)}\frac{g(x) - g(x + h)}{h} \\
&= \lim_{h\to 0}\frac{1}{g(x + h)g(x)}\lim_{h\to 0}\frac{g(x) - g(x + h)}{h} \\
&= -\frac{1}{[g(x)]^2}g'(x).
\end{aligned}$$

This formula is familiar from introductory calculus. We used the product and quotient limit laws, as well as the differentiability, and thus the continuity, of g at x.

For example, using the derivative of $\tan x$, which is $\sec^2 x$, the derivative of the cotangent function, $\cot x = 1/\tan x$, is

$$\frac{d}{dx}\cot x = \frac{d}{dx}\frac{1}{\tan x} = -\frac{1}{\tan^2 x}\frac{d}{dx}\tan x = -\frac{1}{\tan^2 x}\sec^2 x.$$

Therefore the derivative of the cotangent function is $-1/\sin^2 x = -\csc^2 x$. △

The sum of two functions which are differentiable at a is also differentiable at a.

Theorem 5.1.5 (Sum and Difference). *Suppose f and g are real valued functions defined in an interval about a. If f and g are both differentiable at a, then $f \pm g$ is differentiable at a and*

$$(f \pm g)'(a) = f'(a) \pm g'(a).$$

Proof. The proof is left as Exercise 5.1.3. □

If f and g have a left-sided derivative at the point a, that is, $f'_L(a)$ and $g'_L(a)$ both exist, then $f \pm g$ also has a left-sided derivative at a, equal to $f'_L(a) \pm g'_L(a)$. Likewise, if f and g both have a right-sided derivative at the point b, then $f \pm g$ also has a right-sided derivative at b. Variations like this will not be mentioned explicitly from here on.

The rules for differentiation of products and quotients of differentiable functions are stated next.

Theorem 5.1.6 (Product and Quotient). *Suppose f and g are real valued functions defined in an interval about a. If f and g are both differentiable at a, then the following statements are true:*

1. *The product function $(fg)(x) = f(x)g(x)$ is differentiable at a and*

$$(fg)'(a) = f'(a)g(a) + f(a)g'(a).$$

2. *The quotient function $(f/g)(x) = f(x)/g(x)$ is differentiable at a if $g(a) \neq 0$, and*

$$(f/g)'(a) = \frac{f'(a)g(a) - f(a)g'(a)}{[g(a)]^2}.$$

Proof. 1. We first write the difference quotient in a form that allows us to use the hypothesis. In particular we will add and subtract appropriate terms. Thus, we write

$$\begin{aligned}\frac{f(x)g(x) - f(a)g(a)}{x-a} &= \frac{f(x)g(x) - f(a)g(x) + f(a)g(x) - f(a)g(a)}{x-a} \\ &= \frac{f(x)-f(a)}{x-a}g(x) + f(a)\frac{g(x)-g(a)}{x-a}.\end{aligned}$$

Now let $x \to a$ on both sides, to obtain the desired product rule.

2. We use the product rule to compute the derivative at the point a, using the previously computed derivative of $1/g(x)$:

$$\begin{aligned}\frac{d}{dx}\left[\frac{f(x)}{g(x)}\right]\Big|_a &= \frac{d}{dx}\left[f(x)\frac{1}{g(x)}\right]\Big|_a \\ &= f'(a)\frac{1}{g(a)} + f(a)\Big(-\frac{g'(a)}{[g(a)]^2}\Big) \\ &= \frac{f'(a)g(a) - f(a)g'(a)}{[g(a)]^2}.\end{aligned}$$

This is the desired statement of the quotient rule. □

We want to establish the chain rule for the derivative of a composite function. Suppose the composition $h(x) = (g \circ f)(x) = g(f(x))$ is defined in some open interval J about the point a, and suppose that g is differentiable at $f(a)$ and f is differentiable at a. Let $x \in J$ be a point near a, and let $\Delta x = x - a$ and $\Delta f = f(x) - f(a)$, and $\Delta h = g(f(x)) - g(f(a))$. Then we can write the difference quotient for the composition $h(x) = g(f(x))$ as

$$\frac{\Delta h}{\Delta x} = \frac{\Delta h}{\Delta f}\frac{\Delta f}{\Delta x}, \tag{5.1}$$

provided $\Delta f \neq 0$ and $\Delta x \neq 0$. In the limit process, we are allowed to assume the condition $\Delta x \neq 0$; however, what about the assumption $\Delta f \neq 0$? We cannot guarantee the latter condition, since it may be that f takes the value $f(a)$ infinitely often in any open interval about a. We cannot deny that this is possible, so we cannot be sure that $\Delta x \to 0$ implies that $\Delta f \neq 0$. Hence, we cannot argue from (5.1), using the product limit law, that the limit of the difference quotient yields $h'(a) = g'(f(a))f'(a)$. The way to avoid this technical difficulty is to write the derivative definition in an alternative but equivalent way.

If $f'(a)$ exists, then

$$\lim_{\Delta x \to 0} \frac{\Delta f}{\Delta x} = \lim_{x \to a} \frac{f(x) - f(a)}{x - a} = f'(a).$$

By limit laws, this is equivalent to the statement

$$\lim_{\Delta x \to 0} \left[\frac{\Delta f}{\Delta x} - f'(a)\right] = \lim_{x \to a} \left[\frac{f(x) - f(a)}{x - a} - f'(a)\right] = 0,$$

which is equivalent to the statement

$$\lim_{\Delta x \to 0} \left[\frac{\Delta f - f'(a)\Delta x}{\Delta x}\right] = \lim_{x \to a} \left[\frac{f(x) - f(a) - f'(a)(x - a)}{x - a}\right] = 0.$$

When $f'(a)$ exists, we have

$$\lim_{\Delta x \to 0} [\Delta f - f'(a)\Delta x] = \lim_{x \to a} [f(x) - f(a) - f'(a)(x - a)] = 0.$$

But it is important to realize we also have the stronger statement that

$$\lim_{\Delta x \to 0} \frac{\Delta f - f'(a)\Delta x}{\Delta x} = \lim_{x \to a} \frac{f(x) - f(a) - f'(a)(x - a)}{x - a} = 0. \tag{5.2}$$

The expression $\Delta f - f'(a)\Delta x = f(x) - f(a) - f'(a)(x - a)$ is the error when approximating $f(x)$ with the tangent line approximation, $f(a) + f'(a)(x - a)$, at the point $(a, f(a))$. The existence of the derivative $f'(a)$ is equivalent to (5.2), which tells us that *this error goes to zero more rapidly than x approaches a.*

In the definition of a derivative and in other situations we encounter expressions of the form $E(h)/h$ that have zero limit as $h \to 0$. It is convenient to have a notation for this.

Definition 5.1.7. *We say that a function $E(h)$, which is defined for small $|h|$, is* **little-oh** *of h as $h \to 0$, if*

$$\lim_{h\to 0} \frac{E(h)}{h} = 0.$$

We indicate this limit property by writing $E(h) = o(h)$.

Consequently, if $E(h) = o(h)$, then necessarily $\lim_{h\to 0} E(h) = 0$, because $E(h)$ must approach zero faster than h. We may also write $o(h)$ to mean a function of h, perhaps unspecified in detail, which satisfies

$$\lim_{h\to 0} \frac{o(h)}{h} = 0.$$

We may set $h = x - a$ in (5.2) and use the little-oh notation to re-express the definition of differentiability of f at a by saying that there exists a number L such that

$$f(a+h) - f(a) - Lh = o(h). \tag{5.3}$$

We write $L = f'(a)$ for the number so defined. Here is the equivalent definition of a derivative:

Definition 5.1.8. *A function f defined on some interval about a is differentiable at a if there is a number L such that*

$$f(a+h) - f(a) - Lh = o(h) \quad \text{as } h \to 0.$$

This is equivalent to saying that f is differentiable at a if there is a number L such that the quotient

$$\frac{f(a+h) - f(a) - Lh}{h} =: V(h) \ \to 0 \quad \text{as } h \to 0.$$

We can now prove the chain rule for the differentiation of a composite function.

Theorem 5.1.9 (Chain Rule). *Let I and J be open intervals. If $f : J \to I$ is differentiable at $a \in J$ and $g : I \to \mathbf{R}$ is differentiable at $f(a) \in I$, then the composition $(g \circ f)(x) = g(f(x))$ is differentiable at $a \in J$ and*

$$(g \circ f)'(a) = g'(f(a))f'(a).$$

Proof. Let $a \in J$ and let x be near a. In the proof we write $\Delta x = x - a$, $x = a + \Delta x$, $\Delta f = f(x) - f(a)$. We shall use a generic notation, $V(z)$, for a function of a variable z that satisfies $V(z) \to 0$ as $z \to 0$, when such an expression appears in the context of Definition 5.1.8.

Since $g'(f(a))$ exists, we may write

$$g(f(a) + \Delta f) = g(f(a)) + g'(f(a))\Delta f + V(\Delta f)\Delta f. \tag{5.4}$$

Since $f'(a)$ exists, we may write

$$\Delta f = f'(a)\Delta x + V(\Delta x)\Delta x.$$

Since $f(a)+\Delta f = f(x)$, we can substitute the expression for Δf into the right-hand side of (5.4) to obtain

$$\begin{aligned} g(f(x)) = g(f(a+\Delta x)) &= g(f(a)) + g'(f(a))[f'(a)\Delta x + V(\Delta x)\Delta x] \\ &\quad + V(\Delta f)[f'(a)\Delta x + V(\Delta x)\Delta x]. \end{aligned}$$

Rearrange this, making use of the generic V notation to see that

$$g(f(x)) - g(f(a)) - g'(f(a))f'(a)\Delta x = V(\Delta x)\Delta x,$$

where we used the fact that $\Delta f \to 0$ as $\Delta x \to 0$. This expression implies that $(g \circ f)'(a)$ exists and equals $g'(f(a))f'(a)$. □

It follows easily from Theorem 5.1.9 that if f is differentiable on J, $f(J) \subset I$, and g is differentiable on I, then $g \circ f$ is differentiable on J.

Exercises.

Exercise 5.1.1. Let $f : [a, b] \to \mathbf{R}$ and let $a < d < b$. Show that $f'(d)$ exists if and only if $f'_L(d)$ and $f'_R(d)$ both exist and are equal.

Exercise 5.1.2. Define $f : (-1, 1) \to \mathbf{R}$ by $f(x) = x^2 \sin(1/x)$ for $x \neq 0$, and $f(0) = 0$. Show that f is differentiable on $(-1, 1)$.

Exercise 5.1.3. Prove Theorem 5.1.5, the sum and difference rules for derivatives.

Exercise 5.1.4. Let f be defined by $f(x) = x^2$ for $x \in \mathbf{Q}$ and $f(x) = 0$ for $x \in \mathbf{I}$. Show that f is differentiable only at $x = 0$, and $f'(0) = 0$.

5.2. The Mean Value Theorem

We first define the concepts of local maximum value and local minimum value for a real valued function defined on an open interval.

Definition 5.2.1. *Let* $f : (a, b) \to \mathbf{R}$.

1. *The function* f *has a* **relative maximum** *at a point* $x \in (a, b)$ *if there is a* $\delta > 0$ *such that* $f(s) \le f(x)$ *for all* $s \in (x-\delta, x+\delta)$. *The* **relative maximum value** *is then* $f(x)$.
2. *The function* f *has a* **relative minimum** *at a point* $x \in (a, b)$ *if there is a* $\delta > 0$ *such that* $f(s) \ge f(x)$ *for all* $s \in (x-\delta, x+\delta)$. *The* **relative minimum value** *is then* $f(x)$.

Relative extreme points and relative extreme values are also called *local* extreme points and *local* extreme values.

If $f(c)$ is a relative extreme value of f, then the point $(c, f(c))$ on the graph is also called a **relative extreme point** of f. In topographic terms, a relative extreme point might occur on a smooth hilltop or in a smooth valley bottom, or it might occur at a sharp Λ-shaped mountain peak or at the bottom of a **V**-shaped ditch. These different topographic phrasings depend on how smooth, or how jagged, is the peak or the low point. The example $f(x) = |x|$ (or $f(x) = -|x|$) shows that there may be a relative maximum (or minimum) value $f(c)$ at a point where $f'(c)$ does not exist, as at $c = 0$.

The derivative is the mathematical tool for discussing slopes of smooth function graphs. There is a well-defined slope $f'(c)$ at an *extreme point* $(c, f(c))$ of the graph of f only if that slope equals zero.

Theorem 5.2.2. *If $f : (a,b) \to \mathbf{R}$ and f has either a relative maximum or a relative minimum at $c \in (a,b)$, and if $f'(c)$ exists, then $f'(c) = 0$.*

Proof. Suppose f has a relative maximum at $c \in (a,b)$. Then there is a $\delta > 0$ such that if $|x - c| < \delta$, then

$$f(x) \le f(c).$$

Taking $x > c$, we have

$$\frac{f(x) - f(c)}{x - c} \le 0.$$

Letting $x \to c$ with $x > c$, we conclude that

$$f'(c) \le 0.$$

On the other hand, taking $x < c$, we have

$$\frac{f(x) - f(c)}{x - c} \ge 0.$$

Now let $x \to c$ with $x < c$ to get

$$f'(c) \ge 0.$$

Therefore $f'(c) = 0$. If f has a relative minimum at c, then a similar argument shows that $f'(c) = 0$. □

We now use Theorem 5.2.2 and the existence of extreme values for continuous functions on closed intervals to establish a special case of the mean value theorem, often known as *Rolle's theorem.*

Theorem 5.2.3 (Rolle). *Let $f : [a,b] \to \mathbf{R}$ be continuous on $[a,b]$ and differentiable on the open interval (a,b). If $f(a) = f(b)$, then there is a point $c \in (a,b)$ such that $f'(c) = 0$.*

Proof. If f is a constant function on (a,b), then the result is immediate, since the derivative of a constant function is zero.

Thus, suppose that f is not constant on (a,b). If there is an $x \in (a,b)$ such that $f(x) > f(a)$, then f must attain an absolute maximum value $f(c_1)$ for some $c_1 \in (a,b)$, and $f'(c_1) = 0$ by Theorem 5.2.2. On the other hand, if there is an $x \in (a,b)$ such that $f(x) < f(a)$, then f must attain an absolute minimum value $f(c_2)$ for some $c_2 \in (a,b)$, hence, $f'(c_2) = 0$ by Theorem 5.2.2. □

If $f(a) = f(b)$, there can be more than one point at which the derivative of f is zero. Consider the function $f(x) = x^3 - x$ on the interval $[-1,1]$, where $f'(x) = 3x^2 - 1$ equals zero at $c_1 = 1/\sqrt{3}$ and $c_2 = -1/\sqrt{3}$, both in $(-1,1)$.

When it applies, Rolle's theorem states that there is a point $x = c \in (a,b)$ at which the slope $f'(c)$ equals the slope of the chord that joins the points $(a, f(a))$ and $(b, f(b))$, the slope of this chord being zero under the hypothesis that $f(a) = f(b)$.

In the statement of the mean value theorem below, we consider a more general function graph over $[a, b]$ and establish the corresponding statement about equality of derivative and chord slope at some point $c \in (a, b)$, by a reduction to the case of Rolle's theorem.

Theorem 5.2.4 (Mean Value). *If $f : [a, b] \to \mathbf{R}$ is continuous on $[a, b]$ and differentiable on (a, b), then there is a point $c \in (a, b)$ such that*

$$f'(c) = \frac{f(b) - f(a)}{b - a}.$$

Proof. The line passing through the points $(a, f(a))$ and $(b, f(b))$ has equation

$$y = f(a) + \frac{x - a}{b - a}(f(b) - f(a)).$$

We consider the function

$$h(x) := f(x) - f(a) - \frac{x - a}{b - a}(f(b) - f(a)).$$

Then $h(a) = 0 = h(b)$. Since h is continuous on $[a, b]$ and differentiable on (a, b), by Rolle's theorem there is a point $c \in (a, b)$ such that $h'(c) = 0$, hence

$$h'(c) = f'(c) - \frac{f(b) - f(a)}{b - a} = 0,$$

as we wished to show. □

The mean value theorem provides important information about *differences* in function values. Naturally enough, the mean value theorem can be helpful in evaluating limits of expressions that involve differences in function values. See the exercises for some applications. The mean value theorem plays an important role in determining whether certain mappings are contraction mappings in Section 5.5.

Given appropriate information on the sign of the derivative f' over an open interval (a, b), Theorem 5.2.4 allows us to deduce important monotonicity properties of f on (a, b). Recall that a function f on an interval I is **increasing** on I if $f(x) \le f(y)$ whenever $x \le y$. It is **strictly increasing** on I if $f(x) < f(y)$ whenever $x < y$. A function f on an interval I is **decreasing** on I if $f(x) \ge f(y)$ whenever $x \le y$. It is **strictly decreasing** on I if $f(x) > f(y)$ whenever $x < y$.

Theorem 5.2.5. *Suppose $f : [a, b] \to \mathbf{R}$ is continuous on $[a, b]$ and differentiable on (a, b). Then the following implications hold:*

1. *If $|f'(c)| \le M$ for all c in (a, b), then for each $x \in (a, b)$,*

$$|f(x) - f(a)| \le M(x - a) < M(b - a).$$

2. *If $f'(x) = 0$ for all $x \in (a, b)$, then f is a constant function on (a, b).*
3. *If $f'(x) \ge 0$ for all $x \in (a, b)$, then f is increasing on (a, b). If $f'(x) > 0$ for all $x \in (a, b)$, then f is strictly increasing on (a, b).*
4. *If $f'(x) \le 0$ for all $x \in (a, b)$, then f is decreasing on (a, b). If $f'(x) < 0$ for all $x \in (a, b)$, then f is strictly decreasing on (a, b).*

Proof. 1. For any $x \in (a, b)$, by the mean value theorem there is a $c \in (a, b)$ such that

$$|f(x) - f(a)| = |f'(c)(x - a)| \leq M(x - a) < M(b - a).$$

2. Let s and t be points in the interval (a, b) with $s < t$. By the mean value theorem, there is a point c with $s < c < t$ such that

$$f(t) - f(s) = f'(c)(t - s).$$

Under the hypothesis, it follows that $f(t) = f(s)$ for any pair of points s, t, so f is a constant function on (a, b). Proofs of statements 3 and 4 are left to the reader. □

Statement 2 of Theorem 5.2.5 implies another fact which is familiar from elementary calculus. Recall that a function F is an **antiderivative** of f on (a, b) if $F'(x) = f(x)$ for all $x \in (a, b)$.

Theorem 5.2.6. *Let G and F be differentiable on (a, b) and suppose that*

$$G'(x) = F'(x) = f(x) \quad \textit{for all } x \in (a, b).$$

Then $G(x) = F(x) + C$, where C is a constant.

Proof. Since $G'(x) - F'(x) = 0$ for all $x \in (a, b)$, statement 2 of Theorem 5.2.5 implies that $G(x) - F(x) = C$, a constant, that is, $G(x) = F(x) + C$ for all $x \in (a, b)$, where C is a constant. □

Thus, any two antiderivatives of f on (a, b) differ by a constant. The reader is familiar with the indefinite integral notation, $\int f(x)\, dx = F(x) + C$, which denotes the family of antiderivatives of f. The domain being an *interval* is important in statement 2 of Theorem 5.2.5 and Theorem 5.2.6: for example, if $f(x) = 0$ for $x \in (-1, 0)$ and $f(x) = 1$ for $x \in (0, 1)$, then $f'(x) = 0$ for all $x \in (-1, 0) \cup (0, 1)$; similarly, if $g(x) = 2$ for $x \in (-1, 0)$ and $g(x) = 0$ for $x \in (0, 1)$, then $g'(x) = 0$ for all $x \in (-1, 0) \cup (0, 1)$, so $f'(x) = g'(x) = 0$, but neither f nor g is constant on its domain; also, f and g have the same derivative, but do not differ by a constant over the common domain.

According to Theorem 5.2.5, the sign of the derivative f' tells us a great deal over any interval where the sign does not change. However, the existence and knowledge of the derivative $f'(a)$ at a single point a may tell us far less about f than we would like. For example, from the knowledge that $f'(a) > 0$, can we conclude that f is strictly increasing on *some* open interval about the point a? The next example provides a negative answer.

Example 5.2.7. Let $f : \mathbf{R} \to \mathbf{R}$ be the function

$$f(x) = \begin{cases} x^2 \sin(1/x) & \text{if } x \neq 0, \\ 0 & \text{if } x = 0. \end{cases}$$

Then $f'(0)$ exists and $f'(0) = 0$, by the definition of derivative. We have $f'(x) = 2x \sin(1/x) - \cos(1/x)$ for $x \neq 0$. Note that as $x \to 0$, $2x \sin(1/x) \to 0$, while $\cos(1/x)$ oscillates infinitely often between $+1$ and -1. (See Figure 5.1.) Now let $g(x) = f(x) + \frac{1}{2}x$. Then g is differentiable for all x, $g'(x) = f'(x) + \frac{1}{2}$, and $g'(0) = f'(0) + \frac{1}{2} = \frac{1}{2} > 0$, but g is not increasing on any open interval containing 0, because within any interval $(-\delta, \delta)$, $\delta > 0$, there are subintervals on which

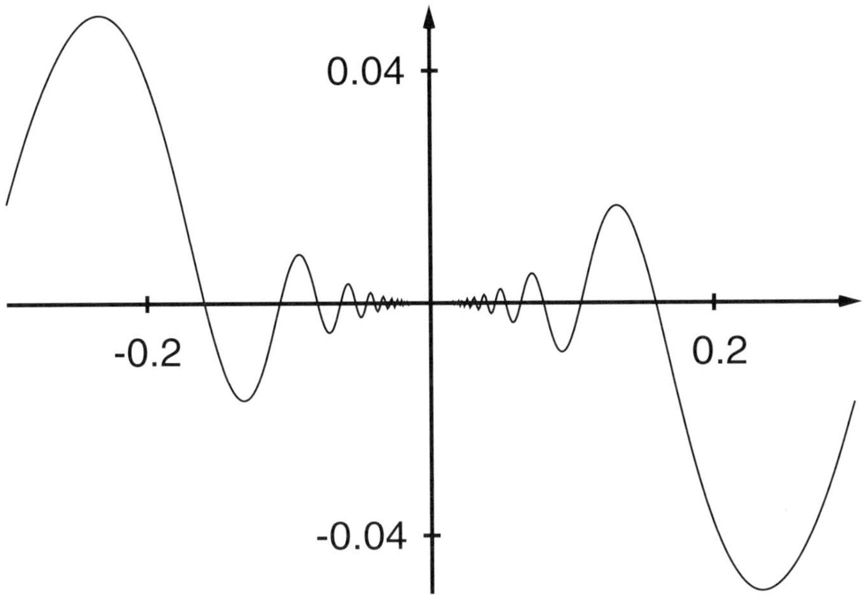

Figure 5.1. The graph of the function f in Example 5.2.7. If $g(x) = f(x) + \frac{1}{2}x$, then $g'(0) = f'(0) + \frac{1}{2} = \frac{1}{2} > 0$, but g is not increasing on any open interval containing 0.

$f'(x) < -\frac{1}{2}$, and thus $g'(x) < 0$. Therefore g is not increasing on any interval containing 0. △

We examine sufficient conditions for the local invertibility of a function in the following section on the inverse function theorem.

Exercises.

Exercise 5.2.1. Prove: If f is differentiable on an open interval I and $|f'|$ is bounded on I, then f is uniformly continuous on I. *Hint*: See Exercise 4.7.2.

Exercise 5.2.2. Use the mean value theorem to find the following limits:

1. $\lim_{x\to\infty}[\sqrt{x+3} - \sqrt{x}\,]$.
2. $\lim_{x\to\infty}[\sqrt{x-5} - \sqrt{x}\,]$.

Exercise 5.2.3. Suppose that f is differentiable on (a, b) and $f'(x) \neq 0$ for all $x \in (a, b)$. Show that f is one-to-one on (a, b).

5.3. The One-Dimensional Inverse Function Theorem

Here we consider the existence of an inverse for a function defined on an interval. Readers who have done Exercise 4.6.5 will recognize the first statement of the next theorem; an outline of the argument for it is included here.

Theorem 5.3.1. *Suppose $f : (a, b) \to \mathbf{R}$.*

1. *If f is continuous on (a, b) and strictly increasing or strictly decreasing, then the inverse function $f^{-1} : f\big((a, b)\big) \to (a, b)$ exists and f^{-1} is continuous.*
2. *If $f : (a, b) \to \mathbf{R}$ is differentiable on (a, b) and f is strictly increasing or strictly decreasing on (a, b), then $(f^{-1})'$ exists for all $y \in f\big((a, b)\big)$ for which $f'\big(f^{-1}(y)\big) \neq 0$, and*

$$(f^{-1})'(y) = \frac{1}{f'\big(f^{-1}(y)\big)}. \tag{5.5}$$

3. *If $f : (a, b) \to \mathbf{R}$ is differentiable on (a, b) and $f'(x) > 0$ for all $x \in (a, b)$ or $f'(x) < 0$ for all $x \in (a, b)$, then $(f^{-1})'$ exists for all $y \in f\big((a, b)\big)$ and (5.5) holds.*

Proof. 1. If f is strictly increasing or strictly decreasing on (a, b), then $f^{-1} : f\big((a, b)\big) \to (a, b)$ exists. It follows from the continuity of f and the intermediate value theorem that f maps open intervals one-to-one and onto open intervals. The same properties show that f^{-1} is continuous on $f\big((a, b)\big)$. We leave the details to the reader.

2. Since f^{-1} is continuous by part 1, $\lim_{y \to y_0} f^{-1}(y) = f^{-1}(y_0)$; that is, $\lim_{y \to y_0} x = x_0$ where $f(x) = y$ and $f(x_0) = y_0$. Hence, if $f'(f^{-1}(y_0)) \neq 0$,

$$(f^{-1})'(y_0) = \lim_{y \to y_0} \frac{f^{-1}(y) - f^{-1}(y_0)}{y - y_0} = \lim_{x \to x_0} \frac{1}{\frac{f(x) - f(x_0)}{x - x_0}} = \frac{1}{f'(x_0)} = \frac{1}{f'\big(f^{-1}(y_0)\big)},$$

as we wished to show.

3. This follows immediately from part 2 since Theorem 5.2.5 implies that f is either strictly increasing or strictly decreasing on (a, b). □

The function $f(x) = x^3$ is differentiable on any interval about 0 and is strictly increasing on any interval, but $f'\big(f^{-1}(0)\big) = 0$ for $y = 0 = f(0)$. This f is an example of a function that satisfies the hypothesis in part 2 but not part 3 of Theorem 5.3.1. The failure of the inverse function $f^{-1}(y) = y^{1/3}$ to be differentiable at $y = 0$ is due to the zero derivative of f at $x = 0$, since this condition gives the inverse function graph a vertical slope at the origin.

If we assume that f has a continuous derivative in an interval about a, and $f'(a) \neq 0$, then the inverse function exists and has a continuous derivative in a neighborhood of $b = f(a)$. This useful sufficient condition for the existence of a continuously differentiable inverse function gives us the following statement known as the *inverse function theorem.*

Theorem 5.3.2. *Let $f : J \to \mathbf{R}$ where J is an interval containing the point a. If f' is continuous on J and $f'(a) \neq 0$, then f is locally invertible on an interval $I \subset J$ with $a \in I$, and $f^{-1} : f(I) \to I$ has a continuous derivative on $f(I)$ given by*

$$(f^{-1})'(y) = \frac{1}{f'\big(f^{-1}(y)\big)}. \tag{5.6}$$

Proof. Let us assume that $f'(a) > 0$; the argument is similar if $f'(a) < 0$. Since $f'(a) > 0$ and f' is continuous on J, then there is an interval $I \subset J$ containing a

such that $f'(x) > 0$ for all $x \in I$. By Theorem 5.2.5, f is strictly increasing, and hence invertible, on I. By part 3 of Theorem 5.3.1, $(f^{-1})'$ exists for all $y \in f(I)$ and (5.6) holds. In view of (5.6), $(f^{-1})'$ is continuous on $f(I)$, since f' is continuous on I and f^{-1} is continuous on $f(I)$. □

It is Theorem 5.3.2 that we will generalize later on for vector valued functions of a vector variable.

Exercises.

Exercise 5.3.1. Let $f(x) = \cos x + 23x^5 - 5x^3 + 16x^2 - x + 1$. Show that f is locally invertible on an interval about 0. Find $(f^{-1})'(2)$.

Exercise 5.3.2. Let $f(x) = x^3 - 3x^2 + 2x$.

1. Find an interval I containing 0 on which f is locally invertible. What is the largest interval containing 0 on which f is invertible?
2. Find $(f^{-1})'(0)$.
3. Investigate the possibility of local inversion of f in an interval about each of these points: $x_1 = 1 - \frac{\sqrt{3}}{3}$, $x_2 = 1$, $x_3 = 1 + \frac{\sqrt{3}}{3}$, and $x_4 = 2$.

Exercise 5.3.3. *Kepler's equation*
If a planet moves along an ellipse, described by $x = a\cos\theta$, $y = b\sin\theta$, $a > b$, with the sun at one focus $((a^2 - b^2)^{1/2}, 0)$, and if t is the time measured from the instant the planet passes through *perihelion* $(a, 0)$, then from Kepler's laws the relation between the position θ and the time t is given by *Kepler's equation*

$$\frac{2\pi}{p}t = \theta - \epsilon\sin\theta,$$

where p is the period of the motion, and $\epsilon \in (0,1)$ is the eccentricity of the elliptic orbit.

1. Show that Kepler's equation can be solved for $\theta = f(t)$.
2. Show that $d\theta/dt = 2\pi/p(1 - \epsilon\cos\theta)$.
3. Conclude that $d\theta/dt$ achieves a maximum at the *perihelion* $(a, 0)$ and a minimum at the *aphelion* $(-a, 0)$.

5.4. Darboux's Theorem

We have seen that derivatives may not be continuous at every point, but any function that is a derivative on an open interval does have a property in common with continuous functions: Any derivative function must have the intermediate value property. This is the assertion of the following theorem, known as *Darboux's theorem.* It is a consequence of Theorem 5.2.2.

Theorem 5.4.1 (Darboux). *Let I be an open interval of the real line, and suppose $f : I \to \mathbf{R}$ is a differentiable function. Then f' has the following intermediate value property on I: If $a, b \in I$ with $a < b$ and $f'(a) \neq f'(b)$, then for any number m between $f'(a)$ and $f'(b)$ there is a point $c \in (a,b) \subset I$ such that $f'(c) = m$.*

Proof. We suppose that $a < b$ and $f'(a) < m < f'(b)$. The function $g(x) = f(x) - m(x-a)$, $x \in I$, is differentiable on I. Since $g'(a) = f'(a) - m < 0$ and $g'(b) = f'(b) - m > 0$, in some small interval about a we have $s_1 < a < s_2$ implies $g(s_1) > g(s_2)$, and in some small interval about b we have $t_1 < b < t_2$ implies $g(t_1) < g(t_2)$. (This follows from the derivative definition.) Therefore g attains its minimum on $[a, b]$ (as it must, since it is continuous there) at some point c in the open interval (a, b). By Theorem 5.2.2, $g'(c) = f'(c) - m = 0$, hence $f'(c) = m$. If, on the other hand, we have $f'(b) < m < f'(a)$, then a similar argument using the same function g shows that g must have its maximum at a point $c \in (a, b)$, leading to the same conclusion. □

Recall the function f from Example 5.2.7:

$$f(x) = \begin{cases} x^2 \sin(1/x) & \text{if } x \neq 0, \\ 0 & \text{if } x = 0. \end{cases}$$

We saw that $f'(0)$ exists, $f'(0) = 0$, and $f'(x) = 2x\sin(1/x) - \cos(1/x)$ for $x \neq 0$. Since $\lim_{x \to 0} f'(x)$ does not exist, f' is not continuous at $x = 0$. Nevertheless, being a derivative, f' must have the intermediate value property, by Darboux's theorem. A comment on the function $g(x) = f(x) + \frac{1}{2}x$ from that example may be helpful here. We had $g'(0) > 0$, but g was not strictly increasing on any interval containing 0; however, we can say (from the derivative definition) that there is some open interval about 0 on which we have $t_1 < 0 < t_2$ implies $g(t_1) < g(t_2)$.

A consequence of the intermediate value property for derivatives is that f' cannot have any jump discontinuities, where both one-sided limits of f' exist at a point but are unequal, since a jump discontinuity precludes the intermediate value property.

Exercise.

Exercise 5.4.1. Is it possible for $g(x) = 1/x$ to be the derivative of a real valued function defined on an open interval containing $x = 0$? Explain.

5.5. Approximations by Contraction Mapping

Suppose we wish to solve an equation that has the form $x = g(x)$. This is known as a *fixed point problem*, since a solution of the equation is a point x mapped to itself by the function g. Suppose we iterate the mapping g starting from a given point x_0, generating the iterates $x_1 = g(x_0)$, $x_2 = g(x_1)$, and so on. In general, we have $x_{n+1} = g(x_n)$ for $n \geq 0$, provided all these iterates are defined. Suppose the sequence (x_n) converges, say $q = \lim_{n\to\infty} x_n$. Then, provided g is continuous and defined at q, we have

$$q = \lim_{n\to\infty} x_n = \lim_{n\to\infty} g(x_{n-1}) = g(\lim_{n\to\infty} x_{n-1}) = g(q),$$

and hence q is a solution of the given equation. Provided the conditions required for the definition and convergence of the sequence of iterates and the continuity of the mapping g hold true, this process is called the *method of successive approximations*, and it yields a solution of the equation $x = g(x)$.

Example 5.5.1. Consider the equation $f(x) = x^3 - x - 1 = 0$. Observe that $f(1) = -1$ and $f(2) = 5$, so the intermediate value theorem for continuous functions implies that there must be a point x_* between 1 and 2 such that $f(x_*) = 0$. There are several ways in which the equation $f(x) = x^3 - x - 1 = 0$ might be expressed as a fixed point problem. By isolating a factor of x in various ways in this equation we can arrive at the following possibilities:

(1) $x = x^3 - 1 = g_1(x)$;

(2) $x = (x+1)^{1/3} = g_2(x)$;

(3) $x = (x^2-1)^{-1} = g_3(x)$;

(4) $x = \sqrt{1 + (1/x)} = g_4(x)$.

In addition, another equation equivalent to $x^3 - x - 1 = 0$ is given by

$$(5) \quad x = x - \frac{x^3 - x - 1}{3x^2 - 1} = g_5(x).$$

This last formulation may appear arbitrary, unless one is already familiar with Newton's method for approximating a root. In any case, Newton's method is examined in more detail in Theorem 5.5.5 below. △

The basis for successive approximations of a solution of an equation $x = g(x)$ by the iteration $x_{n+1} = g(x_n)$ is a theorem known as the Contraction Mapping Theorem. There are more general formulations of the contraction mapping principle than the result considered in this section. A more general version of this principle is considered later in this book.

The definition of our sequence of iterates, and its convergence, is ensured by a contraction condition on the mapping g which is stated next.

Definition 5.5.2. *Let C be a subset of $\mathbf{R}$ and suppose that $g : C \to C$. If there is a real number k with $0 < k < 1$ such that*

$$|g(x) - g(y)| \le k|x - y|$$

for all $x, y \in C$, then g is called a **contraction mapping** *on C with* **contraction constant** *k.*

A contraction mapping $g : C \to C$ is uniformly continuous on C (Exercise 5.5.1).

Contraction mappings are useful for establishing the existence and uniqueness of solutions of equations that take the form $x = g(x)$. If $g : C \to C$ and x_* in C satisfies $g(x_*) = x_*$, then x_* is called a **fixed point** of g. The concept of a contraction mapping helps us to solve such fixed point problems.

The following theorem holds for contraction mappings on *any closed subset* of $\mathbf{R}$, including closed intervals taking any of the forms $[a, b]$, $[a, \infty)$, or $(-\infty, b]$. Notice that the error estimate for the partial sums of a geometric series provides the main estimate needed in this Contraction Mapping Theorem.

Theorem 5.5.3 (Scalar Contraction Mapping). *Let C be a closed subset of $\mathbf{R}$ and $g : C \to C$ a contraction mapping with contraction constant k. Then g has a unique* *fixed point x_* in C. Moreover, given any $\mathbf{x}_0 \in C$, the iteration*

$$x_{n+1} = g(x_n)$$

defines a sequence (x_n) that converges to x_*, and for each n we have

$$|x_n - x_*| \leq \frac{k^n}{1-k}|x_1 - x_0|.$$

Proof. Since $|g(x) - g(y)| \leq k|x - y|$ for any $x, y \in C$, an induction argument shows that

$$|x_{n+1} - x_n| \leq k^n|x_1 - x_0|.$$

If $0 < n < m$, then

$$\begin{aligned} |x_m - x_n| &\leq |x_m - x_{m-1}| + \cdots + |x_{n+1} - x_n| \\ &\leq (k^n + \cdots + k^{m-1})|x_1 - x_0| \\ &< k^n(1 + k + k^2 + \cdots)|x_1 - x_0| \\ &= \frac{k^n}{1-k}|x_1 - x_0|, \end{aligned}$$

where we used the sum of a geometric series in the last line. Since $0 < k < 1$, the sequence (x_n) is a Cauchy sequence which must converge to a limit x_*, and $x_* \in C$ since C is closed. Now hold n fixed and let $m \to \infty$ in the estimate above to yield the estimate in the statement of the theorem. Since a contraction mapping is continuous, we have

$$g(x_*) = \lim_{n\to\infty} g(x_n) = \lim_{n\to\infty} x_{n+1} = x_*,$$

so x_* is a fixed point of g. If there were another fixed point x_{**} of g, we would have

$$|x_* - x_{**}| = |g(x_*) - g(x_{**})| \leq k|x_* - x_{**}|,$$

but since $k < 1$, this implies $x_* = x_{**}$, so the fixed point is unique. □

The application of the contraction mapping theorem requires the reformulation of a given equation $f(x) = 0$ in the form $x = g(x)$ for some suitable choice of function g, as well as the verification that $g : C \to C$ for a suitable subset C. The mean value theorem (Theorem 5.2.4) is often helpful in deciding if a mapping g is a contraction mapping on a particular set C, using an appropriate bound for the derivative of g on C as a contraction constant. Let us try to use the mean value theorem in analyzing the mappings g_1 and g_2 defined in Example 5.5.1.

Example 5.5.4. We continue with Example 5.5.1 and examine three ways in which the equation $f(x) = x^3 - x - 1 = 0$ of that example can be reformulated so as to attempt to apply the contraction mapping theorem.

Rewrite the equation in the equivalent form $x = x^3 - 1 = g_1(x)$. It is difficult to find a closed interval C such that $g_1 : C \to C$. This is because $g_1'(x) = 3x^2$, which is certainly larger than 1 for values of x close to the root q. So g_1 is not a good candidate for a contraction mapping on a closed set containing the desired root q.

Now rewrite the equation in the equivalent form $x = (x+1)^{1/3} = g_2(x)$. Note that $g_2'(x) = \frac{1}{3}(x+1)^{-2/3}$ and the maximum of $|g_2'(x)|$ over the interval $[1, \infty)$ is $k := 1/3(2)^{2/3} < 1$. Also note that $g_2 : [1, \infty) \to [1, \infty)$. By the mean value

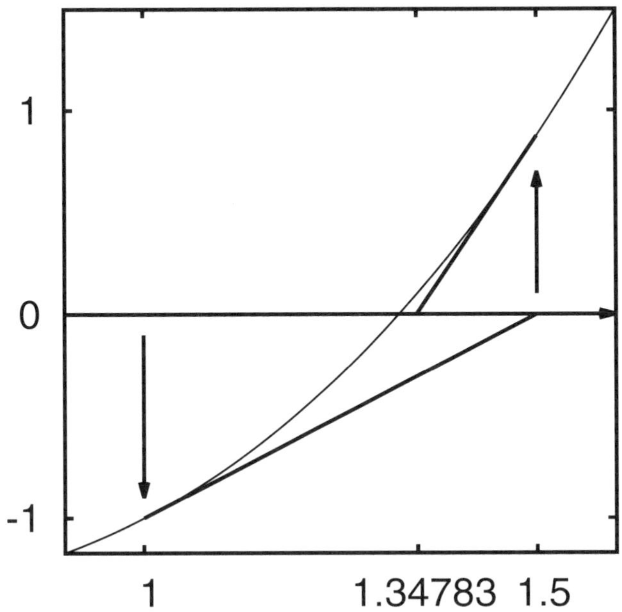

Figure 5.2. Following tangent lines to the next iterate in Newton's method, according to the iteration in equation (5.7). The modified Newton iteration in Theorem 5.5.5 uses known upper and lower bounds on the absolute value of the derivative of f over an interval which encloses an isolated root of f; thus, fewer derivative computations are needed.

theorem (Theorem 5.2.4), g_2 is a contraction mapping on $C = [1, \infty)$. By the contraction mapping theorem, the successive approximations defined by

$$x_{n+1} = g_2(x_n) = (x_n + 1)^{1/3}, \quad n = 0, 1, 2, \ldots,$$

converge to the root $q \in (1, 2)$. $\triangle$

If we rewrite $f(x) = x^3 - x - 1 = 0$ in the equivalent form $x = x - f(x)/f'(x) = g_5(x)$, then we have the iteration

$$(5.7) \qquad x_{n+1} = x_n - \frac{f(x_n)}{f'(x_n)} = x_n - \frac{x_n^3 - x_n - 1}{3x_n^2 - 1} = \frac{2x_n^3 + 1}{3x_n^2 - 1} \quad (x_0 \text{ given}),$$

which is *Newton's method.* The function $g(x) = x - f(x)/f'(x)$ can be shown to be a contraction mapping in some closed interval about the root q, provided $f'(q) \neq 0$. Newton's method is especially valuable because of its rapid convergence to the root.

Let us examine Newton's method in more detail. The geometric idea is to choose a starting approximation x_0 for a root, and then to follow the tangent line approximations for f at each iterate to its intersection with the x axis, which gives the next iterate. (See Figure 5.2). Actually, the next theorem deals with a modified Newton iteration with simpler computations which save on the number of derivative evaluations required.

Theorem 5.5.5 (Simplified Newton Method). *Let $f : [a, b] \to \mathbf{R}$ be differentiable with $f(a)f(b) < 0$, that is, $f(a)$ and $f(b)$ have opposite sign. Suppose $f'(x) \neq 0$ for all $x \in [a, b]$, and let $0 < m \leq |f'(x)| \leq M$ for all $x \in [a, b]$. Then, given any $x_0 \in [a, b]$, the sequence defined by*

$$x_{n+1} = x_n - \frac{f(x_n)}{M}$$

converges to the unique solution x_* of the equation $f(x) = 0$ in $[a, b]$. Moreover, for each positive integer n,

$$|x_n - x_*| \leq \frac{|f(x_0)|}{m}\left(1 - \frac{m}{M}\right)^n.$$

Proof. Since $f(x) = 0$ if and only if $-f(x) = 0$, we may assume that $f(a) < 0 < f(b)$ and $f'(x) > 0$ for all $x \in [a, b]$. (Otherwise, we may work with $-f$.) Thus we have $M \geq f'(x) \geq m > 0$ for all $x \in [a, b]$. Define $g : [a, b] \to \mathbf{R}$ by

$$g(x) = x - \frac{f(x)}{M}.$$

We will show that g maps $[a, b]$ into $[a, b]$, and g is a contraction mapping. Since $M \geq f'(x) \geq m$, we have

$$0 \leq g'(x) = 1 - \frac{f'(x)}{M} \leq 1 - \frac{m}{M} =: k < 1.$$

This shows that g is increasing (not necessarily strictly) on $[a, b]$, and, in view of the mean value theorem, g satisfies

$$|g(x) - g(y)| \leq k|x - y|$$

for all $x, y \in [a, b]$. It remains to show that $g([a, b]) \subset [a, b]$. But $g(a) = a - f(a)/M > a$ since $f(a) < 0$. And $g(b) = b - f(b)/M < b$ since $f(b) > 0$. Since g is increasing on $[a, b]$, it follows that $g([a, b]) \subset [a, b]$. Now we can apply Theorem 5.5.3 to conclude that the sequence $x_{n+1} = x_n - f(x_n)/M$ converges to the unique fixed point x_* of g in $[a, b]$, which must be the unique solution of the equation $f(x) = 0$ in $[a, b]$. Since $k = 1 - m/M$ and $x_1 - x_0 = -f(x_0)/M$, the error estimate from Theorem 5.5.3 yields precisely the estimate

$$\begin{aligned} |x_n - x_*| &\leq \frac{k^n}{1-k}|x_1 - x_0| \\ &= \frac{|f(x_0)|}{m}\left(1 - \frac{m}{M}\right)^n. \end{aligned}$$

This completes the proof. □

As mentioned earlier, there are more general versions of the contraction mapping theorem. Later in this book a version of the contraction theorem for mappings of subsets of n-dimensional Euclidean space $\mathbf{R}^n$ helps to establish the important inverse function theorem. Another extension of the contraction theorem helps to establish the existence and uniqueness of solutions to initial value problems for ordinary differential equations.

Exercises.

Exercise 5.5.1. Show that a contraction mapping $g : C \to C$ is uniformly continuous on C.

Exercise 5.5.2. *Iteration converges, but not to a fixed point*
Let $f : (0, \infty) \to (0, \infty)$ be $f(x) = \frac{1}{2}x$ for $x > 0$. Show that the iteration $x_{n+1} = f(x_n)$ converges for any initial choice x_0, but $x = f(x)$ has no solution.

Exercise 5.5.3. Show that the equation $x^3 - x - 1 = 0$ is equivalent to the fixed point problem $x = g_j(x)$ for each of the functions g_j, $(j = 1, \dots, 5)$ indicated in Example 5.5.1.

Exercise 5.5.4. Investigate the functions g_3 and g_4 of Example 5.5.1 to see if either function is a contraction mapping on a suitable closed interval about the real root x_* between 1 and 2.

Exercise 5.5.5. Set $x_0 = 1$ in the Newton iteration in (5.7) and find the iterates x_1, x_2, x_3 approximating the unique real root of $x^3 - x - 1 = 0$. How many accurate digits do you have at each step? What if $x_0 = 2$?

5.6. Cauchy's Mean Value Theorem

As we have seen, one of the primary uses of the mean value theorem is the approximation represented by statement 1 of Theorem 5.2.5: If $|f'(c)| \le M$ for all c in (a, b), then for each x with $a < x < b$ we have

$$|f(x) - f(a)| \le M(x - a) < M(b - a)$$

or

$$-M(b - a) < -M(x - a) \le f(x) - f(a) \le M(x - a) < M(b - a).$$

This gives a useful approximation to $f(x)$ in terms of the value $f(a)$. By using more derivative information about f, when available, we can obtain better approximations for $f(x)$. As a first step toward that goal, we consider in this section Cauchy's mean value theorem, which will be used to establish Taylor's theorem in the following section.

In the mean value theorem, the graph of f can be viewed as being traced out by parametric equations $x = t$ and $y = f(t)$ for $a \le t \le b$. Now recall the concept of the tangent vector at a point of a parametric plane curve. The mean value theorem states that the line segment joining $(a, f(a))$ and $(b, f(b))$ has slope equal to the slope of the tangent vector $(1, f'(c))$ to this curve at some point $c \in (a, b)$, that is,

$$\frac{f'(c)}{1} = \frac{f(b) - f(a)}{b - a}.$$

Now consider a more general plane curve described by parametric equations $x = g(t)$ and $y = f(t)$, where the functions f and g are continuous on $[a, b]$ and differentiable on (a, b). The curve passes through the points $(g(a), f(a))$ and $(g(b), f(b))$, and the slope of the segment joining these points is

$$\frac{f(b) - f(a)}{g(b) - g(a)},$$

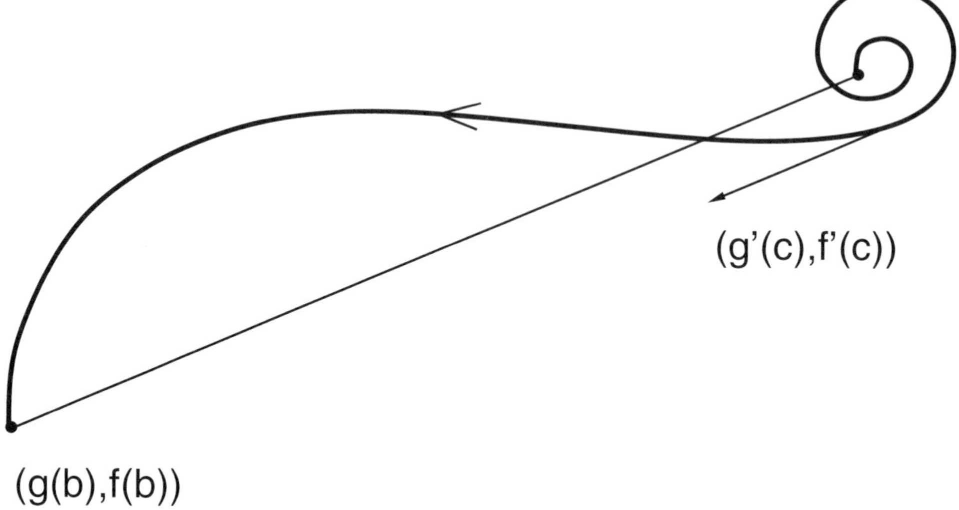

Figure 5.3. Illustrating Cauchy's mean value theorem: The line segment joining $(g(b), f(b))$ and $(g(a), f(a))$ has slope equal to the slope of the tangent vector $(g'(c), f'(c))$, for some c with $a < c < b$.

provided $g(a) \neq g(b)$. Since f and g are differentiable for $t \in (a, b)$, the curve has a tangent vector defined for each value $t \in (a, b)$ by $(g'(t), f'(t)) := g'(t)\mathbf{i} + f'(t)\mathbf{j}$, and this tangent vector varies smoothly with t provided $g'(t)$ and $f'(t)$ are not simultaneously zero at any point t. (See Figure 5.3.) It appears that at some value $t = c$, the smooth curve must have a tangent vector with slope equal to the slope of the segment joining the points $(g(a), f(a))$ and $(g(b), f(b))$. This intuition is confirmed by *Cauchy's mean value theorem.*

Theorem 5.6.1 (Cauchy Mean Value). *Suppose f and g are continuous on $[a, b]$ and differentiable on (a, b). If $f'(x)$ and $g'(x)$ are not both equal to zero at any $x \in (a, b)$ and $g(a) \neq g(b)$, then there is a point $c \in (a, b)$ such that*

$$\frac{f'(c)}{g'(c)} = \frac{f(b) - f(a)}{g(b) - g(a)}. \tag{5.8}$$

Proof. We proceed as in the proof of the mean value theorem. Consider the function

$$h(x) := f(x) - f(a) - \frac{g(x) - g(a)}{g(b) - g(a)}[f(b) - f(a)].$$

Then $h(a) = 0 = h(b)$. Since h is continuous on $[a, b]$ and differentiable on (a, b), by Rolle's theorem there is a point $c \in (a, b)$ such that $h'(c) = 0$, hence

$$h'(c) = f'(c) - g'(c)\frac{f(b) - f(a)}{g(b) - g(a)} = 0,$$

from which (5.8) follows. □

If we take $g(x) = x$ in Cauchy's mean value theorem, we recover the result of the mean value theorem (Theorem 5.2.4) as a special case. Also note that if $g'(x) \neq 0$

for $x \in (a, b)$, then g satisfies the hypotheses of Theorem 5.6.1. In particular, f' and g' cannot simultaneously vanish, and $g(a) \neq g(b)$ because by the mean value theorem, $g(b) - g(a) = g'(c)(b - a)$ for some $c \in (a, b)$, but $g'(c) \neq 0$.

5.6.1. Limits of Indeterminate Forms. We shall explore some applications of Cauchy's mean value theorem to the evaluation of limits of quotients where the numerator and denominator both approach zero, or the numerator and denominator both become unbounded as the variable x approaches a limit point a. The following theorems are usually associated with the name of Guillaume François Marquis de l'Hôpital, 1661-1704.

Theorem 5.6.2 (Rule for 0/0 Forms). *Suppose f and g are defined and continuous on an open interval I containing the point a, f and g are differentiable on $I - \{a\}$, $f(a) = g(a) = 0$, $g'(x) \neq 0$ for $x \in I - \{a\}$, and $\lim_{x \to a} \frac{f'(x)}{g'(x)}$ exists. Then $\lim_{x \to a} \frac{f(x)}{g(x)}$ exists and*

$$\lim_{x \to a} \frac{f(x)}{g(x)} = \lim_{x \to a} \frac{f'(x)}{g'(x)}. \tag{5.9}$$

Proof. Choose $\delta > 0$ such that $B_\delta(a) \subset I$. Let $x \in B_\delta(a)$ so that $|x - a| < \delta$. Cauchy's mean value theorem applies to f and g on an interval containing both x and a. Using the fact that $f(a) = g(a) = 0$, we conclude that for any $x \in I - \{a\}$, there is some c between a and x such that

$$f(x)g'(c) = g(x)f'(c).$$

If $g(x) = 0$ for some $x \in I - \{a\}$, then the mean value theorem applied to g says that

$$g(x) - g(a) = g'(\xi)(x - a) = 0$$

for some $\xi \in I - \{a\}$. But this contradicts the assumption that $g'(x) \neq 0$ for $x \in I - \{a\}$. Therefore $g(x) \neq 0$ for all $x \in I - \{a\}$. Thus, for each $x \in I - \{a\}$, there is some c between a and x such that

$$\frac{f(x)}{g(x)} = \frac{f'(c)}{g'(c)}.$$

As $x \to a$, the point $c \to a$ as well, and the limit statement (5.9) follows. □

Applications of Theorem 5.6.2 can be explored in the exercises.

We recall that the meaning of statements such as $\lim_{x \to a} f(x) = \pm\infty$ is given in Definition 4.4.12.

Theorem 5.6.3 (Rule for ∞/∞ Forms). *Suppose f and g are defined and continuous on $I - \{a\}$ where I is an open interval containing the point a. If f and g are differentiable on $I - \{a\}$, $\lim_{x \to a} f(x) = \pm\infty$, $\lim_{x \to a} g(x) = \pm\infty$, $g'(x) \neq 0$ for $x \in I - \{a\}$, and $\lim_{x \to a} \frac{f'(x)}{g'(x)}$ exists, then $\lim_{x \to a} \frac{f(x)}{g(x)}$ exists and*

$$\lim_{x \to a} \frac{f(x)}{g(x)} = \lim_{x \to a} \frac{f'(x)}{g'(x)}. \tag{5.10}$$

Proof. Let $\delta > 0$ be such that if $0 < |x-a| < \delta$ then $x \in I$. Write $L = \lim_{x\to a} \frac{f'(x)}{g'(x)}$. Then for every $\epsilon > 0$ there is a δ_1 with $0 < \delta_1 < \delta$ such that

$$\left|\frac{f'(x)}{g'(x)} - L\right| < \epsilon \quad \text{for} \quad 0 < |x-a| < \delta_1.$$

We again use Cauchy's mean value theorem. For any x such that $x \in (a-\delta_1, a)$, there is a $c \in (a-\delta_1, x)$ such that

$$\frac{f(x) - f(a-\delta_1)}{g(x) - g(a-\delta_1)} = \frac{f'(c)}{g'(c)}$$

(Also, for any $x \in (a, a+\delta_1)$, there is a $c \in (x, a+\delta_1)$ such that

$$\frac{f(x) - f(a+\delta_1)}{g(x) - g(a+\delta_1)} = \frac{f'(c)}{g'(c)}.)$$

We deal with the first case only, as the second case is handled by a similar argument. Since $g'(x) \neq 0$ for $x \in I - \{a\}$, the mean value theorem guarantees that $g(x) - g(a-\delta_1) \neq 0$. Thus, we may write

$$\left|\frac{f(x) - f(a-\delta_1)}{g(x) - g(a-\delta_1)} - L\right| = \left|\frac{f(x)}{g(x)} \cdot \frac{1 - \frac{f(a-\delta_1)}{f(x)}}{1 - \frac{g(a-\delta_1)}{g(x)}} - L\right| < \epsilon$$

for all $x \in (a-\delta_1, a)$. Letting x approach a, we have

$$\lim_{x\to a} \frac{1 - \frac{f(a-\delta_1)}{f(x)}}{1 - \frac{g(a-\delta_1)}{g(x)}} = 1$$

by the quotient limit law, since $\lim_{x\to a} f(x) = \pm\infty$ and $\lim_{x\to a} g(x) = \pm\infty$. By a straightforward argument using the triangle inequality, it follows that $\lim_{x\to a} \frac{f(x)}{g(x)}$ exists and (5.10) holds. □

Exercises.

Exercise 5.6.1. Find the indicated limits:

$$(1)\ \lim_{x\to 0} \frac{x^n}{e^x - 1} \qquad (2)\ \lim_{x\to 0} x \cot x \qquad (3)\ \lim_{x\to 0} \frac{\tan^{-1}(x^2)}{x \sin x}.$$

Exercise 5.6.2. Find the indicated limit:

$$\lim_{x\to 0} \frac{\sin x \log(\sin^2 x)}{\cos x}.$$

Exercise 5.6.3. Find the indicated limit:

$$\lim_{x\to 0} \frac{\log|\cot x|}{\cot x}.$$

Exercise 5.6.4. Limits as $x \to \pm\infty$ are defined in Definition 4.4.4. Discuss the extension of l'Hôpital's rule to limits at infinity, $\lim_{x\to\pm\infty} f(x)/g(x)$, under appropriate conditions. *Hint*: Set $z = 1/x$ so that $z \to 0$ as $x \to \pm\infty$, and try this approach to find $\lim_{x\to\infty} x(1 - e^{-1/x})$.

5.7. Taylor's Theorem with Lagrange Remainder

Taylor's theorem shows how to use derivative information for a function at a single point in order to approximate the function by a polynomial function in a neighborhood of that point. Taylor's theorem includes an estimate of the error in that approximation. We begin with a result that gives useful local information about a function in the neighborhood of a degenerate critical point.

Lemma 5.7.1. *Let I be an open interval and let n be a nonnegative integer. Suppose that $f : I \to \mathbf{R}$ has $n+1$ derivatives $f', f'', \ldots, f^{(n+1)}$ on I, and that at some point a in I,*

$$f^{(k)}(a) = 0 \quad \text{for } 0 \le k \le n.$$

Then for each $x \neq a$ in I, there is a point c in I between a and x such that

$$f(x) = \frac{f^{(n+1)}(c)}{(n+1)!}(x-a)^{n+1}. \tag{5.11}$$

Proof. Define $g : I \to \mathbf{R}$ by

$$g(x) = \frac{1}{(n+1)!}(x-a)^{n+1}.$$

Then

$$g^{(k)}(a) = 0 \quad \text{for } 0 \le k \le n.$$

If $x \neq a$, then $g'(x) \neq 0$. We assume now without loss of generality that $a < x$. Apply Cauchy's mean value theorem to f and g on the open interval (a, x) to conclude that there is a point $c_1 \in (a, x)$ such that

$$\frac{f(x)}{g(x)} = \frac{f(x) - f(a)}{g(x) - g(a)} = \frac{f'(c_1)}{g'(c_1)}.$$

Given c_1, apply Cauchy's mean value theorem to f' and g' on the interval (a, c_1) to conclude that there is a point $c_2 \in (a, c_1)$ such that

$$\frac{f'(c_1)}{g'(c_1)} = \frac{f'(c_1) - f'(a)}{g'(c_1) - g'(a)} = \frac{f''(c_2)}{g''(c_2)}.$$

We may continue in this way to obtain points $a < c_n < \cdots < c_2 < c_1 < x$ such that

$$\frac{f(x)}{g(x)} = \frac{f'(c_1)}{g'(c_1)} = \frac{f''(c_2)}{g''(c_2)} = \cdots = \frac{f^{(n)}(c_n)}{g^{(n)}(c_n)}.$$

By a final application of Cauchy's mean value theorem to $f^{(n)}$ and $g^{(n)}$ on the interval (a, c_n), there is a point c with $a < c < c_n$ such that

$$\frac{f(x)}{g(x)} = \frac{f^{(n)}(c_n) - f^{(n)}(a)}{g^{(n)}(c_n) - g^{(n)}(a)} = \frac{f^{(n+1)}(c)}{g^{(n+1)}(c)}.$$

By the definition of $g(x)$, $g^{(n+1)}(c) = 1$, and (5.11) follows immediately. □

In general, we define the **Taylor polynomial of degree** n for f at a point a by

$$P_n(x) = f(a) + f'(a)(x-a) + \frac{f''(a)}{2!}(x-a)^2 + \cdots + \frac{f^{(n)}(a)}{n!}(x-a)^n.$$

Notice that the derivatives of P_n and f at a are equal through order n.

The next result is known as *Taylor's theorem* with *Lagrange remainder* term.

Theorem 5.7.2 (Taylor's Theorem with Lagrange Remainder). *Let I be an open interval containing a and let n be a nonnegative integer. If $f : I \to \mathbf{R}$ has $n+1$ derivatives on I, then for any $x \neq a$ in I there is a point c in I between a and x such that*

$$\begin{aligned} f(x) &= f(a) + f'(a)(x-a) + \frac{f''(a)}{2!}(x-a)^2 + \cdots + \frac{f^{(n)}(a)}{n!}(x-a)^n \\ &+ \frac{f^{(n+1)}(c)}{(n+1)!}(x-a)^{n+1}. \end{aligned} \tag{5.12}$$

Proof. Let $P_n(x)$ be the Taylor polynomial of degree n at a,

$$P_n(x) = \sum_{k=0}^{n} \frac{f^{(k)}(a)}{k!}(x-a)^k.$$

As we have noted,

$$f^{(k)}(a) = P_n^{(k)}(a) \quad \text{for } 0 \le k \le n.$$

If $R(x) := f(x) - P_n(x)$, x in I, then

$$R^{(k)}(a) = 0 \quad \text{for } 0 \le k \le n.$$

By Lemma 5.7.1, for any x in I, there is a point c in I such that

$$f(x) - P_n(x) = R(x) = \frac{R^{(n+1)}(c)}{(n+1)!}(x-a)^{n+1}.$$

Since P_n is a polynomial of degree n, $P_n^{(n+1)}(c) = 0$, hence $R^{(n+1)}(c) = f^{(n+1)}(c)$. Thus, by the previous equation,

$$f(x) = P_n(x) + \frac{f^{(n+1)}(c)}{(n+1)!}(x-a)^{n+1},$$

which is the assertion of (5.12). □

Equation (5.12) is known as *Taylor's formula*. Let us indicate the error term, or remainder, in Taylor's formula by writing

$$R_{a,n}(x) = \frac{f^{(n+1)}(c)}{(n+1)!}(x-a)^{n+1}.$$

This expression is called the *Lagrange form of the remainder*, and it gives the error when using the Taylor polynomial of degree n, $P_n(x)$, to approximate f near a. If a bound can be determined for $|f^{(n+1)}(x)|$ on the interval I, let us say

$$|f^{(n+1)}(x)| \le M \quad \text{for } x \in I,$$

then

$$|R_{a,n}(x)| = \left| \frac{f^{(n+1)}(c)}{(n+1)!}(x-a)^{n+1} \right| \le \frac{M}{(n+1)!}(x-a)^{n+1}.$$

According to Taylor's theorem, if the second derivative f'' exists on I, and $a \in I$, then for any $x \in I$ there is a point $c \in I$ between a and x such that

$$f(x) = f(a) + f'(a)(x-a) + \frac{f''(c)}{2!}(x-a)^2.$$

The first degree polynomial $f(a) + f'(a)(x - a)$ is the function that defines the tangent line approximation to f about the point $x = a$. The error when using this tangent line approximation is given by

$$R_{a,1}(x) = \frac{f''(c)}{2!}(x-a)^2,$$

where $c = c(x)$ depends on x. This is a more specific estimate than the one given by the assumption of the existence of the derivative $f'(a)$, which tells us that

$$f(x) = f(a) + f'(a)(x-a) + o(|x-a|) \quad \text{as } x \to a,$$

where the expression $o(|x - a|)$ satisfies the limit property

$$\lim_{x \to a} \frac{o(|x-a|)}{|x-a|} = 0.$$

The assumption of the existence of f'' on I shows that under this additional hypothesis on f, we have

$$o(|x-a|) = \frac{f''(c)}{2!}(x-a)^2,$$

where $c = c(x)$ depends on x.

Exercises.

Exercise 5.7.1. Show that if $p : \mathbf{R} \to \mathbf{R}$ is a polynomial function of degree n, $p(x) = a_n x^n + a_{n-1}x^{n-1} + \cdots + a_1 x + a_0$, then the nth Taylor polynomial for p at $a = 0$ is p itself.

Exercise 5.7.2. Find a bound for the error term when using $P_4(x)$ to approximate the function $f(x) = \cos 2x$ for $|x| \leq \pi/4$, that is, estimate the remainder $R_{0,4}(x)$ for $|x| \leq \pi/4$.

5.8. Extreme Points and Extreme Values

We have seen that if a function f has a relative extremum at x_0 and $f'(x_0)$ exists, then necessarily $f'(x_0) = 0$. The next theorem provides conditions for detecting relative extrema and distinguishing local maxima from local minima, using higher order derivative information.

Theorem 5.8.1. *Let I be an open interval and assume that $f : I \to \mathbf{R}$ has $n+1$ derivatives, $f', f'', \ldots, f^{(n+1)}$ all defined on I, and $f^{(n+1)}$ is continuous on I. Let x_0 be a point in I such that*

$$f'(x_0) = f''(x_0) = \cdots = f^{(n)}(x_0) = 0, \quad \textit{and} \quad f^{(n+1)}(x_0) \neq 0.$$

The following statements are true:

1. *If n is even, then x_0 is not an extreme point for f.*
2. *If n is odd, then x_0 is an extreme point, which is a local minimum point if $f^{(n+1)}(x_0) > 0$ and a local maximum point if $f^{(n+1)}(x_0) < 0$.*

Proof. By the hypothesis of zero derivatives through order n and Taylor's theorem, for any h such that $x_0 + h$ is in I, we have

$$f(x_0+h) - f(x_0) = \frac{f^{(n+1)}(x_0+\theta h)}{(n+1)!}h^{n+1} \tag{5.13}$$

for some $\theta = \theta(h)$ with $0 < \theta < 1$. Since $f^{(n+1)}$ is continuous and $f^{(n+1)}(x_0) \neq 0$, $f^{(n+1)}(x_0 + h) \neq 0$ for all sufficiently small $|h|$, say $|h| \leq h_0$. By continuity, $f^{(n+1)}(x_0 + h)$ cannot change sign for $|h| \leq h_0$ for h_0 sufficiently small.

1. If n is even, then the factor h^{n+1} on the right-hand side of (5.13) takes on both positive and negative values for $|h| \leq h_0$. Therefore x_0 is not an extreme point for f.

2. If n is odd and $f^{(n+1)}(x_0) > 0$, then the right-hand side of (5.13) has constant positive sign for $|h| \leq h_0$, hence $f(x_0 + h) > f(x_0)$ and x_0 is a local minimum point for f. If n is odd and $f^{(n+1)}(x_0) < 0$, then the right-hand side of (5.13) has constant negative sign, hence $f(x_0 + h) < f(x_0)$ for $|h| \leq h_0$ and x_0 is a local maximum point for f. □

Possibly the most commonly used result on extreme points in elementary calculus occurs as case 2 of Theorem 5.8.1 when $n = 1$. This is the case that allows identification of extreme points by means of the second derivative test: $f'(x_0) = 0$ and $f''(x_0) < 0$ imply a local maximum at x_0; while $f'(x_0) = 0$ and $f''(x_0) > 0$ imply a local minimum at x_0.

There are functions to which Theorem 5.8.1 does not apply at all, as in the next example.

Example 5.8.2. Consider the function

$$f(x) = \begin{cases} e^{-1/x^2} & \text{if } x \neq 0, \\ 0 & \text{if } x = 0. \end{cases}$$

It can be shown that f has derivatives of all orders at every real x. (We say that f is of class C^∞ on $\mathbf{R}$.) Moreover, $f^{(n)}(0) = 0$ for every n. Theorem 5.8.1 does not apply to f. However, f has a local minimum at $x_0 = 0$. △

If $n > 1$, and case 1 of Theorem 5.8.1 applies to f, then the function $g(x) = f'(x)$ satisfies

$$g'(x_0) = \cdots = g^{(n-1)}(x_0) = 0, \quad \text{and} \quad g^{(n)}(x_0) \neq 0.$$

Hence g has a local extreme at x_0, by the theorem applied to g, and $g(x_0) = f'(x_0) = 0$. If x_0 is a point where $f'(x_0) = 0$ and x_0 is an extreme point for f', then we call x_0 an **inflection point** for f. For example, $x_0 = 0$ is an inflection point for $f(x) = x^3$; it is a local minimum point for $f'(x) = 3x^2$. On the other hand, $x_0 = 0$ is an inflection point for $f(x) = -x^3$; it is then a local maximum point for $f'(x) = -3x^2$.

Exercises.

Exercise 5.8.1. Verify the statements in Example 5.8.2 about the function f. *Hint*: Show, by induction, that for every positive integer n,

$$f^{(n)}(x) = e^{-1/x^2} q_n\left(\frac{1}{x}\right) \quad \text{for } x \neq 0,$$

where $q_n(t)$ is a polynomial in $t = 1/x$, and $f^{(n)}(0) = 0$.

Exercise 5.8.2. Suppose f is three times continuously differentiable. Let $h = x - a$ for x near a. By Taylor's theorem, for each sufficiently small h, we have

$$f(a+h) - f(a) = f'(a)h + \frac{f''(a)}{2!}h^2 + \frac{f^{(3)}(c_3)}{3!}h^3$$

for some c_3 between a and x. Establish the following using Taylor's theorem.

1. Show that if there is a local minimum of f at $x = a$, then $f'(a) = 0$. (We know this from an earlier theorem, but derive it now from the expression above.)
2. Show that if there is an extremum of f at a and $f''(a) > 0$, then f has a local minimum at a.
3. Show that $f'(a) = 0$ is necessary for a local maximum.
4. Show that the condition $f''(a) < 0$ is sufficient for a local extremum to be a local maximum.

5.9. Notes and References

This chapter was influenced by Folland [**16**], Krantz [**40**] and Sagan [**54**].

Chapter 6

The Riemann Integral

This chapter contains the most fundamental properties of the Riemann integral and consequences of the fundamental theorem of calculus. The primary geometric interpretation of the definite integral in introductory calculus is the computation of the area between the graph of a continuous function and its domain axis. Later in this book we extend this area measure of a planar region to a larger class of sets.

6.1. Partitions and Riemann-Darboux Sums

The Riemann integral is defined for bounded functions on closed intervals. We begin with notation for finite collections of points that subdivide an interval $[a, b]$ into subintervals.

Definition 6.1.1. *Let $x_0, x_1, \ldots, x_n$ be points in the interval $[a, b]$ with $a = x_0 < x_1 < \cdots < x_{n-1} < x_n = b$. Then the set $P = \{x_0, x_1, \ldots, x_n\}$ is called a* **partition** *of $[a, b]$. The* **subintervals of** P *are the intervals $[x_{k-1}, x_k]$ for $k = 1, \ldots, n$. A partition P' is a* **refinement** *of P if $P \subseteq P'$.*

The functions we consider are bounded on $[a, b]$ but not necessarily continuous at every point. So we must consider the supremum and infimum (rather than the maximum and minimum) of the function on each subinterval of a partition when defining our approximating sums, which are called Riemann-Darboux sums.

Definition 6.1.2. *Let $f : [a, b] \to \mathbf{R}$ be bounded and let $P = \{x_0, x_1, \ldots, x_n\}$ be a partition of $[a, b]$. Let*

$$M_k = \sup_{x \in [x_{k-1}, x_k]} f(x) \qquad \text{and} \qquad m_k = \inf_{x \in [x_{k-1}, x_k]} f(x).$$

The **upper and lower Riemann-Darboux sums for f over $[a, b]$ with partition** P *are defined, respectively, by*

$$U(f, P) = \sum_{k=1}^{n} M_k(x_k - x_{k-1}) \qquad \text{and} \qquad L(f, P) = \sum_{k=1}^{n} m_k(x_k - x_{k-1}).$$

For simplicity, the notation for upper and lower sums does not include any reference to the domain interval since the domain is fixed in most of the following discussion.

In the notation of Definition 6.1.2, since $m_k \le M_k$ for each k, it is clear that $L(f, P) \le U(f, P)$ for any given partition P of $[a, b]$. We need to know how these sums behave under refinements of any partition.

Theorem 6.1.3. *Let $f : [a, b] \to \mathbf{R}$ be a bounded function.*

1. *If P is a partition of $[a, b]$ and P' is any refinement of P, then*
$$L(f, P) \le L(f, P') \qquad \text{and} \qquad U(f, P') \le U(f, P).$$
2. *If P_1 and P_2 are any two partitions of $[a, b]$, then*
$$L(f, P_1) \le U(f, P_2).$$

Proof. 1. We prove the inequality regarding the upper sums, $U(f, P') \le U(f, P)$. If $P = P'$, then there is nothing to prove, so we assume that $P \subset P'$ as a proper subset. It is probably easiest to think about the inequality if the refinement P' has only one point more than P, say $P' = P \cup \{x^*\}$, and we prove it for this case first. Since any partition has finitely many points, the general result will follow by induction. Suppose that $I_k = [x_{k-1}, x_k]$ is the subinterval of P that contains x^*. Then the contribution to the upper sums from all the subintervals except I_k are the same under the refinement, and, in the notation of Definition 6.1.2,

$$\begin{aligned} M_k(x_k - x_{k-1}) &= \sup_{x \in [x_{k-1}, x_k]} f(x)(x_k - x_{k-1}) \\ &= \sup_{x \in [x_{k-1}, x_k]} f(x)(x_k - x^*) + \sup_{x \in [x_{k-1}, x_k]} f(x)(x^* - x_{k-1}) \\ &\ge \sup_{x \in [x^*, x_k]} f(x)(x_k - x^*) + \sup_{x \in [x_{k-1}, x^*]} f(x)(x^* - x_{k-1}), \end{aligned}$$

where the inequality holds because each supremum in the last line is taken over a smaller set than the set in the next-to-last line (Exercise 6.1.1). Hence, $U(f, P \cup \{x^*\}) \le U(f, P)$. The inequality for the lower sums is handled in a similar manner.

2. If $P_1 = P_2$, then we have already noted the result. If $P_1 \ne P_2$, then $P_1 \cup P_2$ is a refinement of both P_1 and P_2, and hence

$$L(f, P_1) \le L(f, P_1 \cup P_2) \le U(f, P_1 \cup P_2) \le U(f, P_2)$$

by part 1. □

Exercises.

Exercise 6.1.1. Prove: If $S \subset T$, then $\inf T \le \inf S$ and $\sup S \le \sup T$. Is it possible to have $S \subset T$ and $S \ne T$ with both inequalities being equalities? *Hint*: Consider intervals.

Exercise 6.1.2. Prove the following:

1. If S and T are sets of real numbers such that for every $s \in S$ and $t \in T$, $s < t$, then $\sup S \le \inf T$.

2. If $S = \{L(f, P) : P \text{ partitions } [a, b]\}$ and $T = \{U(f, P) : P \text{ partitions } [a, b]\}$, then $s < t$ for every $s \in S$ and $t \in T$.
3. Let S and T satisfy the hypothesis in 1. Show: If for every $\epsilon > 0$ there exist $s \in S$ and $t \in T$ such that $t - s < \epsilon$, then $\sup S = \inf T$.

6.2. The Integral of a Bounded Function

Let $f : [a, b] \to \mathbf{R}$ be a bounded function. For any sequence of partition refinements of $[a, b]$, say $P_k \subset P_{k+1}$ for $k \in \mathbf{N}$, Theorem 6.1.3 tells us that the corresponding sequence $(U(f, P_k))$ of upper sums for f is bounded below (by any lower sum) and the sequence $(L(f, P_k))$ of lower sums for f is bounded above (by any upper sum). Moreover, we have a nested sequence, $([L(f, P_k), U(f, P_k)])$, of closed, nonempty intervals on the number line. The intersection of those intervals must be nonempty, by the nested interval theorem. But the intersection may not be a single point, because it may not be true that $U(f, P_k) - L(f, P_k) \to 0$ as $k \to \infty$.

Example 6.2.1. Let $f : [0, 1] \to \mathbf{R}$ be defined by

$$f(x) = \begin{cases} 0 & \text{if } x \in [0, 1] \cap \mathbf{Q}, \\ 1 & \text{if } x \in [0, 1] \cap \mathbf{I}. \end{cases}$$

This is called the **Dirichlet[1] function**. It is a very useful example. The Dirichlet function has $U(f, P) = 1$ and $L(f, P) = 0$ for every partition P of $[0, 1]$, hence $U(f, P) - L(f, P) = 1$ for every P. △

Theorem 6.1.3 assures us that the infimum of all possible upper sums and the supremum of all possible lower sums must exist, since the collection of upper sums is bounded below, and the collection of lower sums is bounded above.

Definition 6.2.2. Let $f : [a, b] \to \mathbf{R}$ be a bounded function. The **upper Riemann integral of f over $[a, b]$**, denoted $U_a^b(f)$, is defined by

$$U_a^b(f) := \inf\{U(f, P) : P \text{ partitions } [a, b]\}.$$

The **lower Riemann integral of f over $[a, b]$**, denoted $L_a^b(f)$, is defined by

$$L_a^b(f) := \sup\{L(f, P) : P \text{ partitions } [a, b]\}.$$

In view of item 2 of Theorem 6.1.3 and Exercise 6.1.2, we have

$$L_a^b(f) \le U_a^b(f).$$

As the example of the Dirichlet function shows, there may be an unclosable gap between $L_a^b(f)$ and $U_a^b(f)$ for a given f. We wish to single out those functions for which the upper and lower sums can close the gap.

Definition 6.2.3. *A bounded function $f : [a, b] \to \mathbf{R}$ is* **Riemann integrable** *over $[a, b]$ if $L_a^b(f) = U_a^b(f)$, in which case we denote this common value by*

$$\int_a^b f(x)\, dx.$$

[1] P. G. L. Dirichlet, 1805-1859.

The Dirichlet function in Example 6.2.1 is not Riemann integrable over $[0, 1]$, because $U(f, P) = 1$ and $L(f, P) = 0$ for every partition P of $[0, 1]$, and hence $L_a^b(f) = 0 < 1 = U_a^b(f)$.

We consider two more examples, in each case working from the definition of Riemann integrability.

Example 6.2.4. Any constant function $f(x) = c$ is Riemann integrable on $[a, b]$, since it is straightforward to verify that $U(f, P) = L(f, P) = c(b - a)$ for every partition P. △

Example 6.2.5. The step function

$$f(x) = \begin{cases} 1 & \text{if } 0 \le x < 1, \\ 2 & \text{if } 1 \le x \le 2, \end{cases}$$

having one jump discontinuity, at $x = 1$, is Riemann integrable over $[0, 2]$. This is a piecewise-constant function. Consider any partition P that contains 1 (we can always do so by refinement if necessary). Suppose $x_k = 1$ in the ordering of elements of P. Then $M_k - m_k = 1$ and $M_{k+1} - m_{k+1} = 1$, and all other differences between supremum and infimum of f on subintervals of P are zero. Hence, $U(f, P) - L(f, P) = (x_k - x_{k-1}) + (x_{k+1} - x_k) = x_{k+1} - x_{k-1}$. Given any $\epsilon > 0$, we may choose a partition P containing 1 such that $0 < x_{k+1} - x_{k-1} < \epsilon$. It follows from Exercise 6.1.2 (part 3) that $L_0^2(f) = U_0^2(f)$. Therefore f is Riemann integrable over $[0, 2]$. From this, one can see that $\int_0^2 f(x)\,dx = 3$. △

As seen in this example, the criterion from Exercise 6.1.2 (part 3) for the equality $L_a^b(f) = \sup\{L(f, P)\} = \inf\{U(f, P)\} = U_a^b(f)$ provides the following criterion for Riemann integrability.

Theorem 6.2.6. *A function $f : [a, b] \to \mathbf{R}$ is Riemann integrable if and only if for every $\epsilon > 0$ there is a partition P of $[a, b]$ such that*

$$U(f, P) - L(f, P) < \epsilon.$$

Proof. The sets

$$S = \{L(f, P) : P \text{ partitions } [a, b]\} \quad \text{and} \quad T = \{U(f, P) : P \text{ partitions } [a, b]\}$$

satisfy the conditions in Exercise 6.1.2 (part 1).

If f is Riemann integrable, then $L_a^b(f) = U_a^b(f)$. Thus, given $\epsilon > 0$ there are numbers $s \in S$ and $t \in T$ such that $t - s < \epsilon$. The sums s and t may correspond to different partitions P_1 and P_2, but with the refinement $P_1 \cup P_2$ we obtain a lower sum s' and upper sum t' with $s < s' < t' < t$, and hence $t' - s' < \epsilon$.

Conversely, suppose that for every $\epsilon > 0$ there is a partition P of $[a, b]$ such that $U(f, P) - L(f, P) < \epsilon$. Then we may conclude that

$$L_a^b(f) = \sup S = \inf T = U_a^b(f)$$

by Exercise 6.1.2 (part 3). □

By analogy with the Cauchy criterion for sequential convergence, this theorem might be called a Cauchy criterion for integrability, but the result is due to Riemann, who actually used tagged partitions (choosing a point $c_k \in [x_{k-1}, x_k]$ to form sums) instead of Riemann-Darboux sums, which Darboux introduced. So Theorem 6.2.6 is *Riemann's criterion for integrability.*

Given $f : [a, b] \to \mathbf{R}$ and a partition $P = \{x_0, x_1, \ldots, x_n\}$ of $[a, b]$, we have, in the notation of Definition 6.1.2,

$$U(f, P) - L(f, P) = \sum_{k=1}^{n} (M_k - m_k)(x_k - x_{k-1}).$$

Definition 6.2.7. *Let f be a bounded real function on $[c, d]$ with $c < d$. The* **oscillation of f on** $[c, d]$*, denoted $\omega_f([c, d])$, is defined by*

$$\omega_f([c, d]) := \sup_{x \in [c,d]} f(x) - \inf_{x \in [c,d]} f(x).$$

Since the oscillation $M_k - m_k$ of f on a subinterval $[x_{k-1}, x_k]$ determined by a partition occurs frequently in some arguments, it is convenient to use it in the proofs of some properties of the integral. We restate Riemann's criterion for integrability in terms of the oscillation of f on subintervals.

Theorem 6.2.8. *A function $f : [a, b] \to \mathbf{R}$ is Riemann integrable if and only if for every $\epsilon > 0$ there is a partition $P = \{x_0, x_1, \ldots, x_n\}$ of $[a, b]$ such that*

$$\sum_{k=1}^{n} \omega_f([x_{k-1}, x_k])(x_k - x_{k-1}) < \epsilon.$$

Example 6.2.9. The function $f(x) = x$ is Riemann integrable over $[0, 1]$ and $\int_0^1 x\,dx = 1/2$. We verify both assertions. Let $P_n = \{0, \frac{1}{n}, \frac{2}{n}, \ldots, \frac{n-1}{n}, 1\}$. Let $I_k = [\frac{k-1}{n}, \frac{k}{n}]$ for $k = 1, \ldots, n$. Since f is increasing,

$$M_k = \sup_{I_k} f(x) = \frac{k}{n}, \qquad m_k = \inf_{I_k} f(x) = \frac{k-1}{n},$$

and $\omega_f(I_k) = \frac{k}{n} - \frac{k-1}{n} = \frac{1}{n}$. Then

$$U(f, P_n) = \sum_{k=1}^{n} \frac{k}{n}\frac{1}{n} = \frac{1}{n^2}\sum_{k=1}^{n} k = \frac{1}{n^2}\frac{n(n+1)}{2} = \frac{1}{2} + \frac{1}{2n}$$

and

$$L(f, P_n) = \sum_{k=1}^{n} \frac{k-1}{n}\frac{1}{n} = \frac{1}{n^2}\sum_{k=1}^{n}(k-1) = \frac{1}{n^2}\frac{(n-1)n}{2} = \frac{1}{2} - \frac{1}{2n}.$$

It follows that

$$U(f, P_n) - L(f, P_n) = \sum_{k=1}^{n} \omega_f(I_k)\frac{1}{n} = \sum_{k=1}^{n} \frac{1}{n^2} = \frac{1}{n},$$

as is also seen from the expressions for the upper and lower sum. Given $\epsilon > 0$ there is an n such that $1/n < \epsilon$, so the integrability of f over $[0, 1]$ follows from Theorem 6.2.6. Moreover, we have

$$\frac{1}{2} - \frac{1}{2n} = L(f, P_n) \le L_0^1(f) \le U_0^1(f) \le U(f, P_n) = \frac{1}{2} + \frac{1}{2n}$$

for any $n \in \mathbf{N}$. Letting $n \to \infty$, we conclude $\inf_n\{U(f, P_n)\} = \sup_n\{L(f, P_n)\} = \frac{1}{2}$. Given any partition P of $[0, 1]$, we may consider a partition $P \cup P_n$ which refines both P and P_n. So we conclude that $\inf_P\{U(f,P)\} = \sup_P\{L(f,P)\} = \frac{1}{2}$. Thus, $L_0^1(f) = U_0^1(f) = \frac{1}{2}$, which verifies the integrability of f over $[0,1]$ and the value of the integral. We have $\int_a^b f(x)\,dx = \lim_{n\to\infty} L(f,P_n) = \lim_{n\to\infty} U(f,P_n)$. △

In this example, the conditions $U(f,P_{n+1}) \le U(f,P_n)$ and $L(f,P_n) \le L(f,P_{n+1})$, combined with $U(f,P_n) - L(f,P_n) \to 0$ as $n \to \infty$, enabled us to conclude that $\int_a^b f(x)\,dx = \lim_{n\to\infty} L(f,P_n) = \lim_{n\to\infty} U(f,P_n)$.

Exercises.

Exercise 6.2.1. For the bounded function $f : [0,2] \to \mathbf{R}$ defined by

$$f(x) = \begin{cases} \sqrt{x} & \text{if } 0 \le x \le 1, \\ \sin\big(1/(x-1)\big) & \text{if } 1 < x \le 2, \end{cases}$$

show how to define, for each $\epsilon > 0$, a partition P of $[0,2]$ such that $U_0^2(f) - L_0^2(f) < \epsilon$. *Hint*: The only point of discontinuity is $x = 1$. Given $\epsilon > 0$, let $\delta > 0$ be such that $[1-\delta, 1+\delta] \subset (0,2)$. So f is integrable on $[0, 1-\delta]$ and on $[1+\delta, 2]$. If P_1 and P_2 are partitions of $[0, 1-\delta]$ and $[1+\delta, 2]$, then $P := P_1 \cup P_2$ partitions $[0,2]$. Use boundedness of f to show that $\delta > 0$ can be chosen small enough that $U(f,P) - L(f,P) < \epsilon$.

Exercise 6.2.2. Verify in detail the integrability of $f(x) = x^2$ over $[0,1]$ (as in Example 6.2.9), and verify that $\int_0^1 x^2\,dx = 1/3$.

Exercise 6.2.3. Suppose f is defined on $[0,1]$ and integrable over $[\delta, 1]$ for every $0 < \delta < 1$. Does it follow that f is integrable over $[0,1]$?

6.3. Continuous and Monotone Functions

In this section we show that the continuous functions on $[a,b]$ and the monotone functions on $[a,b]$ are Riemann integrable. We also establish the integral test for infinite series with positive terms.

Theorem 6.3.1. *Every continuous function on $[a,b]$ is Riemann integrable on $[a,b]$.*

Proof. Let $f : [a,b] \to \mathbf{R}$ be continuous on $[a,b]$. By Theorem 4.7.4, f is uniformly continuous on $[a,b]$, so for every $\epsilon > 0$ there is a $\delta(\epsilon) > 0$ such that

$$x, y \in [a,b] \text{ and } |x-y| < \delta(\epsilon) \quad \Longrightarrow \quad |f(x) - f(y)| < \frac{\epsilon}{2(b-a)}.$$

Let $P = \{x_0, x_1, \ldots, x_n\}$ be any partition of $[a,b]$ such that

$$\max\{|x_k - x_{k-1}| : k = 1, \ldots, n\} < \delta(\epsilon).$$

Since f is continuous on the compact set $[a,b]$, f assumes a maximum and a minimum value over each of the compact subintervals $[x_{k-1}, x_k]$, $k = 1, \ldots, n$ (Theorem 4.8.2). By the choice of partition, it follows that

$$\max_{x\in[x_{k-1},x_k]} f(x) - \min_{x\in[x_{k-1},x_k]} f(x) \le \frac{\epsilon}{2(b-a)}.$$

Consequently,

$$\begin{aligned} U(f,P) - L(f,P) &\leq \sum_{k=1}^{n} \frac{\epsilon}{2(b-a)}(x_k - x_{k-1}) \\ &= \frac{\epsilon}{2(b-a)}(b-a) < \epsilon. \end{aligned}$$

By Theorem 6.2.6, f is integrable on $[a, b]$. □

Theorem 6.3.1 says that continuity of f on $[a, b]$ is a *sufficient* condition for Riemann integrability on $[a, b]$. The integrability of the step function in Example 6.2.5 shows that continuity on $[a, b]$ is *not necessary* for Riemann integrability. However, the Dirichlet function on $[0, 1]$, which is discontinuous at *every* point in $[0, 1]$, is not Riemann integrable. These examples suggest that some discontinuity is permissible for an integrable function; a precise description of *how much* discontinuity is permissible will be given later.

Recall the definition of monotone function (Definition 4.9.7).

Theorem 6.3.2. *Every monotone function on $[a, b]$ is Riemann integrable on $[a, b]$.*

Proof. We prove this for monotone decreasing f, and leave the monotone increasing case, which is similar, to Exercise 6.3.1. So let f be monotone decreasing on $[a, b]$. It is possible to have $f(a) = f(b)$, but in that case f must be a constant function and hence integrable, so we assume that $f(a) > f(b)$. Let $P = \{x_0, x_1, \ldots, x_n\}$ be a partition of $[a, b]$. Since f is monotone decreasing, we have

$$\omega_f([x_{k-1}, x_k]) = f(x_{k-1}) - f(x_k).$$

Consequently,

$$\begin{aligned} U(f,P) - L(f,P) &= \sum_{k=1}^{n} \omega_f([x_{k-1}, x_k])(x_k - x_{k-1}) \\ &= \sum_{k=1}^{n} (f(x_{k-1}) - f(x_k))(x_k - x_{k-1}). \end{aligned}$$

Taken by itself, the sum of function differences, $\sum_{k=1}^{n}(f(x_{k-1}) - f(x_k))$, equals $f(a) - f(b)$, by monotonicity. Given $\epsilon > 0$, we may choose a partition P such that

$$0 < x_k - x_{k-1} < \frac{\epsilon}{f(a) - f(b)}, \qquad k = 1, \ldots, n,$$

and then

$$\begin{aligned} U(f,P) - L(f,P) &< \frac{\epsilon}{f(a) - f(b)} \sum_{k=1}^{n} (f(x_{k-1}) - f(x_k)) \\ &= \frac{\epsilon}{f(a) - f(b)}(f(a) - f(b)) = \epsilon. \end{aligned}$$

Hence f is integrable over $[a, b]$ by Theorem 6.2.6 (or Theorem 6.2.8). □

Example 6.3.3. The function f defined below is monotone decreasing on $[0, 1]$ and has a countable infinity of discontinuities:

$$f(x) = \begin{cases} 1, & 0 \le x < 1/2, \\ 1/2, & 1/2 \le x < 3/4, \\ 1/4, & 3/4 \le x < 7/8, \\ \cdots & \cdots \\ 1/2^k, & (2^k - 1)/2^k \le x < (2^{k+1} - 1)/2^{k+1}, \\ \cdots & \cdots \\ 0, & x = 1. \end{cases}$$

Theorem 6.3.2 applies, and f is integrable over $[0, 1]$. △

The reader is invited to construct a monotone increasing function on $[a, b]$ having a countably infinite set of discontinuities (Exercise 6.3.2). Such functions are Riemann integrable by Theorem 6.3.2. We recall that any monotone function on $[a, b]$ has at most countably many discontinuities, by Exercise 4.9.2.

How discontinuous can a function be on $[a, b]$ and still be Riemann integrable? The answer to this question is the subject of the following section.

We end this section with the integral test for infinite series with positive terms, which involves monotone functions. The integral test can be viewed as a test for absolute convergence of a series.

There are many series for which the root test and the ratio test do not apply. For example, the harmonic series, $\sum_{k=1}^{\infty} 1/k$, and the series $\sum_{k=1}^{\infty} 1/k^2$, were noted in Exercise 3.9.3. The integral test addresses series with monotone decreasing positive terms. In stating this test, we assume that our summation index starts with $k = 0$; in practice this condition can be relaxed.

Theorem 6.3.4 (Integral Test). *If $a_k \ge 0$ and the sequence (a_k) is monotone decreasing, then $\sum_{k=0}^{\infty} a_k$ converges if and only if the sequence $(\int_0^k f(x)\,dx)$ converges, where $f : [0, \infty) \to \mathbf{R}$ is any monotone decreasing function for which $f(k) = a_k$ for all integers $k \ge 0$.*

Proof. Since f is decreasing, $f(k) \le f(t)$ for $t \in [k - 1, k]$. Consequently, for any n, we have

$$\sum_{k=1}^{n} f(k) \le \sum_{k=1}^{n} \int_{k-1}^{k} f(t)\,dt \le \sum_{k=0}^{n-1} f(k).$$

We also have

$$\int_0^n f(t)\,dt = \sum_{k=1}^{n} \int_{k-1}^{k} f(t)\,dt.$$

If the sequence $\int_0^n f(x)\,dx$ converges, then the sequence of partial sums $\sum_{k=1}^{n} f(k)$, which is monotone increasing, is bounded and therefore converges. Conversely, if the series $\sum_{k=0}^{\infty} f(k)$ converges, then its partial sums, $\sum_{k=1}^{n} \int_{k-1}^{k} f(t)\,dt$, which increase, are bounded and therefore converge. □

Example 6.3.5. Consider the series $\sum_{k=1}^{\infty} 1/\sqrt{k}$. The root test and ratio test are inconclusive. However, the integral test applies and shows that this series diverges. (We may start the indexing with $k = 1$ and define $f(t) = 1/\sqrt{t}$ for $t \in [1, \infty)$, and

$f(k) = a_k$ for all integers $k \geq 1$.) Of course, the comparison $1/\sqrt{k} \geq 1/k$ with the terms of the harmonic series also shows the divergence. The integral test may also be used to show that the p-series

$$\sum_{k=1}^{\infty} \frac{1}{k^p}$$

converges if $p > 1$ and diverges if $0 < p < 1$. (See Exercise 6.3.3.) △

Exercises.

Exercise 6.3.1. Prove Theorem 6.3.2 for the case of monotone increasing functions.

Exercise 6.3.2. Give an example of a monotone increasing function f on $[0, 1]$ having a countable infinity of discontinuities. Can you find $\int_0^1 f(x)\,dx$ for your example?

Exercise 6.3.3. Use the integral test to show that the p-series $\sum_{k=1}^{\infty} 1/k^p$ converges if $p > 1$ and diverges if $0 < p < 1$.

Exercise 6.3.4. Show that the series $\sum_{k=2}^{\infty} 1/[k(\log k)^p]$ converges if $p > 1$ and diverges if $p \leq 1$.

Exercise 6.3.5. Show that series (a) diverges and series (b) converges:

$$\text{(a)} \quad \sum_{k=3}^{\infty} \frac{1}{k \log k \log(\log k)} \qquad\qquad \text{(b)} \quad \sum_{k=3}^{\infty} \frac{1}{k \log k\, [\log(\log k)]^2}.$$

Exercise 6.3.6. Show that the series $\sum_{n=1}^{\infty} 1/n^x$ converges for $x > 1$ and diverges for $x \leq 1$. The function $\zeta(x) := \sum_{n=1}^{\infty} 1/n^x$, $x > 1$, is called the *Riemann zeta function*.

6.4. Lebesgue Measure Zero and Integrability

If J is an interval with endpoints a and b with $a \leq b$, we define the **length**, or **measure**, of J to be $m(J) = b - a$. The following concept will eventually lead us to a characterization of the Riemann integrable functions on $[a, b]$.

Definition 6.4.1. *A subset S of the real numbers has* **Lebesgue measure zero** *if for every $\epsilon > 0$ there is a sequence of open intervals, J_i, such that $S \subset \bigcup_i J_i$ and*

$$\sum_i m(J_i) < \epsilon.$$

The sum $\sum_i m(J_i)$ is the **total length** *(or total measure) of $\{J_i\}$.*

It is easy to verify from this definition that the empty set has Lebesgue measure zero. Here is a much larger class of sets having measure zero:

Lemma 6.4.2. *Every countable subset of* **R** *has Lebesgue measure zero.*

Proof. We consider finite sets and countably infinite sets separately. If S is finite, write $S = \{x_1, \ldots, x_n\}$. Given $\epsilon > 0$, for each i we may cover x_i by

$(x_1 - \epsilon/3n, x_i + \epsilon/3n)$, an interval of length $2\epsilon/3n$. The union of these n open intervals contains S and has total measure

$$\sum_{i=1}^{n} \frac{2\epsilon}{3n} = \frac{2}{3}\epsilon < \epsilon.$$

If S is countable, let $S = \{x_1, x_2, x_3, \ldots\}$ denote an enumeration of S. Given $\epsilon > 0$, we want to cover S by a union of intervals whose total length is less than ϵ. Think of a series of lengths like $\sum_{i=1}^{\infty} 1/2^i = 1$, or better, $\sum_{i=1}^{\infty} 1/2^{i+1} = 1/2$. For each i, we may choose an open interval of length $\epsilon/2^{i+1}$ centered at x_i, and the total length of these intervals will be

$$\sum_{i=1}^{\infty} \frac{\epsilon}{2^{i+1}} = \epsilon \sum_{i=1}^{\infty} \frac{1}{2^{i+1}} = \frac{\epsilon}{2} < \epsilon.$$

Thus, S has measure zero. □

Since the set $\mathbf{Q}$ of rational numbers is countable, $\mathbf{Q}$ has Lebesgue measure zero. With the simple technique in the proof of Lemma 6.4.2, the reader can show that a union of finitely many sets of measure zero has measure zero (Exercise 6.4.1). In addition, one can show that we obtain the same concept of *measure zero* if we use *closed* intervals instead of *open* intervals in the definition (Exercise 6.4.2).

Even more interesting is that there are uncountable sets that have Lebesgue measure zero. The Cantor set is one of them.

Example 6.4.3. Let us show that the Cantor set $C \subset [0, 1]$ (Definition 3.4.1) has measure zero. For each $k \in \mathbf{N}$, $C \subset C_k = (D_k)^c$ by (3.2), and each closed set C_k is the union of 2^k closed intervals, each interval having length $1/3^k$. The total length of these closed intervals covering C_k is therefore $2^k(1/3^k) = (2/3)^k$. (Or cover C_k with 2^k open intervals, each of length $2/3^k$ if you wish, for a total length of $2^k(2/3^k) = 2(2/3)^k$.) For any $\epsilon > 0$, we can choose a k such that $(2/3)^k < \epsilon$ (or $2(2/3)^k < \epsilon$, if you used the open intervals). Therefore C has measure zero, and we know from Theorem 3.4.3 that C is uncountable. △

We now return to the question of characterizing the Riemann integrable functions on an interval $[a, b]$.

Theorem 6.4.4. *A bounded function $f : [a, b] \to \mathbf{R}$ is Riemann integrable on $[a, b]$ if and only if the set of points at which f is discontinuous is a set of Lebesgue measure zero.*

The proof of Theorem 6.4.4 is postponed to a later section, where a general argument will establish a similar statement for real valued functions defined on a generalized rectangle in $\mathbf{R}^n$. (See Theorem 12.5.1.)

In general, we say that a property holds **almost everywhere (a.e.)** on a set $S \subset \mathbf{R}$ if it holds at all points of S with the possible exception of a set $Z \subset S$ having Lebesgue measure zero. For example, the property of continuity of a function holds (or not) at particular points of the domain. A function $f : [a, b] \to \mathbf{R}$ is said to be **continuous almost everywhere (continuous a.e.)** in (or on) $[a, b]$ if the set of discontinuities of f in $[a, b]$ is a set of Lebesgue measure zero. Similarly, a function

g is differentiable a.e. in $[a, b]$ if g' fails to exist only on a set $Z \subset [a, b]$ having measure zero.

In summary, we have the following characterization of Riemann integrable functions: *A bounded function* $f : [a, b] \to \mathbf{R}$ *is Riemann integrable on* $[a, b]$ *if and only if* f *is continuous a.e. in* $[a, b]$.

Example 6.4.5. Let $f : [0, 1] \to \mathbf{R}$ be defined by $f(x) = 0$ if $x = 1/n$, $n \in \mathbf{N}$, and $f(x) = 1$ otherwise. Then f has countably many points of discontinuity, so f is continuous a.e. on $[0, 1]$. △

Example 6.4.6. If f is integrable on $[a, b]$, then f^2, defined by $f^2(x) = [f(x)]^2$ for $x \in [a, b]$, is also integrable. Notice that the set of discontinuities of f^2 must be contained in the set of discontinuities of f. Since f is integrable, f is continuous a.e. in $[a, b]$, hence f^2 is continuous a.e. in $[a, b]$ and therefore f^2 is integrable. △

Exercises.

Exercise 6.4.1. Show that a finite union of sets of measure zero has measure zero. *Hint*: See the proof of Lemma 6.4.2.

Exercise 6.4.2. Show that a subset $S \subset \mathbf{R}$ has Lebesgue measure zero if and only if for every $\epsilon > 0$ there is a countable collection $\{K_i\}$ of *closed* intervals such that $S \subset \bigcup_i K_i$ and $\sum_i m(K_i) < \epsilon$.

Exercise 6.4.3. Show that the set of all real algebraic numbers (that is, roots of polynomials having rational coefficients) has measure zero.

Exercise 6.4.4. Evaluate $\int_0^1 f(x)\,dx$ for the function f of Example 6.4.5.

Exercise 6.4.5. Prove: If f and g are integrable on $[a, b]$, then fg is integrable on $[a, b]$. *Hint*: $(f + g)^2 = f^2 + 2fg + g^2$.

Exercise 6.4.6. Prove: If f and g are integrable on $[a, b]$ and $g(x) \geq m > 0$ for some m, then f/g is integrable on $[a, b]$.

Exercise 6.4.7. Suppose $f(x) \geq 0$ for $x \in [a, b]$ and f is integrable on $[a, b]$. Show that $\sqrt{f}$ is integrable on $[a, b]$.

Exercise 6.4.8. Give an example of a function $f : [a, b] \to \mathbf{R}$ such that f^2 is integrable but f is not integrable.

6.5. Properties of the Integral

Henceforward in the text, until the Lebesgue integral has been introduced, if a function is Riemann integrable we will simply say that it is integrable. The first properties considered here involve the restriction of integration to a subinterval of a known domain of integrability $[a, b]$, and the splitting of a domain of integration $[a, b]$ into disjoint intervals of integration whose union is $[a, b]$.

Theorem 6.5.1. *If* f *is integrable over* $[a, b]$, *then* f *is integrable over any subinterval* $[c, d] \subset [a, b]$.

Proof. Intuitively, this result is fairly clear: Any partition of $[a, b]$ that contains the points c and d induces a partition (contains a partition) of $[c, d]$. Moreover, the difference between the upper and lower sums over $[a, b]$ is greater than or equal to the difference between upper and lower sums for the induced partition of $[c, d]$. If the former is less than ϵ, then so must be the latter. We write this out in detail.

Given $\epsilon > 0$, there is a partition $P = \{x_0, x_1, \ldots, x_n\}$ of $[a, b]$ such that

$$U(f, P) - L(f, P) < \epsilon.$$

We may assume that $c = x_i$ and $d = x_j$ for some $i < j$; otherwise, we may consider the refinement $P' = P \cup \{c, d\}$, for which we must have $U(f, P') - L(f, P') \leq U(f, P) - L(f, P) < \epsilon$, since $L(f, P) \leq L(f, P') \leq U(f, P') \leq U(f, P)$. With that assumption, we have that

$$c = x_i < x_{i+1} < \cdots < x_j = d$$

is a partition of $[c, d]$. Let us write S and s for the upper and lower sums corresponding to this partition of $[c, d]$. Then

$$\begin{aligned} U(f, P) - L(f, P) &= \sum_{k=1}^{n} \omega_f([x_{k-1}, x_k])(x_k - x_{k-1}) \\ &= S - s + \sum_{k=1}^{i} \omega_f([x_{k-1}, x_k])(x_k - x_{k-1}) \\ &\quad + \sum_{k=j+1}^{n} \omega_f([x_{k-1}, x_k])(x_k - x_{k-1}) \\ &\geq S - s. \end{aligned}$$

Hence, $S - s < \epsilon$, and since ϵ is arbitrary, f is integrable over $[c, d]$. □

Theorem 6.5.2. *If $c \in [a, b]$ and f is integrable over both $[a, c]$ and $[c, b]$, then f is integrable over $[a, b]$, and*

$$\int_a^b f(x)\, dx = \int_a^c f(x)\, dx + \int_c^b f(x)\, dx.$$

Proof. Given $\epsilon > 0$, there is a partition $P^1 = \{x_0 = a, x_1, \ldots, x_j = c\}$ of $[a, c]$ such that

$$\sum_{k=1}^{j} \omega_f([x_{k-1}, x_k])(x_k - x_{k-1}) < \frac{\epsilon}{2},$$

and a partition $P^2 = \{x_j = c, x_{j+1}, \ldots, x_n = b\}$ of $[c, b]$ such that

$$\sum_{k=j+1}^{n} \omega_f([x_{k-1}, x_k])(x_k - x_{k-1}) < \frac{\epsilon}{2}.$$

Then $P^1 \cup P^2 = \{x_0 = a, x_1, \ldots, x_j = c, x_{j+1}, \ldots, x_n = b\}$ is a partition of $[a, b]$, and

$$\sum_{k=1}^{n} \omega_f([x_{k-1}, x_k])(x_k - x_{k-1}) < \frac{\epsilon}{2} + \frac{\epsilon}{2} = \epsilon.$$

This proves that f is integrable over $[a, b]$ by Theorem 6.2.8.

In order to establish the addition formula, we consider a sequence of partitions (P_m^1) for $[a, c]$ and (P_m^2) for $[c, b]$, with notation as above, such that

$$\int_a^c f(x)\,dx = \lim_{m\to\infty} U(f, P_m^1) = \lim_{m\to\infty} L(f, P_m^1)$$

and

$$\int_c^b f(x)\,dx = \lim_{m\to\infty} U(f, P_m^2) = \lim_{m\to\infty} L(f, P_m^2).$$

Then $(P_m^1 \cup P_m^2)$ is a sequence of partitions of $[a, b]$, and $U(f, P_m^1 \cup P_m^2) = U(f, P_m^1) + U(f, P_m^2)$, $L(f, P_m^1 \cup P_m^2) = L(f, P_m^1) + L(f, P_m^2)$. Letting $m \to \infty$ and using the integrability of f over $[a, b]$, we conclude that the addition formula holds. $\square$

Let f be integrable over $[a, b]$. For any subinterval $[\alpha, \beta] \subseteq [a, b]$ we define

$$\int_\beta^\alpha f(x)\,dx = -\int_\alpha^\beta f(x)\,dx.$$

We also define

$$\int_\alpha^\alpha f(x)\,dx = 0$$

for any α. Then for any numbers $\alpha, \beta, \gamma \in [a, b]$, one can verify that the formula

$$\int_\alpha^\beta f(x)\,dx + \int_\beta^\gamma f(x)\,dx = \int_\alpha^\gamma f(x)\,dx \tag{6.1}$$

holds.

Definition 6.5.3. *Let f be Riemann integrable over $[a, b]$ and let c be a point in $[a, b]$. The function*

$$F_c(x) = \int_c^x f(t)\,dt$$

is called an **indefinite integral** *of f on $[a, b]$.*

An indefinite integral F_c of f is an antiderivative of f, since $F_c'(x) = f(x)$ by the fundamental theorem to be established later. For different values of c, we obtain different indefinite integrals F_c of f. For a value $\hat{c}$ we have

$$F_{\hat{c}}(x) = \int_{\hat{c}}^x f(t)\,dt.$$

By the addition formula (6.1), we have

$$F_c(x) - F_{\hat{c}}(x) = \int_c^x f(t)\,dt - \int_{\hat{c}}^x f(t)\,dt = \int_c^x f(t)\,dt + \int_x^{\hat{c}} f(t)\,dt = \int_c^{\hat{c}} f(t)\,dt.$$

Thus, any two indefinite integrals of f differ by a constant.

See Exercise 6.5.1 for some basic monotonicity properties of the integral.

Next, we consider the linearity of the integral.

Theorem 6.5.4. *The integral is a linear function from the set of all Riemann integrable functions on $[a, b]$ into the set of real numbers; that is, the following* *properties hold:*

1. *If f is integrable over $[a,b]$, then so is αf for any real α, and*
$$\int_a^b \alpha f(x)\,dx = \alpha \int_a^b f(x)\,dx.$$
2. *If f and g are integrable over $[a,b]$, then so is $f+g$, and*
$$\int_a^b (f(x)+g(x))\,dx = \int_a^b f(x)\,dx + \int_a^b g(x)\,dx.$$

Proof. If f and g are continuous at x, then $f+g$ is continuous at x. Thus, the set of discontinuities of $f+g$ is contained in the union of the sets of discontinuities of f and g. The set of discontinuities of αf is contained in the set of discontinuities of f. It follows that if f and g are integrable over $[a,b]$, then $f+g$ and αf are continuous a.e. in $[a,b]$, and therefore $\int_a^b (f(x)+g(x))\,dx$ and $\int_a^b \alpha f(x)\,dx$ exist.

For any integrable f and partitions P of $[a,b]$, we have
$$\inf_P U(-f,P) = -\sup_P L(f,P),$$
and it follows immediately that

$$\int_a^b -f(x)\,dx = -\int_a^b f(x)\,dx. \tag{6.2}$$

For any $\alpha > 0$ and any partition $P = \{x_0, x_1, \ldots, x_{n-1}, x_n\}$ of $[a,b]$, we have
(6.3)
$$\sup_{[x_{k-1},x_k]} \alpha f(x) = \alpha \sup_{[x_{k-1},x_k]} f(x) \qquad \text{and} \qquad \inf_{[x_{k-1},x_k]} \alpha f(x) = \alpha \inf_{[x_{k-1},x_k]} f(x)$$
for $k = 1, \ldots, n$. Hence, $U(\alpha f, P) = \alpha U(f,P)$ and $L(\alpha f, P) = \alpha L(f,P)$ for $\alpha > 0$. By (6.2), (6.3) and Theorem 2.2.5, we have
$$\int_a^b \alpha f(x)\,dx = \alpha \int_a^b f(x)\,dx$$
for any real α.

If f and g are integrable over $[a,b]$ and $P = \{x_0, x_1, \ldots, x_{n-1}, x_n\}$ is a partition of $[a,b]$, then by Theorem 2.2.5,
$$\sup_{[x_{k-1},x_k]} (f(x)+g(x)) \le \sup_{[x_{k-1},x_k]} f(x) + \sup_{[x_{k-1},x_k]} g(x), \quad \text{for } k = 1, \ldots, n,$$
and
$$\inf_{[x_{k-1},x_k]} (f(x)+g(x)) \ge \inf_{[x_{k-1},x_k]} f(x) + \inf_{[x_{k-1},x_k]} g(x), \quad \text{for } k = 1, \ldots, n.$$
Thus, for any partition P,
$$U(f+g,P) \le U(f,P) + U(g,P) \tag{6.4}$$
and
$$L(f+g,P) \ge L(f,P) + L(g,P). \tag{6.5}$$

Now for any $\epsilon > 0$, there are partitions P_1 and P_2 such that

$$U(f, P_1) - \frac{\epsilon}{2} < \int_a^b f(x)\,dx \qquad \text{and} \qquad U(g, P_2) - \frac{\epsilon}{2} < \int_a^b g(x)\,dx.$$

If $Q = P_1 \cup P_2$, then (6.4) implies

$$\begin{aligned}\int_a^b (f(x) + g(x))\,dx - \epsilon \quad &< \quad U(f+g, Q) - \epsilon \\ &\leq \quad U(f, Q) + U(g, Q) - \epsilon < \int_a^b f(x)\,dx + \int_a^b g(x)\,dx,\end{aligned}$$

and hence

$$\int_a^b (f(x) + g(x))\,dx < \int_a^b f(x)\,dx + \int_a^b g(x)\,dx + \epsilon. \tag{6.6}$$

A similar argument using (6.5) shows that for every $\epsilon > 0$,

$$\int_a^b (f(x) + g(x))\,dx > \int_a^b f(x)\,dx + \int_a^b g(x)\,dx - \epsilon. \tag{6.7}$$

Since $\epsilon > 0$ is arbitrary in (6.6) and (6.7), this completes the proof of the second statement of the theorem. □

Exercises.

Exercise 6.5.1. Prove the following statements:

1. If f and g are integrable over $[a, b]$ and $f(x) \leq g(x)$ for all $x \in [a, b]$, then $\int_a^b f(x)\,dx \leq \int_a^b g(x)\,dx$.
2. If f is integrable over $[a, b]$, then so is $|f|$, and $|\int_a^b f(x)\,dx| \leq \int_a^b |f(x)|\,dx$.

Exercise 6.5.2. Show that parts 1 and 2 of Theorem 6.5.4, taken together, are equivalent to the statement that if f and g are integrable over $[a, b]$, then so is $\alpha f + \beta g$ for all real α and β, and

$$\int_a^b (\alpha f(x) + \beta g(x))\,dx = \alpha \int_a^b f(x)\,dx + \beta \int_a^b g(x)\,dx.$$

Exercise 6.5.3. Supply the details to establish (6.7) in the proof of additivity of the integral in Theorem 6.5.4.

6.6. Integral Mean Value Theorems

Consider the positive valued function $f(x) = x$ on $[0, 3]$. The area of the triangular region between the graph of f and the interval $[0, 3]$ along the x-axis is determined by geometry to equal

$$\text{area} = \frac{1}{2}(\text{base})\ (\text{height}) = \frac{1}{2}(3)(3) = \frac{9}{2}.$$

This can be viewed as the area of a rectangle of base length 3 and height equal to the function value at a specific point, namely, at $x = c = 3/2$. That is, area $=$ $(3)(f(3/2)) = (3)(3/2) = 9/2$. We generalize this geometrically evident fact in the next theorem.

Theorem 6.6.1 (First Mean Value Theorem for Integrals). *If $f : [a, b] \to \mathbf{R}$ is continuous on $[a, b]$, then there is a point $c \in [a, b]$ such that*

$$\int_a^b f(x)\, dx = f(c)(b - a).$$

Proof. Since f is continuous on the closed interval $[a, b]$, f attains its minimum and maximum values on that interval. Thus there exist numbers $m = \min_{[a,b]} f(x)$ and $M = \max_{[a,b]} f(x)$ such that

$$m \le f(x) \le M \quad \text{for all } x \in [a, b].$$

Integrating from a to b preserves the inequality. Hence,

$$m(b - a) \le \int_a^b f(x)\, dx \le M(b - a).$$

By the intermediate value theorem for continuous functions, f takes on every value between m and M. Hence, there is a point $c \in [a, b]$ such that

$$f(c) = \frac{1}{b - a} \int_a^b f(x)\, dx,$$

as we wished to show. □

We define the **average value** of f on the interval $[a, b]$ by

$$\frac{1}{b - a} \int_a^b f(x)\, dx.$$

Then the first mean value theorem for integrals says that a continuous function assumes its average value on $[a, b]$ at some point $c \in [a, b]$.

The first mean value theorem resulted from an integration of the inequality $m \le f(x) \le M$ from a to b. We preserve this inequality if we multiply it by any nonnegative function. This leads to the second mean value theorem for integrals.

Theorem 6.6.2 (Second Mean Value Theorem for Integrals). *If f is continuous on $[a, b]$ and g is integrable with $g(x) \ge 0$ for $x \in [a, b]$, then there is a point $c \in [a, b]$ such that*

$$\int_a^b f(x)g(x)\, dx = f(c) \int_a^b g(x)\, dx.$$

Proof. Since $g(x) \ge 0$ for $x \in [a, b]$, we may write

$$mg(x) \le f(x)g(x) \le Mg(x) \quad \text{for all } x \in [a, b],$$

where $m = \min_{[a,b]} f(x)$ and $M = \max_{[a,b]} f(x)$. An integration from a to b gives

$$m \int_a^b g(x)\, dx \le \int_a^b f(x)g(x)\, dx \le M \int_a^b g(x)\, dx.$$

If $\int_a^b g(x)\, dx \ne 0$, then

$$m \le \frac{\int_a^b f(x)g(x)\, dx}{\int_a^b g(x)\, dx} \le M,$$

and by the intermediate value theorem applied to f there is a point $c \in [a, b]$ such that

$$f(c) = \frac{\int_a^b f(x)g(x)\,dx}{\int_a^b g(x)\,dx},$$

thus proving the theorem. On the other hand, if $\int_a^b g(x)\,dx = 0$, then we must have $\int_a^b f(x)g(x)\,dx = 0$, in which case the conclusion of the theorem holds for any choice of the point c. □

Exercises.

Exercise 6.6.1. Show: If f is integrable over $[a, b]$ and $m = \inf_{a \le x \le b} f(x)$, $M = \sup_{a \le x \le b} f(x)$, then there exists a number μ with $m \le \mu \le M$ such that

$$\int_a^b f(x)\,dx = \mu(b - a).$$

Exercise 6.6.2. Show: If $f, g : [a, b] \to \mathbf{R}$ are integrable over $[a, b]$, $g(x) \ge 0$ for $x \in [a, b]$, and $m = \inf_{a \le x \le b} f(x)$, $M = \sup_{a \le x \le b} f(x)$, then there is a number $m \le \mu \le M$ such that

$$\int_a^b f(x)g(x)\,dx = \mu \int_a^b g(x)\,dx.$$

Exercise 6.6.3. Show that if g is continuous on $[a, b]$ and $\int_a^b g(x)\,dx = 0$, then $g(x) \equiv 0$ on $[a, b]$. *Hint*: Apply Theorem 6.6.2 with $f = g$. Then argue by contradiction.

6.7. The Fundamental Theorem of Calculus

We begin by showing that every indefinite integral of f on $[a, b]$ is uniformly continuous there.

Theorem 6.7.1. *If f is Riemann integrable over $[a, b]$, then any indefinite integral of f is uniformly continuous on $[a, b]$.*

Proof. An indefinite integral of f on $[a, b]$ has the form $F(x) = \int_c^x f(t)\,dt$ for some point c in $[a, b]$ (Definition 6.5.3). Since f is Riemann integrable, it is bounded, so that for some M, $|f(x)| \le M$ for $x \in [a, b]$. Let x and y be in $[a, b]$. By the definition of F and the addition formula (6.1),

$$|F(x) - F(y)| = \left| \int_c^x f(t)\,dt - \int_c^y f(t)\,dt \right| = \left| \int_y^x f(t)\,dt \right| \le M|x - y|.$$

Given any $\epsilon > 0$, if $x, y \in [a, b]$ with $|x - y| < \delta := \epsilon/M$, then $|F(x) - F(y)| < \epsilon$. This shows that F is uniformly continuous on $[a, b]$. □

With the stronger hypothesis of continuity of f, an indefinite integral is differentiable and its derivative is equal to f.

Theorem 6.7.2 (Fundamental Theorem of Calculus I). *If f is continuous on $[a, b]$, then for any c in $[a, b]$, the indefinite integral*

$$F_c(x) := \int_c^x f(t)\,dt$$

is differentiable on $[a, b]$ and $F_c'(x) = f(x)$ for all $x \in [a, b]$.

Proof. Let x_0 and x be in $[a, b]$. By the first mean value theorem for integrals of continuous functions, there is a point $\xi = \xi_x$ between x_0 and x such that

$$F_c(x) - F_c(x_0) = \int_c^x f(t)\,dt - \int_c^{x_0} f(t)\,dt = \int_{x_0}^x f(t)\,dt = f(\xi)(x - x_0).$$

As x approaches x_0, the point $\xi = \xi_x$ approaches x_0. Since f is continuous at x_0, we have

$$F_c'(x_0) = \lim_{x \to x_0} \frac{F_c(x) - F_c(x_0)}{x - x_0} = \lim_{x \to x_0} \frac{f(\xi)(x - x_0)}{x - x_0} = \lim_{x \to x_0} f(\xi) = f(x_0).$$

This is true for each x_0 in $[a, b]$ (with one-sided limits for the points $x_0 = a$ and $x_0 = b$), and this completes the proof. $\square$

The next example shows that if we relax the continuity hypothesis on f to Riemann integrability, then the conclusion of Theorem 6.7.2 need not hold.

Example 6.7.3. An indefinite integral of a Riemann integrable function need not be differentiable on $[a, b]$. Consider the function

$$f(x) = \begin{cases} 0 & \text{for } 0 \le x \le 1, \\ 1 & \text{for } 1 < x \le 2, \end{cases}$$

with a jump discontinuity at $x = 1$. Then the indefinite integral $F(x) = \int_0^x f(t)\,dt$ is given by

$$F(x) = \begin{cases} 0 & \text{for } 0 \le x \le 1, \\ x - 1 & \text{for } 1 < x \le 2, \end{cases}$$

and F is continuous on $[0, 2]$. However, $F'(1)$ does not exist. $\triangle$

It can happen that an indefinite integral of a Riemann integrable f may be differentiable on $[a, b]$ but its derivative may not equal $f(x)$ for all $x \in [a, b]$.

Example 6.7.4. Consider the function f on $[0, 1]$ defined by $f(x) = 0$ for $x \neq 1/3$ and $f(1/3) = 2$. Then f is discontinuous at $x = 1/3$, but f is integrable and $F(x) = \int_0^x f(t)\,dt$ is differentiable, since $F(x) = 0$ for all x in $[0, 1]$. Hence, $F'(x) = 0$ for all x in $[0, 1]$, but then $F'(1/3) \neq f(1/3)$. Note also that f does not have the intermediate value property, so f cannot equal the derivative of any function on $[0, 1]$. $\triangle$

Theorem 6.7.5 (Fundamental Theorem of Calculus II). *Let $f : [a, b] \to \mathbf{R}$ be Riemann integrable over $[a, b]$. If $F : [a, b] \to \mathbf{R}$ is continuous on $[a, b]$, differentiable on (a, b), and $F'(x) = f(x)$ for all $x \in (a, b)$, then*

$$\int_a^b f(x)\,dx = F(b) - F(a) =: F(x)\Big|_a^b.$$

Proof. Let $P = \{x_0 = a, x_1, x_2, \ldots, x_{n-1}, x_n = b\}$ be a partition of $[a, b]$. Then we may use a telescoping sum to write

$$F(b) - F(a) = \sum_{k=1}^{n} [F(x_k) - F(x_{k-1})].$$

Since F is continuous on $[a, b]$ and differentiable on (a, b), we may apply the mean value theorem on each subinterval $[x_{k-1}, x_k]$, and conclude that there exist points ξ_k with $x_{k-1} < \xi_k < x_k$ such that

$$F(x_k) - F(x_{k-1}) = F'(\xi_k)(x_k - x_{k-1}) = f(\xi_k)(x_k - x_{k-1}),$$

for $1 \le k \le n$. Then

$$F(b) - F(a) = \sum_{k=1}^{n} f(\xi_k)(x_k - x_{k-1}),$$

and consequently, the lower and upper Riemann sums for f associated with P satisfy

$$L(f, P) \le F(b) - F(a) \le U(f, P).$$

Since this is true for each partition P of $[a, b]$, and f is integrable over $[a, b]$, we have

$$\int_a^b f(x)\,dx = \sup_P\{L(f, P)\} = \inf_P\{U(f, P)\} = F(b) - F(a),$$

and the theorem is proved. □

It is possible to have F continuous on $[a, b]$ and differentiable on (a, b), but $f(x) := F'(x)$ is not bounded on (a, b), and thus $f = F'$ is not Riemann integrable. In such a case, the evaluation conclusion of Theorem 6.7.5 cannot hold.

Example 6.7.6. Let $F(x) = x^2 \sin(1/x^2)$ for $x > 0$, and let

$$F(0) = \lim_{x \to 0} \left(x^2 \sin \frac{1}{x^2} \right) = 0.$$

Then F is continuous on $[0, 1]$ and differentiable on $(0, 1)$. However,

$$F'(x) = 2x \sin\left(\frac{1}{x^2}\right) - \frac{2}{x} \cos\left(\frac{1}{x^2}\right) \qquad \text{for } x \neq 0,$$

for $x > 0$. Hence, $f := F'$ is unbounded on $(0, 1)$ and therefore not Riemann integrable. The evaluation conclusion of Theorem 6.7.5 does not hold. △

We needed continuity of F on the closed interval $[a, b]$ in Theorem 6.7.5 so that we could apply the mean value theorem. What if all the other hypotheses hold, except continuity at one of the endpoints? For example, take $f : [0, 1] \to \mathbf{R}$ to be $f(x) \equiv 1$ and define $F : [0, 1] \to \mathbf{R}$ by $F(x) = x$ for $0 \le x < 1$ and $F(1) = 0$. Then $F'(x) = f(x)$ for all $x \in (0, 1)$, but F is not continuous on $[0, 1]$, and we have

$$F(1) - F(0) = 0 \neq \int_0^1 f(x)\,dx = 1.$$

However, $\lim_{x \to 1} F(x) = 1$ and

$$\int_0^1 f(x)\,dx = 1 = \lim_{x \to 1-} F(x) - \lim_{x \to 0+} F(x).$$

This example suggests the slightly stronger version of Theorem 6.7.5 in Exercise 6.7.1.

There are two important methods of integration that follow from the fundamental theorem of calculus. First, we have the *change of variables formula* for integrals, also known as *integration by substitution*.

Theorem 6.7.7 (Change of Variables). *Let $g : [\alpha, \beta] \to [a, b]$ be continuous on $[\alpha, \beta]$ and suppose that g' exists and is continuous on $[\alpha, \beta]$, and $g(\alpha) = a$, $g(\beta) = b$. If f is continuous on $[a, b]$, then*

$$\int_\alpha^\beta f(g(t))g'(t)\,dt = \int_a^b f(u)\,du.$$

Proof. Define

$$h(u) = \int_a^u f(s)\,ds.$$

By Theorem 6.7.2, h is differentiable on $[a, b]$ and $h'(u) = f(u)$ for $u \in [a, b]$. Theorem 6.7.5 applies, yielding

$$h(b) - h(a) = \int_a^b f(u)\,du.$$

The continuous function $F(t) = h(g(t))$ satisfies

$$F'(t) = h'(g(t))g'(t) = f(g(t))g'(t)$$

for $t \in [\alpha, \beta]$. Since f, g and g' are continuous, $F'(t)$ is continuous on $[\alpha, \beta]$, and hence by Theorem 6.7.5,

$$\int_\alpha^\beta f(g(t))g'(t)\,dt = \int_\alpha^\beta F'(t)\,dt = F(\beta) - F(\alpha).$$

Since

$$F(\beta) - F(\alpha) = h(g(\beta)) - h(g(\alpha)) = h(b) - h(a),$$

this completes the proof. □

Next, we have the method of *integration by parts*.

Theorem 6.7.8 (Integration by Parts). *Suppose f and g are differentiable on $[a, b]$ and f', g' are Riemann integrable over $[a, b]$. Then*

$$\int_a^b f(x)g'(x)\,dx = f(x)g(x)\Big|_a^b - \int_a^b g(x)f'(x)\,dx.$$

Proof. By the hypotheses on f and g, both fg' and gf' are Riemann integrable over $[a, b]$. By the product rule for differentiation,

$$\frac{d}{dx}(f(x)g(x)) = f'(x)g(x) + f(x)g'(x),$$

and fg is continuous on $[a, b]$. Hence, Theorem 6.7.5 implies that

$$\int_a^b (f'(x)g(x) + f(x)g'(x))\,dx = f(x)g(x)\Big|_a^b$$

and a rearrangement yields the desired formula. □

The properties developed to this point allow us to define the natural logarithm function, denoted log.

Theorem 6.7.9. *The function* $\log : (0, \infty) \to \mathbf{R}$ *defined by*

$$\log x = \int_1^x \frac{1}{t}\, dt, \qquad x > 0,$$

and called the **natural logarithm**, *satisfies the following properties:*

1. $\log(xy) = \log x + \log y$ *for all* $x, y > 0$;
2. $\log(1/x) = -\log x$ *for all* $x > 0$;
3. $\log$ *is differentiable and* $\log'(x) = 1/x$ *for all* $x > 0$;
4. $\log$ *is strictly increasing and has range* $\mathbf{R}$, *and the inverse function* $\log^{-1}$ *is differentiable, with* $[\log^{-1}]'(y) = \log^{-1}(y)$ *for all* $y \in \mathbf{R}$.

Remark. *We usually write* $\exp = \log^{-1}$, *and thus* $\exp'(y) = \exp(y)$ *for all real* y. *Further properties of the function* exp *are discussed in Section* 7.5.

Proof. We first prove statement 1. If $x, y > 0$ and $x > 1$, let $g : [y, xy] \to [1, x]$ be $g(t) = t/y$. By Theorem 6.7.7, with $f(s) = 1/s$, we have

$$\int_1^x \frac{1}{s}\, ds = \int_y^{xy} f(g(t))g'(t)\, dt = \int_y^{xy} \frac{1}{t/y}\frac{1}{y}\, dt = \int_y^{xy} \frac{1}{t}\, dt,$$

from which it follows that $\log x = \log xy - \log y$, as we wanted. If $x, y > 0$ and $x < 1$, let $g : [xy, y] \to [x, 1]$ be $g(t) = t/y$; then

$$\int_x^1 \frac{1}{s}\, ds = \int_{xy}^{y} f(g(t))g'(t)\, dt = \int_{xy}^{y} \frac{1}{t}\, dt,$$

from which it follows that $-\log x = \log y - \log xy$, again as desired. Thus statement 1 holds. From the definition, $\log 1 = 0$. Now statement 2 follows from 1, since

$$0 = \log 1 = \log[x(1/x)] = \log x + \log 1/x \quad \Longrightarrow \quad \log 1/x = -\log x.$$

Theorem 6.7.2 implies statement 3, and $\log'(x) = 1/x$ for $x > 0$ implies that log is strictly increasing on $(0, \infty)$. For the proof of statement 4, note that on the interval $[k, k+1]$, $1/t \geq 1/(k+1)$, so for integers $n \geq 2$,

$$\int_1^n \frac{1}{t}\, dt \geq \frac{1}{2} + \cdots + \frac{1}{n} = \sum_{k=1}^{n-1} \frac{1}{k+1}.$$

Since the series $\sum_{k=1}^{\infty} 1/(k+1)$ diverges, and since log is increasing and continuous on $(0, \infty)$, it follows that $\log x \to +\infty$ as $x \to +\infty$. Exercise 6.7.6 shows that $\log x \to -\infty$ as $x \to 0+$, and hence the range of log is $\mathbf{R}$. For each $y \in \mathbf{R}$, we have, for $x = \log^{-1}(y)$,

$$[\log^{-1}]'(y) = \frac{1}{\log'(x)} = \frac{1}{1/x} = x = \log^{-1}(y),$$

and this completes the proof of statement 4. $\square$

Later, using the theory of power series, we show that for each real number x, $\log^{-1}(x) = \exp(x)$ is actually the sum of the convergent series

$$\sum_{k=0}^{\infty} \frac{1}{k!} x^k.$$

In view of Theorem 3.5.1, we see that the Euler number e equals $\exp(1)$.

Example 6.7.10. We evaluate $\int_0^{\pi/4} \tan t \log(\cos t)\, dt$. Let $f(u) = u$ and $g(t) = \log(\cos t)$. Then $g'(t) = -\sin t/\cos t = -\tan t$, and therefore

$$\int_0^{\pi/4} \tan t \log(\cos t)\, dt = -\int_0^{\pi/4} f(g(t))g'(t)\, dt = -\int_0^{\pi/4} g(t)g'(t)\, dt.$$

The required antiderivative is $F(t) = [\log(\cos t)]^2/2$, with $F'(t) = g(t)g'(t)$, and F is continuous on $[0, \pi/4]$. Thus,

$$-\int_0^{\pi/4} g(t)g'(t)\, dt = -\Big([\log(1/\sqrt{2})]^2/2 - [\log 1]^2/2\Big) = -\frac{1}{2}[\log\sqrt{2}]^2.$$

Let $u = g(t) = \log(\cos t)$, so that $du = -\tan t\, dt$, and hence

$$\int_0^{\pi/4} \tan t \log(\cos t)\, dt = -\int_0^{\log(1/\sqrt{2})} u\, du = -\frac{u^2}{2}\Big|_0^{\log(1/\sqrt{2})}.$$

The u-substitution is indeed a helpful device. △

Example 6.7.11. We apply Theorem 6.7.8 to evaluate $\int_0^{\pi} e^x \cos x\, dx$. Let $f(x) = e^x$, $g'(x) = \cos x$. Then

$$\int_0^{\pi} e^x \cos x\, dx = e^x \sin x\big|_0^{\pi} - \int_0^{\pi} e^x \sin x\, dx.$$

Applying integration by parts once more to the integral $\int_0^{\pi} e^x(-\sin x)\, dx$ on the right side, with $f(x) = e^x$ and $g'(x) = -\sin x$, we have

$$\int_0^{\pi} e^x \cos x\, dx = e^x \sin x\Big|_0^{\pi} + e^x \cos x\Big|_0^{\pi} - \int_0^{\pi} e^x \cos x\, dx.$$

Let us denote the quantity we want by $I := \int_0^{\pi} e^x \cos x\, dx$, which satisfies

$$I = e^x \sin x\Big|_0^{\pi} + e^x \cos x\Big|_0^{\pi} - I,$$

so we have $2I = -(e^{\pi} + 1)$, and hence $I = -(e^{\pi} + 1)/2$. △

Exercises.

Exercise 6.7.1. Prove: If $f : [a, b] \to \mathbf{R}$ is Riemann integrable over $[a, b]$, and the function $F : (a, b) \to \mathbf{R}$ is differentiable on (a, b), $\lim_{x\to b-} F(x)$ and $\lim_{x\to a+} F(x)$ exist, and $F'(x) = f(x)$ for all $x \in (a, b)$, then

$$\int_a^b f(x)\, dx = \lim_{x\to b-} F(x) - \lim_{x\to a+} F(x).$$

Exercise 6.7.2. Apply Theorem 6.7.7 to evaluate these integrals:

(a) $\displaystyle\int_e^{e^2} \frac{\log(t+1)}{t+1}\, dt$ (b) $\displaystyle\int_0^1 2t\sqrt{2 - t^2}\, dt.$

Exercise 6.7.3. Apply Theorem 6.7.7 to evaluate:

$$\int_1^2 \frac{\cos(\log x)}{x}\,dx.$$

Exercise 6.7.4. Apply Theorem 6.7.8 to evaluate these integrals:

$$\text{(a)} \int_0^1 \sqrt{1+x^2}\,dx \qquad\qquad \text{(b)} \int_0^1 \tan^{-1} x\,dx.$$

Exercise 6.7.5. Apply Theorem 6.7.8 to evaluate these integrals:

$$\text{(a)} \int_0^1 \sin^{-1} x\,dx \qquad\qquad \text{(b)} \int_0^{\pi/2} x\cos x\,dx.$$

Exercise 6.7.6. Complete the proof that the range of log is **R** by showing that $\log x \to -\infty$ as $x \to 0+$.

6.8. Taylor's Theorem with Integral Remainder

Recall that the degree n Taylor polynomial of f at a is defined by

$$\sum_{k=0}^{n} \frac{f^{(k)}(a)}{k!}h^k = f(a) + f'(a)h + \frac{f''(a)}{2!}h^2 + \cdots + \frac{f^{(n)}(a)}{n!}h^n,$$

where $h = x - a$. We have seen already in Taylor's Theorem 5.7.2 that this polynomial approximates $f(x)$ well for $x = a + h$ near a. The error or remainder term is defined by

$$R_{a,n}(h) = f(a+h) - \sum_{k=0}^{n} \frac{f^{(k)}(a)}{k!}h^k. \tag{6.8}$$

Theorem 5.7.2 assumed only that the order $(n+1)$ derivative of f exists on an interval about a, and showed that $R_{a,n}(h) = f^{(n+1)}(c)h^{n+1}/(n+1)!$, where c is some point between a and $a+h$.

The following version of Taylor's theorem has a stronger hypothesis and leads to a more detailed error estimate for the remainder. The proof involves repeated integrations by parts.

Theorem 6.8.1 (Taylor's Theorem with Integral Remainder)**.** *Let $n \geq 0$ and suppose that f is C^{n+1} on an open interval I containing the point a. If h is such that $a + h \in I$, then*

$$\begin{aligned} f(a+h) &= \sum_{j=0}^{n} \frac{f^{(j)}(a)}{j!}h^j + R_{a,n}(h) \\ &= f(a) + f'(a)h + \frac{f''(a)}{2!}h^2 + \cdots + \frac{f^{(n)}(a)}{n!}h^n + R_{a,n}(h), \end{aligned} \tag{6.9}$$

where

$$R_{a,n}(h) = \frac{h^{n+1}}{n!}\int_0^1 f^{(n+1)}(a+th)(1-t)^n\,dt. \tag{6.10}$$

Proof. Let $0 \le t \le 1$. By the chain rule, we have $\frac{d}{dt} f(a+th) = hf'(a+th)$, so by the fundamental theorem of calculus,

$$f(a+h) - f(a) = h \int_0^1 f'(a+th)\, dt.$$

This is exactly the assertion of the theorem in the case where $n = 0$. On the right-hand side, integrate by parts, with $u = f'(a+th)h$ and $dv = dt$, and choose $v = t - 1 = -(1-t)$ to get

$$\begin{aligned} h \int_0^1 f'(a+th)\, dt &= -(1-t)hf'(a+th)|_0^1 + h \int_0^1 (1-t)f''(a+th)h\, dt \\ &= f'(a)h + h^2 \int_0^1 (1-t)f''(a+th)\, dt, \end{aligned}$$

and substitution of this result into the first equation above yields exactly the statement of the theorem for the case $n = 1$. We can continue this process of integration by parts. Suppose that we have obtained

$$f(a+h) = \sum_{j=0}^{k-1} \frac{f^{(j)}(a)}{j!} h^j + \frac{h^k}{(k-1)!} \int_0^1 f^{(k)}(a+th)(1-t)^{k-1}\, dt.$$

(This formula has been shown above to hold for $k = 1$ and $k = 2$, corresponding to $n = 0$ and $n = 1$, respectively.) In the integral on the right-hand side we may integrate again by parts, with $u = h^k f^{(k)}(a+th)$ and $v = -(1-t)^k/k!$, to find that the final term in the previous equation equals

$$\frac{f^{(k)}(a)}{k!} h^k + \frac{h^{k+1}}{k!} \int_0^1 f^{(k+1)}(a+th)(1-t)^k\, dt,$$

and consequently

$$f(a+h) = \sum_{j=0}^{k} \frac{f^{(j)}(a)}{j!} h^j + \frac{h^{k+1}}{k!} \int_0^1 f^{(k+1)}(a+th)(1-t)^k\, dt.$$

Observe that this is the assertion of the theorem if f is C^{k+1} on I. By induction we may continue the process up through the step involving the Taylor polynomial of degree n, plus the remainder involving the integration of $f^{(n+1)}$, yielding exactly (6.9) and (6.10) of the theorem statement. This completes the proof. □

If we know (or assume) a bound on $|f^{(n+1)}(x)|$ for x near a, we can deduce the remainder estimate in the following corollary from the earlier Taylor's Theorem 5.7.2. However, the *continuity* of $f^{(n+1)}$ on some open interval about a guarantees such a bound on a possibly smaller *closed* interval about a. We can then restrict to an open interval *within* that closed interval to get the following result.

Corollary 6.8.2. *If f is C^{n+1} on an open interval I and if $|f^{(n+1)}(x)| \le M$ for all $x \in I$, then for each $a \in I$ and each h such that $a + h \in I$,*

$$|R_{a,n}(h)| \le M \frac{|h|^{n+1}}{(n+1)!}.$$

Proof. Using the remainder formula (6.10), the bound on $|f^{(n+1)}(x)|$ on I gives

$$|R_{a,n}(h)| \le M\frac{|h|^{n+1}}{n!}\int_0^1 (1-t)^n\, dt,$$

and since

$$\int_0^1 (1-t)^n\, dt = \frac{1}{n+1},$$

the result follows. $\square$

Using the second mean value theorem for integrals we can also recover the Lagrange form of the remainder from (6.10). Assuming the hypotheses of Taylor's Theorem 6.8.1, the remainder term is

$$R_{a,n}(h) = \frac{h^{n+1}}{n!}\int_0^1 f^{(n+1)}(a+th)(1-t)^n\, dt.$$

Since $1-t \ge 0$ for $t \in [0,1]$, the second mean value theorem applied to this integral implies that

$$R_{a,n}(h) = \frac{h^{n+1}}{n!} f^{(n+1)}(c)\int_0^1 (1-t)^n\, dt = \frac{f^{(n+1)}(c)}{(n+1)!}h^{n+1}$$

for some number c between a and $a+h$. This is Lagrange's remainder.

It is important to observe that from either the integral remainder or the Lagrange remainder, one may deduce deduce easily deduce that

$$\frac{R_{a,n}(h)}{h^n} \to 0 \quad \text{as} \quad h \to 0. \tag{6.11}$$

This limit statement expresses the idea that the remainder goes to zero *faster than* h^n as $h \to 0$. It gives us a measure of *how good* the degree n Taylor polynomial approximation of f is for small h. Note that when f is C^{n+1}, Corollary 6.8.2 provides a computable bound for $R_{a,n}(h)$ in many cases. It is also possible to achieve the limit behavior (6.11) with a weaker hypothesis on f; see Exercises 6.8.1-6.8.2.

Finally, notice that it is sometimes convenient to write the result of either of the Taylor Theorems 5.7.2, 6.8.1 in the form

$$f(x) = f(a) + f'(a)(x-a) + \cdots + \frac{f^{(n)}(a)}{n!}(x-a)^n + R_{a,n}(x-a)$$

where $x - a = h$ and $R_{a,n}(x-a) = R_{a,n}(h)$.

Exercises.

Exercise 6.8.1. This exercise presents a version of Taylor's theorem in which no derivatives are assumed beyond those appearing in the degree n Taylor polynomial for f at the point a:

Theorem 6.8.3 (C^n Taylor theorem)**.** *Let $n \ge 1$ and suppose that f is C^n on an interval I, $a \in I$ and $a+h \in I$. Then the remainder $R_{a,n}(h)$ defined by (6.8) is given by*

$$R_{a,n}(h) = \frac{h^n}{(n-1)!}\int_0^1 [f^{(n)}(a+th) - f^{(n)}(a)](1-t)^{n-1}\, dt.$$

Prove Theorem 6.8.3 as follows:

1. Apply Theorem 6.8.1 with n there replaced by $n-1$, and write the result in the form $f(a+h) - T_{a,n-1}(h) = R_{a,n-1}(h)$ where $T_{a,n-1}$ is the Taylor polynomial of degree $n-1$ of f at a.
2. Subtract $f^{(n)}(a)h^n/n!$ from both sides of your result from part 1.
3. Use the fact that $\int_0^1 (1-t)^{n-1}\,dt = 1/n$ to deduce the desired expression for $R_{a,n}(h)$.

Exercise 6.8.2. This exercise presents a remainder estimate for the C^n Taylor Theorem 6.8.3 in the previous exercise:

Corollary 6.8.4. *If $n \geq 1$, f is C^n on an interval I, and $a, a+h \in I$, then*

$$\frac{R_{a,n}(h)}{h^n} \to 0 \quad \text{as} \quad h \to 0.$$

Prove Corollary 6.8.4. *Hints*: Use the definition of continuity of $f^{(n)}$ at a to get an upper bound on $|R_{a,n}(h)|$ for small $|h|$, then use statement 3 in the previous exercise, together with the definition of limit of a function.

Exercise 6.8.3. Find the Taylor polynomial of degree 3 about the point $a = 0$ for $f(x) = 1/(x+5)$. Find a constant M such that the remainder $R_{a,3}$ satisfies $|R_{a,3}| \leq M|h|^4$ for $|h| \leq .5$. Sketch f and the Taylor polynomial of degree 3.

Exercise 6.8.4. Find the Taylor polynomial of degree 3 about the point $a = 0$ for $f(x) = \cos x$. Find a constant M such that the remainder $R_{a,3}$ satisfies $|R_{a,3}| \leq M|h|^4$ for $|h| \leq .5$. Sketch f and the Taylor polynomial of degree 3.

6.9. Improper Integrals

The Riemann integral is defined for bounded functions on closed and bounded intervals. A different approach is required for the integration of unbounded functions, or the integration of functions defined on intervals that are either unbounded or not closed. In the discussion to follow it is important to remember that all functions are assumed to be Riemann integrable over any finite closed intervals that appear.

6.9.1. Functions on $[a, \infty)$ or $(-\infty, b]$. We shall call on the definition of limits at infinity from Definition 4.4.4.

Suppose that $f : [a, \infty) \to \mathbf{R}$ is Riemann integrable on $[a, b]$ for each $b > a$. Then the equation

$$F(b) = \int_a^b f(x)\,dx, \quad \text{for } b > a,$$

defines a function $F : [a, \infty) \to \mathbf{R}$. This function F may or may not have a limit at infinity. If $\lim_{b\to\infty} F(b) = \lim_{b\to\infty} \int_a^b f(x)\,dx$ exists, then the limit is called the **improper integral of f on $[a, \infty)$** and is written

$$\int_a^\infty f(x)\,dx = \lim_{b\to\infty} \int_a^b f(x)\,dx.$$

Example 6.9.1. Let $f : [1, \infty) \to \mathbf{R}$ be the function $f(x) = 1/x^2$. Then f is integrable on $[1, b]$ for each $b > 1$, and

$$\int_1^b \frac{1}{x^2}\, dx = -\frac{1}{x}\Big|_1^b = -\frac{1}{b} + 1.$$

Since $\lim_{b\to\infty}(-\frac{1}{b} + 1) = 1$, we have

$$\int_1^\infty \frac{1}{x^2}\, dx = 1$$

for the improper integral of $f(x) = 1/x^2$ on $[1, \infty)$. △

We can extend the technique illustrated above to functions $f : (-\infty, b] \to \mathbf{R}$, Riemann integrable on $[a, b]$ for each $a < b$. Here is an example.

Example 6.9.2. Let $f(x) = xe^x$ on $(-\infty, 0]$. Then f is integrable on $[a, 0]$ for each $a < 0$, and, using integration by parts,

$$\int_a^0 xe^x\, dx = -ae^a - 1 + e^a.$$

Since $\lim_{a\to-\infty}(-ae^a - 1 + e^a) = -1$, we say the improper integral of f on $(-\infty, 0]$ exists and $\int_{-\infty}^0 xe^x = -1$. △

Sometimes a comparison with a simpler integrand can determine the convergence or divergence of an improper integral.

Theorem 6.9.3. *Suppose $f : [a, \infty) \to \mathbf{R}$ and $g : [a, \infty) \to \mathbf{R}$ are Riemann integrable on $[a, b]$ for every $b > a$. If $0 \le f(x) \le g(x)$ for all $x \ge a$ and $\int_a^\infty g(x)\, dx$ exists, then $\int_a^\infty f(x)\, dx$ exists.*

Proof. By the hypotheses, we have, for each $b > a$,

$$0 \le \int_a^b f(x)\, dx \le \int_a^b g(x)\, dx \le \int_a^\infty g < \infty.$$

Therefore the set

$$\Big\{\int_a^b f(x)\, dx : b > a\Big\}$$

is bounded above and thus has a least upper bound. Since the function $F(b) = \int_a^b f(x)\, dx$ is increasing with b, $\lim_{b\to\infty} F(b)$ exists. □

6.9.2. Functions on $(a, b]$ or $[a, b)$. Suppose that $f : (a, b] \to \mathbf{R}$ is Riemann integrable on $[\alpha, b]$ for each $\alpha \in (a, b]$. Then the equation

$$F(\alpha) = \int_\alpha^b f(x)\, dx, \quad \text{for } \alpha \in (a, b]$$

defines a function $F : (a, b] \to \mathbf{R}$. This function F may or may not have a limit as $\alpha \to a$. If $\lim_{\alpha\to a} F(\alpha) = \lim_{\alpha\to a} \int_\alpha^b f(x)\, dx$ exists, then the limit is called the **improper integral of f on $(a, b]$** and is written

$$\int_a^b f(x)\, dx = \lim_{\alpha\to a} \int_\alpha^b f(x)\, dx.$$

Example 6.9.4. If $f : (0, 1] \to \mathbf{R}$ is given by $f(x) = 1/\sqrt{x}$, then the improper integral $\int_0^1 f(x)\,dx$ exists, since

$$\lim_{\alpha \to 0} \int_\alpha^1 \frac{1}{\sqrt{x}}\,dx = \lim_{\alpha \to 0} 2x^{1/2}\Big|_\alpha^1 = \lim_{\alpha \to 0} (2 - 2\alpha^{1/2}) = 2,$$

and we write $\int_0^1 f(x)\,dx = \int_0^1 (1/\sqrt{x})\,dx = 2$. However, if $g : (0, 1] \to \mathbf{R}$ is given by $g(x) = 1/x^2$, then the improper integral $\int_0^1 g(x)\,dx$ does not exist, since

$$\lim_{\alpha \to 0} \int_\alpha^1 \frac{1}{x^2}\,dx = \lim_{\alpha \to 0} -x^{-1}\Big|_\alpha^1 = \lim_{\alpha \to 0} (-1 + \frac{1}{\alpha})$$

does not exist. △

Theorem 6.9.5. *Suppose $f : (a, b] \to \mathbf{R}$ and $g : (a, b] \to \mathbf{R}$ are Riemann integrable on $[\epsilon, b]$ for every $\epsilon \in (a, b]$. If $0 \le f(x) \le g(x)$ for all $x \in (a, b]$ and $\int_a^b g(x)\,dx$ exists, then $\int_a^b f(x)\,dx$ exists.*

The proof of Theorem 6.9.5 is similar to the proof of Theorem 6.9.3 and is left to Exercise 6.9.2.

Improper integrals over $[a, b)$ in which the integrand becomes unbounded as x approaches the right-hand endpoint are handled in a similar manner. See Exercise 6.9.4.

6.9.3. Functions on (a, ∞), $(-\infty, b)$ or $(-\infty, \infty)$. The potential difficulty here is the open intervals, where a function might not be defined at an endpoint or might become unbounded in an approach to the endpoint.

If f is defined on (a, ∞), and if, for some $c > a$, the improper integrals

$$\int_a^c f(x)\,dx \qquad \text{and} \qquad \int_c^\infty f(x)\,dx$$

exist, then we define the improper integral

$$\int_a^\infty f(x)\,dx = \int_a^c f(x)\,dx + \int_c^\infty f(x)\,dx,$$

the sum of two improper integrals. Note that the right-hand side here is

$$\begin{aligned}\int_a^c f(x)\,dx + \int_c^\infty f(x)\,dx &= \lim_{\epsilon \to 0} \int_{a+\epsilon}^c f(x)\,dx + \lim_{N \to \infty} \int_c^N f(x)\,dx \\ &= \lim_{\substack{N \to \infty \\ \epsilon \to 0}} \int_{a+\epsilon}^N f(x)\,dx.\end{aligned}$$

(The reader should verify that the choice of c does not matter; that is, if the improper integrals exist for one such c, then they exist for any such c, and the value given for the improper integral is independent of the choice of c.)

If a function f is defined on $(-\infty, b)$, and if, for some $c < b$, the improper integrals

$$\int_{-\infty}^c f(x)\,dx \qquad \text{and} \qquad \int_c^b f(x)\,dx$$

exist, then we define

$$\int_{-\infty}^{b} f(x)\,dx = \int_{-\infty}^{c} f(x)\,dx + \int_{c}^{b} f(x)\,dx,$$

which is the sum of two improper integrals.

Similarly, if $f : (-\infty, \infty) \to \mathbf{R}$ and if, for some real number c, the improper integrals

$$\int_{-\infty}^{c} f(x)\,dx \qquad \text{and} \qquad \int_{c}^{\infty} f(x)\,dx$$

exist, then

$$\int_{-\infty}^{\infty} f(x)\,dx = \int_{-\infty}^{c} f(x)\,dx + \int_{c}^{\infty} f(x)\,dx. \tag{6.12}$$

Note that (6.12) actually means

$$\begin{aligned}\int_{-\infty}^{\infty} f(x)\,dx &= \lim_{M\to\infty} \int_{-M}^{c} f(x)\,dx + \lim_{N\to\infty} \int_{c}^{N} f(x)\,dx \\ &= \lim_{\substack{N\to\infty \\ M\to\infty}} \int_{-M}^{N} f(x)\,dx.\end{aligned}$$

Again, the existence of these integrals is independent of the number c, provided there exists some c producing convergent integrals. For example, if c produces convergent integrals as indicated, and if $b < c$, then both

$$\int_{-\infty}^{b} f(x)\,dx \qquad \text{and} \qquad \int_{b}^{\infty} f(x)\,dx$$

are convergent, since $\int_b^\infty f(x)\,dx = \int_b^c f(x)\,dx + \int_c^\infty f(x)\,dx$.

6.9.4. Cauchy Principal Value. A concept of improper integral for $\int_{-\infty}^{\infty} f(x)\,dx$ different from (6.12) is the concept of the *Cauchy principal value* of this integral. The Cauchy principal value may exist even in cases where the improper integral as defined by (6.12) does not exist.

To define the Cauchy principal value, we suppose that $f : \mathbf{R} \to \mathbf{R}$ and f is integrable over any interval of the form $[-N, N]$, where N is real. If the limit

$$\lim_{N\to\infty} \int_{-N}^{N} f(x)\,dx$$

exists, then this limit is called the **Cauchy principal value** of $f(x)$ over $(-\infty, \infty)$, and we write

$$\int_{-\infty}^{\infty} f(x)\,dx = \lim_{N\to\infty} \int_{-N}^{N} f(x)\,dx.$$

If the improper integral of f as in (6.12) exists, then the Cauchy principal value of f over $(-\infty, \infty)$ exists (Exercise 6.9.5). For an example where f has a Cauchy principal value over $(-\infty, \infty)$, but the improper integral (6.12) fails to exist, see Exercise 6.9.6.

Another case of the Cauchy principal value occurs if $f : [a, c) \cup (c, b] \to \mathbf{R}$ is Riemann integrable on $[a, c-\epsilon]$ and $[c+\epsilon, b]$ for all ϵ satisfying $0 < \epsilon < \min\{c-a, b-c\}$. Then, provided the limit exists,

$$\lim_{\epsilon\to 0+} \Big(\int_a^{c-\epsilon} f(x)\,dx + \int_{c+\epsilon}^b f(x)\,dx \Big)$$

is called the **Cauchy principal value** of $\int_a^b f(x)\,dx$. For example, the Cauchy principal value of $\int_{-1}^1 1/x\,dx$ exists (Exercise 6.9.7). The Cauchy principal value of $\int_a^b f(x)\,dx$, for f as described in this paragraph, may exist with f being neither Riemann integrable on $[a, b]$ nor improperly integrable on $[a, c]$ and $[c, b]$.

For either type of integral considered here, our discussion shows that it makes sense to attempt to compute the Cauchy principal value.

Exercises.

Exercise 6.9.1. Show that the improper integral $\int_1^\infty \sqrt{x}e^{-x}$ exists. *Hint*: Compare the integrand with xe^{-x}.

Exercise 6.9.2. Prove Theorem 6.9.5.

Exercise 6.9.3. Formulate a theorem analogous to Theorem 6.9.3 that allows one to determine that a given improper integral over $[a, \infty)$ diverges.

Exercise 6.9.4. Determine whether the improper integral $\int_{-1}^0 -x^{-1/3}\,dx$ exists, and find its value if it does exist.

Exercise 6.9.5. Show that if the improper integral of f as in (6.12) exists, then the Cauchy principal value of f over $(-\infty, \infty)$ exists.

Exercise 6.9.6. Define $f : (-\infty, \infty) \to \mathbf{R}$ by

$$f(x) = \begin{cases} 1/(1+x) & \text{if } |x| > 2, \\ 0 & \text{if } |x| \le 2. \end{cases}$$

Show that $\int_{-\infty}^\infty f(x)\,dx$ exists as a Cauchy principal value equal to $-\log 3$, but that the improper integral (6.12) does not exist.

Exercise 6.9.7. Show that the Cauchy principal value of $\int_{-1}^1 1/x\,dx$ exists and evaluate it.

Exercise 6.9.8. Show that $\int_0^2 (\sqrt{|x-1|})^{-1}\,dx$ exists and evaluate it.

Exercise 6.9.9. Show that $\int_0^1 x \log x\,dx$ exists and evaluate it.

Exercise 6.9.10. Let $f : [0, \infty) \to \mathbf{R}$. The **Laplace transform** of f is the function $\mathcal{F}(s)$ defined by $\mathcal{F}(s) = \int_0^\infty f(t)e^{-st}\,dt$, when the improper integral exists. This operation is linear in f and is often written as $\mathcal{L}(f(t)) = \mathcal{F}(s)$. Clearly $\mathcal{L}(0) = 0$. Find the indicated Laplace transforms:

1. $\mathcal{L}(e^{at})$ for a real,
2. $\mathcal{L}(t^k)$ for $k = 0, 1, 2$,
3. $\mathcal{L}(\sin \omega t)$, ω real,

4. $\mathcal{L}(\cos \omega t)$, ω real,
5. $\mathcal{L}(e^{at} f(t))$, if $\mathcal{L}(f(t)) = \mathcal{F}(s)$.

Exercise 6.9.11. See Exercise 6.9.10 for the definition of the Laplace transform of f, $\mathcal{L}(f(t))$. Use integration by parts to show that if $\mathcal{L}(f(t)) = \mathcal{F}(s)$, then $\mathcal{L}(f'(t)) = s\mathcal{F}(s) - f(0)$ and $\mathcal{L}(f''(t)) = s^2\mathcal{F}(s) - f(0)s - f'(0)$.

Exercise 6.9.12. Solve the differential equation $f'(t) + 2f(t) = 0$ with initial condition $f(0) = 1$ by Laplace transforming both sides of the equation to get an algebraic equation for $\mathcal{L}(f(t)) = \mathcal{F}(s)$. Then determine $f(t)$ from $\mathcal{F}(s)$ by an appropriate *inverse Laplace transform* using results from Exercise 6.9.10.

Exercise 6.9.13. Solve the differential equation $f''(t) + 3f'(t) + 2f(t) = 0$ with initial conditions $f(0) = 0$, $f'(0) = 1$ using Laplace transforms. A partial fraction expansion will be helpful in obtaining an expression for $\mathcal{L}(f(t)) = \mathcal{F}(s)$ from which you may recover $f(t)$ by using results from Exercise 6.9.10. (This final step is called finding the **inverse Laplace transform** of $\mathcal{F}(s)$. It is straightforward for this equation.) *Hint*: A partial fraction expansion of the resulting expression for $\mathcal{F}(s)$ will facilitate the determination of $f(t)$ from $\mathcal{F}(s)$.

6.10. Notes and References

The books by Folland [**16**], Krantz [**40**], Lang [**42**], Sagan [**54**] and Schramm [**57**] were all helpful resources on Riemann integration. Folland [**16**] influenced the coverage of Taylor's theorem and the remainder expressions. Our discussion of improper integrals draws on Friedman [**18**].

Chapter 7

Sequences and Series of Functions

In this chapter we extend the analysis of infinite series with a study of sequences and series of functions. We study pointwise convergence and uniform convergence of sequences and series of functions and emphasize the important role of uniform convergence in limit processes. Power series provide the definitions for the exponential and trigonometric functions. The chapter concludes with a statement and proof of the Weierstrass approximation theorem.

7.1. Sequences of Functions: Pointwise and Uniform Convergence

In this section we define two convergence concepts for sequences of functions, pointwise convergence and uniform convergence. Uniform convergence is a stronger condition than pointwise convergence, and it has the advantage that many desirable properties of terms in the series, such as continuity and integrability, are preserved in the limit function.

7.1.1. Pointwise Convergence. The concept of pointwise convergence is the natural starting point for a discussion of convergence of a sequence of functions.

Definition 7.1.1. *Let $f_n : D \subseteq \mathbf{R} \to \mathbf{R}$ for $n \in \mathbf{N}$. The sequence (f_n)* **converges pointwise** *on $S \subseteq D$ to a function $f : S \to \mathbf{R}$ if for every $x \in S$,*

$$\lim_{n\to\infty} f_n(x) = f(x).$$

The resulting function f on S is the **limit function** *of the sequence (f_n).*

We also write "$f_n \to f$ on S" to indicate pointwise convergence of the functions f_n to f on the set S. When the domain S is understood, pointwise convergence of (f_n) to f on S may be indicated simply by writing "$f_n \to f$". Since limits of real sequences are unique, the limit function f is uniquely determined when (f_n) converges pointwise.

According to the definition of limit of a sequence, for a given point x, the sequence $f_n(x)$ has a limit (denoted $f(x)$ in Definition 7.1.1) if for every $\epsilon > 0$ there is an $N = N(\epsilon, x)$ such that if $n \geq N(\epsilon, x)$, then

$$|f_n(x) - f(x)| < \epsilon.$$

We observe the important fact that for a given $\epsilon > 0$ the required $N = N(\epsilon, x)$ may depend in general on the point x as well as on ϵ.

Example 7.1.2. The sequence of functions $f_n(x) = x^n$ for $x \in [0, 1]$ converges pointwise to the function f on $[0, 1]$ where

$$f(x) = \begin{cases} 0 & \text{if } 0 \leq x < 1, \\ 1 & \text{if } x = 1. \end{cases}$$

The verification is simply that $\lim_{n\to\infty} x^n = 0$ for $0 \leq x < 1$, and $f_n(1) = 1$ for all n, so $\lim_{n\to\infty} f_n(1) = 1$. Notice also that for a given x with $0 < x < 1$, if we want to have

$$|x^n - 0| = x^n < \epsilon < 1,$$

or equivalently, $n \log x < \log \epsilon$, we must choose $n \geq N(\epsilon, x) \geq \log \epsilon / \log x$. △

As we will see, the observation that the $N = N(\epsilon, x)$ generally depends on both ϵ and the point x is the key to understanding why a definition as natural as Definition 7.1.1 leads to a concept with some important limitations. The remainder of this section is devoted to illustrating pointwise convergence by several examples. We are especially interested in whether or not the properties of continuity, boundedness, differentiability, and integrability of the terms f_k carries over to the pointwise limit. Some examples are chosen to illustrate certain limitations of pointwise convergence with regard to these issues:

1. The pointwise limit of a sequence of continuous functions need not be continuous. (This is illustrated already by the previous example.)
2. The pointwise limit of a sequence of bounded functions need not be bounded.
3. The pointwise limit of a sequence of Riemann integrable functions need not be Riemann integrable. Even if the limit function f for (f_n) is integrable, it may not be true that the integral of f equals the limit of the integrals of the f_n.
4. The pointwise limit of a sequence of differentiable functions need not be differentiable. Even if the f_n's are differentiable and the limit function f is differentiable, it may not be true that the derivative of f equals the limit of the derivatives of the f_n.

Despite the negative nature of some conclusions we draw from these examples, the concept of pointwise convergence of a sequence of functions is fundamental in the development of other concepts of convergence for sequences.

Example 7.1.3. Let $f_n : (0, 1] \to \mathbf{R}$ be defined by

$$f_n(x) = \begin{cases} n & \text{for } 0 < x \leq 1/n \\ 1/x & \text{for } 1/n < x \leq 1. \end{cases}$$

For any given $x \in (0,1]$ there is an N such that if $n \geq N$, then $1/N < x$, and consequently

$$|f_n(x) - 1/x| = |1/x - 1/x| = 0.$$

Hence, $f_n \to f$ where $f(x) = 1/x$ for $x \in (0,1]$. Observe that each f_n is bounded on $(0,1]$, but the limit function f is not bounded on $(0,1]$. △

Example 7.1.4. In order to give an example of a sequence of integrable functions f_n on $[a,b]$ having a pointwise limit function f that is not integrable, we can simply augment the definition of the f_n's from the previous example. Let $f_n(x)$ be as in Example 7.1.3 for $x \in (0,1]$ and, in addition, we define $f_n(0) = 0$ for each n. Then $f_n \to f$ on the closed interval $[0,1]$, where

$$f(x) = \begin{cases} 0 & \text{for } x = 0, \\ 1/x & \text{for } 0 < x \leq 1. \end{cases}$$

Each f_n is Riemann integrable on $[0,1]$ since it is continuous a.e., while the limit function f is not integrable since it is not bounded on $[0,1]$. △

It is possible to have integrable functions f_n that converge to an integrable limit function f, and yet the integral of the limit f does not equal the limit of the integrals of the f_n. See Exercise 7.1.1.

We consider the behavior of differentiability with respect to pointwise limits in section 7.2.

7.1.2. Uniform Convergence.

We have seen that the concept of pointwise convergence has some serious limitations because important and desirable properties possessed by the functions in a sequence need not be preserved in the pointwise limit function. Uniform convergence is a stronger concept that ensures the preservation of such desirable properties as continuity, boundedness, and integrability.

Definition 7.1.5. *Let $f_n : D \subseteq \mathbf{R} \to \mathbf{R}$ for $n \in \mathbf{N}$. The sequence (f_n)* **converges uniformly** *on $S \subseteq D$ to a function $f : S \to \mathbf{R}$ if for every $\epsilon > 0$ there is an $N = N(\epsilon) \in \mathbf{N}$ such that*

$$|f_n(x) - f(x)| < \epsilon \quad \text{for all } n \geq N(\epsilon) \text{ and all } x \in S.$$

Figure 7.1 shows an ϵ-band about f illustrating uniform convergence of f_n to f. When the domain S is understood, the uniform convergence of (f_n) to f on S may be indicated by writing

$$\text{“}f_n \to f \text{ uniformly”} \qquad \text{or} \qquad \text{“}f_n \overset{\text{unif}}{\to} f\text{”}.$$

It should be clear from Definition 7.1.5 and the definition of limit of a sequence that if $f_n \overset{\text{unif}}{\to} f$, then f_n converges pointwise to f on S.

Proposition 7.1.6. *If f_n converges uniformly to f on S, then f_n converges pointwise to f on S.*

In symbolic language using the quantifiers $\forall$ (for all) and $\exists$ (there exists), the definition of $f_n \overset{\text{unif}}{\to} f$ states that

$$\forall \epsilon > 0 \; \exists N(\epsilon) \; \forall x \in S \; \forall n \geq N(\epsilon) \; (|f_n(x) - f(x)| < \epsilon).$$

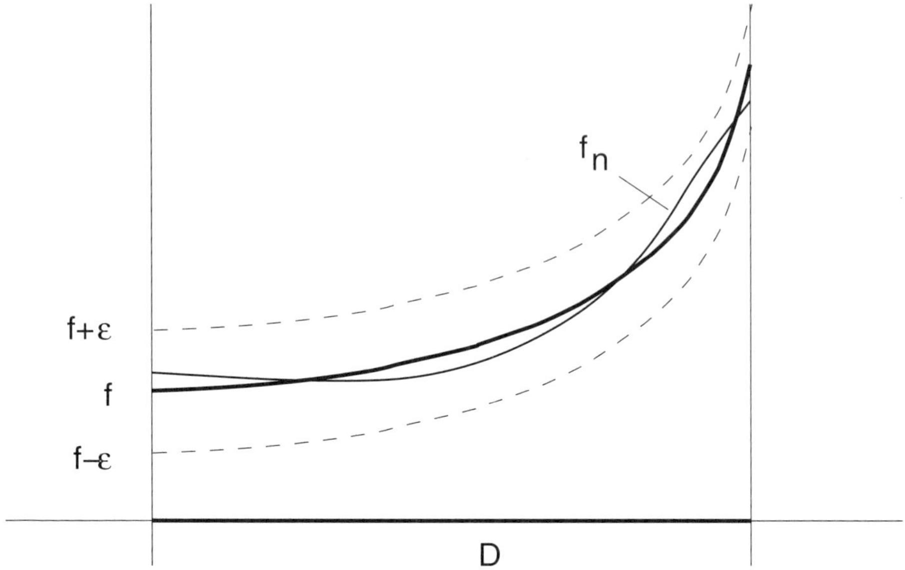

Figure 7.1. Illustrating uniform convergence of the sequence of functions f_n to the function f. The dotted curves bound an ϵ-band about the graph of f.

In order to gain a better understanding of uniform convergence, it is useful to examine the negation of the definition of uniform convergence in the case where the sequence converges pointwise to a function f. The negation of Definition 7.1.5 is

$$\exists \epsilon > 0 \ \forall N \ \exists x \in S \ \exists n \geq N \ (|f_n(x) - f(x)| \geq \epsilon).$$

Thus we can say that a sequence (f_n) fails to converge uniformly to a pointwise limit f if and only if there exists some $\epsilon_0 > 0$ such that for every n there is a point x_n such that

$$|f_n(x_n) - f(x_n)| \geq \epsilon_0.$$

We record this statement as a useful proposition.

Proposition 7.1.7. *A sequence of functions (f_n) defined on S and having the pointwise limit f on S fails to converge uniformly to f if and only if there exists an $\epsilon_0 > 0$ such that for every positive integer n there is a point x_n in S such that*

$$|f_n(x_n) - f(x_n)| \geq \epsilon_0.$$

As noted above, uniform convergence implies pointwise convergence. The converse does not hold, as shown by the next example.

Example 7.1.8. The sequence of functions $f_n(x) = x^n$ for $x \in [0,1]$ converges pointwise. However, (f_n) does not converge uniformly to f on $[0,1]$. In order to see why, notice that the limit function f has a jump discontinuity at $x = 1$. Let $\epsilon_0 = 1/2$ be half the distance of this jump. Then no matter how large we take n, there is some point $0 < x_n < 1$ for which we have

$$|f_n(x_n) - f(x_n)| = |x_n^n - 0| = x_n^n \geq \epsilon_0 = 1/2.$$

For example, we can take $x_n = (\epsilon_0)^{1/n} = (1/2)^{1/n}$ to achieve this inequality. Notice that x_n is an increasing sequence that moves to the right just fast enough as n increases to guarantee that the distance $|f_n(x_n) - f(x_n)|$ is larger than 1/2. Of course, other points might be chosen to achieve an even larger gap. In any case, f_n does not converge to f uniformly on $[0, 1]$. △

The example of $f_n(x) = x^n$ on $[0, 1]$ also shows that a sequence of continuous functions on a set S can converge pointwise to a limit function that is not continuous on S. The next result shows the strength of uniform convergence with regard to continuity.

Theorem 7.1.9. *If the functions $f_n : S \subseteq \mathbf{R} \to \mathbf{R}$ are continuous on S and f_n converges uniformly to f on S, then f is continuous on S.*

Proof. Let x_0 be a point in S. Let us show continuity of the limit function f at x_0. This requires that $f(x)$ can be made as close as we like to $f(x_0)$ by taking x sufficiently close to x_0. In order to motivate the essential estimate that is the key to the argument, notice that the hypotheses say that $f(x) \approx f_n(x)$ for large n, $f_n(x) \approx f_n(x_0)$ by continuity of f_n at x_0, and $f_n(x_0) \approx f(x_0)$ for large n. So we should be able to get $f(x) \approx f(x_0)$ by appropriate estimates. This motivates us to write

$$f(x) - f(x_0) = f(x) - f_n(x) + f_n(x) - f_n(x_0) + f_n(x_0) - f(x_0),$$

which implies

$$|f(x) - f(x_0)| \le |f(x) - f_n(x)| + |f_n(x) - f_n(x_0)| + |f_n(x_0) - f(x_0)|.$$

Let $\epsilon > 0$. By the uniform convergence, we may choose a fixed n sufficiently large so that for all $x \in S$ (including x_0),

$$|f(x) - f_n(x)| < \epsilon/3, \quad \text{and hence also} \quad |f_n(x_0) - f(x_0)| < \epsilon/3.$$

It should be clear that the choice of n places no restriction on x. Now, by continuity of f_n at x_0, there is a $\delta > 0$ such that if $|x - x_0| < \delta$, then

$$|f_n(x) - f_n(x_0)| < \epsilon/3.$$

In summary, then, if $|x - x_0| < \delta$, then

$$|f(x) - f(x_0)| < \epsilon/3 + \epsilon/3 + \epsilon/3 = \epsilon.$$

Therefore f is continuous at x_0. Since the argument applies to any and all x_0 in S, we conclude that f is continuous on S. □

Recall that a pointwise limit of bounded functions need not be bounded; see Example 7.1.4. However, a uniform limit of bounded functions is bounded.

Theorem 7.1.10. *If the functions $f_n : S \subseteq \mathbf{R} \to \mathbf{R}$ are bounded on S and f_n converges uniformly to f on S, then f is bounded on S.*

Proof. Let $\epsilon = 1$. By the uniform convergence, there is an N such that for $n \ge N$ and all $x \in S$,

$$\Big||f(x)| - |f_n(x)|\Big| \le |f_n(x) - f(x)| < 1.$$

Thus, if we let $n = N$, then for all $x \in S$, we have

$$|f(x)| \leq 1 + |f_N(x)| \leq 1 + \sup_{x \in S} |f_N(x)| < \infty,$$

since f_N is bounded. Therefore f is bounded on S. □

In the next theorem, we invoke the result, proved in Theorem 12.4.7 by a unified argument for any $\mathbf{R}^n$, that a countable union of sets of Lebesgue measure zero has Lebesgue measure zero. We have placed the unified argument at that point in the text to avoid duplicating the argument here for the case $n = 1$ alone. Readers might reasonably postpone a reading of that argument; however, it can be read at this point since it applies to the current case of dimension $n = 1$ as written.[1]

Theorem 7.1.11. *If the functions $f_n : [a, b] \to \mathbf{R}$ are Riemann integrable and f_n converges uniformly to f on $[a, b]$, then f is integrable on $[a, b]$.*

Proof. Since the functions f_n are Riemann integrable on $[a, b]$, each f_n is bounded on $[a, b]$, by the definition of integrability. By Theorem 7.1.10, a uniform limit of bounded functions is bounded, so f is bounded on $[a, b]$. In order to show that f is Riemann integrable on $[a, b]$ it only remains to show that f is continuous almost everywhere in $[a, b]$.

Let D_f be the set of points in $[a, b]$ at which f is discontinuous. If $x_0 \in D_f$, then there must be some k for which f_k is discontinuous at x_0, for otherwise the uniform convergence of f_n to f guarantees that f is continuous at x_0. So $D_f \subset D_{f_k}$, where D_{f_k} denotes the set of discontinuities of f_k. Accordingly, for each n, let D_{f_n} be the set of discontinuities of f_n. Then we have

$$D_f \subset \bigcup_{n=1}^{\infty} D_{f_n}.$$

By hypothesis, each f_n is Riemann integrable on $[a, b]$, so D_{f_n} has Lebesgue measure zero. Since D_f is a subset of a countable union of sets of Lebesgue measure zero, D_f has Lebesgue measure zero. Hence, by Theorem 6.4.4, f is Riemann integrable on $[a, b]$. □

Theorem 7.1.12. *If the functions $f_n : [a, b] \to \mathbf{R}$ are Riemann integrable and f_n converges uniformly to f on $[a, b]$, then*

$$\lim_{n\to\infty} \int_a^b f_n(x)\, dx = \int_a^b \lim_{n\to\infty} f_n(x)\, dx = \int_a^b f(x)\, dx.$$

Proof. By Theorem 7.1.11, the limit function f is Riemann integrable. Given $\epsilon > 0$, there exists N such that if $n \geq N$, then for all x in $[a, b]$,

$$|f_n(x) - f(x)| < \epsilon/(b - a).$$

[1] Readers who wish to read Theorem 12.4.7 now should observe the comment on notation in the paragraph immediately preceding the statement of that theorem.

For $n \geq N$, we may estimate the difference of integrals by

$$\begin{aligned} \left| \int_a^b f_n(x)\,dx - \int_a^b f(x)\,dx \right| &= \left| \int_a^b (f_n(x) - f(x))\,dx \right| \\ &\leq \int_a^b |f_n(x) - f(x)|\,dx \\ &< \frac{\epsilon}{(b-a)}(b-a) = \epsilon. \end{aligned}$$

Hence the statement of the theorem follows. ☐

Theorem 7.1.12 states that uniform convergence allows us to interchange the operations of integration and limit of a sequence of functions.

Can we interchange the operations of differentiation and limit of a sequence of functions? That is, if we know that $f_n \to f$ and f_n' exists for each n, is it true that $f_n' \to f'$? The example of the functions $f_n(x) = x^n$ shows that the pointwise limit function f need not be continuous at every point, and hence need not be differentiable at every point. In such a case we do not have pointwise convergence $f_n' \to f' = (\lim_n f_n(x))'$ everywhere since the derivative on the right does not exist at some points.

What if the f_n are differentiable and $f_n \overset{\text{unif}}{\to} f$? Are these conditions sufficient for the limit function f to be differentiable? No, they are not sufficient, and Exercise 7.1.12 outlines an argument for a specific counterexample.

However, using Theorem 7.1.12 we can get a positive result on the limit of a sequence of derivatives f_n', if $f_n \to f$, the f_n' are continuous, and f_n' converges uniformly. In fact, we really only need to assume that the sequence $f_n(x_0)$ converges for *some* point x_0, and that f_n' converges uniformly to *some* function g, and then we may conclude that f_n has a pointwise limit f, and $f' = g$.

Theorem 7.1.13. *Suppose that each f_n is defined on an open interval (a, b) and the derivative function f_n' is continuous on (a, b). If f_n' converges uniformly to a function g on (a, b), and there is a point x_0 in (a, b) such that $\lim_{n\to\infty} f_n(x_0) = L$ exists, then f_n converges pointwise to f, where*

$$f(x) = L + \int_{x_0}^{x} g(s)\,ds, \quad x \in (a, b),$$

and $f'(x) = g(x) = \lim_{n\to\infty} f_n'(x)$ for all x in (a, b).

Proof. For each n, we have

$$f_n(x) = f_n(x_0) + \int_{x_0}^{x} f_n'(s)\,ds, \quad \text{for } x \in (a, b),$$

by the fundamental theorem of calculus. Now fix a point x in (a, b). The hypotheses $\lim_{n\to\infty} f_n(x_0) = L$ and $f_n' \overset{\text{unif}}{\to} g$ imply, using Theorem 7.1.12, that

$$\lim_{n\to\infty} f_n(x) = \lim_{n\to\infty} \left(f_n(x_0) + \int_{x_0}^{x} f_n'(s)\,ds \right) = L + \int_{x_0}^{x} g(s)\,ds.$$

Therefore f_n has the pointwise limit f as stated, and

$$f'(x) = g(x) = \lim_{n\to\infty} f_n'(x), \quad x \in (a,b)$$

follows by the fundamental theorem. $\square$

In Theorem 7.1.13 we needed the convergence of $f_n(x_0)$ at some point x_0 because the uniform convergence of a sequence of functions (f_n' in this case) does not imply the convergence of the sequence of antiderivatives, as shown by this simple example: Let $f_n(x) = n$ for all real x. Then $f_n'(x) = 0$ and $f_n' \overset{\text{unif}}{\to} 0$ on $\mathbf{R}$. But $f_n(x)$ does not converge at any point x.

Theorem 7.1.13 allows us to differentiate functions even in some cases where we possess no explicit representation for the limit function, as in the next example.

Example 7.1.14. Let $0 < b < 1$ and let $f_n : [0,b] \to \mathbf{R}$ be defined by

$$f_n(x) = 1 + x + \frac{x^2}{2} + \cdots + \frac{x^n}{n}.$$

Then $\lim_{n\to\infty} f_n(0) = 1$ exists, and we have

$$f_n'(x) = 1 + x + \cdots + x^{n-1} = \frac{1-x^n}{1-x} \quad \text{for } x \in [0,b].$$

If $m > n$, then

$$|f_m'(x) - f_n'(x)| = \left|\frac{1-x^m}{1-x} - \frac{1-x^n}{1-x}\right| = \frac{|x^n - x^m|}{1-x} \le \frac{|x^n - x^m|}{1-b} < \epsilon$$

if $m, n > N(\epsilon)$ since the sequence x^n converges uniformly on $[0,b]$. Therefore the sequence (f_n') converges uniformly on $[0,b]$. Theorem 7.1.13 applies and we conclude that

$$f' = \lim_{n\to\infty} f_n' = (\lim_{n\to\infty} f_n)',$$

where $f'(x) = \lim_{n\to\infty}(1-x^n)/(1-x) = 1/(1-x)$. We can recover f from f' in this case using the fundamental theorem of calculus and the fact that $f(0) = \lim_{n\to\infty} f_n(0) = 1$. Thus,

$$f(x) = 1 + \int_0^x f'(s)\,ds = 1 + \int_0^x \frac{1}{1-s}\,ds = 1 - \log(1-x),$$

and we conclude that

$$f(x) = \lim_{n\to\infty}\left(1 + x + \frac{x^2}{2} + \cdots + \frac{x^n}{n}\right) = 1 - \log(1-x).$$

This also yields the useful series expansion

$$\log(1-x) = -\lim_{n\to\infty}\left(x + \frac{x^2}{2} + \cdots + \frac{x^n}{n}\right) = -\sum_{k=1}^{\infty} \frac{x^k}{k}$$

for $x \in [0,b]$ where $0 < b < 1$. $\triangle$

A pointwise limit function $f = \lim_n f_n$ may be differentiable without the f_n being differentiable, as in the next example.

Example 7.1.15. Let $f_n(x)$ be defined by

$$f_n(x) = \begin{cases} 1/n & \text{if } x \in \mathbf{I} \cap [0,1], \\ 0 & \text{if } x \in \mathbf{Q} \cap [0,1]. \end{cases}$$

Then each f_n is nowhere continuous on $[0,1]$ so each f_n is nowhere differentiable on $[0,1]$. The limit function is $f(x) = 0$ for all $x \in [0,1]$, so the limit function is differentiable on $[0,1]$. △

Exercises.

Exercise 7.1.1. Consider the functions $f_n : [0,1] \to \mathbf{R}$ defined by

$$f_n(x) = \begin{cases} 0 & \text{for } x = 0, \\ n & \text{for } 0 < x < 1/n, \\ 0 & \text{for } 1/n \le x \le 1. \end{cases}$$

Show that $f_n \to f$ on $[0,1]$, where $f(x) \equiv 0$ on $[0,1]$. Then show that $\int_0^1 f(x)\,dx \neq \lim_{n\to\infty} \int_0^1 f_n(x)\,dx$.

Exercise 7.1.2. Define functions f_n on $\mathbf{R} - \{1\}$ by $f_n(x) = 1 - [x^n/(1-x)]$. Find the largest set S on which the sequence (f_n) converges pointwise, and determine the limit function on S.

Exercise 7.1.3. Let $\{q_1, q_2, q_3, \ldots\} = \mathbf{Q} \cap [0,1]$ be an enumeration of the rationals in the interval $[0,1]$. Define $f_n : [0,1] \to \mathbf{R}$ by

$$f_n(x) = \begin{cases} 0 & \text{if } x \in \{q_1, q_2, q_3, \ldots, q_n\}, \\ 1 & \text{if } x \in [0,1] - \{q_1, q_2, q_3, \ldots, q_n\}. \end{cases}$$

Explain why each f_n is integrable on $[0,1]$. Show that f_n converges pointwise to the Dirichlet function f, where $f(x) = 0$ if $x \in \mathbf{Q} \cap [0,1]$ and $f(x) = 1$ if $x \in [0,1] - \mathbf{Q}$, which we know is not integrable.

Exercise 7.1.4. Consider the series $\sum_{n=1}^{\infty} x(1-x^2)^n$.

1. Show that this series converges to
$$f(x) = \begin{cases} 0 & x = 0, \\ \frac{1}{x} - x & -\sqrt{2} < x < \sqrt{2}, \quad x \neq 0. \end{cases}$$
2. Verify that each term of the series is continuous, differentiable, and integrable on $(-\sqrt{2}, \sqrt{2})$, but the sum f has none of these properties.

Exercise 7.1.5. Let $f_n(x) = 1/x - x/n$ for $x \in (0,1]$. Show that f_n converges uniformly on $(0,1]$ to the limit function $f(x) = 1/x$, $x \in (0,1]$.

Exercise 7.1.6. Define the continuous functions f_n on $[0,1]$ by

$$f_n(x) = \begin{cases} 1 - nx & 0 \le x \le 1/n, \\ 0 & 1/n \le x \le 1. \end{cases}$$

Find the pointwise limit function and show that it is not continuous on $[0,1]$.

Exercise 7.1.7. Define the continuous bounded functions f_n on $(0,1]$ by

$$f_n(x) = \begin{cases} n & 0 < x \le 1/n, \\ 1/x & 1/n < x \le 1. \end{cases}$$

Find the pointwise limit function and show that it is not bounded on $(0,1]$.

Exercise 7.1.8. Let $f_n(x) = \frac{1}{n}\sin nx$ for $x \in \mathbf{R}$. Show that $f_n \overset{\text{unif}}{\to} 0$ on $\mathbf{R}$.

Exercise 7.1.9. Does the sequence $f_n(x) = nx/(1+nx)$ converge uniformly on $\mathbf{R}$? What about on the interval $[1, 2]$?

Exercise 7.1.10. Define the sequence f_n on $[0, 1]$ for $n > 1$ by

$$f_n(x) = \begin{cases} nx & 0 \le x \le 1/n, \\ -\frac{n}{n-1}(x-1) & 1/n \le x \le 1. \end{cases}$$

Show that f_n has a pointwise limit f on $[0, 1]$. Is f continuous? Is the convergence uniform?

Exercise 7.1.11. Show that the sequence of functions defined by

$$f_n(x) = \frac{n+x}{4n+x}, \qquad n = 1, 2, 3, \ldots,$$

converges uniformly on the interval $[0, N]$ for any $N < \infty$, but does not converge uniformly on $[0, \infty)$.

Exercise 7.1.12. This exercise deals with a specific sequence of differentiable functions f_n such that $f_n \overset{\text{unif}}{\to} f$ on $\mathbf{R}$ but the limit function f fails to be differentiable on $\mathbf{R}$. Let $f_n : \mathbf{R} \to \mathbf{R}$ be given by

$$f_n(x) = \cos x + \frac{1}{2}\cos 3x + \frac{1}{4}\cos 9x + \cdots + \frac{1}{2^n}\cos 3^n x.$$

1. Show that (f_n) converges uniformly on $\mathbf{R}$. *Hint*: Show that for $m > n$,

$$|f_m(x) - f_n(x)| < 1/2^n.$$

2. Let f be the pointwise (and uniform) limit function, $f = \lim_{n\to\infty} f_n$. Show that f is not differentiable at 0 as follows. First, consider the sequence $x_k = \pi/3^k \to 0$ as $k \to \infty$. Show that for $n \ge k$, the difference quotient $(f_n(x_k) - f_n(0))/(x_k - 0)$ is given by

$$\frac{1}{\frac{\pi}{3^k}}\left[\cos\frac{\pi}{3^k} + \frac{1}{2}\cos\frac{\pi}{3^{k-1}} + \cdots + \frac{1}{2^{k-1}}\cos\frac{\pi}{3} - \frac{1}{2^k} - \cdots - \frac{1}{2^n} - 1 - \frac{1}{2} - \cdots - \frac{1}{2^n}\right].$$

3. Then use the fact that

$$\frac{f(x_k) - f(0)}{x_k - 0} = \lim_{n\to\infty}\frac{f_n(x_k) - f_n(0)}{x_k - 0}$$

to show that

$$\frac{f(x_k) - f(0)}{x_k - 0} < -\frac{3^k}{\pi}\frac{1}{2^k} = -\frac{1}{\pi}\left(\frac{3}{2}\right)^k,$$

and therefore the difference quotient diverges to $-\infty$. Consequently, $f'(0)$ does not exist. Since f must be periodic with period 2π, this argument shows that f fails to be differentiable at all points $2k\pi$, $k \in \mathbf{Z}$. (In fact, Hardy [**27**] showed in 1909 that f is nowhere differentiable on $\mathbf{R}$.)

7.2. Series of Functions

Recall that a numerical series $\sum_{k=1}^{\infty} a_k$ is defined to be the sequence of partial sums (s_n), where $s_n = \sum_{k=1}^{n} a_k$. In the same way, a series of real functions of a real variable x, denoted

$$\sum_{k=1}^{\infty} f_k(x),$$

is defined to be the sequence of **partial sums**

$$s_n(x) = \sum_{k=1}^{n} f_k(x), \quad n = 1, 2, 3, \ldots.$$

For series of functions we must specify a common domain for the f_k.

Definition 7.2.1. *Let $f_k : D \subseteq \mathbf{R} \to \mathbf{R}$ for all $k \in \mathbf{N}$.*

1. *The series $\sum_{k=1}^{\infty} f_k(x)$* **converges pointwise** *on $S \subseteq D$ if the sequence (s_n) converges pointwise on S.*
2. *The series $\sum_{k=1}^{\infty} f_k(x)$* **converges uniformly** *on $S \subseteq D$ if the sequence (s_n) converges uniformly on S.*
3. *The series $\sum_{k=1}^{\infty} f_k(x)$* **converges absolutely** *on $S \subseteq D$ if $\sum_{k=1}^{\infty} |f_k(x)|$ converges pointwise on S.*
4. *If the series $\sum_{k=1}^{\infty} f_k(x)$ converges pointwise, then the* **sum** *of the series is the function $f(x) := \lim_{n\to\infty} s_n(x)$.*

Remark. *If the series $\sum_{k=1}^{\infty} f_k(x)$ converges pointwise on S with sum f, we may write any of the following phrases to indicate this fact: $f = \sum_{k=1}^{\infty} f_k$ on S, or $\lim_{n\to\infty} s_n = f$ on S, or $f(x) = \sum_{k=1}^{\infty} f_k(x)$ for all $x \in S$, or $s_n \to f$ on S.*

It should be clear that absolute convergence of a series of functions on S implies pointwise convergence of the series on S, and that uniform convergence of a series of functions implies pointwise convergence of the series. Again we are interested in whether or not the properties of continuity, boundedness, differentiability, and integrability of the terms f_k carries over to the sum of a series.

Theorem 7.2.2. *Suppose the series $\sum_{k=1}^{\infty} f_k(x)$ converges uniformly on S to the sum $f : S \to \mathbf{R}$. Then the following statements are true:*

1. *If all the f_k are continuous on S, then the sum f is continuous on S.*
2. *If all the f_k are bounded on S, then the sum f is bounded on S.*

Proof. Let $s_n(x) = \sum_{k=1}^{n} f_k(x)$. If each f_k is continuous on S, then each s_n is continuous on S. If each f_k is bounded on S, then each s_n is bounded on S. Since $f(x) = \lim_{n\to\infty} s_n(x)$, statement 1 on continuity of f follows from the continuity of a uniform sequential limit of continuous functions (Theorem 7.1.9), and statement 2 on boundedness of f follows from the boundedness of a uniform sequential limit of bounded functions (Theorem 7.1.10). $\square$

In practice we may start the indexing of a series at any convenient value.

Example 7.2.3. If $f_k : \mathbf{R} \to \mathbf{R}$ is defined by $f_k(x) = x^k$ for $k = 0, 1, 2, \ldots$, then we have the geometric series

$$\sum_{k=0}^{\infty} f_k(x) = \sum_{k=0}^{\infty} x^k = 1 + x + x^2 + \cdots$$

which converges pointwise on the set $S = \{x : |x| < 1\}$. We agree to make s_n the partial sum through the term x^n, and thus write

$$s_n(x) = \sum_{k=0}^{n} x^k = \frac{1 - x^{n+1}}{1 - x}.$$

The limit function on S is

$$f(x) = \lim_{n\to\infty} s_n = \lim_{n\to\infty} \frac{1 - x^{n+1}}{1 - x} = \frac{1}{1 - x}.$$

We now show the uniform convergence of this series on the closed interval $[-b, b]$ for any $0 < b < 1$, directly from the definition. Given $\epsilon > 0$, if $m > n$ we have

$$|s_m(x) - s_n(x)| = \left|\frac{x^{n+1} - x^{m+1}}{1 - x}\right| \le \frac{|x^{n+1} - x^{m+1}|}{1 - b} < \epsilon/2$$

for all $x \in [-b, b]$ if $m, n > N(\epsilon)$, since the sequence (x^n) converges uniformly on $[-b, b]$. Now let $m \to \infty$ in the last inequality and use the continuity of the absolute value function to conclude that for all $x \in [-b, b]$, we have

$$|f(x) - s_n(x)| \le \epsilon/2 < \epsilon$$

provided $n > N(\epsilon)$. Therefore the geometric series converges uniformly to $f(x) = 1/(1 - x)$ on $[-b, b]$, if $0 < b < 1$. △

7.2.1. Integration and Differentiation of Series.

The sum of a uniformly convergent series of Riemann integrable functions is Riemann integrable, and the series defining the sum can be integrated term-by-term.

Theorem 7.2.4 (Integration of a Series Term-by-Term). *If the functions $f_k : [a, b] \to \mathbf{R}$ are Riemann integrable on $[a, b]$ and $\sum_{k=1}^{\infty} f_k(x)$ converges uniformly on $[a, b]$, then the sum $f(x) = \sum_{k=1}^{\infty} f_k(x)$ is Riemann integrable on $[a, b]$ and*

$$\int_a^b f(x)\,dx = \int_a^b \sum_{k=1}^{\infty} f_k(x)\,dx = \sum_{k=1}^{\infty} \int_a^b f_k(x)\,dx.$$

Proof. Let $s_n = \sum_{k=1}^{n} f_k$ on $[a, b]$. Then Theorem 7.1.11 applied to the s_n implies that $f = \lim_{n\to\infty} s_n$ is Riemann integrable on $[a, b]$. An application of Theorem 7.1.12 to the sequence s_n yields

$$\lim_{n\to\infty} \int_a^b s_n(x)\,dx = \int_a^b \lim_{n\to\infty} s_n(x)\,dx = \int_a^b f(x)\,dx.$$

But

$$\int_a^b s_n(x)\,dx = \int_a^b \sum_{k=1}^{n} f_k(x)\,dx = \sum_{k=1}^{n} \int_a^b f_k(x)\,dx$$

is the n-th partial sum of the series $\sum_{k=1}^{\infty} \int_a^b f_k(x)\,dx$. The result follows. □

We also have the following consequence of Theorem 7.1.13 on term-by-term differentiation of a uniformly convergent series of differentiable functions.

Theorem 7.2.5 (Differentiation of a Series Term-by-Term). *Suppose the functions $f_k : (a,b) \to \mathbf{R}$ are differentiable and f_k' is continuous on (a,b) for each k, $\sum_{k=1}^{\infty} f_k(x_0)$ converges for some $x_0 \in (a,b)$, and $\sum_{k=1}^{\infty} f_k'(x)$ converges uniformly on (a,b). Then the sum $f = \sum_{k=1}^{\infty} f_k$ is differentiable on (a,b) and*

$$f'(x) = \sum_{k=1}^{\infty} f_k'(x) \quad \textit{for all } x \in (a,b).$$

Proof. The result follows by a direct application of Theorem 7.1.13 to the partial sums $s_n = \sum_{k=1}^{n} f_k$, since the derivative $s_n' = \sum_{k=1}^{n} f_k'$ is continuous for each n, and by hypothesis, $(s_n(x_0))$ converges for some x_0 and the sequence (s_n') converges uniformly on (a,b). □

7.2.2. Weierstrass's Test: Uniform Convergence of Series. Uniform convergence of a series of functions can often be determined using the following simple test.

Theorem 7.2.6 (Weierstrass Test). *Suppose the real valued functions f_k, $k \geq 1$, are defined on a common domain $D \subseteq \mathbf{R}$. If each f_k satisfies a bound of the form*

$$|f_k(x)| \leq M_k \quad \textit{for all } x \in D,$$

where the M_k are fixed numbers, and if the series of the M_k converges, that is, $\sum_{k=1}^{\infty} M_k < \infty$, then the series $\sum_{k=1}^{\infty} f_k(x)$ converges uniformly on D.

Proof. Since $\sum_{k=1}^{\infty} M_k$ converges, the series $\sum_{k=1}^{\infty} f_k(x)$ converges absolutely by a direct comparison test, and therefore converges for each $x \in D$, so a pointwise limit function f exists on D. It remains to show that the series converges uniformly to f on D. Define the partial sums s_n of the series by

$$s_n(x) = \sum_{k=1}^{n} f_k(x), \quad x \in D.$$

We want to show that the sequence s_n converges uniformly on D. Let S_n be the sequence of partial sums of the series $\sum_{k=1}^{\infty} M_k$, and note that each $S_n \geq 0$. Given $\epsilon > 0$, there is a number $N(\epsilon) > 0$ such that

$$m > n > N(\epsilon) \quad \Longrightarrow \quad \sum_{k=n+1}^{m} M_k = S_m - S_n < \frac{\epsilon}{2}.$$

Thus, for $m > n > N(\epsilon)$ and all $x \in D$, the partial sums $s_n(x)$ satisfy

$$|s_m(x) - s_n(x)| = \left| \sum_{k=n+1}^{m} f_k(x) \right| \leq \sum_{k=n+1}^{m} M_k < \frac{\epsilon}{2}.$$

Fix $x \in D$ and let $m \to \infty$. By the continuity of the absolute value function, for any fixed $n > N(\epsilon)$ we have

$$\lim_{m\to\infty} |s_m(x) - s_n(x)| = |f(x) - s_n(x)| \leq \frac{\epsilon}{2} < \epsilon.$$

Since x in D was fixed but arbitrary, we conclude that if $n > N(\epsilon)$, then $|f(x) - s_n(x)| < \epsilon$ for all $x \in D$. Thus the sequence s_n converges uniformly to f on D. $\square$

Example 7.2.7. For the geometric series of Example 7.2.3, we may bound the terms by

$$|f_k(x)| = |x^k| \leq b^k$$

when $x \in [-b, b]$. The series $\sum_{k=1}^{\infty} b^k$ converges since it is a geometric series with base $0 < b < 1$. By the Weierstrass test, the geometric series converges uniformly on $[-b, b]$ when $0 < b < 1$. $\triangle$

Exercises.

Exercise 7.2.1. Write out in terms of ϵ and $N(\epsilon)$ the assertions of statements 1-2 of Definition 7.2.1.

Exercise 7.2.2. Show that the series $\sum_{k=0}^{\infty} x^k/k!$ converges uniformly on any bounded interval.

Exercise 7.2.3. Show that the series $\sum_{k=1}^{\infty}(-1)^{k+1}x^k$ converges uniformly on any interval of the form $[-b, b]$, where $0 < b < 1$.

Exercise 7.2.4. Show that the series

$$\sum_{k=1}^{\infty} \frac{1}{k^2} \sin(kx)$$

defines a function f on $\mathbf{R}$. Show that f is continuous on $\mathbf{R}$.

7.3. A Continuous Nowhere Differentiable Function

An example of a continuous nowhere differentiable function can be constructed by superimposing graphs having an increasing number of sharp corner points with slopes on either side of increasing magnitude. Define

$$\psi(x) = \begin{cases} x & \text{for } 0 \leq x \leq 1, \\ 2 - x & \text{for } 1 \leq x \leq 2, \end{cases}$$

and extend ψ to all of $\mathbf{R}$ by the condition that $\psi(x+2) = \psi(x)$ for all real x. Observe that $|\psi(x)| \leq 1$ for all x and $|\psi(x) - \psi(y)| \leq |x - y|$ for all x, y; also, $|\psi(n+1) - \psi(n)| = 1$ for any integer n. The continuous nowhere differentiable function of the following theorem is from Rudin [**52**].

Theorem 7.3.1. *The function $f : \mathbf{R} \to \mathbf{R}$ defined by*

$$f(x) = \sum_{k=0}^{\infty} \left(\frac{3}{4}\right)^k \psi(4^k x)$$

is continuous and nowhere differentiable.

Proof. Since $|(\frac{3}{4})^k \psi(4^k x)| \leq (\frac{3}{4})^k$ and $\sum_{k=0}^{\infty}(\frac{3}{4})^k$ converges, the series converges uniformly on $\mathbf{R}$ by the Weierstrass test, and the sum f is a continuous function on $\mathbf{R}$ by Theorem 7.2.2. Moreover, since

$$\left(\frac{3}{4}\right)^k \psi(4^k(x+2)) = \left(\frac{3}{4}\right)^k \psi(4^k x + 4^k \cdot 2) = \left(\frac{3}{4}\right)^k \psi(4^k x),$$

each of the partial sums has period 2, and consequently the limit function f has period 2 on $\mathbf{R}$. Let us show that f fails to be differentiable at every x. First observe that the graph of the k-th term in the series, $(\frac{3}{4})^k\psi(4^kx)$, is made up of line segments of slope either 3^k or -3^k. Now let x be a fixed real number and let m be any positive integer. Then there exists $n \in \mathbf{Z}$ such that

$$n \leq 4^m x \leq n+1.$$

(Given x, the integer n depends on m, $n = n(m)$, but for simplicity we do not indicate this in the notation.) Define

$$\alpha_m = 4^{-m}n \quad \text{and} \quad \beta_m = 4^{-m}(n+1),$$

and observe that

$$\alpha_m \leq x \leq \beta_m \quad \text{and} \quad \beta_m - \alpha_m = 4^{-m}.$$

Our goal is to show that the difference quotients

$$\frac{f(\beta_m) - f(\alpha_m)}{\beta_m - \alpha_m}$$

become arbitrarily large as $m \to \infty$. (See Exercise 7.3.2.) Consider the numbers $4^k\alpha_m$ and $4^k\beta_m$, $k \in \mathbf{N}$. Then

$$\begin{aligned} k > m \quad &\Longrightarrow \quad 4^k\beta_m - 4^k\alpha_m = 4^{k-m} \quad \text{(an even integer)}, \\ k = m \quad &\Longrightarrow \quad 4^k\beta_m - 4^k\alpha_m = 1, \\ k < m \quad &\Longrightarrow \quad 4^k\beta_m - 4^k\alpha_m = \frac{1}{4^{m-k}} < 1. \end{aligned} \tag{7.1}$$

For $k < m$ there is no integer between $4^k\alpha_m$ and $4^k\beta_m$. By (7.1), properties of ψ we have noted, and triangle inequality arguments, we have

$$\begin{aligned} |f(\beta_m) - f(\alpha_m)| &= \Big|\sum_{k=0}^{m}\left(\frac{3}{4}\right)^k(\psi(4^k\beta_m) - \psi(4^k\alpha_m))\Big| \\ &\geq \left(\frac{3}{4}\right)^m - \sum_{k=0}^{m-1}\Big|\left(\frac{3}{4}\right)^k(\psi(4^k\beta_m) - \psi(4^k\alpha_m))\Big| \\ &\geq \left(\frac{3}{4}\right)^m - \sum_{k=0}^{m-1}\left(\frac{3}{4}\right)^k\left(\frac{1}{4}\right)^{m-k} \\ &= \left(\frac{3}{4}\right)^m - \left(\frac{1}{4}\right)^m\sum_{k=0}^{m-1}3^k \quad \text{(finite geometric series)} \\ &= \left(\frac{3}{4}\right)^m + \frac{1}{2\cdot 4^m} - \frac{1}{2}\left(\frac{3}{4}\right)^m \geq \frac{1}{2}\left(\frac{3}{4}\right)^m. \end{aligned}$$

Consequently,

$$\left|\frac{f(\beta_m) - f(\alpha_m)}{\beta_m - \alpha_m}\right| \geq 4^m\frac{1}{2}\left(\frac{3}{4}\right)^m = \frac{1}{2}3^m \to \infty \quad \text{as} \quad m \to \infty.$$

This shows that $f'(x)$ does not exist (Exercise 7.3.2). Since x was arbitrary, the function f is nowhere differentiable. □

Exercises.

Exercise 7.3.1. Graph some partial sums of the series in Theorem 7.3.1, for example, five terms and ten terms, to see how the sharp corners proliferate.

Exercise 7.3.2. Suppose $f : (a, b) \to \mathbf{R}$, $x \in (a, b)$ and $f'(x)$ exists. Let $a < \alpha_n \leq x \leq \beta_n < b$ for $n \in N$, and suppose that $\alpha_n \to x$ and $\beta_n \to x$ as $n \to \infty$. Show that

$$f'(x) = \lim_{n\to\infty} \frac{f(\beta_n) - f(\alpha_n)}{\beta_n - \alpha_n}.$$

Hint: Verify that

$$\frac{f(\beta_n) - f(\alpha_n)}{\beta_n - \alpha_n} - f'(x)$$

is equivalent to

$$c_n\left[\frac{f(\beta_n) - f(x)}{\beta_n - x} - f'(x)\right] + (1 - c_n)\left[\frac{f(\alpha_n) - f(x)}{\alpha_n - x} - f'(x)\right]$$

where $c_n = (\beta_n - x)/(\beta_n - \alpha_n)$. Note that $0 \leq c_n \leq 1$.

7.4. Power Series; Taylor Series

Power series are a very useful type of infinite series of functions.

Definition 7.4.1. *If x_0 is a fixed real or complex number and a_k is a sequence of real or complex numbers, then a series of the form*

$$\sum_{k=0}^{\infty} a_k(x - x_0)^k$$

is called a **power series**. *The partial sums of such a series are given by*

$$s_n = \sum_{k=0}^{n} a_k(x - x_0)^k \quad \textit{for } n = 0, 1, 2, \ldots.$$

Power series have special properties that distinguish them from more general series of functions. These properties may be deduced in the special case where $x_0 = 0$ and translated easily to the general case where $x_0 \neq 0$; consequently, we assume throughout this section that $x_0 = 0$ and we study power series of the form

$$\sum_{k=0}^{\infty} a_k x^k.$$

Theorem 7.4.2. *If the power series $\sum_{k=0}^{\infty} a_k x^k$ converges for $x = b \neq 0$, then it converges absolutely for all x with $|x| < |b|$.*

Proof. If $\sum_{k=0}^{\infty} a_k b^k$ converges, then the terms are bounded, so there exists an $M > 0$ such that $|a_k b^k| \leq M$ for all k. For any fixed x with $|x| < |b|$, and for any k,

$$|a_k x^k| = \left|a_k b^k \left(\frac{x^k}{b^k}\right)\right| \leq M\left|\frac{x}{b}\right|^k.$$

Since $|x/b| < 1$, we have the convergent geometric series

$$\sum_{k=0}^{\infty} M\left|\frac{x}{b}\right|^k = \frac{M}{1 - |\frac{x}{b}|},$$

and therefore for all $|x| < |b|$,

$$\sum_{k=0}^{n} |a_k x^k| \leq \frac{M}{1 - |\frac{x}{b}|}.$$

Hence $\sum_{k=0}^{\infty} |a_k x^k|$ converges for all $|x| < |b|$, as we wished to show. $\square$

Now let S be the subset of $\mathbf{R}$ defined by

$$S = \left\{x : \sum_{k=0}^{\infty} a_k x^k \text{ converges}\right\}. \tag{7.2}$$

If S is not bounded above, then the power series $\sum_{k=0}^{\infty} a_k x^k$ converges for all x, and hence converges absolutely for all x, by Theorem 7.4.2. If S is bounded above, then it has a least upper bound $r = \sup S \geq 0$. It is an immediate consequence of the definition of $r = \sup S$ that $\sum_{k=0}^{\infty} a_k x^k$ converges absolutely for all $|x| < r$. Thus the next definition makes sense.

Definition 7.4.3. *If the set S in (7.2) is bounded above, then the number*

$$r = \sup S = \sup\left\{x : \sum_{k=0}^{\infty} a_k x^k \text{ converges}\right\}$$

is called the **radius of convergence** *of $\sum_{k=0}^{\infty} a_k x^k$. If S is not bounded above, then we write $r = \infty$ for the radius of convergence.*

It can happen that the radius of convergence is $r = 0$; for example, as in the series

$$\sum_{k=1}^{\infty} (kx)^k = x + 2^2x^2 + 3^3x^3 + \cdots,$$

which must diverge for $x \neq 0$, since then $\lim_{k\to\infty} k^k x^k$ does not exist.

Theorem 7.4.4. *Let r be the radius of convergence of the power series $\sum_{k=0}^{\infty} a_k x^k$, where $r = \infty$ is possible. Then the following statements are true:*

1. *$\sum_{k=0}^{\infty} a_k x^k$ converges absolutely for all x with $|x| < r$.*
2. *$\sum_{k=0}^{\infty} a_k x^k$ diverges for all x with $|x| > r$.*
3. *$\sum_{k=0}^{\infty} a_k x^k$ converges uniformly to a continuous bounded function on every closed real interval $[a, b] \subset (-r, r)$.*
4. *The function $f(x) = \sum_{k=0}^{\infty} a_k x^k$ may be integrated term-by-term from 0 to x for any $x \in (-r, r)$:*

$$\int_0^x f(t)\,dt = \int_0^x \sum_{k=0}^{\infty} a_k t^k\,dt = \sum_{k=0}^{\infty} \int_0^x a_k t^k\,dt = \sum_{k=0}^{\infty} \frac{a_k}{k+1} x^{k+1}.$$

5. *$f(x) = \sum_{k=0}^{\infty} a_k x^k$ may be differentiated term-by-term at any $x \in (-r, r)$:*

$$f'(x) = \sum_{k=0}^{\infty} a_k \frac{d}{dx} x^k = \sum_{k=1}^{\infty} k a_k x^{k-1}.$$

6. *The sum function f of a convergent power series $\sum_{k=0}^{\infty} a_k x^k$ has derivatives of all orders on $(-r, r)$.*

Proof. Statements 1 and 2 follow from the definition of r and Theorem 7.4.2.

3. On the interval $[a, b] \subset (-r, r)$, we have $|a_k x^k| \le |a_k| \max\{|a|, |b|\}^k$, and $\sum |a_k| \max\{|a|, |b|\}^k$ converges since $\max\{|a|, |b|\} < r$. By the Weierstrass test, the series converges absolutely and uniformly on $[a, b]$ to a continuous limit function. The partial sums are continuous and therefore bounded on $[a, b]$, and hence their uniform limit on $[a, b]$ is also bounded.

4. By the uniform convergence on the interval between 0 and x, this follows directly from Theorem 7.2.4 applied to the power series on that interval.

5. If $r > 0$, the series converges at all points of $(-r, r)$, and the termwise differentiated series $\sum_{k=1}^{\infty} k a_k x^{k-1}$ is a power series having the same radius of convergence r (Exercise 7.4.1). Thus the differentiated series converges uniformly on any closed interval about a given point $x \in (-r, r)$. By an application of Theorem 7.2.5, we may conclude that

$$f'(x) = a_1 + 2a_2 x + 3a_3 x^2 + \cdots = \sum_{k=1}^{\infty} k a_k x^{k-1}.$$

Finally, statement 6 follows by an inductive application of statement 5. □

The next example illustrates the term-by-term integration property.

Example 7.4.5. From the definition of the natural logarithm function and a simple substitution, we have

$$\log(1 + x) = \int_1^{1+x} \frac{1}{t}\, dt = \int_0^x \frac{1}{1+t}\, dt.$$

For $|t| < 1$, we have the geometric series sum,

$$\frac{1}{1+t} = 1 - t + t^2 - t^3 + \cdots = \sum_{k=0}^{\infty} (-1)^k t^k,$$

with radius of convergence equal to 1. By Theorem 7.4.4 (statement 4), term-by-term integration yields

$$\log(1 + x) = \sum_{k=0}^{\infty} \int_0^x (-1)^k t^k\, dt = \sum_{k=0}^{\infty} (-1)^k \frac{x^{k+1}}{k+1} = \sum_{k=1}^{\infty} (-1)^{k+1} \frac{x^k}{k}.$$

This power series converges uniformly on any closed subinterval of $(-1, 1)$. Clearly it cannot converge when $x = -1$, but it does converge when $x = 1$ to the value $\log 2$. △

The root test (Theorem 3.11.1) provides an explicit rule for the radius of convergence of a power series.

Theorem 7.4.6 (Cauchy-Hadamard). *If $\sum_{k=0}^{\infty} a_k x^k$ converges for some $x \neq 0$, then the radius of convergence of the power series is given by*

$$r = \frac{1}{\limsup \sqrt[k]{a_k}}, \quad \textit{if } \limsup \sqrt[k]{a_k} > 0,$$

and by $r = \infty$, if $\limsup \sqrt[k]{a_k} = 0$.

Proof. If $x \neq 0$ and $\sum_{k=0}^{\infty} a_k x^k$ converges, then the terms $|a_k x^k|$ are bounded and therefore

$$\limsup \sqrt[k]{|a_k x^k|} = |x| \limsup \sqrt[k]{|a_k|}$$

exists as a finite value. Hence $\limsup \sqrt[k]{|a_k|}$ exists as a finite value.

Suppose $\limsup \sqrt[k]{|a_k|} > 0$. By the root test, $\sum_{k=0}^{\infty} a_k x^k$ converges if

$$|x| \limsup \sqrt[k]{|a_k|} < 1 \quad \text{or} \quad |x| < \frac{1}{\limsup \sqrt[k]{a_k}}.$$

On the other hand, the series diverges if

$$|x| \limsup \sqrt[k]{|a_k|} > 1 \quad \text{or} \quad |x| > \frac{1}{\limsup \sqrt[k]{a_k}}.$$

Hence the radius of convergence equals $(\limsup \sqrt[k]{|a_k|})^{-1}$ as stated.

If $\limsup \sqrt[k]{|a_k|} = 0$, then the root test implies that the series converges absolutely for all x. □

In Section 7.5 the sine and cosine functions are defined by power series, yielding solutions of the differential equation $y''(x)+y(x) = 0$ with initial conditions $y(0) = 0$ and $y'(0) = 1$ ($y = \sin x$), and $y(0) = 1$ and $y'(0) = 0$ ($y = \cos x$). The usual geometric interpretations of these functions can be shown to follow from these definitions.

Clearly, power series can be used to define important functions, and such functions have derivatives of all orders in the interior of the interval of convergence.

On the other hand, if a given function f has derivatives of all orders on an interval, we can define an infinite power series associated with f, using the derivative values of f at a given point a in the interval. The resulting series is called the *Taylor series* of f about a (or centered at a). We now define these Taylor series and consider the convergence of the resulting series to the function f that gives rise to the series.

We say that f is C^∞, or infinitely differentiable, on an interval I if f has derivatives of all orders at each point in I. Suppose that $f : I \to \mathbf{R}$ is infinitely differentiable and $a \in I$. Then it is possible to define the **Taylor series** for f centered at $x = a$ by

$$\sum_{k=0}^{\infty} \frac{f^{(k)}(a)}{k!}(x-a)^k.$$

The Taylor series for f centered at $x = a$ is, by definition, the sequence $(s_n(x))_{n=1}^{\infty}$ of partial sums given by

$$s_n(x) = \sum_{k=0}^{n} \frac{f^{(k)}(a)}{k!}(x-a)^k = f(a) + f'(a)(x-a) + \cdots + \frac{f^{(n)}(a)}{n!}(x-a)^n$$

for $x \in I$. It follows directly from Taylor's Theorem 5.7.2 and the definition of Taylor series that a necessary and sufficient condition for the pointwise convergence of s_n to f on I is that the remainder $R_{a,n}(x)$ has limit zero as $n \to \infty$, for each $x \in I$,

$$\lim_{n\to\infty} R_{a,n}(x) = 0 \quad \text{for each } x \in I.$$

Example 7.4.7. Let $f(x) = \sin x$ and $a = 0$. Since $|f^{(n)}(x)| \leq 1$ for all n, the remainder term $R_{a,n}(x)$, for any choice of x in Taylor's theorem, satisfies

$$|R_{a,n}(x)| \leq \frac{|x|^{n+1}}{(n+1)!}.$$

In particular, by computing derivatives of f through order 5 at $a = 0$, we find

$$\sin x = x - \frac{1}{3!}x^3 + \frac{1}{5!}x^5 + R_{0,5}(x)$$

where $|R_{0,5}(x)| \leq |x|^6/6!$. In fact, it is not difficult to see that for any positive integer n, $|R_{0,n}(x)| \leq |x|^{n+1}/(n+1)!$. It follows that for $|x| \leq b$,

$$\lim_{n\to\infty} R_{0,n}(x) = 0,$$

so the Taylor series for the sine function converges to the value $\sin x$ for $|x| \leq b$. Since this is true for arbitrary $b > 0$, the Taylor series for the sine function converges to $\sin x$ for any real number x. △

The Taylor series for e^x, and a Taylor series for $\log(1+x)$, for $x \in (-1, 1)$, are considered in the exercises.

The existence of derivatives of all orders for a function f on an interval does not suffice to ensure that f has a representation by Taylor series. Consider the next example.

Example 7.4.8. Define $f : \mathbf{R} \to \mathbf{R}$ by

$$f(x) = \begin{cases} e^{-1/x^2} & \text{for } x \neq 0, \\ 0 & \text{for } x = 0. \end{cases}$$

The graph of f appears in Figure 7.2. One can show that $f^{(k)}(0) = 0$ for every positive integer k, and consequently f is C^∞ on $\mathbf{R}$. Thus the Taylor series for f centered at $x = 0$ exists and has all coefficients equal to zero. Thus f cannot be the sum of its Taylor series centered at $x = 0$ over any nontrivial interval containing the origin, and thus we say that f is not an *analytic function*.[2] △

The next result is a uniqueness result for power series representation of f on an interval. The proof is left to Exercise 7.4.6.

[2]The term *analytic function* is a standard term in complex analysis which describes a differentiable complex valued function $f(z)$ of the complex variable z. The existence of the complex derivative is a strong condition, and such functions are in fact represented by their Taylor series expansions.

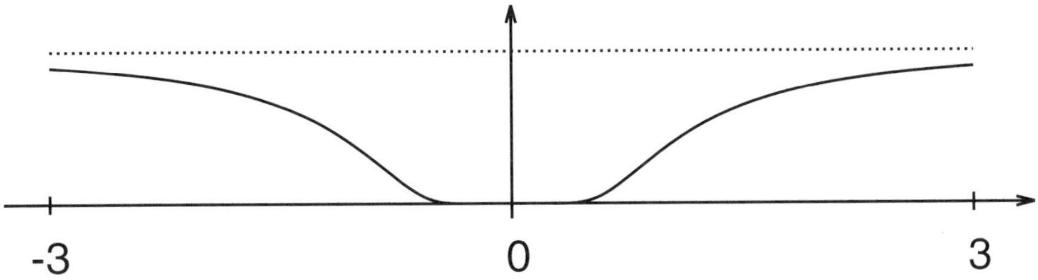

Figure 7.2. The graph of $f(x) = e^{-1/x^2}$. The Taylor series for f centered at the origin has all coefficients zero, since all derivatives of f at the origin equal zero. Thus the Taylor series cannot represent f on any nontrivial interval about the origin, and consequently we say that f is not an analytic function.

Theorem 7.4.9. *Suppose $f(x) = \sum_{k=0}^{\infty} a_k(x-a)^k$ for all $x \in (a-r, a+r)$, where r is the radius of convergence of the power series. Then $\sum_{k=0}^{\infty} a_k(x-a)^k$ is the Taylor series of f centered at $x = a$; that is, $a_k = f^{(k)}(a)/k!$.*

Exercises.

Exercise 7.4.1. Show that the radius of convergence of $\sum_{k=1}^{\infty} k a_k x^{k-1}$ is the same as the radius of convergence of $\sum_{k=0}^{\infty} a_k x^k$. Thus, verify that in Theorem 7.4.4, statement 6 follows from statement 5 by induction.

Exercise 7.4.2. Let $f(x) = e^x$. Show that for any real x and any positive integer n,

$$e^x = 1 + x + \frac{x^2}{2!} + \cdots + \frac{x^n}{n!} + R_{0,n}(x),$$

where

$$|R_{0,n}(x)| \le \begin{cases} e^x |x|^{n+1}/(n+1)! & \text{if } x > 0, \\ |x|^{n+1}/(n+1)! & \text{if } x < 0. \end{cases}$$

Conclude that the Taylor series for $f(x) = e^x$ at $a = 0$ converges to the value e^x for any real x.

Exercise 7.4.3. Let $f(x) = \log(1+x)$, which is defined on the interval $(-1, 1)$ centered at $a = 0$. Show that for $k \ge 1$, $f^{(k)}(x) = (-1)^{k+1}(k-1)!\,(1+x)^{-k}$, and thus f has the Taylor series representation

$$\log(1+x) = \sum_{k=1}^{\infty} \frac{(-1)^{k+1}}{k} x^k = x - \frac{1}{2}x^2 + \frac{1}{3}x^3 - \cdots,$$

which converges for $-1 < x < 1$. Show that the remainder $R_{0,k}(x)$, when $x = 1$, satisfies $\lim_{k\to\infty} R_{0,k}(1) = 0$, and therefore the series

$$\sum_{k=1}^{\infty} \frac{(-1)^{k+1}}{k} = 1 - \frac{1}{2} + \frac{1}{3} - \frac{1}{4} + \cdots$$

converges to $\log 2$.

Exercise 7.4.4. Show that for the function f in Example 7.4.8, $f^{(k)}(0) = 0$ for every positive integer k.

Exercise 7.4.5. Suppose that all derivatives of f at the point $x = a$ are uniformly bounded on an interval I containing the point $x = a$, say $|f^{(k)}(x)| \leq M$ for all $x \in I$ and all $k \in \mathbf{N}$. Show that $\lim_{n\to\infty} R_{a,n}(x) = 0$ for all $x \in I$, and therefore f is the sum of its Taylor series centered at a, that is,

$$f(x) = \sum_{k=0}^{\infty} \frac{f^{(k)}(a)}{k!}(x-a)^k, \quad \text{for all } x \in I.$$

Exercise 7.4.6. Prove Theorem 7.4.9. *Hint*: Use term-by-term differentiation.

Exercise 7.4.7. Define the function F by

$$F(x) := \int_0^x \log(1-t)\, dt, \quad \text{for } x < 1.$$

Find the Taylor series centered at $a = 0$ for F, and determine its interval of convergence. *Hint*: Differentiate to get a known series, then justify term-by-term integration.

Exercise 7.4.8. Consider the function $f(x) = x/(1+x)$, for $-1 < x < \infty$.

1. Find the Taylor series centered at $a = 0$ for f, and determine its interval of convergence.
2. Estimate the remainder $R_{0,5}$, and specify an interval on which the estimate is valid.

7.5. Exponentials, Logarithms, Sine and Cosine

This section applies some earlier results to rigorously establish the most fundamental properties of the elementary transcendental functions: the exponential, logarithm, and trigonometric functions.

In Theorem 6.7.9, we defined the natural logarithm function by the integral

$$\log x = \int_1^x \frac{1}{t}\, dt$$

and established its fundamental properties. In particular, $\log x$ is differentiable for $x > 0$, $\log : (0,\infty) \to \mathbf{R}$, with $\frac{d}{dx}\log x = 1/x$ for all $x > 0$. Since $\log x$ is a strictly increasing function, it is invertible on its range, which is $(-\infty,\infty)$, and

$\log^{-1} : (-\infty, \infty) \to (0, \infty)$. We know from experience in calculus courses that $\log^{-1}(x) = \exp(x)$ and, in fact, we usually write e^x instead of $\exp(x)$. A major goal of the section is to justify this name and notation for the inverse of log.

7.5.1. Exponentials and Logarithms. We begin with a power series definition of the exponential function $\exp(x)$ in Definition 7.5.1 below. We then deduce that $\exp(x)$ really equals

$$\log^{-1} : (-\infty, \infty) \to (0, \infty).$$

First we recall that the Euler number e was defined by the result of Theorem 3.1.5, which established that the limit

$$\lim_{n\to\infty} \left(1 + \frac{1}{n}\right)^n$$

exists. By definition, this limit is the Euler number e. In Theorem 3.5.1, it was shown that e is also the sum of the series

$$\sum_{k=0}^{\infty} \frac{1}{k!}.$$

When b is a positive number and x is a *rational* number, we know what is meant by the expression b^x. For example, since $e > 0$, for a rational number $x = p/q$, with positive $p, q \in \mathbf{Z}$, we have $e^x = e^{p/q} = \sqrt[q]{e^p}$, which is the q-th root of $e^p = e \cdots e$ (p factors). But what about e^x (or, more generally, b^x) for *irrational* x? An exponential operation b^x for all real x that is true to its name should certainly have the properties $b^x b^y = b^{x+y}$ and $(b^x)^{-1} = b^{-x}$ for all real x and y. The key to assigning a meaning to the expression e^x for irrational x is the series definition given here.

Definition 7.5.1. *The function* $\exp : \mathbf{R} \to \mathbf{R}$ *defined by*

$$\exp(x) = \sum_{k=0}^{\infty} \frac{1}{k!} x^k$$

is called the **real exponential function**.

The series $\sum_{k=0}^{\infty} \frac{1}{k!} x^k$ converges absolutely for any real x by the ratio test, and uniformly on any bounded set of real numbers by the Weierstrass test. So exp is defined and continuous on $\mathbf{R}$.

We should note that the **complex exponential function**, $\exp(z)$, $z \in \mathbf{C}$, is defined by the same power series with a complex variable z in place of the real variable x,

$$\exp(z) = \sum_{k=0}^{\infty} \frac{1}{k!} z^k.$$

Again, this series converges absolutely for any complex z, and uniformly on any bounded set in the complex plane.

From the series definition, we have $\exp(0) = 1$ and $\exp(1) = e$, the Euler number. The next proposition shows that $\exp(x)$ has another essential property of an exponentiation operation with base e.

Proposition 7.5.2. *For any real numbers x and y,*

$$\exp(x)\exp(y) = \exp(x+y).$$

Proof. Using the fact that the series converges absolutely for any x, we find by a rearrangement that

$$\begin{aligned}
\exp(x)\exp(y) &= \Big(\sum_{k=0}^{\infty} \frac{1}{k!}x^k\Big)\Big(\sum_{m=0}^{\infty} \frac{1}{m!}y^m\Big) \\
&= \sum_{n=0}^{\infty}\Big(\sum_{k+m=n} \frac{1}{k!}\frac{1}{m!}x^k y^m\Big) \qquad \text{(by rearrangement)} \\
&= \sum_{n=0}^{\infty}\Big(\sum_{k=0}^{n} \frac{1}{k!}\frac{1}{(n-k)!}x^k y^{n-k}\Big) \qquad (\text{set } m = n-k) \\
&= \sum_{n=0}^{\infty}\frac{1}{n!}\Big(\sum_{k=0}^{n} \frac{n!}{k!(n-k)!}x^k y^{n-k}\Big)
\end{aligned}$$

by the introduction of the $n!$ factors in both the numerator and the denominator. The inner sum is the binomial expansion of $(x+y)^n$, so we have $\exp(x)\exp(y) = \sum_{n=0}^{\infty} \frac{1}{n!}(x+y)^n$, which equals $\exp(x+y)$ by the series definition. □

Other important properties of exp follow from Proposition 7.5.2.

Proposition 7.5.3. *The function* $\exp : \mathbf{R} \to \mathbf{R}$ *is differentiable and has the following properties:*

1. $\frac{d}{dx}\exp(x) = \exp(x)$ *for all x;*
2. $\exp(x) > 0$ *for all x;*
3. $\exp(-x) = 1/\exp(x)$ *for all x;*
4. $\exp : \mathbf{R} \to \mathbf{R}^+$ *is one-to-one (it is strictly increasing) and onto* $\mathbf{R}^+ = (0, \infty)$.

Proof. 1 follows by term-by-term differentiation of the defining series for $\exp(x)$, which is valid due to the uniform convergence of the series on bounded intervals. By induction, it follows that $\exp(x)$ is a C^∞ function on $\mathbf{R}$.

2 For any x, $\exp(\frac{1}{2}x)\exp(\frac{1}{2}x) = [\exp(\frac{1}{2}x)]^2 = \exp(x)$, so $\exp(x) \geq 0$. Since we have $\exp(x)\exp(-x) = \exp(0) = 1$, $\exp(x) \neq 0$, and therefore $\exp(x) > 0$ for all x. This also shows that $\exp(-x) = (\exp(x))^{-1} = 1/\exp(x)$, so 3 holds.

4 That exp is strictly increasing follows from 2 and 1. Since $\exp(x)$ is C^∞ on $\mathbf{R}$, it is one-to-one and has a C^∞ inverse by the inverse function theorem. By (2), the range of exp is a subset of $\mathbf{R}^+$. Since $\exp(1) = e$,

$$\exp(n) = \exp(1)\cdots\exp(1) = e\cdots e = e^n,$$

and since $e > 1$, $e^n \to \infty$ as $n \to \infty$. So $\exp(x)$ is not bounded above. Also, $\exp(-1) = 1/\exp(1) = 1/e$, so $\exp(-n) = 1/e^n \to 0$ as $n \to \infty$. (Alternatively, by the fundamental theorem of calculus and the fact that $\exp(x) > 1$ for $x > 0$, we have

$$\exp(x) = \exp(0) + \int_0^x \exp(t)\,dt > 1 + \int_0^x 1\,dt = 1 + x, \qquad \text{for } x > 0,$$

so $\exp(x) \to \infty$ as $x \to \infty$, and since $\exp(-x) = 1/\exp(x)$, $\exp(x) \to 0$ as $x \to -\infty$.) Consequently, the range of exp must be $(0, \infty)$, as we now show. Given any $y > 1$, there is an $n_1 \in \mathbf{N}$ such that $1 < y < e^{n_1}$, and then by continuity of exp and the intermediate value theorem, there is an x such that $0 < x < n_1$ and $\exp(x) = y$. If $0 < y < 1$, there is an $n_2 \in \mathbf{N}$ such that $e^{-n_2} < y < 1$, and the intermediate value theorem implies that there is an x such that $-n_2 < x < 0$ and $\exp(x) = y$. Hence, exp is onto $(0, \infty)$. □

We have $\exp^{-1} : (0, \infty) \to (-\infty, \infty)$. Since the derivative of $\exp^{-1}$ at $y \in (0, \infty)$ is the reciprocal of the derivative of exp at the point x such that $\exp(x) = y$, we have

$$\frac{d}{dy}\exp^{-1}(y) = \frac{1}{\frac{d}{dx}\exp(x)} = \frac{1}{\exp(x)} = \frac{1}{y}, \quad \text{for } y = \exp(x).$$

Thus $\log y$ and $\exp^{-1}(y)$ have the same derivative for all $y > 0$, and $\log 1 = 0 = \exp^{-1}(1)$, as we have seen. Thus,

$$\log = \exp^{-1} \quad \text{and} \quad \exp^{-1} = \log .$$

By the definition of inverse, we have

$$\log(\exp(x)) = x \quad \text{for all real } x$$

and

$$\exp(\log y) = y \quad \text{for all } y > 0.$$

In particular, $\log 1 = 0$ and $\log e = 1$.

The next proposition shows that $\exp : \mathbf{R} \to \mathbf{R}^+$ really can be interpreted as a continuous exponentiation operation that extends the definition of e^x defined for rational x.

Proposition 7.5.4. $\exp(x)$ *is a differentiable (hence continuous) function on* $\mathbf{R}$ *such that* $\exp(x) = e^x$ *for all* $x \in \mathbf{Q}$.

Proof. We only need to show that $e^x = \exp(x)$ for all $x \in \mathbf{Q}$. We have seen already in the proof of Proposition 7.5.3 (item 4) that for any positive integer n, $\exp(n) = \exp(1 + \cdots + 1) = \exp(1) \cdots \exp(1) = e^n$. And by Proposition 7.5.3 (item 3), we have, for positive integer n, $\exp(-n) = 1/\exp(n) = 1/e^n$, which we usually write as e^{-n}, so $\exp(-n) = e^{-n}$. We also have

$$(\exp(1/n))^n = \exp(1/n) \cdots \exp(1/n) = \exp(1/n + \cdots + 1/n) = \exp(1) = e,$$

so $\exp(1/n)$ is an n-th root of e, which we usually write as $\exp(1/n) = \sqrt[n]{e} = e^{1/n}$.

Now let m and n be positive integers. By the facts just given, we have

$$\exp(m/n) = \exp(1/n + \cdots + 1/n) = (\exp(1/n))^m = (\sqrt[n]{e})^m = e^{m/n}.$$

If $m < 0$ and $n > 0$, then we have

$$\exp(m/n) = \exp(-|m|/n) = 1/\exp(|m|/n) = 1/e^{|m|/n} = e^{-|m|/n} = e^{m/n},$$

which completes the proof. □

Proposition 7.5.4 indicates that for a continuous exponentiation operation with base e we must define

$$e^x := \exp(x)$$

when x is irrational, and this completes the definition of the expression e^x for all real x. In particular, this defines power expressions such as e^π and $e^{\sqrt{2}}$.

But what about π^e? or $\sqrt{2}^e$? or 2^π? In other words, for a full range of exponentiation operations with various bases $b > 0$, we should extend the reasoning applied to the base e exponential and its inverse, the base e logarithm, to other bases $b > 0$. (The function $\log = \exp^{-1}$ could be denoted by $\log_e$ (or ln, for **natural logarithm**), to distinguish it from other logarithm functions associated with exponential functions with other base numbers b, to be defined; however, log without a subscript notation will always mean the inverse of exp.) First, we record once more the main properties of log which were derived in Theorem 6.7.9. The interested reader may wish to derive these properties here from the properties of exp and the fact that $\log = \exp^{-1}$.

Proposition 7.5.5. *The function* $\log : \mathbf{R}^+ \to \mathbf{R}$ *has the following properties:*

1. $\log(xy) = \log(x) + \log(y)$ *for every* $x, y > 0$*;*
2. $\log(1/x) = -\log x$*;*
3. $\log$ *is differentiable and* $\frac{d}{dx}\log x = 1/x$ *for every* $x > 0$*;*
4. $\log$ *is one-to-one (it is strictly increasing) and onto* $\mathbf{R}$*.*

We now proceed to consider other bases. Let $b > 0$. For *rational* numbers $x = p/q$, we clearly have

$$b^x = b^{p/q} = (b^p)^{1/q} = [(e^{\log b})^p]^{1/q} = e^{(p/q)\log b} = \exp\left(\frac{p}{q}\log b\right) = \exp(x \log b).$$

In order to have a function b^x which is continuous on $\mathbf{R}$, we define

$$b^x := \exp(x \log b) \qquad \text{for irrational } x.$$

Definition 7.5.6. *The function* $\exp_b : \mathbf{R} \to \mathbf{R}^+$ *defined by*

$$\exp_b(x) = \exp(x \log b)$$

is called the **exponential function with base** b.

Note that if $b = 1$, then $\log b = \log 1 = 0$, so $\exp_b(x) = \exp(0) = 1$ for all x, and from here on we remove this case from consideration.

From the definition, we have $\exp_b(0) = \exp(0) = 1$. The other basic properties of $\exp_b(x)$ follow from the properties of exp.

Proposition 7.5.7. *If* $b > 0$*,* $b \neq 1$*, then the function* $\exp_b : \mathbf{R} \to \mathbf{R}^+$ *has the property that*

$$\exp_b(x)\exp_b(y) = \exp_b(x + y)$$

for any real x *and* y*, and also the following properties:*

1. $\exp_b(x) > 0$ *for all* x*.*
2. $\exp_b(-x) = 1/\exp_b(x)$ *for all* x*.*
3. $\frac{d}{dx}\exp_b(x) = \exp_b(x)\log b$ *for all* x*.*

4. *If* $b > 1$, *then* $\exp_b(x)$ *is strictly increasing on* $\mathbf{R}$, *and*
   a) $\exp_b(x) \to +\infty$ *as* $x \to +\infty$;
   b) $\exp_b(x) \to 0$ *as* $x \to -\infty$.
5. *If* $0 < b < 1$, *then* $\exp_b : \mathbf{R} \to \mathbf{R}^+$ *is strictly decreasing on* $\mathbf{R}$, *and*
   a) $\exp_b(x) \to 0$ *as* $x \to +\infty$;
   b) $\exp_b(x) \to +\infty$ *as* $x \to -\infty$.
6. $\exp_b : \mathbf{R} \to \mathbf{R}^+$ *is one-to-one and onto* $\mathbf{R}^+$.

Proof. For real x and y, using the corresponding property of exp gives

$$\exp_b(x)\exp_b(y) = \exp(x\log b)\exp(y\log b) = \exp((x+y)\log b) = \exp_b(x+y).$$

Properties 1 and 2 follow from the corresponding properties of exp. The derivative formula 3 follows from the known derivative of exp and the chain rule:

$$\frac{d}{dx}\exp_b(x) = \frac{d}{dx}\exp(x\log b) = \exp(x\log b)\log b = \exp_b(x)\log b,$$

and this formula determines the increasing or decreasing property of $\exp_b$, according to whether $b > 1$ or $0 < b < 1$, which determines the sign of $\log b$. The final properties, a) and b) for both parts 4 and 5, as well as part 6, follow readily from property 4 of Proposition 7.5.3 and Definition 7.5.6, and are left as exercises for the reader. □

The function $\exp_b$ provides a continuous exponentiation operation with base $b > 0$ that extends the definition of b^x defined for rational x.

Proposition 7.5.8. $\exp_b(x)$ *is a differentiable (hence continuous) function on* $\mathbf{R}$ *such that* $\exp_b(x) = b^x$ *for all* $x \in \mathbf{Q}$.

The proof of Proposition 7.5.8 is left to the interested reader.

Let $b > 0$, $b \neq 1$. Since $\exp_b : \mathbf{R} \to \mathbf{R}^+$ is a C^∞ one-to-one function onto $\mathbf{R}^+$, by Proposition 7.5.8 and Proposition 7.5.7 (part 6), there is a C^∞ function, denoted $\log_b : \mathbf{R}^+ \to \mathbf{R}$, which is the inverse of $\exp_b$, so that

$$\log_b(\exp_b(x)) = x \quad \text{for all } x \in \mathbf{R}$$

and

$$\exp_b(\log_b(x)) = x \quad \text{for all } x \in \mathbf{R}^+.$$

We have seen that $\frac{d}{dx}\exp_b(x) = \exp_b(x)\log b$ for every real x. Differentiating $\exp_b(\log_b(x)) = x$, we find that

$$\exp_b(\log_b(x))\log b \cdot \frac{d}{dx}\log_b(x) = 1 \quad \text{for all } x > 0,$$

so that

$$\frac{d}{dx}\log_b(x) = \frac{1}{x\log b} \quad \text{for all } x > 0.$$

We now have all the tools needed to write the basic properties for exponentials b^x and their logarithms. Using the basic exp and log, we can obtain the final major properties of exponentiation with base b. For any $b > 0$, $b \neq 1$, we have $\log(b^x) = \log(\exp(x\log b)) = x\log b$, so

$$\log(b^x) = x\log b.$$

Consequently, $(b^x)^y = \exp(y\log(b^x)) = \exp(yx\log(b)) = \exp(xy\log(b)) = b^{xy}$, that is,

$$(b^x)^y = b^{xy}.$$

Here is a summary of the basic properties of exponentiation with base $b > 0$, all of which have now been established.

Proposition 7.5.9. *If $b > 0$, $b \neq 1$, and $x, y \in \mathbf{R}$, then*

1. $b^0 = 1$, *and* $b^x > 0$ *for every* x;
2. $b^x b^y = b^{x+y}$;
3. $(b^x)^y = b^{xy}$.

Here are the corresponding logarithm properties, for $b > 0$ and $b \neq 1$.

Proposition 7.5.10. *If $b > 0$, $b \neq 1$, and $x, y \in \mathbf{R}^+$, then*

1. $\log_b(1) = 0$;
2. $\log_b(xy) = \log_b(x) + \log_b(y)$;
3. $\log_b(x^a) = a\log_b(x)$ *for every real number* a.

Proof. It is similar to the arguments for the basic log function. Property 2 follows from Proposition 7.5.9 (property 2), by writing $x = b^r$ and $y = b^s$ and applying $\log_b$ to the identity $e^r e^s = e^{r+s}$, to get

$$\log_b xy = \log_b(b^r b^s) = \log_b b^{r+s} = r + s = \log_b x + \log_b y.$$

Property 3 follows from Proposition 7.5.9 (property 3), by writing $x = b^r$, so that

$$\log_b(x^a) = \log_b((b^r)^a) = \log_b(b^{ra}) = ra = a\log_b x,$$

as desired. □

See Exercise 7.5.1 for an alternative definition of b^x.

7.5.2. Power Functions.

Power functions x^r, for rational r, are well understood from basic calculus. Our interest here is power functions with irrational exponents, such as x^π or $x^{\sqrt{2}}$. Most of the work has been done in the subsection on exponentials.

If $x > 0$, then x^a is well-defined by $x^a := \exp(a\log x)$. The exponential rule $x^a x^b = x^{a+b}$ follows from Proposition 7.5.9 (property 2). The exponential rule $(x^a)^b = x^{ab}$ follows from Proposition 7.5.9 (property 3). The rule $x^a y^a = (xy)^a$ follows from the calculation

$$x^a y^a = \exp(a\log x)\exp(a\log y) = \exp(a(\log x + \log y)) = \exp(a\log(xy)) = (xy)^a.$$

The derivative of x^a is

$$\frac{d}{dx}x^a = ax^{a-1} \quad \text{for } x > 0,$$

which follows from the chain rule calculation

$$\frac{d}{dx}x^a = \frac{d}{dx}\exp(a\log x) = \exp(a\log x)\frac{a}{x} = ax^{-1}x^a = ax^{a-1},$$

using the first exponential rule in this subsection.

The properties given above assumed that $x > 0$. There are restrictions on powers x^a when $x < 0$. For example, if p and q are integers, we may write

$$x^{p/q} = \sqrt[q]{x^p}$$

for *any* $x < 0$, *provided* p *is even.*

Finally, if $x = 0$, we define $x^a = 0^a = 0$ for any $a > 0$.

7.5.3. Sine and Cosine Functions. We define the sine and cosine functions by infinite series.

Definition 7.5.11. *We define the function* $\sin : \mathbf{R} \to \mathbf{R}$ *by the series*

$$\sin x = \sum_{k=1}^{\infty} (-1)^{k+1} \frac{x^{2k-1}}{(2k-1)!} = x - \frac{x^3}{3!} + \frac{x^5}{5!} - \frac{x^7}{7!} + \cdots ,$$

and call it the **sine** *function. We define the function* $\cos : \mathbf{R} \to \mathbf{R}$ *by the series*

$$\cos x = \sum_{k=0}^{\infty} (-1)^k \frac{x^{2k}}{(2k)!} = 1 - \frac{x^2}{2!} + \frac{x^4}{4!} - \frac{x^6}{6!} - \cdots ,$$

and call it the **cosine** *function.*

Using the ratio test for absolute convergence and Weierstrass' test for uniform convergence, we may deduce that each of these series converges absolutely for each real x and uniformly on bounded intervals.

By direct substitution of $-x$ for x in the defining series, we verify directly that

$$\sin(-x) = -\sin x \qquad \text{and} \qquad \cos(-x) = \cos x$$

for all real x. Thus the sine function is an odd function and the cosine function is an even function.

The derivatives of the sine and cosine functions may be obtained by termwise differentiation of the series, since the termwise differentiated series also converge uniformly on bounded intervals. Thus we obtain the formulas

$$\frac{d}{dx} \sin x = \cos x \qquad \text{and} \qquad \frac{d}{dx} \cos x = -\sin x.$$

From these formulas, one can deduce by induction that $\sin x$ and $\cos x$ have derivatives of all orders, that is, the sine and cosine functions are C^∞ functions on $\mathbf{R}$.

Let us show the Pythagorean identity,

$$\cos^2 x + \sin^2 x = 1$$

for all real x. Indeed, differentiation of $\cos^2 x + \sin^2 x$ gives

$$2\cos x(-\sin x) + 2\sin x \cos x = 0 \quad \Longrightarrow \quad \cos^2 x + \sin^2 x = \text{constant},$$

and from the series definitions, we have $\sin 0 = 0$ and $\cos 0 = 1$, so the constant equals 1. It follows easily that for all real x,

$$|\sin x| \leq 1 \quad \text{and} \quad |\cos x| \leq 1.$$

It is convenient here to establish the uniqueness of the solution pair f, g for the initial value problem consisting of the pair of differential equations

$$f' = g \qquad \text{and} \qquad g' = -f \tag{7.3}$$

and the initial conditions

$$f(0) = 0 \quad \text{and} \quad g(0) = 1. \tag{7.4}$$

We do this here without calling on the general theory of existence and uniqueness for solutions of initial value problems. We already know that $f_1(x) = \sin x$ and $g_1(x) = \cos x$ provide a solution of the initial value problem (7.3), (7.4). Now suppose that f and g satisfy (7.3) and (7.4). Consider the pair of functions

$$-f(x)\cos x + g(x)\sin x$$

and

$$f(x)\sin x + g(x)\cos x.$$

It is straightforward to verify that these functions have zero derivative for all x, by virtue of the differential equations (7.3) satisfied by the pairs f, g and $\sin x$, $\cos x$. Consequently, there are real constants α and β such that for all x,

$$-f(x)\cos x + g(x)\sin x = \alpha, \tag{7.5}$$

$$f(x)\sin x + g(x)\cos x = \beta. \tag{7.6}$$

By setting $x = 0$ and using (7.4) we find that $\alpha = 0$ and $\beta = 1$. For each x, we can view $f(x)$ and $g(x)$ as the unknowns in these equations, written in matrix form as

$$\begin{bmatrix} -\cos x & \sin x \\ \sin x & \cos x \end{bmatrix} \begin{bmatrix} f(x) \\ g(x) \end{bmatrix} = \begin{bmatrix} 0 \\ 1 \end{bmatrix}.$$

We may invert the coefficient matrix, whose determinant is -1 (or use elementary operations), to solve the equations for $f(x)$ and $g(x)$, obtaining

$$\begin{aligned} \begin{bmatrix} f(x) \\ g(x) \end{bmatrix} &= \begin{bmatrix} -\cos x & \sin x \\ \sin x & \cos x \end{bmatrix}^{-1} \begin{bmatrix} 0 \\ 1 \end{bmatrix} \\ &= \begin{bmatrix} -\cos x & \sin x \\ \sin x & \cos x \end{bmatrix} \begin{bmatrix} 0 \\ 1 \end{bmatrix} \\ &= \begin{bmatrix} \sin x \\ \cos x \end{bmatrix} \end{aligned}$$

for all real x. This shows the uniqueness of the solution of (7.3), (7.4).

We now establish the addition formulas for sine and cosine, that is, for all real x and y,

$$\sin(x + y) = \sin x \cos y + \cos x \sin y, \tag{7.7}$$

$$\cos(x + y) = \cos x \cos y - \sin x \sin y. \tag{7.8}$$

Consider the left-hand sides of (7.7) and (7.8) with y considered fixed but arbitrary; they define a pair of functions $f_2 = \sin(x + y)$, $g_2 = \cos(x + y)$ of the variable x, which satisfy

$$f_2'(x) = g_2(x) \quad \text{and} \quad g_2'(x) = -f_2(x),$$

that is, f_2, g_2 satisfy equation (7.3). Their initial conditions at $x = 0$ are $f_2(0) = \sin y$ and $g_2(0) = \cos y$. By an argument similar to the one given earlier, we find that $f = f_2$ and $g = g_2$ satisfy equations (7.5), (7.6), except with $\alpha = -\sin y$ and $\beta = \cos y$. We then solve (7.5), (7.6) with these right-hand sides, to find that

$$\begin{aligned}\begin{bmatrix} f_2(x) \\ g_2(x) \end{bmatrix} &= \begin{bmatrix} -\cos x & \sin x \\ \sin x & \cos x \end{bmatrix} \begin{bmatrix} -\sin y \\ \cos y \end{bmatrix} \\ &= \begin{bmatrix} \sin x \cos y + \cos x \sin y \\ \cos x \cos y - \sin x \sin y \end{bmatrix}\end{aligned}$$

for all real x, which establishes (7.7) and (7.8).

We can now define the number π and establish the periodicity of sine and cosine.

First, note that if $g(x)$ is continuous and satisfies $g(0) > 0$ and $g(x_0) = 0$ for *some* $x_0 > 0$, then there is a smallest positive number z such that $g(z) = 0$ (Exercise 7.5.3). Since $g(x) = \cos x$ is continuous and $g(0) = \cos 0 = 1 > 0$, there is a smallest positive number z such that $\cos z = 0$ provided we show that $\cos x_0 = 0$ for some $x_0 > 0$. Since $\frac{d}{dx} \sin x = \cos x$, by the mean value theorem there is a number c such that

$$\sin 2 - \sin 0 = 2 \cos c \quad \Longrightarrow \quad \sin 2 = 2 \cos c$$

from which it follows that $|\cos c| \leq 1/2$. By the cosine addition formula and the Pythagorean identity,

$$\cos 2c = \cos^2 c - \sin^2 c = 2\cos^2 c - \sin^2 c - \cos^2 c = 2\cos^2 c - 1 < 0.$$

This gives us $\cos 0 > 0$ and $\cos 2c < 0$, so by the intermediate value theorem, there is a number x_0 between 0 and $2c$ such that $\cos x_0 = 0$. Thus, by Exercise 7.5.3, there is a smallest positive number z such that $\cos z = 0$.

Definition 7.5.12. *Let z be the smallest positive number such that $\cos z = 0$. Then the number π is defined by $\pi := 2z$.*

Theorem 7.5.13. *The functions $\sin x$ and $\cos x$ are periodic with period 2π, and 2π is the least positive period of these functions.*

Proof. By Definition 7.5.12, $\cos \pi/2 = 0$ and since $\cos^2 x + \sin^2 x = 1$, $\sin^2 \pi/2 = 1$. By the addition formula (7.8) for cosine,

$$\cos \pi = \cos^2 \pi/2 - \sin^2 \pi/2 = 0 - 1 = -1.$$

Then it follows from the Pythagorean identity that $\sin \pi = 0$, and hence, by (7.8),

$$\cos 2\pi = \cos^2 \pi - \sin^2 \pi = 1 - 0 = 1,$$

and therefore $\sin 2\pi = 0$. Now, by (7.7) and (7.8),

$$\sin(x + 2\pi) = \sin x \quad \text{and} \quad \cos(x + 2\pi) = \cos x$$

for all $x \in \mathbf{R}$. Thus $\sin x$ and $\cos x$ are periodic with period 2π. That 2π is the least positive period of these functions is left to Exercise 7.5.4. □

We can now relate the analytic definition of π with the geometric significance of the number π as the area enclosed by a circle of radius 1. If this circle is centered at the origin in the plane, it is described by the equation $x^2 + y^2 = 1$. We define the areas of planar regions by integration, and thus by symmetry we may define the area enclosed by this circle by stating that

$$\int_0^1 \sqrt{1-x^2}\,dx = \frac{1}{4}(\text{area of the unit circle}).$$

Thus we wish to prove that

$$\int_0^1 \sqrt{1-x^2}\,dx = \frac{\pi}{4}.$$

To do so, we use an integration by substitution. We have $\sin 0 = 0$, and $\sin \pi/2 = 1$, by the Pythagorean identity and the fact that $\cos \pi/2 = 0$. Since $\sin u$ is increasing and differentiable for $u \in [0, \pi/2]$, we may let $x = \sin u$ and use the addition formula (7.8) and Pythagorean identity to get

$$\int_0^1 \sqrt{1-x^2}\,dx = \int_0^{\pi/2} \cos^2 u\,du = \int_0^{\pi/2} \frac{1+\cos 2u}{2}\,du = \frac{\pi}{4}.$$

7.5.4. Some Inverse Trigonometric Functions.

The functions

$$\sin : [-\pi/2, \pi/2] \to [-1, 1] \qquad \text{and} \qquad \cos : [0, \pi] \to [-1, 1]$$

have continuous inverses, considered in Exercises 7.5.5-7.5.6. We focus here on the tangent function and its principal value inverse. Since $\cos x > 0$ for all $x \in (-\pi/2, \pi/2)$, the function $\tan : (-\pi/2, \pi/2) \to \mathbf{R}$ is well defined by

$$\tan x = \frac{\sin x}{\cos x}, \qquad x \in (-\pi/2, \pi/2).$$

Indeed this formula defines $\tan x$ for all real $x \neq \pm k\frac{\pi}{2}$ for odd positive integers k (where $\cos(\pm k\frac{\pi}{2}) = 0$), and $\tan(x + \pi) = \tan x$ where defined. Differentiation gives

$$\frac{d}{dx}[\tan x] = \frac{1}{\cos^2 x}, \qquad x \in (-\pi/2, \pi/2),$$

so $\tan x$ has a positive derivative on $(-\pi/2, \pi/2)$ and is therefore strictly increasing there. The range of tan is all of $\mathbf{R}$. From the Pythagorean identity, it follows that $\cos^2 x = 1/(1 + \tan^2 x)$. Write $y = \tan x$. By Theorem 5.3.1, the inverse $\tan^{-1} : \mathbf{R} \to (-\pi/2, \pi/2)$ has derivative

$$\frac{d}{dy}[\tan^{-1} y] = \cos^2(\tan^{-1} y) = \frac{1}{1+y^2}, \qquad y \in \mathbf{R}.$$

Since $\tan 0 = 0$, we have $0 = \tan^{-1}(0)$, and by the fundamental theorem,

$$\tan^{-1} y = \int_0^y \frac{1}{1+t^2}\,dt, \qquad y \in \mathbf{R}.$$

7.5.5. The Elementary Transcendental Functions.

The functions e^x, $\sin x$, $\cos x$, $\tan x$ are known as the *elementary transcendental functions*. They are called elementary because everybody has known about them for a long time and found them to be very useful. The best reason for the name *transcendental* comes from work of both F. Lindemann and K. Weierstrass in the last two decades of the nineteenth century. In particular, we note the following theorem, quoted from Niven [**49**].

Theorem 7.5.14 (Transcendental values of the elementary functions). *The following statements are true:*

1. *The values* e^x, $\sin x$, $\cos x$, *and* $\tan x$, *as well as*
$$\sinh x = \frac{e^x - e^{-x}}{2}, \quad \cosh x = \frac{e^x + e^{-x}}{2} \quad \text{and} \quad \tanh x = \frac{\sinh x}{\cosh x} = \frac{e^x - e^{-x}}{e^x + e^{-x}},$$
are transcendental numbers for any nonzero algebraic number x.
2. *In addition, the values of* $\log x$, $\sin^{-1} x$, *and, in general, the inverse functions of the functions listed in statement 1, are transcendental for any nonzero algebraic number* $x \neq 1$.

In particular, since 1 is a nonzero algebraic number, $e^1 = e$ is transcendental. What about π? If π were algebraic, then so would be $\pi/4$, and hence, by the theorem, we would have that $\tan \pi/4 = 1$ is transcendental, which is clearly false. So π is not algebraic, and therefore π is transcendental.

> *If one considers that the point set* $(\mathbf{Q} \cup \mathbf{I}_a) \times (\mathbf{Q} \cup \mathbf{I}_a)$ *of "algebraic points" is everywhere dense in* $\mathbf{R}^2$, *it seems truly incredible that the graphs of these functions manage to wind their way through the plane avoiding all but one of these algebraic points.* H. Sagan [**54**] (page 616).

Exercises.

Exercise 7.5.1. Since $\mathbf{Q}$ is dense in $\mathbf{R}$, for any real number x there exists an increasing sequence (q_n) of rational numbers such that $\lim_{n\to\infty} q_n = x$. If $b > 0$, $b \neq 1$, and $x \in \mathbf{R}$, suppose we define $b^x := \lim_{n\to\infty} b^{q_n}$, where (q_n) is an increasing sequence of rationals with $\lim_{n\to\infty} q_n = x$.

1. Show that this definition of b^x does not depend on the specific rational sequence used; that is, show that if (q_n) and (r_n) are increasing sequences of rational numbers such that $\lim_{n\to\infty} q_n = x = \lim_{n\to\infty} r_n$, then
$$\lim_{n\to\infty} a^{q_n} = \lim_{n\to\infty} a^{r_n}.$$
In particular, it is consistent with our previous definition of b^q when q is rational, since the constant rational sequence $q_n = q$ has limit q.
2. Conclude that this definition yields the same functions defined in the text.
3. Show that, under this new definition, the laws of real exponents follow from the corresponding laws of rational exponents by limit arguments.

Exercise 7.5.2. The **complex hyperbolic cosine** and **complex hyperbolic sine** functions are defined in terms of the complex exponential function by $\cosh z = (e^z + e^{-z})/2$ and $\sinh z = (e^z - e^{-z})/2$ for complex z. (Replacing z by real x in these formulas defines the **real hyperbolic cosine** and **real hyperbolic sine** functions.)

1. Show that $\sinh ix = i \sin x$ and $\cosh ix = \cos x$ for all real x.
2. Show that for all complex z and w, $\sinh(z+w) = \sinh z \cosh w + \cosh z \sinh w$, and $\cosh(z + w) = \cosh z \cosh w + \sinh z \sinh w$.
3. Express $\sinh(x + iy)$ and $\cosh(x + iy)$ in terms of real functions of the real variables x and y.

Exercise 7.5.3. Prove: If $g(x)$ is continuous and satisfies $g(0) > 0$ and $g(x_0) = 0$ for *some* $x_0 > 0$, then there is a smallest positive number z such that $g(z) = 0$. *Hint*: Define $z = \inf\{x : g(x) = 0\}$, and show that $z > 0$ and $g(z) = 0$.

Exercise 7.5.4. Prove that 2π is the least positive period of $\sin x$ and $\cos x$.

Exercise 7.5.5. Show that $\sin : [-\pi/2, \pi/2] \to [-1, 1]$ has a continuous inverse on $[-1, 1]$. Denote the inverse by $\sin^{-1} : [-1, 1] \to [-\pi/2, \pi/2]$, and show that

$$\sin^{-1} y = \int_0^y \frac{1}{\sqrt{1-t^2}}\, dt \quad \text{for } |y| < 1.$$

Interpret geometrically.

Exercise 7.5.6. Show that $\cos : [0, \pi] \to [-1, 1]$ has a continuous inverse on $[-1, 1]$. Denote the inverse by $\cos^{-1} : [-1, 1] \to [0, \pi]$, and show that

$$\cos^{-1} y = \frac{\pi}{2} - \int_0^y \frac{1}{\sqrt{1-t^2}}\, dt \quad \text{for } |y| < 1.$$

Interpret geometrically. Show that $\sin^{-1} y + \cos^{-1} y = \pi/2$ for all $y \in [-1, 1]$.

Exercise 7.5.7. *Before the gamma function*
Show that $\int_0^b e^{-x}\, dx = 1 - e^{-b}$ for all $b > 0$, and thus $\int_0^\infty e^{-x}\, dx = 1$.

Exercise 7.5.8. *Towards the gamma function*

1. Use integration by parts to show that

$$\int_0^b x^n e^{-x}\, dx = -b^n e^{-b} + n \int_0^b x^{n-1} e^{-x}\, dx, \quad \text{for } n = 1, 2, \ldots.$$

2. Deduce that for $n \geq 1$,

$$\int_0^\infty x^n e^{-x}\, dx = n \int_0^\infty x^{n-1} e^{-x}\, dx.$$

3. Using the result of Exercise 7.5.7, we obtain $\int_0^\infty x e^{-x}\, dx = \int_0^\infty e^{-x}\, dx = 1$ and $\int_0^\infty x^2 e^{-x}\, dx = 2 \cdot 1 = 2$. Show that

$$\int_0^\infty x^n e^{-x}\, dx = n! \quad \text{for } n = 0, 1, 2, \ldots.$$

Exercise 7.5.9. *The gamma function*
Let $\alpha > 0$. Notice that for $0 < \epsilon < 1 < b$,

$$\int_\epsilon^b x^{\alpha-1} e^{-x}\, dx = \int_\epsilon^1 x^{\alpha-1} e^{-x}\, dx + \int_1^b x^{\alpha-1} e^{-x}\, dx.$$

1. Deduce that $\lim_{\epsilon \to 0+} \int_\epsilon^1 x^{\alpha-1} e^{-x}\, dx$ exists if $0 < \alpha < 1$, and when $\alpha \geq 1$, it is a Riemann integral. *Hint*: For the first statement, note that $0 \leq x^{\alpha-1} e^{-x} \leq x^{\alpha-1}$ for all $x \geq 0$.

2. Show that $\lim_{b \to \infty} \int_1^b x^{\alpha-1} e^{-x}\, dx$ exists for fixed $\alpha > 0$. *Hint*: Divide the integrand $f(x) = x^{\alpha-1} e^{-x}$ by $g(x) = 1/x^2$, and find $\lim_{x \to \infty} x^{\alpha+1} e^{-x} = 0$; then deduce that, given $\epsilon > 0$, there exists $N(\epsilon)$ such that $0 \leq x^{\alpha-1} e^{-x} \leq \epsilon x^{-2}$ for $x > N(\epsilon)$.

3. Conclude that for $\alpha > 0$,

$$\Gamma(\alpha) := \int_0^\infty x^{\alpha-1}e^{-x}\,dx = \lim_{\epsilon\to 0+}\int_\epsilon^1 x^{\alpha-1}e^{-x}\,dx + \lim_{b\to\infty}\int_1^b x^{\alpha-1}e^{-x}\,dx$$

defines a function, called the **gamma function**. Use Exercise 7.5.8 to show that $\Gamma(n+1) = n!$ for $n = 0, 1, 2, 3, \ldots$.

4. Integrate by parts to find that $\Gamma(\alpha+1) = \alpha\Gamma(\alpha)$ for $\alpha > 0$.

7.6. The Weierstrass Approximation Theorem

Polynomial functions on $[a, b]$ are continuous functions, and we know that a uniform limit of polynomials defined on $[a, b]$ is necessarily a continuous function. The Weierstrass approximation theorem asserts that every continuous real valued function on $[a, b]$ can be approximated uniformly by polynomial functions. There is no assumption of derivatives in this statement. This is a stronger result than the uniform convergence of Taylor polynomials on closed intervals within the interval of convergence of a Taylor series.

First, we observe that by a linear mapping from $[0, 1]$ to $[a, b]$ we only need consider a continuous function on the interval $[0, 1]$.

The Russian mathematician S. N. Bernstein used a simple probabilistic idea as the basis of his proof of the Weierstrass approximation theorem. Suppose a coin has the property that the probability of showing heads on a single toss is x, and the probability of showing tails is therefore $1 - x$. The probability of showing exactly k heads after n tosses is

$$\binom{n}{k} x^k(1-x)^{n-k} = \frac{n!}{k!\,(n-k)!}x^k(1-x)^{n-k},$$

because the probabilities of the outcomes of the n independent tosses multiply to give the probability of the final outcome of the sequence of n tosses, and the number of ways that the k heads can occur is the same as the number of combinations of k objects chosen from n objects. In n tosses of the coin, some number k of heads must appear, where $0 \le k \le n$, with probability 1 (certainty in probability theory) so the probabilities for achieving k heads, for $0 \le k \le n$, add to 1:

$$\sum_{k=0}^{n} \binom{n}{k} x^k(1-x)^{n-k} = 1. \tag{7.9}$$

Note that this is the special case of the binomial expansion of $(x+y)^n$ when $y = 1-x$:

$$1 = (x + 1 - x)^n = \sum_{k=0}^{n} \binom{n}{k} x^k(1-x)^{n-k}.$$

Let f be continuous on $[0, 1]$. Since f is uniformly continuous on $[0, 1]$, for a given $x \in (0, 1]$ and n sufficiently large, we can approximate $f(x)$ by the value $f(k/n)$ for some k with $0 \le k \le n$. Consider the polynomial function $B_n(x)$ defined by

$$B_n(x) = \sum_{k=0}^{n} \binom{n}{k} x^k(1-x)^{n-k} f(k/n).$$

These polynomials are called the *Bernstein polynomials* associated with f. Using (7.9), we may write

$$f(x) - B_n(x) = \sum_{k=0}^{n} \binom{n}{k} x^k (1-x)^{n-k} [f(x) - f(k/n)]$$

and consequently

$$|f(x) - B_n(x)| \le \sum_{k=0}^{n} \binom{n}{k} x^k (1-x)^{n-k} |f(x) - f(k/n)|.$$

For fixed n, the expression $x^k(1-x)^{n-k}$ is bounded for $x \in [0,1]$. The difference $|f(x) - f(k/n)|$ can be made small for a given x by choosing n large and k appropriately. However, for a given $\epsilon > 0$ we must be able to see how to choose n large enough so that the entire summation is less than ϵ for all $x \in [0,1]$.

The key to showing that the approximation is uniform is provided by two identities that follow from (7.9) by differentiation with respect to x. Differentiation of (7.9) with respect to x gives

$$0 = \sum_{k=0}^{n} \binom{n}{k} x^{k-1} (1-x)^{n-k-1} (k - nx).$$

Then multiplication by $x(1-x)$ gives

$$0 = \sum_{k=0}^{n} \binom{n}{k} x^k (1-x)^{n-k} (k - nx). \tag{7.10}$$

Now differentiation of (7.10) with respect to x gives

$$0 = \sum_{k=0}^{n} \binom{n}{k} [-nx^k (1-x)^{n-k} + x^{k-1}(1-x)^{n-k-1}(k-nx)^2]. \tag{7.11}$$

An application of (7.9) to (7.11) yields

$$n = \sum_{k=0}^{n} \binom{n}{k} x^{k-1} (1-x)^{n-k-1} (k - nx)^2,$$

and multiplication of this last result by $x(1-x)$ gives

$$nx(1-x) = \sum_{k=0}^{n} \binom{n}{k} x^k (1-x)^{n-k} (k - nx)^2.$$

A final division by n^2 gives

$$\frac{x(1-x)}{n} = \sum_{k=0}^{n} \binom{n}{k} x^k (1-x)^{n-k} (x - k/n)^2. \tag{7.12}$$

The key identities for the proof of Theorem 7.6.1 are (7.9) and (7.12). The knowledge of a definite bound for $x(1-x)$ on $[0,1]$ will imply that the right-hand side of (7.12) can be made arbitrarily small uniformly in x. The proof will be completed by relating the bound for the right-hand side of (7.12) with the bound given above for $|f(x) - B_n(x)|$.

Theorem 7.6.1 (Weierstrass Approximation Theorem). *Suppose $a < b$. Let $f : [a,b] \to \mathbf{R}$ be continuous on $[a,b]$ and let $\epsilon > 0$. Then there exists a polynomial function $p : [a,b] \to \mathbf{R}$ with real coefficients such that for all $x \in [a,b]$,*

$$|f(x) - p(x)| < \epsilon.$$

Proof. We have seen that for all $x \in [0,1]$ and all n,

$$|f(x) - B_n(x)| \le \sum_{k=0}^{n} \binom{n}{k} x^k (1-x)^{n-k} |f(x) - f(k/n)|. \tag{7.13}$$

Let $\epsilon > 0$. By the uniform continuity of f on $[0,1]$, there is a $\delta > 0$ such that

$$|x - k/n| < \delta \quad \Longrightarrow \quad |f(x) - f(k/n)| < \epsilon/2.$$

For any fixed x, the right-hand side of (7.13) can be split into two summations, one of them, labeled $\sum_x$, over those k such that $|x - k/n| < \delta$, and the other, labeled $\sum_x'$, over those k such that $|x - k/n| \ge \delta$. For any fixed x, one can check without difficulty that $\sum_x < \epsilon/2$. The proof will be completed by showing that the sum $\sum_x'$ (independently of x, that is, for all x with $|x - k/n| \ge \delta$) can be made less than $\epsilon/2$ by choosing n sufficiently large. Since f is continuous on $[0,1]$ it is bounded there, so there exists M such that $|f(x)| \le M$ for all $x \in [0,1]$. Thus, for any x,

$$0 \le \sum_x' \le 2M \sum_x'' \binom{n}{k} x^k (1-x)^{n-k}$$

where the summation $\sum_x''$ on the right is over those k such that $|x - k/n| \ge \delta$. So we want to show that the summation on the right can be made less than $\epsilon/4M$ independently of x. Now identity (7.12) provides the needed bound, for it shows that for each x,

$$\begin{aligned} \frac{x(1-x)}{n} &= \sum_{k=0}^{n} \binom{n}{k} x^k (1-x)^{n-k} (x - k/n)^2. \\ &\ge \delta^2 \sum_x'' \binom{n}{k} x^k (1-x)^{n-k}. \end{aligned}$$

Hence, for each x,

$$\sum_x'' \binom{n}{k} x^k (1-x)^{n-k} \le \frac{x(1-x)}{\delta^2 n}.$$

Since $x(1-x)$ is bounded by $1/4$ for $x \in [0,1]$, we have

$$\sum_x'' \binom{n}{k} x^k (1-x)^{n-k} \le \frac{1}{4\delta^2 n}.$$

With an abbreviated notation, the right-hand side of (7.13) is now bounded as follows:

$$\sum_x + \sum_x' < \frac{\epsilon}{2} + 2M \sum_x'' \le \frac{\epsilon}{2} + \frac{M}{2\delta^2 n}.$$

Now choose n such that $n > M/(\delta^2 \epsilon)$. Then the Bernstein polynomial $B_n(x)$ associated with f satisfies

$$|f(x) - B_n(x)| < \frac{\epsilon}{2} + \frac{\epsilon}{2} = \epsilon$$

for all $x \in [0,1]$. ☐

Exercise.

Exercise 7.6.1. Verify the derivation of (7.10), (7.11) and (7.12) in the proof of the Weierstrass Approximation Theorem 7.6.1.

7.7. Notes and References

For additional information on the probabilistic ideas behind Bernstein's approach to the proof of the Weierstrass theorem, see Kazarinoff [**34**]. There is a more abstract view of the Weierstrass approximation theorem in the book by Simmons [**59**], together with a discussion of the important generalization of it known as the Stone-Weierstrass theorem.

Chapter 8

The Metric Space $\mathbf{R}^n$

The most basic concepts of analysis appear in the preceding chapters on real valued functions of a real variable. However, most applied problems deal with more than one variable and more than one real valued function. They often involve vector valued functions of a vector variable. The position, velocity and acceleration of even a single point mass in space are vector functions of time, each vector function having three real component functions. A continuous dynamical model for the populations of n interacting species may involve a system of n nonlinear ordinary differential equations with n dependent variables for the populations of the species (time being the independent variable). Even a nodding acquaintance with problems arising in the sciences or engineering indicates the need for a rigorous look at multivariable calculus. The most basic study of multivariable problems involves vector valued functions of a vector variable in Euclidean space, that is, mappings from $\mathbf{R}^n$ to $\mathbf{R}^m$. For such mappings, we will discuss important analogues of the derivatives, integrals, mean value theorem, and inverse function theorem that are important in single variable analysis. To discuss these analogues, we need the basic theory of normed vector spaces. While our main focus in this chapter is on the n-dimensional vector space $\mathbf{R}^n$, we also give some attention to function spaces, that is, vector spaces whose elements are functions.

8.1. The Vector Space $\mathbf{R}^n$

The Cartesian product of n nonempty sets $S_1, S_2, \ldots, S_n$ is the set

$$S_1 \times S_2 \times \cdots \times S_n = \{(x_1, x_2, \ldots, x_n) : x_j \in S_j \text{ for } j = 1, \ldots, n\}$$

consisting of all ordered n-tuples $(x_1, x_2, \ldots, x_n)$ where the j-th element x_j is an element of S_j for $j = 1, \ldots, n$. These are *ordered* n-tuples because we require that $(x_1, x_2, \ldots, x_n) = (y_1, y_2, \ldots, y_n)$ if and only if $x_j = y_j$ for all $j = 1, \ldots, n$, that is, these n-tuples are equal as finite sequences defined on the set $\{1, 2, \ldots, n\}$.

Definition 8.1.1. *The collection of all ordered n-tuples of real numbers is the n-fold Cartesian product*

$$\underbrace{\mathbf{R} \times \mathbf{R} \times \cdots \times \mathbf{R}}_{n \text{ factors}} = \{(x_1, x_2, \ldots, x_n) : x_j \in \mathbf{R} \quad \text{for} \quad j = 1, \ldots, n\}$$

and we denote this set by $\mathbf{R}^n$.

In introductory multivariable calculus, most of the work is done in the plane $\mathbf{R}^2$ or in space $\mathbf{R}^3$. In those situations, ordered pairs (x_1, x_2) or ordered triples (x_1, x_2, x_3) are visualized as position vectors with terminal point at the point with coordinates (x_1, x_2) (or (x_1, x_2, x_3)) and initial point at the origin $(0, 0)$ (or $(0, 0, 0)$). We shall write elements of $\mathbf{R}^n$ using boldface letters, for example

$$\mathbf{x} = (x_1, x_2, \ldots, x_n) \quad \text{and} \quad \mathbf{y} = (y_1, y_2, \ldots, y_n).$$

By definition, $\mathbf{x} = \mathbf{y}$ if and only if $x_j = y_j$ for $j = 1, \ldots, n$.

The set $\mathbf{R}^n$ is given algebraic structure by the operations of addition of n-tuples and scalar multiplication of an n-tuple by a real number. Here are the definitions.

Definition 8.1.2. *If $\mathbf{x} = (x_1, x_2, \ldots, x_n)$ and $\mathbf{y} = (y_1, y_2, \ldots, y_n)$ are elements of $\mathbf{R}^n$ and α is a real number, we define*

$$\mathbf{x} + \mathbf{y} = (x_1, x_2, \ldots, x_n) + (y_1, y_2, \ldots, y_n) = (x_1 + y_1, x_2 + y_2, \ldots, x_n + y_n)$$

and

$$\alpha\mathbf{x} = \alpha(x_1, x_2, \ldots, x_n) = (\alpha x_1, \alpha x_2, \ldots, \alpha x_n).$$

The set $\mathbf{R}^n$ together with these operations satisfies the axioms of a real vector space, which we state below. First, a remark about notation:

Remark. *Elements of $\mathbf{R}^n$ (for any n) will be denoted by* **boldface** *letters. Discussions about elements in an arbitrary general vector space will* not *use boldface for those elements.*

We assume the reader is familiar with the axioms that define a real vector space, but we recall the definition here.

Definition 8.1.3. *A set V together with an addition operation and a scalar multiplication operation is a real vector space if the following properties hold:*

1. *Closure under addition and multiplication: If $x, y \in V$, then $x + y \in V$. If $x \in V$ and $\alpha \in \mathbf{R}$, then $\alpha x \in V$.*
2. *Addition is commutative: $x + y = y + x$ for all $x, y \in V$.*
3. *Addition is associative: $x + (y + z) = (x + y) + z$ for all $x, y, z \in V$.*
4. *There exists a unique additive identity $0 \in V$ that satisfies $x + 0 = x$ for all $x \in V$.*
5. *Each $x \in V$ has a unique additive inverse y such that $x + y = 0 \in V$.*
6. *$(\alpha\beta)x = \alpha(\beta x)$ for all $x \in V$ and $\alpha, \beta \in \mathbf{R}$.*
7. *$(\alpha + \beta)x = \alpha x + \beta x$ for all $x \in V$ and $\alpha, \beta \in \mathbf{R}$.*
8. *$\alpha(x + y) = \alpha x + \alpha y$ for all $x, y \in V$ and $\alpha \in \mathbf{R}$.*
9. *If $x \in V$, then $0x = 0 \in V$ and $1x = x$.*

Properties 1-9 of a set V with operations of addition (a mapping $V \times V \to V$) and scalar multiplication (a mapping $\mathbf{R} \times V \to V$) are the axioms that define a **real vector space** (or a vector space over the real scalar field $\mathbf{R}$). Notice that these properties only guarantee the *existence* of a single element in V, and that is the additive identity, the zero vector, 0. Indeed, the set $V = \{0\}$ with the operation table $0 + 0 = 0$ and $\alpha 0 = 0$ for all $\alpha \in \mathbf{R}$ (necessarily, by property 9), is a real vector space. It is an important space, describing for example the solution set of the system of linear equations $A\mathbf{x} = \mathbf{0}$, where A is a real $n \times n$ *nonsingular* matrix and $\mathbf{0}$ is the zero vector in $\mathbf{R}^n$.

If properties 1-9 hold for elements of a set V and scalars from the complex field $\mathbf{C}$, then V is a **complex vector space** (or a vector space over the complex scalar field $\mathbf{C}$).

The elements of a space V are often called vectors, but there are vector spaces whose elements are functions, and vector spaces whose elements are matrices. Examples are given below.

Readers with some experience with abstract algebra will recognize the first five axioms as the axioms of an object called an abelian (commutative) group under addition.

In discussions of real vector spaces, real numbers are often called *scalars*. It follows from the definitions of addition and scalar multiplication that if $\mathbf{x}, \mathbf{y} \in \mathbf{R}^n$ and α, β are scalars, then

$$\alpha\mathbf{x} + \beta\mathbf{y} = (\alpha x_1 + \beta y_1, \alpha x_2 + \beta y_2, \ldots, \alpha x_n + \beta y_n).$$

When we said that these two algebraic operations provide $\mathbf{R}^n$ with algebraic structure, we mean precisely that the set $V = \mathbf{R}^n$ supplied with these operations is a real vector space. We record this as the next theorem and encourage the reader to think through a proof of this theorem in Exercise 8.1.1.

Theorem 8.1.4. *The set $\mathbf{R}^n$, with the operations of addition and scalar multiplication in Definition* 8.1.2, *is a real vector space.*

Recall that addition and scalar multiplication of (real or complex valued) functions are defined pointwise: If a set S of functions has common domain D, then for $f, g \in S$,

$$(f + g)(x) = f(x) + g(x), \quad (\alpha f)(x) = \alpha f(x) \quad \text{for all } x \in D.$$

With these definitions of addition and scalar multiplication, a few moments' thought will show that S is a vector space (real or complex) if S is closed under the operations (axiom 1). Note that the additive identity is the zero function whose value at any $x \in D$ is zero (axiom 4). The other axioms (axioms 2-3,5-9) are immediate consequences of the definitions of addition and scalar multiplication of functions.

The following theorem is useful when dealing with subsets W of a known vector space V. A proof is requested in Exercise 8.1.2.

Theorem 8.1.5. *Let V be a real vector space and let W be a subset of V closed under the operations on V, that is, for all $x, y \in W$ and $\alpha \in \mathbf{R}$, we have $x + y \in W$ and $\alpha x \in W$. Then W is a real vector space with the same addition and scalar multiplication as defined on V. We call W a* **subspace** *of the vector space V.*

The variety of vector spaces is immense.

Example 8.1.6. Let $\mathcal{F}[a, b]$ be the set of real valued functions defined on $[a, b]$. It is straightforward to verify that $\mathcal{F}[a, b]$ has the structure of a real vector space if we define pointwise addition and scalar multiplication of functions in the usual way: if $f, g \in \mathcal{F}[a, b]$ and $\alpha \in \mathbf{R}$, then $(f + g)(x) = f(x) + g(x)$, $x \in [a, b]$, and $(\alpha f)(x) = \alpha f(x)$, $x \in [a, b]$. △

The next examples describe some subspaces of the vector space $\mathcal{F}[a, b]$.

Example 8.1.7. Let $B[a, b]$ be the set of real valued functions bounded on $[a, b]$. By definition, $f \in B[a, b]$ if and only if there is some $M > 0$ such that $|f(x)| \leq M$ for all $x \in [a, b]$. Then $B[a, b]$ is a real vector space. For the closure under addition and scalar multiplication, let $f, g \in B[a, b]$ and let M_1, M_2 be real numbers such that $|f(x)| \leq M_1$ and $|g(x)| \leq M_2$ for all $x \in [a, b]$. Then for all $x \in [a, b]$, we have $|f(x) + g(x)| \leq |f(x)| + |g(x)| \leq M_1 + M_2$ and $|(\alpha f)(x)| \leq |\alpha| M_1$. Hence, $f + g, \alpha f \in B[a, b]$. △

Example 8.1.8. The set $C[a, b]$ of real-valued functions continuous on $[a, b]$ is a real vector space. Closure under pointwise addition and scalar multiplication were established in the previous example, since every function continuous on $[a, b]$ is bounded on $[a, b]$. Thus, $C[a, b]$ is a subspace of the vector space $B[a, b]$. △

Example 8.1.9. The set $P[a, b]$ of polynomial functions on $[a, b]$ with real coefficients is a real vector space. It is a subspace of $C[a, b]$, and thus a subspace of $B[a, b]$. △

Example 8.1.10. We assume some basic familiarity with matrices. The set $\mathbf{R}^{mn}$ of $m \times n$ real matrices (m rows and n columns, with real entries) is a vector space if addition and scalar multiplication are defined entrywise:

$$\mathbf{A} = [a_{ij}], \quad \mathbf{B} = [b_{ij}] \quad \Longrightarrow \quad \mathbf{A} + \mathbf{B} = [a_{ij} + b_{ij}],$$

and, if $\alpha \in \mathbf{R}$,

$$\alpha \mathbf{A} = [\alpha a_{ij}],$$

where $1 \leq i \leq m$ and $1 \leq j \leq n$. If the entries are taken to be complex numbers (including real numbers) and scalars are allowed to be complex, then we get the complex vector space $\mathbf{C}^{mn}$ of $m \times n$ complex matrices. △

Let V be a real (or complex) vector space. It follows from the axioms that arbitrary linear combinations of elements of V are also elements in V. That is, if $v_1, v_2, \ldots, v_m \in V$ and $c_1, c_2, \ldots, c_m \in V$, then the finite sum

$$c_1 v_1 + c_2 v_2 + \cdots + c_m v_m$$

is defined unambiguously as an element in V as a consequence of the closure and associativity axioms, and is called a **linear combination** of the v_j.

Definition 8.1.11. *Let V be a real vector space, and X a subset of V. The set X is* **linearly independent** *if whenever $\{x_1, \ldots, x_n\}$ is a finite subset of X and $c_1 x_1 + \cdots + c_n x_n = 0$ for some scalars $c_1, \ldots, c_n$, it follows that $0 = c_1 = \cdots = c_n$. The set X is* **linearly dependent** *if it is not linearly independent, that is, if there is some finite subset $\{x_1, \ldots, x_n\}$ of X and scalars $c_1, \ldots, c_n$, not all zero, such that $c_1 x_1 + \cdots + c_n x_n = 0$.*

A subset $S \subset V$ is a **spanning set** for V if every element of V can be written as a linear combination of elements from S; then we also say that S **spans** V. A **basis** B of V is a linearly independent spanning set. A vector space V is **finite-dimensional** if it contains a finite spanning set; if no finite spanning set exists, then V is **infinite-dimensional**. For a finite-dimensional space, the **dimension** of the space is the minimum number of elements in a spanning set. One can show that, in a finite-dimensional space, a spanning set is a basis if and only if it has minimal size.

There are some special and important vectors in $\mathbf{R}^n$ which we now identify. In $\mathbf{R}^n$, the n vectors

$$\mathbf{e}_1 = (1, 0, \ldots, 0), \quad \mathbf{e}_2 = (0, 1, 0, \ldots, 0), \quad \ldots, \quad \mathbf{e}_n = (0, \ldots, 0, 1)$$

are called the **standard basis vectors** for $\mathbf{R}^n$. These vectors form a basis for $\mathbf{R}^n$. Consequently, every vector $\mathbf{x}$ in $\mathbf{R}^n$ can be written as a linear combination of the $\mathbf{e}_j$ in a unique way; see Exercise 8.1.3.

The spaces $B[a, b]$, $C[a, b]$ and $P[a, b]$ are infinite-dimensional. The space $\mathbf{R}^n$ is finite-dimensional with dimension n (by Exercise 8.1.3). The matrix space $\mathbf{R}^{mn}$ is finite-dimensional and has dimension mn; a basis is given by the matrices $\mathbf{E}_{ij}$, $1 \le i \le m$, $1 \le j \le n$, where $\mathbf{E}_{ij}$ has all entries zero except for the (i, j)-entry, which equals 1.

Vector spaces are present everywhere in mathematical analysis. We give several more examples below. In each of these examples, one should check closure of the relevant set under the operations of addition and scalar multiplication; the other algebraic properties for a vector space are then easily seen to be satisfied.

Example 8.1.12. The rational number field $V = \mathbf{Q}$ is a vector space with the rational number field $\mathbf{Q}$ as the field of scalars. This space is not especially useful for analysis since $\mathbf{Q}$ is not complete. The real field $\mathbf{R} = \mathbf{R}^1$ is a vector space with the real number field $\mathbf{R}$ as the field of scalars. △

Example 8.1.13. Let $C'[a, b]$ be the set of real valued functions defined on $[a, b]$ and differentiable on (a, b). Then $C'[a, b]$ is a real vector space. △

Example 8.1.14. Let $C^1[a, b]$ be the set of real valued functions defined on $[a, b]$ and having a continuous derivative on (a, b). Then $C^1[a, b]$ is a real vector space and it is a proper subspace of $C'[a, b]$. We also note that $C[a, b]$ is a proper subspace of $C'[a, b]$. △

Example 8.1.15. Let $\mathcal{R}[a, b]$ be the set of real valued functions that are integrable on $[a, b]$. Then $\mathcal{R}[a, b]$ is a real vector space. Only closure needs to be checked since $\mathcal{R}[a, b]$ is a subset of the vector space $\mathcal{F}[a, b]$ of real valued functions on $[a, b]$. Clearly, $\mathcal{R}[a, b]$ is closed under real scalar multiplication. If f and g are in $\mathcal{R}[a, b]$, then the discontinuities of $f + g$ are contained in the union of the sets of discontinuities of f and g. Since those sets have Lebesgue measure zero, so does the set of discontinuities of $f + g$, hence $f + g \in \mathcal{R}[a, b]$. △

Additional examples of real vector spaces are considered in the exercises for this section.

Exercises.

Exercise 8.1.1. Prove Theorem 8.1.4. *Hint*: Many properties might be verified without writing them down, but writing them is good practice. Be sure to prove the uniqueness of the additive identity and uniqueness of additive inverses.

Exercise 8.1.2. Prove Theorem 8.1.5.

Exercise 8.1.3. Show that the standard basis vectors $\mathbf{e}_1, \ldots, \mathbf{e}_n$ deserve their name: Show that $\{\mathbf{e}_1, \ldots, \mathbf{e}_n\}$ is a basis of $\mathbf{R}^n$. Then show that every vector $\mathbf{x}$ in $\mathbf{R}^n$ can be written in a unique way as a linear combination of the $\mathbf{e}_j$.

Exercise 8.1.4. Check closure under addition and scalar multiplication for each of the spaces considered in Examples 8.1.13-8.1.14. Also, state the containment relations for all pairs of the spaces $R[a,b]$, $P[a,b]$, $C[a,b]$, $C'[a,b]$, $C^1[a,b]$ and $B[a,b]$.

Exercise 8.1.5. Consider the set S consisting of all real sequences $\xi = (\xi_k)$. Prove: S is a real vector space.

Exercise 8.1.6. Consider the set l^1 consisting of the real sequences $\xi = (\xi_k)$ such that $\sum_{k=1}^{\infty} |\xi_k|$ converges. Prove: l^1 is a real vector space.

Exercise 8.1.7. Consider the set l^∞ consisting of the bounded real sequences $\xi = (\xi_k)$. Prove: l^∞ is a real vector space.

8.2. The Euclidean Inner Product

We are assuming that the reader has some experience from an introductory multivariable calculus course with the dot product

$$\mathbf{x} \cdot \mathbf{y} = x_1 y_1 + x_2 y_2 + x_3 y_3$$

of vectors $\mathbf{x} = (x_1, x_2, x_3)$ and $\mathbf{y} = (y_1, y_2, y_3)$ in $\mathbf{R}^3$. In this book, we denote this product by

$$(\mathbf{x}, \mathbf{y}) = x_1 y_1 + x_2 y_2 + x_3 y_3,$$

and we shall use a similar notation for generalizations of this product in certain other vector spaces.

The next definition is based on the fundamental properties of this product.

Definition 8.2.1. *Let V be a real vector space. A function mapping $V \times V$ into $\mathbf{R}$, with values denoted (x, y), is a real* **inner product** *if it has these properties:*

(i) $(x, x) \geq 0$ *for all* $x \in V$, *and* $(x, x) = 0$ *if and only if* $x = 0$;

(ii) $\alpha(x, y) = (\alpha x, y) = (x, \alpha y)$ *for all* $x, y \in V$ *and* $\alpha \in \mathbf{R}$;

(iii) $(x, y) = (y, x)$ *for all* $x, y \in V$;

(iv) $(x, y + z) = (x, y) + (x, z)$ *for all* $x, y, z \in V$.

Here are some important consequences of these properties. For any $x \in V$, $(x, 0) = 0$; this follows from (ii), since we can take $\alpha = 0$ and any x, y to give

$$0 = 0(x, y) = (x, 0y) = (x, 0).$$

From properties (iii) and (iv), it follows that $(x+y,z) = (x,z)+(y,z)$ for all $x,y,z \in V$. Property (iii) says that a real inner product is *symmetric* in its arguments. Properties (ii), (iii) and (iv) combine to imply that an inner product is linear in both of its arguments, so it is *bilinear*:

$$(\alpha x + \beta y, z) = \alpha(x,z) + \beta(y,z) \quad \text{and} \quad (z, \alpha x + \beta y) = \alpha(z,x) + \beta(z,y)$$

for all $x, y, z \in V$ and real numbers α and β.

The familiar dot product of vectors in $\mathbf{R}^3$ is an inner product by this definition. It is a special case of the next theorem.

Theorem 8.2.2. *The mapping* $\mathbf{R}^n \times \mathbf{R}^n$ *to* $\mathbf{R}$ *defined by*

$$(\mathbf{x}, \mathbf{y}) = x_1 y_1 + x_2 y_2 + \cdots + x_n y_n, \quad \textit{for } \mathbf{x}, \mathbf{y} \in \mathbf{R}^n,$$

is an inner product. This is called the **Euclidean inner product**, *or the* **standard inner product**, *on* $\mathbf{R}^n$.

Proof. We have $(\mathbf{x}, \mathbf{x}) = x_1^2 + x_2^2 + \cdots + x_n^2 \geq 0$ for all $\mathbf{x} \in V$. If $\mathbf{x} \neq \mathbf{0}$, then for some j, $x_j \neq 0$, hence $(\mathbf{x}, \mathbf{x}) > 0$; thus, if $(\mathbf{x}, \mathbf{x}) = 0$, then $x_j = 0$ for each j. Therefore property (i) of the definition holds. For property (ii),

$$\alpha \sum_{j=1}^{n} x_j y_j = \sum_{j=1}^{n} (\alpha x_j) y_j = \sum_{j=1}^{n} x_j (\alpha y_j),$$

by the distributive and associative laws of the real numbers, hence

$$\alpha(\mathbf{x}, \mathbf{y}) = (\alpha \mathbf{x}, \mathbf{y}) = (\mathbf{x}, \alpha \mathbf{y}).$$

Property (iii) follows from commutativity of real number multiplication. Verification of property (iv) of the definition is left to Exercise 8.2.1. □

A real vector space V with a real inner product $(\cdot, \cdot)$ is called a **real inner product space**.

Theorem 8.2.3 (Cauchy-Schwarz Inequality). *If* V *is a real inner product space and* $x, y \in V$, *then*

$$|(x,y)| \leq \sqrt{(x,x)}\sqrt{(y,y)},$$

and equality holds if and only if the vectors are collinear, that is, if and only if $x + t_0 y = 0$ *for some* $t_0 \in \mathbf{R}$.

Proof. If $y = 0$, then $(y, y) = 0$, as we have shown, so the stated inequality simply requires that $(x, 0) = 0$, and we have also shown this above. Now assume that x and $y \neq 0$ are fixed vectors in V and t is real. Properties (ii), (iii) and (iv) of Definition 8.2.1 imply that

$$(x + ty, x + ty) = (x, x) + 2t(x, y) + t^2(y, y).$$

The right-hand side is a quadratic polynomial in t, and by property (i) of Definition 8.2.1,

$$(x + ty, x + ty) \geq 0,$$

hence, for all t,

$$(x, x) + 2t(x, y) + t^2(y, y) \geq 0.$$

The minimum value of this quadratic polynomial must occur at the point

$$t = t_0 = -\frac{(x, y)}{(y, y)}.$$

Substituting t_0 for t in the quadratic implies

$$0 \leq (x + t_0 y, x + t_0 y) = (x, x) - \frac{(x, y)^2}{(y, y)},$$

which is equivalent to

$$(x, y)^2 \leq (x, x)(y, y).$$

Taking the positive square root of each side yields the desired result. Finally, note that equality holds if and only if $0 = (x + t_0 y, x + t_0 y)$, hence if and only if $x + t_0 y = 0$. □

For the Euclidean inner product of vectors $\mathbf{x}, \mathbf{y}$ in $\mathbf{R}^n$, the Cauchy-Schwarz inequality is

$$|(\mathbf{x}, \mathbf{y})| \leq \sqrt{(\mathbf{x}, \mathbf{x})}\sqrt{(\mathbf{y}, \mathbf{y})}.$$

Letting $\mathbf{y} = \mathbf{e}_j$, the j-th standard basis vector, we immediately have

$$|x_j| = |(\mathbf{x}, \mathbf{e}_j)| \leq \sqrt{(\mathbf{x}, \mathbf{x})}, \quad \text{where } \mathbf{x} = (x_1, \ldots, x_n).$$

The next example introduces an important inner product for a space of continuous functions.

Example 8.2.4. Consider the space $C[a, b]$ of continuous functions on $[a, b]$. We define an inner product on $C[a, b]$ by

$$(f, g) = \int_a^b f(x)g(x)\,dx \quad \text{for } f, g \in C[a, b].$$

It is straightforward to check that this is indeed an inner product on $C[a, b]$. Let us verify in particular that $(f, f) = 0$ implies $f = \theta \in C[a, b]$. To see it, notice that if $f(x_0) \neq 0$ for some $x_0 \in [a, b]$, then $[f(x_0)]^2 > 0$, and by continuity, $[f(x)]^2 > 0$ for all x in some neighborhood of x_0. But then $(f, f) = \int_a^b [f(x)]^2\,dx > 0$. Thus, if $(f, f) = 0$, then necessarily $f(x) \equiv 0$, that is, $f = \theta \in C[a, b]$. △

Example 8.2.5. Consider the vector space $\mathcal{R}[a, b]$ of Riemann integrable functions on $[a, b]$ (Example 8.1.15). If we define

$$(f, g) = \int_a^b f(x)g(x)\,dx \quad \text{for } f, g \in \mathcal{R}[a, b],$$

then it can be verified that this product satisfies all the requirements for an inner product in Definition 8.2.1 except condition (i) which states that $(f, f) = 0$ implies $f = \theta \in \mathcal{R}[a, b]$. For example, choose any finite set F of points in $[a, b]$ and let f take the value 1 on those points, and then define $f(x) = 0$ for $x \in [a, b] - F$. Then f is Riemann integrable and $\int_a^b [f(x)]^2\,dx = 0$, but $f \neq \theta \in \mathcal{R}[a, b]$. What we can say is that if $(f, f) = 0$, then $\int_a^b f^2\,dx = 0$, hence $f^2 = 0$ almost everywhere in $[a, b]$, and hence $f = 0$ almost everywhere in $[a, b]$. Since the product (f, g) is so useful, we can modify our thinking about the Riemann integrable functions by identifying two functions if they are equal except on a set of measure zero in $[a, b]$, that is, they are equal a.e. in $[a, b]$. (We can define an equivalence relation $\sim$ on $\mathcal{R}[a, b]$ such that

$f \sim g$ if and only if $f = g$ almost everywhere in $[a, b]$. Then we have well defined operations of addition and scalar multiplication of equivalence classes. See Exercise 8.2.4.) In practice, we can continue to work with individual functions and simply identify functions that are equal almost everywhere in $[a, b]$. By this agreement, we consider $\mathcal{R}[a, b]$ to be an inner product space with inner product (f, g) as defined above. We remark that, in linear algebra terms, we are thus actually working with the quotient space, $\mathcal{R}[a, b]$ modulo the subspace of functions that are equal to zero almost everywhere in $[a, b]$. △

Exercises.

Exercise 8.2.1. Complete the proof of Theorem 8.2.2 by verifying that property (iv) of Definition 8.2.1 holds for the Euclidean inner product $(\mathbf{x}, \mathbf{y})$ on $\mathbf{R}^n$.

Exercise 8.2.2. Show that $(\mathbf{x}, \mathbf{x}) = 0$ if and only if $\mathbf{x} = \mathbf{0}$, using only the definition of inner product (Definition 8.2.1).

Exercise 8.2.3. Prove: If $a_1, a_2, \ldots, a_n \in \mathbf{R}$, then

$$\left(\sum_{i=1}^{n} a_i\right)^2 \le n \sum_{i=1}^{n} a_i^2.$$

Exercise 8.2.4. Consider the equivalence relation $\sim$ on $\mathcal{R}[a, b]$: $f \sim g$ if and only if $f = g$ almost everywhere in $[a, b]$.

1. Show that if $f_1 \sim f_2$, $g_1 \sim g_2$ and $\alpha \in \mathbf{R}$, then $[f_1] + [g_1] = [f_2] + [g_2]$ and $\alpha[f_1] = \alpha[f_2]$. Thus, the addition and scalar multiplication of equivalence classes of Riemann integrable functions on $[a, b]$ is well defined.
2. Verify that the equivalence classes, with addition and scalar multiplication as defined, form a real vector space.
3. Show that $([f], [g]) := \int_a^b f(x)g(x)\, dx$ defines an inner product on the space of equivalence classes. First, show that if $f_1 \sim f_2$ and $g_1 \sim g_2$, then $([f_1], [g_1]) = ([f_2], [g_2])$. Then verify properties (i), (ii), (iii), (iv) of Definition 8.2.1.

8.3. Norms

On the real line we measure the magnitude of a number by its absolute value and the length of an interval $[a, b]$ (or the distance between any two points a and b) by $|a - b|$. In the plane, we measure the length of a line segment from $\mathbf{a}$ to $\mathbf{b}$ by the formula

$$\sqrt{(b_1 - a_1)^2 + (b_2 - a_2)^2},$$

and this is the same as the length of the vector $(b_1 - a_1, b_2 - a_2) = \mathbf{b} - \mathbf{a}$. We have a similar distance formula in three-dimensional space. In this section we begin to think about the general idea of the magnitude, or *norm*, of vectors in a vector space. As we will see, there can be many ways to define this magnitude measure, or norm, and a variety of norms are useful in problems of analysis.

Remark. *We use a single bar notation, $|\cdot|$, for norms in $\mathbf{R}^n$, and different norms may be subscripted differently, for example, $|\mathbf{x}|_1$ or $|\mathbf{x}|_2$. We also use a single bar notation, $|\cdot|$, in some general statements about general normed vector spaces. In contrast, for matrix norms and function space norms, we use the double bar: $\|\cdot\|$.*

Definition 8.3.1. (Norm)
Let V be a real vector space. A function mapping V into $\mathbf{R}$, with values denoted by $|x|$ for $x \in V$, is a **norm** *if it has these properties:*

(i) $|x| \geq 0$ *for all* $x \in V$, *and* $|x| = 0$ *if and only if* $x = 0$.

(ii) $|\alpha x| = |\alpha|\,|x|$ *for all* $x \in V$ *and all* $\alpha \in \mathbf{R}$.

(iii) $|x + y| \leq |x| + |y|$ *for all* $x, y \in V$ *(the triangle inequality).*

A real vector space equipped with a norm is called a **real normed vector space**, or a real normed linear space.

The triangle inequality (iii) implies another useful inequality, as follows. For all x, y in a real normed space, we have

$$\Big||x| - |y|\Big| \leq |x + y|. \tag{8.1}$$

The proof of this **reverse triangle inequality** is the same as the argument for the corresponding statement about absolute values in the vector space of real numbers. (See Exercise 8.3.1.)

An inner product space is always a normed space, by the next result.

Theorem 8.3.2. *If V is a real vector space with an inner product, then the function $|\cdot|$ on V given by $|x| = \sqrt{(x, x)}$ defines a norm on V.*

Proof. We verify directly properties (i), (ii), and (iii) of Definition 8.3.1.

(i) If $x \neq 0$, then $|x| > 0$ since $(x, x) > 0$. If $x = 0$, then $(x, x) = 0$, hence $|x| = 0$.

(ii) $|\alpha x| = \sqrt{(\alpha x, \alpha x)} = \sqrt{\alpha^2 (x, x)}$ using property (ii) of the inner product Definition 8.2.1. Hence, $|\alpha x| = |\alpha||x|$.

(iii) Using first the distributive property and then the commutative property of an inner product, we find that

$$|x + y|^2 = (x + y, x + y) = (x, x) + 2(x, y) + (y, y).$$

By the Cauchy-Schwartz inequality and the definition of $|x|$, we may write $2(x, y) \leq 2|x||y|$, and consequently,

$$|x + y|^2 \leq (|x| + |y|)^2.$$

Taking the positive square root of each side yields the triangle inequality. □

The Euclidean inner product yields the **Euclidean norm** on the real vector space $\mathbf{R}^n$ by the construction in Theorem 8.3.2. We record this in the next corollary.

Corollary 8.3.3. *The function from $\mathbf{R}^n$ to $\mathbf{R}^n$ defined by*

$$|\mathbf{x}|_2 = \sqrt{(\mathbf{x}, \mathbf{x})} = \sqrt{x_1^2 + x_2^2 + \cdots + x_n^2}$$

for $\mathbf{x} = (x_1, x_2, \ldots, x_n) \in \mathbf{R}^n$, is a norm, called the **Euclidean norm**.

The definition of a norm in an inner product space V as in Theorem 8.3.2 allows us to express the Cauchy-Schwarz inequality in terms of the induced norm, by writing

$$|(x, y)| \leq |x|\,|y|.$$

For the Euclidean inner product (the dot product) and the associated Euclidean norm, the Cauchy-Schwarz inequality reads

$$|(\mathbf{x}, \mathbf{y})| \leq |\mathbf{x}|_2 \, |\mathbf{y}|_2.$$

Since $|\mathbf{e}_j|_2 = 1$ for each standard basis vector, the Cauchy-Schwarz inequality implies that

$$|x_j| = |(\mathbf{x}, \mathbf{e}_j)| \leq |\mathbf{x}|_2$$

for $j = 1, 2, \ldots, n$, for any $\mathbf{x} = (x_1, \ldots, x_n)$.

We have seen that an inner product allows us the useful notion of orthogonality (perpendicularity) of two vectors. Recall that the angle θ between two nonzero vectors $\mathbf{x}, \mathbf{y}$ in $\mathbf{R}^n$ is defined by the formula

$$\cos\theta = \frac{(\mathbf{x}, \mathbf{y})}{|\mathbf{x}|_2 |\mathbf{y}|_2}, \quad \text{where } 0 \leq \theta \leq \pi.$$

This formula agrees with the ones in the case of analytic geometry in the plane $n = 2$ and in space $n = 3$. We say that the vectors $\mathbf{x}$ and $\mathbf{y}$ are **orthogonal** if $(\mathbf{x}, \mathbf{y}) = 0$. This definition also gives us the following generalization of the law of cosines from trigonometry: By expansion of the inner products, we have

$$|\mathbf{x} - \mathbf{y}|_2^2 = |\mathbf{x}|_2^2 + |\mathbf{y}|_2^2 - 2(\mathbf{x}, \mathbf{y}),$$

and hence

$$|\mathbf{x} - \mathbf{y}|_2^2 = |\mathbf{x}|_2^2 + |\mathbf{y}|_2^2 - 2|\mathbf{x}|_2 |\mathbf{y}|_2 \cos\theta.$$

Another interesting geometric property of a norm is the **parallelogram law**. See Exercises 8.3.2, 8.3.3.

A norm allows us to define open balls and the notion of an open set in a normed space. The definitions will be familiar as they are direct generalizations of open intervals and open sets in the real line. If V is a normed space with norm $|\cdot|$, then the **open ball** of radius $\delta > 0$ centered at the point $a \in V$ is the set $B_\delta(a) = \{x \in V : |x - a| < \delta\}$. If $S \subset V$, then a point $a \in S$ is an **interior point** of S if there exists some $\delta > 0$ such that $B_\delta(a) \subset S$. A set S is **open** if every point of S is an interior point. A set $F \subset V$ is **closed** if its complement $V - F$ is open. All topological notions and convergence questions in V are based on these concepts, and explored further in later sections of this chapter. We have the following basic definition.

Definition 8.3.4. *Let V be a normed vector space with norm $|\cdot|$.*

1. *A sequence $(\mathbf{a}_k)$ in V is a* **Cauchy sequence** *if for every $\epsilon > 0$ there is a positive integer $N = N(\epsilon)$ such that if $m, n \geq N$, then $|x_m - x_n| < \epsilon$.*
2. *A sequence (x_k)* **converges with limit** *$b \in V$ if for every $\epsilon > 0$ there is an $N = N(\epsilon)$ such that if $k \geq N$, then $|x_k - b| < \epsilon$.*
3. *A normed space V is called* **complete** *if every Cauchy sequence in V converges to a limit that is an element of V.*

If a sequence converges, there is a unique limit, and the proof is formally identical to the proof in the case of convergent real sequences.

Recalling Example 8.2.5, we see that a norm on the space of (equivalence classes of) Riemann integrable functions is induced by the inner product given there. That is,

$$\|f\|_2 := (f,f)^{1/2} = \left(\int_a^b |f(x)|^2\,dx\right)^{1/2}, \tag{8.2}$$

which is called the L^2 **norm**. (The symbolic designation L^2 norm is due to the important role of this norm in the study of the vector space of functions that are square integrable in the Lebesgue sense. This space is discussed in greater detail later in this book after the development of the Lebesgue integral.) To explore this norm now for Riemann integrable functions, see Exercises 8.3.10, 8.3.11, 8.3.12.

Example 8.3.5 (Hilbert sequence space l^2). Here we introduce the interesting set l^2 of real number sequences that are square summable, that is,

$$l^2 = \left\{(\xi_k)_{k=1}^\infty : \sum_{k=1}^\infty \xi_k^2 < \infty\right\}.$$

If $x = (\xi_1, \xi_2, \xi_3, \ldots)$ and $y = (\eta_1, \eta_2, \eta_3, \ldots)$ are in l^2, we shall write

$$(x,y) = \sum_{k=1}^\infty \xi_k \eta_k \tag{8.3}$$

and

$$\|x\|_2 = (x,x)^{1/2} = \left(\sum_{k=1}^\infty \xi_k^2\right)^{1/2}. \tag{8.4}$$

We now justify this notation by showing that l^2 is indeed a real vector space, (8.3) does define an inner product on l^2, and thus (8.4) is indeed a norm on l^2. This space is called the **Hilbert sequence space**. It is studied in more detail at several places later in the book. Let us verify that l^2 is a real vector space under componentwise addition and scalar multiplication, by which we mean that if x and y are in l^2, with

$$x = (\xi_1, \xi_2, \xi_3, \ldots) \quad \text{and} \quad y = (\eta_1, \eta_2, \eta_3, \ldots),$$

and $\alpha \in \mathbf{R}$, then

$$\alpha x = (\alpha\xi_1, \alpha\xi_2, \alpha\xi_3, \ldots) \quad \text{and} \quad x + y = (\xi_1 + \eta_1, \xi_2 + \eta_2, \xi_3 + \eta_3, \ldots).$$

The zero vector in l^2 is the element of l^2 with all entries zero. With these definitions it is clear that if $x \in l^2$, then $\alpha x \in l^2$. Now let us show that $x + y \in l^2$. For each positive integer n, let us write

$$x_n = (\xi_1, \ldots, \xi_n, 0, 0, \ldots) \quad \text{and} \quad y_n = (\eta_1, \ldots, \eta_n, 0, 0, \ldots),$$

where all components beyond the n-th are zero. By the triangle inequality in the finite-dimensional space $\mathbf{R}^n$, we have

$$\begin{aligned}\left(\sum_{k=1}^n (\xi_k + \eta_k)^2\right)^{1/2} &\le \left(\sum_{k=1}^n \xi_k^2\right)^{1/2} + \left(\sum_{k=1}^n \eta_k^2\right)^{1/2}\\ &\le \left(\sum_{k=1}^\infty \xi_k^2\right)^{1/2} + \left(\sum_{k=1}^\infty \eta_k^2\right)^{1/2}.\end{aligned}$$

Since the right-hand side is a finite real number, we may let $n \to \infty$ on the left to see that

$$\sum_{k=1}^{\infty}(\xi_k + \eta_k)^2 < \infty$$

and thus $x + y \in l^2$, and in fact,

$$\|x + y\|_2 \leq \|x\|_2 + \|y\|_2$$

by means of (8.4). It is easy to verify that this directly establishes the norm on l^2. We now want to show that the series (8.3) converges, from which fact we can define the inner product on l^2. Using the same finite truncations of $x, y \in l^2$, the Cauchy-Schwarz inequality in the finite-dimensional space $\mathbf{R}^n$ yields

$$(x_n, y_n) = \sum_{k=1}^{n} \xi_k \eta_k \leq \Big(\sum_{k=1}^{n} \xi_k^2\Big)^{1/2} \Big(\sum_{k=1}^{n} \eta_k^2\Big)^{1/2}.$$

We may let $n \to \infty$ to obtain

$$(x, y) = \sum_{k=1}^{\infty} \xi_k \eta_k \leq \Big(\sum_{k=1}^{\infty} \xi_k^2\Big)^{1/2} \Big(\sum_{k=1}^{\infty} \eta_k^2\Big)^{1/2} = \|x\|_2 \|y\|_2 < \infty$$

which establishes at the same time the convergence of (8.3) and the Cauchy-Schwarz inequality for the inner product. All the defining properties of a real inner product in Definition 8.2.1 are now easy to verify. Therefore the inner product for the vector space l^2 is well defined by (8.3), and the induced norm is given by (8.4). $\triangle$

The next three examples present norms on $\mathbf{R}^n$ different from the Euclidean norm.

Example 8.3.6. Define $|\mathbf{x}|_1$ for $\mathbf{x} \in \mathbf{R}^n$ by

$$|\mathbf{x}|_1 = \sum_{j=1}^{n} |x_j| = |x_1| + |x_2| + \cdots + |x_n|.$$

Again, the triangle inequality follows from the triangle inequality for real numbers. If $\mathbf{x}, \mathbf{y} \in \mathbf{R}^n$, then

$$|\mathbf{x} + \mathbf{y}|_1 = \sum_{j=1}^{n} |x_j + y_j| \leq \sum_{j=1}^{n} |x_j| + |y_j| = |\mathbf{x}|_1 + |\mathbf{y}|_1.$$

Properties (i) and (ii) of Definition 8.3.1 are left as an exercise. $\triangle$

Example 8.3.7. Let p be a real number greater than 1, and $p \neq 2$. Define $|\mathbf{x}|_p$ for $\mathbf{x} \in \mathbf{R}^n$ by

$$|\mathbf{x}|_p = \Big(\sum_{j=1}^{n} |x_j|^p\Big)^{1/p}.$$

This defines a norm on $\mathbf{R}^n$. Properties (i) and (ii) of the norm definition clearly hold. Property (iii) is the triangle inequality, also known as the Minkowski inequality for finite sums in the present instance, and it is proved near the end of Section 9.4. $\triangle$

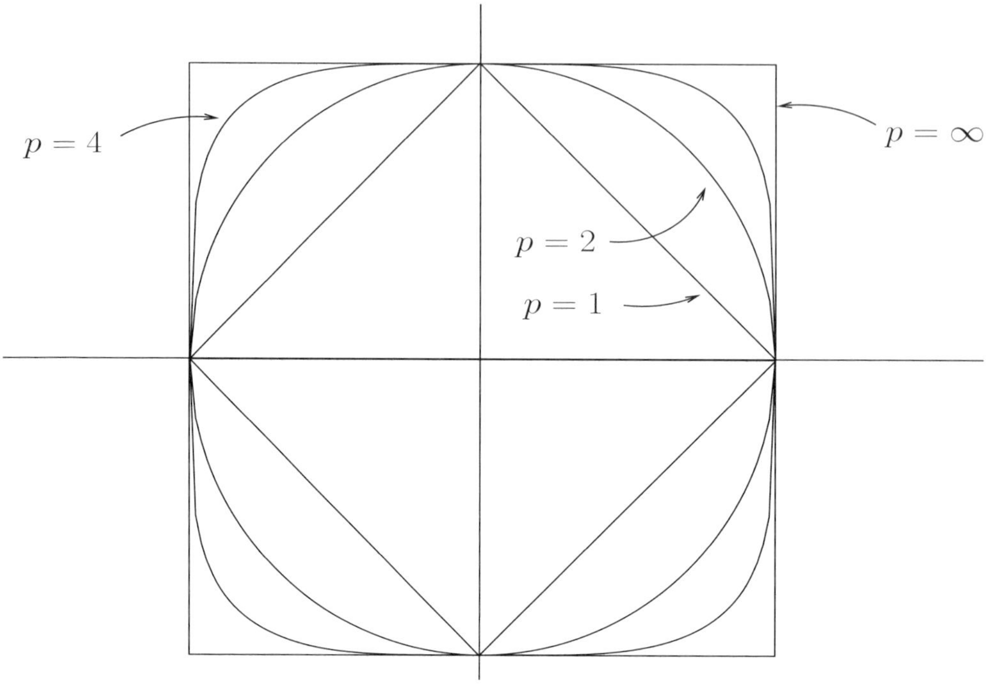

Figure 8.1. The p-norm unit balls in $\mathbf{R}^2$ for $p = 1, 2, 4, \infty$. The arrows point to the boundary (the unit sphere) for each ball.

Example 8.3.8. Define $|\mathbf{x}|_\infty$ for $\mathbf{x} \in \mathbf{R}^n$ by

$$|\mathbf{x}|_\infty = \max_{1 \le j \le n} |x_j|.$$

This can be called the max norm on $\mathbf{R}^n$. The triangle inequality follows from the triangle inequality for real numbers. If $\mathbf{x}, \mathbf{y} \in \mathbf{R}^n$, we have

$$\max_{1 \le j \le n} |x_j + y_j| \le \max_{1 \le j \le n} (|x_j| + |y_j|) \le \max_{1 \le j \le n} |x_j| + \max_{1 \le j \le n} |y_j|,$$

hence $|\mathbf{x} + \mathbf{y}|_\infty \le |\mathbf{x}|_\infty + |\mathbf{y}|_\infty$. Properties (i) and (ii) of Definition 8.3.1 clearly hold for $|\cdot|_\infty$. △

It is interesting to compare the unit balls for the p-norms $|\cdot|_p$, for $p = 1, 2, 4, \infty$. Figure 8.1 shows the unit balls for these norms in $\mathbf{R}^2$. The next proposition shows the relation between the max norm and the p-norms and explains the notation, $|\cdot|_\infty$, for the max norm.

Proposition 8.3.9. *For any fixed* $\mathbf{x} \in \mathbf{R}^n$,

$$\lim_{p \to \infty} |\mathbf{x}|_p = |\mathbf{x}|_\infty.$$

Proof. It is clear for $\mathbf{x} = \mathbf{0}$, so let $\mathbf{x} = (x_1, \ldots, x_n)$ be nonzero, and assume, without loss of generality, that the first component of $\mathbf{x}$ is the one with the maximum absolute value, thus, $|\mathbf{x}|_\infty = \max_{1\le j\le n} |x_j| = |x_1| \neq 0$. Then

$$\begin{aligned} |\mathbf{x}|_p &= \Big(\sum_{j=1}^n |x_j|^p\Big)^{1/p} = \Big(|x_1|^p + |x_2|^p + \cdots + |x_n|^p\Big)^{1/p} \\ &= |x_1|\left[1 + \left|\frac{x_2}{x_1}\right|^p + \cdots + \left|\frac{x_n}{x_1}\right|^p\right]^{1/p} \\ &= |\mathbf{x}|_\infty \left[1 + \left|\frac{x_2}{x_1}\right|^p + \cdots + \left|\frac{x_n}{x_1}\right|^p\right]^{1/p}. \end{aligned}$$

We can write

$$\log\left[1 + \left|\frac{x_2}{x_1}\right|^p + \cdots + \left|\frac{x_n}{x_1}\right|^p\right]^{1/p} = \frac{1}{p}\log\left[1 + \left|\frac{x_2}{x_1}\right|^p + \cdots + \left|\frac{x_n}{x_1}\right|^p\right],$$

and as $p \to \infty$, this quantity has limit zero, since $|x_j/x_1| < 1$ for $2 \le j \le n$ and log is continuous at 1. Therefore by continuity of the exponential function at 0,

$$\lim_{p\to\infty}\left[1+\left|\frac{x_2}{x_1}\right|^p+\cdots+\left|\frac{x_n}{x_1}\right|^p\right]^{1/p} = \lim_{p\to\infty}\exp\left(\log\left[1+\left|\frac{x_2}{x_1}\right|^p+\cdots+\left|\frac{x_n}{x_1}\right|^p\right]^{1/p}\right) = 1.$$

It follows that $\lim_{p\to\infty} |\mathbf{x}|_p = |\mathbf{x}|_\infty$. □

The space $C[a,b]$ can also be normed in different ways.

Example 8.3.10. The definition $\|f\|_{\max} = \max_{a\le x\le b} |f(x)|$ gives a norm on $C[a,b]$, the real vector space of real valued functions continuous on $[a,b]$. A different norm on $C[a,b]$ is defined by $\|f\|_1 = \int_a^b |f(x)|\,dx$. Both statements are straightforward to verify. (See Exercise 8.3.7 and Exercise 8.3.8.) This space is discussed more fully later. In particular, the completeness of $C[a,b]$ with respect to the norm $\|f\|_{\max}$ (that is, the fact that every Cauchy sequence converges) is shown in Theorem 9.3.1. However, $C[a,b]$ is *not* complete with respect to the norm $\|f\|_1$. (See Exercise 9.3.2 or Exercise 9.3.3.) △

Since we have been able to define different norms on $\mathbf{R}^n$, two questions arise. First, does the topology (the collection of open sets and the notion of convergence) change with the norm employed in $\mathbf{R}^n$? We show below that all norms on $\mathbf{R}^n$ are equivalent in the sense that they define the same collection of open sets and the same notion of convergence in $\mathbf{R}^n$. (And in light of the statement about $C[a,b]$ in Example 8.3.10, this norm equivalence in $\mathbf{R}^n$ should be a reassuring statement about Euclidean space.) Second, is it useful to employ more than one norm function on $\mathbf{R}^n$? The answer is Yes, as it is a matter of convenience, and this is illustrated more fully later.

We now proceed to discuss the equivalence of norms and the proof that all norms on $\mathbf{R}^n$ are equivalent. We first require an appropriate notion of equivalence of norms. Recall we use single bars for vector norms.

Definition 8.3.11. *Let V be a vector space (finite- or infinite-dimensional). Two norms $|\cdot|$ and $|\cdot|_0$ on V are called* **equivalent** *if there are numbers $\alpha > 0$ and $\beta > 0$ such that for all $x \in V$,*

$$\alpha|x| \le |x|_0 \le \beta|x|.$$

If the norms $|\cdot|$ and $|\cdot|_0$ on V are equivalent, then every open ball in one norm contains an open ball in the other norm (Exercise 8.3.15). Consequently, the two norms define the same collection of open subsets of V. Also, a sequence converges as measured by $|\cdot|$ if and only if it converges as measured by $|\cdot|_0$ (Exercise 8.3.16).

The next lemma will be used to establish the uniqueness of a norm topology on a finite-dimensional real vector space V. This means that the choice of a norm on V is a matter of convenience, as all possible norms on V generate the same collection of open sets in the space. This is a nice feature of finite-dimensional spaces, since a particular norm sometimes offers advantages in certain arguments. The proof of the lemma uses only the defining properties of a norm and the Bolzano-Weierstrass theorem for sequences of real numbers.

Lemma 8.3.12. *Let V be a finite-dimensional real normed vector space with norm $|\cdot|$ and suppose that $\{v_1, \ldots, v_n\}$ is any basis of V. Then there exists a number $m > 0$ such that if $v = \sum_{i=1}^n a_i v_i$ for real scalars $a_1, \ldots, a_n$, then*

$$|v| = \left|\sum_{i=1}^n a_i v_i\right| \geq m(|a_1| + \cdots + |a_n|). \tag{8.5}$$

Proof. We note first that if all $a_i = 0$, then (8.5) holds for any m. Thus we may assume that $\sum_{i=1}^n |a_i| > 0$. Now observe that the inequality in (8.5) implies that

$$\left|\sum_{i=1}^n b_i v_i\right| \geq m \tag{8.6}$$

for all $(b_1, \ldots, b_n)$ such that $\sum_{i=1}^n |b_i| = 1$. And, conversely, if (8.6) holds for all $(b_1, \ldots, b_n)$ such that $\sum_{i=1}^n |b_i| = 1$, then (8.5) holds for all $\sum_{i=1}^n |a_i| > 0$, as we see by setting $b_i = a_i/(\sum_{i=1}^n |a_i|)$ and dividing (8.5) by $\sum_{i=1}^n |a_i|$ to obtain (8.6), using property (ii) of the norm definition.

Thus, we will prove that there exists $m > 0$ such that (8.6) holds for all $(b_1, \ldots, b_n)$ such that $\sum_{i=1}^n |b_i| = 1$. The proof is by contradiction. Thus, we assume that there is no such $m > 0$. Then there exists a sequence of elements y_k of V,

$$y_k = \sum_{i=1}^n b_i^k v_i = b_1^k v_1 + \cdots + b_n^k v_n,$$

such that $|y_k| \to 0$ as $k \to \infty$. (Note that the *superscript* k is not a power.) And for each k, we have $\sum_{i=1}^n b_i^k = 1$. Consequently, for all k and each $i = 1, \ldots, n$, $|b_i^k| \leq 1$. So each component sequence $(b_i^k)_{k=1}^\infty$, $i = 1, \ldots, n$, is bounded. By the Bolzano-Weierstrass theorem for real sequences, the sequence $(b_1^k)_{k=1}^\infty$ has a convergent subsequence. Let b_1 be the limit of that subsequence, and let us denote by $(y_{1,k})$ the corresponding subsequence of (y_k). Again by the Bolzano-Weierstrass theorem, the sequence $(y_{1,k})$ has a subsequence $(y_{2,k})$ for which the corresponding subsequence of second components (a subsequence of $(b_2^k)_{k=1}^\infty$) converges, with limit b_2. We continue in this way, and after n steps, we obtain a subsequence of (y_k) which we denote in the usual way by (y_{m_k}), with elements expressed by

$$y_{m_k} = \sum_{i=1}^n c_i^k v_i, \qquad \text{where} \quad \sum_{i=1}^n |c_i^k| = 1,$$

and such that for each $i = 1, \ldots, n$, the i-th component sequence $(c_i^k)_{k=1}^{\infty}$ converges to b_i. Letting $y = \sum_{i=1}^{n} b_i v_i$, a triangle inequality estimate shows that $|y_{m_k} - y| \to 0$ as $k \to \infty$. Moreover,

$$1 = \lim_{k\to\infty} \sum_{i=1}^{n} |c_i^k| = \sum_{i=1}^{n} |b_i|.$$

Consequently, not all the b_i are zero, and linear independence of the v_i implies that $y \neq 0$. However, (y_{m_k}) is a subsequence of (y_k) and by hypothesis $|y_k| \to 0$ as $k \to \infty$, so $|y_{m_k}| \to 0$ as $k \to \infty$. Since $y_{m_k} \to y$, we also have $|y_{m_k}| \to |y|$, so $|y| = 0$. Hence, $y = 0$, a contradiction of the previous deduction that $y \neq 0$. This proves (8.6), and hence (8.5). □

The next result reassures us that on a finite-dimensional vector space there is only one possible norm topology and only one possible notion of sequential convergence in norm on V.

Theorem 8.3.13. *Any two norms $|\cdot|$ and $|\cdot|_0$ on a finite-dimensional real vector space are equivalent.*

Proof. Let $n = \dim V$, and let $\{v_1, \ldots, v_n\}$ be any basis for V. Every $v \in V$ has a unique representation

$$v = \sum_{i=1}^{n} \alpha_i v_i = \alpha_1 v_1 + \cdots + \alpha_n v_n$$

with $\alpha_i \in \mathbf{R}$ for $i = 1, \ldots, n$. By Lemma 8.3.12 applied to $|\cdot|$ there is a number $m > 0$ such that

$$|v| = |\alpha_1 v_1 + \cdots + \alpha_n v_n| \geq m(|\alpha_1| + \cdots + |\alpha_n|).$$

Since $|\cdot|_0$ is a norm, the triangle inequality implies that

$$|v|_0 = |\alpha_1 v_1 + \cdots + \alpha_n v_n|_0 \leq \sum_{i=1}^{n} |\alpha_i|\, |v_i|_0 \leq k \sum_{i=1}^{n} |\alpha_i|,$$

where $k = \max_{1\leq i\leq n} |v_i|_0$. Thus, we have for all $v \in V$,

$$\frac{1}{k}|v|_0 \leq \sum_{i=1}^{n} |\alpha_i| \leq \frac{1}{m}|v|,$$

and hence

$$\frac{m}{k}|v|_0 \leq |v|. \tag{8.7}$$

Now we may reverse the roles of $|\cdot|$ and $|\cdot|_0$ in the argument just given, to find numbers $\hat{m}$ and $\hat{k}$ such that for all $v \in V$,

$$\frac{\hat{m}}{\hat{k}}|v| \leq |v|_0. \tag{8.8}$$

Consequently, (8.7) and (8.8) together imply

$$\frac{\hat{m}}{\hat{k}}|v| \leq |v|_0 \leq \frac{k}{m}|v|$$

for all $v \in V$. Therefore $|\cdot|$ and $|\cdot|_0$ are equivalent norms. □

The fact that there is a single equivalence class of norms on a finite-dimensional vector space allows us to choose convenient norms in certain situations with no danger that the results will change with the norm topology, since the norm topology cannot change. For example, estimates using the max norm $|\mathbf{x}|_\infty$ are especially useful in the chapters on multiple Riemann integrals and their transformations by coordinate change.

It is important to realize that not every norm comes from an inner product. Exercises 8.3.5-8.3.6 show that the norms $|\mathbf{x}|_1$ and $|\mathbf{x}|_\infty$ on $\mathbf{R}^n$ are not induced by *any* inner product in $\mathbf{R}^n$.

Exercises.

Exercise 8.3.1. *Reverse triangle inequality*
Show that for all v, w in a real normed space V, we have

$$\Big||v| - |w|\Big| \leq |v + w|.$$

Exercise 8.3.2. *Parallelogram law*
Show that the following identity holds for the Euclidean norm: If $\mathbf{x}, \mathbf{y}$ are any two vectors in $\mathbf{R}^n$, then

$$|\mathbf{x} + \mathbf{y}|_2^2 + |\mathbf{x} - \mathbf{y}|_2^2 = 2(|\mathbf{x}|_2^2 + |\mathbf{y}|_2^2).$$

Make a sketch and explain what this has to do with parallelograms.

Exercise 8.3.3. *Parallelogram law again*
Let V be an inner product space with inner product denoted (v, w) for $v, w \in V$. Show that the following identity holds for the induced norm $|v| = \sqrt{(v, v)}$: If $v, w \in V$, then

$$|v + w|^2 + |v - w|^2 = 2(|v|^2 + |w|^2).$$

Hint: If you completed Exercise 8.3.2, did your argument depend strictly on the Euclidean inner product?

Exercise 8.3.4. *Recovering an inner product from the induced norm*
Verify by direct calculation that in an inner product space V, the inner product can be expressed in terms of the induced norm as follows:

$$(x, y) = \frac{1}{4}\Big(|x + y|^2 - |x - y|^2\Big) \quad \text{for all } x, y \in V.$$

Exercise 8.3.5. Assume $n \geq 2$. Show that the norm $|\mathbf{x}|_1$ in Example 8.3.6 is not induced by *any* inner product in $\mathbf{R}^n$. *Hint*: Use the result of Exercise 8.3.3.

Exercise 8.3.6. Assume $n \geq 2$. Show that the norm $|\mathbf{x}|_\infty$ in Example 8.3.8 is not induced by *any* inner product in $\mathbf{R}^n$.

Exercise 8.3.7. 1. Show that $\|f\| := \max_{a \leq x \leq b} |f(x)|$ is a norm on the real vector space $C[a, b]$ of real valued functions continuous on $[a, b]$.
2. Find the distance between the functions $\phi(t) = t$ and $\psi(t) = t^3$ in the space $C[0, 1]$ with metric given by the maximum norm.

Exercise 8.3.8. Show that $\|f\|_1 := \int_a^b |f(x)|\, dx$ defines a norm on the real vector space $C[a, b]$ of real valued functions continuous on $[a, b]$.

Exercise 8.3.9. Show that $\|f\| := \sup_{a \le x \le b} |f(x)|$ is a norm on the real vector space $B[a,b]$ of real valued functions bounded on $[a,b]$.

Exercise 8.3.10. Consider Theorem 8.3.2 with reference to the space $\mathcal{R}[-\pi,\pi]$ with its inner product and norm $\|\cdot\|_2$ (on equivalence classes) defined by

$$\|f\|_2 = \left(\int_{-\pi}^{\pi} |f(x)|^2 \, dx\right)^{1/2}$$

and called the L^2 norm. We know from Example 8.2.5 that there are specific nonzero integrable functions for which $\|f\|_2 = 0$.

1. Show that if $f \in \mathcal{R}[-\pi,\pi]$ and $\|f\|_2 = 0$, then $f(x) = 0$ at every point x where f is continuous.
2. Show that if $f \in \mathcal{R}[-\pi,\pi]$ and $f(x) = 0$ at every point x where f is continuous, then $\|f\|_2 = 0$.

Exercise 8.3.11. Show that if f, g are Riemann integrable functions on $[-\pi,\pi]$, then

$$\left(\int_{-\pi}^{\pi} |f(x)+g(x)|^2 \, dx\right)^{1/2} \le \left(\int_{-\pi}^{\pi} |f(x)|^2 \, dx\right)^{1/2} + \left(\int_{-\pi}^{\pi} |g(x)|^2 \, dx\right)^{1/2}.$$

Exercise 8.3.12. *$\mathcal{R}[a,b]$ is not complete in the L^2 norm*
This exercise shows that $\mathcal{R}[0,2\pi]$ is not complete in the L^2 norm. Consider the function f defined by

$$f(x) = \begin{cases} 0 & \text{for } x = 0, \\ \log(1/x) & \text{for } 0 < x \le 2\pi. \end{cases}$$

In the space $\mathcal{R}[0,2\pi]$, define a sequence $(f_k)_{k=1}^{\infty}$ by

$$f_k(x) = \begin{cases} 0 & \text{for } 0 \le x \le 1/k, \\ f(x) & \text{for } 1/k < x \le 2\pi. \end{cases}$$

Observe that f is not in $\mathcal{R}[0,2\pi]$ since f is not bounded on $[0,2\pi]$, but $f_k \in \mathcal{R}[0,2\pi]$ for each k.

1. Show that $(f_k)_{k=1}^{\infty}$ is a Cauchy sequence in $\mathcal{R}[0,2\pi]$. *Hint*: Note that for $m > n$,
$$\|f_m - f_n\|_2^2 = \int_{1/m}^{1/n} |f_m(x) - f_n(x)|^2 \, dx = \int_{1/m}^{1/n} |\log(1/x)|^2 \, dx = \int_{1/m}^{1/n} (\log x)^2 \, dx$$
and use the fact that $\frac{d}{dx}[x(\log x)^2 - 2x\log x + 2x] = (\log x)^2$.
2. Verify that $\lim_{k\to\infty} \int_0^{2\pi} (f_k(x) - f(x))^2 \, dx = 0$, but $f \notin \mathcal{R}[0,2\pi]$.

Exercise 8.3.13. Consider the vector space l^1 from Exercise 8.1.6. Prove that, with $\xi = (\xi_k) \in l^1$, $|\xi|_1 := \sum_{k=1}^{\infty} |\xi_k|$ defines a norm on l^1.

Exercise 8.3.14. Consider the vector space l^∞ from Exercise 8.1.7. Prove that, with $\xi = (\xi_k) \in l^\infty$, $|\xi|_\infty := \sup_k |\xi_k|$ defines a norm on l^∞.

Exercise 8.3.15. Suppose $|\cdot|$ and $|\cdot|_0$ are equivalent norms on V. Show that every open ball in one norm contains an open ball in the other norm. Conclude that both norms define the same collection of open subsets of V.

Exercise 8.3.16. Suppose $|\cdot|$ and $|\cdot|_0$ are equivalent norms on V. Show that a sequence in V converges with respect to $|\cdot|$ if and only if it converges with respect to $|\cdot|_0$.

Exercise 8.3.17. Prove the following statements:

1. If S is a complete subspace of a normed space Y, then S is a closed set in Y.
2. If Y is a complete normed space and S is a subspace that is a closed set in Y, then S is complete.
3. Every *finite-dimensional* subspace S of a normed space Y is complete, hence closed. As a corollary, every finite-dimensional normed space is complete. *Hint*: Choose a basis of S and a Cauchy sequence in S, and apply Lemma 8.3.12 to the elements of the sequence.

Exercise 8.3.18. Consider the space l^2 with the orthonormal set $\{e_k\}_{k=1}^{\infty}$. Establish the Cauchy-Schwarz inequality as follows:

1. Suppose $\|y\|_2 = 1$, and expand the right-hand side of the inequality $0 \leq \|x - (x,y)y\|_2^2$.
2. For $\|y\|_2 \neq 1$, rescale y and use the result of part 1.

Exercise 8.3.19. Consider the vector space $\mathcal{R}[-\pi,\pi]$ with its integral inner product and norm

$$\|f\| = \left(\int_{-\pi}^{\pi} |f(x)|^2\,dx\right)^{1/2}.$$

1. Show that there exist nonzero integrable functions f for which $\|f\| = 0$.
2. However, show that if $f \in \mathcal{R}[-\pi,\pi]$ with $\|f\| = 0$, then $f(x) = 0$ whenever f is continuous at x.
3. Conversely, show that if $f \in \mathcal{R}[-\pi,\pi]$, and f is zero at all of its points of continuity, then $\|f\| = 0$.

8.4. Fourier Expansion in $\mathbf{R}^n$

In this section we consider some fundamental facts about orthogonal expansion in $\mathbf{R}^n$. The properties are extended to other inner product spaces, including infinite-dimensional spaces, in later developments in the book. Recall that the Euclidean inner product $(\mathbf{x}, \mathbf{y})$ in $\mathbf{R}^3$ is a linear function of each of its arguments and that $\sqrt{(\mathbf{x}, \mathbf{x})}$ is the Euclidean length of the vector $\mathbf{x}$.

An inner product provides the important concept of orthogonality (perpendicularity) of two vectors. The angle θ between two nonzero vectors $\mathbf{x}, \mathbf{y}$ in $\mathbf{R}^n$ is given by the formula

$$\cos\theta = \frac{(\mathbf{x}, \mathbf{y})}{|\mathbf{x}|_2 |\mathbf{y}|_2}, \quad \text{where } 0 \leq \theta \leq \pi.$$

We say that the vectors $\mathbf{x}$ and $\mathbf{y}$ are **orthogonal** if $(\mathbf{x}, \mathbf{y}) = 0$.

The first results of the section are vector expressions of the Pythagorean theorem in $\mathbf{R}^3$.

Theorem 8.4.1. *If* $\mathbf{x}, \mathbf{y} \in \mathbf{R}^3$ *and* $(\mathbf{x}, \mathbf{y}) = \mathbf{0}$, *then* $|\mathbf{x} + \mathbf{y}|_2^2 = |\mathbf{x}|_2^2 + |\mathbf{y}|_2^2$.

Proof. Expanding $|\mathbf{x}+\mathbf{y}|_2^2$ using the properties of the inner product, we have

$$|\mathbf{x}+\mathbf{y}|_2^2 = \big((\mathbf{x}+\mathbf{y}),(\mathbf{x}+\mathbf{y})\big) = (\mathbf{x},\mathbf{x}) + 2(\mathbf{x},\mathbf{y}) + (\mathbf{y},\mathbf{y}) = |\mathbf{x}|_2^2 + |\mathbf{y}|_2^2$$

since $(\mathbf{x},\mathbf{y}) = 0$. □

In the space $\mathbf{R}^3$ there is no trouble in defining sets of three pairwise orthogonal vectors. For example, recall that if $\mathbf{v}$ and $\mathbf{w}$ are nonzero orthogonal vectors in $\mathbf{R}^3$, then $\mathbf{v}\times\mathbf{w}$ is nonzero and orthogonal to both $\mathbf{v}$ and $\mathbf{w}$. We have the following *Pythagorean theorem.*

Theorem 8.4.2 (Pythagorean Theorem). *If $\mathbf{v}_1, \mathbf{v}_2, \mathbf{v}_3 \in \mathbf{R}^3$ are pairwise orthogonal, then*

$$|\mathbf{v}_1+\mathbf{v}_2+\mathbf{v}_3|_2^2 = |\mathbf{v}_1|_2^2 + |\mathbf{v}_2|_2^2 + |\mathbf{v}_3|_2^2.$$

Consequently, for any $c_1, c_2, c_3 \in \mathbf{R}$, we have

$$|c_1\mathbf{v}_1 + c_2\mathbf{v}_2 + c_3\mathbf{v}_3|_2^2 = |c_1|^2|\mathbf{v}_1|_2^2 + |c_2|^2|\mathbf{v}_2|_2^2 + |c_3|^2|\mathbf{v}_3|_2^2.$$

Proof. By the pairwise orthogonality of $\mathbf{v}_1, \mathbf{v}_2, \mathbf{v}_3$, we have

$$\begin{aligned}|\mathbf{v}_1+\mathbf{v}_2+\mathbf{v}_3|_2^2 &= \sum_{i=1}^{3}\sum_{j=1}^{3}(\mathbf{v}_i,\mathbf{v}_j)\\ &= \sum_{j=1}^{3}(\mathbf{v}_j,\mathbf{v}_j) = |\mathbf{v}_1|_2^2 + |\mathbf{v}_2|_2^2 + |\mathbf{v}_3|_2^2.\end{aligned}$$

Now observe that pairwise orthogonality of $\mathbf{v}_1, \mathbf{v}_2, \mathbf{v}_3$ implies pairwise orthogonality of $c_1\mathbf{v}_1, c_2\mathbf{v}_2, c_3\mathbf{v}_3$ for any scalars c_1, c_2, c_3. From this, the second statement of the theorem follows easily, since $|c_j\mathbf{v}_j|_2^2 = |c_j|^2|\mathbf{v}_j|_2^2$ for $j = 1, 2, 3$. □

The standard basis vectors in $\mathbf{R}^3$, defined by $\mathbf{e}_1 = (1,0,0)$, $\mathbf{e}_2 = (0,1,0)$, and $\mathbf{e}_3 = (0,0,1)$, are important because every $\mathbf{x} = (x_1, x_2, x_3) \in \mathbf{R}^3$ can be written in a unique way as a linear combination

$$\mathbf{x} = x_1\mathbf{e}_1 + x_2\mathbf{e}_2 + x_3\mathbf{e}_3.$$

One can use pairwise orthogonality of the $\mathbf{e}_j$ to obtain a useful representation of the coefficients x_j, which are the components of $\mathbf{x}$ with respect to the basis $\{\mathbf{e}_1, \mathbf{e}_2, \mathbf{e}_3\}$. We have

$$(\mathbf{x},\mathbf{e}_1) = x_1, \qquad (\mathbf{x},\mathbf{e}_2) = x_2, \qquad (\mathbf{x},\mathbf{e}_3) = x_3.$$

Notice that each $\mathbf{e}_j$ has unit norm, $|\mathbf{e}_j|_2 = 1$.

Every basis of $\mathbf{R}^3$ consists of three linearly independent vectors $\mathbf{v}_1, \mathbf{v}_2, \mathbf{v}_3$. If the $\mathbf{v}_j$ are pairwise orthogonal, then we obtain an **orthogonal basis** for $\mathbf{R}^3$. The justification for this statement is that if $\mathbf{v}_1, \mathbf{v}_2, \mathbf{v}_3$ are pairwise orthogonal, then $\{\mathbf{v}_1, \mathbf{v}_2, \mathbf{v}_3\}$ is a linearly independent set, and thus a basis for $\mathbf{R}^3$. In order to see this, suppose that

$$c_1\mathbf{v}_1 + c_2\mathbf{v}_2 + c_3\mathbf{v}_3 = \mathbf{0}.$$

Take the dot product of both sides with $\mathbf{v}_j$ to verify that $c_j(\mathbf{v}_j,\mathbf{v}_j) = 0$, so $c_j = 0$ for $j = 1, 2, 3$ since each $\mathbf{v}_j \neq 0$. If for each $j = 1, 2, 3$ we also have $|\mathbf{v}_j|_2 = 1$, then the orthogonal basis $\{\mathbf{v}_1, \mathbf{v}_2, \mathbf{v}_3\}$ is called an **orthonormal basis** for $\mathbf{R}^3$. The standard basis $\{\mathbf{e}_1, \mathbf{e}_2, \mathbf{e}_3\}$ is an orthonormal basis for $\mathbf{R}^3$.

Theorem 8.4.3 (Fourier Expansion). *If $\{\mathbf{v}_1, \mathbf{v}_2, \mathbf{v}_3\}$ is an orthogonal basis for $\mathbf{R}^3$ and $\mathbf{x} \in \mathbf{R}^3$ has the representation $\mathbf{x} = c_1\mathbf{v}_1 + c_2\mathbf{v}_2 + c_3\mathbf{v}_3$, then*

$$c_j = \frac{(\mathbf{x}, \mathbf{v}_j)}{(\mathbf{v}_j, \mathbf{v}_j)} = \frac{(\mathbf{x}, \mathbf{v}_j)}{|\mathbf{v}_j|_2^2} \quad \text{for } j = 1, 2, 3.$$

If the basis is orthonormal, then $c_j = (\mathbf{x}, \mathbf{v}_j)$ for $j = 1, 2, 3$, and hence

$$\mathbf{x} = (\mathbf{x}, \mathbf{v}_1)\mathbf{v}_1 + (\mathbf{x}, \mathbf{v}_2)\mathbf{v}_2 + (\mathbf{x}, \mathbf{v}_3)\mathbf{v}_3.$$

Proof. Take the inner product of each side of $\mathbf{x} = c_1\mathbf{v}_1 + c_2\mathbf{v}_2 + c_3\mathbf{v}_3$ with $\mathbf{v}_j$ and use orthogonality. We find that $(\mathbf{x}, \mathbf{v}_j) = c_j(\mathbf{v}_j, \mathbf{v}_j)$. Since each $\mathbf{v}_j$ is a basis vector, we have $|\mathbf{v}_j|_2^2 \neq 0$, hence we can solve for each c_j, $j = 1, 2, 3$, and we have the first result. If the basis is orthonormal, then for each j we have $|\mathbf{v}_j|_2^2 = 1$ and the second statement follows. □

In either of the cases covered by Theorem 8.4.3, the coefficients c_j of the element $\mathbf{x}$ with respect to the basis $\{\mathbf{v}_1, \mathbf{v}_2, \mathbf{v}_3\}$ are called the **Fourier coefficients** of $\mathbf{x}$. Given $\mathbf{x}$, the linear combination

$$\sum_{j=1}^{3} c_j\mathbf{v}_j = c_1\mathbf{v}_1 + c_2\mathbf{v}_2 + c_3\mathbf{v}_3$$

is called the **Fourier expansion** or the **Fourier representation** of $\mathbf{x}$. Theorem 8.4.3 states that each vector $\mathbf{x}$ in $\mathbf{R}^3$ equals the sum of its Fourier expansion.

Example 8.4.4. An important application of orthogonal bases occurs in the problem of diagonalizing a real symmetric matrix. We consider the case of the real symmetric 3×3 matrix A given by

$$A = \begin{bmatrix} 2 & 1 & 0 \\ 1 & 2 & 0 \\ 0 & 0 & 2 \end{bmatrix}.$$

The eigenvalues of A are $\lambda_1 = 1$, $\lambda_2 = 2$, and $\lambda_3 = 3$. Corresponding eigenvectors satisfying $A\mathbf{v}_1 = \mathbf{v}_1$, $A\mathbf{v}_2 = 2\mathbf{v}_2$ and $A\mathbf{v}_3 = 3\mathbf{v}_3$ are given by

$$\mathbf{v}_1 = \begin{bmatrix} 1 \\ -1 \\ 0 \end{bmatrix} \quad \mathbf{v}_2 = \begin{bmatrix} 0 \\ 0 \\ 1 \end{bmatrix} \quad \mathbf{v}_3 = \begin{bmatrix} 1 \\ 1 \\ 0 \end{bmatrix}.$$

It should be clear that $\{\mathbf{v}_1, \mathbf{v}_2, \mathbf{v}_3\}$ is an orthogonal basis of $\mathbf{R}^3$. The action of the linear transformation $T(\mathbf{x}) = A\mathbf{x}$ is especially simple on the basis $\{\mathbf{v}_1, \mathbf{v}_2, \mathbf{v}_3\}$, since $A\mathbf{v}_1 = \mathbf{v}_1$, $A\mathbf{v}_2 = 2\mathbf{v}_2$, and $A\mathbf{v}_3 = 3\mathbf{v}_3$. These three equations are equivalent to the statement that

$$A\begin{bmatrix} \mathbf{v}_1 & \mathbf{v}_2 & \mathbf{v}_3 \end{bmatrix} = \begin{bmatrix} \mathbf{v}_1 & \mathbf{v}_2 & \mathbf{v}_3 \end{bmatrix} \begin{bmatrix} 1 & 0 & 0 \\ 0 & 2 & 0 \\ 0 & 0 & 3 \end{bmatrix}.$$

Thus the matrix S defined by

$$S = \begin{bmatrix} \mathbf{v}_1 & \mathbf{v}_2 & \mathbf{v}_3 \end{bmatrix} = \begin{bmatrix} 1 & 0 & 1 \\ -1 & 0 & 1 \\ 0 & 1 & 0 \end{bmatrix}$$

has the property that $AS = SD$, where D is diagonal with the eigenvalues of A along its main diagonal. Equivalently, we have $S^{-1}AS = D$. △

Exercises 8.4.1-8.4.2 explore the application of Fourier coefficients to some basic approximation problems.

The Fourier expansion of $\mathbf{x}$ with respect to an orthonormal basis leads to an expression of the Pythagorean theorem called *Parseval's theorem.*

Theorem 8.4.5 (Parseval's Theorem). *If* $\{\mathbf{v}_1, \mathbf{v}_2, \mathbf{v}_3\}$ *is an orthonormal basis for* $\mathbf{R}^3$, *then for every* $\mathbf{x} \in \mathbf{R}^3$,

$$|\mathbf{x}|_2^2 = \sum_{j=1}^{3} (\mathbf{x}, \mathbf{v}_j)^2 = (\mathbf{x}, \mathbf{v}_1)^2 + (\mathbf{x}, \mathbf{v}_2)^2 + (\mathbf{x}, \mathbf{v}_3)^2.$$

Proof. Given $\mathbf{x}$, we have $\mathbf{x} = c_1\mathbf{v}_1 + c_2\mathbf{v}_2 + c_3\mathbf{v}_3$, where $c_j = (\mathbf{x}, \mathbf{v}_j)$ for $j = 1, 2, 3$. By the Fourier expansion of Theorem 8.4.3 and orthogonality of the basis,

$$\begin{aligned} |\mathbf{x}|_2^2 &= \left|(\mathbf{x}, \mathbf{v}_1)\mathbf{v}_1 + (\mathbf{x}, \mathbf{v}_2)\mathbf{v}_2 + (\mathbf{x}, \mathbf{v}_3)\mathbf{v}_3\right|_2^2 \\ &= (\mathbf{x}, \mathbf{v}_1)^2|\mathbf{v}_1|_2^2 + (\mathbf{x}, \mathbf{v}_2)^2|\mathbf{v}_2|_2^2 + (\mathbf{x}, \mathbf{v}_3)^2|\mathbf{v}_3|_2^2. \end{aligned}$$

Since $|\mathbf{v}_j|_2^2 = 1$ for each j, $|\mathbf{x}|_2^2 = (\mathbf{x}, \mathbf{v}_1)^2 + (\mathbf{x}, \mathbf{v}_2)^2 + (\mathbf{x}, \mathbf{v}_3)^2$. □

It is possible to state and prove straightforward generalizations of Theorem 8.4.3 and Theorem 8.4.5 in the space $\mathbf{R}^n$, where the Euclidean inner product of $\mathbf{x} = (x_1, \ldots, x_n)$ and $\mathbf{y} = (y_1, \ldots, y_n)$ in $\mathbf{R}^n$ is given by

$$(\mathbf{x}, \mathbf{y}) = \sum_{j=1}^{n} x_j y_j = x_1 y_1 + \cdots + x_n y_n,$$

and $|\mathbf{x}|_2 = \sqrt{(\mathbf{x}, \mathbf{x})}$. The extensions for $\mathbf{R}^n$ are considered in Exercise 8.4.3. Later in this book we extend the ideas and results of this section to other inner product spaces, such as the sequence space l^2 and some other function spaces.

Exercises.

Exercise 8.4.1. *Best approximation from a line*
Let $\mathbf{v}$ be a vector in $\mathbf{R}^2$ with unit norm, and $V = \{c\mathbf{v} : c \in \mathbf{R}\}$ the line spanned by $\mathbf{v}$. Let $\mathbf{x}$ be a vector not in the subspace V, so that $\mathbf{x} \neq c\mathbf{v}$ for any choice of scalar c. Show that the closest approximation to $\mathbf{x}$ by a vector in V is given by $\mathbf{p}_\mathbf{x} = (\mathbf{x}, \mathbf{v})\mathbf{v}$. *Hint*: Think geometrically.

Exercise 8.4.2. *Best approximation from a plane*
Let $\mathbf{v}_1$ and $\mathbf{v}_2$ be orthogonal vectors in $\mathbf{R}^3$ with unit norm, and $V = \{c_1\mathbf{v}_1 + c_2\mathbf{v}_2 : c_1, c_2 \in \mathbf{R}\}$ the plane spanned by $\mathbf{v}_1$ and $\mathbf{v}_2$. Let $\mathbf{x}$ be a vector not in the subspace V, so that $\mathbf{x} \neq c_1\mathbf{v}_1 + c_2\mathbf{v}_2$ for any choice of scalars c_1 and c_2. Show that the closest approximation to $\mathbf{x}$ by a vector in V is given by $\mathbf{p}_\mathbf{x} = (\mathbf{x}, \mathbf{v}_1)\mathbf{v}_1 + (\mathbf{x}, \mathbf{v}_2)\mathbf{v}_2$. *Hint*: Think geometrically.

Exercise 8.4.3. *Fourier expansion and Parseval's theorem in* $\mathbf{R}^n$
Formulate and prove an extension of the Fourier expansion Theorem 8.4.3 for $\mathbf{R}^n$. Do the same for Parseval's Theorem 8.4.5.

8.5. Real Symmetric Matrices

In this section we present the spectral theorem for real symmetric matrices, which states that every real symmetric matrix can be diagonalized by means of an orthogonal transformation matrix. The discussion provides some useful review of important linear algebra and matrix theory and employs the idea of an orthonormal basis for subspaces of $\mathbf{R}^n$.

8.5.1. Definitions and Preliminary Results. We will be implicitly using the fundamental theorem of algebra which states that every polynomial of degree n with complex number coefficients has exactly n roots, taking account of multiplicities for repeated roots.

Recall that the transpose A^T of an $n \times n$ matrix A is formed by making row j of A^T the row vector with the same entries left-to-right in the same order as the column j entries of U read top-to-bottom, for $1 \le j \le n$. For example,

$$A = \begin{bmatrix} 0 & 1 & 2 \\ 3 & 4 & 5 \\ 6 & 7 & 8 \end{bmatrix} \quad \Longrightarrow \quad A^T = \begin{bmatrix} 0 & 3 & 6 \\ 1 & 4 & 7 \\ 2 & 5 & 8 \end{bmatrix}.$$

Definition 8.5.1. *A real $n \times n$ matrix A is called* **symmetric** *if $A^T = A$.*

We also need the concept of an orthogonal matrix.

Definition 8.5.2. *A real $n \times n$ matrix U is called* **orthogonal** *if its columns form an orthonormal basis of $\mathbf{R}^n$, or, what is equivalent, U satisfies the condition $U^T U = I$, where I is the $n \times n$ identity matrix.*

Remark. *It follows from this definition that an orthogonal matrix U has orthonormal columns and is invertible, with $U^{-1} = U^T$.*

An arbitrary real $n \times n$ matrix can have complex eigenvalues and associated eigenvectors with complex number components. For this reason, we study a real symmetric matrix initially as a linear transformation of the space $\mathbf{C}^n$ whose elements are vectors with n components that can be any complex numbers. The space $\mathbf{C}^n$ is a vector space over the complex number field, which means that it has properties 1-9 of the vector space definition (Definition 8.1.3) with scalars from the complex field $\mathbf{C}$. We will also use an inner product on $\mathbf{C}^n$. First, we define what is meant by a complex inner product on a vector space with complex scalars, such as $\mathbf{C}^n$. Recall that if $z = x + iy \in \mathbf{C}$, then $\bar{z} = x - iy$ is the complex conjugate of z. If z and w are complex numbers, then $\overline{zw} = \bar{z}\,\bar{w}$ and $\overline{z + w} = \bar{z} + \bar{w}$.

Definition 8.5.3 (Complex Inner Product)**.** *Let V be a vector space over the complex number field. A function mapping $V \times V$ into $\mathbf{C}$, with values denoted (v, w), is a* **complex inner product** *if it has these properties:*

(i) $(v_1 + v_2, w) = (v_1, w) + (v_2, w)$ *for all $v_1, v_2, w \in V$.*

(ii) $(\alpha v, w) = \alpha(v, w)$ *for all $v, w \in V$ and $\alpha \in \mathbf{C}$;*

(iii) $(v, w) = \overline{(w, v)}$ *for all $v, w \in V$;*

(iv) $(v, v) \ge 0$ *for all $v \in V$, and $(v, v) = 0$ if and only if $v = 0$.*

For the vector space $\mathbf{C}^n$, we can define an inner product as follows: If $\mathbf{v}, \mathbf{w} \in \mathbf{C}^n$, written as $\mathbf{v} = (v_1, \ldots, v_n)$, $\mathbf{w} = (w_1, \ldots, w_n)$, we define

$$(\mathbf{v}, \mathbf{w}) = v_1\overline{w_1} + v_2\overline{w_2} + \cdots + v_n\overline{w_n}.$$

Properties (i) and (ii) are easily verified for this product (Exercise 8.5.1). Consider (iii): We have

$$(\mathbf{w}, \mathbf{v}) = w_1\overline{v_1} + w_2\overline{v_2} + \cdots + w_n\overline{v_n},$$

and conjugation yields

$$\overline{(\mathbf{w}, \mathbf{v})} = \overline{w_1}v_1 + \overline{w_2}v_2 + \cdots + \overline{w_n}v_n,$$

which equals $(\mathbf{v}, \mathbf{w})$ since complex number multiplication is commutative. Now (iii) implies that $(\mathbf{v}, \mathbf{v})$ is real, and (iv) says

$$(\mathbf{v}, \mathbf{v}) = v_1\overline{v_1} + \cdots + v_n\overline{v_n} = \sum_{j=1}^{n} |v_j|^2,$$

and this sum is positive if and only if $\mathbf{v} \neq \mathbf{0}$. By combining properties (iii) and (i), we deduce that

$$(\mathbf{v}, \mathbf{w}_1 + \mathbf{w}_2) = (\mathbf{v}, \mathbf{w}_1) + (\mathbf{v}, \mathbf{w}_2),$$

and by combining (ii) and (iii), we deduce that

$$(\mathbf{v}, \alpha\mathbf{w}) = \overline{\alpha}(\mathbf{v}, \mathbf{w}).$$

(The latter two properties hold, of course, for any complex inner product.)

The norm on $\mathbf{C}^n$ induced by the inner product we have just defined is given by

$$|\mathbf{v}| = (\mathbf{v}, \mathbf{v})^{1/2} = \Big(\sum_{j=1}^{n} |v_j|^2\Big)^{1/2}.$$

The Cauchy-Schwarz inequality is $\big|(\mathbf{v}, \mathbf{w})\big| \leq |\mathbf{v}|\,|\mathbf{w}|$. Since we need neither the norm nor the inequality in this section, we leave the verification of these facts to the interested reader.

In the computations that follow, we think of a complex vector as a column vector, and its transpose as a row vector, and then the complex inner product $(\mathbf{v}, \mathbf{w})$ may also be written as $\mathbf{v}^T\overline{\mathbf{w}}$, with the definition

$$\overline{\mathbf{w}} = \begin{bmatrix} \overline{w_1} \\ \cdots \\ \overline{w_n} \end{bmatrix}.$$

In order to establish the spectral theorem we need three preliminary results, all centered around a given eigenvalue of a symmetric matrix and the associated eigenspace. Recall that λ is an eigenvalue of A if there is a nonzero vector $\mathbf{v}$ such that $A\mathbf{v} = \lambda\mathbf{v}$. Then $\mathbf{v}$ is said to be an **eigenvector** for λ. By definition, eigenvectors of A associated with λ are *nonzero* vectors.

Lemma 8.5.4. *Every eigenvalue of an $n \times n$ real symmetric matrix is real and has a corresponding eigenvector in $\mathbf{R}^n$.*

Proof. Let A be an $n \times n$ real symmetric matrix, and let λ be an eigenvalue of A with corresponding eigenvector $\mathbf{v}$. Since $A\mathbf{v} = \lambda\mathbf{v}$ and A is real, complex conjugation gives $A\overline{\mathbf{v}} = \overline{\lambda\mathbf{v}} = \overline{\lambda}\overline{\mathbf{v}}$. Using the complex inner product $(\mathbf{u}, \mathbf{w}) = \mathbf{u}^T\overline{\mathbf{w}}$ on $\mathbf{C}^n$, we have

$$(A\mathbf{v}, \mathbf{v}) = (A\mathbf{v})^T\overline{\mathbf{v}} = \mathbf{v}^T A^T\overline{\mathbf{v}} = \mathbf{v}^T A\overline{\mathbf{v}} = \mathbf{v}^T(\overline{\lambda\mathbf{v}}) = \overline{\lambda}\,\mathbf{v}^T\overline{\mathbf{v}}.$$

Since we also have

$$(A\mathbf{v}, \mathbf{v}) = (\lambda\mathbf{v}, \mathbf{v}) = \lambda\mathbf{v}^T\overline{\mathbf{v}},$$

it follows that $\overline{\lambda}\,\mathbf{v}^T\overline{\mathbf{v}} = \lambda\mathbf{v}^T\overline{\mathbf{v}}$. Since $\mathbf{v}^T\overline{\mathbf{v}} > 0$ for an eigenvector $\mathbf{v}$, we have $\lambda = \overline{\lambda}$, so λ is real. To see that λ has a corresponding eigenvector in $\mathbf{R}^n$, write $\mathbf{v} = \mathbf{u} + i\mathbf{w}$ where $\mathbf{u}, \mathbf{w} \in \mathbf{R}^n$. Then $A\mathbf{v} = A\mathbf{u} + iA\mathbf{w}$, and both $A\mathbf{u}$ and $A\mathbf{w}$ are real since A is real. Since $A\mathbf{v} = \lambda\mathbf{v} = \lambda\mathbf{u} + i\lambda\mathbf{w}$ and λ is real, we have $A\mathbf{u} = \lambda\mathbf{u}$ and $A\mathbf{w} = \lambda\mathbf{w}$, and at least one of $\mathbf{u}, \mathbf{w}$ is nonzero since $\mathbf{v}$ is nonzero. This proves the lemma. □

We define the **eigenspace** for an eigenvalue λ of a real symmetric matrix A to be the null space

$$\mathcal{N}(\lambda I - A) = \{\mathbf{v} \in \mathbf{R}^n : A\mathbf{v} = \lambda\mathbf{v}\},$$

the set of all solutions of the linear system of equations, $(\lambda I - A)\mathbf{v} = \mathbf{0}$. This null space is closed under vector addition and scalar multiplication by real numbers, so $\mathcal{N}(\lambda I - A)$ is a subspace of $\mathbf{R}^n$.

We introduced the complex inner product on $\mathbf{C}^n$ for the immediate purpose of establishing Lemma 8.5.4. It is certainly worth knowing about for advanced work in linear algebra and numerical analysis. Since we can regard the real symmetric matrix A as a linear transformation of $\mathbf{R}^n$, we can now revert to using the Euclidean inner product for real vectors in the remaining results of this section.

Lemma 8.5.5. *If A is an $n \times n$ real symmetric matrix, λ is an eigenvalue of A, and $W = \mathcal{N}(\lambda I - A)$ is the corresponding eigenspace for λ, then W has an orthonormal basis.*

Proof. Suppose W is a subspace of $\mathbf{R}^n$ of dimension k. Then W has a basis $\{\mathbf{x}_1, \ldots, \mathbf{x}_k\}$. From this set of k linearly independent vectors, an orthogonal basis can be constructed as follows: Let $\mathbf{v}_1 = \mathbf{x}_1$, and for $j = 2, \ldots, k$, define

$$\mathbf{v}_j = \mathbf{x}_j - \sum_{i=1}^{j-1} \frac{(\mathbf{v}_j, \mathbf{v}_i)}{\|\mathbf{v}_i\|^2}\mathbf{v}_i.$$

This procedure is called the **Gram-Schmidt process**, and it produces an orthogonal set that spans the subspace W, hence an orthogonal basis $\{\mathbf{v}_1, \ldots, \mathbf{v}_k\}$ for W. (See Exercise 8.5.2). We can normalize the vectors $\mathbf{v}_j$ by setting $\mathbf{u}_j = \mathbf{v}_j/\|\mathbf{v}_j\|$ for $1 \le j \le k$, and thus obtain an orthonormal basis of W. □

Let V be a subspace of $\mathbf{R}^n$. We say that V is an **invariant subspace** for A if $\mathbf{v} \in V$ implies $A\mathbf{v} \in V$. We express this invariance of V by writing $A(V) \subseteq V$. It is easy to see that the eigenspace $W = \mathcal{N}(\lambda I - A)$ of an eigenvalue λ is invariant for A, since $A\mathbf{v} = \lambda\mathbf{v} \in W$ for every $\mathbf{v} \in W$.

If V is a subspace of $\mathbf{R}^n$, the **orthogonal complement** of V in $\mathbf{R}^n$ is the subspace

$$V^\perp := \{\mathbf{w} \in \mathbf{R}^n : (\mathbf{v}, \mathbf{w}) = 0 \text{ for all } \mathbf{v} \in V\}.$$

It is easily verified that this set is closed under vector addition and scalar multiplication, so $V^\perp$ is indeed a subspace of $\mathbf{R}^n$. (We read $V^\perp$ as "V perp".)

We need one more fact about symmetric matrices before addressing the spectral theorem. Note that we can still write $(\mathbf{v}, \mathbf{w}) = \mathbf{v}^T\mathbf{w}$ when computing with the Euclidean inner product for real vectors.

Lemma 8.5.6. *If A is an $n \times n$ real symmetric matrix and $W \subset \mathbf{R}^n$ is the eigenspace for an eigenvalue λ of A, then $W^\perp$ is an invariant subspace for A, that is, $A(W^\perp) \subseteq W^\perp$.*

Proof. Let $\mathbf{v} \in W$, so that $A\mathbf{v} = \lambda\mathbf{v}$, and let $\mathbf{w} \in W^\perp$. We have

$$(A\mathbf{v}, \mathbf{w}) = (\lambda\mathbf{v}, \mathbf{w}) = \lambda(\mathbf{v}, \mathbf{w}) = 0,$$

and

$$(A\mathbf{v}, \mathbf{w}) = (A\mathbf{v})^T\mathbf{w} = \mathbf{v}^T A^T\mathbf{w} = \mathbf{v}^T A\mathbf{w} = (\mathbf{v}, A\mathbf{w}),$$

since $A^T = A$. Hence, $(\mathbf{v}, A\mathbf{w}) = 0$. This is true for every $\mathbf{v} \in W$ and $\mathbf{w} \in W^\perp$. Thus $\mathbf{w} \in W^\perp$ implies $A\mathbf{w} \in W^\perp$. Hence $W^\perp$ is invariant under A. □

8.5.2. The Spectral Theorem for Real Symmetric Matrices.

We are ready to prove the spectral theorem for real symmetric matrices, which says that every real symmetric matrix can be diagonalized by means of an orthogonal transformation matrix.

Theorem 8.5.7 (Spectral Theorem). *If A is an $n \times n$ real symmetric matrix, then there exists an orthonormal basis $\{\mathbf{u}_1, \dots, \mathbf{u}_n\}$ of $\mathbf{R}^n$ consisting of eigenvectors of A. If*

$$U = \begin{bmatrix} \mathbf{u}_1 & \cdots & \mathbf{u}_n \end{bmatrix},$$

then U is an orthogonal matrix and $U^{-1}AU = U^T AU = \Lambda$, where Λ is a diagonal matrix with the eigenvalues of A on the main diagonal.

Proof. The proof is by induction on the dimension n. For $n = 1$, a real 1×1 matrix $A = [\lambda]$ is symmetric and has λ as its sole eigenvalue. Then the set $\{1\}$ (consisting of an eigenvector for A) is an orthonormal basis for $\mathbf{R}$ where the Euclidean inner product is real number multiplication. To proceed to the induction step, we assume that A is a real $n \times n$ symmetric matrix and that the theorem holds for all real symmetric matrices of size less than n. (Thus we are using the principle of induction in Theorem 1.3.4.) Let λ_1 be an eigenvalue for A and let W be the corresponding eigenspace. Suppose $\dim W = k$. Then W has an orthonormal basis $\{\mathbf{u}_1, \dots, \mathbf{u}_k\}$. If $k = n$, then we are done. If $k < n$, then we may augment this basis for W with an orthonormal basis $\{\mathbf{v}_{k+1}, \dots, \mathbf{v}_n\}$ for $W^\perp$. Then $\{\mathbf{u}_1, \dots, \mathbf{u}_k, \mathbf{v}_{k+1}, \dots, \mathbf{v}_n\}$ is an orthonormal basis for $\mathbf{R}^n$. Note that the vectors $\mathbf{v}_{k+1}, \dots, \mathbf{v}_n$ need not be eigenvectors for A; they merely serve to fill out an orthonormal basis of $\mathbf{R}^n$, and thus ensure that the matrix Q_1 defined by

$$Q_1 = [\mathbf{u}_1 \quad \cdots \quad \mathbf{u}_k \quad \mathbf{v}_{k+1} \quad \cdots \quad \mathbf{v}_n]$$

is orthogonal, that is, $Q_1^T Q_1 = I$, and hence $Q_1^{-1} = Q_1^T$. Since both W and $W^\perp$ are invariant under A, the matrix representation of the linear transformation $\mathcal{L}\mathbf{x} = A\mathbf{x}$

in the basis $\{\mathbf{u}_1, \ldots, \mathbf{u}_k, \mathbf{v}_{k+1}, \ldots, \mathbf{v}_n\}$ is block diagonal, since we have

$$Q_1^{-1}AQ_1 = Q_1^TAQ_1 = \begin{bmatrix} \lambda_1 I_k & 0 \\ 0 & \hat{A} \end{bmatrix}$$

where I_k is a $k \times k$ identity matrix and $\hat{A}$ is $(n-k) \times (n-k)$. Moreover, $\hat{A}$ is symmetric, since $Q_1^TAQ_1$ is symmetric by the symmetry of A:

$$(Q_1^TAQ_1)^T = Q_1^TA^TQ_1 = Q_1^TAQ_1.$$

The similarity transformation preserves the eigenvalues of A, so the eigenvalues of $\hat{A}$ are eigenvalues of A. Now we invoke the induction hypothesis: There exists an orthonormal set $\{\hat{\mathbf{u}}_{k+1}, \ldots, \hat{\mathbf{u}}_n\}$ in $\mathbf{R}^{n-k}$ consisting of eigenvectors of $\hat{A}$. If we set

$$Q_2 = \begin{bmatrix} I_k & & 0 & \\ 0 & [\hat{\mathbf{u}}_{k+1} & \cdots & \hat{\mathbf{u}}_n] \end{bmatrix},$$

then we have

$$Q_2^T(Q_1^TAQ_1)Q_2 = \Lambda$$

where Λ is an $n \times n$ diagonal matrix with the eigenvalues of A along the diagonal, with each eigenvalue appearing a number of times equal to its geometric multiplicity, which is the number of linearly independent eigenvectors for that eigenvalue. Set $U = Q_1Q_2$. Then $U^TU = I$ and $U^{-1}AU = U^TAU = \Lambda$, which is equivalent to $AU = U\Lambda$. Therefore the columns of U are eigenvectors of A and form an orthonormal basis of $\mathbf{R}^n$. $\square$

The spectral theorem allows a convenient characterization of positive definite and negative definite quadratic forms. First, the definitions.

Definition 8.5.8. *Let A be an $n \times n$ real symmetric matrix. The function $Q : \mathbf{R}^n \to \mathbf{R}$ given by*

$$Q(\mathbf{x}) = (A\mathbf{x}, \mathbf{x}) = \mathbf{x}^TA\mathbf{x}$$

is the **quadratic form** *determined by A. We say that A is* **positive definite** *if $\mathbf{x}^TA\mathbf{x} > 0$ for all nonzero $\mathbf{x} \in \mathbf{R}^n$, and that A is* **negative definite** *if $\mathbf{x}^TA\mathbf{x} < 0$ for all nonzero $\mathbf{x} \in \mathbf{R}^n$. Otherwise, A is said to be* **indefinite**. *These terms are also applied to the quadratic form itself.*

Theorem 8.5.9. *Let A be an $n \times n$ real symmetric matrix. Then A is positive definite if and only if all eigenvalues of A are positive, and A is negative definite if and only if all eigenvalues of A are negative.*

Proof. Suppose A is positive definite. Let $A\mathbf{v} = \lambda\mathbf{v}$. Then

$$\mathbf{v}^TA\mathbf{v} = \mathbf{v}^T(\lambda\mathbf{v}) = \lambda\mathbf{v}^T\mathbf{v} > 0.$$

Since $\mathbf{v}^T\mathbf{v} > 0$ for an eigenvector $\mathbf{v}$, we have $\lambda > 0$. Therefore all eigenvalues of A are positive. On the other hand, suppose that all eigenvalues of A are positive. Let $\mathbf{x}$ be a nonzero vector, and write

$$\mathbf{x} = \sum_{i=1}^{n} c_i\mathbf{u}_i$$

where $\{\mathbf{u}_1, \ldots, \mathbf{u}_n\}$ is an orthonormal basis of eigenvectors of A, with $A\mathbf{u}_i = \lambda_i \mathbf{u}_i$ for $i = 1, \ldots, n$. There is at least one nonzero c_i, and by orthonormality of the $\mathbf{u}_i$ we have

$$\mathbf{x}^T A\mathbf{x} = \sum_{i=1} \lambda_i c_i^2 > 0,$$

since all $\lambda_i > 0$. Therefore A is positive definite.

The proof of the statement about negative definite A is similar (Exercise 8.5.3). □

The spectral theorem also implies that a real symmetric matrix A is positive semidefinite, that is, $\mathbf{x}^T A\mathbf{x} \geq 0$ for all $\mathbf{x}$, if and only if all eigenvalues of A are nonnegative. Similarly, a real symmetric matrix A is negative semidefinite, that is, $\mathbf{x}^T A\mathbf{x} \leq 0$ for all $\mathbf{x}$, if and only if all eigenvalues of A are less than or equal to zero.

Exercises.

Exercise 8.5.1. Show that if z and w are complex numbers, then $\overline{zw} = \overline{z}\,\overline{w}$ and $\overline{z+w} = \overline{z} + \overline{w}$. Then check the details to show that the complex inner product for $\mathbf{C}^n$ satisfies properties (i)-(iv) of Definition 8.5.3.

Exercise 8.5.2. *Gram-Schmidt orthogonalization in* $\mathbf{R}^n$

1. Suppose $\{\mathbf{x}_1, \mathbf{x}_2, \mathbf{x}_3\}$ is a linearly independent set in $\mathbf{R}^3$. Sketch these vectors as nonorthogonal vectors in $\mathbf{R}^3$. Define $\mathbf{v}_1 = \mathbf{x}_1$, and then define

$$\mathbf{v}_2 = \mathbf{x}_2 - \frac{(\mathbf{x}_2, \mathbf{v}_1)}{|\mathbf{v}_1|_2^2}\mathbf{v}_1 \qquad \text{and} \qquad \mathbf{v}_3 = \mathbf{x}_3 - \frac{(\mathbf{x}_3, \mathbf{v}_1)}{|\mathbf{v}_1|_2^2}\mathbf{v}_1 - \frac{(\mathbf{x}_3, \mathbf{v}_2)}{|\mathbf{v}_2|_2^2}\mathbf{v}_2.$$

Sketch the vectors $\mathbf{v}_1, \mathbf{v}_2, \mathbf{v}_3$. Show that $\text{span}\,\{\mathbf{v}_1, \mathbf{v}_2\} = \text{span}\,\{\mathbf{x}_1, \mathbf{x}_2\}$. Show that $\{\mathbf{v}_1, \mathbf{v}_2, \mathbf{v}_3\}$ is an orthogonal set, and hence a basis for $\mathbf{R}^3$. By normalizing the vectors $\mathbf{v}_j$, we obtain an orthonormal basis of $\mathbf{R}^3$.

2. Suppose $\{\mathbf{x}_1, \ldots, \mathbf{x}_n\}$ is a linearly independent set in $\mathbf{R}^n$. Let $\mathbf{v}_1 = \mathbf{x}_1$, and for $2 \leq j \leq n$, define

$$\mathbf{v}_j = \mathbf{x}_j - \sum_{i=1}^{j-1} \frac{(\mathbf{v}_j, \mathbf{v}_i)}{|\mathbf{v}_i|_2^2}\mathbf{v}_i.$$

Show that $\text{span}\,\{\mathbf{v}_1, \ldots, \mathbf{v}_k\} = \text{span}\,\{\mathbf{x}_1, \ldots, \mathbf{x}_k\}$ for $1 \leq k \leq n$, and thus $\{\mathbf{v}_1, \ldots, \mathbf{v}_n\}$ is an orthogonal basis for $\mathbf{R}^n$. By normalizing the vectors $\mathbf{v}_j$, we obtain an orthonormal basis of $\mathbf{R}^n$.

Exercise 8.5.3. Complete the details to show that an $n \times n$ real symmetric matrix is negative definite if and only if all eigenvalues of A are negative.

Exercise 8.5.4. If B is any $n \times n$ real matrix, not necessarily symmetric, then $\mathbf{x}^T B\mathbf{x}$ is the quadratic form determined by B. Show that there is a symmetric matrix P such that $\mathbf{x}^T B\mathbf{x} = \mathbf{x}^T P\mathbf{x}$ for all $\mathbf{x} \in \mathbf{R}^n$. *Hint*: Write $B = \frac{1}{2}(B + B^T) + \frac{1}{2}(B - B^T)$.

Exercise 8.5.5. Find an orthogonal matrix P such that $P^{-1}AP$ is diagonal, if

$$A = \begin{bmatrix} -2 & 1 & 0 \\ 1 & -2 & 1 \\ 0 & 1 & -2 \end{bmatrix}.$$

8.6. The Euclidean Metric Space $\mathbf{R}^n$

The length of vectors in the plane and in space was abstracted to provide the definition of a norm. And a norm can be used to define the distance between two points. If $\mathbf{x}, \mathbf{y} \in \mathbf{R}^3$, then the Euclidean distance between $\mathbf{x} = (x_1, x_2, x_3)$ and $\mathbf{y} = (y_1, y_2, y_3)$ is defined by

$$d(\mathbf{x}, \mathbf{y}) = \sqrt{(x_1 - y_1)^2 + (x_2 - y_2)^2 + (x_3 - y_3)^2}.$$

Note that this is exactly the Euclidean norm $|\mathbf{x} - \mathbf{y}|_2$ for $\mathbf{x}, \mathbf{y} \in \mathbf{R}^3$.

Often it is useful to have a distance function defined for pairs of points in sets that are not vector spaces, including the case of a proper subset of a vector space which is not a subspace. The next definition is stated for an arbitrary nonempty set and lists the essential properties of a distance function, or *metric*.

Definition 8.6.1 (Metric Space). *Let X be a nonempty set. A function $d : X \times X \to \mathbf{R}$ with values denoted $d(x, y)$ is a* **metric** *in X if the following properties hold for $x, y, z \in X$:*

1. $d(x, y) > 0$ *if* $x \neq y$.
2. $d(x, x) = 0$.
3. $d(x, y) = d(y, x)$.
4. $d(x, y) \leq d(x, z) + d(z, y)$.

A nonempty set X with a metric is called a **metric space**.

We see that a metric in $\mathbf{R}^3$ is determined by the Euclidean norm. More generally, we have the following theorem:

Theorem 8.6.2. *If V is a normed space with norm $|\cdot|$, then the function $d(x, y) = |x - y|$ is a metric for V, hence V is a metric space.*

Proof. 1. $d(x, y) = |x - y| > 0$ if $x - y \neq 0$.

2. $d(x, x) = |x - x| = |0| = 0$.

3. $d(x, y) = |x - y| = |(-1)(y - x)| = |y - x| = d(y, x)$.

4. $d(x, y) = |x - y| = |x - z + z - y| \leq |x - z| + |z - y| = d(x, z) + d(z, y)$. □

If $\mathbf{x}, \mathbf{y} \in \mathbf{R}^n$, then the distance between $\mathbf{x} = (x_1, \ldots, x_n)$ and $\mathbf{y} = (y_1, \ldots, y_n)$ as determined by the Euclidean norm is

$$d(\mathbf{x}, \mathbf{y}) = \sqrt{(x_1 - y_1)^2 + \cdots + (x_n - y_n)^2},$$

and we naturally call this the **Euclidean metric** in $\mathbf{R}^n$. If $\mathbf{x} \in \mathbf{R}^n$, then in the Euclidean norm, the set of all points $\mathbf{y}$ such that $d(\mathbf{x}, \mathbf{y}) < r$ is usually denoted $B_r(\mathbf{x}) = \{\mathbf{y} : |\mathbf{y} - \mathbf{x}|_2 < r\}$, the open ball of radius r about $\mathbf{x}$.

Example 8.6.3. We identify the product space $\mathbf{R}^n \times \mathbf{R}^p$ with $\mathbf{R}^{n+p}$. The Euclidean metric on the product space $\mathbf{R}^n \times \mathbf{R}^p$ is the Euclidean metric on $\mathbf{R}^{n+p}$ under the identification. △

We proceed to define sequential convergence, Cauchy sequences, and completeness, for general metric spaces.

Definition 8.6.4. *Let X be a metric space with metric d. A sequence $(x_k)_{k=1}^{\infty}$ in X* **converges** *to a point x in X if for every $\epsilon > 0$ there is a positive integer N such that if $k \geq N$, then $d(x_k, x) < \epsilon$. Then x is called the* **limit** *of the sequence (x_k).*

This definition implies that a sequence (x_k) converges with limit x if and only if $\lim_{k\to\infty} d(x_k, x) = 0$. It is straightforward to show that the limit of a convergent sequence is unique.

Definition 8.6.5. *A sequence $(x_k)_{k=1}^{\infty}$ in X is a* **Cauchy sequence** *if for every $\epsilon > 0$ there is a positive integer N such that if $m, n \geq N$, then $d(x_m, x_n) < \epsilon$.*

A sequence that converges is necessarily a Cauchy sequence.

Definition 8.6.6. *A metric space X is said to be* **complete** *if every Cauchy sequence in X has a limit in X.*

A space X may be complete with respect to one metric but not complete with respect to a different metric.

Example 8.6.7. On the real vector space $C[a, b]$ of real valued functions continuous on $[a, b]$,

$$d(f, g) := \max_{a \leq x \leq b} |f(x) - g(x)|$$

is a metric. (See also Example 8.3.10.) An alternative choice of distance function,

$$d(f, g) := \int_a^b |f(x) - g(x)|\, dx,$$

provides another metric on $C[a, b]$. (See Example 8.3.10, Exercise 8.3.7 and Exercise 8.3.8.) We will see in Section 9.3 that $C[a, b]$ is complete with respect to the first of these metrics, but $C[a, b]$ is not complete with respect to the second, integral metric. These facts will be examined more fully in Theorem 9.3.1 and Exercise 9.3.2. △

We end this section with an example of a metric on a vector space of sequences. This example serves to give an impression of how general the concept of a metric space is, as the metric in this space does not arise from a norm.

Example 8.6.8. Let S denote the set of all real sequences. Then S is a vector space with componentwise operations of addition and scalar multiplication. Define a metric on the sequence space s as follows: If $\sigma = (\sigma_k), \mu = (\mu_k) \in S$, define the distance between σ and μ by

$$d(\sigma, \mu) := \sum_{k=1}^{\infty} \frac{1}{2^k} \frac{|\sigma_k - \mu_k|}{1 + |\sigma_k - \mu_k|}.$$

Then properties (i), (ii), (iii) of Definition 8.6.1 hold. The triangle inequality (iv) also holds, as we now show. Let

$$g(t) = \frac{t}{1+t}, \quad \text{for } t \text{ real.}$$

Then $g'(t) = (1+t)^{-2} > 0$ for all t, so g is an increasing function of t. Hence,

$$|a + b| \leq |a| + |b| \quad \Longrightarrow \quad g(|a + b|) \leq g(|a| + |b|).$$

Consequently, the triangle inequality for numbers implies

$$\begin{aligned} \frac{|a+b|}{1+|a+b|} &\leq \frac{|a|+|b|}{1+|a|+|b|} \\ &= \frac{|a|}{1+|a|+|b|} + \frac{|b|}{1+|a|+|b|} \\ &\leq \frac{|a|}{1+|a|} + \frac{|b|}{1+|b|}. \end{aligned}$$

If $\xi = (\xi_k)$, $\zeta = (\zeta_k)$, and $\eta = (\eta_k)$ are in S, then, letting $a = \xi_k - \zeta_k$ and $b = \zeta_k - \eta_k$, we have $a + b = \xi_k - \eta_k$, and

$$\frac{|\xi_k - \eta_k|}{1+|\xi_k - \eta_k|} \leq \frac{|\xi_k - \zeta_k|}{1+|\xi_k - \zeta_k|} + \frac{|\zeta_k - \eta_k|}{1+|\zeta_k - \eta_k|}.$$

Now multiply both sides of this inequality by $1/2^k$ and sum from $k = 1$ to ∞ to obtain

$$d(\xi, \eta) \leq d(\xi, \zeta) + d(\zeta, \eta),$$

which is the triangle inequality for the space S. △

It is important to realize that not every metric comes from a norm. Metrics induced by a norm according to $d(x, y) = |x - y|$ also have the properties given in the following theorem.

Theorem 8.6.9. *If the metric d is induced by a norm $|\cdot|$ on a normed space X, according to $d(x, y) = |x-y|$, then for all $x, y, a \in X$ and all scalars α, the following properties hold:*

1. $d(x + a, y + a) = d(x, y)$;
2. $d(\alpha x, \alpha y) = |\alpha| d(x, y)$.

Proof. By definition, $d(x + a, y + a) = |(x + a) - (y + a)| = |x - y| = d(x, y)$ and $d(\alpha x, \alpha y) = |\alpha x - \alpha y| = |\alpha||x - y| = |\alpha| d(x, y)$. □

We can see that the metric for the sequence space S in Example 8.6.8 is not induced by any norm on S because the metric there does not satisfy property 2 of Theorem 8.6.9.

In summary, an inner product on a vector space defines a norm, and a norm defines a metric distance. However, not every norm is derived from an inner product, and not every metric is derived from a norm.

Exercise.

Exercise 8.6.1. Show that in the metric space S of Example 8.6.8, the distance between any two elements of S is less than 1.

8.7. Sequences and the Completeness of $\mathbf{R}^n$

By the equivalence of norms on $\mathbf{R}^n$ (Theorem 8.3.13), we may discuss convergence of sequences using any norm we wish. Let us work with the Euclidean norm $|\cdot|_2$ on $\mathbf{R}^n$. A sequence in $\mathbf{R}^n$ is a function $\mathbf{a} : \mathbf{N} \to \mathbf{R}^n$ where $\mathbf{a}(k) = \mathbf{a}_k$, and we denote the sequence by $(\mathbf{a}_k)$. Definition 8.3.4 includes the definition of Cauchy sequence and convergent sequence for a normed space such as $\mathbf{R}^n$. The uniqueness of sequential limits in $\mathbf{R}^n$ follows by an argument similar to the one that shows uniqueness of real number sequential limits (Theorem 2.4.2). Convergent sequences are bounded in norm, and every subsequence of a convergent sequence must converge to the same limit. The formulation and proof of these facts is left to Exercise 8.7.1. If $(\mathbf{a}_k)$ converges with limit L, we may write

$$\lim_{k\to\infty} \mathbf{a}_k = L \qquad \text{or} \qquad \mathbf{a}_k \to L \quad (\text{as } k \to \infty).$$

Any sequence in $\mathbf{R}^n$ determines an ordered n-tuple of real number sequences. Consider the following example.

Example 8.7.1. Consider the sequence in $\mathbf{R}^3$ defined by

$$\mathbf{a}_k = \left(\frac{1}{k^2},\ \sin\frac{k\pi}{2},\ \frac{3k-2}{4k+5}\right).$$

The first and third component sequences have limits 0 and 3/4, respectively, as $k \to \infty$. However, $\sin(k\pi/2)$ has no limit, since

$$\limsup \sin(k\pi/2) = 1 \qquad \text{and} \qquad \liminf \sin(k\pi/2) = -1.$$

Thus there is no point L in $\mathbf{R}^3$ which can be the limit of $(\mathbf{a}_k)$. △

The next lemma helps us to reduce convergence questions for sequences in $\mathbf{R}^n$ to questions about convergence of the real component sequences.

Lemma 8.7.2. *If $\mathbf{x} = (x_1, x_2, \ldots, x_n) \in \mathbf{R}^n$, then*

$$|x_j| \le |\mathbf{x}|_2 \le \sqrt{n}\,\max\{|x_1|, |x_2|, \ldots, |x_n|\}$$

for $j = 1, 2, \ldots, n$.

Proof. We noted earlier that $|x_j| \le |\mathbf{x}|_2$ for $j = 1, 2, \ldots, n$. Since

$$|\mathbf{x}|_2^2 = \sum_{j=1}^{n} |x_j|^2 \le \sum_{j=1}^{n} [\max\{|x_1|, \ldots, |x_n|\}]^2 = n[\max\{|x_1|, \ldots, |x_n|\}]^2,$$

taking positive square roots on both sides yields the second inequality. □

Theorem 8.7.3. *A sequence in $\mathbf{R}^n$ converges with limit $\mathbf{L}$ if and only if each real component sequence converges to the corresponding component of $\mathbf{L}$.*

Proof. Suppose $\lim_{k\to\infty} \mathbf{a}_k = \mathbf{L} = (L_1, \ldots, L_n)$. Let j be a fixed integer in $\{1, 2, \ldots, n\}$, and denote the real j-th component sequence by (a_{kj}). By Lemma 8.7.2,

$$|a_{kj} - L_j| \le |\mathbf{a}_k - \mathbf{L}|_2 \le \sqrt{n}\,\max\{|a_{k1} - L_1|, |a_{k2} - L_2|, \ldots, |a_{kn} - L_n|\}.$$

Given $\epsilon > 0$, there is an $N = N(\epsilon)$ such that $|\mathbf{a}_k - \mathbf{L}|_2 < \epsilon$ for all $k \ge N$, hence $|a_{kj} - L_j| < \epsilon$ for all $k \ge N$. This shows that the j-th component sequence converges with limit L_j. Since j was arbitrary in this argument, each component sequence converges to the corresponding component of $\mathbf{L}$.

Conversely, suppose that each component sequence converges to the corresponding component of $\mathbf{L} = (L_1, \dots, L_n) \in \mathbf{R}^n$. Since there are finitely many components, given $\epsilon > 0$ there is an N such that

$$\sqrt{n}\max\{|a_{k1} - L_1|, |a_{k2} - L_2|, \dots, |a_{kn} - L_n|\} < \epsilon$$

for all $k \geq N$. Then also $|\mathbf{a}_k - \mathbf{L}|_2 < \epsilon$ for all $k \geq N$. Therefore $(\mathbf{a}_k)$ converges to $\mathbf{L} = (L_1, \dots, L_n)$. □

Notice that Theorem 8.7.3 establishes without any doubt that the sequence in Example 8.7.1 diverges.

Example 8.7.4. Consider the sequence in $\mathbf{R}^2$ defined by

$$\mathbf{a}_k = \left(\frac{1}{k^2}, k^{101}e^{-k}\right).$$

The first component sequence $(1/k^2)$ clearly converges, so $(\mathbf{a}_k)$ converges if and only if $\lim_{k\to\infty} k^{101}e^{-k}$ exists. But this limit exists and equals zero, by repeated application of l'Hôpital's rule. △

We are now ready to discuss the completeness of the space $\mathbf{R}^n$. As in the case of sequences of real numbers, a convergent sequence in $\mathbf{R}^n$ must be a Cauchy sequence, and the reader is encouraged to write this out in Exercise 8.7.2. It is now possible to address the converse question: Does every Cauchy sequence in $\mathbf{R}^n$ converge to a limit in $\mathbf{R}^n$?

It follows from the fundamental inequality in Lemma 8.7.2 that the sequence $(\mathbf{a}_k)$ in $\mathbf{R}^n$ is a Cauchy sequence if and only if each of its component sequences is a Cauchy sequence of real numbers. The proof of this fact follows directly from the inequality

$$|a_{j1} - a_{j1}| \leq |\mathbf{a}_l - \mathbf{a}_k|_2 \leq \sqrt{n}\max\{|a_{l1} - a_{k1}|, |a_{l2} - a_{k2}|, \dots, |a_{ln} - a_{kn}|\},$$

which holds for $j = 1, \dots, n$. (See also Exercise 8.7.3.)

Theorem 8.7.5. *A sequence in $\mathbf{R}^n$ converges to a limit in $\mathbf{R}^n$ if and only if it is a Cauchy sequence.*

Proof. If the sequence $(\mathbf{a}_k)$ is Cauchy, then each of its component sequences is Cauchy, by Lemma 8.7.2 and Exercise 8.7.3, and hence each component sequence converges to a real number limit. Thus, $(\mathbf{a}_k)$ itself converges, by Theorem 8.7.3.

For the converse statement (the only if part), see Exercise 8.7.2. □

Thus the completeness of $\mathbf{R}^n$ is a consequence of the completeness of the real number field $\mathbf{R}$.

Exercises.

Exercise 8.7.1. Suppose that the sequence $(\mathbf{a}_k)$ in $\mathbf{R}^n$ converges with limit $\mathbf{L}$. Prove the following:

1. The limit is unique.

2. The sequence is bounded: There exists M such that $|\mathbf{a}_k| \leq M$ for all k.
3. If $(\mathbf{a}_{n_k})$ is any subsequence of $(\mathbf{a}_k)$, then $\lim_{k\to\infty} \mathbf{a}_{n_k} = \mathbf{L}$.

Exercise 8.7.2. Prove: If a sequence $(\mathbf{a}_k)$ in $\mathbf{R}^n$ converges to a limit in $\mathbf{R}^n$, then it is a Cauchy sequence.

Exercise 8.7.3. Write an $\epsilon, N(\epsilon)$ proof of the following statement:
A sequence $(\mathbf{a}_k)$ in $\mathbf{R}^n$ is a Cauchy sequence if and only if each of its component sequences is a Cauchy sequence of real numbers.

8.8. Topological Concepts for $\mathbf{R}^n$

This section presents the basic definitions of topological concepts in $\mathbf{R}^n$.

8.8.1. Topology of $\mathbf{R}^n$.

The concepts and definitions in this subsection flow from the pattern set down earlier for topological notions for subsets of the real numbers. The accumulation of definitions is not meant to intimidate but to encourage some review of the earlier material from Chapter 4, since the pattern of definition and argument for the facts presented here already appears in the earlier work, and it is important to observe the similarities.

We begin with the basic topological definitions describing points in relation to a given subset of $\mathbf{R}^n$. The statements are essentially the same as in the case of subsets of real numbers.

Definition 8.8.1. *Let S be a subset of $\mathbf{R}^n$.*

1. *A point $\mathbf{x} \in S$ is an* **interior point** *of S if there exists an $\epsilon > 0$ such that the open ball $B_\epsilon(\mathbf{x})$ is contained in S. The set of all interior points of S, sometimes denoted* Int S, *is called the* **interior** *of S.*
2. *A point $\mathbf{x}$ is a* **boundary point** *of S if for every $\epsilon > 0$ the open ball $B_\epsilon(\mathbf{x})$ has nonempty intersection with both S and its complement S^c. The set of all boundary points of S, denoted ∂S, is called the* **boundary** *of S. A boundary point of S need not be a point of S.*
3. *A point $\mathbf{x}$ is a* **cluster point** *(or,* **accumulation point***) of S if for every $\epsilon > 0$ the open ball $B_\epsilon(\mathbf{x})$ contains infinitely many points of S distinct from* $\mathbf{x}$.
4. *If $\mathbf{x} \in S$ and $\mathbf{x}$ is not a cluster point of S, then it is an* **isolated point**.

A subset S of $\mathbf{R}^n$ is **open** if every point of S is an interior point. A subset F of $\mathbf{R}^n$ is **closed** if its complement $\mathbf{R}^n - F$ is open. As in the case of subsets of the real numbers, a set $F \subseteq \mathbf{R}$ is closed if and only if F contains all its cluster points, if and only if $\overline{F} = F$.

A set $S \subset \mathbf{R}^n$ is **dense** in $\mathbf{R}^n$ if $\overline{S} = \mathbf{R}^n$. For example, $\mathbf{Q}^n$, the Cartesian product of n copies of $\mathbf{Q}$, is dense in $\mathbf{R}^n$. More generally, a set S is **dense in an open set** U if $U \subset \overline{S}$. A set $S \subset \mathbf{R}^n$ is **nowhere dense** if its closure $\overline{S}$ has no interior point.

We record several fundamental theorems whose proofs are straightforward translations of the proofs of earlier results.

Theorem 8.8.2. *In $\mathbf{R}^n$, the following statements hold:*

1. *The union of any collection of open sets in $\mathbf{R}^n$ is an open set in $\mathbf{R}^n$.*
2. *The intersection of a finite collection of open sets in $\mathbf{R}^n$ is open in $\mathbf{R}^n$.*

The proof of part 1 of Theorem 8.8.2 is essentially the same as the proof of Theorem 4.1.4; the proof of part 2 is essentially the same as the argument for Theorem 4.1.5. By essentially the same, we mean that only the notation of the earlier argument might need adaptation to reflect the new setting of $\mathbf{R}^n$.

Theorem 8.8.3. *In $\mathbf{R}^n$, the following statements hold:*

1. *The intersection of any collection of closed sets in $\mathbf{R}^n$ is a closed set in $\mathbf{R}^n$.*
2. *The union of any finite collection of closed sets in $\mathbf{R}^n$ is closed in $\mathbf{R}^n$.*

The proof of part 1 of Theorem 8.8.3 is essentially the same as the proof of Theorem 4.1.9; the proof of part 2 is essentially the same as the argument for Theorem 4.1.10.

Theorem 8.8.4. *Let $S \subset \mathbf{R}^n$. A point $\mathbf{x}_0 \in S$ is a cluster point of S if and only if there is a nonconstant sequence $(\mathbf{x}_n)$ of points of S distinct from $\mathbf{x}_0$ such that $\mathbf{x}_n \to \mathbf{x}_0$ as $n \to \infty$.*

To prove Theorem 8.8.4, translate the proof of Theorem 4.1.12 to $\mathbf{R}^n$.

8.8.2. Relative Topology of a Subset. Each subset of $\mathbf{R}^n$ inherits a topology, a designated collection of open subsets, from $\mathbf{R}^n$ itself. This inherited topology is called the relative topology on the subset. This section presents some global properties of continuous functions, described conveniently in the language of relative topology. Definition 8.8.5 (Definition 8.8.8) describes the sets that are open (closed) relative to a given subset E. The basic properties of the relative topology are set out in Theorem 8.8.7, Theorem 8.8.10 and Theorem 8.8.11. The concept of the relative topology on a subset of the reals provides a convenient language for a topological characterization of continuous functions on a domain $E \subset \mathbf{R}$ (Theorem 8.10.9). We also discuss continuous images of connected sets for real valued functions of a real variable.

Definition 8.8.5. *Let $E \subset \mathbf{R}^n$. A subset S of E is* **open relative to** *E if $S = E \cap O$ for some open subset O of $\mathbf{R}^n$.*

If S is open relative to $E \subset \mathbf{R}^n$, we also say that S is open in E. For every set $E \subset \mathbf{R}^n$, $E = E \cap \mathbf{R}^n$ and $\mathbf{R}^n$ is open, so every set E is open in itself. The empty set is also open in E.

If $E = \mathbf{R}^n$, then Definition 8.8.5 agrees with the definition of open subset of $\mathbf{R}^n$. In fact, if E is any open set, then the subsets of E that are open relative to E are exactly the subsets of E that are open in $\mathbf{R}^n$. We now consider a set $E \subset \mathbf{R}$ which is not open, and some of its subsets that are open in E.

Example 8.8.6. Let $E = [0, 2)$. For any b with $0 < b < 2$, the subset $S_b = [0, b)$ is open in E since, for example, we have

$$S_b = [0, b) = E \cap (-b, b)$$

and $(-b, b)$ is an open subset of the reals. △

In the general study of topology, the collection of open sets in a space defines what we call the **topology** of the space, and the set of open sets in S determined by Definition 8.8.5 defines the **relative topology** of the subset S, meaning the topology that S inherits from the ambient space ($\mathbf{R}^n$, in this case). The relative topology on E provides a topology for E in the same sense that the open subsets of $\mathbf{R}^n$ provide a topology for $\mathbf{R}^n$. This is the content of the next theorem, whose proof is left to Exercise 8.8.1.

Theorem 8.8.7. *The collection of sets that are open relative to $E \subset \mathbf{R}^n$ contains the empty set and E itself, and is closed under arbitrary unions and finite intersections.*

An equally useful concept is that of the collection of subsets of S that are closed relative to S.

Definition 8.8.8. *Let $E \subset \mathbf{R}^n$. A subset S of E is* **closed relative to** *E if $S = E \cap F$ for some closed subset F of $\mathbf{R}^n$.*

If S is closed relative to $E \subset \mathbf{R}^n$, we also say that S is closed in E.

For every set E, the empty set is closed in E. Also, for every set $E \subset \mathbf{R}^n$, $E = E \cap \mathbf{R}^n$ and $\mathbf{R}^n$ is closed, so every set E is closed in itself. For this reason, if $E \subset \mathbf{R}^n$ and $S \subset E$, we say that S is **dense in** E if the closure of S intersected with E equals E, that is, $\overline{S} \cap E = E$. This definition recognizes that E is the universal set for the purpose of discussing the denseness of S, and is equivalent to forming the closure of S using only those cluster points of S that are elements of E.

Example 8.8.9. The set $S = \mathbf{I} \cap (0,1)$ of irrational numbers in $E = (0,1)$ is dense in $(0,1)$, since $\overline{S} \cap (0,1) = (0,1)$. The rational numbers in $(0,1)$ are also dense in $(0,1)$. △

If $E = \mathbf{R}^n$, then Definition 8.8.8 agrees with the earlier definition of closed subset of $\mathbf{R}^n$. If E is a closed subset of $\mathbf{R}^n$, then the subsets of E that are closed relative to E are exactly the subsets of E that are closed in $\mathbf{R}^n$.

Theorem 8.8.10. *The collection of sets that are closed relative to $E \subset \mathbf{R}^n$ contains the empty set and E itself, and is closed under finite unions and arbitrary intersections.*

The next theorem shows that relatively closed sets are just as important as relatively open sets.

Theorem 8.8.11. *Let $E \subset \mathbf{R}^n$. A subset S of E is open relative to E if and only if its complement in E, $E - S = S^c \cap E$, is closed relative to E.*

Exercises.

Exercise 8.8.1. Prove Theorem 8.8.7.

Exercise 8.8.2. Prove Theorem 8.8.10.

Exercise 8.8.3. Prove Theorem 8.8.11.

8.9. Nested Intervals and the Bolzano-Weierstrass Theorem

We have seen that the real number field has the least upper bound property if and only if it has the nested interval property. Since $\mathbf{R}^n$ is not a field it makes no sense to speak of a least upper bound property for this space. However, we can extend the notion of interval to $\mathbf{R}^n$, and prove a nested interval property for this space.

First, we generalize real intervals with the concept of intervals in $\mathbf{R}^n$. An **open interval** in $\mathbf{R}^n$ $(n \geq 2)$ is a Cartesian product of n real intervals,

$$B = (a_1, b_1) \times \cdots \times (a_n, b_n),$$

where $a_i < b_i$ for $1 \leq i \leq n$. A **closed interval** in $\mathbf{R}^n$ $(n \geq 2)$ has the form

$$B = [a_1, b_1] \times \cdots \times [a_n, b_n],$$

where $a_i \leq b_i$ for $1 \leq i \leq n$. (Note that the interior of a closed interval is the open interval having the same endpoints for each interval factor.)

A set is open in $\mathbf{R}^n$ if every point of the set is an interior point; a set is closed if its complement in $\mathbf{R}^n$ is open. It is an exercise to show that a closed interval in $\mathbf{R}^n$ is a closed set, and an open interval in $\mathbf{R}^n$ is an open set.

Theorem 8.9.1 (Nested Interval Property). *If $I_k = [a_1^{(k)}, b_1^{(k)}] \times \cdots \times [a_n^{(k)}, b_n^{(k)}]$ is a nested sequence of closed nonempty intervals in $\mathbf{R}^n$, that is, if*

$$I_{k+1} \subseteq I_k \quad \textit{for each } k \in \mathbf{N}, \tag{8.9}$$

then the intersection $\bigcap_{k=1}^{\infty} I_k$ is nonempty. If $\lim_{k\to\infty} |b_j^{(k)} - a_j^{(k)}| = 0$ for each $j = 1, 2 \ldots, n$, then the intersection consists of a single point.

Proof. By the hypothesis (8.9), we have for each $j = 1, 2, \ldots, n$ a nested sequence of real intervals

$$[a_j^{(k+1)}, b_j^{(k+1)}] \subseteq [a_j^{(k)}, b_j^{(k)}], \quad k \in \mathbf{N}.$$

By the nested interval Theorem 2.5.1, there are real numbers $c_j \in [a_j^{(k)}, b_j^{(k)}]$ for all $k \in \mathbf{N}$, $j = 1, 2, \ldots, n$. Thus, $\mathbf{c} = (c_1, c_2, \ldots, c_n) \in I_k$ for all $k \in \mathbf{N}$. If, in addition, we have $\lim_{k\to\infty} |b_j^{(k)} - a_j^{(k)}| = 0$ for each $j = 1, 2, \ldots, n$, then again by Theorem 2.5.1, there is exactly one number $c_j \in [a_j^{(k)}, b_j^{(k)}]$ for all $k \in \mathbf{N}$, for each $j = 1, 2, \ldots, n$, and thus a single point in the intersection $\bigcap_k I_k$. □

We can now prove a Bolzano-Weierstrass theorem for $\mathbf{R}^n$. First we need the definition of bounded set in $\mathbf{R}^n$.

Definition 8.9.2. *A set $S \subset \mathbf{R}^n$ is* **bounded** *if there is a number $M > 0$ such that $|\mathbf{x}|_2 \leq M$ for all $\mathbf{x} \in S$.*

We have used the Euclidean norm in this definition of bounded set, but it does not matter which norm on $\mathbf{R}^n$ we use (Exercise 8.9.3).

Theorem 8.9.3 (Bolzano-Weierstrass). *A bounded infinite set S in $\mathbf{R}^n$ has at least one cluster point, which need not be an element of S.*

Proof. See Exercise 8.9.4. □

Exercises.

Exercise 8.9.1. Prove that an open interval in $\mathbf{R}^n$ is an open set.

Exercise 8.9.2. Prove that a closed interval in $\mathbf{R}^n$ is a closed set.

Exercise 8.9.3. Explain why we could use any norm on $\mathbf{R}^n$ in Definition 8.9.2. To be precise, if $|\cdot|_0$ denotes any norm on $\mathbf{R}^n$, explain why there is a number $M > 0$ such that $|\mathbf{x}|_2 \leq M$ for all $\mathbf{x} \in S$ if and only if there is a number $M_0 > 0$ such that $|\mathbf{x}|_0 \leq M_0$ for all $\mathbf{x} \in S$.

Exercise 8.9.4. Write a detailed proof of Theorem 8.9.3. *Hint*: Follow the idea of the proof of Theorem 2.6.3, observing that now an enclosing interval B for S must be subdivided into 2^n subintervals, and the process continued by such subdivisions at each step.

8.10. Mappings of Euclidean Spaces

This section presents the basic definitions and results for mappings of Euclidean spaces. Many of the results of this section have a global character.

8.10.1. Limits of Functions and Continuity.

The definitions of limit of a function $\mathbf{F} : D \to \mathbf{R}^m$ at a cluster point of the domain $D \subseteq \mathbf{R}^n$ and continuity of $\mathbf{F}$ at a point $\mathbf{a} \in D$ take the same form as in the single variable case.

Definition 8.10.1. *Let $D \subseteq \mathbf{R}^n$ and $\mathbf{F} : D \to \mathbf{R}^m$. Let $\mathbf{a}$ be a cluster point of D, and let $\mathbf{L} \in \mathbf{R}^m$. We say that $\mathbf{F}$ has* **limit** $\mathbf{L}$ *as $\mathbf{x}$ approaches $\mathbf{a}$, and write*

$$\lim_{\mathbf{x}\to\mathbf{a}} \mathbf{F}(\mathbf{x}) = \mathbf{L}$$

if for every $\epsilon > 0$ there is a $\delta > 0$ such that

$$\mathbf{x} \in D \text{ and } 0 < |\mathbf{x} - \mathbf{a}| < \delta \quad \Longrightarrow \quad |\mathbf{F}(\mathbf{x}) - \mathbf{L}| < \epsilon.$$

As expected, limits are unique, and the proof is exactly the same as the proof of Theorem 4.4.6 with the absolute value replaced by a norm.

Theorem 8.10.2. *Let $\mathbf{F} : D \to \mathbf{R}^n$ and let $\mathbf{a}$ be a cluster point of D. If $\lim_{\mathbf{x}\to\mathbf{a}} \mathbf{F}(\mathbf{x}) = \mathbf{L}_1$ and $\lim_{\mathbf{x}\to\mathbf{a}} \mathbf{F}(\mathbf{x}) = \mathbf{L}_2$ according to Definition 8.10.1, then $\mathbf{L}_1 = \mathbf{L}_2$.*

For functions mapping a subset of $\mathbf{R}^n$ into the real numbers, we usually employ a lowercase f instead of the boldfaced capital $\mathbf{F}$. Limits may then be indicated by $\lim_{\mathbf{x}\to\mathbf{a}} f(\mathbf{x}) = L \in \mathbf{R}$ (that is, L is not boldfaced).

Definition 8.10.3. *Let $D \subseteq \mathbf{R}^n$ and $\mathbf{F} : D \to \mathbf{R}^m$. If $\mathbf{a} \in D$, then $\mathbf{F}$ is* **continuous** *at $\mathbf{a}$ if*

$$\lim_{\mathbf{x}\to\mathbf{a}} \mathbf{F}(\mathbf{x}) = \mathbf{F}(\mathbf{a}).$$

We have the following characterization of limits and continuity in terms of sequential convergence. The proof mimics the proof in the single variable case. (See Theorem 4.4.8.)

Theorem 8.10.4. *Let $D \subseteq \mathbf{R}^n$ and $\mathbf{F} : D \to \mathbf{R}^m$. Let $\mathbf{a}$ be a cluster point of D. Then the following are true:*

1. $\lim_{\mathbf{x}\to\mathbf{a}} \mathbf{F}(\mathbf{x}) = \mathbf{L}$ *if and only if for every sequence $(\mathbf{x}_n)$ such that* $\lim_{n\to\infty} \mathbf{x}_n = \mathbf{a}$, *we have* $\lim_{n\to\infty} \mathbf{F}(\mathbf{x}_n) = \mathbf{L}$.
2. *In particular, $\mathbf{F}$ is continuous at $\mathbf{a}$ if and only if for every sequence $(\mathbf{x}_n)$ in D such that* $\lim_{n\to\infty} \mathbf{x}_n = \mathbf{a}$, *we have* $\lim_{n\to\infty} \mathbf{F}(\mathbf{x}_n) = \mathbf{F}(\mathbf{a})$.

Two of the properties of limits in Theorem 4.4.7 have direct analogues here. In particular, $\lim_{\mathbf{x}\to\mathbf{a}} \mathbf{F}(\mathbf{x}) = \mathbf{L} \in \mathbf{R}^n$ if and only if $\lim_{\mathbf{x}\to\mathbf{a}} |\mathbf{F}(\mathbf{x}) - \mathbf{L}| = 0$. And if $\lim_{\mathbf{x}\to\mathbf{a}} \mathbf{F}(\mathbf{x}) = \mathbf{L}$, then for any scalar $c \in \mathbf{R}$, $\lim_{\mathbf{x}\to\mathbf{a}} c\mathbf{F}(\mathbf{x}) = c\mathbf{L}$. The squeeze theorem (Theorem 4.4.7(c)) has a direct extension only for *real valued* functions of $\mathbf{x} \in \mathbf{R}^n$, and the formulation and proof of this extension is an exercise.

Sums, differences, products, and quotients of functions are defined pointwise as usual. The results of Theorem 4.4.9 on limits of sums and differences extend to the case of mappings from $D \subseteq \mathbf{R}^n$ into $\mathbf{R}^m$ and follow from the sequential characterization in Theorem 8.10.4. For real valued functions, the product and quotient limit laws hold as well, and these are summarized in the following theorem.

Theorem 8.10.5. *Let $\mathbf{F}$ and $\mathbf{G}$ be functions defined on a common domain D, and let $\mathbf{a}$ be a cluster point of D.*

1. *If* $\lim_{\mathbf{x}\to\mathbf{a}} \mathbf{F}(\mathbf{x}) = \mathbf{L}_1$ *and* $\lim_{\mathbf{x}\to\mathbf{a}} \mathbf{G}(\mathbf{x}) = \mathbf{L}_2$, *then*
$$\lim_{\mathbf{x}\to\mathbf{a}} (\mathbf{F} \pm \mathbf{G})(\mathbf{x}) = \mathbf{L}_1 \pm \mathbf{L}_2.$$
2. *If f and g are real valued on D, and if* $\lim_{\mathbf{x}\to\mathbf{a}} f(\mathbf{x}) = L_1$ *and* $\lim_{\mathbf{x}\to\mathbf{a}} g(\mathbf{x}) = L_2$, *then*
$$\lim_{\mathbf{x}\to\mathbf{a}} f(\mathbf{x})g(\mathbf{x}) = L_1L_2, \quad \textit{and} \quad \lim_{\mathbf{x}\to\mathbf{a}} f(\mathbf{x})/g(\mathbf{x}) = L_1/L_2 \quad \textit{if } L_2 \neq 0.$$

The next result is an immediate consequence of Theorem 8.10.5.

Theorem 8.10.6. *Suppose $\mathbf{F}$ and $\mathbf{G}$ are vector valued, and f and g are real valued, functions defined in an open set $D \subseteq \mathbf{R}^n$ that contains the point $\mathbf{a}$.*

1. *If $\mathbf{F}$ and $\mathbf{G}$ are both continuous at $\mathbf{a}$, then $\mathbf{F} \pm \mathbf{G}$ is continuous at a.*
2. *If f and g are both continuous at $\mathbf{a}$, then fg is continuous at $\mathbf{a}$, and, if $g(\mathbf{a}) \neq 0$, then f/g is continuous at $\mathbf{a}$.*

Let $U \subseteq \mathbf{R}^n$, $V \subseteq \mathbf{R}^m$ and suppose that $\mathbf{G} : U \to \mathbf{R}^m$ and $\mathbf{F} : V \to \mathbf{R}^p$. Then the composition $\mathbf{F} \circ \mathbf{G}$ is the function defined by

$$(\mathbf{F} \circ \mathbf{G})(\mathbf{x}) = \mathbf{F}(\mathbf{G}(\mathbf{x})), \quad \text{for } \mathbf{x} \in \mathbf{G}^{-1}(V) \cap U,$$

where $\mathbf{G}^{-1}(V) = \{\mathbf{x} : \mathbf{G}(\mathbf{x}) \in V\}$. We have the following theorem on the continuity of the composite mapping at a point.

Theorem 8.10.7 (Continuity of Composition). *Let $U \subseteq \mathbf{R}^n$ and $V \subseteq \mathbf{R}^m$, and suppose that $\mathbf{G} : U \to \mathbf{R}^m$ and $\mathbf{F} : V \to \mathbf{R}^p$. If $\mathbf{G}$ is continuous at $\mathbf{a} \in U$ and $\mathbf{F}$ is continuous at $\mathbf{G}(\mathbf{a}) \in V$, then $\mathbf{F} \circ \mathbf{G}$ is continuous at $\mathbf{a}$.*

Proof. We use the sequential characterization of continuity. Let $(\mathbf{x}_j)_{j=1}^{\infty}$ be any sequence such that $\mathbf{x}_j \to a$ as $j \to \infty$. Then, by continuity of $\mathbf{G}$ at $\mathbf{a}$, $\mathbf{G}(\mathbf{x}_j) \to \mathbf{G}(\mathbf{a})$, and by continuity of $\mathbf{F}$ at $\mathbf{G}(\mathbf{a})$, $\mathbf{F}(\mathbf{G}(\mathbf{x}_j)) \to \mathbf{F}(\mathbf{G}(\mathbf{a}))$. □

We record the following definitions for continuity and uniform continuity of functions $\mathbf{F} : D \subseteq \mathbf{R}^n \to \mathbf{R}^m$ over their domain D.

Definition 8.10.8. *We say that a function* $\mathbf{F} : D \subseteq \mathbf{R}^n \to \mathbf{R}^m$ *is* **continuous on** D *if it is continuous at each* $\mathbf{x} \in D$. *We say that* F *is* **uniformly continuous on** D *if for every* $\epsilon > 0$ *there is a* $\delta = \delta(\epsilon) > 0$ *such that if* $x, y \in D$ *and* $|x - y| < \delta(\epsilon)$, *then* $|f(x) - f(y)| < \epsilon$.

Exercises.

Exercise 8.10.1. Let $f(x, y) = x^2y/(x^4 + y^2)$ if $(x, y) \neq (0, 0)$, and $f(0, 0) = 0$. Show that $\lim_{(x,y)\to(0,0)} f(x, y)$ does not exist. *Hint*: Consider different paths of approach to $(0, 0)$.

Exercise 8.10.2. Show that the function

$$f(x, y) = \begin{cases} xy/(x^2 + y^2) & \text{if } (x, y) \neq (0, 0), \\ 0 & \text{if } (x, y) = (0, 0) \end{cases}$$

is not continuous at $(0, 0)$.

Exercise 8.10.3. Show that the function

$$f(x, y) = \begin{cases} xy/\sqrt{x^2 + y^2} & \text{if } (x, y) \neq (0, 0), \\ 0 & \text{if } (x, y) = (0, 0) \end{cases}$$

is continuous on $\mathbf{R}^2$. Use the limit definition to show that f is continuous at $(0, 0)$.

Exercise 8.10.4. Formulate and prove a squeeze theorem (as in Theorem 4.4.7 (c)) for real valued functions of $\mathbf{x}$ in $\mathbf{R}^n$.

Exercise 8.10.5. Prove Theorem 8.10.2.

Exercise 8.10.6. Prove Theorem 8.10.4.

Exercise 8.10.7. Let $f : B \subset \mathbf{R}^n \to \mathbf{R}$ be a bounded function, where B is an interval in $\mathbf{R}^n$, and let $\mathbf{p} \in B$. For each open set U containing $\mathbf{p}$, let us define $o(f, U)$ by

$$o(f, U) = \sup\Big\{|f(\mathbf{x}_1) - f(\mathbf{x}_2)| : \mathbf{x}_1, \mathbf{x}_2 \in U \cap B\Big\}.$$

Then define the **oscillation of** f **at** $\mathbf{p} \in B$ by

$$o(f, \mathbf{p}) = \inf\Big\{o(f, U) : U \text{ open and } \mathbf{p} \in U\Big\}.$$

Show the following:

1. $o(f, \mathbf{p})$ exists for any $p \in B$, and $o(f, \mathbf{p}) \geq 0$.
2. f is continuous at $\mathbf{p}$ if and only if $o(f, \mathbf{p}) = 0$.
3. If $D = \{\mathbf{x} \in B : f \text{ is discontinuous at } \mathbf{x}\}$, then $D = \bigcup_{n=1}^{\infty} D_{1/n}$, where $D_{1/n} = \{\mathbf{x} \in B : o(f, \mathbf{x}) \geq 1/n\}$.
4. For any $n \in \mathbf{N}$, the set $D_{1/n} = \{\mathbf{x} \in B : o(f, \mathbf{x}) \geq 1/n\}$ is a closed set.

8.10.2. Continuity on a Domain. We can now give a global characterization of continuity on a domain. We consider real valued functions f and leave the extension to vector valued functions as an exercise. Let $f : E \subset \mathbf{R}^n \to \mathbf{R}$. Recall that if V is any set of real numbers, then the set $f^{-1}(V) = \{x \in E : f(x) \in V\}$ is called the *inverse image* of V under f.

Theorem 8.10.9. *A function $f : E \subset \mathbf{R}^n \to \mathbf{R}$ is continuous on E if and only if the inverse image $f^{-1}(V)$ of every open set V is open relative to E. If the domain E is an open set in $\mathbf{R}^n$, then f is continuous on E if and only if the inverse image $f^{-1}(V)$ of every open set V is open.*

Proof. Suppose f is continuous on E. Let O be an open set in $\mathbf{R}$. If $f^{-1}(O)$ is empty, then $f^{-1}(O)$ is open (the empty set is open) and we are done, so we assume now that $f^{-1}(O)$ is nonempty. If $\mathbf{a} \in f^{-1}(O)$, then $f(\mathbf{a}) \in O$. Since O is open, there is an $\epsilon = \epsilon(\mathbf{a}) > 0$ such that $B_\epsilon(f(\mathbf{a})) \subset O$. Since f is continuous at $\mathbf{a}$, there is a $\delta = \delta(\epsilon, \mathbf{a})$ such that

$$f(B_\delta(\mathbf{a})) \subset B_\epsilon(f(\mathbf{a})) \subset O,$$

that is to say,

$$E \cap B_\delta(\mathbf{a}) \subset f^{-1}(O).$$

This construction can be carried out for each $\mathbf{a} \in f^{-1}(O)$. Now let

$$O_1 = \bigcup_{\mathbf{a} \in f^{-1}(O)} B_{\delta(\epsilon,\mathbf{a})}(\mathbf{a}).$$

Then O_1 is open since it is a union of open sets. If $\mathbf{x} \in f^{-1}(O)$, then clearly $x \in O_1$ (since $\mathbf{x} \in B_{\delta(\epsilon,\mathbf{x})}(\mathbf{x})$). So $f^{-1}(O) \subset O_1$. And since $f^{-1}(O) \subset E$, we have

$$f^{-1}(O) \subset E \cap O_1.$$

But if $\mathbf{x} \in E \cap O_1$, then there is some $\mathbf{a} \in f^{-1}(O)$ for which

$$\mathbf{x} \in E \cap B_{\delta(\epsilon,\mathbf{a})}(\mathbf{a}) \subset f^{-1}(O),$$

and therefore

$$E \cap O_1 \subset f^{-1}(O).$$

Hence $f^{-1}(O) = E \cap O_1$.

Now suppose that for every open set O in $\mathbf{R}$ there is an open set O_1 in $\mathbf{R}^n$ such that

$$f^{-1}(O) = E \cap O_1.$$

Let $\mathbf{a} \in E$. We want to show that f is continuous at $\mathbf{a}$. For every $\epsilon > 0$, $B_\epsilon(f(\mathbf{a}))$ is open in $\mathbf{R}$, and $\mathbf{a} \in f^{-1}(B_\epsilon(f(\mathbf{a})))$. There is an open set O_1 in $\mathbf{R}^n$ such that

$$f^{-1}(B_\epsilon(f(\mathbf{a}))) = E \cap O_1.$$

Since $\mathbf{a} \in O_1$, there is a $\delta = \delta(\epsilon) > 0$ such that $B_\delta(\mathbf{a}) \subset O_1$. Since

$$E \cap B_\delta(\mathbf{a}) \subset E \cap O_1 \subset f^{-1}(B_\epsilon(f(\mathbf{a}))),$$

we have

$$f(E \cap B_\delta(\mathbf{a})) \subset B_\epsilon(f(\mathbf{a})).$$

Since ϵ was arbitrary, this proves that f is continuous at $\mathbf{a}$. Since this is true for each $\mathbf{a} \in E$, f is continuous on E. □

The second statement of this theorem covers the case where f is defined on all of $\mathbf{R}^n$, and then f is continuous on $\mathbf{R}^n$ if and only if the inverse image $f^{-1}(O)$ of every open set O is open in $\mathbf{R}^n$.

The proof of Theorem 8.10.9 extends with minor modifications to cover the case of functions $\mathbf{F} : E \subset \mathbf{R}^n \to \mathbf{R}^m$. Thus, a mapping $\mathbf{F} : E \subseteq \mathbf{R}^n \to \mathbf{R}^m$ is continuous on E if and only if for every open set $O \subset \mathbf{R}^m$ there is an open set $O_1 \subset \mathbf{R}^n$ such that $\mathbf{F}^{-1}(O) = E \cap O_1$.

A closed set is the complement of an open set. The continuity of a function f on a set E may also be characterized using inverse images of closed sets. Recall that for any set W, we have

$$f^{-1}(W^c) = (f^{-1}(W))^c.$$

Theorem 8.10.10. *Let $f : E \subset \mathbf{R}^n \to \mathbf{R}$. The function f is continuous on E if and only if the inverse image $f^{-1}(W)$ of every closed set W is closed relative to E. If the domain E is a closed set in $\mathbf{R}^n$, then f is continuous on E if and only if the inverse image $f^{-1}(W)$ of every closed set W is closed.*

Proof. Suppose f is continuous on E. If W is closed, then W^c is open, and thus $f^{-1}(W^c) = (f^{-1}(W))^c = E \cap O$ where O is an open set, by Theorem 8.10.9. Then $f^{-1}(W) = (E \cap O)^c = E^c \cup O^c$, and by the definition of inverse image, we must have $f^{-1}(W) = E \cap O^c$.

Conversely, if for any closed set W we have $f^{-1}(W) = E \cap O^c$ where O is open, then $(f^{-1}(W))^c = (E \cap O^c)^c = E^c \cup O$ and therefore $f^{-1}(W^c) = E^c \cup O$. By the definition of inverse image, we must have $f^{-1}(W^c) = E \cap O$. Since this is true for any closed set W, it follows that the inverse image under f of any open set is open relative to E, so f is continuous on E by Theorem 8.10.9. □

Note that if f is defined on all of $\mathbf{R}^n$, then f is continuous on $\mathbf{R}^n$ if and only if the inverse image $f^{-1}(W)$ of every closed set W is closed in $\mathbf{R}^n$.

We now discuss the continuous image of a connected set, with the emphasis here on real valued functions of a real variable. Recall the characterization of real intervals as the connected subsets of $\mathbf{R}$, proved in Theorem 4.3.2.

Theorem 8.10.11. *If J is a real interval and $f : J \to \mathbf{R}$ is continuous, then $f(J)$ is connected, hence an interval.*

Proof. Suppose, for the purpose of reaching a contradiction, that U and V are open sets in $f(J)$ that disconnect $f(J)$. Since f is continuous on J, the sets $f^{-1}(U)$ and $f^{-1}(V)$ are open in J, they are disjoint since $U \cap V = \emptyset$, and their union is J since $U \cup V = f(J)$. This gives a disconnection of J, which is a contradiction since we know that every interval is connected. □

An extension of Theorem 8.10.11 is considered in Theorem 8.10.31.

The intermediate value theorem, proved earlier in Theorem 4.6.3 by a different argument, follows now as a corollary of Theorem 8.10.11:

Corollary 8.10.12. *If $f : [a, b] \to \mathbf{R}$ is a continuous function, and c is any real number between $f(a)$ and $f(b)$, then there exists a point $x \in (a, b)$ such that $f(x) = c$.*

Proof. The image $f([a,b])$ is connected, by Theorem 8.10.11, and hence is an interval by Theorem 4.3.2. If c lies between $f(a)$ and $f(b)$, then c must belong to $f([a,b])$, so $f(x) = c$ for some $x \in (a,b)$. □

8.10.3. Open Mappings. A continuous function $f : \mathbf{R} \to \mathbf{R}$ need not map open sets onto open sets. For example, consider the function in Exercise 8.10.8. However, mappings that send open sets in the domain onto open sets in the range are of interest, as we will see shortly.

Definition 8.10.13. *The function $\mathbf{F} : E \subset \mathbf{R}^n \to \mathbf{R}^m$ is called an* **open mapping** *(or, an* **open map***) on E if for every open set $O \subset \mathbf{R}^n$ there is an open set $O_1 \subseteq \mathbf{R}^m$ such that*

$$\mathbf{F}(E \cap O) = \mathbf{F}(E) \cap O_1. \tag{8.10}$$

Thus $\mathbf{F}$ is an open mapping on E if and only if the image of every open set in E is open in $\mathbf{F}(E)$.

Theorem 8.10.14. *If $\mathbf{F} : E \subset \mathbf{R}^n \to \mathbf{R}^m$ is a one-to-one open mapping, then the inverse function $\mathbf{F}^{-1} : \mathbf{F}(E) \to E$ is continuous on $\mathbf{F}(E)$.*

Proof. Let $\mathbf{G} = \mathbf{F}^{-1}$, which is defined on $\mathbf{F}(E)$ since $\mathbf{F}$ is one-to-one. For any $S \subset \mathbf{R}^n$, clearly we have $\mathbf{G}^{-1}(S) = \mathbf{F}(S)$. Since $\mathbf{F}$ is an open mapping on E, if O is open in $\mathbf{R}^n$, then there is an open set O_1 in $\mathbf{R}^m$ such that

$$\mathbf{F}(E \cap O) = \mathbf{F}(E) \cap O_1,$$

and therefore

$$\mathbf{G}^{-1}(O) = \mathbf{G}^{-1}(E \cap O) = \mathbf{F}(E \cap O) = \mathbf{F}(E) \cap O_1.$$

Since O was an arbitrary open set in $\mathbf{R}^n$, $\mathbf{G} = \mathbf{F}^{-1}$ is continuous on its domain $\mathbf{F}(E)$ by a straightforward extension of Theorem 8.10.9 to functions taking values in $\mathbf{R}^m$. □

Exercises.

Exercise 8.10.8. Define $f(x) = -x$ for $x \in (-\infty, 0)$, $f(x) = 0$ for $x \in [0,1]$, and $f(x) = x - 1$ for $x \in (1, \infty)$. Show that f does not generally map open sets onto open sets.

Exercise 8.10.9. A continuous function $g : \mathbf{R} \to \mathbf{R}$ need not map closed sets onto closed sets. Show that the function $g(x) = x/(1+x^2)$ maps the closed set $[1, \infty)$ onto the set $(0, 1/2]$, which is not closed.

8.10.4. Continuous Images of Compact Sets. We first define open covers by extending Definition 4.2.1.

Definition 8.10.15 (Open Cover). *Let S be a subset of $\mathbf{R}^n$ and let $O_\gamma \subseteq \mathbf{R}^n$ be an open set for each γ in some index set Γ. If $S \subseteq \bigcup_\gamma O_\gamma$, then the collection $\{O_\gamma\}_{\gamma \in \Gamma}$ is called an* **open cover** *of S. If $\{O_\gamma\}_{\gamma \in \Gamma}$ is an open cover of S, and if $\Gamma_0 \subset \Gamma$ and $S \subseteq \bigcup_{\gamma \in \Gamma_0} O_\gamma$, then the collection $\{O_\gamma\}_{\gamma \in \Gamma}$ contains the* **subcover** *$\{O_\gamma\}_{\gamma \in \Gamma_0}$ of S. If a subcover has only finitely many elements it is a* **finite subcover** *of S.*

Definition 8.10.16 (Compact Set). *A set K in $\mathbf{R}^n$ is* **compact** *if every open cover of K contains a finite subcover of K.*

As in the case of compact subsets of the real line, any compact set K in $\mathbf{R}^n$ must be bounded, because K is covered by the open cover consisting of the open cubes centered at the origin and having edge length j, $j \in \mathbf{N}$. There is a finite subcover, so K is contained within one of these cubes, hence K is bounded.

We have the following generalization of Theorem 4.2.9.

Theorem 8.10.17 (Heine-Borel). *A subset K of $\mathbf{R}^n$ is compact if and only if it is closed and bounded.*

Theorem 8.10.17 can be proved in much the same way as for Theorem 4.2.9, with appropriate adaptations for the n-dimensional case; the details are left to Exercise 8.10.10.

We have the following generalization of Theorem 4.2.10.

Theorem 8.10.18. *A set K in $\mathbf{R}^n$ is compact if and only if every infinite subset of K contains a nonconstant convergent sequence with limit in K.*

We leave the proof as an exercise.

Theorem 4.8.1 on the continuous image of a compact set has the following extension.

Theorem 8.10.19. *If $\mathbf{F} : K \subset \mathbf{R}^n \to \mathbf{R}^m$ is continuous on a compact set K, then $f(K)$ is compact.*

Proof. Let $\Omega = \{O_\alpha : \alpha \in \mathcal{A}\}$ be an open cover of the image $\mathbf{F}(K)$. Since $\mathbf{F}$ is continuous on K, for each $O_\alpha \in \Omega$ there is an open set G_α in $\mathbf{R}^n$ such that

$$\mathbf{F}^{-1}(O_\alpha) = G_\alpha \cap K.$$

Then $\mathcal{G} = \{G_\alpha : \alpha \in \mathcal{A}\}$ is an open cover of K, since for any $\mathbf{x} \in K$, we have $\mathbf{F}(\mathbf{x}) \in \mathbf{F}(K)$, and hence $\mathbf{F}(\mathbf{x}) \in O_\alpha$ for some $\alpha \in \mathcal{A}$. Since K is compact, there is a finite subcover, say

$$\mathcal{G}' = \{G_{\alpha_1}, G_{\alpha_2}, \dots, G_{\alpha_m}\},$$

so that $K \subset \bigcup_{k=1}^m G_{\alpha_k}$. Given any point $\mathbf{F}(\mathbf{x}) \in \mathbf{F}(K)$, we have $\mathbf{x} \in G_{\alpha_k}$ for some k with $1 \le k \le m$, and thus $\mathbf{F}(\mathbf{x}) \in O_{\alpha_k}$. Hence the collection $\Omega' = \{O_{\alpha_k} : 1 \le k \le m\}$ is a finite subcover of the open cover Ω of $\mathbf{F}(K)$. This proves that $\mathbf{F}(K)$ is compact. $\square$

We defined uniform continuity of a function $\mathbf{F} : D \subseteq \mathbf{R}^n \to \mathbf{R}^m$ in Definition 8.10.8. We have the following theorem, whose proof is suggested in Exercise 8.10.18.

Theorem 8.10.20. *If $\mathbf{F} : K \subseteq \mathbf{R}^n \to \mathbf{R}^m$ and K is compact, then $\mathbf{F}$ is uniformly continuous on K.*

We have an *extreme value theorem* for real valued continuous functions on compact domains:

Theorem 8.10.21 (Extreme Value Theorem). *A real valued continuous function $f : K \subset \mathbf{R}^n \to \mathbf{R}$ on a compact set K achieves its absolute maximum and absolute minimum value on K; that is, there exist points $\mathbf{x}_M$ and $\mathbf{x}_m$ in K such that $f(\mathbf{x}_m) \leq f\mathbf{x})$ for all $\mathbf{x} \in K$, and $f(\mathbf{x}_M) \geq f(\mathbf{x})$ for all $\mathbf{x} \in K$.*

Proof. The image set $f(K)$ is a compact subset of $\mathbf{R}$, by Theorem 8.10.19, so $f(K)$ is bounded and closed, and hence $F(K)$ contains its limit points $\inf f(K)$ and $\sup f(K)$. □

There is an easy corollary to the extreme value theorem.

Theorem 8.10.22. *If $\mathbf{F} : K \subset \mathbf{R}^n \to \mathbf{R}^m$ is continuous on a compact set K, then for any norm $|\cdot|$ on $\mathbf{R}^m$, the function $|\mathbf{F}| : K \to \mathbf{R}$, defined by $|\mathbf{F}|(\mathbf{x}) = |\mathbf{F}(\mathbf{x})|$, $\mathbf{x} \in K$, achieves its maximum and minimum value on K.*

Proof. We only need to know that the composite function $|\mathbf{F}|$ is continuous on K. For any fixed $\mathbf{a} \in K$, the reverse triangle inequality for the norm implies that

$$\Big| |\mathbf{F}(\mathbf{x})| - |\mathbf{F}(\mathbf{a})| \Big| \leq |\mathbf{F}(\mathbf{x}) - \mathbf{F}(\mathbf{a})|$$

for $\mathbf{x} \in K$. Therefore the continuity of $\mathbf{F}$ on K implies continuity of $|\mathbf{F}|$. □

An important special case of Theorem 8.10.22 leads to the next result.

Theorem 8.10.23. *Let $C_n[a,b] = \{\psi : [a,b] \to \mathbf{R}^n : \psi \text{ is continuous}\}$ be the set of continuous curves in $\mathbf{R}^n$ defined on the interval $[a,b]$. With pointwise addition and scalar multiplication,*

$$(\phi + \psi)(t) = \phi(t) + \psi(t), \quad (\alpha\psi)(t) = \alpha\psi(t), \quad \alpha \in \mathbf{R}, \ t \in [a,b],$$

$C_n[a,b]$ is a real vector space. If $|\cdot|$ is a fixed norm on $\mathbf{R}^n$, then $C_n[a,b]$ is normed by

$$\|\psi\| := \max_{t \in [a,b]} |\psi(t)|. \tag{8.11}$$

Proof. The existence of the maximum follows from Theorem 8.10.22 since $[a,b]$ is compact and $|\psi(t)|$ is continuous on $[a,b]$. That (8.11) defines a norm is left to Exercise 8.10.16. □

Exercises.

Exercise 8.10.10. Prove the following statements:

1. If K is a compact set in $\mathbf{R}^n$, then K is closed and bounded. *Hint*: Follow the arguments leading to Theorem 4.2.5.
2. If $K \subset \mathbf{R}^n$ is closed and bounded, then K is compact. *Hint*: Follow the arguments leading to Theorem 4.2.8.

Exercise 8.10.11. Suppose $f : \mathbf{R} \to \mathbf{R}$ is continuous and $|f(x)| \to \infty$ as $|x| \to \infty$.

1. Show that $f^{-1}(K)$ is compact for every compact set $K \subset \mathbf{R}$. Compare this with the situation presented by the function $g(x) = 1/(1+x^2)$; in particular, consider $g^{-1}([-1/2, 1/2])$.
2. Let $p : \mathbf{R} \to \mathbf{R}$ be a nonconstant polynomial function. Show that $p^{-1}(K)$ is compact for every compact set K.

Exercise 8.10.12. Prove Theorem 8.10.18. *Hint*: Follow the approach of Theorem 4.2.10, modifying statements appropriately for the case of $\mathbf{R}^n$.

Exercise 8.10.13. Show that $K \subset \mathbf{R}^n$ is compact if and only if every infinite sequence in K has a convergent subsequence with limit in K. (See also Exercise 4.2.5.)

Exercise 8.10.14. Give a proof of Theorem 8.10.19 following the approach taken in Theorem 4.8.1, and modifying statements appropriately for the case of $\mathbf{R}^n$.

Exercise 8.10.15. Prove: If $K \subset \mathbf{R}$ is compact and $f : K \to \mathbf{R}$ is continuous and one-to-one, then $f^{-1} : \mathcal{R}(f) \to K$ is continuous on $\mathcal{R}(f)$, where $\mathcal{R}(f)$ is the range of f.

Exercise 8.10.16. Let $|\cdot|$ be a norm on $\mathbf{R}^n$. Show that $\|\psi\| = \max_{t\in[a,b]} |\psi(t)|$, for $\psi \in C_n[a, b]$, defines a norm on the vector space $C_n[a, b]$.

Exercise 8.10.17. *Equivalence of norms revisited*
This exercise gives an alternative approach to proving the equivalence of any two norms on $\mathbf{R}^n$ (Theorem 8.3.13). Let $|\cdot|_2$ denote the Euclidean norm on $\mathbf{R}^n$ and let $|\cdot|$ denote any other norm on $\mathbf{R}^n$.

1. Write $\mathbf{x} = (x_1, \dots, x_n) = \sum_{k=1}^n x_k \mathbf{e}_k$. Use the triangle inequality to show that $|\mathbf{x}| \le \sum_{k=1}^n |x_k|\,|\mathbf{e}_k|$.
2. Then use the Cauchy-Schwarz inequality to show that $\sum_{k=1}^n |x_k|\,|\mathbf{e}_k| \le M|\mathbf{x}|_2$, where $M = (\sum_{k=1}^n |\mathbf{e}_k|^2)^{1/2}$. Note that this proves one inequality required of an equivalence, and shows that $f(\mathbf{x}) = |\mathbf{x}|$ is continuous in the Euclidean norm.
3. Use the continuity established in part 2 to conclude that $f(\mathbf{x}) = |\mathbf{x}|$ attains a positive minimum value m on the compact unit sphere where $|\mathbf{x}|_2 = 1$. Conclude that $|\mathbf{x}| \ge m|\mathbf{x}|_2$ for all $\mathbf{x} \ne \mathbf{0}$.
4. Conclude by transitivity of norm equivalence that any two norms on $\mathbf{R}^n$ are equivalent.

Exercise 8.10.18. Prove Theorem 8.10.20. *Suggestion*: Prove this by contradiction with the help of Theorem 8.10.18. By the negation of the definition of uniform continuity, the function $F : K \subseteq \mathbf{R}^n \to \mathbf{R}$ is not uniformly continuous on K if and only if there is an $\epsilon > 0$ and sequences $\mathbf{x}_n$, $\mathbf{z}_n$ in K such that $\lim_{n\to\infty} |\mathbf{x}_n - \mathbf{z}_n| = 0$ but $|F(\mathbf{x}_n) - F(\mathbf{z}_n)| \ge \epsilon$.

8.10.5. Differentiation under the Integral. This section provides an opportunity to discuss certain situations where a differentiation follows an integration, and the issue is whether it is legitimate to differentiate a definite integral by *differentiating under the integral sign*. We give a precise statement below in Theorem 8.10.24.

The theorem employs the compactness of the Cartesian product $[a, b] \times [c, d]$ of two finite closed intervals and the boundedness and uniform continuity of a continuous function on a compact set, all from the previous section. (Theorem 8.10.21 covers the boundedness, and Exercise 8.10.18 the uniform continuity, used in this section.) We also need a partial derivative.

We consider a function $f : [a, b] \times [c, d] \to \mathbf{R}$, and we define the partial derivative of f with respect to x at a point (t_0, x_0) in $[a, b] \times [c, d]$ by

$$D_2 f(t_0, x_0) := \frac{\partial f}{\partial x}(t_0, x_0) = \lim_{h \to 0} \frac{f(t_0, x_0 + h) - f(t_0, x_0)}{h}.$$

If $x_0 = d$, then the limit is taken for $h < 0$, and if $x_0 = c$, then the limit is taken for $h > 0$.

If the mapping $t \to f(t, x)$ is Riemann integrable in t for each fixed x in $[c, d]$, then

$$g(x) = \int_a^b f(t, x)\, dt \tag{8.12}$$

is defined for each x in $[c, d]$. Under certain conditions, we can show that the function $g : [c, d] \to \mathbf{R}$ is differentiable at every x in $[c, d]$, and compute its derivative.

Theorem 8.10.24. *Suppose $f(t, x)$ and $D_2 f(t, x)$ are defined and continuous for (t, x) in $[a, b] \times [c, d]$. Then the function g in (8.12) is differentiable on $[c, d]$, and*

$$g'(x) = \int_a^b D_2 f(t, x)\, dt$$

for every $x \in [c, d]$.

Proof. Fix x in $[c, d]$. For any fixed t in $[a, b]$, and for real h, the chain rule gives

$$\frac{d}{ds} f(t, x + sh) = D_2 f(t, x + sh) h.$$

Then the fundamental theorem of calculus (integration of a derivative) implies

$$f(t, x + h) - f(t, x) = \int_0^1 D_2 f(t, x + sh) h\, ds.$$

Let $\lambda(x) = \int_a^b D_2 f(t, x)\, dt$. We wish to show that $\lambda(x) = g'(x)$. By definition of g and λ, we have

$$\begin{aligned} g(x+h) - g(x) - \lambda(x)h &= \int_a^b [f(t, x+h) - f(t, x) - D_2 f(t, x)h]\, dt \\ &= \int_a^b \Big[\int_0^1 D_2 f(t, x + sh) h\, ds - D_2 f(t, x) h \Big]\, dt \\ &= \int_a^b \Big\{ \int_0^1 [D_2 f(t, x + sh) - D_2 f(t, x)] h\, ds \Big\}\, dt. \end{aligned}$$

Let

$$M_x(h) = \max_{0 \le s \le 1,\, a \le t \le b} \big| D_2 f(t, x + sh) - D_2 f(t, x) \big|,$$

which exists since $D_2f(t, x + sh) - D_2f(t, x)$ is continuous for $(t, s) \in [a, b] \times [0, 1]$. Then

$$\left|g(x + h) - g(x) - \lambda(x)h\right| \le M_x(h)\,|h|\,(b - a).$$

By the uniform continuity of D_2f on the compact set $[a, b] \times [c, d]$, given any $\epsilon > 0$ there is a $\delta > 0$ such that if $|h| < \delta$, then $M_x(h) < \epsilon/(b - a)$, and hence

$$\left|g(x + h) - g(x) - \lambda(x)h\right| < \epsilon|h|.$$

Thus, by the definition of derivative, $\lambda(x) = g'(x)$. □

As we noted above, the result of Theorem 8.10.24 is often called *differentiation under the integral*, since it asserts that under the assumed continuity conditions, we have

$$\frac{d}{dx}\left[\int_a^b f(t, x)\,dt\right] = \int_a^b D_2f(t, x)\,dt = \int_a^b \frac{\partial f}{\partial x}(t, x)\,dt$$

for each $x \in [c, d]$. By definition of $g'(x)$, this is the same as saying that

$$\begin{aligned}\lim_{h\to 0} \frac{g(x + h) - g(x)}{h} &= \lim_{h\to 0} \frac{1}{h}\Big(\int_a^b [f(t, x + h) - f(t, x)]\,dt\Big)\\ &= \int_a^b \lim_{h\to 0}[f(t, x + h) - f(t, x)]\,dt\\ &= \int_a^b \frac{\partial}{\partial x} f(t, x)\,dt\end{aligned}$$

for $(t, x) \in [a, b] \times [c, d]$. Thus, Theorem 8.10.24 gives sufficient conditions under which the limit as $h \to 0$ can be passed inside the integral operation, yielding the integrand $\frac{\partial}{\partial x} f(t, x)$. Note that a direct application of the definition of $g'(x)$ requires that the integration be done for each h, and then the limit of the resulting function of h be found. The continuity conditions of Theorem 8.10.24 allow the interchange of order of the limit and integral operations on the right-hand side.

Exercises.

Exercise 8.10.19. Let $f(t, x) = \sin(tx)/t$ for $(t, x) \in [a, b] \times (-\infty, \infty)$, where $[a, b]$ does not contain 0. Use Theorem 8.10.24 to show that if x is real and $g(x)$ is defined by (8.12), then

$$g'(x) = \int_a^b D_2f(t, x)\,dt = \int_a^b \cos(tx)\,dt.$$

Exercise 8.10.20. Find $g'(x)$ if

$$g(x) = \int_0^1 \frac{x}{\sqrt{1 - x^2t^2}}\,dt, \quad 0 \le x \le 1/2.$$

Exercise 8.10.21. Find $g'(x)$ if $g(x) = \int_0^1 e^{-tx^2}\,dt$ for $x \ge 1$.

8.10.6. Continuous Images of Connected Sets. We now consider the property of connectedness for subsets of $\mathbf{R}^n$. Earlier we defined connectedness for subsets of the reals and we saw in Theorem 4.3.2 that a subset of $\mathbf{R}$ is connected if and only if it is an interval. In particular, $\mathbf{R}$ itself is connected. The next definition is a straightforward extension of Definition 4.3.1.

Definition 8.10.25. *A subset $S \subseteq \mathbf{R}^n$ is* **disconnected** *if there exist open sets U and V in $\mathbf{R}^n$ such that*

1. $U \cap S \neq \emptyset, \quad V \cap S \neq \emptyset,$
2. $(U \cap S) \cap (V \cap S) = \emptyset,$
3. $(U \cap S) \cup (V \cap S) = S.$

Then the pair U, V are said to disconnect S. A set S is **connected** *if it is not disconnected.*

Using the language of relative topology, an equivalent definition of disconnected set is that $S \subseteq \mathbf{R}^n$ is disconnected if there are relatively open sets U_1 and V_1 in S such that

(i) $U_1 \neq \emptyset, \quad V_1 \neq \emptyset,$

(ii) $U_1 \cap V_1 = \emptyset,$

(iii) $U_1 \cup V_1 = S.$

It is not always easy to decide whether a subset of $\mathbf{R}^n$ is connected or disconnected. For example, let $a_j \leq b_j$ for $j = 1, \ldots, n$, and consider the **closed interval** defined by

$$J = \{(x_1, \ldots, x_n) \in \mathbf{R}^n : a_j \leq x_j \leq b_j\}.$$

If we specify instead that $a_j < x_j < b_j$ and $a_j < b_j$ for $j = 1, \ldots, n$, then we get an **open interval** in $\mathbf{R}^n$. If we have any mixture of strict and nonstrict inequalities, then we still get an **interval**, one that is neither open nor closed. It is not immediate from the definition of connected set that an interval in $\mathbf{R}^n$ is connected.

Let us first prove that $\mathbf{R}^n$ is connected, since the proof provides a useful idea.

Theorem 8.10.26. $\mathbf{R}^n$ *is a connected set.*

Proof. Suppose not, and let U and V be open sets that disconnect $\mathbf{R}^n$. Let $\mathbf{a} \in U$ and let $\mathbf{b} \in V$. Let $\mathbf{r}(t) = \mathbf{a} + t\mathbf{b}$ for $t \in \mathbf{R}$. Since $\mathbf{r} : \mathbf{R} \to \mathbf{R}^n$ is continuous, $\mathbf{r}^{-1}(U)$ and $\mathbf{r}^{-1}(V)$ are open sets in $\mathbf{R}$. Since U and V are nonempty and disjoint, the same is true of $\mathbf{r}^{-1}(U)$ and $\mathbf{r}^{-1}(V)$. Since $U \cup V = \mathbf{R}^n$, we have $\mathbf{r}^{-1}(U) \cup \mathbf{r}^{-1}(V) = \mathbf{R}$. Therefore $\mathbf{r}^{-1}(U)$ and $\mathbf{r}^{-1}(V)$ are open sets that disconnect $\mathbf{R}$, contradicting Theorem 4.3.2. Hence $\mathbf{R}^n$ is connected. $\square$

Corollary 8.10.27. *The only subsets of* $\mathbf{R}^n$ *that are both open and closed are* $\mathbf{R}^n$ *and the empty set.*

Proof. Suppose $A \subset \mathbf{R}^n$, $A \neq \mathbf{R}^n$, $A \neq \emptyset$, and A is both open and closed. Then A^c is open since A is closed, $A^c \neq \emptyset$, $A \cap A^c = \emptyset$ and $A \cup A^c = \mathbf{R}^n$. Thus, A and A^c disconnect $\mathbf{R}^n$, which contradicts Theorem 8.10.26. $\square$

The next concept is quite helpful in determining whether certain subsets are connected or disconnected. Here we make further use of the idea in the proof of Theorem 8.10.26.

Definition 8.10.28. *A set $S \subseteq \mathbf{R}^n$ is* **path connected** *if for every pair of points* $\mathbf{a}$ *and* $\mathbf{b}$ *in S, there is a continuous path* $\mathbf{p} : [0,1] \to S$ *such that* $\mathbf{p}(0) = \mathbf{a}$ *and* $\mathbf{p}(1) = \mathbf{b}$. *(Thus for every* $\mathbf{a}$ *and* $\mathbf{b}$ *in S there is a continuous path from* $\mathbf{a}$ *to* $\mathbf{b}$ *that lies entirely within S.)*

Theorem 8.10.29. *If $S \subseteq \mathbf{R}^n$ is path connected, then S is connected.*

A proof of Theorem 8.10.29 is requested in Exercise 8.10.22, and the proof can be patterned on the argument in the proof of Theorem 8.10.26. Using the concept of path connectedness, it is relatively easy to show that intervals in $\mathbf{R}^n$ are connected (Exercise 8.10.23). A set $S \subseteq \mathbf{R}^n$ is **convex** if for any pair of points $\mathbf{a}$ and $\mathbf{b}$ in S, the line segment joining them, which is the image of the path $\mathbf{r}(t) = \mathbf{a} + t\mathbf{b}$, $0 \le t \le 1$, lies entirely within S. Since $\mathbf{r}(t) = \mathbf{a} + t\mathbf{b}$ is continuous on $[0,1]$, if S is convex, then S is path connected, and thus Theorem 8.10.29 implies that every convex set is connected.

The converse of Theorem 8.10.29 is not true. There are connected sets that are not path connected, as in the next example.

Example 8.10.30. Let $S \subset \mathbf{R}^2$ be the set

$$S = \big\{(x,y) : x = 0 \text{ and } -1 \le y \le 1\big\} \cup \big\{(x,y) : x > 0, y = \sin(1/x)\big\},$$

that is, S is the union of the line segment $\{0\} \times [-1,1]$ on the y axis and the graph of $y = \sin(1/x)$ for $x > 0$. Then S is connected, but not path connected (Exercise 8.10.24). △

For open sets, however, connectedness and path connectedness are equivalent; this follows from the result of Exercise 8.10.25 and Theorem 8.10.29. Observe that the set S in Example 8.10.30 is not open.

Theorem 8.10.11 showed that if f is a continuous real-valued mapping of an interval $J \subseteq \mathbf{R}$, then $f(J)$ is connected. We can now prove a general result which covers the earlier statement.

Theorem 8.10.31. *If $\mathbf{F} : D \subseteq \mathbf{R}^n \to \mathbf{R}^m$ is continuous and D is a connected set in $\mathbf{R}^n$, then $\mathbf{F}(D)$ is a connected set in $\mathbf{R}^m$.*

Proof. Assume $\mathbf{F}$ is continuous on the connected set D. Suppose $\mathbf{F}(D)$ is disconnected, so that there are relatively open sets U_1 and V_1 in $\mathbf{F}(D)$ such that

$$U_1 \neq \emptyset, \quad V_1 \neq \emptyset, \quad U_1 \cap V_1 = \emptyset, \quad U_1 \cup V_1 = \mathbf{F}(D).$$

By continuity of $\mathbf{F}$, $\mathbf{F}^{-1}(U_1)$ and $\mathbf{F}^{-1}(V_1)$ are relatively open in D. Moreover, these sets are nonempty and disjoint since U_1 and V_1 are nonempty and disjoint, and $\mathbf{F}^{-1}(U_1) \cup \mathbf{F}^{-1}(V_1) = D$ since $U_1 \cup V_1 = \mathbf{F}(D)$. This shows that D is disconnected, the contradiction we were seeking. □

Exercises.

Exercise 8.10.22. Prove Theorem 8.10.29.

Exercise 8.10.23. Show that every interval in $\mathbf{R}^n$ is a connected set.

Exercise 8.10.24. Show that the set S of Example 8.10.30 is connected but not path connected.

Exercise 8.10.25. Prove: If $S \subset \mathbf{R}^n$ is open and connected, then S is path connected.

Exercise 8.10.26. Prove: If $S = [0, 1]$ and $f : S \to S$ is continuous, then f has at least one fixed point; that is, there exists some point $x_0 \in S$ such that $f(x_0) = x_0$.

8.11. Notes and References

Much of this chapter is influenced by Sagan [**54**]. Much of the material on norms is influenced by Kreyszig [**41**].

Exercise 8.10.26 is the simplest case of a more general fixed point theorem known as the *Brouwer fixed point theorem*; see Simmons [**59**].

Chapter 9

Metric Spaces and Completeness

This chapter introduces basic topological concepts in metric spaces, proves the widely applicable contraction mapping theorem for complete metric spaces, and establishes completeness for some important spaces used later in the book. The modest level of abstraction required in this chapter provides a basic foundation for moving beyond Euclidean spaces in the study of analysis. The concept of a complete metric space arises even when we consider a closed subset of a normed space when the subset is not a vector space.

9.1. Basic Topology in Metric Spaces

Let X be a metric space with metric d (Definition 8.6.1). The definition does not require X to be a vector space, although most metric spaces considered in this book are vector spaces (or subsets of a vector space) and the metric distance function is defined in terms of a norm. We refer to the elements of X as points, and in general statements about metric spaces we use lowercase letters to indicate the points.

The following definitions are direct generalizations of definitions in the case of the metric space $\mathbf{R}$, where $d(x, y) = |x-y|$ is defined by the absolute value function, and the metric space $\mathbf{R}^n$, where $d(\mathbf{x}, \mathbf{y}) = |\mathbf{x} - \mathbf{y}|$, where $|\cdot|$ is any norm on $\mathbf{R}^n$. In the definitions which follow we assume that X is a metric space with metric d.

Definition 9.1.1. *If X is a metric space, $x \in X$, and $\delta > 0$, then the set of points*

$$B_\delta(x) = \{y \in X : d(y, x) < \delta\}$$

is called an **open ball** *about x. The number δ is called the radius of the ball.*

Definition 9.1.2. *A subset S of a metric space (X, d) is* **bounded** *if there is a positive number M such that $d(x, y) \leq M$ for all $x, y \in S$.*

Definition 9.1.3. *If $S \subset X$, then an element x is called an* **interior point** *of S if there exists a positive number $\delta > 0$ such that the open ball $B_\delta(x) \subset S$.*

Definition 9.1.4. *A set $S \subseteq X$ is called an* **open set** *if every element of S is an interior point of S. A set $S \subseteq X$ is called a* **closed set** *if its complement, $X - S$, is an open set.*

By this definition, both X and the empty set are open sets. Thus, by complementation, X and the empty set are also closed sets. Also, any open ball is an open set (Exercise 9.1.1).

Definition 9.1.5. *If $S \subset X$, a point x is called a* **boundary point** *of S if every open ball about x contains at least one point of S and at least one point of $X - S$. The* **boundary** *of S, denoted ∂S, is the set of boundary points of S.*

Definition 9.1.6. *If $S \subset X$, a point x is called a* **cluster point** *(or* **accumulation point***) of S if every open ball about x contains infinitely many points of S. A point $\mathbf{x} \in S$ which is not a cluster point of S is called an* **isolated point** *of S. The* **closure** *of S, denoted $\overline{S}$, is the union of S and the set of all cluster points of S.*

Theorem 9.1.7. *A set $S \subset X$ is closed if and only if it contains all its cluster points. For any set S, the closure $\overline{S}$ is a closed set.*

Proof. Suppose S is closed. If $x \in X$ and x does not belong to S, then $x \in X - S$, and since $X - S$ is open, there is an open ball about x that contains only points of $X - S$. Therefore x is not a cluster point of S. Hence, every cluster point of S is in S.

Suppose every cluster point of S belongs to S. If $x \in X - S$, then x is not a cluster point of S, so there is an open ball about x that is contained in $X - S$. Thus, every point of $X - S$ is an interior point, so $X - S$ is open, and hence S is closed.

For any set S, the closure $\overline{S}$ is the union of S and its set of cluster points, hence $\overline{S}$ is a closed set. □

The definition of a set being *dense in an open set* and the definition of a set being *nowhere dense* are the same as for sets in Euclidean space.

Definition 9.1.8. *Let X be a metric space. A subset $S \subset X$ is* **dense in an open set** *U of X if $U \subset \overline{S}$. A set S is defined to be* **nowhere dense** *if $\overline{S}$ has no interior points.*

For example, the Weierstrass approximation theorem (Theorem 7.6.1) implies that the set of polynomial functions on the interval $[a, b]$ is dense in the metric space $C[a, b]$ of continuous functions on $[a, b]$ with the metric determined by the uniform norm:

$$\|f - g\| = \max_{x \in [a,b]} |f(x) - g(x)|.$$

All vector spaces described earlier for which a norm was defined also provide examples of metric spaces. The space $C[a, b]$ with the max norm (Theorem 8.10.23) is a complete normed space, and hence a complete metric space; its completeness is proved later in Theorem 9.3.1.

The behavior of open sets and closed sets under unions and intersections will seem familiar from Section 4.1. The proofs of the next two theorems are essentially the same as the corresponding results in the case of open sets and closed sets in the real line, and are left as exercises.

Theorem 9.1.9. *In a metric space, the following statements are true:*

1. *The union of any collection of open sets is open.*
2. *The intersection of any finite collection of open sets is open.*

Theorem 9.1.10. *In a metric space, the following statements are true:*

1. *The intersection of any collection of closed sets is closed.*
2. *The union of any finite collection of closed sets is closed.*

In Section 8.6 we stated the definitions of convergent sequence, Cauchy sequence, and completeness, for general metric spaces. Readers might wish to review those definitions before proceeding. (See Definitions 8.6.4, 8.6.5, 8.6.6.)

Theorem 9.1.11. *If X is a complete metric space with respect to the metric d, and Y is a nonempty closed subset of X, then Y is a complete metric space with the same metric d.*

Proof. If X is a metric space with metric d, then $Y \subset X$ is also a metric space with metric d. Supposing that X is complete, let (x_k) be a Cauchy sequence in Y. Then (x_k) is Cauchy in X, and by completeness of X there is a limit $x \in X$ for (x_k). For any $\epsilon > 0$, there is an N such that if $k \geq N$, then $d(x_k, x) < \epsilon$. So every open ball about x contains a point of Y. Thus, either $x \in Y$ or x is a cluster point of Y. Since Y is nonempty and closed, x must belong to Y. Therefore every Cauchy sequence in Y has a limit in Y, and Y is complete. □

Example 9.1.12. We have asserted that the space $C[a, b]$ with the max norm is a complete normed space, and hence a complete metric space (Theorem 9.3.1). Let $t_0 \in [a, b]$, let $x_0 \in \mathbf{R}$, and define

$$Y = \{\psi \in C[a, b] : \psi(t_0) = x_0\}.$$

Then Y is a closed subset of $C[a, b]$, and hence Y is a complete metric space in the metric defined by the max norm. To see that Y is closed, notice that convergence of a sequence (ψ_k) with respect to the max norm on $C[a, b]$ means uniform convergence of the ψ_k on $[a, b]$. If (ψ_k) has limit ψ in $C[a, b]$, then, since $\psi_k(t_0) = x_0$ for each k, we have $\psi(t_0) = x_0$, and thus $\psi \in Y$. △

The next definition is the natural generalization of the sequential characterization of continuity at a point for mappings from $\mathbf{R}$ to $\mathbf{R}$ or from $\mathbf{R}^n$ to $\mathbf{R}^m$.

Definition 9.1.13. *Let (X, d) and (Y, ρ) be metric spaces. A mapping $f : X \to Y$ is continuous at $a \in X$ if for any sequence (a_k) in X such that $d(a_k, a) \to 0$ as $k \to \infty$, it is true that $\rho(f(a_k), f(a)) \to 0$ as $k \to \infty$.*

As usual, we say that a mapping $f : X \to Y$ is **continuous on** X if for every $x \in X$, f is continuous at x. With Theorem 8.10.9 in view, the next result will not be surprising; its proof is left to Exercise 9.1.8.

Theorem 9.1.14. *Let (X, d) and (Y, ρ) be metric spaces. A mapping $f : X \to Y$ is continuous on X if and only if $f^{-1}(O)$ is an open set in X for every open set O in Y.*

In a metric space, the definition of compact sets is familiar.

Definition 9.1.15. *A subset K of a metric space X is compact if every open cover of K contains a finite subcover of K.*

Based on some experience with compact sets in $\mathbf{R}$ and $\mathbf{R}^n$ the next two theorems will look familiar, and their proofs are left as exercises, though we mention here that both results follow from the characterization of compactness in Theorem 9.1.21 below.

Theorem 9.1.16. *If K is a compact subset of a metric space X, then K is closed and bounded.*

Theorem 9.1.17. *Let K be a closed subset of a compact metric space X. Then K is compact.*

Theorem 9.1.18. *If K is a compact subset of a metric space X, then every sequence in K has a subsequence that converges to a point in K.*

Proof. Every sequence that is eventually constant, that is, every sequence with finite range, converges; thus, we need to consider only sequences with infinite range.

We shall prove the contrapositive of the theorem statement. We suppose that (x_k) is a sequence in K with no convergent subsequence. The sequence therefore has infinitely many distinct elements. For each element x in K, there exists a ball B_x about x which contains only finitely many elements of (x_k). (Otherwise, there is a point p such that every ball about p contains infinitely many elements of the sequence, and then some subsequence converges to p.) Then $\{B_x : x \in K\}$ is an open cover of K, but there is no finite subcover, since any finite subcover can contain only finitely many elements of (x_k). Hence, K is not compact. $\square$

In general metric spaces, the property that a set is closed and bounded does not generally imply compactness. For example, the unit ball in l^2 is closed and bounded, but it is not compact: Theorem 9.1.18 can be used to establish this fact (Exercise 9.1.12).

The next definition formalizes two properties that we have seen to be equivalent in $\mathbf{R}$ and $\mathbf{R}^n$.

Definition 9.1.19. *Let X be a metric space. A subset $A \subseteq X$ is* **sequentially compact** *if every sequence in A has a subsequence that converges to a point in A. The set A has the* **Bolzano-Weierstrass property** *if every infinite subset of A has a limit point (cluster point) that belongs to A.*

We shall use both properties in Definition 9.1.19 in Theorem 9.1.21 below. We note that Theorem 9.1.18 establishes that compactness implies sequential compactness.

In order to obtain sufficient conditions for a subset of a general metric space to be compact, the boundedness property must be strengthened to the concept of *total boundedness*.

Definition 9.1.20. *A subset S of a metric space (X, d) is* **totally bounded** *if, for every $\epsilon > 0$, S can be covered by finitely many open balls of radius ϵ.*

It is clear that total boundedness is necessary for a set to be compact: If K is compact, then the open cover of K given by $\{B_\epsilon(x) : x \in K\}$, for arbitrary $\epsilon > 0$, must have a finite subcover. Completeness is also necessary for compactness: A Cauchy sequence in a compact set K has a subsequence that converges to a point p in K, by Theorem 9.1.18, and thus the Cauchy sequence itself must converge to p; hence, K is complete. The statement *complete and totally bounded implies compact* requires some more work, but this is accomplished with the following theorem, which summarizes the essential characterizations of compactness in a metric space.

Theorem 9.1.21. *Let A be a subset of a metric space (X, d). The following are equivalent:*

1. *A is compact.*
2. *A is sequentially compact.*
3. *A has the Bolzano-Weierstrass property.*
4. *A is complete and totally bounded.*

Proof. The plan of the proof is as follows: We shall prove that 1 implies 2 and 4; then prove that 2 and 4 are equivalent; and then, that 2 and 4 together imply 1. This will establish the equivalence of 1, 2, and 4. Finally, we show that 2 and 3 are equivalent. This will establish the equivalence of the four statements.

Item 1 implies 2 and 4: The implication 1 implies 2 was established in Theorem 9.1.18. That 1 implies 4 was established in our comments after Definition 9.1.20.

Item 2 implies 4: Suppose A is sequentially compact. If A is not complete, then there exists a Cauchy sequence in A which does not converge to an element of A, and hence no subsequence converges to an element of A, for if a subsequence of a Cauchy sequence converges, then the Cauchy sequence itself converges, and to the same limit as the subsequence. But no convergent subsequence contradicts the hypothesis that A is sequentially compact. Thus, if A is sequentially compact, then A is complete. Now suppose A is not totally bounded. Then there exists $\epsilon > 0$ such that A cannot be covered by finitely many balls of radius ϵ. Define a sequence inductively as follows: Select $x_1 \in A$. Then $A - B_\epsilon(x_1)$ is nonempty by our hypothesis. Choose $x_2 \in A - B_\epsilon(x_1)$. Then $A - (B_\epsilon(x_1) \cup B_\epsilon(x_2))$ is nonempty and $d(x_1, x_2) \geq \epsilon$. If $x_1, \ldots, x_{k-1}$ have been chosen such that $d(x_i, x_j) \geq \epsilon$ when $1 \leq i < j \leq k-1$, then we can select

$$x_k \in A - \bigcup_{i=1}^{k-1} B_\epsilon(x_i),$$

since A is assumed not totally bounded. Thus, we have defined a sequence (x_k) with $d(x_i, x_j) \geq \epsilon$ when $i \neq j$. Then (x_k) has no convergent subsequence, a contradiction of the sequential compactness hypothesis. Thus, if A is sequentially compact, then A is totally bounded.

Statement 4 implies 2: Suppose A is complete and totally bounded. Let (x_k) be a sequence in A. We may assume that this sequence has infinitely many elements.

Since A is totally bounded, A can be covered by finitely many open balls of radius $1/2$, and at least one of these balls contains infinitely many elements of the sequence. Choose such a ball B_1; then there is an infinite subset N_1 of positive integers such that $x_k \in B_1$ for $k \in N_1$. Now $A \cap B_1$ can be covered by finitely many balls of radius $1/2^2$; we choose one of them, say B_2, such that $x_k \in A \cap B_1 \cap B_2$ if $k \in N_2$, where N_2 is an infinite subset of N_1. Continuing in this way, we inductively define a sequence of balls B_j of radius $1/2^j$ and a nested sequence $N_{j+1} \subset N_j$ of infinite subsets of positive integers, such that

$$x_k \in A \cap \left(\bigcap_{i=1}^{j} B_j \right) \quad \text{for } k \in N_j.$$

Now choose indices $n_1 < n_2 < n_3 < \cdots$, with $n_k \in N_k$ for each k. Then the subsequence (x_{n_k}) of (x_k) has the property that

$$d(x_{n_j}, x_{n_k}) < \frac{2}{2^j} = \frac{1}{2^{j-1}} \quad \text{for } k > j,$$

since $k > j$ implies $n_k > n_j$, hence $x_{n_k}, x_{n_j} \in B_j$ and B_j has radius $1/2^j$. Therefore (x_{n_k}) is a Cauchy sequence in A, and since A is complete, (x_{n_k}) converges to an element of A. This shows that A is sequentially compact.

Statements 2 and 4 together imply 1: Now assume that A is sequentially compact as well as complete and totally bounded. Let $\{O_\beta\}$ be an open cover of A, where β belongs to an arbitrary index set. We claim that there exists an $\epsilon > 0$ such that every ball of radius ϵ that intersects A is contained in some open set O_β of the given cover. This property will prove that A is compact, since the total boundedness of A implies that A is contained in the union of finitely many of these ϵ balls, and hence there are finitely many of the open sets O_β, say $O_{\beta_1}, \ldots, O_{\beta_m}$, such that

$$A \subset \bigcup_{i=1}^{m} O_{\beta_i}.$$

We prove the claim by contradiction. Thus, suppose there is no such $\epsilon > 0$. Then for each positive integer k, there is a ball B_k of radius $1/k$ such that $B_k \cap A$ is nonempty and B_k is not contained in any of the open sets O_β. For each k, we can select a point $x_k \in B_k \cap A$. Since A is sequentially compact, the sequence (x_k) has a subsequence (x_{n_k}) that converges to a point x in A. The limit x is in O_β for some β. Since O_β is open, there exists $\epsilon > 0$ such that the ball $B_\epsilon(x) \subset O_\beta$. Since $x_{n_k} \to x$ as $k \to \infty$, there is an $N(\epsilon)$ such that if $k \geq N(\epsilon)$, then $1/n_k < \epsilon/4$ and $d(x_{n_k}, x) < \epsilon/2$. It follows that $B_{n_k} \subset B_\epsilon(x) \subset O_\beta$, by the following estimate: If $y \in B_{n_k}$, which has radius $1/n_k$, then $d(y, x_{n_k}) < 2/n_k$, and

$$\begin{aligned} d(y,x) &\leq d(y, x_{n_k}) + d(x_{n_k}, x) \\ &< \frac{2}{n_k} + \frac{\epsilon}{2} \\ &< \frac{2\epsilon}{4} + \frac{\epsilon}{2} = \epsilon, \end{aligned}$$

and hence $y \in B_\epsilon(x) \subset O_\beta$. This contradicts the hypothesis that B_{n_k} is not contained in any of the open sets O_β. This proves the claim above, and the compactness of A.

At this point, we have established that statements 1, 2, and 4 are equivalent.

Statement 2 is equivalent to 3: That 3 implies 2 is easy to see, since every sequence in A which is not eventually constant has an infinite set in A as its range; hence, by 3, there is a limit point of the range that belongs to A, and this limit point is the limit of a subsequence of the given sequence. Now assume 2 and suppose that A contains an infinite subset, and hence there is a sequence in A with infinitely many elements of that subset, none of which are repeated. By 2 there is a subsequence that converges to a point of A, which is a limit point (cluster point) of A. Thus every infinite subset of A has a limit point that belongs to A. □

There is an easy corollary of the preceding theorem.

Corollary 9.1.22. *A closed subset of a complete metric space is compact if and only if it is totally bounded.*

Proof. A closed subset of a complete metric space is complete, and therefore by Theorem 9.1.21, it is compact if and only if it is totally bounded. □

Although the language of total boundedness is relatively easy to understand, it can be difficult to verify for specific closed subsets of a metric space. For example, closed subsets of the space $C[a, b]$ of continuous real valued functions on $[a, b]$ often play a role in applications where it is important to know whether every sequence in the closed subset has a convergent subsequence. A criterion for compactness is needed which is expressed in terms of the individual functions in the subset. Given a compact metric space $\mathcal{M}$ (such as the interval $[a, b]$) and the space $C(\mathcal{M})$ of real valued continuous functions on $\mathcal{M}$, a theorem known as the Arzelà-Ascoli Theorem characterizes the compact subsets of $C(\mathcal{M})$. References for this theorem are given in the Notes and References for this chapter.

Exercises.

Exercise 9.1.1. Prove: In a metric space X, any open ball $B_r(x)$ is an open set.

Exercise 9.1.2. Let X be a metric space and $S \subset X$. Prove the following statements:

1. If b is a boundary point of S, then either $b \in S$ or b is a cluster point of S.
2. If y is a cluster point of S, then either $y \in S$ or $y \in \partial S$.
3. S contains all its cluster points if and only if it contains all its boundary points.
4. S is closed if and only if S contains all its boundary points.

Exercise 9.1.3. Prove Theorem 9.1.9.

Exercise 9.1.4. Prove Theorem 9.1.10.

Exercise 9.1.5. Show that the set Y defined by

$$Y = \{\phi \in C[a, b] : \phi(t_0) = x_0 \text{ and } |\phi(t) - x_0| \leq r \text{ for all } t \in [a, b]\},$$

where $t_0 \in [a, b]$, x_0 and $r > 0$ are fixed real numbers, is a closed set in $C[a, b]$, and thus Y is a complete metric space in the metric induced by the sup norm.

Exercise 9.1.6. Let $L : C[0,1] \to \mathbf{R}$ be defined by $L(f) = f(0)$.

1. Is L continuous in the norm $\|f\| = \max_{0\le t\le 1} |f(t)|$ on $C[0,1]$?
2. Is L continuous in the norm $\|f\|_1 = \int_0^1 |f(t)|\, dt$ on $C[0,1]$?

Exercise 9.1.7. Let $S : C[0,1] \to C[0,1]$ be defined by $S(f) = f^2$.

1. Is S continuous in the norm $\|f\| = \max_{0\le t\le 1} |f(t)|$ on $C[0,1]$?
2. Is S continuous in the norm $\|f\|_1 = \int_0^1 |f(t)|\, dt$ on $C[0,1]$?

Exercise 9.1.8. Write a detailed proof of Theorem 9.1.14.

Exercise 9.1.9. Let S be a connected subset of a metric space (X, d), with connectedness defined as in the case of connected sets in $\mathbf{R}^n$. Prove: If $f : S \subset X \to Y$ is a continuous mapping into a metric space (Y, ρ), then $f(S)$ is a connected set in Y.

Exercise 9.1.10. Prove Theorem 9.1.16. *Hint*: Follow the argument for Theorem 4.2.5.

Exercise 9.1.11. Prove Theorem 9.1.17. *Hint*: Follow the argument for Theorem 4.2.6.

Exercise 9.1.12. Show that the unit ball in l^2,

$$B = \left\{ (\xi_k) \in l^2 : \|(\xi_k)\|_2^2 = \sum_{k=1}^{\infty} \xi_k^2 = 1 \right\},$$

is closed and bounded. Show that B is not compact by giving an example of an infinite sequence in B which has no convergent subsequence.

9.2. The Contraction Mapping Theorem

Let X be a complete metric space with metric d. We wish to prove a generalization of the scalar contraction mapping theorem (Theorem 5.5.3).

Definition 9.2.1. *A point $x \in X$ is a* **fixed point** *of a mapping $T : X \to X$ if $T(x) = x$.*

Definition 9.2.2. *A mapping $T : X \to X$ of a metric space X with metric d is a* **contraction mapping** *if there is a number $0 < r < 1$ such that*

$$d(T(x), T(y)) \le r\, d(x, y)$$

for all $x, y \in X$. Such a constant r is called a **contraction constant** *for T.*

From this definition, we immediately deduce that a contraction mapping on X must be continuous at every point in X. Indeed, if $d(x_k, y) \to 0$ as $k \to \infty$, then $d(T(x_k), T(y)) \to 0$ as $k \to \infty$ by the contraction condition.

We now prove the contraction mapping theorem for complete metric spaces. The reader should observe that the proof of Theorem 9.2.3 is the same as the proof of the earlier scalar result in Theorem 5.5.3, except for the change in notation needed to express the more general situation of a complete metric space X with metric function d.

Theorem 9.2.3 (Contraction Mapping Theorem). *A contraction mapping $T : X \to X$ of a complete metric space X has a unique fixed point x^*. Moreover, if r is a contraction constant for T, then given any $x_0 \in X$, the iteration*

$$x_{k+1} = T(x_k), \quad k = 0, 1, 2, 3, \ldots,$$

defines a sequence (x_k) that converges to x^*, and for each k we have

$$d(x_k, x^*) \le \frac{r^k}{1-r} d(x_1, x_0). \tag{9.1}$$

Proof. Let x_0 be an arbitrary initial point in X. Since $d(T(x), T(y)) \le r d(x, y)$ for all x, y, it follows by induction that

$$d(x_{k+1}, x_k) \le r^k d(x_1, x_0)$$

for every positive integer k. If $0 < n < m$, then

$$\begin{aligned} d(x_m, x_n) &\le d(x_m, x_{m-1}) + d(x_{m-1}, x_{m-2}) + \cdots + d(x_{n+1}, x_n) \\ &\le (r^n + \cdots + r^{m-1}) d(x_1, x_0) \\ &= r^n(1 + r + r^2 + \cdots + r^{m-1-n}) d(x_1, x_0) \\ &< r^n(1 + r + r^2 + \cdots) d(x_1, x_0) \\ &= \frac{r^n}{1-r} d(x_1, x_0), \end{aligned} \tag{9.2}$$

where we used the sum of a geometric series in the last estimate. Thus the sequence (x_k) is a Cauchy sequence which converges to a limit x^* in X since X is complete. Since a contraction mapping is continuous, we have

$$T(x^*) = \lim_{k\to\infty} T(x_k) = \lim_{k\to\infty} x_{k+1} = x^*,$$

so x^* is a fixed point of T. If there were another fixed point x^{**} of T, we would have

$$d(x^*, x^{**}) = d(T(x^*), T(x^{**})) \le r d(x^*, x^{**}),$$

but since $r < 1$, this implies $x^* = x^{**}$, so the fixed point is unique.

Finally, letting $m \to \infty$ in the estimate (9.2) yields the estimate (9.1) in the statement of the theorem. □

The following corollary is useful when we have a mapping $T : Y \to Y$ of a closed subset Y in a complete metric space X.

Corollary 9.2.4. *Let Y be a closed, nonempty subset of a complete metric space X. A contraction mapping $T : Y \to Y$ has a unique fixed point.*

Proof. A closed subset of a complete metric space is a complete metric space in its own right using the metric function d of X (Theorem 9.1.11). The corollary follows immediately from Theorem 9.2.3. □

The contraction mapping theorem has some far-reaching consequences for the existence and uniqueness of solutions of equations of various kinds. Important applications of the contraction theorem in this book occur in the proof of the inverse function theorem for mappings $\mathbf{F} : \mathbf{R}^n \to \mathbf{R}^m$ and in the proof of existence and uniqueness for solutions of initial value problems for systems of ordinary differential equations. There are many other applications.

Exercises.

Exercise 9.2.1. Let $f : [1, \infty) \to [1, \infty)$ be defined by $f(x) = x + e^{1-x}$.

(a) Show that $|f(x) - f(y)| < |x - y|$ for all $x \neq y$ in $[0, \infty)$.

(b) Show that f has no fixed point in $[1, \infty)$.

Exercise 9.2.2. Let $f : (1, \infty) \to (1, \infty)$ be defined by $f(x) = \sqrt{x}$.

(a) Show that f is a contraction mapping on $(1, \infty)$.

(b) Show that f has no fixed point in $(1, \infty)$.

Exercise 9.2.3. Let S be a closed subset of $\mathbf{R}^n$. A mapping $T : S \to \mathbf{R}^n$ such that $T(S) \supseteq S$ is an **expansion mapping** with respect to a norm $|\cdot|$ on $\mathbf{R}^n$, if there is a constant $L > 1$ such that $|T(\mathbf{x}) - T(\mathbf{y})| \geq L|\mathbf{x} - \mathbf{y}|$ for all $\mathbf{x}$, $\mathbf{y}$ in S. Show that an expansion mapping has a unique fixed point.

9.3. The Completeness of $C[a, b]$ and l^2

In this section we establish the completeness of two of the normed spaces that play important roles later in the book, the spaces $C[a, b]$ and l^2.

First, we prove the completeness of the space $C[a, b]$ of continuous real valued functions on the interval $[a, b]$ with respect to the norm $\|x\|_{\max} = \max_{t \in [a,b]} |x(t)|$. Recall that by Theorem 7.1.9, the limit of a uniformly convergent sequence of real valued continuous functions on $[a, b]$ is a continuous function on $[a, b]$. As we will see, convergence of a sequence in the space $C[a, b]$ with the maximum norm is precisely uniform convergence of that sequence on the interval $[a, b]$.

Theorem 9.3.1. *The vector space $C[a, b]$ with the maximum norm is a complete normed space.*

Proof. Let (x_m) be a Cauchy sequence in $C[a, b]$. Then for any $\epsilon > 0$, there is an $N = N(\epsilon)$ such that if $m, n \geq N$, then

$$\|x_m - x_n\|_{\max} = \max_{t \in [a,b]} |x_m(t) - x_n(t)| < \epsilon. \tag{9.3}$$

For any fixed $t_0 \in [a, b]$,

$$|x_m(t_0) - x_n(t_0)| < \epsilon,$$

provided $m, n \geq N$. Hence the sequence $(x_1(t_0), x_2(t_0), x_3(t_0), \ldots)$ is a Cauchy sequence of real numbers, and since $\mathbf{R}$ is complete, this sequence converges, and we denote the limit by $x(t_0) = \lim_{m \to \infty} x_m(t_0)$. Since t_0 was an arbitrary point in $[a, b]$, we have for any $t \in [a, b]$ a number $x(t)$ such that

$$x(t) = \lim_{m \to \infty} x_m(t).$$

Thus (x_m) converges pointwise to the function x. It remains to show that x is continuous on $[a, b]$ and that $x_m \to x$ in the norm on $C[a, b]$. Observe that in (9.3) we may let $n \to \infty$ to obtain, for any $m \geq N$,

$$\max_{t \in [a,b]} |x_m(t) - x(t)| \leq \epsilon.$$

Then for any $t \in [a, b]$ we have

$$(9.4) \qquad |x_m(t) - x(t)| \le \epsilon \quad \text{for } m \ge N.$$

By (9.4), the sequence x_m converges uniformly to x on $[a, b]$. By Theorem 7.1.9, the limit function x is continuous on $[a, b]$, that is, $x \in C[a, b]$. Now that we know $x \in C[a, b]$, we can write $\|x_m - x\|_{\max} \le \epsilon$ if $m \ge N(\epsilon)$. This proves that $x_m \to x$ in the norm on $C[a, b]$. Since (x_m) was an arbitrary Cauchy sequence, $C[a, b]$ is complete in the maximum norm. □

Since convergence of a sequence in the maximum norm in $C[a, b]$ means uniform convergence of that sequence on $[a, b]$, the maximum norm is also called the *uniform norm* on $C[a, b]$. The resulting metric $d(x, y) = \|x - y\|_{\max} = \max_{t \in [a,b]} |x(t) - y(t)|$ is also called the *uniform metric* on $C[a, b]$. By Theorem 9.1.11, any closed subset of $C[a, b]$ is a complete metric space in the uniform metric. Exercise 9.3.4 is an application of the contraction theorem to the question of existence and uniqueness of solutions to an integral equation.

As we noted earlier in Example 8.3.10, $C[a, b]$ is not complete with respect to the norm $\|x\|_1 = \int_a^b |x(t)|\, dt$, $x \in C[a, b]$. (See Exercise 9.3.2.)

Definition 9.3.2. *A* **Banach space** *is a complete normed vector space.*

Thus $C[a, b]$ with the maximum norm is a Banach space.

The l^2 sequence space was introduced in Example 8.3.5. It is an inner product space with inner product given by

$$((\xi_k), (\eta_k)) = \sum_{k=1}^{\infty} \xi_k \eta_k$$

and norm given by

$$\|(\xi_k)\|_2 = \Big(\sum_{k=1}^{\infty} \xi_k^2 \Big)^{1/2}.$$

The space l^2 is of great importance. The mathematician D. Hilbert employed this sequence space in an influential study of integral equations early in the twentieth century. We have the following important definition.

Definition 9.3.3. *A* **Hilbert space** *is an inner product space that is complete in the norm induced by the inner product.*

We note that every Hilbert space is a Banach space, but not conversely, since there are norms that are not induced by any inner product.

We now show that l^2 is a Hilbert space.

Theorem 9.3.4. *The inner product space l^2 is a Hilbert space.*

Proof. Let (x_m) be a Cauchy sequence in the space l^2, and let us write $x_m = (\xi_1^m, \xi_2^m, \xi_3^m, \ldots)$. Then for any $\epsilon > 0$, there is an $N = N(\epsilon)$ such that if $m, n \ge N$, then

$$(9.5) \qquad \|x_m - x_n\|_2 = \Big(\sum_{j=1}^{\infty} |\xi_j^m - \xi_j^n|^2 \Big)^{1/2} < \epsilon.$$

For any fixed j_0, we have $|\xi_{j_0}^m - \xi_{j_0}^n|^2 \le \sum_{j=1}^{\infty} |\xi_j^m - \xi_j^n|^2$, and therefore (9.5) implies that for $m, n \ge N$,

$$|\xi_j^m - \xi_j^n| < \epsilon \quad \text{for all } j \in \mathbf{N}. \tag{9.6}$$

If we fix j, then (9.6) implies that the sequence $(\xi_j^1, \xi_j^2, \xi_j^3, \ldots)$ is a Cauchy sequence of real numbers. By the completeness of $\mathbf{R}$, this sequence converges; let us write

$$\xi_j = \lim_{n \to \infty} \xi_j^n.$$

These limits define a sequence

$$x = (\xi_1, \xi_2, \xi_3, \ldots).$$

We wish to show that $x \in l^2$ and $x_m \to x$ in the norm on l^2. By (9.5), for each $k \in \mathbf{N}$, and for $m, n \ge N(\epsilon)$, we have

$$\sum_{j=1}^{k} |\xi_j^m - \xi_j^n|^2 < \epsilon^2.$$

With k arbitrary but fixed, and $m \ge N(\epsilon)$, we may let $n \to \infty$ to obtain

$$\sum_{j=1}^{k} |\xi_j^m - \xi_j|^2 \le \epsilon^2,$$

where ξ_j is the j-th entry in x. Then let $k \to \infty$ to obtain for $m \ge N(\epsilon)$,

$$\sum_{j=1}^{\infty} |\xi_j^m - \xi_j|^2 \le \epsilon^2. \tag{9.7}$$

Estimate (9.7) says that for $m \ge N(\epsilon)$, $\|x_m - x\|_2^2 \le \epsilon^2$, and it also proves that $x_m - x \in l^2$. Since $x_m \in l^2$ by hypothesis, the triangle inequality for l^2 implies that

$$x = x_m + (x - x_m) \in l^2.$$

Now that we know $x \in l^2$, the estimate (9.7) shows that $x_m \to x$ in the norm on l^2. Since (x_m) was an arbitrary Cauchy sequence in l^2, this proves that l^2 is complete. □

Exercises.

Exercise 9.3.1. Show that the vector space $P[a, b]$ of real valued polynomial functions on $[a, b]$ is not complete in the maximum norm this space inherits from $C[a, b]$. For example, consider $[a, b] = [-\delta, \delta]$ where $0 < \delta < 1$, and let $p_k(t) = \sum_{j=0}^{k} t^j = 1 + t + t^2 + \cdots + t^k$, for $k \in \mathbf{N}$. Show that the uniform limit of the p_k is not an element of $P[-\delta, \delta]$.

Exercise 9.3.2. Show that $C[0, 1]$ is not complete with respect to the norm $\|x\|_1 = \int_0^1 |x(t)|\, dt$, as follows:

1. For $k \ge 2$, define $x_k(t) = 0$ if $0 \le t \le 1/2 - 1/k$; $x_k(t) = kt + 1 - k/2$ if $1/2 - 1/k < t < 1/2$; and $x_k(t) = 1$ if $1/2 \le t \le 1$. Show that the sequence (x_k) is a Cauchy sequence with respect to the norm $\|\cdot\|_1$. *Hint*: A sketch of x_k and x_n with $n > k$ will suggest a quick geometric argument.

2. Show that (x_k) does not converge to an element of $C[0,1]$ with respect to the norm $\|\cdot\|_1$. *Hint*: Assuming $\|x_k - x\|_1 \to 0$ as $k \to \infty$, express $\|x_k - x\|_1$ as the sum of three integrals: from 0 to $1/2 - 1/k$; from $1/2 - 1/k$ to $1/2$; and from $1/2$ to 1. Deduce that x cannot be continuous at $t = 1/2$.

Exercise 9.3.3. Show that $C[0,1]$ is not complete with respect to the norm $\|x\|_1 = \int_0^1 |x(t)|\,dt$, as follows: Consider the sequence (x_k) given by

$$x_k(t) = \begin{cases} k & \text{if } 0 \le t \le 1/k^2, \\ 1/\sqrt{t} & \text{if } 1/k^2 \le t \le 1. \end{cases}$$

Show that (x_k) is Cauchy with respect to $\|\cdot\|_1$, but that (x_k) does not converge in the norm $\|\cdot\|_1$ to an element of $C[0,1]$.

Exercise 9.3.4. An integral equation of the form

$$f(x) - \int_a^b k(x,y) f(y)\,dy = g(x), \tag{9.8}$$

where g is given, is a *linear Fredholm equation of the second kind*.[1] We may write (9.8) as a fixed point problem, $Tf = f$, where

$$(Tf)(x) := g(x) + \int_a^b k(x,y) f(y)\,dy.$$

Suppose $k : [a,b] \times [a,b] \to \mathbf{R}$ is a continuous function such that

$$\sup_{a \le x \le b} \left\{ \int_a^b |k(x,y)|\,dy \right\} < 1,$$

and $g : [a,b] \to \mathbf{R}$ is a continuous function. Show that there is a unique continuous function $f : [a,b] \to \mathbf{R}$ that satisfies (9.8). *Hint*: Show that T is a contraction mapping on the space $C[a,b]$ with the uniform norm (max norm).

9.4. The l^p Sequence Spaces

This section will be of interest to students who are interested in studying functional analysis. The main goal of the section is to establish the Minkowski inequality, which allows us to define the sequence spaces l^p, for $p > 1$, and also establishes the triangle inequality for the norm on l^p. The Hölder and Minkowski inequalities are two classical inequalities at the foundation of the study of function spaces in functional analysis.

It is appropriate to recall here the definitions of certain sequence spaces already defined. Exercise 8.1.6 and Exercise 8.3.13 deal with the sequence space l^1 consisting of all real sequences $\xi = (\xi_k)$ such that $\sum_{k=1}^{\infty} |\xi_k|$ converges. Exercise 8.1.6 establishes that l^1 is a vector space and Exercise 8.3.13 shows that l^1 is normed by

$$\|\xi\|_1 := \sum_{k=1}^{\infty} |\xi_k|, \quad \xi = (\xi_k) \in l^1.$$

The triangle inequality for the norm in l^1 follows rather easily from the triangle inequality for the real numbers and basic properties of convergent series. Also, we

[1] A linear Fredholm equation of the *first kind* takes the form $\int_a^b k(x,y) f(y)\,dy = g(x)$.

introduced the Hilbert sequence space l^2 in Example 8.3.5 and established there that l^2 is a real inner product space. Finally, Exercise 8.1.7 and Exercise 8.3.14 established a normed space structure for the sequence space l^∞ consisting of bounded real sequences.

Now let p be a real number with $p > 1$, and let l^p denote the *set* of all real sequences $\xi = (\xi_k)$ such that $\sum_{k=1}^\infty |\xi_k|^p$ converges. We will establish that l^p is actually a vector space under componentwise addition and scalar multiplication. We define $\|\xi\|_p$ for $\xi \in l^p$ by

$$\|\xi\|_p := \Big(\sum_{k=1}^\infty |\xi_k|^p\Big)^{1/p}, \quad \xi = (\xi_k) \in l^p.$$

Then $\|\xi\|_p$ is well defined for each $\xi \in l^p$. We have $\|\xi\|_p \geq 0$, and $\|\xi\|_p = 0$ implies $\xi = (0, 0, 0, \ldots) = \theta \in l^p$. If α is a real scalar, then $\|\alpha\xi\|_p = |\alpha|\|\xi\|_p$.

It remains to show that the sum of two elements of l^p is an element of l^p and that the triangle inequality

$$\|\xi + \eta\|_p \leq \|\xi\|_p + \|\eta\|_p$$

holds. In what follows, let $p > 1$ and suppose q is defined by

$$\frac{1}{p} + \frac{1}{q} = 1.$$

Then p and q are called *conjugate exponents*. For conjugate exponents, we have

$$\frac{p+q}{pq} = 1 \quad \Longrightarrow \quad pq = p + q \quad \Longrightarrow \quad (p-1)(q-1) = 1,$$

and hence $1/(p-1) = q - 1$. Thus, if $y = x^{p-1}$, then $x = y^{q-1}$, so these power functions are inverses of each other.

In order to establish the triangle inequality, we establish the following important inequalities, where $p > 1$ and q are conjugate exponents:

1. **Young's inequality**. If a and b are any positive numbers, then
$$ab \leq \frac{a^p}{p} + \frac{b^q}{q}.$$
2. **Hölder's inequality**. For any nonzero $\xi = (\xi_k) \in l^p$ and $\eta = (\eta_k) \in l^q$,
$$\sum_{k=1}^\infty |\xi_k\eta_k| \leq \Big(\sum_{i=1}^\infty |\xi_i|^p\Big)^{1/p}\Big(\sum_{j=1}^\infty |\eta_j|^q\Big)^{1/q}.$$
3. **Minkowski's inequality**. For any $\xi = (\xi_k), \eta = (\eta_k) \in l^p$, $p > 1$,
$$\Big(\sum_{k=1}^\infty |\xi_k + \eta_k|^p\Big)^{1/p} \leq \Big(\sum_{i=1}^\infty |\xi_i|^p\Big)^{1/p} + \Big(\sum_{j=1}^\infty |\eta_j|^p\Big)^{1/p}.$$
This is precisely the triangle inequality $\|\xi + \eta\|_p \leq \|\xi\|_p + \|\eta\|_p$ for ξ, η in l^p, $p > 1$.

We first prove Young's inequality. Then we show that Young's inequality implies Hölder's inequality, and finally that Hölder's inequality implies Minkowski's inequality.

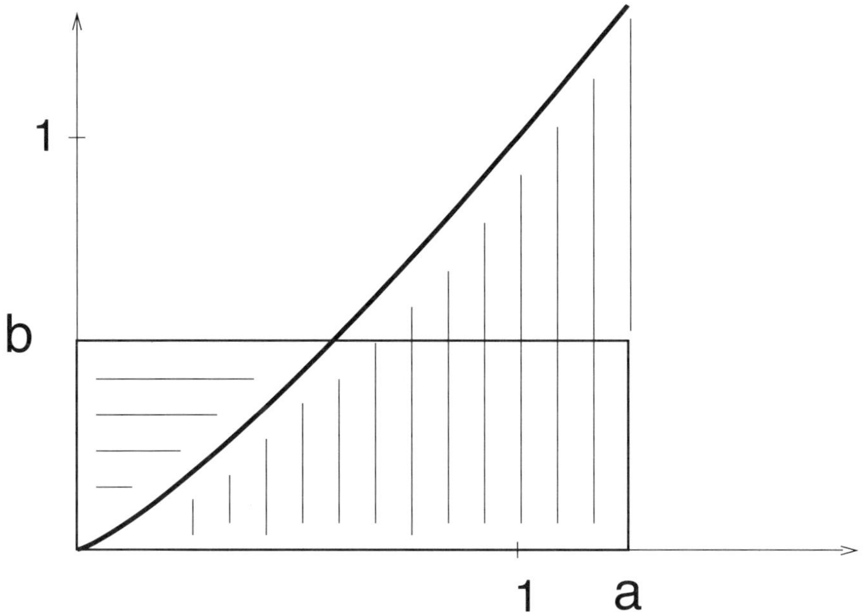

Figure 9.1. Illustrating Young's inequality geometrically: The graph of $y = x^{p-1}$, which is also the graph of $x = y^{q-1}$, is shown, and $ab \leq \int_0^a x^{p-1}\,dx + \int_0^b y^{q-1}\,dy = \frac{a^p}{p} + \frac{b^q}{q}$. There are also cases where the graph of $y = x^{p-1}$ intersects the right-hand side of the rectangle at or below the level of b, with the same inequality relating the sum of the two integrals and the product ab.

Young's inequality. Consider the rectangle in the xy-plane with vertices at the points $(0,0)$, $(a,0)$, $(0,b)$ and (a,b). The area of this rectangle is ab. Consider also the graph of the function $y = x^{p-1}$, which is also the graph of the inverse relation $x = y^{q-1}$. Now the area between this graph and the x-axis from $x = 0$ to $x = a$, added to the area between this graph and the y-axis from $y = 0$ to $y = b$, must be greater than or equal to the rectangle area ab. That is,

$$ab \leq \int_0^a x^{p-1}\,dx + \int_0^b y^{q-1}\,dy.$$

See Figure 9.1. Computation of the integrals yields Young's inequality

$$ab \leq \frac{a^p}{p} + \frac{b^q}{q}.$$

Young's inequality implies Hölder's inequality. We first prove Hölder's inequality for $\xi \in l^p$ and $\eta \in l^q$ with $\|\xi\|_p = 1$ and $\|\eta\|_q = 1$. Afterwards, we scale general elements of these spaces to get the general result. Thus, assume that $\xi = (\xi_k) \in l^p$ and $\eta = (\eta_k) \in l^q$ satisfy

$$\sum_{k=1}^{\infty} |\xi_k|^p = 1 \qquad \text{and} \qquad \sum_{k=1}^{\infty} |\eta_k|^q = 1.$$

For each k we have, by Young's inequality,

$$|\xi_k \eta_k| = |\xi_k||\eta_k| \le \frac{|\xi_k|^p}{p} + \frac{|\eta_k|^q}{q}.$$

Now sum over k, using a direct comparison of convergent series and the relation $1/p + 1/q = 1$ to obtain

$$\sum_{k=1}^{\infty} |\xi_k \eta_k| \le \frac{1}{p} + \frac{1}{q} = 1.$$

This is Hölder's inequality for the case where $\|\xi\|_p = 1$ and $\|\eta\|_q = 1$. Now let $\xi \in l^p$ and $\eta \in l^q$ be any nonzero elements and define

$$\hat{\xi} = (\hat{\xi}_k) \qquad \text{and} \qquad \hat{\eta} = (\hat{\eta}_k)$$

by

$$\hat{\xi}_k = \frac{\xi_k}{(\sum_{k=1}^{\infty} |\xi_k|^p)^{1/p}} \qquad \text{and} \qquad \hat{\eta_k} = \frac{\eta_k}{(\sum_{k=1}^{\infty} |\eta_k|^q)^{1/q}}.$$

Then $\hat{\xi} \in l^p$ with $\|\hat{\xi}\|_p = 1$, and $\hat{\eta} \in l^q$ with $\|\hat{\eta}\|_q = 1$. Hence

$$\sum_{k=1}^{\infty} |\hat{\xi}_k \hat{\eta}_k| \le 1,$$

which is equivalent to Holder's inequality

$$\sum_{k=1}^{\infty} |\xi_k \eta_k| \le \Big(\sum_{i=1}^{\infty} |\xi_i|^p \Big)^{1/p} \Big(\sum_{j=1}^{\infty} |\eta_j|^q \Big)^{1/q}.$$

Observe that the special case where $p = 2$ and $q = 2$ gives the Cauchy-Schwarz inequality

$$\sum_{k=1}^{\infty} |\xi_k \eta_k| \le \Big(\sum_{i=1}^{\infty} |\xi_i|^2 \Big)^{1/2} \Big(\sum_{j=1}^{\infty} |\eta_j|^2 \Big)^{1/2},$$

established earlier for l^2 in Example 8.3.5.

Hölder's inequality implies Minkowski's inequality. We begin by noting that Hölder's inequality certainly holds for finite sums. Let $\xi = (\xi_k) \in l^p$ and $\eta = (\eta_k) \in l^p$ with $p > 1$. By the triangle inequality for numbers,

$$|\xi_k + \eta_k|^p = |\xi_k + \eta_k||\xi_k + \eta_k|^{p-1} \le (|\xi_k| + |\eta_k|)|\xi_k + \eta_k|^{p-1}.$$

For any positive integer n,

$$\sum_{k=1}^{n} |\xi_k + \eta_k|^p \le \sum_{k=1}^{n} |\xi_k||\xi_k + \eta_k|^{p-1} + \sum_{k=1}^{n} |\eta_k||\xi_k + \eta_k|^{p-1}.$$

Now apply Hölder's inequality to each of the sums on the right side. For the first sum on the right side, Hölder's inequality yields

$$\sum_{k=1}^{n} |\xi_k||\xi_k + \eta_k|^{p-1} \le \Big(\sum_{i=1}^{n} |\xi_i|^p \Big)^{1/p} \Big(\sum_{j=1}^{n} |\xi_j + \eta_j|^{(p-1)q} \Big)^{1/q},$$

and $(p-1)q = p$ since $pq = p + q$. Hölder's inequality for the second sum yields

$$\sum_{k=1}^{n} |\eta_k||\xi_k + \eta_k|^{p-1} \le \Big(\sum_{i=1}^{n} |\eta_i|^p \Big)^{1/p} \Big(\sum_{j=1}^{n} |\xi_j + \eta_j|^{(p-1)q} \Big)^{1/q},$$

where again $(p-1)q = p$. Combining these results, we have

$$\sum_{k=1}^{n} |\xi_k + \eta_k|^p \le \Big(\sum_{i=1}^{n} |\xi_i|^p\Big)^{1/p} \Big(\sum_{j=1}^{n} |\xi_j + \eta_j|^p\Big)^{1/q} + \Big(\sum_{i=1}^{n} |\eta_i|^p\Big)^{1/p} \Big(\sum_{j=1}^{n} |\xi_j + \eta_j|^p\Big)^{1/q},$$

and hence

$$\sum_{k=1}^{n} |\xi_k + \eta_k|^p \le \Big[\Big(\sum_{i=1}^{n} |\xi_i|^p\Big)^{1/p} + \Big(\sum_{i=1}^{n} |\eta_i|^p\Big)^{1/p}\Big] \Big(\sum_{j=1}^{n} |\xi_j + \eta_j|^p\Big)^{1/q}.$$

Divide by the last factor on the right side and use the fact that $1 - 1/q = 1/p$ to find

$$\Big(\sum_{k=1}^{n} |\xi_k + \eta_k|^p\Big)^{1/p} \le \Big(\sum_{i=1}^{n} |\xi_i|^p\Big)^{1/p} + \Big(\sum_{i=1}^{n} |\eta_i|^p\Big)^{1/p}.$$

This is the Minkowski inequality for finite sums. (Note that this proves the triangle inequality for the p-norms on $\mathbf{R}^n$.) Let $n \to \infty$; the resulting series on the right-hand side both converge since $\xi, \eta \in l^p$. Hence the series on the left-hand side also converges, yielding Minkowski's inequality for elements of l^p, $p > 1$.

As a consequence of Minkowski's inequality, the sum of two elements of l^p is also an element of l^p, and the triangle inequality holds for $\|\cdot\|_p$. Thus we have defined the sequence space l^p for $p > 1$.

By following the pattern of proof in Theorem 9.3.4, one can show that the sequence spaces l^p, for all $p \ge 1$, are complete (Exercise 9.4.1). Thus, the l^p sequence spaces are Banach spaces. However, l^p, for $p \ne 2$, is not an inner product space (Exercise 9.4.2).

Exercises.

Exercise 9.4.1. Write a detailed proof that l^p, $p \ge 1$, is complete with respect to the p-norm, by following the pattern of the argument in the proof of Theorem 9.3.4.

Exercise 9.4.2. Show that l^p, for $p \ne 2$, is not an inner product space. *Hint*: The parallelogram law of Exercise 8.3.3 holds in any inner product space. Show that this law does not hold for the norm on l^p, $p \ne 2$.

9.5. Matrix Norms and Completeness

Matrix norms are an essential tool in analysis. In this section we present several convenient ways to norm the space $\mathbf{R}^{n\times n}$ of $n \times n$ real matrices.

9.5.1. Matrix Norms. This section discusses matrix norms and the complete normed space of $n \times n$ real matrices, and includes some basic matrix analysis that is useful later in the book. In addition to the background in linear algebra from Chapter 8, we also assume some familiarity with linear transformations $L : \mathbf{R}^n \to \mathbf{R}^m$ defined by an $m \times n$ matrix A,

$$L\mathbf{x} = A\mathbf{x}, \quad \mathbf{x} \in \mathbf{R}^n,$$

and familiarity with the concepts of eigenvalues and eigenvectors for A. We usually think of vectors in Euclidean space as column vectors. However, sometimes it is convenient to write $\mathbf{x} = (x_1, \ldots, x_n)$ for points of $\mathbf{R}^n$ and $\mathbf{y} = (y_1, \ldots, y_m)$ for points in $\mathbf{R}^m$ to indicate the coordinates of these vectors, without having to display a column vector or the transposed (row) vector, for example $\mathbf{x}^T = [x_1 \cdots x_n]$.

For convenience we recall the definition of a linear transformation.

Definition 9.5.1. *Let V and W be real vector spaces. A mapping $L : V \to W$ is* **linear** *if*

(a) $L(u + v) = L(u) + L(v)$ *for all* $u, v \in V$ *and*

(b) $L(\alpha u) = \alpha L(u)$ *for all* $\alpha \in \mathbf{R}$ *and* $u \in V$.

The conjunction of (a) and (b) is equivalent to the statement that for all $u, v \in V$ and $\alpha, \beta \in \mathbf{R}$, $L(\alpha u + \beta v) = \alpha L(u) + \beta L(v)$. Suppose that V and W have dimension n and m, respectively. Recall that a choice of bases in V and W determines an $m \times n$ matrix representation A of the linear transformation L with respect to the specified bases, as follows. Let V have basis $\{\mathbf{v}_1, \ldots, \mathbf{v}_n\}$ and W have basis $\{\mathbf{w}_1, \ldots, \mathbf{w}_m\}$. For each $\mathbf{x} \in V$, we have

$$\mathbf{x} = \sum_{j=1}^{n} x_j \mathbf{v}_j$$

for a unique choice of coefficients x_j. For each j with $1 \leq j \leq n$, we have

$$L\mathbf{v}_j = \sum_{i=1}^{m} a_{ij} \mathbf{w}_i$$

for a unique choice of coefficients a_{ij}. Thus,

$$L\mathbf{x} = \sum_{j=1}^{n} x_j L\mathbf{v}_j = \sum_{j=1}^{n} x_j \sum_{i=1}^{m} a_{ij} \mathbf{w}_i = \sum_{i=1}^{m} \Big(\sum_{j=1}^{n} a_{ij} x_j \Big) \mathbf{w}_i.$$

The linear mapping $L : V \to W$ is thus represented by the matrix $A = [a_{ij}]$, which maps the coordinate vector $(x_1, \ldots, x_n)$ of $\mathbf{x}$ with respect to the basis $\{\mathbf{v}_1, \ldots, \mathbf{v}_n\}$ to the coordinate vector of $L\mathbf{x}$ with respect to the basis $\{\mathbf{w}_1, \ldots, \mathbf{w}_m\}$.

Let us write $\mathbf{R}^{n \times n}$ for the vector space of $n \times n$ matrices with real entries. Notice that the spaces $\mathbf{R}^{n \times n}$ and $\mathbf{R}^{n^2}$ are isomorphic, since they both have dimension n^2. (See Exercise 9.5.2.)

Definition 9.5.2. *A* **matrix norm** *is a function $\| \cdot \|$ on the vector space $\mathbf{R}^{n \times n}$ of $n \times n$ real matrices which satisfies, in addition to the properties* (i), (ii), *and* (iii) *of a norm in Definition* 8.3.1, *the following property:*

(iv) $\|AB\| \leq \|A\| \, \|B\|$ *for any two $n \times n$ matrices A and B.*

A matrix norm $\| \cdot \|$ is **compatible** *with a given vector norm $| \cdot |$ if it satisfies the inequality*

(v) $|A\mathbf{x}| \leq \|A\| \, |\mathbf{x}|$ *for any $n \times n$ matrix A and any vector $\mathbf{x} \in \mathbf{R}^n$.*

For any matrix norm on real $n \times n$ matrices, it is straightforward to show by induction that $\|A^k\| \leq \|A\|^k$ for every positive integer k.

Example 9.5.3 (Absolute sum matrix norm)**.** For any $n \times n$ matrix A, let us define

$$\|A\|_{as} = \sum_{i=1}^{n} \sum_{j=1}^{n} |a_{ij}|, \quad \text{where} \quad A = [a_{ij}], \tag{9.9}$$

where the subscript "as" stands for the **absolute sum** matrix norm. Then $\|\cdot\|_{as}$ is a matrix norm, as the reader may verify in Exercise 9.5.3. We will now show that this matrix norm is compatible with the vector norm

$$|\mathbf{x}|_1 = \sum_{j=1}^{n} |x_j| = |x_1| + \cdots + |x_n|. \tag{9.10}$$

Writing $(A\mathbf{x})_i$ for the i-th component of $A\mathbf{x}$, we have

$$|A\mathbf{x}|_1 = \sum_{i=1}^{n} |(A\mathbf{x})_i| = \sum_{i=1}^{n} \Big| \sum_{j=1}^{n} a_{ij} x_j \Big| \le \sum_{i=1}^{n} \sum_{j=1}^{n} |a_{ij}| \, |x_j|.$$

Since $|x_j| \le |\mathbf{x}|_1$ for each j, we have $|A\mathbf{x}|_1 \le \|A\|_{as} \, |\mathbf{x}|_1$, as desired. △

Theorem 9.5.4. *Let $|\cdot|$ denote a vector norm on $\mathbf{R}^n$. This vector norm induces a matrix norm on $\mathbf{R}^{n\times n}$, given by*

$$\|A\| := \max_{|\mathbf{x}|=1} |A\mathbf{x}|, \tag{9.11}$$

where $|A\mathbf{x}|$ is the vector norm of the image vector $A\mathbf{x}$. Moreover, this matrix norm is compatible with the vector norm $|\cdot|$; that is, $|A\mathbf{x}| \le \|A\| \, |\mathbf{x}|$ for any $n \times n$ matrix A and all $\mathbf{x} \in \mathbf{R}^n$.

Proof. Since the mapping $\mathbf{x} \mapsto |A\mathbf{x}|$ is continuous and the unit sphere $\{\mathbf{x} : |\mathbf{x}| = 1\}$ is compact, the maximum in (9.11) exists. Clearly $\|A\| \ge 0$ and $\|A\| = 0$ if and only if $A\mathbf{x} = \mathbf{0}$ for all $\mathbf{x} \in \mathbf{R}^n$ with $|\mathbf{x}| = 1$. Since $A(\frac{\mathbf{x}}{|\mathbf{x}|}) = \frac{1}{|\mathbf{x}|} A\mathbf{x}$ for any nonzero $\mathbf{x}$, this is equivalent to saying $\|A\| = 0$ if and only if $A\mathbf{x} = \mathbf{0}$ for all $\mathbf{x} \in \mathbf{R}^n$, that is, A is the zero matrix. If α is a real scalar, then

$$\|\alpha A\| = \max_{|\mathbf{x}|=1} \big|\alpha A\mathbf{x}\big| = \max_{|\mathbf{x}|=1} |\alpha|\big|A\mathbf{x}\big| = |\alpha| \max_{|\mathbf{x}|=1} \big|A\mathbf{x}\big| = |\alpha| \, \|A\|.$$

Finally, for the triangle inequality, we have

$$\begin{aligned} \|A + B\| &= \max_{|\mathbf{x}|=1} \big|(A+B)\mathbf{x}\big| \le \max_{|\mathbf{x}|=1} (\big|A\mathbf{x}\big| + \big|B\mathbf{x}\big|) \\ &\le \max_{|\mathbf{x}|=1} |A\mathbf{x}| + \max_{|\mathbf{x}|=1} |B\mathbf{x}| = \|A\| + \|B\|. \end{aligned}$$

Therefore (9.11) does define a norm on the space $\mathbf{R}^{n\times n}$. If $\mathbf{x} \in \mathbf{R}^n$ is any nonzero vector, then $\mathbf{x}/|\mathbf{x}|$ has unit norm, and therefore

$$\frac{1}{|\mathbf{x}|} \big|A\mathbf{x}\big| = \Big|A\Big(\frac{1}{|\mathbf{x}|}\mathbf{x}\Big)\Big| \le \|A\| \quad \Longrightarrow \quad \big|A\mathbf{x}\big| \le \|A\| \, |\mathbf{x}|.$$

Since the last inequality is also satisfied automatically by $\mathbf{x} = \mathbf{0}$, the matrix norm defined by (9.11) is compatible with the given vector norm $|\cdot|$. □

We consider two examples to illustrate Theorem 9.5.4.

Example 9.5.5 (Matrix norm induced by vector norm $|\cdot|_\infty$). The vector norm $|\mathbf{x}|_\infty = \max_{1\le j\le n}|a_j|$ on $\mathbf{R}^n$ (Example 8.3.8) is especially useful later in this book. We now compute the induced matrix norm $\|A\|_\infty$ according to (9.11). In fact we will show that

$$\|A\|_\infty = \max_{1\le i\le n}\sum_{j=1}^n |a_{ij}|,$$

that is, $\|A\|_\infty$ is the **maximum absolute row sum** of A. Fix an index i, and consider the i-th row of A, with entries a_{ij}, $j=1,\dots,n$. Define a vector $\mathbf{v}$ as follows. Let $v_j = 1$ if $a_{ij}\ge 0$ and let $v_j=-1$ if $a_{ij}<0$. Then $|\mathbf{v}|_\infty = 1$, and the i-th component of $A\mathbf{v}$ equals $\sum_{j=1}^n |a_{ij}|$, the i-th absolute row sum of A. Thus, since we can realize each of the absolute row sums of A as a component in an image vector $A\mathbf{v}$ for some $\mathbf{v}$ with unit norm, it follows from (9.11) that

$$\|A\|_\infty \ge \max_{1\le i\le n}\sum_{j=1}^n |a_{ij}|.$$

To show the reverse inequality, let $|\mathbf{x}|_\infty = 1$. Then $|x_j|\le 1$ for $j=1,\dots,n$. For each $i=1,\dots,n$, the i-th component of $A\mathbf{x}$ is $\sum_{j=1}^n a_{ij}x_j$, and hence

$$\left|A\mathbf{x}\right|_\infty = \max_{1\le i\le n}\left|\sum_{j=1}^n a_{ij}x_j\right|.$$

For each i,

$$\left|\sum_{j=1}^n a_{ij}x_j\right| \le \sum_{j=1}^n |a_{ij}|\,|x_j| \le \sum_{j=1}^n |a_{ij}|,$$

the right-hand side being the ith absolute row sum of A. Hence,

$$\left|A\mathbf{x}\right|_\infty \le \max_{1\le i\le n}\sum_{j=1}^n |a_{ij}|$$

as we wanted to show. △

Example 9.5.6 (Matrix norm induced by the Euclidean norm $|\cdot|_2$). We apply some facts from Section 8.5 and the spectral theorem for real symmetric matrices to discuss the matrix norm $\|A\|_2$ induced by the Euclidean vector norm $|\mathbf{x}|_2$. Let A be an $n\times n$ real matrix. Then A^TA is symmetric and positive semidefinite, that is, $\mathbf{x}^TA^TA\mathbf{x}\ge 0$ for all $\mathbf{x}$, and hence all its eigenvalues are nonnegative. Let

$$\lambda_{\max} := \text{the maximum eigenvalue of } A^TA.$$

We will show that

$$\|A\|_2 = (\lambda_{\max})^{1/2}.$$

By definition, $\|A\|_2 = \max_{|\mathbf{x}|_2=1}|A\mathbf{x}|_2$. Let λ be an eigenvalue of A^TA and let $\mathbf{x}$ be a corresponding eigenvector with $|\mathbf{x}|_2=1$. Then $A^TA\mathbf{x}=\lambda\mathbf{x}$ implies

$$|A\mathbf{x}|_2^2 = \mathbf{x}^TA^TA\mathbf{x} = \mathbf{x}^T(\lambda\mathbf{x}) = \lambda|\mathbf{x}|_2^2 = \lambda,$$

and hence $|A\mathbf{x}|_2 = (\lambda)^{1/2}$. Since this is true for any eigenvalue λ of A, we have

$$\|A\|_2 = \max_{|\mathbf{x}|_2=1}\left|A\mathbf{x}\right|_2 \ge (\lambda_{\max})^{1/2}.$$

To get the reverse inequality, let us list the eigenvalues of $A^T A$, including multiplicities, as $\lambda_1, \dots, \lambda_n$. Let $\mathbf{x} \in \mathbf{R}^n$. Then we may write

$$\mathbf{x} = \mathbf{x}_1 + \mathbf{x}_2 + \cdots + \mathbf{x}_n$$

where for each k, $\mathbf{x}_k$ is an eigenvector for λ_k, and these eigenvectors are pairwise orthogonal and form a basis of $\mathbf{R}^n$. Then we have

$$\mathbf{x}^T A^T A\mathbf{x} = \mathbf{x}^T \sum_{k=1}^{n} \lambda_k \mathbf{x}_k = \sum_{k=1}^{n} \lambda_k |\mathbf{x}_k|_2^2 \leq \lambda_{\max} \sum_{k=1}^{n} |\mathbf{x}_k|_2^2 = \lambda_{\max} |\mathbf{x}|_2^2.$$

Hence, $|A\mathbf{x}|_2^2 \leq \lambda_{\max}|\mathbf{x}|_2^2$. Thus, $|A\mathbf{x}|_2 \leq (\lambda_{\max})^{1/2}|\mathbf{x}|_2$ for all $\mathbf{x}$, and it follows immediately from the definition that $\|A\|_2 \leq (\lambda_{\max})^{1/2}$. △

We need a norm for linear mappings from one Euclidean space to another. We extend the definition of matrix norms given for square matrices in Definition 9.5.2 by defining a **matrix norm** for $\mathbf{R}^{m\times n}$, the space of $m \times n$ real matrices, to be a function $\|\cdot\| : \mathbf{R}^{m\times n} \to \mathbf{R}$ that satisfies properties (i), (ii), and (iii) of the norm definition (Definition 8.3.1). Given fixed vector norms on $\mathbf{R}^n$ and $\mathbf{R}^m$, there is a standard way of defining a matrix norm on $\mathbf{R}^{m\times n}$ that is **compatible** with the given vector norms. This is the content of the following definition.

Definition 9.5.7. *Let $\mathbf{L} : \mathbf{R}^n \to \mathbf{R}^m$ be a linear transformation or an $m \times n$ real matrix. Given norms on $\mathbf{R}^n$ and $\mathbf{R}^m$ (both denoted here by $|\cdot|$) we define the associated norm for $\mathbf{L}$ by*

$$\|\mathbf{L}\| = \max_{|\mathbf{x}|=1} \big|\mathbf{L}\mathbf{x}\big|.$$

Thus $\|\mathbf{L}\|$ is the maximum of the image vector norms $|\mathbf{L}\mathbf{x}|$, for $\mathbf{x}$ in the unit ball in $\mathbf{R}^n$ defined by the given norm in the domain space. We note that every linear transformation of Euclidean spaces is continuous, since each component function of $\mathbf{L}$ takes the form of a linear combination, with real scalars, of the components of the vector $\mathbf{x}$. Since the unit sphere defined by $|\mathbf{x}| = 1$ is compact, the maximum indicated in the definition exists. If $|\mathbf{x}| = 1$, then $|\mathbf{L}\mathbf{x}| \leq \|\mathbf{L}\|$. For any $\mathbf{x} \neq \mathbf{0}$, the vector $\frac{1}{|\mathbf{x}|}\mathbf{x}$ has unit norm, hence

$$\frac{1}{|\mathbf{x}|}\big|\mathbf{L}\mathbf{x}\big| = \Big|\frac{1}{|\mathbf{x}|}\mathbf{L}\mathbf{x}\Big| = \Big|\mathbf{L}\Big(\frac{1}{|\mathbf{x}|}\mathbf{x}\Big)\Big| \leq \|\mathbf{L}\|,$$

and therefore $|\mathbf{L}\mathbf{x}| \leq \|\mathbf{L}\|\,|\mathbf{x}|$, which is also true when $\mathbf{x} = \mathbf{0}$. For example, if we use the Euclidean norm, denoted $|\cdot|_2$, for the domain $\mathbf{R}^2$ and range $\mathbf{R}^3$, then for convenience and identification we naturally denote the compatible matrix norm of Definition 9.5.7 by $\|\cdot\|_2$, and thus for any real 3×2 matrix A, we have the inequality

$$\big|\mathbf{A}\mathbf{x}\big|_2 \leq \|\mathbf{A}\|_2\,|\mathbf{x}|_2,$$

which holds for all $\mathbf{x} \in \mathbf{R}^2$ and corresponding image $\mathbf{A}\mathbf{x} \in \mathbf{R}^3$.

Theorem 9.5.8. *If $\mathbf{L} : \mathbf{R}^n \to \mathbf{R}^m$ is a linear transformation, then $\mathbf{L}$ is uniformly continuous on $\mathbf{R}^n$.*

Proof. Assume that norms are given on domain and range space, both indicated here by the symbol $|\cdot|$. Given $\mathbf{u}$, $\mathbf{v}$ in the domain,

$$\big|\mathbf{L}\mathbf{u} - \mathbf{L}\mathbf{v}\big| = \big|\mathbf{L}(\mathbf{u} - \mathbf{v})\big| \leq \|\mathbf{L}\|\,|\mathbf{u} - \mathbf{v}|,$$

where $\|\mathbf{L}\|$ denotes the associated norm in Definition 9.5.7. Uniform continuity follows immediately: Given $\epsilon > 0$, the choice of $|\mathbf{u}-\mathbf{v}| < \epsilon/\|\mathbf{L}\| = \delta(\epsilon)$ implies that $|\mathbf{Lu} - \mathbf{Lv}| < \epsilon$. □

9.5.2. Completeness of $\mathbf{R}^{n\times n}$. We now show that the normed vector space $\mathbf{R}^{n\times n}$ of $n \times n$ real matrices is complete, that is, a Banach space.

Theorem 9.5.9. *The vector space $\mathbf{R}^{n\times n}$ of $n \times n$ real matrices is a Banach space, complete with respect to any matrix norm.*

Proof. By the equivalence of norms on a finite-dimensional space, we may use any norm on $\mathbf{R}^{n\times n}$. We choose the matrix norm $\|\cdot\|_{as}$, the absolute sum norm: $\|A\|_{as} = \sum_{i=1}^{n}\sum_{j=1}^{n}|a_{ij}|$. Every Cauchy sequence of matrices A_k in $\mathbf{R}^{n\times n}$ must converge to a matrix $A \in \mathbf{R}^{n\times n}$. To see this, note that if $A_k \in \mathbf{R}^{n\times n}$ defines a Cauchy sequence, then for each index pair i, j, the sequence of ij entries A_k^{ij} is a real number Cauchy sequence, since $|A_n^{ij} - A_k^{ij}| \leq \|A_n - A_k\|_{as}$. Therefore (A_k^{ij}) converges to a real number, denoted a_{ij}. Then the matrix $A = [a_{ij}]$ is such that $\lim_{k\to\infty} A_k = A$, that is, $\lim_{k\to\infty}\|A_k - A\|_{as} = 0$. Therefore $\mathbf{R}^{n\times n}$ is complete. □

The following concept is useful in later discussions in the book on matrix exponential series and matrix geometric series.

Definition 9.5.10. *Let V be a normed vector space with norm $\|\cdot\|$. The infinite series $\sum_{k=1}^{\infty} a_k$, $a_k \in V$, is* **absolutely convergent** *if the real numerical series $\sum_{k=1}^{\infty}\|a_k\|$ converges.*

A simple but remarkable theorem asserts that absolute convergence in a complete normed space implies convergence.

Theorem 9.5.11 (Absolute Convergence Implies Convergence)**.** *Let V be a complete normed vector space. If the infinite series $\sum_{k=1}^{\infty} a_k$, $a_k \in V$, is absolutely convergent, then it converges in the norm on V to a limit $s \in V$, that is,*

$$\sum_{k=1}^{\infty} a_k = \lim_{n\to\infty}\sum_{k=1}^{n} a_k = s.$$

Proof. Let us write

$$s_n = \sum_{k=1}^{n} a_k \quad \text{and} \quad S_n = \sum_{k=1}^{n}\|a_k\|$$

for the partial sums of the series in V and the real numerical series of norms, respectively. Since V is complete, the proof hinges on showing that if (S_n) is Cauchy, then (s_n) is Cauchy in the norm on V. If (S_n) is Cauchy, then given any $\epsilon > 0$ there exists an $N = N(\epsilon)$ such that if $m > n \geq N$, then

$$|S_m - S_n| = S_m - S_n = \sum_{k=n+1}^{m}\|a_k\| < \epsilon.$$

Now we can estimate that for $m > n \geq N$,

$$\|s_m - s_n\| = \left\|\sum_{k=n+1}^{m} a_k\right\| \leq \sum_{k=n+1}^{m}\|a_k\| = S_m - S_n < \epsilon.$$

Thus the sequence (s_n) of partial sums of the series in V is a Cauchy sequence. Since V is complete, $s = \lim_{n\to\infty} s_n = \sum_{k=1}^{\infty} a_k$ exists. □

Example 9.5.12 (Matrix geometric series). Suppose A is an $n \times n$ real matrix such that $\|A\| < 1$ for some matrix norm. Then $I - A$ is invertible, and

$$(I - A)^{-1} = \sum_{k=0}^{\infty} A^k = I + A + A^2 + A^3 + \cdots,$$

where I is the $n \times n$ identity matrix. In fact, we will prove this result using the real numerical geometric series result along with the completeness of $\mathbf{R}^{n\times n}$ and Theorem 9.5.11. First, the series of norms $\sum_{k=0}^{\infty} \|A^k\|$ converges since we have $\|A^k\| \le \|A\|^k$ for each k, and the real geometric series

$$\sum_{k=0}^{\infty} \|A\|^k$$

converges since $\|A\| < 1$. Thus, $\sum_{k=0}^{\infty} \|A^k\|$ converges by the direct comparison test. Since $\mathbf{R}^{n\times n}$ is complete, we may now apply Theorem 9.5.11 to conclude that the matrix series $\sum_{k=0}^{\infty} A^k$ converges to a uniquely defined limit matrix X. We wish to show that $X = (I - A)^{-1}$. We follow the proof of the numerical geometric series result. Write

$$S_n = \sum_{k=0}^{n-1} A^k = I + A + A^2 + \cdots + A^{n-1}.$$

Then $X = \lim_{n\to\infty} S_n$. We have a telescoping sum for the product

$$(I - A)S_n = I - A^n.$$

Now let $n \to \infty$ in this equation to get

$$(I - A)X = I,$$

since $\|A\| < 1$ implies that $\lim_{n\to\infty} A^n$ is the zero matrix. Therefore X is a right inverse for the square matrix $I - A$, hence $I - A$ is invertible and $X = (I - A)^{-1}$. This proves what we wanted. (See also Exercises 9.5.6, 9.5.7.) △

Exercises.

Exercise 9.5.1. Let V and W be real vector spaces. Prove that a one-to-one linear mapping $L : V \to W$ maps linearly independent sets of vectors to linearly independent sets.

Exercise 9.5.2. Verify that the set $\mathbf{R}^{n\times n}$ of $n \times n$ matrices with real entries is a real vector space, and identify a basis for this space.

Exercise 9.5.3. Show that $\|A\|_{as} = \sum_{i=1}^{n} \sum_{j=1}^{n} |a_{ij}|$ defines a matrix norm on the space $\mathbf{R}^{n\times n}$ of $n \times n$ matrices.

Exercise 9.5.4. Define $R : l^2 \to l^2$ by $R(\xi_1, \xi_2, \xi_3, \ldots) = (\xi_1, \xi_2, \xi_3, \ldots)$ for $\xi = (\xi_k)$ in l^2. Show that R is a linear transformation. Show that R is one-to-one and has no eigenvalues. *Hint*: Recall that eigenvectors must be nonzero, by definition.

Exercise 9.5.5. *Matrix norm induced by* $|\cdot|_1$
This exercise shows that the absolute sum matrix norm, $\|\cdot\|_{as}$, is not induced by the vector norm $|\mathbf{x}|_1 = \sum_{k=1}^n |x_k|$. Show that the matrix norm induced by the norm $|\mathbf{x}|_1$, according to Theorem 9.5.4, is

$$\|A\|_1 = \max_{1\le j\le n} \sum_{i=1}^n |a_{ij}|,$$

the **maximum absolute column sum** of A.

Exercise 9.5.6. Let $A \in \mathbf{R}^{n\times n}$ and let I be the $n \times n$ identity matrix. Show that if $\|I - A\| < 1$ for some matrix norm, then A is invertible and

$$A^{-1} = \sum_{k=0}^{\infty} (I - A)^k.$$

Exercise 9.5.7. Suppose $A \in \mathbf{R}^{n\times n}$ and A is invertible. Find a number $r > 0$ such that if $B \in \mathbf{R}^{n\times n}$ and $\|A - B\| < r$ for some matrix norm, then B is invertible. (This shows that the set of invertible matrices is an open set in the space $\mathbf{R}^{n\times n}$ of real $n \times n$ matrices.) *Hint*: Write $B = A-(A-B) = A(I-A^{-1}(A-B))$ and then consider the matrix $X := A^{-1}(A - B)$.

Exercise 9.5.8. *Jacobi Iteration*
Let $A = [a_{ij}]$ be $n \times n$ and invertible. Write $A = D - L - U$, where $D = \operatorname{diag}[a_{11}, a_{22}, \dots, a_{nn}]$ is the **diagonal part** of A, $-L$ is the **lower triangular part** of A, and $-U$ is the **upper triangular part** of A. For example, if A is 3×3, then

$$D = \begin{bmatrix} a_{11} & 0 & 0 \\ 0 & a_{22} & 0 \\ 0 & 0 & a_{33} \end{bmatrix}, \quad -L = \begin{bmatrix} 0 & 0 & 0 \\ a_{21} & 0 & 0 \\ a_{31} & a_{32} & 0 \end{bmatrix}, \quad -U = \begin{bmatrix} 0 & a_{12} & a_{13} \\ 0 & 0 & a_{23} \\ 0 & 0 & 0 \end{bmatrix}.$$

Suppose A is **diagonally dominant**, that is,

$$|a_{ii}| > \sum_{j=1, j\ne i}^{n} |a_{ij}|, \quad \text{for } i = 1, \dots, n. \tag{9.12}$$

Thus D is invertible.

1. Show that the equation $A\mathbf{x} = \mathbf{b}$ may be written as $\mathbf{x} = D^{-1}(L+U)\mathbf{x} + D^{-1}\mathbf{b}$.
2. Show that the resulting iteration, $\mathbf{x}_{k+1} = D^{-1}(L + U)\mathbf{x}_k + D^{-1}\mathbf{b}$, converges to the unique solution of $A\mathbf{x} = \mathbf{b}$ for any starting value $\mathbf{x}_0$.

Exercise 9.5.9. *Gauss-Seidel Iteration*
Let $A = [a_{ij}]$ be invertible and 3×3, for simplicity. Write $A = D - L - U$ as in the exercise on Jacobi iteration, and assume that D is invertible.

1. Show that the equation $A\mathbf{x} = \mathbf{b}$ may be written as
$$\mathbf{x} = (D - L)^{-1}U\mathbf{x} + (D - L)^{-1}\mathbf{b}.$$
2. State a condition which guarantees that the resulting iteration,
$$\mathbf{x}_{k+1} = (D - L)^{-1}U\mathbf{x}_k + (D - L)^{-1}\mathbf{b},$$
converges to the unique solution of $A\mathbf{x} = \mathbf{b}$ for any starting value $\mathbf{x}_0$.

3. Show that the Gauss-Seidel iteration in part 2 converges if

$$A = \begin{bmatrix} 4 & -1 & 0 \\ -1 & 4 & -1 \\ 0 & -1 & 4 \end{bmatrix}.$$

Gauss-Seidel iteration is helpful in the sparse linear systems that arise from finite difference approximations to certain partial differential equations. It can be shown that if A is diagonally dominant, that is, (9.12) holds, then the Gauss-Seidel iteration converges.

Exercise 9.5.10. Let X be a Banach space with norm $|\cdot|$, and let $K : X \to X$ be a linear transformation. We say that K is **bounded** if there is a number $M > 0$ such that $|K(x)| \leq M|x|$ for all $x \in X$. Let $\mathcal{B}(X)$ denote the set of bounded linear transformations from X to X.

1. Show that $\mathcal{B}(X)$ is a vector space.
2. Verify that we define a norm on $\mathcal{B}(X)$ by setting
$$\|K\| = \sup_{|x|=1} |K(x)|, \quad K \in \mathcal{B}(X),$$
and show that $\|K_1 K_2\| \leq \|K_1\| \|K_2\|$ when $K_1, K_2 \in \mathcal{B}(X)$.
3. Extend the result of Example 9.5.12 by proving that if $\|K\| < 1$, then $I - K$ is invertible and
$$(I - K)^{-1} = I + K + K^2 + K^3 + \cdots$$
where K^m is the m-fold composition of K with itself, and the series converges absolutely in the norm defined above.
4. Revisit (or visit) Exercise 9.3.4, and show that the unique solution of the Fredholm integral equation (9.8) is given by
$$f = \sum_{m=0}^{\infty} K^m g,$$
where $K : C[a, b] \to C[a, b]$ is given by $(Kh)(x) = \int_a^b k(x, y)h(y)\, dy$ for $h \in C[a, b]$.

9.6. Notes and References

The presentation of the l^p spaces is drawn from Kreyszig [**41**].

For further discussion of contraction theorem applications, see Kreyszig [**41**], which includes a discussion of the Jacobi and Gauss-Seidel iterations of Exercises 9.5.8-9.5.9. Gauss-Seidel iteration is helpful in the sparse linear systems that arise from finite difference approximations to certain partial differential equations; see for example Smith [**60**]. It can be shown that if A is diagonally dominant, that is, (9.12) holds, then the Gauss-Seidel iteration converges. An introductory efficiency comparison of the Jacobi and Gauss-Seidel iterations is available in Strang [**62**].

Friedman [**17**] and Simmons [**59**] cover the Arzelà-Ascoli (or Ascoli-Arzelà) theorem, mentioned at the end of Section 9.1.

Chapter 10

Differentiation in $\mathbf{R}^n$

In this chapter we study differentiation of functions that map one Euclidean space into another Euclidean space. The basic results on limits and continuity for these functions appears in Chapter 8. We denote real valued functions by lowercase letters, and **vector valued** functions by **boldface** letters (either uppercase or lowercase).

10.1. Partial Derivatives

Let D be an open subset of $\mathbf{R}^n$, and $f : D \to \mathbf{R}$ a function on D with values in $\mathbf{R}$. We write $\mathbf{x} = (x_1, x_2, \ldots, x_n)$ for points of $\mathbf{R}^n$ and $f(\mathbf{x}) = f(x_1, x_2, \ldots, x_n)$ for the image point in $\mathbf{R}$. For a function f of two or three real variables we will often write $f(x, y)$ or $f(x, y, z)$.

Definition 10.1.1. *The partial derivative of $f : D \to \mathbf{R}$ with respect to x_j at the point $\mathbf{a} \in D$ is defined by*

$$D_j f(\mathbf{a}) = \lim_{h\to 0} \frac{f(\mathbf{a} + h\mathbf{e}_j) - f(\mathbf{a})}{h},$$

if the limit exists.

The Leibniz notation $\frac{\partial f}{\partial x_j}(\mathbf{a})$ for this defining limit will also be used at times; in particular, for some specific calculations when f is a function of two or three variables and we write $f(x, y)$ or $f(x, y, z)$. In component detail, we have

$$D_j f(\mathbf{a}) = \lim_{h\to 0} \frac{f(a_1, \ldots, a_j + h, \ldots, a_n) - f(a_1, \ldots, a_j, \ldots, a_n))}{h},$$

so this limit is the derivative of the real valued function

$$x_j \to f(a_1, \ldots, x_j, \ldots, a_n)$$

at the point a_j, with a_k, for $k \neq j$, held constant. Thus the usual derivative rules apply when computing partial derivatives. We need only hold x_k constant for $k \neq j$ and differentiate with respect to x_j, then evaluate as required, to obtain $D_j f(\mathbf{a})$.

In contrast to the case for functions of a single variable, the existence of all partial derivatives of f at $\mathbf{a}$,

$$D_1 f(\mathbf{a}), \quad D_2 f(\mathbf{a}), \quad \ldots, \quad D_{n-1} f(\mathbf{a}), \quad D_n f(\mathbf{a}),$$

does not guarantee that f is continuous at $\mathbf{a}$.

Example 10.1.2. Let $f : \mathbf{R}^2 \to \mathbf{R}$ be the function defined by

$$f(x, y) = \begin{cases} 1 & \text{if } xy = 0, \\ 0 & \text{if } xy \neq 0. \end{cases}$$

At $(a_1, a_2) = (0, 0)$, the definition of partial derivative yields the results that

$$\frac{\partial f}{\partial x}(0, 0) = 0 \quad \text{and} \quad \frac{\partial f}{\partial y}(0, 0) = 0.$$

Thus f has partial derivatives at $(a_1, a_2) = (0, 0)$. However, f is not continuous at $(0, 0)$. Observe that the sequence $(1/n, 1/n)$ converges to $(0, 0)$, with $f(1/n, 1/n) = 0 \to 0 \neq 1 = f(0, 0)$. Also notice that $\partial f/\partial y$ does not exist on the x-axis, except at $(0, 0)$, and $\partial f/\partial x$ does not exist on the y-axis, except at $(0, 0)$. △

See also Exercise 10.1.1, where the function has partial derivatives at every point of $\mathbf{R}^2$, the partial derivatives are continuous at every point except the origin, and the function fails to be continuous at the origin.

Recall that if a real function of a single variable is differentiable at a (that is, if $f'(a)$ exists), then f is continuous at a. For the functions in Example 10.1.2 and Exercise 10.1.1, the existence of partial derivatives at a point $\mathbf{a}$, and even the existence of partial derivatives throughout a neighborhood of $\mathbf{a}$, is seen to be *insufficient* to guarantee continuity of the function at $\mathbf{a}$. The way to understand this phenomenon in those examples is to realize that the function graph did not have a well-defined tangent plane approximation at the point $(\mathbf{0}, f(\mathbf{0}))$. The existence of a tangent plane approximation at $(\mathbf{0}, f(\mathbf{0}))$ would imply continuity there. We will see this more precisely when we have defined *differentiability* for multivariable functions in the appropriate way. For the moment, the key idea is to realize that *differentiability* of f at $\mathbf{a}$ should mean a well-defined linear approximation to the graph of f at the point $(\mathbf{a}, f(\mathbf{a}))$. Under that definition, differentiability at $\mathbf{a}$ does imply continuity at $\mathbf{a}$. What conditions are sufficient for the graph of f to have a well-defined linear approximation at $(\mathbf{a}, f(\mathbf{a}))$? Intuitively, it appears that continuity of the partial derivatives at all points in some neighborhood of $\mathbf{a}$ should be sufficient, since this guarantees a smoothly varying surface as the graph of f for $\mathbf{x}$ near $\mathbf{a}$. This intuition can be justified with the definition of *differentiability* and its consequences.

The definition of second-order (and higher-order) partial derivatives of a real function f is covered by inductive application of Definition 10.1.1. For example, given the existence of the real function $D_j f(\mathbf{x})$ for all $\mathbf{x}$ in some ball about $\mathbf{a}$, the second-order partial derivative of f indicated by $D_i D_j f(\mathbf{a})$ is defined by

$$D_i D_j f(\mathbf{a}) = \lim_{h \to 0} \frac{D_j f(\mathbf{a} + h\mathbf{e}_i) - D_j f(\mathbf{a})}{h},$$

when the limit exists. This is the partial derivative of $D_j f(\mathbf{x})$ with respect to the variable x_i at $\mathbf{a}$.

Some comments are in order on notations for partial derivatives. Readers may already be familiar with another subscript notation for partial derivatives, and recognize that if f is a function of $\mathbf{x} = (x_1, x_2, \dots, x_n) \in \mathbf{R}^n$, then

$$D_j f(\mathbf{a}) = \frac{\partial f}{\partial x_j}(\mathbf{a}) = f_{x_j}(\mathbf{a}).$$

Without the evaluation notation at some point $\mathbf{a}$, the notations $D_j f$, $\frac{\partial f}{\partial x_j}$ and f_{x_j} are usually taken to mean the derivative function itself rather than the evaluation of the derivative function at any particular point. Care is always needed with partial derivative notation. Recall that the ordering of subscripts in the subscript notation for derivatives is the opposite of the ordering of the indicated derivatives in the D or ∂ notation. For example, the second-order partial derivative $D_i D_j f(\mathbf{a})$ can be indicated in any of these ways:

$$D_i D_j f(\mathbf{a}) = \frac{\partial^2 f}{\partial x_i \partial x_j}(\mathbf{a}) = f_{x_j x_i}(\mathbf{a}).$$

Care is needed when using parentheses.[1] The variety of partial derivative notations proves convenient in various circumstances, as clarity is sometimes enhanced by using one notation rather than another, and clarity takes priority.

We now turn to proving the equality of mixed partial derivatives such as f_{xy} and f_{yx}. The essential issue is a two-dimensional phenomenon, so we begin with the following lemma.

Lemma 10.1.3. *Let $f(x, y)$ be defined on $D \subset \mathbf{R}^2$ and let (a, b) be an interior point of D. If the partial derivatives $f_x(x, y)$, $f_y(x, y)$ and $f_{yx}(x, y)$ exist in the open ball $B_\delta(a, b)$ for some $\delta > 0$ and if f_{yx} is continuous at (a, b), then $f_{xy}(a, b)$ exists and*

$$f_{xy}(a, b) = f_{yx}(a, b).$$

Proof. We want to show that $f_{xy}(a, b)$ exists and $f_{xy}(a, b) = f_{yx}(a, b)$. By defining

$$\Delta f(x, y) = f(x + h, y) - f(x, y),$$

we may write

$$f_x(x, y) = \lim_{h \to 0} \frac{\Delta f(x, y)}{h}.$$

Thus we want to show that

$$\begin{aligned} f_{xy}(a, b) &= \lim_{k \to 0} \frac{f_x(a, b + k) - f_x(a, b)}{k} \\ &= \lim_{k \to 0} \frac{1}{k} \Big(\lim_{h \to 0} \frac{\Delta f(a, b + k) - \Delta f(a, b)}{h} \Big) \end{aligned} \tag{10.1}$$

exists and equals $f_{yx}(a, b)$. It may be helpful to visualize certain domain points in the argument that follows. (See Figure 10.1.)

By hypothesis we may differentiate with respect to y to see that

$$(\Delta f)_y(x, y) = f_y(x + h, y) - f_y(x, y) \tag{10.2}$$

[1]For example, the expression $\frac{\partial}{\partial x_i}\big(\frac{\partial f}{\partial x_j}(\mathbf{a})\big)$ has the value zero, but might not have been intended; if the intention is to indicate a second-order derivative at the point $\mathbf{a}$, then $\frac{\partial}{\partial x_i}\big(\frac{\partial f}{\partial x_j}\big)(\mathbf{a})$ is fine, and $\big(\frac{\partial}{\partial x_i}\big(\frac{\partial f}{\partial x_j}\big)\big)(\mathbf{a}) = \big(\frac{\partial}{\partial x_i}\frac{\partial f}{\partial x_j}\big)(\mathbf{a}) = \frac{\partial^2 f}{\partial x_i \partial x_j}(\mathbf{a})$ are also clear.

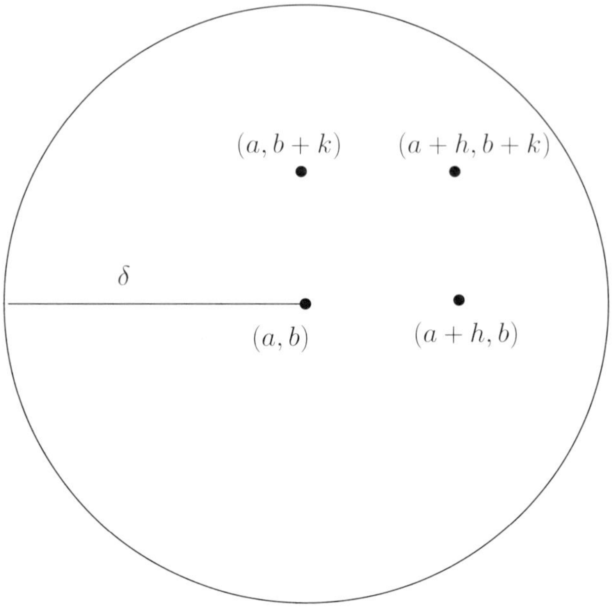

Figure 10.1. Proving the equality of mixed partial derivatives: Points near (a, b) in the proof of Lemma 10.1.3.

exists for all points (x, y) on the line segment joining (a, b) and $(a, b+k)$ if $|k| < \delta$. By the mean value theorem (Theorem 5.2.4) there is a θ_1 depending on h, k with $0 < \theta_1 < 1$ such that

$$\Delta f(a, b+k) - \Delta f(a, b) = (\Delta f)_y(a, b+\theta_1 k)\, k.$$

By (10.1), we have

$$f_{xy}(a, b) = \lim_{k\to 0}\Big(\lim_{h\to 0}\frac{(\Delta f)_y(a, b+\theta_1 k)}{h}\Big). \tag{10.3}$$

Now by (10.2),

$$(\Delta f)_y(a, b+\theta_1 k) = f_y(a+h, b+\theta_1 k) - f_y(a, b+\theta_1 k).$$

By choice of h and k, we have $(a, b+\theta_1 k) \in B_\delta(a, b)$, and $f_{yx}(x, y)$ exists for (x, y) in $B_\delta(a, b)$, hence

$$\begin{aligned}\lim_{h\to 0}\frac{1}{h}[(\Delta f)_y(a, b+\theta_1 k)] &= \lim_{h\to 0}\frac{1}{h}[f_y(a+h, b+\theta_1 k) - f_y(a, b+\theta_1 k)]\\ &= f_{yx}(a, b+\theta_1 k).\end{aligned}$$

Now (10.3) and continuity of f_{yx} at (a, b) imply that

$$f_{xy}(a, b) = \lim_{k\to 0} f_{yx}(a, b+\theta_1 k) = f_{yx}(a, b),$$

and this completes the proof. $\square$

The hypothesis of continuity of f_{yx} at $\mathbf{a}$ was used in the last step of the proof of Lemma 10.1.3. See Exercise 10.1.2 for an example where this continuity hypothesis does not hold, and neither does the conclusion of the lemma.

If we changed the hypotheses in the lemma to read "if $f_x(x,y)$, $f_y(x,y)$ and $f_{xy}(x,y)$ exist in the open ball $B_\delta(a,b)$ for some $\delta > 0$ and if f_{xy} is continuous at (a,b)", then a similar proof allows us to conclude that $f_{yx}(a,b)$ exists and $f_{yx}(a,b) = f_{xy}(a,b)$.

The general result on equality of mixed partials for real functions now follows.

Theorem 10.1.4. *Let $f : D \subseteq \mathbf{R}^n \to \mathbf{R}$ and let $\mathbf{a}$ be an interior point of D. If the partial derivatives $D_i f(\mathbf{x})$, $D_j f(\mathbf{x})$ and $D_i D_j f(\mathbf{x})$ exist in an open ball $B_\delta(\mathbf{a})$ for some $\delta > 0$ and if $D_i D_j f$ is continuous at $\mathbf{a}$, then $D_j D_i f(\mathbf{a})$ exists and*

$$D_j D_i f(\mathbf{a}) = D_i D_j f(\mathbf{a}).$$

Proof. We may assume without loss of generality that $i < j$. Apply Lemma 10.1.3 to the function $\phi(x,y)$ defined by

$$\phi(x,y) = f(a_1, \ldots, a_{i-1}, x, a_{i+1}, \ldots, a_{j-1}, y, a_{j+1}, \ldots, a_n),$$

to conclude that $\phi_{xy}(a,b) = \phi_{yx}(a,b)$; that is, $D_j D_i f(\mathbf{a}) = D_i D_j f(\mathbf{a})$. □

Here it is convenient to introduce some terminology for orders of continuous differentiability.

Definition 10.1.5. *The mapping $\mathbf{F} : U \subseteq \mathbf{R}^n \to \mathbf{R}^m$ is of* **class C^1** *(or simply,* **is C^1***) at an interior point $\mathbf{a} \in U$ if the partial derivatives of each component function of $\mathbf{F}$ exist throughout some open ball containing $\mathbf{a}$ and are continuous at $\mathbf{a}$.*

Definition 10.1.6. *Let U be an open subset of $\mathbf{R}^n$. The mapping $\mathbf{F} : U \subset \mathbf{R}^n \to \mathbf{R}^m$ is said to be*

1. **of class C^1 on** *U, or more simply,* **$\mathbf{F}$ is C^1 on** *U, if the first-order partial derivatives of each component function of $\mathbf{F}$ exist and are continuous on U;*
2. **of class C^k on** *U, or* **is C^k on** *U, if all the partial derivatives of each component function of $\mathbf{F}$ through order k exist and are continuous on U.*

With Theorem 10.1.4 in mind, observe that if the derivatives $D_i f(\mathbf{x})$, $D_j f(\mathbf{x})$ and $D_i D_j f(\mathbf{x})$ exist and are continuous throughout some open ball $B_\delta(\mathbf{a})$, then the hypotheses of the lemma are satisfied at each point of that ball, and therefore

$$D_j D_i f(\mathbf{x}) = D_i D_j f(\mathbf{x}) \quad \text{for all } \mathbf{x} \in B_\delta(\mathbf{a}).$$

Therefore, in view of Definition 10.1.6, we can say that if f is of class C^2 on U, then the order of differentiation in mixed second-order partial derivatives does not matter.

From introductory multivariable calculus, the reader has some experience with the differential operators of vector analysis.

If $f : D \subset \mathbf{R}^n \to \mathbf{R}$ has partial derivatives at the point $\mathbf{x} \in D$, then we write

$$\operatorname{grad} f(\mathbf{x}) = \Big(\frac{\partial f}{\partial x_1}(\mathbf{x}), \ldots, \frac{\partial f}{\partial x_n}(\mathbf{x})\Big)$$

and call $\operatorname{grad} f(\mathbf{x})$ the **gradient** of f at $\mathbf{x}$.

A mapping $\mathbf{F} : D \subseteq \mathbf{R}^n \to \mathbf{R}^n$ defines a **vector field** on D. For $n = 3$, we get vector fields in space. For a vector field in space, write

$$\mathbf{F}(\mathbf{x}) = (F_1(\mathbf{x}), F_2(\mathbf{x}), F_3(\mathbf{x})).$$

If the indicated partial derivatives exist, then

$$\operatorname{div} \mathbf{F}(\mathbf{x}) = \frac{\partial F_1}{\partial x_1}(\mathbf{x}) + \frac{\partial F_2}{\partial x_2}(\mathbf{x}) + \frac{\partial F_3}{\partial x_3}(\mathbf{x})$$

is called the **divergence** of $\mathbf{F}$ at $\mathbf{x}$. For a vector field $\mathbf{F}$ on $D \subset \mathbf{R}^n$, the divergence of $\mathbf{F}$ at $\mathbf{x}$ is

$$\operatorname{div} \mathbf{F}(\mathbf{x}) = \frac{\partial F_1}{\partial x_1}(\mathbf{x}) + \cdots + \frac{\partial F_n}{\partial x_n}(\mathbf{x}).$$

If $\mathbf{F} : D \subset \mathbf{R}^3 \to \mathbf{R}^3$, written $\mathbf{F}(\mathbf{x}) = (M(\mathbf{x}), N(\mathbf{x}), P(\mathbf{x}))$ with $\mathbf{x} = (x, y, z)$, and the partial derivatives exist, then the vector

$$\operatorname{curl} \mathbf{F}(\mathbf{x}) = \Big(\frac{\partial P}{\partial y}(\mathbf{x}) - \frac{\partial N}{\partial z}(\mathbf{x}), \frac{\partial M}{\partial z}(\mathbf{x}) - \frac{\partial P}{\partial x}(\mathbf{x}), \frac{\partial N}{\partial x}(\mathbf{x}) - \frac{\partial M}{\partial y}(\mathbf{x})\Big)$$

is called the **curl** of $\mathbf{F}$ at $\mathbf{x}$.

Now consider functions f and $\mathbf{F}$ defined in space with domain $D \subset \mathbf{R}^3$ and points $\mathbf{x} \in D$ written $\mathbf{x} = (x, y, z)$. There is some classical notation associated with these constructions using the differential operator $\nabla = (\frac{\partial}{\partial x}, \frac{\partial}{\partial y}, \frac{\partial}{\partial z})$ treated as a vector. Thus,

$$\operatorname{grad} f(\mathbf{x}) = \nabla f(\mathbf{x}) = \Big[\Big(\frac{\partial}{\partial x}, \frac{\partial}{\partial y}, \frac{\partial}{\partial z}\Big) f\Big](x) = \Big(\frac{\partial f}{\partial x}(\mathbf{x}), \frac{\partial f}{\partial y}(\mathbf{x}), \frac{\partial f}{\partial z}(\mathbf{x})\Big).$$

If $\mathbf{F}(\mathbf{x}) = (M(\mathbf{x}), N(\mathbf{x}), P(\mathbf{x}))$, then

$$\operatorname{curl} \mathbf{F}(\mathbf{x}) = \nabla \times \mathbf{F}(\mathbf{x}) = \begin{bmatrix} \mathbf{e}_1 & \mathbf{e}_2 & \mathbf{e}_3 \\ \frac{\partial}{\partial x} & \frac{\partial}{\partial y} & \frac{\partial}{\partial z} \\ M(\mathbf{x}) & N(\mathbf{x}) & P(\mathbf{x}) \end{bmatrix}$$

yields $\operatorname{curl} \mathbf{F}(\mathbf{x})$ by a symbolic cross product-determinant calculation. Also,

$$\operatorname{div} \mathbf{F}(\mathbf{x}) = \nabla \cdot \mathbf{F}(\mathbf{x}) = \Big(\frac{\partial}{\partial x}, \frac{\partial}{\partial y}, \frac{\partial}{\partial z}\Big) \cdot (M, N, P) = \frac{\partial M}{\partial x}(\mathbf{x}) + \frac{\partial N}{\partial y}(\mathbf{x}) + \frac{\partial P}{\partial z}(\mathbf{x})$$

yields $\operatorname{div} \mathbf{F}(\mathbf{x})$ as a symbolic dot product calculation.

Theorem 10.1.7. *Suppose $f : U \subset \mathbf{R}^3 \to \mathbf{R}$ is a C^2 function on U and $\mathbf{F} : U \subset \mathbf{R}^3 \to \mathbf{R}^3$ is a C^2 vector field on U. Then the following identities hold:*

1. $\operatorname{curl} \operatorname{grad} f(\mathbf{x}) = \nabla \times (\nabla f)(\mathbf{x}) = \mathbf{0}$ *for all* $\mathbf{x} \in U$*;*
2. $\operatorname{div} \operatorname{curl} \mathbf{F}(\mathbf{x}) = \nabla \cdot (\nabla \times \mathbf{F})(\mathbf{x}) = \mathbf{0}$ *for all* $\mathbf{x} \in U$*.*

Proof. In each case the proof is a direct consequence of the equality of mixed partial derivatives, and the verification is left to Exercise 10.1.4. □

The important construction, div grad $f(\mathbf{x})$, for a function $f : D \subset \mathbf{R}^3 \to \mathbf{R}$, may be written

$$\begin{aligned} \operatorname{div} \operatorname{grad} f(\mathbf{x}) &= \nabla \cdot \nabla f(\mathbf{x}) = \left(\frac{\partial}{\partial x}, \frac{\partial}{\partial y}, \frac{\partial}{\partial z}\right) \cdot \left(\frac{\partial f}{\partial x}, \frac{\partial f}{\partial y}, \frac{\partial f}{\partial z}\right) \\ &= \frac{\partial^2 f}{\partial x^2}(\mathbf{x}) + \frac{\partial^2 f}{\partial y^2}(\mathbf{x}) + \frac{\partial^2 f}{\partial z^2}(\mathbf{x}). \end{aligned}$$

This construction is often written

$$\nabla^2 f(\mathbf{x}) = \operatorname{div} \operatorname{grad} f(\mathbf{x}) = \frac{\partial^2 f}{\partial x^2}(\mathbf{x}) + \frac{\partial^2 f}{\partial y^2}(\mathbf{x}) + \frac{\partial^2 f}{\partial z^2}(\mathbf{x}).$$

The operator $\nabla^2 = \operatorname{div} \operatorname{grad}$ is called the **Laplacian operator**, and $\nabla^2 f(\mathbf{x})$ is the **Laplacian** of f. The Laplacian is also frequently denoted by the simpler notation

$$\Delta f(\mathbf{x}) = \frac{\partial^2 f}{\partial x^2}(\mathbf{x}) + \frac{\partial^2 f}{\partial y^2}(\mathbf{x}) + \frac{\partial^2 f}{\partial z^2}(\mathbf{x}).$$

The symbol Δ (called simply "del") denotes the Laplacian operator. The partial differential equation

$$\Delta u(\mathbf{x}) = \operatorname{div} \operatorname{grad} u(\mathbf{x}) = \frac{\partial^2 u}{\partial x^2}(\mathbf{x}) + \frac{\partial^2 u}{\partial y^2}(\mathbf{x}) + \frac{\partial^2 u}{\partial z^2}(\mathbf{x}) = 0$$

is **Laplace's equation** for the unknown $u(x, y, z)$, and it plays an important role in most of the fundamental differential equations of mathematical physics. In two space dimensions, for functions $u : U \subset \mathbf{R}^2 \to \mathbf{R}$, written as $u(x, y)$, Laplace's equation reads

$$\frac{\partial^2 u}{\partial x^2}(\mathbf{x}) + \frac{\partial^2 u}{\partial y^2}(\mathbf{x}) = 0,$$

and its C^2 solutions can be viewed as the building blocks of the theory of functions of a complex variable.[2] Laplace's equation in n dimensions reads

$$\frac{\partial^2 u}{\partial x_1^2}(\mathbf{x}) + \frac{\partial^2 u}{\partial x_2^2}(\mathbf{x}) + \cdots + \frac{\partial^2 u}{\partial x_n^2}(\mathbf{x}) = 0.$$

In all cases, the C^2 solutions of Laplace's equation are called **harmonic functions**.

Exercises.

Exercise 10.1.1. Let $f : \mathbf{R}^2 \to \mathbf{R}$ be the function defined by

$$f(x, y) = \begin{cases} xy/(x^2 + y^2) & \text{if } (x, y) \neq (0, 0), \\ 0 & \text{if } (x, y) = (0, 0). \end{cases}$$

1. Show that for $(x, y) \neq (0, 0)$, the partial derivatives are computed by the usual differentiation rules to obtain

$$f_x(x, y) = \frac{y^3 - yx^2}{(x^2 + y^2)^2} \quad \text{and} \quad f_y(x, y) = \frac{x^3 - xy^2}{(x^2 + y^2)^2}.$$

2. Show that at $(x, y) = (0, 0)$, the definition of partial derivative yields $f_x(0, 0) = 0$ and $f_y(0, 0) = 0$, so f has partial derivatives at every point.

[2] Both the real and imaginary parts of a differentiable function of a complex variable, $f : \mathbf{C} \to \mathbf{C}$, are harmonic functions.

3. Show that f is not continuous at $(0,0)$.
4. Show that the partial derivatives $\partial f/\partial x$, $\partial f/\partial y$ are discontinuous at $(0,0)$.

Exercise 10.1.2. Let $\tan^{-1}$ be the inverse tangent function taking values in the interval $(-\pi/2, \pi/2)$, and let $f : \mathbf{R}^2 \to \mathbf{R}$ be

$$f(x,y) = \begin{cases} x^2 \tan^{-1}(y/x) - y^2 \tan^{-1}(x/y) & \text{if } xy \neq 0, \\ 0 & \text{if } xy = 0. \end{cases}$$

Show that $D_1 D_2 f(0,0) \neq D_2 D_1 f(0,0)$.

Exercise 10.1.3. Find the partial derivative $D_1 D_2 f = f_{x_2 x_1}$ for the function

$$f(\mathbf{x}) = x_1 x_2 + \sqrt{1 + x_2^2}\, \cos\left(x_2 / \sqrt{1 + x_2^2}\right)$$

using the least amount of computation, and justify your computation.

Exercise 10.1.4. Prove Theorem 10.1.7.

Exercise 10.1.5. Show that each of these functions is harmonic:

(*a*) $u(x,y) = \cosh x \sin y$; (*b*) $v(x,y) = \sinh x \cos y$; (*c*) $w(x,y) = e^y \sin x$.

Exercise 10.1.6. Show that $\Phi(\mathbf{x}) = \Phi(x,y,z) = (x^2 + y^2 + z^2)^{-1/2}$ satisfies Laplace's equation for $\mathbf{x} \neq \mathbf{0}$.

Exercise 10.1.7. *(This exercise presents preliminary versions of a general mean value theorem proved later using the chain rule. We do not use the chain rule here, as it has not yet been introduced.)*

1. Suppose that $f(x_1, x_2)$ has first-order partial derivatives at every point of an open ball $B_r(\mathbf{a}) = \{\mathbf{x} : |\mathbf{x} - \mathbf{a}|_2 < r\}$ in the plane. Let h be a fixed real number (positive or negative) with $|h| < r$, so that for $j = 1$ and $j = 2$, the line segment joining $\mathbf{a}$ and $\mathbf{a} + h\mathbf{e}_j$ lies in $B_r(\mathbf{a})$. Prove that there are numbers θ_1 and θ_2 with $0 < \theta_j < 1$ for $j = 1, 2$, such that
$$f(\mathbf{a} + h\mathbf{e}_j) - f(\mathbf{a}) = \frac{\partial f}{\partial x_j}(\mathbf{a} + \theta_j h \mathbf{e}_j)\, h, \quad \text{for } j = 1, 2.$$
Hint: Apply the single variable mean value theorem to $\phi_j(t) = f(\mathbf{a} + t\mathbf{e}_j)$ where I is an interval of real numbers containing both 0 and h. Use only the *definition* of $\phi_j'(t)$ to find that $\phi_j'(t) = \frac{\partial f}{\partial x_j}(\mathbf{a} + t\mathbf{e}_j)$. You should find the argument is the same for $j = 1, 2$.

2. Extend the result of part 1 by going from $\mathbf{a}$ to $\mathbf{a} + \mathbf{h}$ for *any* increment $\mathbf{h}$. That is, suppose $f(x_1, x_2)$ has first-order partial derivatives at every point of the open ball $B_r(\mathbf{a}) = \{\mathbf{x} : |\mathbf{x} - \mathbf{a}|_2 < r\}$ in the plane. Let $|\mathbf{h}|_2 < r$. Prove that there exist points $\mathbf{c}_1, \mathbf{c}_2$ in $B_r(\mathbf{a})$ with $|\mathbf{a} - \mathbf{c}_j|_2 < |\mathbf{h}|_2$ for $j = 1, 2$, such that
$$f(\mathbf{a} + \mathbf{h}) - f(\mathbf{a}) = \frac{\partial f}{\partial x_1}(\mathbf{c}_1)\, h_1 + \frac{\partial f}{\partial x_2}(\mathbf{c}_2)\, h_2.$$
Hint: Write the difference $f(\mathbf{a} + \mathbf{h}) - f(\mathbf{a})$ as follows:
$$\begin{aligned} f(\mathbf{a} + \mathbf{h}) - f(\mathbf{a}) &= f(a_1 + h_1, a_2 + h_2) - f(a_1, a_2) \\ &= f(a_1 + h_1, a_2 + h_2) - f(a_1 + h_1, a_2) \\ &\quad + f(a_1 + h_1, a_2) - f(a_1, a_2). \end{aligned}$$

View the first difference as involving an increment in the second variable only, and the second difference as involving an increment in the first variable only. Apply part 1 to each of these differences, noting that $B_r(\mathbf{x})$ is convex.

10.2. Differentiability: Real Functions and Vector Functions

In order to define differentiability of a function at an interior point of its domain, we take our cue from Definition 5.1.8 in the single variable case. We may use any specific norm for vectors in $\mathbf{R}^n$ and $\mathbf{R}^m$, so we do not subscript the norms here. We indicate vector norms by single bars and norms of matrices or linear transformations by double bars.

Definition 10.2.1. *Let* $\mathbf{F} : U \to \mathbf{R}^m$, $U \subseteq \mathbf{R}^n$, *and let* $\mathbf{a}$ *be an interior point of* U. *Then* $\mathbf{F}$ *is* **differentiable** *at* $\mathbf{a}$ *if there is a linear mapping* $\mathbf{T} : \mathbf{R}^n \to \mathbf{R}^m$ *such that*

$$\lim_{|\mathbf{h}|\to 0} \frac{|\mathbf{F}(\mathbf{a}+\mathbf{h}) - \mathbf{F}(\mathbf{a}) - \mathbf{T}\mathbf{h}|}{|\mathbf{h}|} = 0.$$

The linear mapping $\mathbf{T}$ *is called the* **derivative** *of* $\mathbf{F}$ *at* $\mathbf{a}$, *which we write henceforward as* $D\mathbf{F}(\mathbf{a})$.

Write $\mathbf{x} = \mathbf{a} + \mathbf{h}$. By the definition of limit we have that $\mathbf{F}$ is differentiable at $\mathbf{a}$ if for every $\epsilon > 0$ there is a $\delta > 0$ such that if $|\mathbf{x} - \mathbf{a}| < \delta$, then

$$\frac{|\mathbf{F}(\mathbf{x}) - \mathbf{F}(\mathbf{a}) - D\mathbf{F}(\mathbf{a})(\mathbf{x}-\mathbf{a})|}{|\mathbf{x}-\mathbf{a}|} < \epsilon,$$

that is,

$$|\mathbf{F}(\mathbf{x}) - \mathbf{F}(\mathbf{a}) - D\mathbf{F}(\mathbf{a})(\mathbf{x}-\mathbf{a})| < \epsilon|\mathbf{x}-\mathbf{a}|.$$

This limit statement expresses a precise sense in which the tangent estimate,

$$\mathbf{F}(\mathbf{x}) \approx \mathbf{F}(\mathbf{a}) + D\mathbf{F}(\mathbf{a})(\mathbf{x}-\mathbf{a}) \qquad (\mathbf{x} \text{ near } \mathbf{a}),$$

or, alternatively,

$$\mathbf{F}(\mathbf{a}+\mathbf{h}) - \mathbf{F}(\mathbf{a}) \approx D\mathbf{F}(\mathbf{a})\mathbf{h} \qquad (\mathbf{h} \text{ near } \mathbf{0}),$$

is valid. The next theorem gives assurance that the derivative of $\mathbf{F}$ at $\mathbf{a}$, if such a linear mapping exists, is uniquely determined by Definition 10.2.1.

Theorem 10.2.2. *If* $\mathbf{F} : U \to \mathbf{R}^m$ *is differentiable at the interior point* $\mathbf{a} \in U$, *then the derivative* $D\mathbf{F}(\mathbf{a})$ *is uniquely determined.*

Proof. Suppose $\mathbf{T}_1$ and $\mathbf{T}_2$ are linear mappings from $\mathbf{R}^n$ to $\mathbf{R}^m$ that satisfy Definition 10.2.1. Then for every $\epsilon > 0$ there is a $\delta > 0$ such that if $|\mathbf{h}| < \delta$, then

$$|\mathbf{F}(\mathbf{a}+\mathbf{h}) - \mathbf{F}(\mathbf{a}) - \mathbf{T}_1\mathbf{h}| < \frac{\epsilon}{2}|\mathbf{h}|$$

and

$$|\mathbf{F}(\mathbf{a}+\mathbf{h}) - \mathbf{F}(\mathbf{a}) - \mathbf{T}_2\mathbf{h}| < \frac{\epsilon}{2}|\mathbf{h}|.$$

By the triangle inequality, for all $|\mathbf{h}| < \delta$ we have

$$|(\mathbf{T}_1 - \mathbf{T}_2)\mathbf{h}| = |\mathbf{T}_1\mathbf{h} - \mathbf{T}_2\mathbf{h}| < \epsilon|\mathbf{h}|.$$

But this implies that for all $\mathbf{u}$ with $|\mathbf{u}| = 1$, we have $|(\mathbf{T}_1 - \mathbf{T}_2)\mathbf{u}| < \epsilon$, and therefore $\|\mathbf{T}_1 - \mathbf{T}_2\| < \epsilon$. Since $\epsilon > 0$ was arbitrary, $\mathbf{T}_1 = \mathbf{T}_2$. $\square$

Example 10.2.3. Using the uniqueness result of Theorem 10.2.2, one verifies easily from Definition 10.2.1 that the function $f(x_1, x_2) = x_1^2 + x_2^2$ on $\mathbf{R}^2$ has derivative at $(0, 0)$ equal to the zero linear transformation, that is,

$$Df(0, 0)(h_1, h_2) = 0$$

for all (h_1, h_2). △

Example 10.2.4. A differentiable curve $\mathbf{r} : J \to \mathbf{R}^n$ from an open interval J into $\mathbf{R}^n$, given by $\mathbf{r}(t) = (x_1(t), \ldots, x_n(t))$, has derivative at $t_0 \in J$ given by the linear mapping $D\mathbf{r}(t_0)h = h\mathbf{r}'(t_0)$, where

$$\mathbf{r}'(t_0) = (x_1'(t_0), \ldots, x_n'(t_0)).$$

See Exercise 10.2.1. Recall that the tangent line to the curve at the point $\mathbf{r}(t_0)$ has equation $\mathbf{R}(h) = \mathbf{r}(t_0) + h\mathbf{r}'(t_0)$, as $\mathbf{r}'(t_0)$ is a direction vector for the line. △

If $\mathbf{a}$ is not an interior point of the domain of $\mathbf{F}$, then, even if there is a linear mapping $\mathbf{T}$ satisfying Definition 10.2.1, there may not be a unique such linear mapping, so the derivative of $\mathbf{F}$ at $\mathbf{a}$ may not be uniquely or well defined in that situation.

Theorem 10.2.5. *If $\mathbf{F} : U \subset \mathbf{R}^n \to \mathbf{R}^m$ is differentiable at an interior point $\mathbf{a} \in U$, then $\mathbf{F}$ is continuous at $\mathbf{a}$.*

Proof. The existence of the derivative at $\mathbf{a}$ implies that for every $\epsilon > 0$ there is a $\delta > 0$ such that if $|\mathbf{x} - \mathbf{a}| < \delta$, then

$$|\mathbf{F}(\mathbf{x}) - \mathbf{F}(\mathbf{a}) - D\mathbf{F}(\mathbf{a})(\mathbf{x} - \mathbf{a})| < \epsilon|\mathbf{x} - \mathbf{a}|.$$

Letting $\epsilon = 1$, there is a $\delta_1 > 0$ such that $|\mathbf{x} - \mathbf{a}| < \delta_1$ implies that

$$|\mathbf{F}(\mathbf{x}) - \mathbf{F}(\mathbf{a}) - D\mathbf{F}(\mathbf{a})(\mathbf{x} - \mathbf{a})| < |\mathbf{x} - \mathbf{a}|,$$

and hence, by a reverse triangle inequality argument, that

$$|\mathbf{F}(\mathbf{x}) - \mathbf{F}(\mathbf{a})| < |D\mathbf{F}(\mathbf{a})(\mathbf{x} - \mathbf{a})| + |\mathbf{x} - \mathbf{a}| \le (\|D\mathbf{F}(\mathbf{a})\| + 1)\,|\mathbf{x} - \mathbf{a}|.$$

Now let $\mathbf{x} \to \mathbf{a}$ and use the fact that a linear transformation from $\mathbf{R}^n$ to $\mathbf{R}^m$ is continuous at $\mathbf{a}$, to conclude that $\lim_{\mathbf{x} \to \mathbf{a}} \mathbf{F}(\mathbf{x}) = \mathbf{F}(\mathbf{a})$. □

Exercises.

Exercise 10.2.1. Carry out the required limit calculation from Definition 10.2.1 to establish the derivatives asserted in Example 10.2.3 and Example 10.2.4.

Exercise 10.2.2. Let $f : \mathbf{R}^2 \to \mathbf{R}^2$ be

$$f(x_1, x_2) = \begin{cases} x_1 + x_2 & \text{if } x_1 x_2 = 0, \\ 1 & \text{if } x_1 x_2 \neq 0. \end{cases}$$

Show that $D_1 f(0, 0) = 1 = D_2 f(0, 0)$, but that f is not continuous at $(0, 0)$. Is f differentiable at $(0, 0)$?

Exercise 10.2.3. Let $\mathbf{F} : U \to \mathbf{R}^n$ be the identity mapping on $U \subseteq \mathbf{R}^n$. Show that at each interior point $\mathbf{a} \in U$, $D\mathbf{F}(\mathbf{a})$ is the identity linear mapping.

Exercise 10.2.4. Prove: If $\mathbf{F} : \mathbf{R}^n \to \mathbf{R}^m$ is a linear mapping, then at each point $\mathbf{a} \in \mathbf{R}^n$, $D\mathbf{F}(\mathbf{a}) = \mathbf{F}$.

Exercise 10.2.5. *C^k but not C^{k+1}*
Consider functions $f : \mathbf{R} \to \mathbf{R}$.

1. Give an example of a function f that is C^1 but not C^2. *Hint*: Start with $g(x) = |x|$ and integrate.
2. Give an example of a function f that is C^k but not C^{k+1}.

10.3. Matrix Representation of the Derivative

When we defined partial derivatives of a real function $f : U \subseteq \mathbf{R}^n \to \mathbf{R}$, we assumed that points (vectors) in the domain were expressed relative to the standard basis of $\mathbf{R}^n$ given by $\{\mathbf{e}_1, \ldots, \mathbf{e}_n\}$. On the other hand, the definition of the derivative made no reference to any basis of the space of vectors.

Theorem 10.3.1. *If U is open in $\mathbf{R}^n$, $\mathbf{a} \in U$, and $\mathbf{F} : U \to \mathbf{R}^m$ is differentiable at $\mathbf{a}$, then the linear transformation $D\mathbf{F}(\mathbf{a})$ is represented in the standard bases of $\mathbf{R}^n$ and $\mathbf{R}^m$ by the* **Jacobian matrix**

$$A = J_{\mathbf{F}}(\mathbf{a}) = \begin{bmatrix} D_1 f_1(\mathbf{a}) & \cdots & D_n f_1(\mathbf{a}) \\ \vdots & \vdots & \vdots \\ D_1 f_m(\mathbf{a}) & \cdots & D_n f_m(\mathbf{a}) \end{bmatrix} = \begin{bmatrix} \frac{\partial f_1}{\partial x_1}(\mathbf{a}) & \cdots & \frac{\partial f_1}{\partial x_n}(\mathbf{a}) \\ \vdots & \vdots & \vdots \\ \frac{\partial f_m}{\partial x_1}(\mathbf{a}) & \cdots & \frac{\partial f_m}{\partial x_n}(\mathbf{a}) \end{bmatrix}$$

where $\mathbf{F} = (f_1, \ldots, f_m)$ is the component expression of $\mathbf{F}$.

Proof. We must show that if $\mathbf{F} = (f_1, f_2, \ldots, f_m)$ is differentiable at $\mathbf{a}$, then all the partial derivatives of each component function f_j exist at $\mathbf{a}$ and that the matrix representation of the linear mapping $D\mathbf{F}(\mathbf{a})$ with respect to the standard bases in $\mathbf{R}^n$ and $\mathbf{R}^m$ is the Jacobian matrix of $\mathbf{F}$ at $\mathbf{a}$. To see this, let $A = [a_{ij}]$ be the matrix of $D\mathbf{F}(\mathbf{a})$ with respect to the standard bases, $\{\mathbf{e}_1, \ldots, \mathbf{e}_n\}$ for $\mathbf{R}^n$ and $\{\mathbf{e}_1, \ldots, \mathbf{e}_m\}$ for $\mathbf{R}^m$. We must show that

$$a_{ij} = \frac{\partial f_i}{\partial x_j}(\mathbf{a}) = \lim_{h \to 0} \frac{f_i(\mathbf{a} + h\mathbf{e}_j) - f_i(\mathbf{a})}{h}$$

for $1 \le j \le n$ and $1 \le i \le m$. To be specific, we shall work with Euclidean norms in domain and range.

For any vector $\mathbf{z} = (z_1, \ldots, z_m)$, we have $|z_i| \le |\mathbf{z}|_2$ for $i = 1, \ldots, m$. Thus for $i = 1, \ldots, m$ and $j = 1, \ldots, n$ we have

$$\begin{aligned} 0 &\le \left| \frac{f_i(\mathbf{a} + h\mathbf{e}_j) - f_i(\mathbf{a})}{h} - a_{ij} \right| \\ &\le \left| \frac{\mathbf{F}(\mathbf{a} + h\mathbf{e}_j) - \mathbf{F}(\mathbf{a})}{h} - A\mathbf{e}_j \right|_2 \\ &= \frac{|\mathbf{F}(\mathbf{a} + h\mathbf{e}_j) - \mathbf{F}(\mathbf{a}) - A(h\mathbf{e}_j)|_2}{|h\mathbf{e}_j|_2}, \end{aligned}$$

where the last step used the linearity of A and the fact that $|\mathbf{e}_j| = 1$. Since $D\mathbf{F}(\mathbf{a})$ exists, the limit of this last quantity as $h \to 0$ exists and equals zero. Consequently,

$$\lim_{h \to 0} \frac{f_i(\mathbf{a} + h\mathbf{e}_j) - f_i(\mathbf{a})}{h} = a_{ij}$$

for $1 \le j \le n$ and $1 \le i \le m$, as desired. This proves that each derivative $\frac{\partial f_i}{\partial x_j}(a)$ exists, so the Jacobian matrix $A = J_{\mathbf{F}}(\mathbf{a})$ is defined. Our argument also implies that

$$\lim_{|\mathbf{h}|_2 \to 0} \frac{|\mathbf{F}(\mathbf{a}+\mathbf{h}) - \mathbf{F}(\mathbf{a}) - A\mathbf{h}|_2}{|\mathbf{h}|_2} = 0,$$

so $A = J_{\mathbf{F}}(\mathbf{a})$ is the matrix representation of $D\mathbf{F}(\mathbf{a})$ with respect to the standard bases in $\mathbf{R}^n$ and $\mathbf{R}^m$. □

Unless specifically indicated otherwise, we will assume henceforward that our vectors, in both domain and range space, are expressed in terms of the standard bases. There may be a tendency to identify $D\mathbf{F}(\mathbf{x})$ with the Jacobian matrix of $\mathbf{F}$ at $\mathbf{x}$, but we continue to write $D\mathbf{F}(\mathbf{x})$ for the derivative (the linear transformation) and $J_{\mathbf{F}}(\mathbf{x})$ for the matrix representation of $D\mathbf{F}(\mathbf{x})$ in the standard bases. The Jacobian matrix

$$J_{\mathbf{F}}(\mathbf{x}) = \begin{bmatrix} \partial f_1/\partial x_1(\mathbf{x}) & \partial f_1/\partial x_2(\mathbf{x}) & \cdots & \partial f_1/\partial x_n(\mathbf{x}) \\ \partial f_2/\partial x_1(\mathbf{x}) & \partial f_2/\partial x_2(\mathbf{x}) & \cdots & \partial f_2/\partial x_n(\mathbf{x}) \\ \vdots & \vdots & & \vdots \\ \partial f_m/\partial x_1(\mathbf{x}) & \partial f_m/\partial x_2(\mathbf{x}) & \cdots & \partial f_m/\partial x_n(\mathbf{x}) \end{bmatrix}$$

may also be viewed as

$$J_{\mathbf{F}}(\mathbf{x}) = \begin{bmatrix} \nabla^T f_1(\mathbf{x}) \\ \nabla^T f_2(\mathbf{x}) \\ \vdots \\ \nabla^T f_m(\mathbf{x}) \end{bmatrix}$$

using the row gradients of the component functions, indicated by $\nabla^T f_i(\mathbf{x})$.

If f is a real valued function of n real variables, then the derivative is usually written with a lowercase d, and thus

$$df(\mathbf{x})\mathbf{h} = \nabla^T f(\mathbf{x})\mathbf{h} = \nabla f(\mathbf{x}) \cdot \mathbf{h}$$

for all $\mathbf{h} \in \mathbf{R}^n$, since the standard matrix representation of the derivative $df(\mathbf{x})$ is the row gradient $\nabla^T f(\mathbf{x})$.

Exercise.

Exercise 10.3.1. Define $\mathbf{F} : \mathbf{R}^2 \to \mathbf{R}^2$ by its component functions

$$\begin{aligned} f_1(x_1, x_2) &= x_1^2 x_2, \\ f_2(x_1, x_2) &= x_1 - 3x_2. \end{aligned}$$

1. Show that $D\mathbf{F}(\mathbf{0})$ is the linear mapping

$$D\mathbf{F}(\mathbf{0})(h_1, h_2) = (0, h_1 - 3h_2).$$

2. Find $D\mathbf{F}(1, 2)(2, 1)$.

10.4. Existence of the Derivative

We begin with a result that should not be too surprising.

Theorem 10.4.1. *Let* $\mathbf{F} : U \subseteq \mathbf{R}^n \to \mathbf{R}^m$ *with* $\mathbf{F} = (f_1, \ldots, f_m)$*, and let* $\mathbf{a}$ *be an interior point of* U*. Then* $D\mathbf{F}(\mathbf{a})$ *exists if and only if* $df_j(\mathbf{a})$ *exists for each component function* f_j*,* $j = 1, \ldots, m$*.*

Proof. Suppose $D\mathbf{F}(\mathbf{a})$ exists. Then for every $\epsilon > 0$, there is a $\delta = \delta(\epsilon) > 0$ such that if $|\mathbf{x} - \mathbf{a}| < \delta$, then

$$|\mathbf{F}(\mathbf{x}) - \mathbf{F}(\mathbf{a}) - D\mathbf{F}(\mathbf{a})(\mathbf{x} - \mathbf{a})|_2 < \epsilon|\mathbf{x} - \mathbf{a}|_2.$$

If $\mathbf{F}(\mathbf{x}) = (f_1(\mathbf{x}), f_2(\mathbf{x}), \ldots, f_m(\mathbf{x}))$, then we want to show that for each j, $df_j(\mathbf{a})$ exists and equals the j-th component function of $D\mathbf{F}(\mathbf{a})$. Since $D\mathbf{F}(\mathbf{a})$ is a linear mapping each of its components l_j is a linear mapping of $\mathbf{R}^n$ into $\mathbf{R}$. Thus, the j-th component of $\mathbf{F}(\mathbf{x}) - \mathbf{F}(\mathbf{a}) - D\mathbf{F}(\mathbf{a})(\mathbf{x} - \mathbf{a})$ can be written as

$$f_j(\mathbf{x}) - f_j(\mathbf{a}) - l_j(\mathbf{x} - \mathbf{a}).$$

Since we have $|x_j| \leq |\mathbf{x}|_2$ for any $\mathbf{x} = (x_1, \ldots, x_n)$ in $\mathbf{R}^n$, we obtain

$$|f_j(\mathbf{x}) - f_j(\mathbf{a}) - l_j(\mathbf{x} - \mathbf{a})| < \epsilon|\mathbf{x} - \mathbf{a}|_2$$

if $|\mathbf{x} - \mathbf{a}|_2 < \delta$. Since this is true for every $\epsilon > 0$, we conclude that $df_j(\mathbf{a})$ exists and must be the linear mapping l_j, the j-th component function of $D\mathbf{F}(\mathbf{a})$.

Suppose $df_j(\mathbf{a})$ exists for each component function f_j, $j = 1, \ldots, m$, of $\mathbf{F}$. Given $\epsilon > 0$, there are numbers $\delta_j = \delta_j(\epsilon) > 0$, $j = 1, \ldots, m$, such that if $|\mathbf{x} - \mathbf{a}|_2 < \delta_j$, then

$$|f_j(\mathbf{x}) - f_j(\mathbf{a}) - df_j(\mathbf{a})(\mathbf{x} - \mathbf{a})| < \frac{\epsilon}{\sqrt{m}}|\mathbf{x} - \mathbf{a}|_2.$$

The function defined by

$$L(\mathbf{h}) = (df_1(\mathbf{a}), df_2(\mathbf{a}), \ldots, df_m(\mathbf{a}))$$

is linear since each of its components is linear. Let $\delta := \min\{\delta_1, \delta_2, \ldots, \delta_m\}$. If $|\mathbf{x} - \mathbf{a}|_2 < \delta$, then each of our component estimates above holds, and we conclude that

$$|\mathbf{F}(\mathbf{x}) - \mathbf{F}(\mathbf{a}) - L(\mathbf{x} - \mathbf{a})|_2 < \left[m\left(\frac{\epsilon^2}{m}|\mathbf{x} - \mathbf{a}|_2^2\right)\right]^{1/2} = \epsilon|\mathbf{x} - \mathbf{a}|_2.$$

Since $\epsilon > 0$ was arbitrary, we conclude that $D\mathbf{F}(\mathbf{a})$ exists and equals L. □

Theorem 10.4.1 allows us to reduce the argument for vector valued functions in the next result to the basic case of the real valued component functions.

Theorem 10.4.2. *Let* $\mathbf{F} : U \subseteq \mathbf{R}^n \to \mathbf{R}^m$ *and let* $\mathbf{a}$ *be an interior point of* U*. If the partial derivatives of each component function of* $\mathbf{F}$ *exist on a ball* $B_r(\mathbf{a})$ *for some* $r > 0$*, and the partial derivatives are continuous at* $\mathbf{a}$*, then* $D\mathbf{F}(\mathbf{a})$ *exists.*

Proof. By Theorem 10.4.1, it suffices to prove this result for the case of a single real component function, denoted $f : U \subseteq \mathbf{R}^n \to \mathbf{R}$. Thus we assume that all partial derivatives of f exist on a ball $B_r(\mathbf{a})$ for some $r > 0$, and all partial derivatives are continuous at $\mathbf{a}$. We wish to show that $df(\mathbf{a})$ exists. Let $\mathbf{h} = (h_1, \ldots, h_n)$ be such that $\mathbf{a} + \mathbf{h} \in B_r(\mathbf{a})$. We may write the increment $f(\mathbf{a} + \mathbf{h}) - f(\mathbf{a})$ as a sum of increments, each of which is taken parallel to a single coordinate axis, as follows. Write $\mathbf{h}_j = (h_1, \ldots, h_j, 0, \ldots, 0)$ for $j = 1, \ldots, n$, and $\mathbf{h}_0 = \mathbf{0}$. Then we have

$$f(\mathbf{a}+\mathbf{h}) - f(\mathbf{a}) = \sum_{j=1}^{n} [f(\mathbf{a}+\mathbf{h}_j) - f(\mathbf{a}+\mathbf{h}_{j-1})].$$

Now consider the j-th summand as the increment of the function

$$g_j(x) = f(a_1 + h_1, \ldots, a_{j-1} + h_{j-1}, x, a_{j+1}, \ldots, a_n)$$

of x in $(a_j, a_j + h_j)$. Apply the mean value theorem to each of these summands to obtain real numbers $b_j \in (a_j, a_j + h_j)$, such that

$$f(\mathbf{a}+\mathbf{h}_j) - f(\mathbf{a}+\mathbf{h}_{j-1}) = D_j f(a_1 + h_1, \ldots, a_{j-1} + h_{j-1}, b_j, a_{j+1}, \ldots, a_n)\, h_j,$$

where $D_j = \partial/\partial x_j$, for $j = 1, \ldots, n$. Hence, $f(\mathbf{a}+\mathbf{h}_j) - f(\mathbf{a}+\mathbf{h}_{j-1}) = D_j f(\mathbf{b}_j) h_j$ where the points $\mathbf{b}_j \to \mathbf{a}$ as $\mathbf{h} \to \mathbf{0}$, and

$$f(\mathbf{a}+\mathbf{h}) - f(\mathbf{a}) = \sum_{j=1}^{n} D_j f(\mathbf{b}_j)\, h_j.$$

Theorem 10.3.1 asserts that if the derivative $df(\mathbf{a})$ exists, then its matrix representation must be $\nabla f(\mathbf{a})$. Observe that $L : \mathbf{R}^n \to \mathbf{R}$ defined by

$$L\,\mathbf{h} = \sum_{j=1}^{n} D_j f(\mathbf{a}) h_j$$

determines a linear mapping with $L\,\mathbf{h} = \nabla f(\mathbf{a}) \cdot \mathbf{h}$, and we estimate

$$\begin{aligned}
\frac{|f(\mathbf{a}+\mathbf{h}) - f(\mathbf{a}) - L\,\mathbf{h}|}{|\mathbf{h}|_2} &= \frac{|f(\mathbf{a}+\mathbf{h}) - f(\mathbf{a}) - \sum_{j=1}^{n} D_j f(\mathbf{a}) h_j|}{|\mathbf{h}|_2} \\
&= \frac{|\sum_{j=1}^{n} [D_j f(\mathbf{b}_j) - D_j f(\mathbf{a})] h_j|}{|\mathbf{h}|_2} \\
&\le \sum_{j=1}^{n} |D_j f(\mathbf{b}_j) - D_j f(\mathbf{a})| \frac{|h_j|}{|\mathbf{h}|_2} \\
&\le \sum_{j=1}^{n} |D_j f(\mathbf{b}_j) - D_j f(\mathbf{a})|,
\end{aligned}$$

by the triangle inequality for real numbers and the fact that $|h_j| \le |\mathbf{h}|_2$. As $\mathbf{h} \to \mathbf{0}$, $\mathbf{b}_j \to \mathbf{a}$, and the continuity of $D_j f$ at $\mathbf{a}$ for each j implies that $df(\mathbf{a})$ exists and $df(\mathbf{a}) = L$. □

Theorem 10.4.2 often allows an easy determination of the existence of $D\mathbf{F}(\mathbf{a})$ for $\mathbf{a} \in U$, as in many cases the continuity of the partial derivatives is clear from a recognition that they belong to known classes of continuous functions. In view of Definition 10.1.5 (Section 10.1), Theorem 10.4.2 states that if $\mathbf{F}$ is C^1 at $\mathbf{a}$, then $D\mathbf{F}(\mathbf{a})$ exists. In practice, we are more likely to want this C^1 property to hold over some open set contained in the domain of the mapping, and consequently we often assume that $\mathbf{F} : U \subseteq \mathbf{R}^n \to \mathbf{R}^m$ is C^1 on U (Definition 10.1.6 in Section 10.1). Under this stronger hypothesis, we easily deduce from Theorem 10.4.2 that if the Jacobian matrix $J_{\mathbf{F}}(\mathbf{x})$ is defined for all $\mathbf{x}$ in U and all entries of the Jacobian matrix are continuous functions on U, then the linear mapping $D\mathbf{F}(\mathbf{x})$ exists for all $\mathbf{x}$ in U and is represented with respect to the standard bases in $\mathbf{R}^n$ and $\mathbf{R}^m$ by $J_{\mathbf{F}}(\mathbf{a})$.

Example 10.4.3. Here is an example to which Theorem 10.4.2 does not apply, since one of the partial derivatives is not continuous at the point in question. Consider the function $\mathbf{F} = (f_1, f_2)$, from $\mathbf{R}^2$ to $\mathbf{R}^2$, where

$$f_1(x_1, x_2) = \begin{cases} x_1 + x_1^2 \sin(1/x_1) & \text{if } x_1 \neq 0, \\ 0 & \text{if } x_1 = 0, \end{cases}$$

and $f_2(x_1, x_2) = x_2$. Clearly, f_2 has partial derivatives continuous at every point. And $(f_1)_{x_2}(x_1, x_2) = 0$ for all (x_1, x_2). By the usual differentiation rules, for $x_1 \neq 0$,

$$(f_1)_{x_1}(x_1, x_2) = 1 + 2x_1 \sin(1/x_1) - \cos(1/x_1).$$

If $x_1 = 0$, then

$$(f_1)_{x_1}(0, x_2) = \lim_{h \to 0} \frac{f(h, x_2) - 0}{h} = \lim_{h \to 0} \frac{h + h^2 \sin(1/h)}{h} = 1.$$

Thus the partial derivatives of f_1 exist at every point. Observe that $(f_1)_{x_1}$ is not continuous at any point where $x_1 = 0$, due to the $\cos(1/x_1)$ term. But Definition 10.2.1 applies and shows that $(df_1)(0, x_2)$ exists for any x_2 (Exercise 10.4.1). △

Example 10.4.4. The function

$$f(\mathbf{x}) = \begin{cases} (x_1^2 + x_2^2) \sin(1/(x_1^2 + x_2^2)) & \text{if } \mathbf{x} \neq \mathbf{0}, \\ 0 & \text{if } \mathbf{x} = \mathbf{0} \end{cases}$$

clearly has partial derivatives continuous at every point $(x_1, x_2) \neq (0, 0)$ in $\mathbf{R}^2$, given by the usual differentiation rules. And both partial derivatives exist at the origin. Observe that

$$D_1 f(0,0) = \lim_{h \to 0} \frac{f(h,0) - f(0,0)}{h} = \lim_{h \to 0} \frac{h^2 \sin(1/h^2)}{h} = 0$$

and that a similar calculation gives $D_2 f(0,0) = 0$. The partial derivatives fail to be continuous only at the origin (Exercise 10.4.2). Nevertheless, $Df(\mathbf{0})$ exists, because the tangency estimate of Definition 10.2.1 holds for the linear mapping $\mathbf{h} \mapsto \nabla f(\mathbf{0})\mathbf{h} = [f_{x_1}(\mathbf{0}) \quad f_{x_2}(\mathbf{0})]\mathbf{h} = [0 \quad 0]\mathbf{h} = 0$. △

Perhaps the most important differentiation rule is the chain rule for the derivative of a composition of functions, and the following section is devoted to it. We end this section with a few differentiation rules we state without proof.

Theorem 10.4.5. *Let* $\mathbf{a}$ *be an interior point of* $D \subseteq \mathbf{R}^n$. *If* $\mathbf{F}, \mathbf{G} : D \to \mathbf{R}^m$ *and* $D\mathbf{F}(\mathbf{a})$ *and* $D\mathbf{G}(\mathbf{a})$ *exist, then* $D(\mathbf{F} \pm \mathbf{G})(\mathbf{a})$ *and* $D(\mathbf{F} \cdot \mathbf{G})(\mathbf{a})$ *exist, and*

1. $D(\mathbf{F} \pm \mathbf{G})(\mathbf{a}) = D\mathbf{F}(\mathbf{a}) \pm D\mathbf{G}(\mathbf{a})$;
2. $D(\mathbf{F} \cdot \mathbf{G})(\mathbf{a})\mathbf{h} = \mathbf{F}(\mathbf{a}) \cdot D\mathbf{G}(\mathbf{a})\mathbf{h} + \mathbf{G}(\mathbf{a}) \cdot D\mathbf{F}(\mathbf{a})\mathbf{h}$, *for all* $\mathbf{h}$ *in* $\mathbf{R}^n$;
3. $D(\lambda\mathbf{F})(\mathbf{a})\mathbf{h} = \lambda(\mathbf{a}) D\mathbf{F}(\mathbf{a})\mathbf{h} + [d\lambda(\mathbf{a})\mathbf{h}]\, \mathbf{F}(\mathbf{a})$, *for all* $\mathbf{h}$ *in* $\mathbf{R}^n$ *and real* λ.

Exercises.

Exercise 10.4.1. In Example 10.4.3, use Definition 10.2.1 of derivative to show that $(df_1)(0, x_2)$ exists for any x_2. Thus, conclude that $D\mathbf{F}(0,0)$ exists.

Exercise 10.4.2. Refer to Example 10.4.4.

1. Show that the partial derivatives $D_1 f$ and $D_2 f$ fail to be continuous at the origin. *Hint*: Consider the sequence $\mathbf{x}_k = (1/k, 0)$ for $D_1 f$ and the sequence $\mathbf{y}_k = (0, 1/k)$ for $D_2 f$, or use the symmetry of f in x_1, x_2.
2. Verify that $Df(\mathbf{0})$ equals the zero mapping, since the tangency estimate of Definition 10.2.1 holds for the linear mapping $\mathbf{h} \mapsto \nabla f(\mathbf{0})\mathbf{h} = \mathbf{0}$.

Exercise 10.4.3. Prove Theorem 10.4.5.

10.5. The Chain Rule

We shall use the notation $\mathcal{D}(\mathbf{H})$ for the domain of a mapping $\mathbf{H}$. We shall also call on the following facts involving direct and inverse images of any mapping $\mathbf{H}$: If S is a set, then

$$S \cap \mathcal{D}(\mathbf{H}) \subseteq \mathbf{H}^{-1}(\mathbf{H}(S)),$$

and if $B_1 \subseteq B_2$, then

$$\mathbf{H}^{-1}(B_1) \subseteq \mathbf{H}^{-1}(B_2).$$

Lemma 10.5.1. *Let $\mathbf{G} : \mathcal{D}(\mathbf{G}) \to \mathbf{R}^m$ and let $\mathbf{a}$ be an interior point of $\mathcal{D}(\mathbf{G})$. If $\mathbf{G}$ is continuous at $\mathbf{a}$ and if $\mathbf{G}(\mathbf{a}) = \mathbf{b}$ is an interior point of $A \subseteq \mathbf{R}^m$, then $\mathbf{a}$ is also an interior point of $\mathbf{G}^{-1}(A)$.*

Proof. Since $\mathbf{b} = \mathbf{G}(\mathbf{a})$ is an interior point of A, there is an $\epsilon > 0$ such that $B_\epsilon(\mathbf{G}(\mathbf{a})) \subset A$. Since $\mathbf{G}$ is continuous at $\mathbf{a}$ and $\mathbf{a}$ is an interior point of $\mathcal{D}(\mathbf{G})$, there is a $\delta > 0$ such that $\mathbf{G}(B_\delta(\mathbf{a})) \subset B_\epsilon(\mathbf{G}(\mathbf{a}))$ and $B_\delta(\mathbf{a}) \subset \mathcal{D}(\mathbf{G})$. Thus we have, for the set $S = B_\delta(\mathbf{a})$,

$$\begin{aligned} B_\delta(\mathbf{a}) &= B_\delta(\mathbf{a}) \cap \mathcal{D}(\mathbf{G}) \\ &\subseteq \mathbf{G}^{-1}(\mathbf{G}(B_\delta(\mathbf{a}))) \\ &\subseteq \mathbf{G}^{-1}(B_\epsilon(\mathbf{G}(\mathbf{a}))) \\ &\subseteq \mathbf{G}^{-1}(A). \end{aligned}$$

This proves that $\mathbf{a}$ is an interior point of $\mathbf{G}^{-1}(A)$. □

A proof of the chain rule in the multivariable case can be based on an appropriate extension of the argument for the single variable case in the proof of Theorem 5.1.9, using the little-oh ($o(|\mathbf{h}|)$) concept. Here we shall give a different proof to provide practice of a different kind.

Theorem 10.5.2. *Let $\mathbf{G} : U \subset \mathbf{R}^n \to \mathbf{R}^m$ and $\mathbf{F} : V \subset \mathbf{R}^m \to \mathbf{R}^p$. Let $\mathbf{a}$ be an interior point of U and let $\mathbf{b} = \mathbf{G}(\mathbf{a})$ be an interior point of V. If $D\mathbf{G}(\mathbf{a})$ and $D\mathbf{F}(\mathbf{b})$ exist, then the composition $\mathbf{F} \circ \mathbf{G}$ is differentiable at $\mathbf{a}$, and*

$$D(\mathbf{F} \circ \mathbf{G})(\mathbf{a}) = D\mathbf{F}(\mathbf{b}) \circ D\mathbf{G}(\mathbf{a}).$$

Proof. First, by Lemma 10.5.1 with $A = \mathcal{D}(\mathbf{F})$, it follows that $\mathbf{a}$ is an interior point of $\mathbf{G}^{-1}(\mathcal{D}(\mathbf{F})) = \mathcal{D}(\mathbf{F} \circ \mathbf{G})$, so it makes sense to consider the existence of $D(\mathbf{F} \circ \mathbf{G})(\mathbf{a})$.

By reference to the definition of derivative, we must show that, for every $\epsilon > 0$ there is a $\delta > 0$ such that $B_\delta(\mathbf{a}) \subset \mathcal{D}(\mathbf{F} \circ \mathbf{G})$ and

$$|(\mathbf{F} \circ \mathbf{G})(\mathbf{x}) - (\mathbf{F} \circ \mathbf{G})(\mathbf{a}) - D\mathbf{F}(\mathbf{G}(\mathbf{a})) \circ D\mathbf{G}(\mathbf{a})(\mathbf{x} - \mathbf{a})| \le \epsilon |\mathbf{x} - \mathbf{a}| \tag{10.4}$$

for all $\mathbf{x} \in B_\delta(\mathbf{a})$. This will be sufficient to prove the theorem, since the composition of linear mappings is indeed a linear mapping.

Write $\mathbf{y} = \mathbf{G}(\mathbf{x})$ and $\mathbf{b} = \mathbf{G}(\mathbf{a})$. Then, for the vector quantity within the norm symbols on the left-hand side of (10.4), we may write

$$\mathbf{F}(\mathbf{y}) - \mathbf{F}(\mathbf{b}) - D\mathbf{F}(\mathbf{b}) \circ D\mathbf{G}(\mathbf{a})(\mathbf{x} - \mathbf{a}),$$

and then adding and subtracting $D\mathbf{F}(\mathbf{b})(\mathbf{y} - \mathbf{b})$ gives

$$\begin{aligned}&\mathbf{F}(\mathbf{y}) - \mathbf{F}(\mathbf{b}) - D\mathbf{F}(\mathbf{b})(\mathbf{y} - \mathbf{b}) + D\mathbf{F}(\mathbf{b})(\mathbf{y} - \mathbf{b}) - D\mathbf{F}(\mathbf{b}) \circ D\mathbf{G}(\mathbf{a})(\mathbf{x} - \mathbf{a}) \\ &\quad = \mathbf{F}(\mathbf{y}) - \mathbf{F}(\mathbf{b}) - D\mathbf{F}(\mathbf{b})(\mathbf{y} - \mathbf{b}) + D\mathbf{F}(\mathbf{b})[\mathbf{G}(\mathbf{x}) - \mathbf{G}(\mathbf{a}) - D\mathbf{G}(\mathbf{a})(\mathbf{x} - \mathbf{a})].\end{aligned}$$

Since $\mathbf{G}$ is differentiable at $\mathbf{a}$, there is a $\delta_1 > 0$ and a constant $M > 0$ such that if $\mathbf{x} \in B_{\delta_1}(\mathbf{a})$, then

$$|\mathbf{G}(\mathbf{x}) - \mathbf{G}(\mathbf{a})| \le M|\mathbf{x} - \mathbf{a}|.$$

(This follows from an argument in the proof of Theorem 10.2.5; see also Exercise 10.5.3.) And we may choose δ_1 such that $B_{\delta_1}(\mathbf{a}) \subset \mathcal{D}(\mathbf{F} \circ \mathbf{G})$ since $\mathbf{a}$ is an interior point of this domain.

Since $\mathbf{F}$ is differentiable at $\mathbf{b}$, there is a $\delta_2 > 0$ such that

$$|\mathbf{F}(\mathbf{y}) - \mathbf{F}(\mathbf{b}) - D\mathbf{F}(\mathbf{b})(\mathbf{y} - \mathbf{b})| \le \frac{\epsilon}{2M}|\mathbf{y} - \mathbf{b}|$$

for all $\mathbf{y} \in B_{\delta_2}(\mathbf{b})$, and we may choose δ_2 so that $B_{\delta_2}(\mathbf{b}) \subset \mathcal{D}(\mathbf{F})$.

Now choose $\delta_3 = \min\{\delta_1, \delta_2/M\}$ and take $\mathbf{x} \in B_{\delta_3}(\mathbf{a})$. Then

$$|\mathbf{y} - \mathbf{b}| = |\mathbf{G}(\mathbf{x}) - \mathbf{G}(\mathbf{a})| \le M|\mathbf{x} - \mathbf{a}| \le M\frac{\delta_2}{M} = \delta_2,$$

and we then obtain, for $\mathbf{x} \in B_{\delta_3}(\mathbf{a})$,

$$\begin{aligned}|\mathbf{F}(\mathbf{y}) - \mathbf{F}(\mathbf{b}) - D\mathbf{F}(\mathbf{b})(\mathbf{y} - \mathbf{b})| &\le \frac{\epsilon}{2M}|\mathbf{y} - \mathbf{b}| \\ &\le \frac{\epsilon}{2M}M|\mathbf{x} - \mathbf{a}| \\ &\le \frac{\epsilon}{2}|\mathbf{x} - \mathbf{a}|. \qquad (10.5)\end{aligned}$$

Since $D\mathbf{F}(\mathbf{b})$ is a linear mapping, there is a constant $K > 0$ such that

$$|D\mathbf{F}(\mathbf{b})\mathbf{u}| \le K|\mathbf{u}|$$

for all $\mathbf{u} \in \mathbf{R}^m$. Since $\mathbf{G}$ is differentiable at $\mathbf{a}$, there is a $\delta_4 > 0$ such that $B_{\delta_4}(\mathbf{a}) \subset \mathcal{D}(\mathbf{F} \circ \mathbf{G})$ and

$$|\mathbf{G}(\mathbf{x}) - \mathbf{G}(\mathbf{a}) - D\mathbf{G}(\mathbf{a})(\mathbf{x} - \mathbf{a})| \le \frac{\epsilon}{2K}|\mathbf{x} - \mathbf{a}|$$

for all $\mathbf{x} \in B_{\delta_4}(\mathbf{a})$. Now, by the last two inequalities, if $\mathbf{x} \in B_{\delta_4}(\mathbf{a})$ then

$$\begin{aligned} \left|D\mathbf{F}(\mathbf{b})[\mathbf{G}(\mathbf{x}) - \mathbf{G}(\mathbf{a}) - D\mathbf{G}(\mathbf{a})(\mathbf{x} - \mathbf{a})]\right| &\leq K|\mathbf{G}(\mathbf{x}) - \mathbf{G}(\mathbf{a}) - D\mathbf{G}(\mathbf{a})(\mathbf{x} - \mathbf{a})| \\ &\leq \frac{\epsilon}{2}|\mathbf{x} - \mathbf{a}|. \end{aligned} \tag{10.6}$$

Thus, by (10.5), (10.6) and the triangle inequality, we have that if $\delta = \min\{\delta_3, \delta_4\}$ and if $\mathbf{x} \in B_\delta(\mathbf{a})$, then (10.4) holds, as we wished to show. □

Example 10.5.3. Let $\mathbf{G} : \mathbf{R}^2 \to \mathbf{R}^3$ be the mapping with component functions

$$\begin{aligned} g_1(x_1, x_2) &= x_1^3 + x_2, \\ g_2(x_1, x_2) &= 2x_2, \\ g_3(x_1, x_2) &= x_1 + 3x_2, \end{aligned}$$

and let $\mathbf{F} : \mathbf{R}^3 \to \mathbf{R}$ be given by

$$\mathbf{F}(y_1, y_2, y_3) = 3y_1 + y_2^3 + y_3.$$

Then $\mathcal{D}(\mathbf{F} \circ \mathbf{G}) = \mathbf{G}^{-1}(\mathcal{D}(\mathbf{F})) = \mathbf{G}^{-1}(\mathbf{R}^3) = \mathbf{R}^2$, and $\mathbf{F} \circ \mathbf{G} : \mathbf{R}^2 \to \mathbf{R}$. Since all the partial derivatives of $\mathbf{F}$ and $\mathbf{G}$ exist and are continuous everywhere, these functions are differentiable everywhere. In the standard bases assumed by our descriptions of these functions, the derivative $D\mathbf{G}(\mathbf{a})$, where $\mathbf{a} = (a_1, a_2)$, is represented by its Jacobian matrix

$$J_{\mathbf{G}}(\mathbf{a}) = \begin{bmatrix} 3a_1^2 & 1 \\ 0 & 2 \\ 1 & 3 \end{bmatrix}$$

and $D\mathbf{F}(\mathbf{b})$, where $\mathbf{b} = (b_1, b_2, b_3)$, is represented by its Jacobian matrix

$$J_{\mathbf{F}}(\mathbf{b}) = \begin{bmatrix} 3 & 3b_2^2 & 1 \end{bmatrix}.$$

Therefore, by Theorem 10.5.2, $D(\mathbf{F} \circ \mathbf{G})(\mathbf{a})$ is represented with respect to the standard bases in $\mathbf{R}^2$ and $\mathbf{R}$ by the matrix product

$$\begin{aligned} J_{\mathbf{F}\circ\mathbf{G}}(\mathbf{a}) = J_{\mathbf{F}}(\mathbf{G}(\mathbf{a}))J_{\mathbf{G}}(\mathbf{a}) &= \begin{bmatrix} 3 & 3(2a_2)^2 & 1 \end{bmatrix} \begin{bmatrix} 3a_1^2 & 1 \\ 0 & 2 \\ 1 & 3 \end{bmatrix} \\ &= \begin{bmatrix} (9a_1^2 + 1) & (24a_2^2 + 6) \end{bmatrix}. \end{aligned}$$

Since the explicit composite function is given by

$$\mathbf{F} \circ \mathbf{G}(\mathbf{x}) = 3(x_1^3 + x_2) + (2x_2)^3 + (x_1 + 3x_2) = 3x_1^3 + x_1 + 6x_2 + 8x_2^3,$$

we can also directly verify the Jacobian matrix $J_{\mathbf{F}\circ\mathbf{G}}(\mathbf{a})$ above for the composite function. △

Suppose $\mathcal{D}(\mathbf{F}) \subseteq \mathbf{R}^n$ and $\mathbf{F} : \mathcal{D}(\mathbf{F}) \to \mathbf{R}^n$. If $\mathbf{F}^{-1}$ exists in a neighborhood of $\mathbf{F}(\mathbf{a}) = \mathbf{b}$, and if $D\mathbf{F}^{-1}(\mathbf{b})$ exists, then the identity $\mathbf{F}^{-1}(\mathbf{F}(\mathbf{x})) = \mathbf{x}$, valid in some ball about $\mathbf{a}$, implies via Theorem 10.5.2 that, in particular,

$$D\mathbf{F}^{-1}(\mathbf{b}) \circ D\mathbf{F}(\mathbf{a}) = I,$$

and hence that

$$D\mathbf{F}^{-1}(\mathbf{F}(\mathbf{a})) = [D\mathbf{F}(\mathbf{a})]^{-1}. \tag{10.7}$$

The inverse function theorem, discussed later, explains precisely under what conditions equation (10.7) can be extended to an identity that holds throughout some neighborhood of the point $\mathbf{a}$. In this connection it may be useful to note that the function $f(x) = x^3$ is invertible on the real line, but $f^{-1}(y) = y^{1/3}$ is not differentiable at $y = 0$; thus, although $f'(0)$ exists, it is not invertible (as a linear mapping from $\mathbf{R}$ to $\mathbf{R}$), so the mere existence of $f'(a)$ is not sufficient for (10.7) because it does not guarantee the existence of the derivative of the inverse function. The inverse function theorem (Theorem 11.2.2) will show that the extension of (10.7) to a neighborhood of $\mathbf{a}$ is guaranteed if $\mathbf{F}$ is continuously differentiable in a neighborhood of $\mathbf{a}$ and if $D\mathbf{F}(\mathbf{a})$ is invertible; in fact, these conditions guarantee that $\mathbf{F}$ is locally invertible in a neighborhood of $\mathbf{a}$, and that $D\mathbf{F}^{-1}(\mathbf{y})$ exists and is continuous for $\mathbf{y}$ in some neighborhood of $\mathbf{b} = \mathbf{F}(\mathbf{a})$.

Exercises.

Exercise 10.5.1. Write a clear and concise description of why the multivariable chain rule statement in Theorem 10.5.2 is really an extension of the single variable chain rule statement in Theorem 5.1.9.

Exercise 10.5.2. Define $\mathbf{F} : \mathbf{R}^3 \to \mathbf{R}^3$ by

$$\mathbf{F}(\mathbf{y}) = (y_1 - y_3,\, y_2 + 4y_3,\, y_1 - 2y_2 + 3y_3),$$

and $\mathbf{G} : \mathbf{R}^3 \to \mathbf{R}^3$ by

$$\mathbf{G}(\mathbf{y}) = \big(\cosh(y_1) - \sinh(y_3),\, y_2 + 4y_3,\, \log\big(1 + 2y_2^2 + 3y_3^2\big)\big).$$

Find $D(\mathbf{F} \circ \mathbf{G})(\mathbf{0})$ and its matrix representation.

Exercise 10.5.3. Show that if $D\mathbf{G}(\mathbf{a})$ exists, then there is a $\delta_1 > 0$ and a constant $M > 0$ such that $|\mathbf{G}(\mathbf{x}) - \mathbf{G}(\mathbf{a})| \le M|\mathbf{x} - \mathbf{a}|$ for all $\mathbf{x} \in B_{\delta_1}(\mathbf{a})$. *Hint*: See the proof of Theorem 10.2.5.

10.6. The Mean Value Theorem: Real Functions

The single variable mean value theorem (Theorem 5.2.4) does not have a direct generalization to mappings $\mathbf{r} : [a, b] \to \mathbf{R}^m$ where I is an interval and $m \ge 2$. A direct generalization would assert that if $\mathbf{r}$ is continuous on $[a, b]$ and differentiable on (a, b), then there exists a $t_0 \in (a, b)$ such that

$$\mathbf{r}(b) - \mathbf{r}(a) = D\mathbf{r}(t_0)(b - a) = (b - a)\mathbf{r}'(t_0).$$

Consider the following counterexample to that assertion.

Example 10.6.1. Let $\mathbf{r} : [0, \pi] \to \mathbf{R}^2$ be the curve given by $\mathbf{r}(t) = (\cos t, \sin t)$ for $0 \le t \le \pi$. If we have

$$\mathbf{r}(\pi) - \mathbf{r}(0) = (-1, 0) - (1, 0) = D\mathbf{r}(t_0)(\pi - 0) = (-\pi \sin t_0, \pi \cos t_0)$$

for some $t_0 \in (0, \pi)$, then $\sin t_0 = 2/\pi$ and $\cos t_0 = 0$. The second condition holds only for $t_0 = \pi/2 \in (0, \pi)$, but then $\sin \pi/2 = 1 \ne 2/\pi$. (See Exercise 10.6.1). △

There is, however, a direct generalization of the single variable mean value theorem to the case of differentiable functions $f : U \subset \mathbf{R}^n \to \mathbf{R}$. The extension is really a deduction from the single variable result (Theorem 5.2.4) and it has many important consequences for the analysis of functions of several variables. In presenting this generalization, we will need to describe the line segment l joining two points $\mathbf{a}$ and $\mathbf{b}$ in $\mathbf{R}^n$; for this purpose, we may write

$$l = l_{\mathbf{ab}} = \{(1-t)\mathbf{a} + t\mathbf{b} : 0 \le t \le 1\} = \{\mathbf{a} + t(\mathbf{b} - \mathbf{a}) : 0 \le t \le 1\}.$$

The next result is the mean value theorem for real functions of several real variables.

Theorem 10.6.2 (Mean Value Theorem for Real Functions). *Let $f : U \subseteq \mathbf{R}^n \to \mathbf{R}$ and suppose that $\mathbf{a}$ and $\mathbf{b}$ are interior points of U. If the line segment $l_{\mathbf{ab}}$ is contained in the interior of U and f is continuous on $l_{\mathbf{ab}}$ and differentiable at all points of $l_{\mathbf{ab}}$ (except possibly at its endpoints $\mathbf{a}$ and $\mathbf{b}$), then there is a point $\mathbf{c} \in l_{\mathbf{ab}}$ such that*

$$f(\mathbf{b}) - f(\mathbf{a}) = \nabla f(\mathbf{c}) \cdot (\mathbf{b} - \mathbf{a}). \tag{10.8}$$

Proof. The curve $\mathbf{r} : [0,1] \to \mathbf{R}^n$ given by $\mathbf{r}(t) = \mathbf{a} + t(\mathbf{b} - \mathbf{a})$ is continuous on $[0,1]$ and differentiable on $(0,1)$, and $\mathbf{r}'(t) = \mathbf{b} - \mathbf{a}$. Define the function $\phi : [0,1] \to \mathbf{R}^n$ by $\phi(t) = f(\mathbf{r}(t)) = f(\mathbf{a} + t(\mathbf{b} - \mathbf{a}))$. Then ϕ is continuous on $[0,1]$ and differentiable on $(0,1)$, and $\phi(0) = f(\mathbf{a})$, $\phi(1) = f(\mathbf{b})$. Since f is differentiable at all points of $l_{\mathbf{ab}}$, the chain rule (Theorem 10.5.2) applies, and we have

$$\phi'(t) = \nabla f(\mathbf{a} + t(\mathbf{b} - \mathbf{a})) \cdot (\mathbf{b} - \mathbf{a}).$$

By the single variable mean value theorem (Theorem 5.2.4), there is a $t_0 \in (0,1)$ such that $\phi(1) - \phi(0) = \phi'(t_0)(1-0) = \phi'(t_0)$, hence (10.8) holds with $\mathbf{c} = \mathbf{a} + t_0(\mathbf{b} - \mathbf{a})$. □

The real importance of the mean value Theorem 5.2.4 and its generalizations, the feature that makes these results most useful, is that they provide *approximations to a difference in function values* for nearby points.

There is an easy estimate for the absolute value of the difference in function values in (10.8). Using the Euclidean vector norm for both $\nabla f(\mathbf{c})$ and $\mathbf{b} - \mathbf{a}$, the Cauchy-Schwarz inequality implies that

$$|f(\mathbf{b}) - f(\mathbf{a})| = |\nabla f(\mathbf{c}) \cdot (\mathbf{b} - \mathbf{a})| \le |\nabla f(\mathbf{c})|_2 \, |\mathbf{b} - \mathbf{a}|_2.$$

The remaining corollaries in this section extend this approximation theme using derivative information.

A set $U \subseteq \mathbf{R}^n$ is **convex** if for any pair of points $\mathbf{a}, \mathbf{b} \in U$ the line segment $l_{\mathbf{ab}}$ joining $\mathbf{a}$ and $\mathbf{b}$ is contained in U. Clearly, $\mathbf{R}^n$ itself is convex, as is any open ball $B_r(\mathbf{x})$ (defined with respect to any norm on $\mathbf{R}^n$).

Corollary 10.6.3. *Let $f : U \subseteq \mathbf{R}^n \to \mathbf{R}$ be a C^1 mapping on an open set U containing the points $\mathbf{a}$ and $\mathbf{a} + \mathbf{h}$ and the entire line segment joining them. Then*

$$|f(\mathbf{a} + \mathbf{h}) - f(\mathbf{a})| \le \Big(\max_{0 \le t \le 1} |\nabla f(\mathbf{a} + t\mathbf{h})|_2 \Big) |\mathbf{h}|_2. \tag{10.9}$$

If U is open and convex and $|\nabla f(\mathbf{x})|_2 \le M$ for each $\mathbf{x} \in U$, then

$$|f(\mathbf{x}) - f(\mathbf{y})| \le M |\mathbf{x} - \mathbf{y}|_2 \tag{10.10}$$

for all $\mathbf{x}, \mathbf{y} \in U$.

Proof. The segment joining $\mathbf{a}$ and $\mathbf{a} + \mathbf{h}$ is $l = \{\mathbf{a} + t\mathbf{h} : 0 \le t \le 1\}$. Since the interval $[0, 1]$ is compact and the function $t \mapsto |\nabla f(\mathbf{a} + t\mathbf{h})|_2$ is continuous on $[0, 1]$, the maximum on the right-hand side of (10.9) exists. The inequality (10.9) follows directly from an application of Theorem 10.6.2 to f on the line segment l, and an application of the Cauchy-Schwarz inequality.

If U is open and convex and $|\nabla f(\mathbf{x})|_2$ is bounded by M on U, then for any two points $\mathbf{x}$, $\mathbf{y}$ in U, we can let $\mathbf{a} = \mathbf{x}$, $\mathbf{h} = \mathbf{y} - \mathbf{x}$ and use M in place of the maximum in (10.9), and (10.10) follows. □

In the corollary, we assume U is open so that we can talk about the differentiability of f on U. On any *closed and bounded subset* of U, we will certainly have a bound on $|\nabla f(\mathbf{x})|_2$, so the second statement of the corollary always applies on compact subsets of U.

This is a convenient place to introduce the idea of directional derivatives.

Definition 10.6.4 (Directional Derivative). *Suppose $f : B_r(\mathbf{x}) \to \mathbf{R}$ where $r > 0$ and $\mathbf{x}$ is in $\mathbf{R}^n$, and let $\mathbf{u} \in \mathbf{R}^n$ be a unit vector in the Euclidean norm, so that $|\mathbf{u}|_2 = 1$. When the limit exists, the number $D_\mathbf{u} f(\mathbf{x})$ defined by*

$$D_\mathbf{u} f(\mathbf{x}) = \lim_{h \to 0} \frac{f(\mathbf{x} + h\mathbf{u}) - f(\mathbf{x})}{h}$$

is called the **directional derivative of f at $\mathbf{x}$ in the direction $\mathbf{u}$**.

When $|\mathbf{u}|_2 = 1$, it is reasonable to interpret the directional derivative $D_\mathbf{u} f(\mathbf{x})$ as a *rate of change* of f at $\mathbf{x}$ in the direction $\mathbf{u}$, because then the difference quotient compares the change in function values at points that differ by an increment $h\mathbf{u}$ having norm equal to $|h|$. Directional derivatives generalize the idea of partial derivatives, and, as we will see shortly, they allow us to compute the rate of change of f at $\mathbf{x}$ in any direction in space. In particular, we have

$$D_{\mathbf{e}_j} f(\mathbf{x}) = \lim_{h \to 0} \frac{f(\mathbf{x} + h\mathbf{e}_j) - f(\mathbf{x})}{h} = \frac{\partial f}{\partial x_j}(\mathbf{x}).$$

On the other hand, the mere existence of partial derivatives does not imply existence of other directional derivatives, as the next example shows.

Example 10.6.5. Let $f : \mathbf{R}^2 \to \mathbf{R}$ be defined by

$$f(x, y) = \begin{cases} 1 & \text{if } xy = 0, \\ 0 & \text{if } x \ne 0 \text{ and } y \ne 0. \end{cases}$$

Then the partial derivatives, $f_x(0, 0)$ and $f_y(0, 0)$, exist and equal zero. However, if $|\mathbf{u}|_2 = 1$ and $\mathbf{u} \ne \mathbf{e}_j$ for $j = 1, 2$, then $D_\mathbf{u} f(0, 0)$ does not exist. (See Exercise 10.6.4.) We note that f is not continuous at $(0, 0)$, and hence f is not differentiable at $(0, 0)$. △

The next result shows that if f is differentiable at $\mathbf{a}$, then all directional derivatives of f at $\mathbf{a}$ exist.

Theorem 10.6.6. *If f is differentiable at $\mathbf{a}$, then all directional derivatives of f at $\mathbf{a}$ exist, and*

$$D_\mathbf{u} f(\mathbf{a}) = \sum_{j=1}^{n} \frac{\partial f}{\partial x_j}(\mathbf{a}) u_j = \nabla f(\mathbf{a}) \cdot \mathbf{u}.$$

Proof. Since f is differentiable at $\mathbf{a}$,

$$\frac{|f(\mathbf{a}+\mathbf{h}) - f(\mathbf{a}) - \nabla f(\mathbf{a}) \cdot \mathbf{h}|}{|\mathbf{h}|_2} \to 0 \qquad \text{as} \quad |h|_2 \to 0.$$

It follows that for a vector $\mathbf{u}$ with $|\mathbf{u}|_2 = 1$, and real h,

$$\frac{|f(\mathbf{a}+h\mathbf{u}) - f(\mathbf{a}) - \nabla f(\mathbf{a}) \cdot (h\mathbf{u})|}{|h\mathbf{u}|_2} \to 0 \qquad \text{as} \quad |h| \to 0.$$

This says that

$$\frac{1}{h}[f(\mathbf{a}+h\mathbf{u}) - f(\mathbf{a}) - h\nabla f(\mathbf{a}) \cdot \mathbf{u}] \to 0 \qquad \text{as} \quad h \to 0,$$

which is equivalent to the statement

$$\lim_{h\to 0} \frac{f(\mathbf{a}+h\mathbf{u}) - f(\mathbf{a})}{h} = \nabla f(\mathbf{a}) \cdot \mathbf{u},$$

as we desired. □

By this result, for a differentiable real valued function $f : \mathbf{R}^n \to \mathbf{R}$, we can view $df(\mathbf{x}) : \mathbf{R}^n \to \mathbf{R}$ as a linear mapping (or *linear functional*) that determines directional derivatives of f at $\mathbf{x}$, in the direction of any unit vector $\mathbf{u}$, according to the chain rule:

$$\frac{d}{dt} f(\mathbf{x} + t\mathbf{u})|_{t=0} = df(\mathbf{x})\mathbf{u} = \nabla f(\mathbf{x}) \cdot \mathbf{u} = D_{\mathbf{u}} f(\mathbf{x}). \tag{10.11}$$

We now return to the idea of $D_{\mathbf{u}} f(\mathbf{x})$ as a rate of change of f when $|\mathbf{u}|_2 = 1$. By the Cauchy-Schwarz inequality,

$$|D_{\mathbf{u}} f(\mathbf{x})| = |\nabla f(\mathbf{x}) \cdot \mathbf{u}| \leq |\nabla f(\mathbf{x})|_2 \, |\mathbf{u}|_2 = |\nabla f(\mathbf{x})|_2,$$

which is equivalent to the inequality

$$-|\nabla f(\mathbf{x})|_2 \leq D_{\mathbf{u}} f(\mathbf{x}) \leq |\nabla f(\mathbf{x})|_2.$$

Moreover, the maximum value of $D_{\mathbf{u}} f(\mathbf{x})$ is $|\nabla f(\mathbf{x})|_2$, and this maximum value is achieved when $\mathbf{u} = \frac{\nabla f(\mathbf{x})}{|\nabla f(\mathbf{x})|_2}$. The minimum value of $D_{\mathbf{u}} f(\mathbf{x})$ is $-|\nabla f(\mathbf{x})|_2$, and this minimum value is achieved when $\mathbf{u} = -\frac{\nabla f(\mathbf{x})}{|\nabla f(\mathbf{x})|_2}$.

Exercises.

Exercise 10.6.1. Consider a particle in motion along the helical curve $\mathbf{r}(t) = (\cos t, \sin t, t)$ for $0 \leq t \leq \pi$. Show that $\mathbf{r}(\pi) - \mathbf{r}(0) = (\pi - 0)\mathbf{r}'(t_0)$ is not possible for any $t_0 \in (0, \pi)$. Explain this result geometrically, without computations.

Exercise 10.6.2. Define $\phi : \mathbf{R} \to \mathbf{R}^2$ by $\phi(t) = (t - t^2, t - t^5)$ so that $\phi(0) = (0, 0)$ and $\phi(1) = (0, 0)$. Show that there is no T such that $0 < T < 1$ with $\phi(1) - \phi(0) = (1)\phi'(T)$.

Exercise 10.6.3. Let $r > 0$ and $\mathbf{a} \in \mathbf{R}^n$. Show that the open ball $B_r(\mathbf{a})$, for any norm on $\mathbf{R}^n$, is convex.

Exercise 10.6.4. In Example 10.6.5, verify that if $|\mathbf{u}|_2 = 1$ and $\mathbf{u} \neq \mathbf{e}_j$ for $j = 1, 2$, then $D_{\mathbf{u}} f(0, 0)$ does not exist.

Exercise 10.6.5. Let $U \subseteq \mathbf{R}^n$ be an open convex set. Show that if $g : U \to \mathbf{R}$ is differentiable on U and $dg(\mathbf{x})$ is zero for all $\mathbf{x} \in U$, that is, for each $\mathbf{x} \in U$, $dg(\mathbf{x})\mathbf{h} = 0$ for all $\mathbf{h} \in \mathbf{R}^n$, then there exists a real constant c such that $g(\mathbf{x}) = c$ for all $\mathbf{x} \in U$.

10.7. The Two-Dimensional Implicit Function Theorem

The two-dimensional implicit function theorem deals with the solution of equations of the form

$$f(x, y) = 0$$

for one of the variables in terms of the other variable, in a neighborhood of a known solution point (x_0, y_0). Here f is a real valued function defined on an open subset of the plane containing the point (x_0, y_0).

It is important to emphasize the local nature of the solution provided by the implicit function theorem. Consider the simple equation

$$x^2 + y^2 - 1 = 0$$

defined by the function $f(x, y) = x^2 + y^2 - 1$. Evidently the full solution set is the circle of radius one in $\mathbf{R}^2$. Suppose (a, b) satisfies $a^2 + b^2 - 1 = 0$. We ask whether the solution set in a neighborhood of the point (a, b) can be written as the graph of a function of x; that is, can we solve the equation for y in terms of x in a neighborhood of the point (a, b)? If $\frac{\partial f}{\partial y}(a, b) = 2b \neq 0$, we may do so, since $\frac{\partial f}{\partial y}(a, b) = 2b \neq 0$ if and only if $b \neq 0$. If $b > 0$, then we may solve for y as $y = \sqrt{1 - x^2}$ for x in an interval $(a - \delta, a + \delta)$ for some $\delta > 0$. (See Figure 10.2.) If $b < 0$, then we may solve for $y = -\sqrt{1 - x^2}$, for x in some interval about a.

The two-dimensional implicit function theorem is the simplest version of the more general implicit function theorem presented later in Theorem 11.3.1. Theorem 11.3.1 is proved as an application of the general inverse function theorem (Theorem 11.2.2) for mappings of subsets of $\mathbf{R}^n$ into $\mathbf{R}^n$.

The argument for the special case of the implicit function theorem considered here is worth studying because it provides a nice workout using several previous results.[3] Specifically, we shall use the following results: the single variable mean value theorem; the fact that differentiability of $f(x, y)$ implies f is continuous (Theorem 10.2.5); the intermediate value theorem; the mean value theorem for real functions of two variables (Theorem 10.6.2); and the boundedness of continuous functions on a compact set. This important result is also known as *Dini's implicit function theorem*.

Theorem 10.7.1 (Dini's Implicit Function Theorem). *Let D be an open subset of $\mathbf{R}^2$ and suppose that $f : D \to \mathbf{R}$ has continuous first order partial derivatives in D. If $(x_0, y_0) \in D$ with $f(x_0, y_0) = 0$ and $\frac{\partial f}{\partial y}(x_0, y_0) \neq 0$, then there is an $r > 0$ and a continuously differentiable $g : (x_0 - r, x_0 + r) \to \mathbf{R}$ such that*

$$f(x, g(x)) = 0 \quad \textit{for all } x \in (x_0 - r, x_0 + r),$$

[3]Up to this point in the text, we only have the one-dimensional inverse function theorem (Theorem 5.3.2), but not the two-dimensional version, so we cannot call on the proof of the general implicit function Theorem 11.3.1 at this point.

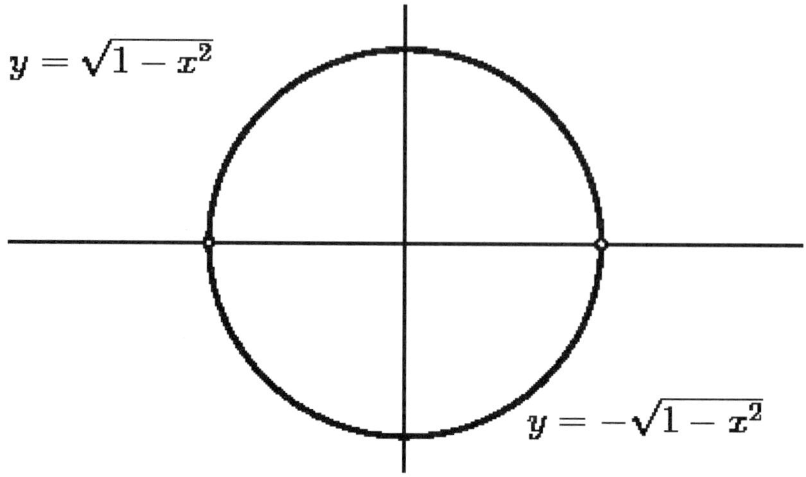

Figure 10.2. We can solve the equation $x^2 + y^2 - 1 = 0$ uniquely for y in terms of x, except around $(\pm 1, 0)$.

and if $|x - x_0| < r$ *and* $|y - y_0| < r$ *with* $f(x, y) = 0$*, then* $y = g(x)$*. Moreover,*

$$\frac{\partial f}{\partial x}(x, g(x)) + \frac{\partial f}{\partial y}(x, g(x))g'(x) = 0 \quad \textit{for all } x \in (x_0 - r, x_0 + r).$$

Proof. Assume that $\frac{\partial f}{\partial y}(x_0, y_0) > 0$. The argument in the case where $\frac{\partial f}{\partial y}(x_0, y_0) < 0$ is similar. Write $\frac{\partial f}{\partial y}(x_0, y_0) =: c > 0$. Since D is open and $\partial f / \partial y$ is continuous at (x_0, y_0), there exists $\delta > 0$ such that

$$R := \{(x, y) : |x - x_0| \leq \delta, |y - y_0| \leq \delta\} \subset D$$

and

$$\frac{\partial f}{\partial y}(x, y) > \frac{c}{2} > 0 \quad \text{for all } (x, y) \in R.$$

It follows from the mean value theorem that for each fixed x with $|x - x_0| < \delta$, the mapping $y \mapsto f(x, y)$ is increasing on the interval $|y - y_0| < \delta$. Then, since $f(x_0, y_0) = 0$, we must have

$$f(x_0, y_0 - \delta) < 0 < f(x_0, y_0 + \delta).$$

Since f is differentiable on D, f is continuous on D by Theorem 10.2.5. By continuity of f, there is a number r such that $0 < r < \delta$ such that

$$f(x, y_0 - \delta) < 0 < f(x, y_0 + \delta) \quad \text{for all } x \in (x_0 - r, x_0 + r).$$

Let $I := (x_0 - r, x_0 + r)$, and let x be a point of I. By the intermediate value theorem, there is some point y between $y_0 - \delta$ and $y_0 + \delta$ such that $f(x, y) = 0$. By the monotone increasing property of f in y, there is only one such point y for a given $x \in I$. So given $x \in I$, let $g(x)$ be this unique point y. Then $g : I \to \mathbf{R}$ and g has the properties that $f(x, g(x)) = 0$ for all $x \in I$ and, if $|x - x_0| < r$ and $|y - y_0| < r$ with $f(x, y) = 0$, then $y = g(x)$.

We proceed to show that g is differentiable at every $x \in I$. We show this first for x_0. Thus, let h be such that $x_0 + h \in I$. Then, by definition of g,

$$f(x_0 + h, g(x_0 + h)) = 0 \quad \text{and} \quad f(x_0, g(x_0)) = 0.$$

By the mean value theorem for real functions of two variables (covered by Theorem 10.6.2), there is a point $\mathbf{c}(h)$ on the line segment joining $(x_0, g(x_0))$ and $(x_0 + h, g(x_0 + h))$ such that

$$\begin{aligned} 0 &= \nabla f(\mathbf{c}(h)) \cdot (h, g(x_0 + h) - g(x_0)) \\ &= \frac{\partial f}{\partial x}(\mathbf{c}(h))h + \frac{\partial f}{\partial y}(\mathbf{c}(h))\,(g(x_0 + h) - g(x_0)), \end{aligned}$$

and hence

$$g(x_0 + h) - g(x_0) = -\frac{\frac{\partial f}{\partial x}(\mathbf{c}(h))}{\frac{\partial f}{\partial y}(\mathbf{c}(h))}\, h.$$

Since R is compact, there is an $M > 0$ such that

$$\left|\frac{\partial f}{\partial x}(x, y)\right| \le M \quad \text{for all } (x, y) \in R.$$

Since $\frac{\partial f}{\partial y}(x, y) > \frac{c}{2}$ for all $(x, y) \in R$, it follows that

$$\Big|g(x_0 + h) - g(x_0)\Big| \le \frac{M}{c/2}|h| \quad \text{if } x_0 + h \in I.$$

Hence, $g(x_0 + h) \to g(x_0)$ as $h \to 0$, and g is continuous at x_0. It follows from the continuity of g at x_0 that $\mathbf{c}(h) \to (x_0, y_0) = (x_0, g(x_0))$ as $h \to 0$. For $h \neq 0$,

$$\frac{g(x_0 + h) - g(x_0)}{h} = -\frac{\frac{\partial f}{\partial x}(\mathbf{c}(h))}{\frac{\partial f}{\partial y}(\mathbf{c}(h))}.$$

Letting $h \to 0$, and using the continuity of $\partial f/\partial x$ and $\partial f/\partial y$, we see that

$$\lim_{h\to 0} \frac{g(x_0 + h) - g(x_0)}{h} = -\lim_{h\to 0} \frac{\frac{\partial f}{\partial x}(\mathbf{c}(h))}{\frac{\partial f}{\partial y}(\mathbf{c}(h))} = -\frac{\frac{\partial f}{\partial x}(x_0, y_0)}{\frac{\partial f}{\partial y}(x_0, y_0)}.$$

Hence, $g'(x_0)$ exists and

$$g'(x_0) = -\frac{\frac{\partial f}{\partial x}(x_0, y_0)}{\frac{\partial f}{\partial y}(x_0, y_0)}.$$

Finally, note that any other point $x \in I = (x_0 - r, x_0 + r)$ satisfies the same conditions as x_0 in the argument above for the existence of the derivative of g and its value. Thus, $g'(x)$ exists for each $x \in I$, and

$$g'(x) = -\frac{\frac{\partial f}{\partial x}(x, y)}{\frac{\partial f}{\partial y}(x, y)}, \qquad x \in I. \tag{10.12}$$

Thus, $\frac{\partial f}{\partial x}(x, g(x)) + \frac{\partial f}{\partial y}(x, g(x))g'(x) = 0$ holds for all $x \in I$. The continuity of g' on I follows from (10.12) and the continuity of $\partial f/\partial x$ and $\partial f/\partial y$. □

In the proof of Theorem 10.7.1 we obtained a formula for $g'(x)$ as part of the argument for its existence, with the result that

$$\frac{\partial f}{\partial x}(x, g(x)) + \frac{\partial f}{\partial y}(x, g(x))g'(x) = 0, \qquad x \in I.$$

If we knew from the start that $g'(x)$ exists for $x \in I$, then the chain rule applied to the identity $f(x, g(x)) = 0$, $x \in I$, would yield the same formula.

The implicit function theorem is a useful, even essential, tool in the investigation of many perturbation problems. As simple examples, we include exercises for some regular perturbation problems for simple algebraic equations. See Exercises 10.7.2, 10.7.3.

Exercises.

Exercise 10.7.1. Suppose $f(x, y)$ is twice continuously differentiable in an open neighborhood of (x_0, y_0), and $f(x_0, y_0) = 0$, $\frac{\partial f}{\partial y}(x_0, y_0) \neq 0$. Show that the solution function $g(x)$ of Theorem 10.7.1 is also twice continuously differentiable in a neighborhood of x_0, and give a formula for $g''(x)$.

Exercise 10.7.2. The equation $\phi(x, \epsilon) = x^2 + 2\epsilon x - 3 = 0$, for small ϵ, is a perturbation of the so-called reduced problem, $\phi(x, 0) = x^2 - 3 = 0$. The reduced problem has two solutions, $x = \pm\sqrt{3}$.

1. Forgetting the quadratic formula for a moment, investigate the solution of $\phi(x, \epsilon) = 0$ for x as a function of ϵ near each of the solutions $(\sqrt{3}, 0)$ and $(-\sqrt{3}, 0)$. Thus find two solution branches, $x = g_1(\epsilon)$ and $x = g_2(\epsilon)$, for ϵ near 0.
2. For each solution branch, give the Taylor expansion of the solution function about $\epsilon = 0$ through second-order terms.
3. Remembering the quadratic formula now, compare your Taylor approximations with the exact solutions.

Exercise 10.7.3. The equation $\phi(x, \epsilon) = x^3 + x^2 - (2 + \epsilon)x + 2\epsilon = 0$, for small nonzero ϵ, is a perturbation of the reduced problem, $\phi(x, 0) = x^3 + x^2 - 2x = 0$, which has the solutions, $x_1 = -2$, $x_2 = 0$, $x_3 = 1$.

1. Show that for sufficiently small ϵ, it is possible to solve for a root $g_j(\epsilon)$ of $\phi(x, \epsilon) = 0$ near the root x_j of the reduced problem, for $j = 1, 2, 3$.
2. Find a Taylor expansion for each of the root branches, $g_j(\epsilon)$, $j = 1, 2, 3$, through second-order terms in ϵ.

10.8. The Mean Value Theorem: Vector Functions

Our next goal is to extend the mean value theorem to the case of mappings defined on subsets of $\mathbf{R}^n$ and taking values in $\mathbf{R}^m$. It is convenient to use the max norm on vectors in the domain and range spaces, and the induced matrix norm on $m \times n$ matrices. These norms are given by

$$|\mathbf{x}|_\infty = \max_{1 \le i \le n} |x_i|, \quad \mathbf{x} = (x_1, \dots, x_n) \in \mathbf{R}^n$$

(and similarly in $\mathbf{R}^m$) and

$$\|L\|_\infty = \max_{|\mathbf{x}|_\infty = 1} |L\mathbf{x}|_\infty, \quad L \in \mathbf{R}^{m \times n}.$$

If L is an $m \times n$ matrix, then

$$|L\,\mathbf{x}|_\infty \le \|L\|_\infty \, |\mathbf{x}|_\infty \tag{10.13}$$

for every $\mathbf{x} \in \mathbf{R}^n$. Write $L\mathbf{x} = (L_1\mathbf{x}, \ldots, L_m\mathbf{x})$ where L_i is the i-th row of L. Then the i-th component of $L\mathbf{x}$ is the i-th row of L times $\mathbf{x}$, that is,

$$(L\mathbf{x})_i = L_i\,\mathbf{x}.$$

Since the matrix norm $\|\cdot\|_\infty$ is the maximum absolute row sum of a matrix, we have that for each $i = 1, \ldots, m$,

$$\|L_i\|_\infty \le \max_{1 \le k \le m} \|L_k\|_\infty = \|L\|_\infty. \tag{10.14}$$

We can now establish the mean value theorem for C^1 mappings $\mathbf{F} : U \subseteq \mathbf{R}^n \to \mathbf{R}^m$.

Theorem 10.8.1 (Mean Value Theorem for Vector Functions). *Let U be an open set in $\mathbf{R}^n$ containing the line segment $l_{\mathbf{ab}}$ joining the points $\mathbf{a}$ and $\mathbf{b}$, and let $\mathbf{F} : U \to \mathbf{R}^m$ be C^1 on U. Then*

$$|\mathbf{F}(\mathbf{b}) - \mathbf{F}(\mathbf{a})|_\infty \le \Big(\max_{\mathbf{x} \in l_{\mathbf{ab}}} \|D\mathbf{F}(\mathbf{x})\|_\infty\Big)|\mathbf{b} - \mathbf{a}|_\infty\ . \tag{10.15}$$

Proof. Let $\mathbf{h} = \mathbf{b} - \mathbf{a}$. Define a curve $\phi : [0,1] \to \mathbf{R}^m$ by

$$\phi(t) = \mathbf{F}(\mathbf{a} + t\mathbf{h}), \quad t \in [0,1].$$

Then ϕ is C^1. If $\mathbf{F}(\mathbf{x}) = (f_1(\mathbf{x}), \ldots, f_m(\mathbf{x}))$, that is, the component functions of $\mathbf{F}$ are denoted by f_i, $i = 1, \ldots, m$, then the i-th component of ϕ is

$$\phi_i(t) = f_i(\mathbf{a} + t\mathbf{h}), \quad t \in [0,1].$$

Note that $\phi_i(0) = f_i(\mathbf{a})$ and $\phi_i(1) = f_i(\mathbf{b})$ for each i. By the chain rule, we have

$$\phi_i'(t) = df_i(\mathbf{a} + t\mathbf{h})\mathbf{h}, \quad t \in [0,1].$$

For the given $\mathbf{a}$ and $\mathbf{b}$, there is a positive integer $j \in \{1, \ldots, m\}$ such that

$$|\mathbf{F}(\mathbf{b}) - \mathbf{F}(\mathbf{a})|_\infty = |f_j(\mathbf{b}) - f_j(\mathbf{a})| = |\phi_j(1) - \phi_j(0)|.$$

So we can estimate the difference we are interested in by concentrating on the j-th component. By the fundamental theorem of calculus,

$$|\mathbf{F}(\mathbf{b}) - \mathbf{F}(\mathbf{a})|_\infty = |\phi_j(1) - \phi_j(0)| = \Big|\int_0^1 \phi_j'(s)\,ds\Big|. \tag{10.16}$$

Now, we estimate

$$\begin{aligned}
\Big|\int_0^1 \phi_j'(s)\,ds\Big| &\le \int_0^1 |\phi_j'(s)|\,ds \\
&= \int_0^1 |df_j(\mathbf{a} + s\mathbf{h})\mathbf{h}|\,ds \\
&\le \max_{0 \le s \le 1} |df_j(\mathbf{a} + s\mathbf{h})\mathbf{h}| \\
&\le \Big(\max_{0 \le s \le 1} \|df_j(\mathbf{a} + s\mathbf{h})\|_\infty\Big)\,|\mathbf{h}|_\infty,
\end{aligned}$$

where the last line follows from (10.13). Letting $\tau \in [0,1]$ be the point at which $\max_{0 \le s \le 1} \|df_j(\mathbf{a} + s\mathbf{h})\|_\infty$ occurs, we then have

$$\Big|\int_0^1 \phi_j'(s)\,ds\Big| \le \|df_j(\mathbf{a} + \tau\mathbf{h})\|_\infty\,|\mathbf{h}|_\infty \le \|D\mathbf{F}(\mathbf{a} + \tau\mathbf{h})\|_\infty\,|\mathbf{h}|_\infty \tag{10.17}$$

by (10.14). Now (10.15) is an immediate consequence of (10.16) and (10.17), since $\mathbf{h} = \mathbf{b} - \mathbf{a}$ and $\mathbf{a} + \tau\mathbf{h} \in l_{\mathbf{ab}}$. $\square$

The mean value Theorem 10.8.1 has many important consequences for the local behavior of mappings, as we will see in the remainder of this section. Theorem 10.8.1 estimates the difference in function values at two given points using an upper bound on the norm of the derivative along the line segment joining the two points. The next corollary shows how close a linear estimate of the function difference can be, using any linear mapping $\mathbf{L}$ and estimating $\mathbf{F}(\mathbf{a}+\mathbf{h}) - \mathbf{F}(\mathbf{a})$ by $\mathbf{Lh}$.

Corollary 10.8.2. *Let U be an open set in $\mathbf{R}^n$ and let $\mathbf{F} : U \to \mathbf{R}^m$ be C^1 on U. If the line segment l joining the points $\mathbf{a}$ and $\mathbf{a}+\mathbf{h}$ is contained in U, then for any linear mapping $\mathbf{L} : \mathbf{R}^n \to \mathbf{R}^m$,*

$$|\,\mathbf{F}(\mathbf{a}+\mathbf{h}) - \mathbf{F}(\mathbf{a}) - \mathbf{Lh}\,|_\infty \le \Big(\max_{\mathbf{x}\in l} \|D\mathbf{F}(\mathbf{x}) - \mathbf{L}\|_\infty\Big)|\mathbf{h}|_\infty.$$

Proof. Define $\mathbf{G} : U \to \mathbf{R}^m$ by $\mathbf{G}(\mathbf{x}) = \mathbf{F}(\mathbf{x}) - \mathbf{Lx}$. Since $\mathbf{L}$ is linear, for each $\mathbf{x} \in U$, we have $D\mathbf{G}(\mathbf{x}) = D\mathbf{F}(\mathbf{x}) - \mathbf{L}$, and again by linearity of $\mathbf{L}$,

$$\mathbf{G}(\mathbf{a}+\mathbf{h}) - \mathbf{G}(\mathbf{a}) = \mathbf{F}(\mathbf{a}+\mathbf{h}) - \mathbf{F}(\mathbf{a}) - \mathbf{Lh}.$$

We may apply Theorem 10.8.1 to get

$$\begin{aligned} |\mathbf{F}(\mathbf{a}+\mathbf{h}) - \mathbf{F}(\mathbf{a}) - \mathbf{Lh}|_\infty &= |\mathbf{G}(\mathbf{a}+\mathbf{h}) - \mathbf{G}(\mathbf{a})|_\infty \\ &\le \Big(\max_{\mathbf{x}\in l} \|D\mathbf{G}(\mathbf{x})\|_\infty\Big)|\mathbf{h}|_\infty \\ &= \Big(\max_{\mathbf{x}\in l} \|D\mathbf{F}(\mathbf{x}) - \mathbf{L}\|_\infty\Big)|\mathbf{h}|_\infty, \end{aligned}$$

which proves the result. $\square$

Corollary 10.8.2 provides an upper bound for the error in using a linear mapping $\mathbf{L}$ to approximate the difference in function values at two points, the upper bound depending on the norm of the difference between $\mathbf{L}$ and the derivatives of $\mathbf{F}$ along the line segment joining the two points. This is not too surprising, given the estimate of the mean value theorem, where the derivatives of $\mathbf{F}$ along the line segment gave the upper bound for the difference in function values. If we use $\mathbf{L} = D\mathbf{F}(\mathbf{a})$, then we have the estimate

$$|\mathbf{F}(\mathbf{a}+\mathbf{h}) - \mathbf{F}(\mathbf{a}) - D\mathbf{F}(\mathbf{a})\mathbf{h}|_\infty \le \Big(\max_{\mathbf{x}\in l} \|D\mathbf{F}(\mathbf{x}) - D\mathbf{F}(\mathbf{a})\|_\infty\Big)|\mathbf{h}|_\infty \tag{10.18}$$

where l is the segment joining $\mathbf{a}$ and $\mathbf{a}+\mathbf{h}$.

With respect to the max norm on $\mathbf{R}^n$, the closed ball of radius $r > 0$, usually denoted by $\bar{B}_r(\mathbf{0})$, is actually a *closed cube of side length* $2r$, which we shall also denote by C_r. Thus,

$$\begin{aligned} C_r &= \{\mathbf{x} \in \mathbf{R}^n : \|\mathbf{x}\|_\infty \le r\} \\ &= \bar{B}_r(\mathbf{0}) \quad \text{for the max norm} \\ &= [-r, r] \times \cdots \times [-r, r]. \end{aligned}$$

We call r the **radius** of the cube C_r and observe that this radius is half the common side length $2r$.

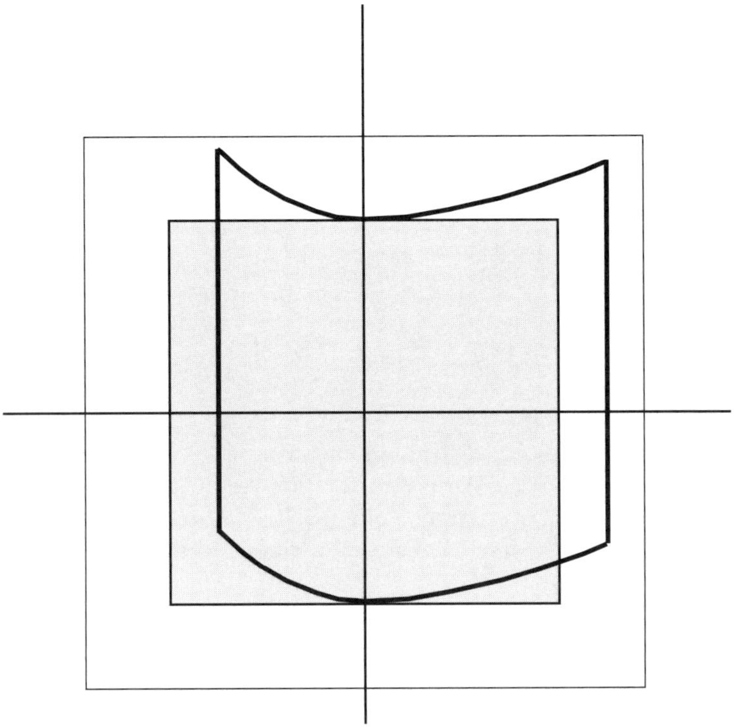

Figure 10.3. Under a C^1 mapping $\mathbf{F}$ such that $\mathbf{F}(\mathbf{0}) = \mathbf{0}$ and $D\mathbf{F}(\mathbf{0}) = I$, the shaded cube C_r centered at the origin is mapped to an image $\mathbf{F}(C_r)$ (with bold boundary curve) which is contained in a slightly larger cube $C_{(1+\epsilon)r}$. See Corollary 10.8.3.

Closed cubes play an important role in the development of results on the transformation of multiple Riemann integrals. The special type of mapping $\mathbf{F} : U \subseteq \mathbf{R}^n \to \mathbf{R}^n$ considered in the next result has important consequences in that context, and also contributes to the development of the inverse function theorem. To provide some intuition for this result, consider that a C^1 mapping $\mathbf{F}$ defined on an open neighborhood of a cube C_r will distort that cube in a smooth way. If the mapping fixes the origin, and if the derivative of the mapping at the origin is the identity, then the image of the cube, $\mathbf{F}(C_r)$, will be contained in another cube of slightly larger radius. (See Figure 10.3.) The next corollary gives a precise statement.

Corollary 10.8.3. *Let U be an open set in $\mathbf{R}^n$ containing the cube*

$$C_r = \{\mathbf{x} \in \mathbf{R}^n : \|\mathbf{x}\|_\infty \leq r\}$$

and let $\mathbf{F} : U \to \mathbf{R}^n$ be C^1 on U with $\mathbf{F}(\mathbf{0}) = \mathbf{0}$ and $D\mathbf{F}(\mathbf{0}) = I$. If

$$\|D\mathbf{F}(\mathbf{x}) - I\|_\infty < \epsilon \quad \textit{for all } \mathbf{x} \in C_r,$$

then $\mathbf{F}(C_r) \subset C_{(1+\epsilon)r}$.

Proof. Observe that C_r, being a closed ball for the max norm, is a convex set. We may apply the previous Corollary 10.8.2 by setting $\mathbf{a} = \mathbf{0}$, $\mathbf{h} = \mathbf{x} \in C_r$, and

$\mathbf{L} = D\mathbf{F}(\mathbf{0}) = I$, and letting l be the line segment joining $\mathbf{0}$ and $\mathbf{x}$. (See Figure 10.8.2.) Then, writing $\mathbf{z}$ for arbitrary points on l, we obtain

$$\begin{aligned} |\,\mathbf{F}(\mathbf{x}) - \mathbf{x}\,|_\infty &\leq \left(\max_{\mathbf{z}\in l} \|D\mathbf{F}(\mathbf{z}) - I\|_\infty\right)|\mathbf{x}|_\infty \\ &< \epsilon\,|\mathbf{x}|_\infty. \end{aligned}$$

The reverse triangle inequality,

$$\Big|\,|\,\mathbf{F}(\mathbf{x})|_\infty - |\mathbf{x}\,|_\infty\Big| \leq |\,\mathbf{F}(\mathbf{x}) - \mathbf{x}\,|_\infty,$$

implies that for all $\mathbf{x} \in C_r$,

$$|\mathbf{F}(\mathbf{x})|_\infty < \epsilon\,|\mathbf{x}|_\infty + |\mathbf{x}|_\infty = (1+\epsilon)|\mathbf{x}|_\infty \leq (1+\epsilon)r,$$

hence $\mathbf{F}(C_r) \subset C_{(1+\epsilon)r}$. □

Corollary 10.8.3 is restricted to mappings from $U \subseteq \mathbf{R}^n$ to $\mathbf{R}^n$ because of the assumption that $D\mathbf{F}(\mathbf{0}) = I$. We note that this result states only that $\mathbf{F}(C_r)$ is *contained in* a cube of slightly larger volume. It is a legitimate question whether or not the image set $\mathbf{F}(C_r)$ itself has a well-defined volume measure. We address this issue later in the book.

We now return to the more general case of mappings from subsets of $\mathbf{R}^n$ to $\mathbf{R}^m$. The next result is a good example of how a property of the derivative $D\mathbf{F}(\mathbf{a})$ reflects a local property of $\mathbf{F}$ itself near $\mathbf{a}$.

Corollary 10.8.4. *Let U be an open set in $\mathbf{R}^n$, let $\mathbf{a} \in U$, and suppose that $\mathbf{F} : U \to \mathbf{R}^m$ is C^1 on some open neighborhood of $\mathbf{a}$. If $D\mathbf{F}(\mathbf{a}) : \mathbf{R}^n \to \mathbf{R}^m$ is one-to-one, then $\mathbf{F}$ is one-to-one on an open neighborhood of $\mathbf{a}$.*

Proof. Consider first what we know about the derivative $D\mathbf{F}(\mathbf{a})$. Since $D\mathbf{F}(\mathbf{a}) : \mathbf{R}^n \to \mathbf{R}^m$ is one-to-one, necessarily $n \leq m$. Let $C_1 = \{\mathbf{x} \in \mathbf{R}^n : |\mathbf{x}|_\infty \leq 1\}$ be the closed cube centered at the origin in $\mathbf{R}^n$ with radius 1. The boundary of C_1 is

$$\partial C_1 = \{\mathbf{x} \in \mathbf{R}^n : |\mathbf{x}|_\infty = 1\},$$

which is a compact set. Since the linear mapping $D\mathbf{F}(\mathbf{a})$ is continuous on ∂C_1,

$$\eta = \min_{|\mathbf{h}|_\infty = 1} |D\mathbf{F}(\mathbf{a})\mathbf{h}|_\infty \tag{10.19}$$

exists, and $\eta > 0$ since $D\mathbf{F}(\mathbf{a})$ is one-to-one.

Let U_1 be an open neighborhood of $\mathbf{a}$ on which $\mathbf{F}$ is C^1. Let $0 < \epsilon < \eta$. Then there is an $r = r(\epsilon) > 0$ such that if

$$V = \{\mathbf{x} \in \mathbf{R}^n : |\mathbf{x} - \mathbf{a}|_\infty < r\},$$

then $V \subseteq U_1$ and

$$\|D\mathbf{F}(\mathbf{x}) - D\mathbf{F}(\mathbf{a})\|_\infty < \epsilon$$

for all $\mathbf{x} \in V$. (The set V is the interior of a closed cube centered at $\mathbf{a}$, that is, $V = \operatorname{Int} C_r(\mathbf{a})$.) If $\mathbf{x}, \mathbf{y} \in V$ and $\mathbf{x} \neq \mathbf{y}$, then by Corollary 10.8.2 with $\mathbf{L} = D\mathbf{F}(\mathbf{a})$ and l being the line segment joining $\mathbf{x}$ and $\mathbf{y}$,

$$\begin{aligned} |\,\mathbf{F}(\mathbf{x}) - \mathbf{F}(\mathbf{y}) - D\mathbf{F}(\mathbf{a})(\mathbf{x}-\mathbf{y})\,|_\infty &\leq \left(\max_{\mathbf{z}\in l} \|D\mathbf{F}(\mathbf{z}) - D\mathbf{F}(\mathbf{a})\|_\infty\right)|\mathbf{x}-\mathbf{y}|_\infty \\ &< \epsilon\,|\mathbf{x}-\mathbf{y}|_\infty. \end{aligned} \tag{10.20}$$

Now an application of the reverse triangle inequality and (10.19) gives

$$\begin{aligned} |\mathbf{F}(\mathbf{x}) - \mathbf{F}(\mathbf{y})|_\infty &> |D\mathbf{F}(\mathbf{a})(\mathbf{x}-\mathbf{y})|_\infty - \epsilon\,|\mathbf{x}-\mathbf{y}|_\infty \\ &\geq \eta\,|\mathbf{x}-\mathbf{y}|_\infty - \epsilon\,|\mathbf{x}-\mathbf{y}|_\infty \\ &= (\eta-\epsilon)|\mathbf{x}-\mathbf{y}|_\infty. \end{aligned}$$

Since $\eta - \epsilon > 0$, $\mathbf{x} \neq \mathbf{y}$ implies $\mathbf{F}(\mathbf{x}) \neq \mathbf{F}(\mathbf{y})$. So $\mathbf{F}$ is one-to-one on V. □

From the proof of Corollary 10.8.4, we have that $\mathbf{F}|_V : V \to \mathbf{F}(V)$ is one-to-one and onto, so the inverse mapping $\mathbf{G} : \mathbf{F}(V) \to V$ exists. Moreover, $\mathbf{G}$ is continuous, because for any $\mathbf{y}_1, \mathbf{y}_2 \in \mathbf{F}(V)$, the final estimate of the proof gives

$$|\mathbf{G}(\mathbf{y}_1) - \mathbf{G}(\mathbf{y}_2)|_\infty < \frac{1}{\eta-\epsilon}|\mathbf{y}_1 - \mathbf{y}_2|_\infty,$$

so $\mathbf{G}$ satisfies a Lipschitz condition on $\mathbf{F}(V)$. If $n = m$, then we can say more about $\mathbf{G}$; in particular, $\mathbf{G}$ is differentiable at $\mathbf{b} = \mathbf{F}(\mathbf{a})$ and $D\mathbf{G}(\mathbf{b}) = [D\mathbf{F}(\mathbf{a})]^{-1}$. To follow up on this suggestion, see Exercise 10.8.1, which can serve as a useful introduction to some of the ideas of the inverse function theorem in Chapter 11.

We wrap up this section with a reminder that Corollary 10.8.4 is a local result. If $\mathbf{F} : U \to \mathbf{R}^m$ is C^1 on U and $D\mathbf{F}(\mathbf{x})$ is one-to-one for each $\mathbf{x} \in U$, we cannot conclude that $\mathbf{F}$ is one-to-one on U, only that $\mathbf{F}$ is locally one-to-one on some open neighborhood of each point of U. Consider the following example.

Example 10.8.5. The mapping $\mathbf{F} : \mathbf{R}^2 \to \mathbf{R}^2$ defined by

$$\mathbf{F}(x, y) = (x^2 - y^2, 2xy)$$

is locally one-to-one in a neighborhood of each point $(x, y) \neq (0, 0)$, since

$$\det J_{\mathbf{F}}(x,y) = \det \begin{bmatrix} 2x & -2y \\ 2y & 2x \end{bmatrix} = 4x^2 + 4y^2 \neq 0$$

for $(x, y) \neq (0, 0)$. But $\mathbf{F}$ is not globally one-to-one on the set $\mathbf{R}^2 - \{(0, 0)\}$. For example, $\mathbf{F}(1, 0) = \mathbf{F}(-1, 0) = (1, 0)$. See also Exercise 10.8.2. △

Exercises.

Exercise 10.8.1. Under the hypotheses of Corollary 10.8.4, suppose in addition that $n = m$ and hence that $D\mathbf{F}(\mathbf{a})$ is invertible. Let $\mathbf{G}$ be the inverse of $\mathbf{F}|_V$. Let $\mathbf{b} = F(\mathbf{a})$ and write $\mathbf{y} = \mathbf{F}(\mathbf{x})$ for $\mathbf{x} \in V$. This exercise shows that $\mathbf{G}$ is differentiable at $\mathbf{b}$ and $D\mathbf{G}(\mathbf{b}) = [D\mathbf{F}(\mathbf{a})]^{-1}$.

1. Write the difference quotient defining $D\mathbf{G}(\mathbf{b})$, inserting the candidate $[D\mathbf{F}(\mathbf{a})]^{-1}$ for $D\mathbf{G}(\mathbf{b})$ and using increment $\mathbf{h} = \mathbf{y} - \mathbf{b}$.
2. Factor out $[D\mathbf{F}(\mathbf{a})]^{-1}$ from the entire vector expression in the numerator of the quotient in part 1.
3. Use the fact, proven in Corollary 10.8.4, that $|\mathbf{F}(\mathbf{x}) - \mathbf{F}(\mathbf{y})|_\infty > (\eta-\epsilon)|\mathbf{x}-\mathbf{y}|_\infty$ for $\mathbf{x}$ and $\mathbf{y}$ in V. (This also proved that the inverse mapping $\mathbf{G}$ is continuous.)
4. Complete the argument to show that $D\mathbf{G}(\mathbf{b}) = [D\mathbf{F}(\mathbf{a})]^{-1}$.

Exercise 10.8.2. Show that the component functions of the mapping $\mathbf{F}$ in Example 10.8.5 are the real and imaginary parts of the complex mapping $f : \mathbf{C} \to \mathbf{C}$, $f(z) = z^2$. Use polar coordinates to show that f (and hence $\mathbf{F}$) maps each circle centered at the origin with positive radius r *twice* onto the circle centered at the origin with radius r^2.

Exercise 10.8.3. If other norms, $|\cdot|_n$ on $\mathbf{R}^n$ and $|\cdot|_m$ on $\mathbf{R}^m$, are used, and we employ the operator norm

$$\|\mathbf{L}\|_{mn} := \max_{|\mathbf{v}|_n=1} |\mathbf{L}\mathbf{v}|_m$$

for linear mappings $\mathbf{L}$ from $\mathbf{R}^n$ to $\mathbf{R}^m$, then the result of the mean value theorem (Theorem 10.8.1) can be stated as

$$|\mathbf{F}(\mathbf{b}) - \mathbf{F}(\mathbf{a})|_m \leq \Big(\max_{\mathbf{x}\in l_{\mathbf{ab}}} \|D\mathbf{F}(\mathbf{x})\|_{mn} \Big) |\mathbf{b} - \mathbf{a}|_n .$$

Prove this by the following two steps:

1. Let $\phi : [0, 1] \to \mathbf{R}^m$ be a continuously differentiable curve with $|\phi'(t)|_m \leq M$ for all $t \in [0, 1]$, where $|\cdot|_m$ is any norm on $\mathbf{R}^m$. Show that
$$|\phi(1) - \phi(0)|_m \leq M.$$
Hint: If $\epsilon > 0$, denote by S_ϵ the set of points $x \in [0, 1]$ such that
$$|\phi(t) - \phi(0)|_m \leq (M + \epsilon)\, t + \epsilon$$
for all $t \leq x$. Verify that S_ϵ is nonempty. Let $b = \sup S_\epsilon$, and verify that $b \in S_\epsilon$. Show that if $b < 1$, then for sufficiently small $|h|$, say $|h| \leq \delta$, where $\delta > 0$, we have $b + h$ in $[0, 1]$ and
$$\left| \frac{\phi(b + h) - \phi(b)}{h} \right| < M + \epsilon.$$
Conclude that $b + \delta \in S_\epsilon$, a contradiction. Hence, $b = 1$. Thus, for every $\epsilon > 0$,
$$|\phi(1) - \phi(0)|_m \leq (M + \epsilon) + \epsilon.$$
2. Let $\mathbf{h} = \mathbf{b} - \mathbf{a}$. Apply part 1 to the curve $\phi(t) = \mathbf{F}(\mathbf{a} + t\mathbf{h})$ to deduce the stated result when $\mathbf{F}$ satisfies the hypotheses of the mean value Theorem 10.8.1.

10.9. Taylor's Theorem

In this section we use the letter r to denote an order of continuous differentiability of our functions. Thus, suppose f is of class C^{r+1} on an open set U in $\mathbf{R}^n$ containing the point $\mathbf{a}$. Then there is some open ball about $\mathbf{a}$ on which $f(\mathbf{a} + t\mathbf{h})$ is defined for $-1 \leq t \leq 1$. The goal of Taylor's theorem, as in the single variable case, is to express the function values $f(\mathbf{a} + \mathbf{h})$ for small $|\mathbf{h}|$ by an expression of the form

$$f(\mathbf{a} + \mathbf{h}) = P(\mathbf{h}) + R_{\mathbf{a},r}(\mathbf{h})$$

where $P(\mathbf{h})$ is a degree r polynomial whose derivatives of order $k \leq r$ at $\mathbf{a}$ all agree with the corresponding derivatives of f, and $R_{\mathbf{a},r}(\mathbf{h})$ is the remainder term, or error, in the approximation of $f(\mathbf{a}+\mathbf{h})$ by $P(\mathbf{h})$.[4] Instead of stating the formula for $P(\mathbf{h})$ outright, let us see how its expression arises along with the remainder term from the known single variable result (Theorem 6.8.1).

[4] Of course, $P(h)$ also depends on $\mathbf{a}$ and r, but $R_{\mathbf{a},r}$ carries this information, P and $R_{\mathbf{a},r}$ are most likely to appear together, and $R_{\mathbf{a},r}$ is the object that requires a reference most often, so we avoid a more complicated notation for P.

With $\mathbf{a}$ and $\mathbf{h}$ fixed, define

$$g(t) = f(\mathbf{a} + t\mathbf{h})$$

for $0 \le t \le 1$. Since $g(1) = f(\mathbf{a}+\mathbf{h})$ and $g(0) = f(\mathbf{a})$, we are interested in the Taylor expansion

$$\begin{aligned} g(1) &= g(0) + g'(0) + \cdots + \frac{1}{r!} g^{(r)}(0) + (\text{remainder}) \\ &= \sum_{k=0}^{r} \frac{1}{k!} g^{(k)}(0) + (\text{remainder}), \end{aligned}$$

where the remainder is determined by (6.10) (with f there replaced by g, a replaced by 0, and h replaced by 1). This expansion exists by Theorem 6.8.1 since g is of class C^{r+1} on an open interval containing $[-1, 1]$, so we need only compute the coefficients. We have $g(0) = f(\mathbf{a})$. By the chain rule,

$$g'(t) = \nabla f(\mathbf{a} + t\mathbf{h}) \cdot \mathbf{h} = \mathbf{h} \cdot \nabla f(\mathbf{a} + t\mathbf{h}),$$

so $g'(0) = \nabla f(\mathbf{a}) \cdot \mathbf{h}$. To help with higher derivatives, note that we may write

$$\begin{aligned} g'(t) &= \mathbf{h} \cdot \nabla f(\mathbf{a} + t\mathbf{h}) \\ &= \sum_{i=1}^{n} h_i \frac{\partial f}{\partial x_i}(\mathbf{a} + t\mathbf{h}) \qquad (10.21) \\ &= \Big[\Big(\sum_{i=1}^{n} h_i \frac{\partial}{\partial x_i}\Big) f\Big](\mathbf{a} + t\mathbf{h}) \\ &= [\mathcal{L} f](\mathbf{a} + t\mathbf{h}), \end{aligned}$$

where we have written

$$\mathcal{L} := \sum_{i=1}^{n} h_i \frac{\partial}{\partial x_i} \qquad (10.22)$$

as an operator (a linear differential operator) applied to f.[5] Since $\mathcal{L}f$ is a new function, one can evaluate it at a point $\mathbf{z}$, indicated by $[\mathcal{L}f](\mathbf{z})$. By (10.21), observe that we will apply the same differential operator $\mathcal{L}$ to each term $\frac{\partial f}{\partial x_i}(\mathbf{a} + t\mathbf{h})$ when we differentiate to obtain the second derivative of g. Thus,

$$\begin{aligned} g''(t) &= \sum_{i=1}^{n} h_i \Big(\sum_{j=1}^{n} h_j \frac{\partial^2 f}{\partial x_i \partial x_j}(\mathbf{a} + t\mathbf{h}) \Big) \qquad (10.23) \\ &= \sum_{i=1}^{n} h_i \Big[\mathcal{L} \frac{\partial f}{\partial x_i}\Big](\mathbf{a} + t\mathbf{h}) \qquad (\text{definition of } \mathcal{L}) \\ &= \Big[\mathcal{L}\Big(\sum_{i=1}^{n} h_i \frac{\partial f}{\partial x_i}\Big)\Big](\mathbf{a} + t\mathbf{h}) \qquad (\text{linearity of } \mathcal{L}) \\ &= [\mathcal{L}(\mathcal{L}f)](\mathbf{a} + t\mathbf{h}) \\ &= [\mathcal{L}^2 f](\mathbf{a} + t\mathbf{h}). \end{aligned}$$

[5] Some texts use, and some readers may like, the symbolic notation $\mathbf{h} \cdot \nabla$ for the operator $\mathcal{L}$, since $[\mathcal{L}f](\mathbf{z})$ is the dot product of the vectors $\mathbf{h}$ and $\nabla f(\mathbf{z})$. We opt for the simpler notation $\mathcal{L}$, since it is easy to remember the operator formula.

Since $\mathcal{L}$ is an operator on functions (that is, a function or mapping applied to functions) the notation $\mathcal{L}^k f$ simply means the iterated application of $\mathcal{L}$ starting from f, defined inductively by $\mathcal{L}^k f = \mathcal{L}(\mathcal{L}^{k-1} f)$. Thus $\mathcal{L}^k f$ will make sense as long as $\mathcal{L}^{k-1} f$ has partial derivatives on the domain of interest for f.

Now back to our function g, which is of class C^{r+1}. Assuming that $g^{(k)}(t) = [\mathcal{L}^k f](\mathbf{a} + t\mathbf{h})$ for some positive integer k, we have $g^{(k+1)}(t) = [\mathcal{L}^{k+1} f](\mathbf{a} + t\mathbf{h})$ if f is C^{k+1}. So we have $g^{(k)}(0) = [\mathcal{L}^k f](\mathbf{a})$ for $0 \le k \le r$ if f is at least C^r. Thus, Theorem 6.8.1 applied to the function $g(t) = f(\mathbf{a} + t\mathbf{h})$ yields *Taylor's theorem.*

Theorem 10.9.1 (Taylor's Theorem). *Suppose $f : U \subseteq \mathbf{R}^n \to \mathbf{R}$ where U is an open set containing* $\mathbf{a}$, *and suppose that f is C^{r+1} on U. Let $\mathcal{L}$ be the operator defined in (10.22). Then for sufficiently small $|\mathbf{h}|$ we have the expansion*

$$f(\mathbf{a} + \mathbf{h}) = P(\mathbf{h}) + R_{\mathbf{a},r}(\mathbf{h}) \tag{10.24}$$

where

$$P(\mathbf{h}) = \sum_{k=0}^{r} \frac{1}{k!} [\mathcal{L}^k f](\mathbf{a}) = \sum_{k=0}^{r} \frac{1}{k!} \Big[\Big(\sum_{i=1}^{n} h_i \frac{\partial}{\partial x_i} \Big)^k f \Big] (\mathbf{a})$$

is the **Taylor polynomial of degree** r **for** f **at** $\mathbf{a}$, *and the* **remainder** $R_{\mathbf{a},r}(\mathbf{h})$ *is defined by $R_{\mathbf{a},r}(\mathbf{h}) = f(\mathbf{a} + \mathbf{h}) - P(\mathbf{h})$.*

A word on notation is in order. For a function of two variables, we can deal with the notational issue by using a binomial expansion of the operator $\mathcal{L} = h_1 \frac{\partial}{\partial x_1} + h_2 \frac{\partial}{\partial x_2}$. This binomial expansion only requires that the operator of multiplication by a constant h_i and the operator $\frac{\partial}{\partial x_j}$ commute for any i, j, as they do: $h_i(\frac{\partial}{\partial x_j} f) = \frac{\partial}{\partial x_j}(h_i f)$ for all differentiable f. Thus, the binomial theorem gives

$$\begin{aligned} \mathcal{L}^k f(\mathbf{a}) &= \Big[\Big(h_1 \frac{\partial}{\partial x_1} + h_2 \frac{\partial}{\partial x_2} \Big)^k f \Big] (\mathbf{a}) \\ &= \sum_{j=0}^{k} \frac{k!}{j!(k-j)!} \frac{\partial^k f}{\partial x_1^j \partial x_2^{k-j}} (\mathbf{a})\, h_1^j\, h_2^{k-j}. \end{aligned} \tag{10.25}$$

See Exercises 10.9.1, 10.9.2 for calculations with specific functions of two variables. For functions of three variables, (10.24) applies with an expansion of $\mathcal{L}^k$ where $\mathcal{L} = \sum_{i=1}^{3} h_i \frac{\partial}{\partial x_i}$.

If an expansion for $f(\mathbf{a} + \mathbf{h})$ is needed only through quadratic terms in the components of $\mathbf{h}$, as in the next section on relative extrema at nondegenerate critical points of f, then from (10.23) and $g(0)$, $g'(0)$ determined earlier, we may write

$$\begin{aligned} f(\mathbf{a} + \mathbf{h}) &= f(\mathbf{a}) + \sum_{i=1}^{n} \frac{\partial f}{\partial x_i} (\mathbf{a})\, h_i \\ &\quad + \frac{1}{2!} \sum_{i=1}^{n} \sum_{j=1}^{n} \frac{\partial^2 f}{\partial x_i \partial x_j} (\mathbf{a})\, h_i\, h_j \, + \, R_{\mathbf{a},2}(\mathbf{h}). \end{aligned}$$

As noted above, this is the remainder for the approximation of $g(1) = f(\mathbf{a}+\mathbf{h})$ by second-order terms, given by (6.10) (with f, a, and h there, replaced by g, 0, and 1, respectively). Thus, we have

$$R_{\mathbf{a},2}(\mathbf{h}) = \frac{1}{2!}\int_0^1 g^{(3)}(t)(1-t)^2\,dt, \qquad \big(g(t) = f(\mathbf{a}+t\mathbf{h})\big).$$

By its definition, the derivative $g^{(k)}(t)$ is k-multilinear in the components of $\mathbf{h} = (h_1, h_2, \ldots, h_n)$. In the case of $R_{\mathbf{a},2}(\mathbf{h})$, $g^{(3)}(t)$ is trilinear in the components of $\mathbf{h}$. As noted just prior to Corollary 6.8.2, if we know a bound for all the partial derivatives of f through third order in a given neighborhood of $\mathbf{a}$, then we have, for all points $\mathbf{a}+\mathbf{h}$ in that neighborhood, an error estimate of the form

$$|R_{\mathbf{a},2}(\mathbf{h})| \le M\frac{|\mathbf{h}|_\infty^3}{3!}, \qquad \text{where } |\mathbf{h}|_\infty = \max_{1\le i\le n} |h_i|.$$

Exercises.

Exercise 10.9.1. Compute the Taylor polynomial $P(\mathbf{h})$ of degree 2 if $\mathbf{a} = (1,1)$ for the function $f(x,y) = 1/(x^2+y^2)$, using (10.24) and (10.25).

Exercise 10.9.2. Compute the Taylor polynomial $P(\mathbf{h})$ of degree 3 if $\mathbf{a} = (-1,1)$ for the function $f(x,y) = xy + \frac{1}{x} - \frac{1}{y}$, using (10.24) and (10.25). We know that the degree 3 Taylor polynomial gives a better approximation to $f(\mathbf{x})$ for $\mathbf{x}$ near $\mathbf{a}$ than does the degree 2 Taylor polynomial, but the next section shows that degree 2 suffices to determine the nature of the critical point at $\mathbf{a} = (-1,1)$.

Exercise 10.9.3. What is the remainder $R_{\mathbf{0},3}$ for the function $f(x,y,z) = xy + yz + x^2z - xyz - yz^2 + z^3$? And for $g(x,y,z) = xy + x^2z + z^3 - y^{10}$?

10.10. Relative Extrema without Constraints

In this section we use a dot notation, $\mathbf{v}\cdot\mathbf{w}$, for the Euclidean inner product $(\mathbf{v},\mathbf{w})$ of vectors $\mathbf{v}$ and $\mathbf{w}$ in $\mathbf{R}^n$, in the belief that this will serve to reduce some of the notational burden.

We begin with the definitions of relative minimum, relative maximum, and saddle points.

Definition 10.10.1. *Let U be an open set in $\mathbf{R}^n$ and let $f: U \to \mathbf{R}$. Then f has a* **relative minimum** *at $\mathbf{a}$ if $f(\mathbf{x}) \ge f(\mathbf{a})$ for all $\mathbf{x}$ in some neighborhood of $\mathbf{a}$, and f has a* **relative maximum** *at $\mathbf{a}$ if $f(\mathbf{x}) \le f(\mathbf{a})$ for all $\mathbf{x}$ in some neighborhood of $\mathbf{a}$. If in every neighborhood of $\mathbf{a}$ there are points $\mathbf{x}$ and $\mathbf{z}$ with $f(\mathbf{x}) > f(\mathbf{a})$ and $f(\mathbf{z}) < f(\mathbf{a})$, then f has a* **saddle point** *at $\mathbf{a}$.*

If f has either a relative minimum or a relative maximum at $\mathbf{a}$, then we say that f has a **relative extremum** at $\mathbf{a}$. Note that by definition relative extrema occur at interior points of the domain of a function.

Theorem 10.10.2. *If $f: U \to \mathbf{R}$ has a relative extremum at $\mathbf{a} \in U$ and if f is differentiable at $\mathbf{a}$, then*

$$\frac{\partial f}{\partial x_j}(\mathbf{a}) = 0 \quad \text{for } j = 1, \ldots, n.$$

Proof. If f has a relative extremum at $\mathbf{a}$, then the function $g_j(t) = f(\mathbf{a} + t\mathbf{e}_j)$, defined for t near 0, must also have a relative extremum at $t = 0$. Since f is differentiable at $\mathbf{a}$, g_j is differentiable at $t = 0$, by the chain rule. By the result for the single variable case (Theorem 5.2.2), necessarily $g_j'(0) = 0$, and the chain rule gives

$$0 = g_j'(0) = \nabla f(\mathbf{a}) \cdot \mathbf{e}_j = \frac{\partial f}{\partial x_j}(\mathbf{a}).$$

Since this is true for each $j = 1, \ldots, n$, the proof is complete. □

We say that $\mathbf{a}$ is a **critical point** for f if $\nabla f(\mathbf{a}) = \mathbf{0}$. Theorem 10.10.2 says that if f has a relative extremum at $\mathbf{a}$ and f is differentiable at $\mathbf{a}$, then $\mathbf{a}$ is a critical point for f.

The single variable result in Theorem 5.8.1 (part 2) implies that a real function f of a real variable has a relative minimum at a critical point t_0 if $f'(t_0) = 0$ and $f''(t_0) > 0$, and f has a relative maximum at t_0 if $f'(t_0) = 0$ and $f''(t_0) < 0$. These deductions are based on the single variable Taylor's theorem. By the multivariable Taylor's Theorem 10.9.1, if f is C^3 near $\mathbf{a}$, then

$$f(\mathbf{a}+\mathbf{h}) = f(\mathbf{a}) + \nabla f(\mathbf{a}) \cdot \mathbf{h} + \frac{1}{2!}\sum_{i=1}^{n}\sum_{j=1}^{n} \frac{\partial^2 f}{\partial x_i \partial x_j}(\mathbf{a})\, h_i\, h_j + R_{\mathbf{a},2}(\mathbf{h})$$

for small $|\mathbf{h}|$. The Hessian matrix of f at $\mathbf{a}$ is

$$H_f(\mathbf{a}) = \begin{bmatrix} f_{x_1x_1}(\mathbf{a}) & f_{x_1x_2}(\mathbf{a}) & \cdots & f_{x_1x_n}(\mathbf{a}) \\ f_{x_2x_1}(\mathbf{a}) & f_{x_2x_2}(\mathbf{a}) & \cdots & f_{x_2x_n}(\mathbf{a}) \\ \cdots & & & \cdots \\ f_{x_nx_1}(\mathbf{a}) & f_{x_nx_2}(\mathbf{a}) & \cdots & f_{x_nx_n}(\mathbf{a}) \end{bmatrix}.$$

The i-th row of $H_f(\mathbf{a})$ is the row gradient of $\partial f/\partial x_i$. Recall that an $n \times n$ real matrix A is **symmetric** if $A^T = A$. The Hessian matrix $H_f(\mathbf{a})$ is a real symmetric matrix. The quadratic terms in the Taylor expansion may be written as

$$\frac{1}{2}\sum_{i=1}^{n}\Big(\sum_{j=1}^{n} f_{x_ix_j}(\mathbf{a})h_j\Big)h_i = \frac{1}{2}H_f(\mathbf{a})\mathbf{h}\cdot\mathbf{h}.$$

If $\mathbf{a}$ is a critical point for f, then

$$f(\mathbf{a}+\mathbf{h}) = f(\mathbf{a}) + \frac{1}{2}H_f(\mathbf{a})\mathbf{h}\cdot\mathbf{h} + R_{\mathbf{a},2}(\mathbf{h}).$$

The local behavior of f near $\mathbf{a}$ can be described completely in terms of the properties of the matrix $H_f(\mathbf{a})$ when $H_f(\mathbf{a})$ is nonsingular. To describe this result, we make use of the spectral theorem for real symmetric matrices (Theorem 8.5.7) and further facts from Section 8.5; thus, a reading or review of that material is beneficial here.

The main result of the section is a generalization of the single-variable second derivative test for local extrema.

Theorem 10.10.3. *Suppose that f is C^3 on an open set $U \subseteq \mathbf{R}^n$, $\mathbf{a} \in U$, $\nabla f(\mathbf{a}) = \mathbf{0}$ and $H_f(\mathbf{a})$ is nonsingular. Then the following statements are true:*

1. *If $H_f(\mathbf{a})$ is positive definite, then f has a relative minimum at $\mathbf{a}$.*
2. *If $H_f(\mathbf{a})$ is negative definite, then f has a relative maximum at $\mathbf{a}$.*
3. *If $H_f(\mathbf{a})$ is indefinite, then f has a saddle point at $\mathbf{a}$.*

Proof. Let $\lambda_1, \ldots, \lambda_n$ be the eigenvalues of $H_f(\mathbf{a})$, which are all real. By the spectral theorem for real symmetric matrices (Theorem 8.5.7), there is an orthonormal basis $\{\mathbf{u}_1, \ldots, \mathbf{u}_n\}$ of $\mathbf{R}^n$ consisting of corresponding orthonormal eigenvectors for the λ_i.

1. If $H_f(\mathbf{a})$ is positive definite then all $\lambda_i > 0$ (Theorem 8.5.9). Let $m = \min\{\lambda_1, \ldots, \lambda_n\}$. If $\mathbf{h} = \sum_{i=1}^n c_i \mathbf{u}_i \neq \mathbf{0}$ then

$$\frac{1}{2} H_f(\mathbf{a})\mathbf{h} \cdot \mathbf{h} = \frac{1}{2} \sum_{i=1}^n \lambda_i c_i^2 \geq \frac{1}{2} m \sum_{i=1}^n c_i^2 = \frac{1}{2} m \mathbf{h} \cdot \mathbf{h}.$$

Since f is C^3 we know that for $|\mathbf{h}|$ sufficiently small,

$$|R_{\mathbf{a},2}(\mathbf{h})| < \frac{1}{4} m \mathbf{h} \cdot \mathbf{h},$$

and consequently for these $\mathbf{h}$ we have

$$\begin{aligned} f(\mathbf{a}+\mathbf{h}) - f(\mathbf{a}) &= \frac{1}{2} H_f(\mathbf{a})\mathbf{h} \cdot \mathbf{h} + R_{\mathbf{a},2}(\mathbf{h}) \\ &\geq \frac{1}{4} m\, \mathbf{h} \cdot \mathbf{h} > 0. \end{aligned}$$

Therefore f has a relative minimum at $\mathbf{a}$.

2. The proof is similar to part 1 and is left to Exercise 10.10.1.

3. As noted earlier, as a consequence of Theorem 8.5.9, since $H_f(\mathbf{a})$ is nonsingular and indefinite, $H_f(\mathbf{a})$ has a positive eigenvalue λ_j and a negative eigenvalue λ_k. Let $\mathbf{u}_j$ and $\mathbf{u}_k$ be corresponding unit eigenvectors. Since $\nabla f(\mathbf{a}) = \mathbf{0}$, the Taylor expansion of the function $g_j(t) = f(\mathbf{a} + t\mathbf{u}_j)$ about $t_0 = 0$ is

$$\begin{aligned} g_j(t) &= g_j(0) + g_j'(0)t + \frac{1}{2} g_j''(0)t^2 + R_{0,2}(t) \\ &= f(\mathbf{a}) + \frac{1}{2} t^2 H_f(\mathbf{a})\mathbf{u}_j \cdot \mathbf{u}_j + R_{0,2}(t) \\ &= f(\mathbf{a}) + \frac{1}{2} \lambda_j t^2 + R_{0,2}(t). \end{aligned}$$

(See Exercise 10.10.2.) Since $\lambda_j > 0$, it follows that $f(\mathbf{a}+t\mathbf{u}_j) > f(\mathbf{a})$ for sufficiently small nonzero $|t|$. A similar argument using the negative eigenvalue λ_k and the function $g_k(t) = f(\mathbf{a} + t\mathbf{u}_k)$ shows that $f(\mathbf{a} + t\mathbf{u}_k) < f(\mathbf{a})$ for sufficiently small nonzero $|t|$. We conclude that f has a saddle point at $\mathbf{a}$. □

Theorem 10.10.3 provides a reasonably quick practical test of the nature of a critical point for moderate size problems, as software can compute $\det H_f(\mathbf{a})$ and the eigenvalues. There are also *necessary* conditions for relative extrema in terms of a semidefinite property of the Hessian of f at the critical point; see Exercise 10.10.3.

For functions of two real variables the following special case of Theorem 10.10.3 is often useful.

Corollary 10.10.4. *Suppose that f is C^3 on an open set $U \subset \mathbf{R}^2$, $\mathbf{a} \in U$ and $\nabla f(\mathbf{a}) = \mathbf{0}$. Write $f(x, y)$ for $(x, y) \in U$, set*

$$\alpha = f_{xx}(\mathbf{a}), \qquad \beta = f_{xy}(\mathbf{a}), \qquad \gamma = f_{yy}(\mathbf{a}),$$

and assume $\alpha\gamma - \beta^2 \neq 0$. The following statements are true.

1. *If $\alpha\gamma - \beta^2 > 0$ and $\alpha > 0$, then f has a relative minimum at $\mathbf{a}$.*
2. *If $\alpha\gamma - \beta^2 > 0$ and $\alpha < 0$, then f has a relative maximum at $\mathbf{a}$.*
3. *If $\alpha\gamma - \beta^2 < 0$, then f has a saddle point at $\mathbf{a}$.*

Proof. We have

$$H_f(\mathbf{a}) = \begin{bmatrix} \alpha & \beta \\ \beta & \gamma \end{bmatrix}$$

and $\det H_f(\mathbf{a}) = \alpha\gamma - \beta^2 \neq 0$, so this is indeed a special case of Theorem 10.10.3. If λ_1 and λ_2 are the eigenvalues of $H_f(\mathbf{a})$, then $\alpha\gamma - \beta^2 = \lambda_1\lambda_2$. We observe also that entry $\alpha = H_f(\mathbf{a})\mathbf{e}_1 \cdot \mathbf{e}_1$.

1. If $\alpha\gamma - \beta^2 > 0$, the eigenvalues have the same sign. If we also have $\alpha > 0$, then, since $\alpha = H_f(\mathbf{a})\mathbf{e}_1 \cdot \mathbf{e}_1 > 0$, we cannot have two negative eigenvalues, so in fact both eigenvalues are positive. Thus $H_f(\mathbf{a})$ is positive definite, and the result follows from Theorem 10.10.3 (part 1).

2. Again $\alpha\gamma - \beta^2 > 0$ implies that the eigenvalues have the same sign. If $\alpha < 0$, then, since $\alpha = H_f(\mathbf{a})\mathbf{e}_1 \cdot \mathbf{e}_1 < 0$, we cannot have two positive eigenvalues, so both eigenvalues are negative. Thus $H_f(\mathbf{a})$ is negative definite, and the result follows from Theorem 10.10.3 (part 2).

3. $H_f(\mathbf{a})$ is indefinite if and only if one eigenvalue is positive and the other negative, if and only if $\alpha\gamma - \beta^2 = \lambda_1\lambda_2 < 0$. So statement 3 follows from Theorem 10.10.3 (part 3). □

The case where $H_f(\mathbf{a})$ is singular, or $\det H_f(\mathbf{a}) = 0$, is the case not covered by either Theorem 10.10.3 or Corollary 10.10.4. In that case, no conclusion can be made based on $H_f(\mathbf{a})$ alone, and the remainder term $R_{\mathbf{a},2}(\mathbf{h})$ becomes significant in determining the local behavior of f near $\mathbf{a}$.

Exercises.

Exercise 10.10.1. Prove: If $H_f(\mathbf{a})$ is negative definite, then f has a relative maximum at $\mathbf{a}$.

Exercise 10.10.2. Show that if $g(t) = f(\mathbf{a} + t\mathbf{u})$ where $\mathbf{u}$ is a unit vector such that $H_f(\mathbf{a})\mathbf{u} = \lambda\mathbf{u}$, then $g''(0) = H_f(\mathbf{a})\mathbf{u} \cdot \mathbf{u} = \lambda$.

Exercise 10.10.3. A real $n \times n$ symmetric matrix A is defined to be **positive semidefinite** if $A\mathbf{h} \cdot \mathbf{h} \geq 0$ for all $\mathbf{h} \in \mathbf{R}^n$, and A is defined to be **negative semidefinite** if $A\mathbf{h} \cdot \mathbf{h} \leq 0$ for all $\mathbf{h} \in \mathbf{R}^n$. Prove the following for a C^3 function f:

1. If f has a relative minimum at a critical point $\mathbf{a}$, then $\lambda_i \geq 0$ for every eigenvalue λ_i of $H_f(\mathbf{a})$, and $H_f(\mathbf{a})$ is positive semidefinite.
2. If f has a relative maximum at a critical point $\mathbf{a}$, then $\lambda_i \leq 0$ for every eigenvalue λ_i of $H_f(\mathbf{a})$, and $H_f(\mathbf{a})$ is negative semidefinite.

Exercise 10.10.4. Determine the type of the critical point at the origin for each of these functions:

1. $f(x, y) = x^2 - 3xy + y^2$,
2. $f(x, y) = x^2 + xy + y^2$,
3. $f(x, y, z) = x^2 + y^2 + z^2 + xz + xyz$,
4. $f(x, y, z, w) = x^2 + y^2 + z^2 + xz + yw + w^2$.

Exercise 10.10.5. Show that

$$f(x, y) = xy + \frac{1}{x} - \frac{1}{y}$$

has a single critical point and f has a local maximum value there.

Exercise 10.10.6. Determine the type of each critical point of

$$f(x, y) = xy - xy^2.$$

Exercise 10.10.7. In the proof of Corollary 10.10.4, we needed and established the *if* part of each statement below:

1. $H_f(\mathbf{a})$ is positive definite if and only if $\alpha\gamma - \beta^2 > 0$ and $\alpha > 0$.
2. $H_f(\mathbf{a})$ is negative definite if and only if $\alpha\gamma - \beta^2 > 0$ and $\alpha < 0$.

Prove the *only if* part of each of these statements.

Exercise 10.10.8. Suppose $f(x, y)$ is C^2 and $H_f(\mathbf{a})$ is written as in Corollary 10.10.4. Complete the square to express the quadratic form $H_f(\mathbf{a})\mathbf{h} \cdot \mathbf{h}$ with $\mathbf{h} = (h_1, h_2)$ as a sum of squares, under appropriate assumptions. Show that this leads to a different proof of parts 1 and 2 of Corollary 10.10.4.

10.11. Notes and References

The books by Sagan [**54**] and Edwards [**10**] were helpful in the preparation of this chapter. In particular, the presentation of the mean value theorem for vector functions and its consequences was influenced by Edwards [**10**].

Chapter 11

The Inverse and Implicit Function Theorems

The inverse function theorem and the implicit function theorem are concerned with the solution of systems of equations. These theorems are equally important, and, in fact, each of these results can be derived from the other. We begin in Section 1 with a result showing that matrix inversion is a continuous mapping on the space of invertible square matrices. We apply this result in Section 2, where we derive the inverse function theorem from the contraction mapping theorem by an iterative procedure closely related to Newton's method for finding a root of an equation. The continuous differentiability of the local inverse function follows using the continuity of matrix inversion. In Section 3 we derive the implicit function theorem from the inverse function theorem. Section 4 covers the Lagrange multiplier theorem and Section 5 presents the Morse lemma as applications of implicit function and inverse function arguments.

11.1. Matrix Geometric Series and Inversion

The numerical geometric series $\sum_{k=0}^{\infty} x^k$ converges absolutely if $|x| < 1$. The proof of this fact, given earlier, extends to the case of a matrix geometric series and provides a useful result on matrix series and matrix invertibility. The matrix norm in the following theorem can be any matrix norm.

Theorem 11.1.1. *If T is an $n \times n$ matrix with $\|T\| < 1$, then the matrix series $\sum_{k=0}^{\infty} T^k$ converges absolutely, $I - T$ is invertible, and*

$$(I - T)^{-1} = \sum_{k=0}^{\infty} T^k. \tag{11.1}$$

Consequently, if $\|I - T\| < 1$*, then* T *is invertible and*

$$T^{-1} = \sum_{k=0}^{\infty} (I - T)^k.$$

Proof. Suppose $\|T\| < 1$. For each k, $\|T^k\| \le \|T\|^k$, so the real series $\sum_{k=0}^{\infty} \|T^k\|$ is dominated termwise by the real geometric series $\sum_{k=0}^{\infty} \|T\|^k$. The latter series converges since $\|T\| < 1$, so $\sum_{k=0}^{\infty} \|T^k\|$ converges by the direct comparison test for series with positive terms. Therefore the matrix series $\sum_{k=0}^{\infty} T^k$ converges absolutely, and hence $\sum_{k=0}^{\infty} T^k$ is a well-defined $n \times n$ matrix. Let $S_n = \sum_{k=0}^{n} T^k$; then $S := \lim_{n\to\infty} S_n = \sum_{k=0}^{\infty} T^k$ exists. For each n,

$$(I - T)S_n = I - T^{n+1},$$

since the product on the left produces a telescoping sum that yields the right-hand side. We have

$$\lim_{n\to\infty} (I - T)S_n = (I - T) \lim_{n\to\infty} S_n = (I - T)S = \lim_{n\to\infty} (I - T^{n+1}) = I,$$

where $\lim_{n\to\infty} T^{n+1} = 0_{n\times n}$ since $\|T\| < 1$. Since $I - T$ and S are $n \times n$, and $I - T$ has right inverse S, we conclude that $I - T$ is invertible and $S = (I - T)^{-1}$. This completes the proof of the first statement of the theorem.

For the second statement of the theorem, if $\|I - T\| < 1$, we have that $T = I - (I - T)$ is invertible, and $T^{-1} = \sum_{k=0}^{\infty} (I - T)^k$. □

Theorem 11.1.1 has applications to the invertibility of linear transformations. A linear transformation $\mathbf{L} : \mathbf{R}^n \to \mathbf{R}^n$ is represented by a matrix $A = [a_{ij}]$ with respect to a given basis of $\mathbf{R}^n$, and $\mathbf{L}$ is invertible if and only if A is invertible. Let $\mathrm{Inv}(\mathbf{R}^n, \mathbf{R}^n)$ be the set of invertible elements in $L(\mathbf{R}^n, \mathbf{R}^n)$, the normed space of linear transformations of $\mathbf{R}^n$. Exercises 11.1.1 and 11.1.2 give two different views of the fact that $\mathrm{Inv}(\mathbf{R}^n, \mathbf{R}^n)$ is an open set in $L(\mathbf{R}^n, \mathbf{R}^n)$.

We consider some useful facts about matrix inversion. Recall that the **cofactor** of the entry a_{ij} of an $n \times n$ matrix A is $(-1)^{i+j}$ times the determinant of the $(n-1) \times (n-1)$ submatrix that remains after deleting row i and column j, $1 \le i, j \le n$.

Theorem 11.1.2. *A real* $n \times n$ *matrix* A *is invertible if and only if* $\det A \neq 0$*, and when* A *is invertible the unique inverse of* A *is given by*

$$A^{-1} = \frac{1}{\det A} \operatorname{adj} A, \tag{11.2}$$

where $\operatorname{adj} A$ *is the* **classical adjoint** *of* A*, defined as the transpose of the matrix of cofactors of* A*.*

We note in passing that the result called Cramer's rule for the solution of linear algebraic equations $A\mathbf{x} = \mathbf{y}$ when A is invertible follows from (11.2). For a complete discussion of Theorem 11.1.2 as well as Cramer's rule, see Hoffman and Kunze [**31**] (Section 5.4).

Example 11.1.3. We illustrate the 2×2 case of Theorem 11.1.2. If A is 2×2, given by

$$A = \begin{bmatrix} a & b \\ c & d \end{bmatrix},$$

then the classical adjoint of A is

$$\operatorname{adj} A = \begin{bmatrix} d & -c \\ -b & a \end{bmatrix}^T = \begin{bmatrix} d & -b \\ -c & a \end{bmatrix}.$$

It is easy to verify that $(\operatorname{adj} A)A = A(\operatorname{adj} A) = (\det A)I = (ad - bc)I$. Thus, if $\det A \neq 0$, then

$$A^{-1} = \frac{1}{\det A} \operatorname{adj} A = \frac{1}{ad - bc} \begin{bmatrix} d & -b \\ -c & a \end{bmatrix},$$

a formula familiar from a first course in linear algebra. △

For our purposes in this discussion, the important fact about the cofactors of A, which give the entries of $\operatorname{adj} A$, is that they are polynomials in the entries of A. Consequently, each of the entries of A^{-1} is a continuous real valued function of the entries of A on the set $\{A : \det A \neq 0\}$.

Another result of interest concerning the operation of inversion is that the entries of the inverse of an invertible matrix function are as smooth as the entries of the original function. This result is used in the proof of the inverse function theorem in the next section, and we now describe this result in more detail. We first prove the continuity of inversion for linear transformations, and it follows for matrix representations since $\|\mathbf{L}\| = \|A\|$ where A is the matrix representation of the transformation $\mathbf{L}$ with respect to a given basis.

Theorem 11.1.4. *The inversion mapping* $\phi : \operatorname{Inv}(\mathbf{R}^n, \mathbf{R}^n) \to \operatorname{Inv}(\mathbf{R}^n, \mathbf{R}^n)$ *defined by*

$$\phi(\mathbf{L}) = \mathbf{L}^{-1}$$

is continuous.

Proof. Let $\mathbf{L}_0 \in \operatorname{Inv}(\mathbf{R}^n, \mathbf{R}^n)$, so $\mathbf{L}_0$ is invertible. By Exercise 11.1.1 the set $\operatorname{Inv}(\mathbf{R}^n, \mathbf{R}^n)$ is open, so if $\mathbf{L}$ is sufficiently near $\mathbf{L}_0$ in norm, then $\mathbf{L}$ is also invertible. We may write

$$\mathbf{L}^{-1} - \mathbf{L}_0^{-1} = \mathbf{L}^{-1}(\mathbf{L}_0 - \mathbf{L})\mathbf{L}_0^{-1}, \tag{11.3}$$

and consequently we may estimate $\phi(\mathbf{L}) - \phi(\mathbf{L}_0)$ in norm by

$$\|\mathbf{L}^{-1} - \mathbf{L}_0^{-1}\| \leq \|\mathbf{L}^{-1}\| \, \|\mathbf{L}_0 - \mathbf{L}\| \, \|\mathbf{L}_0^{-1}\|. \tag{11.4}$$

We want to let $\|\mathbf{L} - \mathbf{L}_0\| \to 0$ and conclude that $\|\mathbf{L}^{-1} - \mathbf{L}_0^{-1}\| \to 0$. But we need to know that $\|\mathbf{L}^{-1}\|$ remains bounded. A rearrangement of (11.3) gives

$$\mathbf{L}^{-1}[I - (\mathbf{L}_0 - \mathbf{L})\mathbf{L}_0^{-1}] = \mathbf{L}_0^{-1},$$

and by Theorem 11.1.1, the quantity in brackets, $\mathbf{Q}_\mathbf{L} := I - (\mathbf{L}_0 - \mathbf{L})\mathbf{L}_0^{-1}$, is invertible for $\|\mathbf{L}_0 - \mathbf{L}\| < 1/[2\|\mathbf{L}_0^{-1}\|]$. With that condition on $\mathbf{L}$, (11.1) implies that

$$\|\mathbf{Q}_\mathbf{L}^{-1}\| = \|[I - (\mathbf{L}_0 - \mathbf{L})\mathbf{L}_0^{-1}]^{-1}\| \leq \sum_{k=0}^{\infty} \|(\mathbf{L}_0 - \mathbf{L})\mathbf{L}_0^{-1}\|^k \leq \sum_{k=0}^{\infty} \|\mathbf{L}_0 - \mathbf{L}\|^k \|\mathbf{L}_0^{-1}\|^k.$$

Consequently, since $\mathbf{L}^{-1} = \mathbf{L}_0^{-1}\mathbf{Q}_\mathbf{L}^{-1}$, we have

$$\|\mathbf{L}^{-1}\| \le \|\mathbf{L}_0^{-1}\|\,\|\mathbf{Q}_\mathbf{L}^{-1}\| \le \|\mathbf{L}_0^{-1}\| \sum_{k=0}^{\infty} \left(\|\mathbf{L}_0 - \mathbf{L}\|\,\|\mathbf{L}_0^{-1}\|\right)^k$$

so $\|\mathbf{L}^{-1}\|$ remains bounded by $2\|\mathbf{L}_0^{-1}\|$. Now if $\|\mathbf{L} - \mathbf{L}_0\| \to 0$, then (11.4) implies that $\|\phi(\mathbf{L}) - \phi(\mathbf{L}_0)\| = \|\mathbf{L}^{-1} - \mathbf{L}_0^{-1}\| \to 0$, showing continuity of ϕ at $\mathbf{L}_0$. Since $\mathbf{L}_0$ is an arbitrary invertible element, the proof is complete. □

We now consider the smoothness of an inverse matrix.

Theorem 11.1.5. *Suppose $A(s)$ is an $n \times n$ matrix function of a real variable s, and $A(s)$ is C^1 in s and invertible for each s in some open interval. Then the inverse $A^{-1}(s)$ is C^1 and*

$$\frac{d}{ds}A^{-1}(s) = -A^{-1}(s)\Big[\frac{d}{ds}A(s)\Big]A^{-1}(s)$$

for each s. If $A(s)$ is C^k in s, then $A^{-1}(s)$ is also C^k in s.

Proof. By hypothesis, all entries of $A(s)$ are continuously differentiable. By Theorem 11.1.2, each entry of $A^{-1}(s)$ is a rational function of the entries of $A(s)$, with denominator $\det A(s) \neq 0$, and therefore a continuously differentiable function of s. Since each entry of $A^{-1}(s)$ is C^1, $A^{-1}(s)$ is a C^1 matrix function of s. Since $\frac{d}{ds}A^{-1}(s)$ exists, we may differentiate the identity $A(s)A^{-1}(s) = I$ to find

$$\Big[\frac{d}{ds}A(s)\Big]A^{-1}(s) + A(s)\Big[\frac{d}{ds}A^{-1}(s)\Big] = 0_{n\times n},$$

and hence

$$\frac{d}{ds}A^{-1}(s) = -A^{-1}(s)\Big[\frac{d}{ds}A(s)\Big]A^{-1}(s).$$

Finally, if $A(s)$ is C^k in s, then from the form of the entries in $A^{-1}(s)$ we conclude that each of those entries is also C^k in s, so $A^{-1}(s)$ is a C^k matrix function of s. □

If $A(\mathbf{x})$ is a C^1 and invertible matrix function of $\mathbf{x}$ on an open set in $\mathbf{R}^n$, then letting $s = x_j$, $1 \le j \le n$, we have

$$\frac{\partial}{\partial x_j}A^{-1}(\mathbf{x}) = -A^{-1}(\mathbf{x})\Big[\frac{\partial}{\partial x_j}A(\mathbf{x})\Big]A^{-1}(\mathbf{x}), \quad 1 \le j \le n.$$

From this relation it follows that if $A(\mathbf{x})$ is C^k in $\mathbf{x}$, then $A^{-1}(\mathbf{x})$ is also C^k in $\mathbf{x}$.

In this section we have highlighted two very different aspects of inversion. First, equation (11.2) of Theorem 11.1.2 shows that if A is invertible, then the entries of A^{-1} are rational functions of the entries of A, with numerators being linear combinations of $(n-1)$-fold products of the entries of A. On the other hand, Theorem 11.1.1 shows that if $\|A - I\| < 1$, then A^{-1} is represented by an *infinite matrix series* involving all powers of A. Note Exercise 11.1.3, regarding the Cayley-Hamilton theorem from linear algebra.

Exercises.

Exercise 11.1.1. Let A and B be elements of $L(\mathbf{R}^n, \mathbf{R}^n)$ (or matrices in $\mathbf{R}^{n\times n}$). Show that if A is invertible and B satisfies $\|A-B\| < 1/\|A^{-1}\|$, then B is invertible. *Hint*: Write $B = A - (A-B) = A(I - A^{-1}(A-B))$, and let $X = A^{-1}(A-B)$.

Exercise 11.1.2. Let $\det : \mathbf{R}^{n\times n} \to \mathbf{R}$ be the determinant function which maps the set of $n \times n$ real matrices into the real numbers. Use the fact that $\det A$ is a degree n polynomial in the entries of A to show that the set of $n \times n$ invertible real matrices is an open set in the space $\mathbf{R}^{n\times n}$.

Exercise 11.1.3. Let

$$p(\lambda) = \det(\lambda I - A) = \lambda^n + a_{n-1}\lambda^{n-1} + \cdots + a_1\lambda_1 + a_0$$

be the characteristic polynomial of an $n \times n$ matrix A. The Cayley-Hamilton theorem states that every $n \times n$ matrix A satisfies its own characteristic polynomial, in the sense that

$$p(A) = A^n + a_{n-1}A^{n-1} + \cdots + a_1 A + a_0 I_{n\times n} = I_{n\times n}.$$

Consequently, for any $n \times n$ matrix A, the powers A^k for $k \geq n$ can be expressed as linear combinations of the powers $I, A, A^2, \ldots, A^{n-1}$. As a simple example, show that if A is the matrix in Example 11.1.3, then $p(\lambda) = \det(\lambda I - A) = \lambda^2 - (a + d)\lambda + (ad - bc)$ and

$$A^2 - (a+d)A + (ad - bc)I = 0.$$

Then write A^3 as a linear combination of I and A.

11.2. The Inverse Function Theorem

In many applications it is important to know whether the inverse of a given mapping F has the same degree of smoothness as F itself. Recall that even if a mapping F is differentiable to all orders, its inverse, if it exists, need not be differentiable everywhere. A simple example of this situation is the mapping $F : \mathbf{R} \to \mathbf{R}$ given by $F(x) = x^3$. Then F is infinitely differentiable, but its inverse mapping $F^{-1}(x) = \sqrt[3]{x}$ is not differentiable at $x = 0$.

The notion of **local C^1-invertibility**, defined next, is an important case illustrating the concept of a mapping and its inverse having the same degree of smoothness.

Definition 11.2.1. *Let $\mathbf{F} : O \to \mathbf{R}^n$, let $\mathbf{a}$ be an interior point of O, and suppose that $\mathbf{F}$ is C^1 on an open neighborhood of $\mathbf{a}$. Then $\mathbf{F}$ is* **locally C^1-invertible at $\mathbf{a}$** *if there exists an open set $U \subset O$ with $\mathbf{a} \in U$, an open set V with $\mathbf{b} = \mathbf{F}(\mathbf{a}) \in V$, and a C^1 function $\mathbf{G} : V \to U$ such that*

$$\mathbf{G} \circ \mathbf{F} = \mathbf{Id}_U, \qquad \mathbf{F} \circ \mathbf{G} = \mathbf{Id}_V,$$

where $\mathbf{Id}_U$ and $\mathbf{Id}_V$ are the identity mappings on U and V. (We use the notation $\mathbf{Id}$ here to avoid potential confusion with an identity matrix I.)

If $\mathbf{F}$ is locally C^1-invertible at $\mathbf{a}$, then of course $\mathbf{F}$ maps U one-to-one and onto V, and the local inverse $\mathbf{G}$ maps V one-to-one and onto U, and Definition 11.2.1 says that

$$\mathbf{G}(\mathbf{F}(\mathbf{x})) = \mathbf{x} \quad \text{for all } \mathbf{x} \in U \qquad \text{and} \qquad \mathbf{F}(\mathbf{G}(\mathbf{y})) = \mathbf{y} \quad \text{for all } \mathbf{y} \in V.$$

The main goal of this section is to prove the inverse function theorem, which gives a sufficient condition for a mapping to be locally C^1-invertible at a point. At the end of Section 10.8, we had developed essentially all the tools needed to prove the next theorem.

Theorem 11.2.2 (Inverse Function Theorem). *Let $\mathbf{F} : \Omega \to \mathbf{R}^n$ be C^1 on the open set $\Omega \subseteq \mathbf{R}^n$, and suppose that $\mathbf{a}$ is in Ω with $\mathbf{F}(\mathbf{a}) = \mathbf{b}$. If $D\mathbf{F}(\mathbf{a})$ is invertible, then $\mathbf{F}$ is locally C^1-invertible at $\mathbf{a}$. If $\mathbf{y}$ is in the domain of the local C^1 inverse and $\mathbf{F}(\mathbf{x}) = \mathbf{y}$, then*

$$D\mathbf{F}^{-1}(\mathbf{y}) = [D\mathbf{F}(\mathbf{x})]^{-1}.$$

Moreover, if $\mathbf{F}$ is C^k, then the local inverse of $\mathbf{F}$ at $\mathbf{a}$ is also C^k.

As in Section 10.8, we will use the convenient max norm on vectors and the associated matrix norm. Recall that

$$\{\mathbf{x} \in \mathbf{R}^n : |\mathbf{x}|_\infty < r\}$$

is the open ball $B_r(\mathbf{0})$ for the max norm, also called an open cube with side r.

We first consider a useful simplification of the problem. Let Ω be an open cube centered at $\mathbf{a}$, that is, let $r > 0$ and let $\Omega = \{\mathbf{x} \in \mathbf{R}^n : |\mathbf{x} - \mathbf{a}|_\infty < r\}$. Suppose $\mathbf{F}$ satisfies the hypotheses of Theorem 11.2.2, so that $\mathbf{F} : \Omega \to \mathbf{R}^n$ is C^1 on Ω, $\mathbf{a} \in \Omega$ with $\mathbf{F}(\mathbf{a}) = \mathbf{b}$, and $D\mathbf{F}(\mathbf{a})$ is invertible. Define a translation mapping on $\mathbf{R}^n$ by $\tau_{\mathbf{z}}(\mathbf{x}) = \mathbf{x} + \mathbf{z}$ for all $\mathbf{x} \in \mathbf{R}^n$, so that $\tau_{\mathbf{z}}$ translates each point of $\mathbf{R}^n$ by the vector $\mathbf{z}$. Then we have

$$\tau_{\mathbf{a}}(\mathbf{x}) = \mathbf{x} + \mathbf{a} \qquad \text{and} \qquad \tau_{-\mathbf{b}}(\mathbf{y}) = \mathbf{y} - \mathbf{b}.$$

Note that $\tau_{-\mathbf{b}} = \tau_{\mathbf{b}}^{-1}$. Let us define the mapping

$$\mathbf{\Psi} = [D\mathbf{F}(\mathbf{a})]^{-1} \circ \tau_{-\mathbf{b}} \circ \mathbf{F} \circ \tau_{\mathbf{a}}. \tag{11.5}$$

(See Figure 11.1.) It should be clear that $\mathbf{F}$ is locally invertible near $\mathbf{a}$ if and only if $\mathbf{\Psi}$ is locally invertible near the origin.

Now $\mathbf{\Psi}$ is defined on the open neighborhood $\tau_{\mathbf{a}}^{-1}(\Omega)$ of the origin and maps the origin to the origin, $\mathbf{\Psi}(\mathbf{0}) = \mathbf{0}$. Using the fact that the derivative of a translation at any point is the identity, the chain rule implies that the derivative of $\mathbf{\Psi}$ at the origin is $D\mathbf{\Psi}(\mathbf{0}) = [D\mathbf{F}(\mathbf{a})]^{-1} D\mathbf{F}(\mathbf{a}) = I$. Thus we have reduced the question of local invertibility to the case of a mapping $\mathbf{\Psi}$ such that $\mathbf{\Psi}(\mathbf{0}) = \mathbf{0}$ and $D\mathbf{\Psi}(\mathbf{0}) = I$.

Given the reductions of the preceding paragraph, we shall prove Theorem 11.2.3 below, and then complete the proof of Theorem 11.2.2 by a continuous differentiability argument.

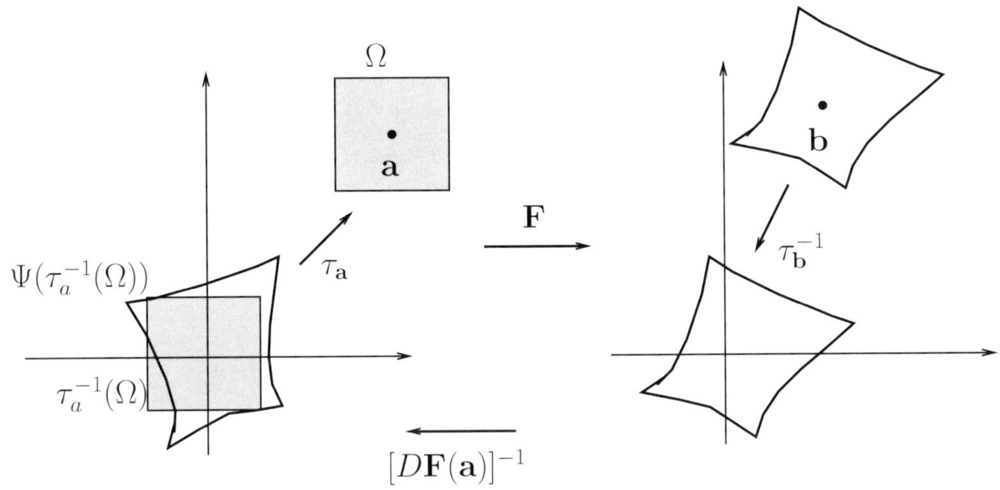

Figure 11.1. Mapping cubes to illustrate the inverse function theorem: Ω is an open cube about $\mathbf{a}$, and the mapping $\Psi = [D\mathbf{F}(\mathbf{a})]^{-1} \circ \tau_b^{-1} \circ \mathbf{F} \circ \tau_{\mathbf{a}}$ takes the open cube $\tau_{\mathbf{a}}^{-1}(\Omega)$ onto $\Psi(\tau_{\mathbf{a}}^{-1}(\Omega))$. The mapping $\mathbf{F}$ is locally invertible near $\mathbf{a}$ if and only if $\mathbf{\Psi}$ is locally invertible near the origin.

Theorem 11.2.3. *Let O be an open set in $\mathbf{R}^n$ containing the origin, let $\mathbf{F} : O \to \mathbf{R}^n$ be C^1 on O, and suppose that $\mathbf{F}(\mathbf{0}) = \mathbf{0}$ and $D\mathbf{F}(\mathbf{0}) = I$. Then the following statements hold:*

1. *Given $0 < \epsilon < 1$, there exists $r = r(\epsilon) > 0$ such that $\|D\mathbf{F}(\mathbf{x}) - I\|_\infty < \epsilon$ for all $\mathbf{x} \in C_r$, and*
$$C_{(1-\epsilon)r} \subseteq \mathbf{F}(C_r) \subseteq C_{(1+\epsilon)r}\ .$$
2. *With $r = r(\epsilon)$ as above, let $V = \operatorname{Int} C_{(1-\epsilon)r}$ and $U = \mathbf{F}^{-1}(V) \cap \operatorname{Int} C_r$. Then $\mathbf{F} : U \to V$ is one-to-one and onto with a continuous inverse.*
3. *The local inverse $\mathbf{G} : V \to U$ is continuously differentiable, and $D\mathbf{G}(\mathbf{0}) = I$.*

Proof. 1. Since $\mathbf{F}$ is C^1 on O, given $0 < \epsilon < 1$ there exists $r = r(\epsilon) > 0$ such that

$$\|D\mathbf{F}(\mathbf{x}) - I\|_\infty < \epsilon \quad \text{for all } \mathbf{x} \in C_r. \tag{11.6}$$

By Corollary 10.8.3, we know already that $\mathbf{F}(C_r) \subseteq C_{(1+\epsilon)r}$, but this fact is also derived here. By Corollary 10.8.2, with $\mathbf{L} = D\mathbf{F}(\mathbf{0}) = I$, and l the segment joining any two points $\mathbf{x}_1, \mathbf{x}_2 \in C_r$, we have

$$\begin{aligned} \big|\, \mathbf{F}(\mathbf{x}_1) - \mathbf{F}(\mathbf{x}_2) - (\mathbf{x}_1 - \mathbf{x}_2) \,\big|_\infty &\le \Big(\max_{\mathbf{z} \in l} \|D\mathbf{F}(\mathbf{z}) - I\|_\infty \Big) |\mathbf{x}_1 - \mathbf{x}_2|_\infty \\ &\le \epsilon\, |\mathbf{x}_1 - \mathbf{x}_2|_\infty. \end{aligned} \tag{11.7}$$

Then reverse triangle inequality arguments from (11.6) and (11.7) together with the mean value theorem applied to $\mathbf{F}$ on C_r yield

$$(1-\epsilon)|\mathbf{x}_1 - \mathbf{x}_2|_\infty \le |\mathbf{F}(\mathbf{x}_1) - \mathbf{F}(\mathbf{x}_2)|_\infty \le (1+\epsilon)|\mathbf{x}_1 - \mathbf{x}_2|_\infty, \tag{11.8}$$

and this holds for all $\mathbf{x}_1, \mathbf{x}_2 \in C_r$. Observe that the left-hand inequality shows that $\mathbf{F}$ is one-to-one on C_r. By setting $\mathbf{x}_2 = \mathbf{0}$ in the right-hand inequality, we see that if $\mathbf{x} = \mathbf{x}_1 \in C_r$, then

$$|\mathbf{F}(\mathbf{x})|_\infty \leq (1+\epsilon)|\mathbf{x}|_\infty \leq (1+\epsilon)r,$$

and hence $\mathbf{F}(C_r) \subseteq C_{(1+\epsilon)r}$. We now show that $C_{(1-\epsilon)r} \subseteq \mathbf{F}(C_r)$, using the contraction mapping theorem. Now fix $\mathbf{y} \in C_{(1-\epsilon)r}$, and consider the mapping

$$\Phi_\mathbf{y}(\mathbf{x}) = \mathbf{x} - \mathbf{F}(\mathbf{x}) + \mathbf{y}$$

for $\mathbf{x} \in C_r$. Using (11.6) in the mean value Theorem 10.8.1 applied to $\mathbf{x} - \mathbf{F}(\mathbf{x})$, we find

$$|\Phi_\mathbf{y}(\mathbf{x})|_\infty \leq |\mathbf{x} - \mathbf{F}(\mathbf{x})|_\infty + |\mathbf{y}|_\infty \leq \epsilon|\mathbf{x}|_\infty + |\mathbf{y}|_\infty \leq \epsilon r + (1-\epsilon)r = r,$$

hence $\Phi_\mathbf{y}$ maps C_r into C_r. Observe carefully that the same estimate shows that if $\mathbf{y} \in \operatorname{Int} C_{(1-\epsilon)r}$, then $\Phi_\mathbf{y}(\mathbf{x}) \in \operatorname{Int} C_r$. That is,

$$\mathbf{y} \in \operatorname{Int} C_{(1-\epsilon)r} \quad \Longrightarrow \quad \Phi_\mathbf{y}(C_r) \subseteq \operatorname{Int} C_r. \tag{11.9}$$

It follows from the estimate (11.7) that $\Phi_\mathbf{y}$ is a contraction on C_r, since $0 < \epsilon < 1$, and for $\mathbf{x}_1, \mathbf{x}_2 \in C_r$, we have

$$\big|\Phi_\mathbf{y}(\mathbf{x}_1) - \Phi_\mathbf{y}(\mathbf{x}_2)\big|_\infty = \big|\mathbf{F}(\mathbf{x_2}) - \mathbf{F}(\mathbf{x_1}) - (\mathbf{x_2} - \mathbf{x_1})\big|_\infty \leq \epsilon|\mathbf{x}_2 - \mathbf{x}_1|_\infty.$$

We have now shown that for each $\mathbf{y} \in C_{(1-\epsilon)r}$ there is a unique $\mathbf{x} \in C_r$ such that $\mathbf{F}(\mathbf{x}) = \mathbf{y}$. Thus $C_{(1-\epsilon)r} \subseteq \mathbf{F}(C_r)$, and this completes the proof of statement 1.

2. We have observed that by (11.9), if $\mathbf{y} \in \operatorname{Int} C_{(1-\epsilon)r}$, then the unique solution of $\mathbf{F}(\mathbf{x}) = \mathbf{y}$ satisfies $\mathbf{x} \in \operatorname{Int} C_r$. We need open sets for the domain and range of the local inverse for $\mathbf{F}$. Given what we know, a natural choice for the range is to let $V = \operatorname{Int} C_{(1-\epsilon)r}$, and then we define the domain to be $U = \mathbf{F}^{-1}(V) \cap \operatorname{Int} C_r$. (Since $\mathbf{F}$ need not be globally one-to-one, we intersect $\mathbf{F}^{-1}(V)$ with $\operatorname{Int} C_r$, on which $\mathbf{F}$ *is* one-to-one.) Thus $\mathbf{F} : U \to V$ is one-to-one and onto, and we denote the local inverse by $\mathbf{G} : V \to U$. The continuity of the local inverse follows from the left-hand inequality of (11.8), since we have

$$(1-\epsilon)|\mathbf{x}_1 - \mathbf{x}_2|_\infty \leq \big|\mathbf{F}(\mathbf{x}_1) - \mathbf{F}(\mathbf{x}_2)\big|_\infty \quad \text{for all } \mathbf{x}_1, \mathbf{x}_2 \in U \subset C_r$$

and hence

$$(1-\epsilon)\big|\mathbf{G}(\mathbf{y}_1) - \mathbf{G}(\mathbf{y}_2)\big|_\infty \leq |\mathbf{y}_1 - \mathbf{y}_2|_\infty \quad \text{for all } \mathbf{y}_1, \mathbf{y}_2 \in V.$$

3. Now we prove continuous differentiability of the local inverse $\mathbf{G}$ on V. By the matrix geometric series Theorem 11.1.1, we know that $D\mathbf{F}(\mathbf{x}_1)$ is invertible for all $\mathbf{x}_1 \in \operatorname{Int} C_r$. From the definition of derivative, we have

$$\mathbf{F}(\mathbf{x}) - \mathbf{F}(\mathbf{x}_1) = D\mathbf{F}(\mathbf{x}_1)(\mathbf{x} - \mathbf{x}_1) + |\mathbf{x} - \mathbf{x}_1|_\infty \psi(\mathbf{x} - \mathbf{x}_1) \tag{11.10}$$

where $\lim_{\mathbf{x}\to\mathbf{x}_1} \psi(\mathbf{x} - \mathbf{x}_1) = \mathbf{0}$. Let $\mathbf{y}_1 = \mathbf{F}(\mathbf{x}_1)$ and $\mathbf{y} = \mathbf{F}(\mathbf{x})$ be in V, and let $\mathbf{x}_1 = \mathbf{G}(\mathbf{y}_1)$, $\mathbf{x} = \mathbf{G}(\mathbf{y})$. We want to establish the appropriate tangent estimate for the expression

$$\mathbf{G}(\mathbf{y}) - \mathbf{G}(\mathbf{y}_1) - [D\mathbf{F}(\mathbf{x}_1)]^{-1}(\mathbf{y} - \mathbf{y}_1). \tag{11.11}$$

Using (11.10), we have

$$\mathbf{G}(\mathbf{y}) - \mathbf{G}(\mathbf{y}_1) - [D\mathbf{F}(\mathbf{x}_1)]^{-1}(\mathbf{y} - \mathbf{y}_1) = -|\mathbf{x} - \mathbf{x}_1|_\infty [D\mathbf{F}(\mathbf{x}_1)]^{-1}(\psi(\mathbf{x} - \mathbf{x}_1)).$$

Letting $\|[D\mathbf{F}(\mathbf{x}_1)]^{-1}\| = \beta$, the left-hand inequality of (11.8) yields

$$|\mathbf{G}(\mathbf{y}) - \mathbf{G}(\mathbf{y}_1) - [D\mathbf{F}(\mathbf{x}_1)]^{-1}(\mathbf{y} - \mathbf{y}_1)|_\infty \le \frac{\beta}{1-\epsilon}|\mathbf{y} - \mathbf{y}_1|_\infty |\psi(\mathbf{G}(\mathbf{y}) - \mathbf{G}(\mathbf{y}_1))|_\infty.$$

By the continuity of the local inverse $\mathbf{G}$ established in part 2, we have $\mathbf{x} \to \mathbf{x}_1$ as $\mathbf{y} \to \mathbf{y}_1$. Consequently, $\lim_{\mathbf{y}\to\mathbf{y}_1} \psi(\mathbf{G}(\mathbf{y}) - \mathbf{G}(\mathbf{y}_1)) = 0$, and hence

$$\lim_{\mathbf{y}\to\mathbf{y}_1} \frac{\left|\mathbf{G}(\mathbf{y}) - \mathbf{G}(\mathbf{y}_1) - [D\mathbf{F}(\mathbf{x}_1)]^{-1}(\mathbf{y} - \mathbf{y}_1)\right|_\infty}{|\mathbf{y} - \mathbf{y}_1|_\infty} = 0.$$

Thus by the definition of derivative, $D\mathbf{G}(\mathbf{y}_1) = [D\mathbf{F}(\mathbf{x}_1)]^{-1}$. In particular, with $\mathbf{x}_1 = \mathbf{0}$ and $\mathbf{y}_1 = \mathbf{0}$, we have $D\mathbf{G}(\mathbf{0}) = I$, since $D\mathbf{F}(\mathbf{0}) = I$. Since $\mathbf{y}_1$ was arbitrary in V, this shows that $\mathbf{G}$ is differentiable on V. Finally, to see that $D\mathbf{G}$ is continuous on V, we observe that $D\mathbf{G}$ is the composition of three continuous mappings:

$$\mathbf{y}_1 \overset{\mathbf{G}}{\mapsto} \mathbf{G}\mathbf{y}_1 = \mathbf{x}_1 \overset{D\mathbf{F}}{\mapsto} D\mathbf{F}(\mathbf{x}_1) \overset{\phi}{\mapsto} [D\mathbf{F}(\mathbf{x}_1)]^{-1} \tag{11.12}$$

where ϕ is the inversion map $\mathbf{L} \mapsto \mathbf{L}^{-1}$ on the set of invertible linear maps from $\mathbf{R}^n$ to $\mathbf{R}^n$, shown to be continuous in Theorem 11.1.4. Hence, $D\mathbf{G}$ is continuous on V. □

We proceed now to complete the proof of Theorem 11.2.2.

Completion of the proof of Theorem 11.2.2. We summarize what we know about the mapping $\mathbf{\Psi}$ in (11.5). Theorem 11.2.3 applies to it. By Theorem 11.2.3 (statement 1), given $0 < \epsilon < 1$ there exists $r = r(\epsilon) > 0$ such that $\|D\mathbf{\Psi}(\mathbf{x}) - I\|_\infty < \epsilon$ for all $\mathbf{x} \in C_r$. By the matrix geometric series Theorem 11.1.1, we also know that $D\mathbf{\Psi}(\mathbf{x})$ is invertible for all $\mathbf{x} \in \operatorname{Int} C_r$. By Theorem 11.2.3 (statement 2), there are open sets $V = \operatorname{Int} C_{(1-\epsilon)r}$ and $U = \mathbf{\Psi}^{-1}(V) \cap \operatorname{Int} C_r$, each containing the origin, such that $\mathbf{\Psi}$ maps U one-to-one and onto V. It follows from (11.5) that our function $\mathbf{F}$ (in Theorem 11.2.2) is locally invertible on the open set $U_1 = \tau_{\mathbf{a}}(U)$ and maps U_1 one-to-one and onto the open set $V_1 = \tau_{\mathbf{b}}(A(V))$. By (11.5), this local inverse $\mathbf{F}^{-1} : V_1 \to U_1$ is given by

$$[\mathbf{F}|_{V_1}]^{-1} = (\tau_{\mathbf{b}} \circ A \circ \mathbf{\Psi} \circ \tau_{-\mathbf{a}})^{-1}|_{V_1} = \tau_{\mathbf{a}} \circ \mathbf{\Psi}^{-1} \circ A^{-1} \circ \tau_{-\mathbf{b}}. \tag{11.13}$$

By the chain rule, we deduce that

$$D\mathbf{F}^{-1}(\mathbf{b}) = A^{-1} = [D\mathbf{F}(\mathbf{a})]^{-1}.$$

Now we want to see that, for any $\mathbf{y} = \mathbf{F}(\mathbf{x}) \in V_1$ with $\mathbf{x} \in U_1$, we have $D\mathbf{F}^{-1}(\mathbf{y}) = [D\mathbf{F}(\mathbf{x})]^{-1}$. This follows from an argument like that in the proof of part (3) of Theorem 11.2.3, since $D\mathbf{F}(\mathbf{x})$ is invertible for $\mathbf{x}$ in U_1 and the local inverse of $\mathbf{F}$ is continuous. (An argument is also outlined earlier in Exercise 10.8.1.) Thus the local inverse $\mathbf{F}^{-1} : V_1 \to U_1$ is differentiable, and its derivative at any point $\mathbf{y} = \mathbf{F}(\mathbf{x}) \in V_1$ is the inverse of the derivative of $\mathbf{F}$ at $\mathbf{x}$.

Continuity of $D\mathbf{F}^{-1}$ on V_1 follows, as in (11.12), from the continuity of inversion of linear mappings, the continuity of $D\mathbf{F}$ on U_1, and the continuity of $\mathbf{F}^{-1}$.

Finally, if the mapping $\mathbf{F}$ is C^k, then the local inverse $\mathbf{F}^{-1}$ is also C^k, as follows from viewing $D\mathbf{F}^{-1}(\mathbf{y})$ as the composition (again as in (11.12)) of C^k mappings, using the smoothness of inversion from Theorem 11.1.5. □

The next example serves to emphasize the local nature of the conclusion of the inverse function theorem.

Example 11.2.4. Define a mapping $\mathbf{F} : \mathbf{R}^2 \to \mathbf{R}^2$ by means of its component functions

$$\begin{aligned} f_1(x, y) &= x^2 - y^2, \\ f_2(x, y) &= 2xy. \end{aligned}$$

The determinant of the Jacobian matrix of $\mathbf{F}$ at (x, y) is given by

$$\det J_{\mathbf{F}}(x, y) = \det \begin{bmatrix} 2x & -2y \\ 2y & 2x \end{bmatrix} = 4(x^2 + y^2),$$

and thus $\mathbf{F}$ is locally C^1-invertible in some neighborhood of every point except the origin. However, $\mathbf{F}$ is not a globally one-to-one function; we observe that $\mathbf{F}(1, 1) = (0, 2) = \mathbf{F}(-1, -1)$. Note that $D\mathbf{F}(0, 0)$ is not invertible; in fact, it can be shown that there is no open neighborhood of the origin on which $\mathbf{F}$ is one-to-one (Exercise 11.2.1). △

Exercises.

Exercise 11.2.1. Consider the function of Example 11.2.4 given by $\mathbf{F}(x, y) = (x^2 - y^2, 2xy)$. Show that there is no open neighborhood of the origin on which $\mathbf{F}$ is one-to-one.

Exercise 11.2.2. Let $\mathbf{F} : \mathbf{R}^2 \to \mathbf{R}^2$ be defined by $\mathbf{F}(x, y) = (x - y, xy)$.

1. At which points does the inverse function Theorem 11.2.2 apply?
2. On a sketch, indicate the points at which $D\mathbf{F}$ fails to be invertible. At these points, the row gradients for the components of $\mathbf{F}$ are linearly dependent. Interpret this in terms of a tangency condition satisfied by the level curves of the component functions. Sketch a few of these level curves in all four quadrants, showing this tangency condition.
3. Note that $\mathbf{F}(2, 1) = (1, 2)$. Let $\mathbf{G}$ be the local inverse of $\mathbf{F}$ such that $\mathbf{G}(1, 2) = (2, 1)$. Find $D\mathbf{G}(1, 2)$.
4. Consider the behavior of $\mathbf{F}$ under the mapping $(x, y) \mapsto (-y, -x)$. Conclude that $\mathbf{F}$ is not locally invertible at any of the points you found in part 2.

Exercise 11.2.3. Let $\mathbf{F} : \mathbf{R}^2 \to \mathbf{R}^2$ be defined by its component functions $f_1(x, y)$ and $f_2(x, y)$, where $f_1(x, y) = x + x^2 \sin(1/x)$ if $x \neq 0$ and $f_1(x, y) = 0$ if $x = 0$, and $f_2(x, y) = y$. Show that $D\mathbf{F}(0, 0)$ is invertible, but the inverse function Theorem 11.2.2 does not apply. (This is the function of Example 10.4.3.)

Exercise 11.2.4. Let $\mathbf{F} : \mathbf{R}^2 \to \mathbf{R}^2$ be defined by $\mathbf{F}(x, y) = (e^x \cos y, e^x \sin y)$.

1. Show that $\mathbf{F}$ is locally C^1-invertible at very point.
2. Show that $\mathbf{F}$ is not a one-to-one mapping.

3. Find a domain on which $\mathbf{F}$ is locally C^1-invertible about $(0,0)$.
4. Show that the range of $\mathbf{F}$ equals $\mathbf{R}^2 - \{(0,0)\}$.

Exercise 11.2.5. Let $D = \{\rho, \phi, \theta : \rho > 0,\ 0 < \phi < \pi, -\pi < \theta \leq \pi\} \subset \mathbf{R}^3$, and let $\mathbf{F} : D \to \mathbf{R}^3$ be the spherical coordinate mapping given by

$$(x, y, z) = \mathbf{F}(\rho, \phi, \theta) = (\rho \sin\phi \cos\theta,\ \rho \sin\phi \sin\theta,\ \rho \cos\phi).$$

We observe that D is $\mathbf{R}^3$ minus the z-axis. Variable ρ is the distance from (x, y, z) to the origin, ϕ is the angle to (x, y, z) from the positive z-axis, and θ is the longitude (the polar coordinate angle θ of the projected point $(x, y, 0)$).

1. Describe the surfaces determined by these equations: (i) $\rho =$ constant, (ii) $\phi =$ constant, (iii) $\theta =$ constant.
2. Show that $\det J_{\mathbf{F}}(\rho, \phi, \theta) = \rho^2 \sin\phi$. Conclude that $\mathbf{F}$ is C^1-invertible in a neighborhood of each point in D.
3. Is $\mathbf{F}$ globally one-to-one on D? *Hint*: Use the fact that $\rho^2 = x^2 + y^2 + z^2$.

11.3. The Implicit Function Theorem

The implicit function theorem is one of the most useful results of basic analysis, as it deals with the solvability of systems of equations. The lowest-dimensional case of the theorem was considered in Theorem 10.7.1, and a review of that material may be beneficial before proceeding. This section extends Theorem 10.7.1 to higher-dimensional cases, using the inverse function theorem of the previous section.

Let W be an open set in $\mathbf{R}^{n+m}$ and let $\mathbf{F} : W \to \mathbf{R}^m$. Then the equation

$$\mathbf{F}(\mathbf{u}) = \mathbf{0}$$

represents m equations in $n+m$ variables. The implicit function theorem deals with the solvability of such an equation for m of the variables in terms of the remaining n variables. We identify $\mathbf{R}^{n+m}$ with $\mathbf{R}^n \times \mathbf{R}^m$ and thus write elements of $\mathbf{R}^{n+m}$ as $\mathbf{u} = (\mathbf{x}, \mathbf{y})$ with $\mathbf{x} \in \mathbf{R}^n$ and $\mathbf{y} \in \mathbf{R}^m$. The equation is then written as

$$\mathbf{F}(\mathbf{x}, \mathbf{y}) = \mathbf{0}.$$

It will aid our intuition to consider the case of a linear mapping $\mathbf{F} : \mathbf{R}^n \times \mathbf{R}^m \to \mathbf{R}^m$. Consider an equation of the form

$$\mathbf{F}(\mathbf{x}, \mathbf{y}) = A\mathbf{x} + B\mathbf{y} = \mathbf{0}$$

where A and B are matrices of size $m \times n$ and $m \times m$, respectively. If the $m \times m$ matrix B is invertible, then we may solve the equation $F(\mathbf{x}, \mathbf{y}) = \mathbf{0}$ uniquely for $\mathbf{y}$ in terms of $\mathbf{x}$, since $\mathbf{y} = -B^{-1}A\mathbf{x}$. Since

$$B = \frac{\partial \mathbf{F}}{\partial \mathbf{y}}$$

is the Jacobian matrix of $\mathbf{F}$ with respect to the $\mathbf{y}$ components, this sufficient condition for solvability can be expressed as the invertibility of the Jacobian matrix $\partial\mathbf{F}/\partial\mathbf{y}$. For linear equations, the relevant Jacobian is a constant matrix. For nonlinear equations, the implicit function theorem provides a local solvability result based on the invertibility of a Jacobian matrix evaluated at a *known* solution of the equations $\mathbf{F}(\mathbf{x}, \mathbf{y}) = \mathbf{0}$. In the proof of the theorem it will be convenient to write

$$\mathbf{F}_{\mathbf{x}}(\mathbf{x}, \mathbf{y}) = \frac{\partial \mathbf{F}}{\partial \mathbf{x}}(\mathbf{x}, \mathbf{y}) \quad \text{and} \quad \mathbf{F}_{\mathbf{y}}(\mathbf{x}, \mathbf{y}) = \frac{\partial \mathbf{F}}{\partial \mathbf{y}}(\mathbf{x}, \mathbf{y}),$$

where $\mathbf{x} \in \mathbf{R}^n$ and $\mathbf{y} \in \mathbf{R}^m$. Note that $\frac{\partial \mathbf{F}}{\partial \mathbf{x}}(\mathbf{x}, \mathbf{y})$ is $m \times n$, and $\frac{\partial \mathbf{F}}{\partial \mathbf{y}}(\mathbf{x}, \mathbf{y})$ is $m \times m$.

Theorem 11.3.1 (Implicit Function Theorem). *Let $W \subset \mathbf{R}^n \times \mathbf{R}^m$ be an open set. Let $\mathbf{F} : W \to \mathbf{R}^m$ be a C^1 mapping. If $(\mathbf{x}_0, \mathbf{y}_0)$ is in W and satisfies $\mathbf{F}(\mathbf{x}_0, \mathbf{y}_0) = \mathbf{0}$, and the linear mapping*

$$D_{\mathbf{y}}\mathbf{F}(\mathbf{x}_0, \mathbf{y}_0) = \frac{\partial \mathbf{F}}{\partial \mathbf{y}}(\mathbf{x}_0, \mathbf{y}_0) : \mathbf{R}^m \to \mathbf{R}^m$$

is invertible, then there exist open sets $U \subset \mathbf{R}^n$ and $V \subset \mathbf{R}^m$ such that

$$(\mathbf{x}_0, \mathbf{y}_0) \in U \times V \subset W,$$

and a unique C^1 mapping

$$\mathbf{g} : U \to V$$

such that

$$\mathbf{F}(\mathbf{x}, \mathbf{g}(\mathbf{x})) = \mathbf{0}$$

for all $\mathbf{x} \in U$. Moreover, $\mathbf{F}(\mathbf{x}, \mathbf{y}) \neq \mathbf{0}$ if $(\mathbf{x}, \mathbf{y}) \in U \times V$ and $\mathbf{y} \neq \mathbf{g}(\mathbf{x})$; thus the graph of $\mathbf{g}$ is exactly the set $\mathbf{F}^{-1}(\mathbf{0}) \cap (U \times V)$.

Proof. We shall apply the inverse function theorem to the mapping $\mathbf{H} : W \to \mathbf{R}^n \times \mathbf{R}^m$ defined by

$$\mathbf{H}(\mathbf{x}, \mathbf{y}) = (\mathbf{x}, \mathbf{F}(\mathbf{x}, \mathbf{y})).$$

This mapping will be used to map the local solution set of the equation $\mathbf{F}(\mathbf{x}, \mathbf{y}) = \mathbf{0}$ in an appropriate neighborhood of $(\mathbf{x}_0, \mathbf{y}_0)$ into an open portion of the $\mathbf{x}$ coordinate space. (See Figure 11.2.)

The derivative of $\mathbf{H}$ at $(\mathbf{a}, \mathbf{b})$, $D\mathbf{H}(\mathbf{a}, \mathbf{b}) : \mathbf{R}^n \times \mathbf{R}^m \to \mathbf{R}^n \times \mathbf{R}^m$, is given by

$$(\mathbf{h}_1, \mathbf{h}_2) \to \left(\mathbf{h}_1, \frac{\partial \mathbf{F}}{\partial \mathbf{x}}(\mathbf{a}, \mathbf{b})\mathbf{h}_1 + \frac{\partial \mathbf{F}}{\partial \mathbf{y}}(\mathbf{a}, \mathbf{b})\mathbf{h}_2\right)$$

or, in matrix form,

$$D\mathbf{H}(\mathbf{a}, \mathbf{b}) \begin{bmatrix} \mathbf{h}_1 \\ \mathbf{h}_2 \end{bmatrix} = \begin{bmatrix} I_n & 0 \\ \mathbf{F}_{\mathbf{x}}(\mathbf{a}, \mathbf{b}) & \mathbf{F}_{\mathbf{y}}(\mathbf{a}, \mathbf{b}) \end{bmatrix} \begin{bmatrix} \mathbf{h}_1 \\ \mathbf{h}_2 \end{bmatrix}.$$

From the structure of the matrix $D\mathbf{H}(\mathbf{a}, \mathbf{b})$, it should be clear that $D\mathbf{H}(\mathbf{a}, \mathbf{b})$ is invertible if and only if $\mathbf{F}_{\mathbf{y}}(\mathbf{a}, \mathbf{b})$ is invertible (Exercise 11.3.1). By hypothesis, $\mathbf{F}_{\mathbf{y}}(\mathbf{x}_0, \mathbf{y}_0)$ is invertible, hence $D\mathbf{H}(\mathbf{x}_0, \mathbf{y}_0)$ is invertible. By the inverse function theorem there is an open set $U_0 \times V \subset W$ containing $(\mathbf{x}_0, \mathbf{y}_0)$ such that $\mathbf{H}$ restricts to a C^1-invertible mapping of $U_0 \times V$ onto an open set $Z \subset \mathbf{R}^n \times \mathbf{R}^m$. Observe that the inverse of $\mathbf{H} : U_0 \times V \to Z$ leaves the first coordinate unchanged, as $\mathbf{H}$ does. Hence, the C^1 mapping $\mathbf{H}^{-1} : Z \to U_0 \times V$ takes the form

$$\mathbf{H}^{-1}(\mathbf{x}, \mathbf{w}) = (\mathbf{x}, \phi(\mathbf{x}, \mathbf{w})), \tag{11.14}$$

where $\phi : Z \to V$ and ϕ is C^1 on Z. In order to define the mapping $\mathbf{g}$ asserted by the theorem, let us choose an open subset of Z of the form $U \times Y$; thus, we choose open sets $U \subset U_0 \subset \mathbf{R}^n$ and $Y \subset \mathbf{R}^m$ such that $\mathbf{x}_0 \in U$, $\mathbf{0} \in Y$, and $U \times Y \subset Z$.

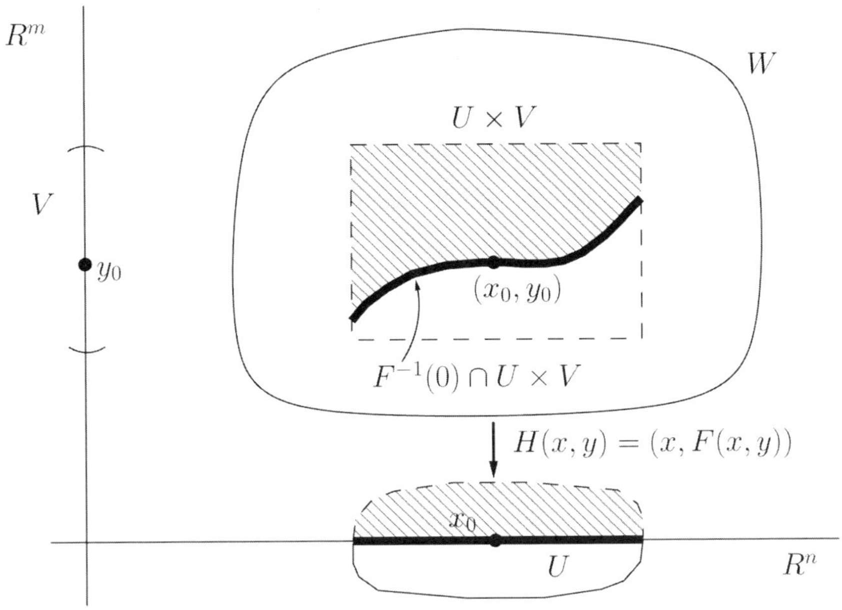

Figure 11.2. The implicit function theorem realizes the inverse image $\mathbf{F}^{-1}(\mathbf{0})$ as a function graph over the $\mathbf{x}$-space: We project the local solution set of the equation $\mathbf{F}(\mathbf{x}, \mathbf{y}) = \mathbf{0}$ to a copy of $\mathbf{R}^n$ (the $\mathbf{x}$-space) by the mapping $\mathbf{H}(\mathbf{x}, \mathbf{y})$ in the proof of Theorem 11.3.1.

Then the restriction of $(\mathbf{H}|U_0 \times V)^{-1}$ to the set $U \times Y$ is a C^1 mapping given by (11.14) except that now we consider $\phi : U \times Y \to V$.

Define a mapping $\mathbf{g} : U \to V$ by

$$\mathbf{g}(\mathbf{x}) = \phi(\mathbf{x}, \mathbf{0}).$$

Then $\mathbf{g}$ is C^1 on U since ϕ is C^1 on Z (and hence on $U \times Y$). From the relation $\mathbf{H} \circ \mathbf{H}^{-1} = \mathbf{Id}|_{U \times Y}$ we obtain, for $(\mathbf{x}, \mathbf{0}) \in U \times Y$:

$$\begin{aligned} (\mathbf{x}, \mathbf{0}) &= \mathbf{H} \circ \mathbf{H}^{-1}(\mathbf{x}, \mathbf{0}) \\ &= \mathbf{H}(\mathbf{x}, \phi(\mathbf{x}, \mathbf{0})) \\ &= (\mathbf{x}, \mathbf{F}(\mathbf{x}, \phi(\mathbf{x}, \mathbf{0}))) \\ &= (\mathbf{x}, \mathbf{F}(\mathbf{x}, \mathbf{g}(\mathbf{x}))). \end{aligned}$$

Therefore $\mathbf{F}(\mathbf{x}, \mathbf{g}(\mathbf{x})) = \mathbf{0}$ for all $\mathbf{x} \in U$.

For the final statement of the theorem, observe that $\mathbf{H}$ is one-to-one on $U \times V$. If $(\mathbf{x}, \mathbf{y}) \in U \times V$ and $\mathbf{y} \neq \mathbf{g}(\mathbf{x})$, then

$$\mathbf{H}(\mathbf{x}, \mathbf{y}) \neq \mathbf{H}(\mathbf{x}, \mathbf{g}(\mathbf{x})),$$

and hence, by definition of $\mathbf{H}$,

$$(\mathbf{x}, \mathbf{F}(\mathbf{x}, \mathbf{y})) \neq (\mathbf{x}, \mathbf{F}(\mathbf{x}, \mathbf{g}(\mathbf{x}))) = (\mathbf{x}, \mathbf{0}),$$

so $\mathbf{F}(\mathbf{x}, \mathbf{y}) \neq \mathbf{0}$. This completes the proof. □

From the identity $\mathbf{F}(\mathbf{x}, \mathbf{g}(\mathbf{x})) = \mathbf{0}$ for $\mathbf{x} \in U$, the chain rule implies that for all $\mathbf{x} \in U$, we have

$$\frac{\partial \mathbf{F}}{\partial \mathbf{x}}(\mathbf{x}, \mathbf{g}(\mathbf{x})) + \frac{\partial \mathbf{F}}{\partial \mathbf{y}}(\mathbf{x}, \mathbf{g}(\mathbf{x})) D\mathbf{g}(\mathbf{x}) = \mathbf{0}_{m \times n}.$$

This yields the formula

$$D\mathbf{g}(\mathbf{x}) = -\left[\frac{\partial \mathbf{F}}{\partial \mathbf{y}}(\mathbf{x}, \mathbf{g}(\mathbf{x}))\right]^{-1} \frac{\partial \mathbf{F}}{\partial \mathbf{x}}(\mathbf{x}, \mathbf{g}(\mathbf{x})).$$

We observe from this formula that if $\mathbf{F}$ is of class C^r, then $\mathbf{g}$ is also C^r.

Exercises.

Exercise 11.3.1. Verify as stated in the proof of Theorem 11.3.1 that $D\mathbf{H}(\mathbf{a}, \mathbf{b})$ is invertible if and only if $\mathbf{F}_{\mathbf{y}}(\mathbf{a}, \mathbf{b})$ is invertible.

Exercise 11.3.2. Derive the inverse function theorem from the implicit function theorem. *Hint*: Consider $\mathbf{G}(\mathbf{x}, \mathbf{y}) = \mathbf{y} - \mathbf{F}(\mathbf{x}) = \mathbf{0}$.

Exercise 11.3.3. Is there a unique solution $y = h(x)$ of the equation $y^2 - x^2 + y = 0$ in a neighborhood of the solution $(x_0, y_0) = (0, 0)$? Find it explicitly.

Exercise 11.3.4. Consider the system of equations

$$\begin{aligned} ye^x - 2yz + 3xz &= 0, \\ xyz - x + 2e^y &= 2. \end{aligned}$$

Investigate the possibilities for solving the system for any two of the variables in terms of the remaining variable near the point $(0, 0, 0)$.

Exercise 11.3.5. Consider the equation $y^3 - x + e^{1-x} = 0$.

1. Show that for each x there is a unique y satisfying the equation, and for each y there is a unique x satisfying the equation.
2. Write the solution for y in terms of x explicitly. Is this solution a C^1 function of x everywhere?
3. Observe that the solution for x in terms of y cannot be written in terms of elementary functions. Is the solution for x a C^1 function of y? Where?

Exercise 11.3.6. Suppose $f : \mathbf{R}^2 \to \mathbf{R}$ is defined by $f(x, y) = x - y^2$.

1. Does there exist a real valued function g defined on a neighborhood of $x = 0$ such that $f(x, g(x)) = 0$ on that neighborhood?
2. Show that there is a unique function g defined on a neighborhood of $x = 1$ such that $f(x, g(x)) = 0$ on that neighborhood and $g(1) = -1$.

Exercise 11.3.7. Suppose $f : \mathbf{R}^2 \to \mathbf{R}$ is defined by $f(x, y) = x^2 - y^2$.

1. Find two distinct real valued continuous functions g defined on a neighborhood of $x = 0$ such that $f(x, g(x)) = 0$ on that neighborhood.
2. Find two more real-valued continuous functions g defined on a neighborhood of $x = 0$ such that $f(x, g(x)) = 0$ on that neighborhood.

11.4. Constrained Extrema and Lagrange Multipliers

Earlier we considered the classification of relative extrema for a real valued function using the Hessian matrix at an interior critical point of the domain. This section deals with extrema of real functions subject to one or more constraint equations.

Definition 11.4.1 (Constrained Extrema). *Let U be an open set in $\mathbf{R}^n$, let $f : U \to \mathbf{R}$, and suppose that $S \subset U$. Let $\mathbf{v}_0$ be a point in S which is an extreme point for the restriction $f|_S : S \to \mathbf{R}$. Then $\mathbf{v}_0$ is a* **constrained extremum** *for f.*

If the constraint set $S \subset U$ is a closed and bounded set, then constrained extrema for f do indeed exist in S.

The following simple example illustrates a key geometric idea behind the method of the Lagrange multiplier theorem.

Example 11.4.2. Suppose we wish to maximize the function $f(x, y) = x + y$ subject to the constraint $g(x, y) = x^2+y^2-1 = 0$. A sketch and some thought about the geometry of the level sets of f and g leads to the conjecture that the maximum should occur at $(1, 1)$ and the maximum value of f subject to the constraint $g = 0$ is 2. (See Figure 11.3.) Here is a geometric argument. Suppose that f has a relative extremum at the point $\mathbf{a}$ in the set where $g = 0$. Let $\gamma(t)$, $|t| < \delta$ for some $\delta > 0$, be a path in the set where $g = 0$ satisfying $\gamma(0) = \mathbf{a}$. Then the function $f(\gamma(t))$ also has a relative extremum, at $t = 0$, hence, by the chain rule, $\nabla f(\mathbf{a}) \cdot \gamma'(0) = 0$. But $\gamma'(0)$ is the tangent vector to the curve $\gamma(t)$ at the point $\mathbf{a}$, so $\gamma'(0)$ is tangent to the level curve $g = 0$, hence $\gamma'(0) \perp \nabla g(\mathbf{a})$, since the gradient of g at $\mathbf{a}$ must be perpendicular to the level curve $g = 0$ at that point. Consequently, we necessarily have $\nabla f(\mathbf{a}) = \lambda \nabla g(\mathbf{a})$, that is, these gradients must be parallel. Thus, $(1, 1) = \lambda(2x, 2y)$. Since we cannot have $\lambda = 0$, it must be that $x = y$. This gives two possibilities where f might be maximized when restricted to the set where $g = 0$: the points $(1, 1)$ and $(-1, -1)$. Since $f(-1, -1) = -2$ and $f(1, 1) = 2$, clearly $f(1, 1) = 2$ is a relative maximum for f subject to the constraint $g = 0$. In this case it is also a global maximum for f over the constraint set. △

A real constraint equation of the form $g(\mathbf{x}_0) = y_0$ can always be rewritten in the form $\tilde{g}(\mathbf{x}_0) = 0$ by defining $\tilde{g}(\mathbf{x}) = g(\mathbf{x}) - y_0$, with $\nabla \tilde{g}(\mathbf{x}) = \nabla g(\mathbf{x})$. For this reason the theorems of this section are formulated with the constraints written as zero level sets.

Theorem 11.4.3. *Let U be an open set in $\mathbf{R}^n$ and let $f, g : U \to \mathbf{R}$ be C^1 functions on U. Let $\mathbf{v}_0 \in U$ with $g(\mathbf{v}_0) = 0$ and $\nabla g(\mathbf{v}_0) \neq \mathbf{0}$, and let*

$$S = \{\mathbf{x} \in U : g(\mathbf{x}) = 0\}.$$

If f, subjected to the constraint $g(\mathbf{x}) = 0$, has a relative extremum at $\mathbf{v}_0$, that is, if $f|_S$ has a relative extremum at $\mathbf{v}_0$, then there is a number λ such that

$$\nabla f(\mathbf{v}_0) = \lambda \nabla g(\mathbf{v}_0).$$

Proof. By a relabeling of the variables, if necessary, we may assume that

$$\frac{\partial g}{\partial x_n}(\mathbf{v}_0) \neq 0.$$

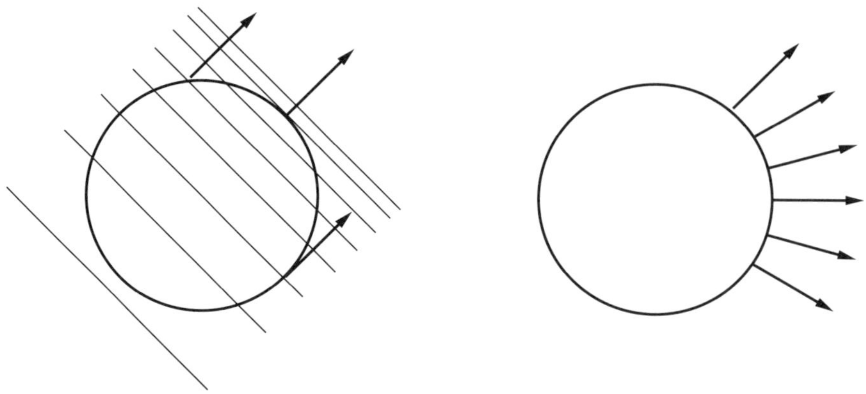

Figure 11.3. Geometry of optimization by the Lagrange multiplier method: We wish to maximize $f(x,y) = x+y$ subject to the constraint $g(x,y) = x^2+y^2-1=0$. Right: Gradient vectors of g at several points on the constraint circle. Left: The three arrows represent gradient vectors for f which are perpendicular to level sets of f. The maximum for f occurs at the base point of the middle arrow, and at this point $\mathbf{a}$, we have $\nabla f(\mathbf{a}) = \lambda \nabla g(\mathbf{a})$.

Write $\mathbf{v}_0 = (v_1, v_2, \ldots, v_{n-1}, v_n)$. By the implicit function theorem, there is an open set U_1 in $\mathbf{R}^{n-1}$ containing the point $(v_1, v_2, \ldots, v_{n-1})$, and a C^1 function $\phi : U_1 \to \mathbf{R}$ such that $\phi(v_1, v_2, \ldots, v_{n-1}) = v_n$ and

$$g(x_1, x_2, \ldots, x_{n-1}, \phi(x_1, x_2, \ldots, x_{n-1})) = 0$$

for all $(x_1, x_2, \ldots, x_{n-1})$ in U_1. Consider the function $h : U_1 \to \mathbf{R}$ defined by

$$h(x_1, x_2, \ldots, x_{n-1}) = f(x_1, x_2, \ldots, x_{n-1}, \phi(x_1, x_2, \ldots, x_{n-1})).$$

Since f has a relative extremum on the constraint set at $\mathbf{v}_0$, the function h has a relative extremum at $(v_1, v_2, \ldots, v_{n-1})$. Consequently, for $i = 1, 2, \ldots, n-1$,

$$\frac{\partial h}{\partial x_i}(\mathbf{v}_0) = \frac{\partial f}{\partial x_i}(\mathbf{v}_0) + \frac{\partial f}{\partial x_n}(\mathbf{v}_0)\frac{\partial \phi}{\partial x_i}(\mathbf{v}_0) = 0.$$

Differentiation of the constraint equation as a function of $(x_1, x_2, \ldots, x_{n-1})$ and evaluation at $\mathbf{v}_0$ gives, for $i = 1, 2, \ldots, n-1$,

$$\frac{\partial g}{\partial x_i}(\mathbf{v}_0) + \frac{\partial g}{\partial x_n}(\mathbf{v}_0)\frac{\partial \phi}{\partial x_i}(\mathbf{v}_0) = 0.$$

Since $\frac{\partial g}{\partial x_n}(\mathbf{v}_0) \neq 0$, we may define

$$\lambda := \frac{\partial f/\partial x_n(\mathbf{v}_0)}{\partial g/\partial x_n(\mathbf{v}_0)}.$$

Then from the previous two equalities it follows that

$$\frac{\partial f}{\partial x_i}(\mathbf{v}_0) = \lambda \frac{\partial g}{\partial x_i}(\mathbf{v}_0), \quad \text{for } i = 1, 2, \ldots, n-1.$$

Since $\partial f/\partial x_n(\mathbf{v}_0) = \lambda \partial g/\partial x_n(\mathbf{v}_0)$ by definition of λ, we are done. □

The number λ in Theorem 11.4.3 is called a *Lagrange multiplier*. As we saw in Example 11.4.2, the multiplier is a tool that can help in locating possible extrema.

In the case of a function f of n variables subject to k independent constraint equations, where $0 < k < n$, we have the more general result in Theorem 11.4.4 below. Its proof is similar to the argument in the example: if we write the constraint set as the graph of a function of $n-k$ variables, then the problem of describing a constrained extremum resolves itself into the problem of describing an unconstrained extremum of f over that graph. The implicit function theorem ensures that the constraint set can be written locally as such a graph.

Theorem 11.4.4 (Lagrange Multiplier Theorem). *Let U be an open set in $\mathbf{R}^n$, and let $f : U \to \mathbf{R}$ be C^1. Suppose $0 < k < n$, and let $\mathbf{g} = (g_1, \dots, g_k) : U \to \mathbf{R}^k$ be C^1. Define*

$$S = \{\mathbf{x} \in U : \mathbf{g}(\mathbf{x}) = \mathbf{0}\} = \bigcap_{i=1}^{k} \{\mathbf{x} \in U : g_i(\mathbf{x}) = 0\}.$$

If $f|_S$ has a relative extremum at the point $\mathbf{v}_0$ and the $k \times n$ matrix $J_{\mathbf{g}}(\mathbf{v}_0)$ has rank k, then there are numbers $\lambda_1, \dots, \lambda_k$ such that

$$\nabla f(\mathbf{v}_0) = \lambda_1 \nabla g_1(\mathbf{v}_0) + \cdots + \lambda_k \nabla g_k(\mathbf{v}_0). \tag{11.15}$$

Proof. Write $n = m + k$. We shall write points of $\mathbf{R}^n$ in the form $(\mathbf{x}, \mathbf{y})$, where $\mathbf{x} \in \mathbf{R}^m$ and $\mathbf{y} \in \mathbf{R}^k$. In particular, let the relative extreme point $\mathbf{v}_0 = (\mathbf{x}_0, \mathbf{y}_0)$.

Since the matrix $D\mathbf{g}(\mathbf{v}_0)$ has rank k, we may relabel components if necessary, and thus permute columns as necessary, so that the $k \times k$ submatrix $D_{\mathbf{y}}\mathbf{g}(\mathbf{v}_0) = D_{\mathbf{y}}\mathbf{g}(\mathbf{x}_0, \mathbf{y}_0)$ is invertible. Then by the implicit function theorem, there is an open neighborhood U_1 of $\mathbf{x}_0$ in $\mathbf{R}^m$ and a mapping $\phi : U_1 \to \mathbf{R}^k$ such that $\mathbf{y}_0 = \phi(\mathbf{x}_0)$ and

$$\mathbf{g}(\mathbf{x}, \phi(\mathbf{x})) = \mathbf{0} \quad \text{for } \mathbf{x} \in U_1. \tag{11.16}$$

Therefore the graph of the mapping $\phi : U_1 \to \mathbf{R}^k$ lies within the constraint set S. Now consider the mapping f restricted to this graph. Define $h : U_1 \to \mathbf{R}$ by

$$h(\mathbf{x}) = f(\mathbf{x}, \phi(\mathbf{x})) \quad \text{for } \mathbf{x} \in U_1.$$

Since U_1 is open in $\mathbf{R}^m$, the point $\mathbf{x}_0$ (the first block component of $\mathbf{v}_0$) must be an unconstrained extremum of the function $h : U_1 \to \mathbf{R}$, since $\mathbf{v}_0$ is an extremum for f on S. Consequently, $\nabla h(\mathbf{x}_0) = \mathbf{0}$. By the definition of h and the chain rule, we have

$$dh(\mathbf{x}_0) = d_{\mathbf{x}} f(\mathbf{x}_0, \mathbf{y}_0) + d_{\mathbf{y}} f(\mathbf{x}_0, \mathbf{y}_0) D\phi(\mathbf{x}_0) = \mathbf{0}. \tag{11.17}$$

On differentiating (11.16), we find, on evaluation at $\mathbf{x}_0$,

$$D_{\mathbf{x}}\mathbf{g}(\mathbf{x}_0, \mathbf{y}_0) + D_{\mathbf{y}}\mathbf{g}(\mathbf{x}_0, \mathbf{y}_0) D\phi(\mathbf{x}_0) = \mathbf{0}.$$

By the invertibility of $D_{\mathbf{y}}\mathbf{g}(\mathbf{x}_0, \mathbf{y}_0)$, we have

$$D\phi(\mathbf{x}_0) = -[D_{\mathbf{y}}\mathbf{g}(\mathbf{x}_0, \mathbf{y}_0)]^{-1} D_{\mathbf{x}}\mathbf{g}(\mathbf{x}_0, \mathbf{y}_0).$$

We may substitute this result into (11.17) to obtain

$$d_{\mathbf{x}} f(\mathbf{x}_0, \mathbf{y}_0) = d_{\mathbf{y}} f(\mathbf{x}_0, \mathbf{y}_0) [D_{\mathbf{y}}\mathbf{g}(\mathbf{x}_0, \mathbf{y}_0)]^{-1} D_{\mathbf{x}}\mathbf{g}(\mathbf{x}_0, \mathbf{y}_0). \tag{11.18}$$

Now observe that the desired identity (11.15) is equivalent to the identity

$$df(\mathbf{x}_0, \mathbf{y}_0) = [\,\lambda_1 \ \cdots \ \lambda_k\,] D\mathbf{g}(\mathbf{x}_0, \mathbf{y}_0),$$

where we read the differentials of real functions as row gradients, and this equation may be split into two block component equations, as follows:

$$d_{\mathbf{x}} f(\mathbf{x}_0, \mathbf{y}_0) = [\,\lambda_1 \;\cdots\; \lambda_k\,] D_{\mathbf{x}}\mathbf{g}(\mathbf{x}_0, \mathbf{y}_0), \tag{11.19}$$

$$d_{\mathbf{y}} f(\mathbf{x}_0, \mathbf{y}_0) = [\,\lambda_1 \;\cdots\; \lambda_k\,] D_{\mathbf{y}}\mathbf{g}(\mathbf{x}_0, \mathbf{y}_0), \tag{11.20}$$

where $D_{\mathbf{x}}\mathbf{g}(\mathbf{x}_0, \mathbf{y}_0)$ is the first m columns of $D\mathbf{g}(\mathbf{x}_0, \mathbf{y}_0)$, and $D_{\mathbf{y}}\mathbf{g}(\mathbf{x}_0, \mathbf{y}_0)$ is the last k columns of $D\mathbf{g}(\mathbf{x}_0, \mathbf{y}_0)$. Again using the invertibility of $D_{\mathbf{y}}\mathbf{g}(\mathbf{x}_0, \mathbf{y}_0)$, we may now *define* the row vector $[\,\lambda_1 \;\cdots\; \lambda_k\,]$ by

$$[\,\lambda_1 \;\cdots\; \lambda_k\,] := d_{\mathbf{y}} f(\mathbf{x}_0, \mathbf{y}_0)\,[D_{\mathbf{y}}\mathbf{g}(\mathbf{x}_0, \mathbf{y}_0)]^{-1},$$

and this definition ensures that (11.20) is satisfied. Finally, note that (11.19) is also satisfied since it is equivalent to (11.18) by our definition of $[\,\lambda_1 \;\cdots\; \lambda_k\,]$. This establishes (11.15) and proves the theorem. □

The proof of Theorem 11.4.4 used the implicit function theorem, by virtue of the rank condition on the Jacobian matrix $J_{\mathbf{g}}(\mathbf{v}_0)$. In practice, given a function f subject to constraints $\mathbf{g} = \mathbf{0}$, a solution of the constraint equations in closed form for some unknowns in terms of the others may be quite difficult, even if possible according to the implicit function theorem. Thus, expressing the given function in terms of a reduced number of variables and optimizing it as in Section 10.10 may not be a viable option. The Lagrange multiplier theorem provides the alternative approach of supplying a necessary condition for constrained extrema that does not require the knowledge of an explicit solution ahead of time. The multiplier equations implied by (11.15) introduce k parameters to help solve the combined system of constraints and multiplier equations for the constrained extrema. Since Theorem 11.4.4 provides necessary conditions for constrained extrema, one must verify the nature of any candidates found. The multiplier method of Theorem 11.4.4 may be difficult to apply unless the functions f and $\mathbf{g}$ are relatively simple.

Exercises.

Exercise 11.4.1. Minimize $f(x, y) = x^2 + 4y^2$ subject to $x + y - 1 = 0$. Can f be maximized subject to this constraint?

Exercise 11.4.2. Minimize $f(x, y, z) = x^2 + y^2 + z^2$ subject to the constraint $g(x, y, z) = x + y + z - 1 = 0$.

Exercise 11.4.3. Maximize and minimize $f(x, y) = x + y$ subject to the constraint $x^2 + 4y^2 = 1$.

Exercise 11.4.4. Find the maximum and minimum values of $f(x, y, z) = x + y + z$, subject to the constraints $x^2 + y^2 + z^2 = 1$ and $x^2 + y^2 + (z - 1)^2 = 1$.

Exercise 11.4.5. Recall that a real matrix A is *symmetric* if $A^T = A$.

1. Give an example of a real 2×2 matrix that has no real eigenvalues.
2. Show that every real symmetric matrix A has a real eigenvalue, that is, there is a real number λ and a nonzero vector $\mathbf{x}$ such that $A\mathbf{x} = \lambda\mathbf{x}$. *Hint*: Define $f(\mathbf{x}) = \mathbf{x}^T A\mathbf{x}$ and $g(\mathbf{x}) = \mathbf{x}^T\mathbf{x} - 1$, and then maximize (or minimize) f subject to the constraint $g(\mathbf{x}) = 0$; in particular, show that the gradient of the quadratic form $\mathbf{x}^T A\mathbf{x}$ equals $2A\mathbf{x}$.

Exercise 11.4.6. Minimize $f(x, y) = \frac{x^2}{4} + y^2$ subject to $x + y = 3$.

Exercise 11.4.7. Find the minimum amount of material required to make a rectangular box, enclosed on all six sides, if the volume of the box is to be 9 cubic feet. *Hint*: Write x (width), y (length) and z (height) for the dimensions of the box, set $z = 9/(xy)$ and write the amount of material as a function of x and y, unconstrained except for $x > 0$ and $y > 0$.

Exercise 11.4.8. Revisit Hölder's inequality using the Lagrange multiplier method.

1. For fixed positive p and q, let
$$f(x, y) = \frac{x^p}{p} + \frac{y^q}{q}, \qquad x > 0,\ y > 0.$$
Show that the minimum of f subject to $g(x, y) = xy - 1 = 0$ is $1/p + 1/q$.
2. (Hölder's inequality) Use the result of part 1 to show that if $p > 1$ and $q > 1$ with $1/p + 1/q = 1$, and $a \geq 0$, $b \geq 0$, then
$$ab \leq \frac{a^p}{p} + \frac{b^q}{q}.$$
Hint: Consider f in part 1 and the constraint $g(x, y) = xy - ab = 0$.

11.5. The Morse Lemma

Suppose f is a real valued function of class C^2 that has a critical point at $\mathbf{x}_0$. Taylor's theorem tells us that the second-order terms in the Taylor expansion of f about $\mathbf{x}_0$ approximate the difference $f(\mathbf{x}) - f(\mathbf{x}_0)$ for $\mathbf{x}$ near $\mathbf{x}_0$. In the case of a nonsingular Hessian matrix at $\mathbf{x}_0$, this approximation provides especially useful qualitative information about f. The Morse lemma states that if the Hessian at $\mathbf{x}_0$ is nonsingular, then there exist local coordinates in a neighborhood of $\mathbf{x}_0$ in which f equals an especially simple quadratic function: a sum of squares.

The argument for this result uses the inverse function theorem, the idea of completing the square, and an induction argument. Given a quadratic form
$$\alpha x_1^2 + 2\beta x_1 x_2 + \gamma x_2^2,$$
if $\alpha \neq 0$, then we can complete a square of the x_1 terms,
$$\begin{aligned}\alpha x_1^2 + 2\beta x_1 x_2 + \gamma x_2^2 &= \alpha\Big(x_1 + \frac{\beta}{\alpha}x_2\Big)^2 - \frac{\beta^2}{\alpha}x_2^2 + \gamma x_2^2 \\ &= \alpha\Big(x_1 + \frac{\beta}{\alpha}x_2\Big)^2 + \Big(\gamma - \frac{\beta^2}{\alpha}\Big)x_2^2.\end{aligned}$$
Similarly, if $\gamma \neq 0$, we can complete a square of the x_2 terms. If both α and γ are zero, but $\beta \neq 0$, then the transformation $x_1 = y_1 - y_2$, $x_2 = y_1 + y_2$ transforms $2\beta x_1 x_2$ into
$$2\beta x_1 x_2 = 2\beta(y_1 - y_2)(y_1 + y_2) = 2\beta y_1^2 - 2\beta y_2^2.$$
Thus, in all cases, a nonzero quadratic form is transformed into the sum of squared terms. This is the basic algebraic idea behind the Morse lemma.

We assume that f is differentiable of class C^p, with p to be determined.

Theorem 11.5.1 (The Morse Lemma). *Let O be an open set in $\mathbf{R}^n$ and $f : O \to \mathbf{R}$ a class C^p function with a nondegenerate critical point at $\mathbf{x}_0 \in O$, that is, $\nabla f(\mathbf{x}_0) = \mathbf{0}$, and the Hessian matrix of f at $\mathbf{x}_0$,*

$$H_f(\mathbf{x}_0) = \begin{bmatrix} f_{x_1x_1}(\mathbf{x}_0) & f_{x_1x_2}(\mathbf{x}_0) & \cdots & f_{x_1x_n}(\mathbf{x}_0) \\ f_{x_2x_1}(\mathbf{x}_0) & f_{x_2x_2}(\mathbf{x}_0) & \cdots & f_{x_2x_n}(\mathbf{x}_0) \\ \cdots & \cdots & \cdots & \cdots \\ f_{x_nx_1}(\mathbf{x}_0) & f_{x_nx_2}(\mathbf{x}_0) & \cdots & f_{x_nx_n}(\mathbf{x}_0) \end{bmatrix},$$

is nonsingular. Then there are open sets $U \subset O$ about $\mathbf{x}_0$ and V about $\mathbf{0} \in \mathbf{R}^n$ and a C^{p-2} change of coordinates $\mathbf{g} : V \to U$, denoted by $\mathbf{x} = \mathbf{g}(\mathbf{u})$, such that for all $\mathbf{u} \in V$,

$$f(\mathbf{g}(\mathbf{u})) = f(\mathbf{x}_0) - u_1^2 - \cdots - u_k^2 + u_{k+1}^2 + \cdots + u_n^2 \tag{11.21}$$

where k is a fixed nonnegative integer, called the **index** *of the critical point $\mathbf{x}_0$.*

Remark. *The index k is also called the index of the quadratic form determined by the fixed matrix $H_f(\mathbf{x}_0)$. The index of a quadratic form is usually defined as the dimension of the largest subspace on which the quadratic form is negative definite; this is the same as the number of negative eigenvalues of the symmetric matrix of the form, counted with multiplicity* [**4**]. *In the present context, if $f(\mathbf{x}_0)$ is a nondegenerate, and hence isolated, relative minimum of f, then the index is $k = 0$, and the level sets of f in a neighborhood of $\mathbf{x}_0$ are topological spheres, expressed in the $\mathbf{u}$ coordinates by the equations*

$$f(\mathbf{x}_0) + \sum_{k=1}^{n} u_k^2 = \textit{constant}.$$

If $f(\mathbf{x}_0)$ is a nondegenerate relative maximum of f, then the index is $k = n$, and the level sets of f in a neighborhood of $\mathbf{x}_0$ are the topological spheres expressed in the $\mathbf{u}$ coordinates by the equations

$$f(\mathbf{x}_0) - \sum_{k=1}^{n} u_k^2 = \textit{constant}.$$

Proof of the Morse Lemma. By translation mappings in the domain and range, we may assume that $\mathbf{x}_0 = \mathbf{0}$ and $f(\mathbf{x}_0) = f(\mathbf{0}) = 0$. By the fundamental theorem of calculus and the chain rule, we may write

$$\begin{aligned} f(x_1, \ldots, x_n) &= \int_0^1 \frac{d}{dt} f(tx_1, \ldots, tx_n)\, dt \\ &= \int_0^1 \sum_{i=1}^{n} x_i \frac{\partial f}{\partial x_i}(tx_1, \ldots, tx_n)\, dt. \end{aligned}$$

Define functions g_i for $1 \le i \le n$ by

$$g_i(x_1, \ldots, x_n) = \int_0^1 \frac{\partial f}{\partial x_i}(tx_1, \ldots, tx_n)\, dt.$$

Then

$$f(x_1, \ldots, x_n) = \sum_{i=1}^{n} x_i\, g_i(x_1, \ldots, x_n).$$

Since f is C^p, the functions g_i are C^{p-1}. As long as the resulting derivatives are continuous, we may differentiate g_i to any order by differentiating under the integral (Theorem 8.10.24). Since $\frac{\partial f}{\partial x_i}(\mathbf{0}) = 0$ at the critical point $\mathbf{0}$, an easy computation shows that $g_i(\mathbf{0}) = 0$. Thus, we may apply the same argument to each g_i that we applied to f, so there are C^{p-2} functions h_{ij} such that

$$g_i(x_1, \ldots, x_n) = \sum_{j=1}^{n} x_j h_{ij}(x_1, \ldots, x_n), \quad 1 \le i \le n.$$

This allows us to write

$$\begin{aligned} f(x_1, \ldots, x_n) &= \sum_{i=1}^{n} x_i \sum_{j=1}^{n} x_j h_{ij}(x_1, \ldots, x_n) \\ &= \sum_{i=1}^{n} \sum_{j=1}^{n} x_i x_j h_{ij}(x_1, \ldots, x_n) \end{aligned} \tag{11.22}$$

where $h_{ij}(\mathbf{0}) = 0$ for all i, j. If we have $h_{ij}(\mathbf{x}) \neq h_{ji}(\mathbf{x})$, then we can symmetrize these coefficient functions by replacing each $h_{ij}(\mathbf{x})$ by $\frac{1}{2}(h_{ij}(\mathbf{x}) + h_{ji}(\mathbf{x}))$, $1 \le i, j \le n$, since this replacement does not change the value of the summation at any point $\mathbf{x}$. We assume from here on that $h_{ij}(\mathbf{x}) = h_{ji}(\mathbf{x})$ for all $\mathbf{x}$. Then we can write

$$f(x_1, \ldots, x_n) = \sum_{i=1}^{n} x_i^2 h_{ii}(\mathbf{x}) + \sum_{1 \le i < j \le n} 2 x_i x_j h_{ij}(\mathbf{x}).$$

Two differentiations of f and evaluation at $\mathbf{0}$ gives

$$\frac{\partial^2 f}{\partial x_i \partial x_j}(\mathbf{0}) = 2 h_{ij}(\mathbf{0})$$

for any i, j. Now let us write (11.22) in the form

$$f(\mathbf{x}) = \mathbf{x}^T H(\mathbf{x}) \mathbf{x}$$

where $H(\mathbf{x})$ is a real symmetric $n \times n$ matrix for each $\mathbf{x}$, and by the nondegeneracy hypothesis on $\mathbf{x}_0 = \mathbf{0}$, $H(\mathbf{0})$ is nonsingular. Hence, by continuity, $H(\mathbf{x})$ is nonsingular for $\mathbf{x}$ in some open neighborhood of the origin.

If necessary, we can make a transformation of coordinates to get a leading coefficient $h_{11}(\mathbf{0}) \neq 0$. The argument for this follows:

Since f is not identically zero near the origin, there is at least one h_{ij} with $h_{ij}(\mathbf{0}) \neq 0$. If some diagonal term $h_{ii}(\mathbf{0}) \neq 0$, then we can get a nonzero $h_{11}(\mathbf{0})$ by interchanging the variables x_1 and x_i, which is accomplished by a permutation matrix transformation. Then $f(\mathbf{x}) = h_{11}(\mathbf{x}) x_1^2 + \cdots$, and $h_{11}(\mathbf{x})$ is nonzero in a neighborhood of $\mathbf{0}$. If all diagonal terms $h_{ii}(\mathbf{0}) = 0$, then there is some pair of indices $i \neq j$ with $h_{ij}(\mathbf{0}) \neq 0$. By permuting the variables according to $x_1 \leftrightarrow x_i$ and $x_2 \leftrightarrow x_j$, we obtain $h_{12}(\mathbf{0}) \neq 0$. By the symmetry, $h_{21}(\mathbf{0}) = h_{12}(\mathbf{0})$. Then f has the form

$$f(x_1, \ldots, x_n) = h_{12}(\mathbf{x}) x_1 x_2 + h_{21}(\mathbf{x}) x_2 x_1 + \cdots = 2 h_{12}(\mathbf{x}) + \cdots .$$

This leading term can be transformed to the form $2h_{12}(\mathbf{x})(y_1^2 - y_2^2)$, with nonzero coefficient $2h_{12}(\mathbf{x})$ near $\mathbf{0}$, by the transformation

$$x_1 = y_1 - y_2, \qquad x_2 = y_1 + y_2, \qquad x_3 = y_3, \quad \ldots \quad , x_n = y_n.$$

This transformation is invertible. This proves the first claim above. Now f has a leading term given by either

$$f(\mathbf{x}) = h_{11}(\mathbf{x})x_1^2 + \cdots$$

or

$$f(\mathbf{x}) = 2h_{12}(\mathbf{x})x_1^2 - 2h_{12}(\mathbf{x})x_2^2 + \cdots,$$

where $h_{11}(\mathbf{0}) \neq 0$ or $h_{12}(\mathbf{0}) \neq 0$, as the case may be. In either case, the important point is that the leading coefficient of x_1^2 is nonzero in some neighborhood of the critical point. Let us call this coefficient function $h_{11}(\mathbf{x})$ to cover either case.

Now we have

$$f(\mathbf{x}) = h_{11}(\mathbf{x})\Big(\sum_{i,j=1}^{n} x_i x_j k_{ij}(\mathbf{x})\Big), \tag{11.23}$$

where now $k_{ij}(\mathbf{x}) = h_{ij}(\mathbf{x})/h_{11}(\mathbf{x})$ for $\mathbf{x}$ near the origin, and $k_{11}(\mathbf{x}) \equiv 1$. By the symmetry $h_{ij} = h_{ji}$, the terms in parentheses in (11.23) that involve x_1 are as follows, on completing a square:

$$\begin{aligned} x_1^2 + 2\sum_{j=2}^{n} x_1 x_j k_{1j}(\mathbf{x}) &= x_1^2 + \Big(2\sum_{j=2}^{n} x_j k_{1j}(\mathbf{x})\Big)\, x_1 \\ &= \Big(x_1 + \sum_{j=2}^{n} x_j k_{1j}(\mathbf{x})\Big)^2 - \Big(\sum_{j=2}^{n} x_j k_{1j}(\mathbf{x})\Big)^2. \end{aligned}$$

Now the coordinate transformation

$$y_1 = x_1 + \sum_{j=2}^{n} x_j k_{1j}(\mathbf{x}), \quad y_2 = x_2, \quad \ldots \quad , \quad y_n = x_n,$$

ensures that the new variable y_1 appears only as y_1^2. This transformation is easily seen to be invertible, since

$$\frac{\partial y_1}{\partial x_1}(\mathbf{x}) = 1 + \sum_{j=2}^{n} x_j \frac{\partial k_{1j}}{\partial x_1}(\mathbf{x}) \approx 1,$$

for $\mathbf{x}$ near the origin. Our function f now has the form

$$h_{11}(\mathbf{y})y_1^2 + \sum_{j,k=2}^{n} y_j y_k q_{jk}(\mathbf{y}),$$

where the sum on the right is a quadratic form in $(n-1)$ variables $y_2, \ldots, y_n$, and the functions q_{jk} are class C^{p-2}. We can redefine y_1, by rescaling it, to be $y_1 = \pm\sqrt{|h_{11}(\mathbf{y})|}\, y_1$, using the plus sign if $h_{11}(\mathbf{0}) > 0$ and the minus sign if $h_{11}(\mathbf{0}) < 0$. (Note that $h_{11}(\mathbf{y})$ maintains the same sign over some neighborhood of the origin.) Then f takes the form

$$f = \pm y_1^2 + \sum_{j,k=2}^{n} y_j y_k q_{jk}(\mathbf{y}).$$

The step just completed starts the induction argument.

Now suppose there exist coordinates $u_1, \dots, u_n$ in a neighborhood U_1 of the origin such that

$$f(\mathbf{u}) = \pm u_1^2 \pm \cdots \pm u_{k-1}^2 + \sum_{i,j \geq k}^{n} u_i u_j H_{ij}(\mathbf{u}) \tag{11.24}$$

for $\mathbf{u} = (u_1, \dots, u_n)$ in U_1, where $H_{ij}(\mathbf{u}) = H_{ji}(\mathbf{u})$ for all i and j, and $k-1 < n$. As we argued above, we can carry out a nonsingular transformation of the last $n-k+1$ variables, if necessary, to ensure that $H_{kk}(\mathbf{0}) \neq 0$. (Otherwise, we contradict the hypothesis that our critical point is nondegenerate, since f is then a sum of $k-1 < n$ squared terms for only $k-1$ independent variables.) Thus we have $H_{kk}(\mathbf{u}) \neq 0$ on a possibly smaller neighborhood $U_2 \subset U_1$, where $H_{kk}(\mathbf{u})$ maintains either a positive or a negative sign.

Now we want to complete another square. To do so, let us separate out from the residual sum in (11.24) only those terms that involve u_k in front of the H_{ij}'s. These are the only terms needed to form the new squared term. Using the symmetry of the H_{ij}, these terms follow, where we complete the square on the u_k terms:

$$\begin{aligned}
u_k^2 H_{kk}(\mathbf{u}) + 2\sum_{i>k} u_k u_i H_{ik}(\mathbf{u}) &= H_{kk}(\mathbf{u})\Big[u_k^2 + 2u_k \sum_{i>k}^{n} u_i \frac{H_{ik}(\mathbf{u})}{H_{kk}(\mathbf{u})}\Big] \\
&= H_{kk}(\mathbf{u})\Big[u_k^2 + \Big(2u_k \sum_{i>k}^{n} u_i \frac{H_{ik}(\mathbf{u})}{H_{kk}(\mathbf{u})}\Big) + \Big(\sum_{i>k}^{n} u_i \frac{H_{ik}(\mathbf{u})}{H_{kk}(\mathbf{u})}\Big)^2\Big] \\
&\qquad - H_{kk}(\mathbf{u})\Big(\sum_{i>k}^{n} u_i \frac{H_{ik}(\mathbf{u})}{H_{kk}(\mathbf{u})}\Big)^2 \\
&= H_{kk}(\mathbf{u})\Big[u_k + \sum_{i>k}^{n} u_i \frac{H_{ik}(\mathbf{u})}{H_{kk}(\mathbf{u})}\Big]^2 - H_{kk}(\mathbf{u})\Big(\sum_{i>k}^{n} u_i \frac{H_{ik}(\mathbf{u})}{H_{kk}(\mathbf{u})}\Big)^2.
\end{aligned}$$

We want the first term on the right to be the next squared term, and the last term on the right will go into the new residual. Therefore we define

$$v_k = |H_{kk}(\mathbf{u})|^{1/2}\Big[u_k + \sum_{i>k}^{n} u_i \frac{H_{ik}(\mathbf{u})}{H_{kk}(\mathbf{u})}\Big] \tag{11.25}$$

to accomplish this. The factor $|H_{kk}(\mathbf{u})|^{1/2}$ is a class C^{p-2} nonzero function defined on U_2. Our new coordinates $v_1, \dots, v_n$ are given by $v_i = u_i$ for $i \neq k$, and $v_k = v_k(\mathbf{u})$ in (11.25). We note that

$$\frac{\partial v_k}{\partial u_k}(\mathbf{0}) = |H_{kk}(\mathbf{0})|^{1/2} \neq 0,$$

and by continuity of all the H_{ik} (if $p \geq 2$), $\partial v_k / \partial u_k$ is nonzero throughout some possibly smaller neighborhood $U_3 \subset U_2$ about the origin. Thus, by the inverse function theorem, $v_1, \dots, v_n$ serve as coordinates in U_3. If we also rescale v_k as $\sqrt{|H_{kk}(\mathbf{u})|}\, v_k$ and relabel it again as v_k, then, in the coordinates $v_1, \dots, v_n$, f takes the form

$$f(\mathbf{v}) = \Big(\sum_{i=1}^{k-1} \pm v_i^2\Big) \pm v_k^2 + \sum_{i,j>k}^{n} v_i v_j H'_{ij}(\mathbf{v}) = \sum_{i=1}^{k} \pm v_i^2 + \sum_{i,j>k}^{n} v_i v_j H'_{ij}(\mathbf{v})$$

for $\mathbf{v}$ in U_3 and C^{p-2} functions H'_{ij} on U_3. This process continues until there are no nonzero terms in the residual sum, which can occur only after exactly n steps, by the nondegeneracy hypothesis. Then a rearrangement and relabeling of the squared terms, if necessary, yields (11.21). □

The proof of the Morse lemma shows that if f is at least class C^2, then the overall coordinate transformation constructed in the proof is at least class C^0, that is, continuous.

In some applications the Morse lemma provides useful information about the topological nature of the level sets of a function. To give one indication of this, in the study of stability of equilibria for systems of ordinary differential equations, certain nonnegative functions $V(\mathbf{x})$ are employed as energy functions (also called Lyapunov functions) that may decrease along solution curves starting near the equilibrium point. This decrease along solutions, if it occurs, may be detected without solving the system explicitly, since the chain rule implies that $V(\mathbf{x})$ decreases along solutions of the system $d\mathbf{x}/dt = \mathbf{f}(\mathbf{x})$ when $\nabla V(\mathbf{x}) \cdot \mathbf{f}(\mathbf{x}) < 0$, a condition that requires knowledge only of the differential equations and V itself. If the equilibrium point is a nondegenerate minimum of a C^2 function V, then the result of the Morse lemma shows that solutions near equilibrium become trapped inside the nested topological spheres that constitute the level sets of V. Thus, such functions help to determine the stability or asymptotic stability of equilibria. These energy function ideas are explored for some low-dimensional examples in subsection 14.4.2; however, the ideas there are not restricted to the use of C^2 functions.

Exercises.

Exercise 11.5.1. Does the Morse lemma apply to any critical point of the function $f(x_1, x_2) = x_1^2 - 6x_1x_2 + 9x_2^2$?

Exercise 11.5.2. Use the Morse lemma to describe the shape of the graph of $f(x, y) = x^2 - 4xy - 6y^2 + xy^3$ for (x, y) in a neighborhood of the origin.

Exercise 11.5.3. Suppose $f(x, y) = x^2 + y^2 + x^5 + 3x \cos x \sin y + 4$. Show that there are local coordinates $\xi = \xi(x, y)$, $\eta = \eta(x, y)$ in a neighborhood V of the origin such that $f(x, y) = 4 - \xi^2 + \eta^2$ for (ξ, η) in V.

11.6. Notes and References

See Halmos [**25**] or Hoffman and Kunze [**31**] for comprehensive linear algebra background. The books of Edwards [**10**] and Lang [**42**] influenced this chapter. The Morse lemma is from [**47**].

The inverse function theorem and implicit function theorem play a fundamental role in the development of the theory of smooth manifolds; see Boothby [**6**], Lee [**44**], or Munkres [**48**].

Both the inverse function theorem and the implicit function theorem are fundamental in the development of many results for ordinary differential equations (see Hale [**24**]) and partial differential equations, see Renardy and Rogers [**50**].

Chapter 12

The Riemann Integral in Euclidean Space

In this chapter we extend the Riemann integral of real valued and vector valued functions to certain subsets of $\mathbf{R}^n$. In outline form, the main goals of the chapter develop as follows:

In Section 1, we define closed intervals in $\mathbf{R}^n$ and extend the Riemann integral to bounded real valued functions defined on these closed intervals.

Section 2 defines integrability for bounded functions on bounded sets by requiring the function to be integrable on a closed interval containing the bounded set. The integral thus defined allows us to define a volume measure known as *Jordan measure* for certain bounded domains in $\mathbf{R}^n$.

In Section 3 we characterize the Jordan measurable sets; these are the sets that have volume measure.

In Section 4, we define subsets of Lebesgue measure zero in $\mathbf{R}^n$. In Section 5 this concept helps us to characterize integrability on a closed interval.

Section 6 covers properties of the integral, and Section 7 addresses the computation of multiple integrals by Fubini's theorem.

12.1. Bounded Functions on Closed Intervals

The development of the Riemann integral for a function of several variables is similar to the development in the case of a function of a single variable. Although the pattern of the development should seem familiar, we present the theory in detail. Instead of closed intervals in $\mathbf{R}$, the basic domains are the closed intervals in $\mathbf{R}^n$, which we now define.

Intervals and Partitions. An **open interval** in $\mathbf{R}^n$ $(n \geq 2)$ is a Cartesian product of n real intervals,

$$B = (a_1, b_1) \times \cdots \times (a_n, b_n),$$

where $a_i < b_i$ for $1 \leq i \leq n$. A **closed interval** in $\mathbf{R}^n$ ($n \geq 2$) has the form

$$B = [a_1, b_1] \times \cdots \times [a_n, b_n],$$

where $a_i \leq b_i$ for $1 \leq i \leq n$. (Note that if $a_i < b_i$, $1 \leq i \leq n$, then the interior of a closed interval is the open interval having the same endpoints for each interval factor.) The **volume** of either of these types of intervals, described by the Cartesian product of real intervals, is defined to be $\nu(B) = \prod_{i=1}^{n}(b_i - a_i)$.[1] We also define the volume of a union of finitely many intervals, any two of which intersect (if at all) only along boundary segments, to be the sum (finite) of the volumes of the intervals.

The most direct extension of the Riemann integral concept is to the case of bounded functions defined on a closed interval. (A closed interval in $\mathbf{R}^1$ is a closed interval $[a, b]$.)

A **partition** P of the closed interval $B = [a_1, b_1] \times \cdots \times [a_n, b_n]$ is obtained by choosing a partition for each of the interval factors $[a_i, b_i]$, say by points $x_0^i, x_1^i, \ldots,$ $x_{m_i}^i$, with $x_0^i = a_i$ and $x_{m_i}^i = b_i$. Given these partition points in each of the intervals $[a_i, b_i]$, the resulting partition P of B is precisely the set of lattice points determined by the partitions of each factor; thus

$$P = \{(x_{j_1}^1, x_{j_2}^2, \ldots, x_{j_n}^n) \in \mathbf{R}^n : 0 \leq j_i \leq m_i, \text{ for } i = 1, \ldots, n\}.$$

Thus a partition of B yields, for $1 \leq i \leq n$, m_i subintervals for $[a_i, b_i]$, and we obtain in this case a total of $m_1 m_2 \cdots m_n$ **intervals of the partition** P, given by all the Cartesian products

$$[x_{j_1}^1, x_{j_1+1}^1] \times [x_{j_2}^2, x_{j_2+1}^2] \times \cdots \times [x_{j_n}^n, x_{j_n+1}^n]$$

of the subintervals within the $[a_i, b_i]$, where $0 \leq j_i \leq m_i - 1$ for $1 \leq i \leq n$.

Figure 12.1 indicates a partition of $[0, 1] \times [0, 1]$ with $(4)(4) = 16$ lattice points and $(3)(3) = 9$ intervals of the partition, and displays the boundary of each subinterval.

Upper and Lower Sums. Let $f : B \to \mathbf{R}$ be a bounded function on an interval B in $\mathbf{R}^n$. We denote lower and upper sums for f associated with the partition P as before; thus, for an interval S of the partition P, we define

$$m_S(f) = \inf_{\mathbf{x} \in S} f(\mathbf{x}) \quad \text{and} \quad M_S(f) = \sup_{\mathbf{x} \in S} f(\mathbf{x}).$$

We then define the **lower sum** of f for P,

$$L(f, P) = \sum_S m_S(f) \nu(S),$$

and the **upper sum** of f for P,

$$U(f, P) = \sum_S M_S(f) \nu(S),$$

where each summation is over all the intervals S of the partition P. Clearly, we have $L(f, P) \leq U(f, P)$ for any partition P.

[1] We could go ahead and define the volume of intervals having some interval factors closed and some open, or some factors half-open, to be the product of the lengths of the interval factors. There is no problem in doing this; however, it is a consequence of the general definition of volume in Definition 12.3.1.

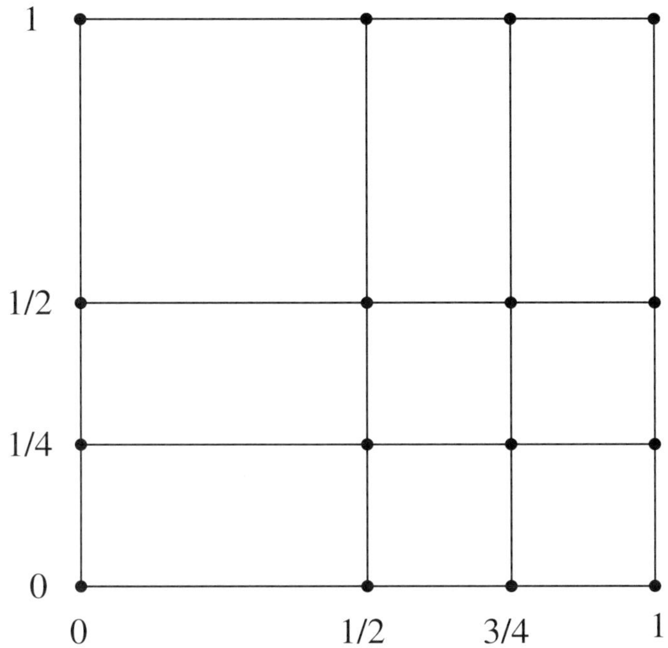

Figure 12.1. A partition of $[0,1] \times [0,1]$ with four mesh points in each factor for a total of 16 lattice points and nine intervals of the partition. The interval boundary segments are shown.

Upper and Lower Sums under Refinement. We define a partition P' to be a **refinement** of a partition P if each interval of P' is contained in some interval of P.

If P' is a refinement of P, then $L(f,P) \leq L(f,P')$ and $U(f,P') \leq U(f,P)$. This is so because on any interval of P', the infimum of f is greater than or equal to the infimum on any containing interval from P, and the supremum of f is less than or equal to the supremum on any containing interval from P. An important consequence of this monotonic behavior of upper and lower sums under refinement is that for any two partitions P_1 and P_2 of $B = [a_1, b_1] \times \cdots \times [a_n, b_n]$, we have

$$L(f, P_1) \leq U(f, P_2). \tag{12.1}$$

Indeed, we have for any partition P that is a refinement of both P_1 and P_2,

$$L(f, P_1) \leq L(f, P) \leq U(f, P) \leq U(f, P_2),$$

which proves (12.1). Observe that we may always choose $P = P_1 \cup P_2$ as a refinement of both P_1 and P_2.

Integrability of a Real Function on a Closed Interval. If $f : B \to \mathbf{R}$ is a bounded function, then the set of lower sums

$$\{L(f,P) : P \text{ is a partition of } B\}$$

and the set of upper sums

$$\{U(f,P) : P \text{ is a partition of } B\}$$

are bounded sets. In particular, each of these sets is contained in the real interval $[(\inf_B f)\nu(B), (\sup_B f)\nu(B)]$. By (12.1), we have

$$\sup\{L(f, P)\} \leq \inf\{U(f, P)\}.$$

Definition 12.1.1. *If $f : B \to \mathbf{R}$ is a bounded function on the closed interval B in $\mathbf{R}^n$, then f is* **Riemann integrable** *on B if*

$$\sup_P L(f, P) = \inf_P U(f, P),$$

where the supremum and infimum are taken over all partitions P of B. If f is Riemann integrable on B, then the Riemann integral of f on B is the common value, denoted by $\int_B f$.

Other notations for the Riemann integral of f on B, instead of the simplest notation $\int_B f$, are $\int_B f(\mathbf{x})\, d\mathbf{x}$ or $\int \cdots \int_B f(x_1, \ldots, x_n)\, dx_1 \cdots dx_n$, although these more complicated notations appear most often as a mnemonic device in the computation of iterated integrals or possibly in some change of variable formulas. For $n = 1$ and $B = [a, b]$, a real closed interval, the simplest notation is either $\int_{[a,b]} f$ or $\int_a^b f$.

Integrability of Vector Valued Functions of a Vector Variable. A direct generalization of Definition 12.1.1 for vector valued functions of a vector variable is as follows.

Definition 12.1.2. *Let $\mathbf{F} : B \to \mathbf{R}^m$ be a function bounded on the closed interval B in $\mathbf{R}^n$, and let us write $\mathbf{F} = (f_1, \ldots, f_m)$ where the f_j are the real valued component functions. We say that $\mathbf{F}$ is* **Riemann integrable** *on B if and only if each component function $f_j : B \to \mathbf{R}$, $1 \leq j \leq m$, is Riemann integrable on B. Then the vector*

$$\int_B \mathbf{F} = \left(\int_B f_1, \ldots, \int_B f_m \right)$$

is called the **Riemann integral** *of $\mathbf{F}$ on B.*

Riemann's Criterion for Integrability on a Closed Interval. The following criterion for integrability of a function over a closed interval will be useful at several places in the development of this chapter.

Theorem 12.1.3. *Let B be a closed interval in $\mathbf{R}^n$. A function $f : B \to \mathbf{R}$ is Riemann integrable on B if and only if for every $\epsilon > 0$ there is a partition P_ϵ of B such that*

$$U(f, P_\epsilon) - L(f, P_\epsilon) < \epsilon.$$

Proof. The proof is similar to that of Theorem 6.2.6.

If f is integrable on B, then $\sup_P L(f, P) = \inf_P U(f, P) = \int_B f$, the supremum and infimum being taken over all partitions P of B. Given $\epsilon > 0$, by the definition of supremum and infimum, there are partitions P_1 and P_2 such that

$$U(f, P_1) < \left(\int_B f \right) + \epsilon/2 \quad \text{and} \quad L(f, P_2) > \left(\int_B f \right) - \epsilon/2.$$

Let $P_\epsilon = P_1 \cup P_2$. Then P_ϵ refines P_1 and P_2, so $U(f, P_\epsilon) \leq U(f, P_1)$ and $L(f, P_\epsilon) \geq L(f, P_2)$. Hence,

$$U(f, P_\epsilon) - L(f, P_\epsilon) \leq U(f, P_1) - L(f, P_2) < \epsilon/2 + \epsilon/2 = \epsilon.$$

Conversely, suppose that for every $\epsilon > 0$ there is a partition P_ϵ of B such that

$$U(f, P_\epsilon) - L(f, P_\epsilon) < \epsilon.$$

Given the partition P_ϵ, we have

$$L(f, P_\epsilon) \leq \sup_P L(f, P) \leq \inf_P U(f, P) \leq U(f, P_\epsilon),$$

and hence

$$0 \leq \inf_P U(f, P) - \sup_P L(f, P) \leq U(f, P_\epsilon) - L(f, P_\epsilon) < \epsilon.$$

Since this is true for every $\epsilon > 0$, we have $\inf_P U(f, P) = \sup_P L(f, P)$, and f is integrable on B. □

Exercises.

Exercise 12.1.1. Let $B = [0, 1] \times [0, 1]$.

1. Partition B by partitioning the first factor using the points $0, 1/2, 3/4, 1$, and partitioning the second factor using the points $0, 1/4, 1/2, 1$. Sketch B together with the complete set of lattice points that comprise the resulting partition P. Also sketch the boundary segments of each interval of the partition, clearly outlining these intervals.
2. Refine the partition P of part 1 by adding points $1/6, 1/3$ to the partition of the first factor, and adding points $2/3, 5/6$ to the partition of the second factor. This defines the refinement P' of P. Sketch B together with the complete set of lattice points that comprise the new partition P'. Also sketch the boundary segments of each interval of P', clearly outlining these intervals.

Exercise 12.1.2. Let $B = [0, 1] \times [0, 1]$ and define $\mathbf{F} : B \to \mathbf{R}^2$ by $\mathbf{F} = (f_1, f_2)$, where

$$f_1(x, y) = 1 \quad \text{for all } (x, y) \in B$$

and

$$f_2(x, y) = \begin{cases} 0 & \text{if } (x, y) \in (\mathbf{Q} \times \mathbf{Q}) \cap B, \\ 1 & \text{if } (x, y) \in (\mathbf{Q} \times \mathbf{Q})^c \cap B. \end{cases}$$

Show that $\mathbf{F}$ is not Riemann integrable on B.

12.2. Bounded Functions on Bounded Sets

We have defined the integral of a bounded function over a closed interval in $\mathbf{R}^n$. We now want to extend the definition of the integral to bounded functions on more general bounded domains.

Now let $S \subset \mathbf{R}^n$ be a bounded set, and $f : S \to \mathbf{R}$ a bounded function. We may extend f to all of $\mathbf{R}^n$ by defining

$$f_S(\mathbf{x}) = \begin{cases} f(\mathbf{x}) & \text{if } \mathbf{x} \in S, \\ 0 & \text{if } \mathbf{x} \notin S. \end{cases}$$

This is called the **extension of** f **by zero**. Let B be a closed interval in $\mathbf{R}^n$ that contains the bounded set S. We want to say that f is integrable on S if f_S is integrable on B, that is, if the integral $\int_B f_S$ exists. However, we have to show that the existence of the integral, and its value, is independent of the enclosing interval B.

Lemma 12.2.1. *Let S be a bounded subset of $\mathbf{R}^n$ and $f : S \to \mathbf{R}$ a bounded function such that $\int_B f_S$ exists for some closed interval B containing S. Then*

$$\int_B f_S = \int_{B'} f_S$$

for any other closed interval B' in $\mathbf{R}^n$ containing S.

Proof. We are assuming that $\int_B f_S$ exists for some closed interval B containing S, and we let B' be any other closed interval containing S. Let B'' be a closed interval containing $B \cup B'$. We propose to show that $\int_B f_S = \int_{B''} f_S = \int_{B'} f_S$, where the existence of the integrals is part of this statement. Let us argue first that $\int_B f_S = \int_{B''} f_S$; we will see that the same argument applies to B'.

Any partition P'' of B'' induces a partition P of B by restriction: we use the partition points in the factors of B'' that are contained in B, and we include the endpoints of the factors of B as needed (if these endpoints are not already partition points included in the description of the partition P''). On the other hand, any partition P of B induces a partition P'' of B'' by extension: in this case, we use exactly the partition points in each factor of B specified in the description of P as the partition points for P'' in the factors of B''. Now let P and P'' be corresponding partitions of B and B'', respectively; that is, either P is the restriction of P'' to B or P'' is the extension of P to B''. Since $f_S(\mathbf{x}) = 0$ for all $\mathbf{x} \in B'' - B$, we have

$$L(f_S, P) = L(f_S, P'') \qquad \text{and} \qquad U(f_S, P) = U(f_S, P'').$$

Hence,

$$\sup_P L(f_S, P) = \sup_{P''} L(f_S, P'') \qquad \text{and} \qquad \inf_P U(f_S, P) = \inf_{P''} U(f_S, P'').$$

Since $\int_B f_S$ exists, we have shown that $\int_{B''} f_S$ exists and $\int_B f_S = \int_{B''} f_S$. Now since $\int_{B''} f_S$ exists, by a similar argument we find that $\int_{B'} f_S$ exists and $\int_{B''} f_S = \int_{B'} f_S$. Hence, $\int_B f_S = \int_{B'} f_S$ since both equal $\int_{B''} f_S$. □

Lemma 12.2.1 justifies the following definition.

Definition 12.2.2. *If $S \subset \mathbf{R}^n$ is a bounded set and $f : S \to \mathbf{R}$ is a bounded function for which $\int_B f_S$ exists for some closed interval B containing S, then f is* **integrable on** S *and*

$$\int_S f = \int_B f_S$$

is the **integral of** f **on** S.

Thus the existence of $\int_S f$, and its value, are independent of the enclosing interval B.

We have now defined integrability over bounded domains. We want to be sure that we can integrate (at the very least) continuous functions over a large class of domains that are likely to arise in practice. A major question concerns the identification of bounded sets S that are reasonable domains of integration, and a major issue concerns the boundary of S. We begin the effort to identify appropriate domains by introducing the concept of volume, or Jordan measure, in the next section.

Exercise.

Exercise 12.2.1. Let $D = \{\mathbf{x} \in \mathbf{R}^3 : \mathbf{x} = (1/k, 1/k^2, 1/k^3), k \in \mathbf{N}\}$. Define $\chi_D : \mathbf{R}^3 \to \mathbf{R}$ by $\chi_D(\mathbf{x}) = 1$ if $\mathbf{x} \in D$ and $\chi_D(\mathbf{x}) = 0$ if $\mathbf{x} \notin D$. (This function χ_D is called the *characteristic function* of the set D.) Find $\int_D \chi_D$.

12.3. Jordan Measurable Sets; Sets with Volume

In introductory calculus, the reader learned how to compute the lengths of curves, areas of planar regions, and volumes of solids. Often the calculation involved some ingenuity in describing the boundary of the set in question. The volume of certain solid objects was computed by integrating the constant function 1 over the set in question.

This last idea is the one pursued here for developing the concept of volume. We ask whether it is possible to integrate the simple constant function 1 over the set. We will see that our definition does lead us to a useful characterization of the boundary of sets having volume. The precise definition follows.

Definition 12.3.1. *If $A \subset \mathbf{R}^n$ is a bounded set, the* **characteristic function** *of A is the mapping $\chi_A : \mathbf{R}^n \to \mathbf{R}$ defined by*

$$\chi_A(\mathbf{x}) = \begin{cases} 1 & \text{if } \mathbf{x} \in A, \\ 0 & \text{if } \mathbf{x} \notin A. \end{cases}$$

We say that the set A is **Jordan measurable** *or that A* **has volume** *if χ_A is integrable on A, that is, $\int_A \chi_A$ exists. The* **volume** *of A, denoted $\nu(A)$, is defined by*

$$\nu(A) = \int_A \chi_A.$$

For open intervals, $S = (a_1, b_1) \times \cdots \times (a_n, b_n)$, and their closure, $\overline{S} = [a_1, b_1] \times \cdots \times [a_n, b_n]$, whose volumes equal $\prod_{i=1}^n (b_i - a_i)$ by axiom, we define, for consistency,

$$\int_S \chi_S = \int_{\overline{S}} \chi_{\overline{S}} = \prod_{i=1}^{n} (b_i - a_i).$$

The volume of A, when it exists, is also called the **Jordan measure**, *or* **Jordan content**, *of A.*

For a subset A of the two-dimensional plane, the volume is the **area** of the region, and this area is numerically equal to the (three-dimensional) volume of the solid lying between the graph of χ_A and the region A in the plane. For an interval $A = [a, b]$ of real numbers, the volume is the **length** of the interval, and this length is numerically the same as the area of the region between the graph of $\chi_{[a,b]}$ and the interval $A = [a, b]$ on the real line.

A set A with volume such that $\nu(A) = 0$ is said to have **volume zero**.

Remark. *The concept of volume zero is also called* **Jordan measure zero** *or* **Jordan content zero**.

It follows from the definition of integrability of χ_A that a set A has volume zero if and only if for every $\epsilon > 0$ there is a finite collection of closed intervals $S_1, \dots, S_N$ such that $A \subseteq \bigcup_{i=1}^{N} S_i$ and

$$\sum_{i=1}^{N} \nu(S_i) < \epsilon.$$

(See Exercise 12.3.2.)

The volume of the open interval $S = (a_1, b_1) \times \cdots \times (a_n, b_n)$ equals the volume of its closure, $\overline{S}$, the closed interval $\overline{S} = [a_1, b_1] \times \cdots \times [a_n, b_n]$. After we learn more about integrability, we will see that the volume is the same, as a consequence of Definition 12.3.1, if some factors of the interval are open and some closed, and/or some factors are half-open.

One of the weaknesses of the volume concept as given in Definition 12.3.1 is that it does not apply to unbounded sets. Another weakness is that a countable union of sets having volume is not necessarily a set having volume, even in some cases where we think it probably should be. Consider the next example.

Example 12.3.2. On the real line, the open set $\bigcup_{k=1}^{\infty}(k - 1/2^k, k + 1/2^k)$ has what we call finite total length, given by

$$\sum_{k=1}^{\infty} \frac{2}{2^k} = \sum_{k=1}^{\infty} \frac{1}{2^{k-1}} = 2,$$

but it does not have volume since it is an unbounded set. △

Part of the problem in the last example is that the set is unbounded. But consider the next example.

Example 12.3.3. Any single point, that is, a singleton set $\{\mathbf{x}\}$, has volume zero, since it can be covered by a single closed interval of arbitrarily small volume (Exercise 12.3.2). On the real line, consider the rational numbers in $[0, 1]$, that is, $S = \mathbf{Q} \cap [0, 1]$. Then S is bounded, and it is the union of countably many (singleton) sets of volume zero, but S does not have volume, much less volume zero, since χ_S is not integrable. There are similar examples in the plane and in higher dimensions. For example, the rational points (the points with rational coordinates) in the unit square in the plane, $\mathbf{Q} \times \mathbf{Q} \cap [0, 1] \times [0, 1]$, is a countable union of sets with volume zero, but it does not have volume, since its characteristic function is not integrable (Exercise 12.3.3). △

We have seen that there are open sets that do not have volume. Since open sets play a fundamental role in analysis, this must be seen as a weakness in the theory of Jordan measure we are discussing, and the weakness is tied to the Riemann integral concept (through Definition 12.3.1). The central issue that prevents some bounded sets S from having volume is that the boundary ∂S may be too complicated to allow integrability of the characteristic function χ_S. This issue about the boundary ∂S is discussed more fully in the next section.

Exercises.

Exercise 12.3.1. Verify that for any two sets $S_1, S_2 \subseteq \mathbf{R}^n$, the sets

$$\partial(S_2 - S_1) = \partial(S_2 \cap (S_1)^c), \qquad \partial(S_1 \cap S_2), \qquad \text{and} \qquad \partial(S_1 \cup S_2)$$

are contained in $\partial S_1 \cup \partial S_2$.

Exercise 12.3.2. Use the definition of integrability to show that a bounded set $A \subset \mathbf{R}^n$ has volume zero if and only if for every $\epsilon > 0$ there is a finite collection of closed intervals $S_1, \ldots, S_N$ such that $A \subseteq \bigcup_{i=1}^N S_i$ and $\sum_{i=1}^N \nu(S_i) < \epsilon$. *Hint*: For each implication, think of the intervals S_i as intervals of a partition P, involved in defining an upper sum $U(\chi_A, P)$ for that partition.

Exercise 12.3.3. Show that $A = \mathbf{Q} \times \mathbf{Q} \cap [0,1] \times [0,1]$ does not have volume, that is, χ_A is not integrable.

12.4. Lebesgue Measure Zero

Henri Lebesgue (1875-1941) made some far-reaching advances in measure and integration theory, and the following concept is central to the success of his approach. Lebesgue's approach to integration is discussed later in this book. For now, we need his concept of *measure zero* to help characterize the functions that are integrable in the Riemann sense. This, in itself, is a major success of Lebesgue's work.

Definition 12.4.1. *Let $S \subset \mathbf{R}^n$, bounded or unbounded. We say that S has n-dimensional* **Lebesgue measure zero** *(or simply* **measure zero***) if for every $\epsilon > 0$ there is a sequence of open intervals, J_i, in $\mathbf{R}^n$ such that $S \subseteq \bigcup_i J_i$ and*

$$\sum_i \nu(J_i) < \epsilon.$$

The concepts of measure zero and volume zero depend on the dimension, and one can write $m_n(S) = 0$ and $\nu_n(S) = 0$ to indicate n-dimensional Lebesgue measure zero and n-dimensional volume zero, respectively, if needed.

Example 12.4.2. Let S be the set of rational numbers in the unit interval, $S = \mathbf{Q} \cap [0,1]$. Then S has Lebesgue measure zero. We enumerate these rationals by the listing $\{r_1, r_2, r_3, \ldots\}$, and then cover the numbers individually by open intervals whose lengths sum to less than a given $\epsilon > 0$. For example, cover r_1 by an open interval of length $\epsilon/2$, r_2 by an open interval of length $\epsilon/2^2$; in general, cover r_k by an open interval of length $\epsilon/2^k$. Then the countable collection of these open intervals covers S and has total length less than $\sum_{k=1}^\infty \epsilon/2^k = \epsilon$. Therefore S has Lebesgue measure zero. △

Example 12.4.3. The set $S = \{(x,0) : 0 \le x \le 1\}$ has 2-dimensional Lebesgue measure zero. To verify this, observe that T can be covered by the single closed interval $[0,1] \times [0,\delta]$ for any $\delta > 0$. Since this interval has volume δ, we conclude that T has volume zero, and hence T has measure zero. Alternatively, given $0 < \epsilon < 1$, S can be covered by a single open interval, for example,

$$R = \left\{(x,y) : -\frac{\epsilon}{4} < x < 1 + \frac{\epsilon}{4},\ -\frac{\epsilon}{4} < y < \frac{\epsilon}{4}\right\}$$

which has volume $\nu(R) = (1 + \epsilon/2)(\epsilon/2) < \epsilon$. Therefore S has measure zero. △

If S has n-dimensional volume zero, then it has n-dimensional Lebesgue measure zero. For if S has volume zero, then for any $\epsilon > 0$, S can be covered by a finite collection of closed intervals I_i, $1 \le i \le N$, such that $\sum_{i=1}^N \nu(I_i) < \epsilon/2$. For each i, we can cover I_i with an open interval J_i of volume $\nu(I_i) + \epsilon/2^{i+1}$, and $\sum_{i=1}^N \nu(J_i) = \sum_{i=1}^N \nu(I_i) + \sum_{i=1}^N \epsilon/2^{i+1} < \epsilon/2 + \epsilon/2 = \epsilon$. Since countable means finite or countably infinite, S has Lebesgue measure zero. On the other hand, there are sets having Lebesgue measure zero that do not have volume, as we see in the next example.

Example 12.4.4. Let S be the set of points in the unit square $[0,1] \times [0,1]$ having rational coordinates, that is,

$$S = \mathbf{Q} \times \mathbf{Q} \cap [0,1] \times [0,1].$$

Then S has Lebesgue measure zero, since S is countable (Exercise 12.4.2). However, S does not have volume, because the characteristic function of S is not integrable, by Exercise 12.3.3. △

We have seen that volume zero implies Lebesgue measure zero; however, the converse does not generally hold. An exception is described in the next proposition.

Proposition 12.4.5. *A compact set in $\mathbf{R}^n$ that has Lebesgue measure zero also has volume zero.*

Proof. Suppose $A \subset \mathbf{R}^n$ is compact (that is, closed and bounded) and has Lebesgue measure zero. Let $\epsilon > 0$. Since A has Lebesgue measure zero, there is a sequence of open intervals, J_i, in $\mathbf{R}^n$ such that $A \subseteq \bigcup_i J_i$ and $\sum_i \nu(J_i) < \epsilon$. Since A is compact, there is a finite subcover $\{J_{i_1}, J_{i_2}, \dots, J_{i_M}\}$ of A. By taking the closure of each of these M open intervals, we have the collection $\{\overline{J}_{i_1}, \overline{J}_{i_2}, \dots, \overline{J}_{i_M}\}$ of closed intervals, which covers A, and

$$\sum_{j=1}^{M} \nu(\overline{J}_{i_j}) \le \sum_i \nu(J_i) < \epsilon.$$

This argument holds for every $\epsilon > 0$, and therefore A has volume zero. □

Observe that if $J_1 \times \cdots \times J_n$ is an interval in $\mathbf{R}^n$, then its boundary is given by

$$\bigcup_{k=1}^{n} J_1 \times \cdots \times J_{k-1} \times (\partial J_k) \times J_{k+1} \times \cdots \times J_n.$$

It is not difficult to see that the boundary of an interval in $\mathbf{R}^n$ has volume zero, and thus the boundary has n-dimensional Lebesgue measure zero. On the other hand, it is not difficult to directly see that the boundary of an interval in $\mathbf{R}^n$ has Lebesgue measure zero, and that the boundary is compact (since it is closed and bounded); hence, the boundary of an interval has volume zero by the proposition.

It follows directly from Definition 12.4.1 that every subset of a set of measure zero has measure zero (Exercise 12.4.1). In particular, since the empty set is a subset of every set, it is covered by a single interval of arbitrarily small volume, and hence has Lebesgue measure zero. By a similar argument, any singleton set $\{\mathbf{x}\}$ has measure zero, and thus every finite set in $\mathbf{R}^n$ has Lebesgue measure zero.

The reader can verify that we could have chosen closed intervals instead of open intervals in the definition of Lebesgue measure zero and obtained the same concept (Exercise 12.4.3).

Example 12.4.6. Let us show that the graph of a continuous function $f : [a, b] \to \mathbf{R}$ has 2-dimensional Lebesgue measure zero. It suffices to show that the graph has 2-dimensional volume zero. Let $G = \{(x, f(x)) : x \in [a, b]\}$ be the graph. Since f is uniformly continuous on $[a, b]$, for every $\epsilon > 0$ there is a $\delta > 0$ such that $|x_1 - x_2| < \delta$ implies $|f(x_1) - f(x_2)| < \epsilon/(b - a)$. Thus, for every $\epsilon > 0$, we can find a finite cover of G by closed rectangles having height $\epsilon/(b - a)$ and nonoverlapping interiors. Thus, G is covered by finitely many closed intervals in $\mathbf{R}^2$ whose total volume is less than or equal to ϵ, so $\nu(G) = 0$. △

Despite the result of this last example, a continuous image of a set with n-dimensional volume zero need not have n-dimensional volume zero. This fact is demonstrated by the existence of space-filling curves, one of which is presented at the beginning of the next chapter.

An advantage of the concept of measure zero over that of volume zero is that the union of a countable infinity of sets, each having measure zero, is also a set of measure zero. The case of dimension $n = 1$ in the next result was invoked earlier in the proof of Theorem 7.1.11 to show that a uniform limit of Riemann integrable functions on $[a, b]$ is Riemann integrable on $[a, b]$. (The proof here applies as written to the case $n = 1$, with the understanding that the volume $\nu(J_j^k)$ of interval J_j^k means the length of that interval.)

Theorem 12.4.7. *A countably infinite union of sets of n-dimensional Lebesgue measure zero is a set of n-dimensional Lebesgue measure zero.*

Proof. Let $\{E_k\}$ be a countable collection of subsets $E_k \subset \mathbf{R}^n$, each having Lebesgue measure zero. Given $\epsilon > 0$, there exists a doubly indexed collection of open intervals $\{J_j^k\}$ in $\mathbf{R}^n$ such that for each k,

$$\sum_j \nu(J_j^k) < \frac{\epsilon}{2^{k+1}}$$

and $\bigcup_j J_j^k$ covers E_k. Thus, $\bigcup_{k,j} J_j^k$ covers $\bigcup_k E_k$. The problem now is to arrange this doubly indexed collection of intervals into a sequence and then sum the volumes according to that definite sequence.

We arrange the volumes $\nu(J_j^k)$ of these intervals in a matrix with $\nu(J_1^1)$ as the upper left entry and, using k as the row index, j as the column index, we can list the volumes, by tracing the diagonals of slope one in our matrix, starting from the upper left entry, in the order

(12.2)
$$\nu(J_1^1), \quad \nu(J_1^2), \nu(J_2^1), \quad \nu(J_1^3), \nu(J_2^2), \nu(J_3^1), \quad \nu(J_1^4), \nu(J_2^3), \nu(J_3^2), \nu(J_4^1),$$

and so on. Let σ_n denote the n-th partial sum based on this ordering of the volumes; thus,

$$\sigma_1 = \nu(J_1^1), \quad \sigma_2 = \nu(J_1^1) + \nu(J_1^2), \quad \sigma_3 = \nu(J_1^1) + \nu(J_1^2) + \nu(J_2^1), \quad \ldots .$$

With the diagonals of slope one in our matrix in view, we can form the *triangular* partial sums

$$s_n = \sum_{k+j\le n} \nu(J_j^k),$$

and notice that we have

$$s_1 = \sigma_1, \quad s_2 = \sigma_3, \quad s_3 = \sigma_6, \quad \dots, \quad s_n = \sigma_{(n(n+1))/2}, \quad \dots .$$

Thus (σ_n) is an increasing sequence with subsequence (s_n). By (12.2), we may write

$$s_n = \sigma_{(n(n+1))/2} = \sum_{k=1}^{n} \Big(\nu(J_1^k) + \nu(J_2^{k-1}) + \cdots + \nu(J_k^1)\Big).$$

Since $\nu(J_j^k) \ge 0$ for all $k, j \in \mathbf{N}$, we have

$$\begin{aligned} s_n = \sigma_{(n(n+1))/2} &= \sum_{k=1}^{n} \Big(\nu(J_1^k) + \nu(J_2^{k-1}) + \cdots + \nu(J_k^1)\Big) \le \sum_{k=1}^{n} \Big(\sum_{j=1}^{n} \nu(J_j^k)\Big) \\ &\le \sum_{k=1}^{n} \Big(\sum_{j=1}^{\infty} \nu(J_j^k)\Big) < \sum_{k=1}^{n} \frac{\epsilon}{2^{k+1}} < \frac{\epsilon}{2} < \epsilon. \end{aligned}$$

Therefore the sequence $(s_n) = (\sigma_{(n(n+1))/2})$ is increasing and bounded above by ϵ, and it converges to a limit $s < \epsilon$. Since it is a subsequence of the increasing sequence (σ_n), we conclude from Theorem 2.4.17 that (σ_n) itself converges to the same limit, hence $\lim_{n\to\infty} \sigma_n = s < \epsilon$. We conclude that the sum of the volumes of the intervals J_j^k, listed in (12.2), is less than ϵ. Since ϵ is arbitrary, this shows that $\bigcup_k E_k$ has Lebesgue measure zero. □

Example 12.4.8. The real numbers of the form $a+b\sqrt{2}$, $a, b \in \mathbf{Q}$, are all irrational. The set of such numbers is a countable union of countable sets, and hence has measure zero. △

We noted in Example 12.4.3 that the real interval $[0,1]$, considered as a subset of the plane, has 2-dimensional Lebesgue measure zero. Let us show that the entire real line, considered as a subset of the plane, has measure zero.

Example 12.4.9. The entire real line, considered as a subset of the plane, has measure zero. The proof depends on showing that increasingly larger chunks of the embedded line can be covered by smaller and smaller 2-dimensional interval volumes. Given $0 < \epsilon < 1$, we must find open intervals J_j in $\mathbf{R}^2$ such that $\sum_j \nu(J_j) < \epsilon$. For example, we may choose

$$J_j = \Big(-\frac{j}{2}, \frac{j}{2}\Big) \times \Big(-\frac{\epsilon}{2j(2^j)}, \frac{\epsilon}{2j(2^j)}\Big).$$

The 2-dimensional volume of J_j is $\nu(J_j) = \epsilon/2^j$, and thus $\sum_j \nu(J_j) < \epsilon$. Moreover, it is clear that the entire real line, considered as a subset of the plane, that is, the set $\{(x,0) : x \in \mathbf{R}\}$, is contained in the union of the J_j. Thus, the embedded real line has measure zero in $\mathbf{R}^2$. △

Exercises.

Exercise 12.4.1. Show that if $A \subset B \subset \mathbf{R}^n$ and B has Lebesgue measure zero, then so does A.

Exercise 12.4.2. Show that every countably infinite set in $\mathbf{R}^n$ has Lebesgue measure zero.

Exercise 12.4.3. Show that a set $A \subset \mathbf{R}^n$ has Lebesgue measure zero if and only if for every $\epsilon > 0$ there is a sequence of *closed* intervals I_i such that $A \subseteq \bigcup_i I_i$ and $\sum_i \nu(I_i) < \epsilon$.

Exercise 12.4.4. Show that if $Z \subset \mathbf{R}^n$ has Lebesgue measure zero, then for any $\mathbf{a} \in \mathbf{R}^n$ the translated set $\mathbf{a}+Z = \{\mathbf{a}+\mathbf{x} : \mathbf{x} \in Z\}$ also has Lebesgue measure zero.

Exercise 12.4.5. *Volume measure is invariant under translations*
Show that if $Z \subset \mathbf{R}^n$ has volume zero, then for any $\mathbf{a} \in \mathbf{R}^n$ the translated set $\mathbf{a}+Z = \{\mathbf{a}+\mathbf{x} : \mathbf{x} \in Z\}$ also has volume zero. Show that, if S has volume, then for any $\mathbf{a} \in \mathbf{R}^n$, the translated set $\mathbf{a}+S = \{\mathbf{a}+\mathbf{x} : \mathbf{x} \in S\}$ also has volume, and $\nu(\mathbf{a}+S) = \nu(S)$.

Exercise 12.4.6. Give an example of a bounded set $A \subset \mathbf{R}^n$ (for some n) such that ∂A does not have Lebesgue measure zero.

Exercise 12.4.7. Give an example of a set $A \subset \mathbf{R}^n$ (for some n) having measure zero, such that ∂A does not have measure zero.

12.5. A Criterion for Riemann Integrability

The purpose of this section is to prove the following criterion for the integrability of a real valued function on a closed interval in $\mathbf{R}^n$.

Theorem 12.5.1. *Let B be a closed interval in $\mathbf{R}^n$. A bounded function $f : B \to \mathbf{R}$ is Riemann integrable on B if and only if the set of points where f is discontinuous has Lebesgue measure zero.*

With $n = 1$, the theorem includes the case when $B = [a,b]$, a closed interval, and $f : [a,b] \to \mathbf{R}$, in which case a reading of the proof adapts itself to that special case with no difficulty, providing a proof of Theorem 6.4.4.

Let B be an interval in $\mathbf{R}^n$ and $f : B \to \mathbf{R}$ a bounded function. We begin by recalling the definitions and result of Exercise 8.10.7 which involves the concept of the oscillation $o(f, \mathbf{p})$ of f at $\mathbf{p} \in B$. For each open set U containing $\mathbf{p}$, the **oscillation of f on U** is defined by

$$o(f,U) = \sup\Big\{|f(x_1) - f(x_2)| : x_1, x_2 \in U \cap B\Big\}.$$

The **oscillation of f at $\mathbf{p} \in B$** is defined by

$$o(f,\mathbf{p}) = \inf\Big\{o(f,U) : U \text{ open and } \mathbf{p} \in U\Big\}.$$

For any $\mathbf{p} \in B$, we have $o(f,\mathbf{p}) \geq 0$. By Exercise 8.10.7, f is discontinuous at $\mathbf{p}$ if and only if $o(f,\mathbf{p}) > 0$. Thus, f is discontinuous at $\mathbf{p}$ if and only if there is an $m \in \mathbf{N}$ such that $o(f,\mathbf{p}) > 1/m$. The main advantage of the oscillation concept is that it allows us to write the entire set of points of discontinuity of f as a countable union. Let D be the set of points in B at which f is discontinuous. For each $m \in \mathbf{N}$, let

$$D_{1/m} = \{\mathbf{p} \in B : o(f, \mathbf{p}) \geq 1/m\}.$$

Then $D_{1/m} \subseteq D$ and

$$D = \bigcup_{m=1}^{\infty} D_{1/m}.$$

From Exercise 8.10.7, we have that each set $D_{1/m}$ is a closed set, but we shall prove this here. If $\mathbf{y}$ is a cluster point of $D_{1/m}$, then every open neighborhood of $\mathbf{y}$ contains points of $D_{1/m}$. So every open neighborhood U of $\mathbf{y}$ is an open neighborhood of a point $\mathbf{p}$ of $D_{1/m}$; hence, by definition of $D_{1/m}$, we have

$$o(f, U) = \sup \Big\{ |f(\mathbf{x}_1) - f(\mathbf{x}_2)| : \mathbf{x}_1, \mathbf{x}_2 \in U \cap B \Big\} \geq \frac{1}{m}.$$

Therefore $o(f, \mathbf{y}) \geq 1/m$, so $\mathbf{y} \in D_{1/m}$. So $D_{1/m}$ contains all its cluster points and is therefore closed.

Let P be a partition of B, and let $m \in \mathbf{N}$ be a fixed positive integer. The subintervals S of the partition P are of two types:

(i) $S \cap D_{1/m} \neq \emptyset$, so there exist points in S such that the oscillation of f in every open neighborhood of those points is $\geq 1/m$.

(ii) $S \cap D_{1/m} = \emptyset$, which means there are no points in S such that the oscillation of f in every open neighborhood is $\geq 1/m$. Equivalently, for each $\mathbf{p} \in S$, we have $o(f, \mathbf{p}) < 1/m$.

Proposition 12.5.2. *Let B be a closed interval in $\mathbf{R}^n$. If $f : B \to \mathbf{R}$ is a bounded function and the set D of discontinuities of f has Lebesgue measure zero, then f is Riemann integrable on B.*

Proof. We are assuming that $\mu(D) = 0$. Let $\epsilon > 0$, and fix a positive integer $m \in \mathbf{N}$ such that $1/m < \epsilon$. Since $D_{1/m} \subset D \subset B$, $D_{1/m}$ is bounded, and since $D_{1/m}$ is closed, $D_{1/m}$ is compact, and since $\mu(D) = 0$, $\mu(D_{1/m}) = 0$. So there is a sequence of open intervals $\{J_k\}$ such that $D_{1/m} \subseteq \bigcup_k J_k$ and

$$\sum_{k=1}^{\infty} \nu(J_k) < \epsilon.$$

Since $D_{1/m}$ is compact, a finite number of the J_k cover $D_{1/m}$; suppose $J_1, \ldots, J_N$ cover $D_{1/m}$. Then clearly

$$\sum_{k=1}^{N} \nu(J_k) < \epsilon. \tag{12.3}$$

Let P be a partition of B. By refining the partition if necessary, we may assume that each interval of the partition is either disjoint from $D_{1/m}$ or contained in one of the intervals $J_1, \ldots, J_N$ that cover $D_{1/m}$. Thus the intervals of the partition are split into two classes, $C1$ and $C2$, not necessarily disjoint:

$C1$: the collection of intervals of the partition contained in one of the open intervals $J_1, \ldots, J_N$, which cover $D_{1/m}$;

$C2$: the collection of intervals of the partition that do not intersect $D_{1/m}$.

For each interval S in $C2$, since S does not intersect $D_{1/m}$, the oscillation of f at each point of S is $< 1/m$. So for each $\mathbf{x} \in S$ we can find an open neighborhood $U_\mathbf{x}$ of $\mathbf{x}$ such that

$$M_{U_\mathbf{x}}(f) - m_{U_\mathbf{x}}(f) < \frac{1}{m},$$

where we have employed our usual notation for the supremum (M) and infimum (m) of f over a set. Since S is compact, a finite number of the open sets $U_\mathbf{x}$ covers S. Moreover, we may refine the partition P such that within S, each interval of the new partition is contained in one of the finitely many open sets $U_\mathbf{x}$ that covers S. We may do this for each of the intervals S in $C2$, yielding a partition P' of B. There is an M such that $|f(\mathbf{x})| \leq M$ for $\mathbf{x} \in B$. And for the new partition P' and classes $C1$, $C2$ relative to P', we have

$$U(f,P') - L(f,P') \leq \sum_{S \in \mathbf{C1}} [M_S(f) - m_S(f)]\nu(S) + \sum_{S \in \mathbf{C2}} [M_S(f) - m_S(f)]\nu(S),$$

and thus

$$U(f,P') - L(f,P') \leq \sum_{S \in \mathbf{C1}} 2M\nu(S) + \frac{1}{m}\nu(B).$$

Since $S \subset \bigcup J_k$ for S in $\mathbf{C1}$, (12.3) and the fact that $1/m < \epsilon$ imply

$$U(f,P') - L(f,P') \leq 2M\epsilon + \epsilon\,\nu(B) = [2M + \nu(B)]\,\epsilon.$$

Since ϵ is arbitrary, Riemann's criterion (Theorem 12.1.3) implies that f is integrable on B. □

We wish to prove the converse of Proposition 12.5.2.

Proposition 12.5.3. *Let B be a closed interval in $\mathbf{R}^n$. If $f : B \to \mathbf{R}$ is Riemann integrable on B, then the set D of discontinuities of f has Lebesgue measure zero.*

Proof. We have $D = \bigcup_{m=1}^{\infty} D_{1/m}$ where $D_{1/m} = \{\mathbf{p} \in B : o(f,\mathbf{p}) \geq 1/m\}$. We wish to show that for each $m \in \mathbf{N}$, $D_{1/m}$ has measure zero, for then D is a countable union of sets of measure zero, and hence D has measure zero by Theorem 12.4.7.

Let $\epsilon > 0$. By integrability of f, there is a partition P of B such that

$$U(f,P) - L(f,P) = \sum_S [M_S(f) - m_S(f)]\nu(S) < \epsilon, \tag{12.4}$$

where the sum is over all intervals S of the partition P. Let $n \in \mathbf{N}$ be a fixed positive integer. We write $D_{1/n}$ as a disjoint union as follows. Let

$$S_1 := \{\mathbf{x} \in D_{1/n} : \mathbf{x} \in \partial S \text{ for some rectangle } S \text{ of } P\}$$

and

$$S_2 := \{\mathbf{x} \in D_{1/n} : \mathbf{x} \in \operatorname{Int} S \text{ for some rectangle } S \text{ of } P\}.$$

Then $D_{1/n} = S_1 \cup S_2$, and $\mu(S_1) = 0$, since the boundary of any interval has measure zero and there are finitely many intervals of the partition.

Since S_1 has measure zero, we can refine the partition P to produce partition P', if necessary, and find a finite collection C' of intervals of P' that covers S_1 such that

$$\sum_{S\in C'} \nu(S) < \epsilon.$$

Let C denote the collection of intervals S of the partition P' that have an element of $D_{1/n}$ in their interior. Thus, if S is in C, then

$$M_S(f) - m_S(f) \geq \frac{1}{n},$$

and therefore

$$\frac{1}{n}\sum_{S\in C} \nu(S) \leq \sum_{S\in C}[M_S(f) - m_S(f)]\nu(S) \leq \sum_{S}[M_S(f) - m_S(f)]\nu(S) < \epsilon,$$

by (12.4). Thus, C is a collection of intervals that cover S_2, and we have

$$\sum_{S\in C} \nu(S) \leq n\,\epsilon.$$

Then $C \cup C'$ covers $S_1 \cup S_2 = D_{1/n}$ and

$$\sum_{S\in C\cup C'} \nu(S) < n\epsilon + \epsilon = (n+1)\epsilon.$$

Since ϵ is arbitrary (and n fixed), $D_{1/n}$ has measure zero. Since n was arbitrary, $D = \bigcup_{m=1}^{\infty} D_{1/m}$ has measure zero by Theorem 12.4.7. □

Taken together, Propositions 12.5.2 and 12.5.3 complete the proof of Theorem 12.5.1.

The next corollary is an immediate consequence of Theorem 12.5.1 and the definition of Riemann integrability for a bounded real function on a bounded set S in $\mathbf{R}^n$ (Definition 12.2.2).

Corollary 12.5.4. *Let S be a bounded set in $\mathbf{R}^n$ and B any closed interval containing S. A bounded function $f : S \to \mathbf{R}$ is integrable on S if and only if the set of discontinuities of f_S, the extension of f by zero to B, has Lebesgue measure zero.*

If a property or statement involving points of B holds for all points except on a subset of B of Lebesgue measure zero, we may say that the property holds **almost everywhere (a.e.)** in B.

Corollary 12.5.5. *A bounded set S in $\mathbf{R}^n$ has volume if and only if ∂S has Lebesgue measure zero.*

Proof. A bounded set S in $\mathbf{R}^n$ has volume if and only if the characteristic function χ_S is integrable on S. For any set S, the set of discontinuities of χ_S coincides with ∂S (Exercise 12.5.1). Thus, by Theorem 12.5.1, χ_S is integrable if and only if ∂S has measure zero. □

The next result summarizes the situation with integrability in a convenient manner for later reference.

Corollary 12.5.6. *Let S be a bounded set that has volume. A bounded function $f : S \to \mathbf{R}$ is integrable on S if and only if f is continuous a.e. in the interior of S.*

Proof. Extend f by zero to f_S on a closed interval B that contains S. The set D of discontinuities of f_S on B is

$$D = (D \cap \operatorname{Int} S) \cup (D \cap \partial S),$$

since $f_S \equiv 0$ outside the closed set $\overline{S} = \operatorname{Int} S \cup \partial S$. Since S has volume, ∂S has Lebesgue measure zero. So $D \cap \partial S$, being a subset of ∂S, has Lebesgue measure zero. Thus, f_S is integrable on B if and only if $D \cap \operatorname{Int} S$ has measure zero. Therefore f is integrable on S if and only if f is continuous a.e. in the interior of S. □

Example 12.5.7. Let f be defined on the closed unit square $B = [0,1] \times [0,1]$ in the plane by $f(x,y) = 1$ for (x,y) in the interior and for points with rational coordinates on the boundary, and $f(x,y) = 0$ for all other points on the boundary. The boundary has measure zero in $\mathbf{R}^2$ and f is continuous everywhere in the interior, which is the open unit square $(0,1) \times (0,1)$, so f is integrable on B. Observe that f is discontinuous everywhere on the boundary, an uncountable set. △

Exercise.

Exercise 12.5.1. Prove: For any set $S \subset \mathbf{R}^n$, the set of discontinuities of χ_S coincides with ∂S.

12.6. Properties of Volume and Integrals

Suppose S has volume in $\mathbf{R}^n$. We wish to prove the linearity of the integral over S. By Corollary 12.5.5 and Corollary 12.5.6 at the end of the previous section, and Definition 12.2.2, it suffices to prove linearity of the integral on closed intervals B.

Theorem 12.6.1. *If B is a closed interval in $\mathbf{R}^n$, $f, g : B \to \mathbf{R}$ are integrable on B, and α and β are any real numbers, then $\alpha f + \beta g$ is integrable on B and*

$$\int_B (\alpha f + \beta g) = \alpha \int_B f + \beta \int_B g.$$

Proof. If f and g are continuous at $\mathbf{p}$, then so is $\alpha f + \beta g$. It follows that the set of discontinuities of $\alpha f + \beta g$ is contained in the union of the set of discontinuities of f with the set of discontinuities of g. If f and g are integrable on B, then they are continuous a.e. on $\operatorname{Int} B$, so it follows that $\alpha f + \beta g$ is continuous a.e. on $\operatorname{Int} B$, hence $\alpha f + \beta g$ is integrable on B.

(i) Prove that $\int_B \alpha f = \alpha \int_B f$ for $\alpha > 0$: We have $U(\alpha f, P) = \alpha U(f, P)$ and $L(\alpha f, P) = \alpha L(f, P)$ for any partition P. Since we have established integrability of αf, it suffices to consider lower sums only. By the integrability of f, we have

$$\sup L(\alpha f, P) = \sup \alpha L(f, P) = \alpha \sup L(f, P) = \alpha \int_B f,$$

so the integrability of αf implies that $\int_B \alpha f = \alpha \int_B f$.

Recall that if S is a set which is bounded above and bounded below, then $\inf(-S) = -\sup S$ and $\sup(-S) = -\inf S$.

(ii) Prove that $\int_B(-f) = -\int_B f$: We have established that $-f$ is integrable. If P is a partition of B, then $L(-f,P) = -U(f,P)$. With

$$S = \{L(-f,P) : P \text{ a partition of } B\},$$

we have $-S = \{U(f,P) : P \text{ a partition of } B\}$, so we have

$$\inf\{U(f,P)\} = \inf(-S) = -\sup S = -\sup\{L(-f,P)\}.$$

By the integrability of $-f$, it follows that $\sup\{L(-f,P)\} = \int_B(-f)$, and by the integrability of f, we have

$$\int_B (-f) = \sup\{L(-f,P)\} = -\inf\{U(f,P)\} = -\int_B f.$$

(iii) Prove that $\int_B \alpha f = \alpha \int_B f$ for $\alpha < 0$: This follows immediately from (i) and (ii) above, since we have

$$\int_B \alpha f = \int_B -|\alpha| f = -\int_B |\alpha| f = -|\alpha| \int_B f = \alpha \int_B f.$$

(iv) Prove that $\int_B(f+g) = \int_B f + \int_B g$: Let P be a partition of B and let S be any rectangle of the partition P. Since

$$M_S(f+g) = \sup_S(f(\mathbf{x}) + g(\mathbf{x})) \le \sup_S f(\mathbf{x}) + \sup_S g(\mathbf{x}) = M_S(f) + M_S(g)$$

and

$$m_S(f+g) = \inf_S(f(\mathbf{x}) + g(\mathbf{x})) \ge \inf_S f(\mathbf{x}) + \inf_S g(\mathbf{x}) = m_S(f) + m_S(g),$$

it follows that

$$U(f+g,P) \le U(f,P) + U(g,P) \tag{12.5}$$

and

$$L(f+g,P) \ge L(f,P) + L(g,P). \tag{12.6}$$

Given $\epsilon > 0$, there exist partitions P_1 and P_2 such that

$$U(f,P_1) < \frac{\epsilon}{2} + \int_B f \quad \text{and} \quad U(g,P_2) < \frac{\epsilon}{2} + \int_B g.$$

Let $P_3 = P_1 \cup P_2$; then P_3 is a refinement of P_1 and P_2. By (12.5),

$$U(f+g,P_3) \le U(f,P_3) + U(g,P_3) \le U(f,P_1) + U(g,P_2) < \epsilon + \int_B f + \int_B g,$$

hence

$$\int_B (f+g) = \inf\{U(f+g,P) : P \text{ a partition of } B\} \le \epsilon + \int_B f + \int_B g.$$

By a similar argument using lower sums, (12.6) yields the other inequality,

$$\int_B (f+g) = \sup\{L(f+g,P) : P \text{ a partition of } B\} \ge \epsilon + \int_B f + \int_B g.$$

Since ϵ is arbitrary, we conclude that $\int_B(f+g) = \int_B f + \int_B$. This completes the proof of linearity of the integral on B. □

As we indicated at the beginning of the section, Theorem 12.6.1 establishes the linearity of the integral over any bounded set S that has volume, for functions f and g integrable on S and real numbers α and β, since for any closed interval B containing S,

$$\int_S (\alpha f + \beta g) = \int_B (\alpha f_S + \beta g_S) = \alpha \int_B f_S + \beta \int_B g_S = \alpha \int_S f + \beta \int_S g.$$

We can now list the basic properties of sets that have volume, that is, the Jordan measurable sets. The reader might wish to read or work out Exercise 12.3.1 before reading the proofs of these properties.

Theorem 12.6.2. *Let S_1 and S_2 be subsets of $\mathbf{R}^n$ that have volume. Then the following statements are true:*

1. *$S_1 \cup S_2$ and $S_1 \cap S_2$ have volume, and*
$$\nu(S_1 \cup S_2) = \nu(S_1) + \nu(S_2) - \nu(S_1 \cap S_2).$$
2. *If $\operatorname{Int} S_1 \cap \operatorname{Int} S_2$ is the empty set, then*
$$\nu(S_1 \cup S_2) = \nu(S_1) + \nu(S_2).$$
3. *If $S_1 \subseteq S_2$, then $S_2 - S_1 = S_2 \cap S_1^c$ has volume and*
$$\nu(S_2 \cap S_1^c) = \nu(S_2) - \nu(S_1).$$
4. *If $S_1 \subseteq S_2$, then*
$$\nu(S_1) \leq \nu(S_2).$$

Proof. 1. Since S_1 and S_2 have volume, they are bounded and both ∂S_1, ∂S_2 have volume zero, so $\partial S_1 \cup \partial S_2$ has volume zero. Then $S_1 \cup S_2$ and $S_1 \cap S_2$ are also bounded, and both $\partial(S_1 \cup S_2)$ and $\partial(S_1 \cap S_2)$ are contained in $\partial S_1 \cup \partial S_2$, and thus have volume zero, so we may conclude that both $S_1 \cup S_2$ and $S_1 \cap S_2$ have volume. It remains to show the formula for the volume of the union. For the characteristic functions of the four sets S_1, S_2, $S_1 \cup S_2$ and $S_1 \cap S_2$, it is straightforward to show that

$$\chi_{S_1} + \chi_{S_2} = \chi_{S_1 \cup S_2} + \chi_{S_1 \cap S_2} \tag{12.7}$$

(Exercise 12.6.2). Since χ_{S_1} and χ_{S_2} are integrable on any closed interval B that contains S_1 and S_2, we have by the linearity of the integral and the definition of volume,

$$\int_B (\chi_{S_1} + \chi_{S_2}) = \int_B \chi_{S_1} + \int_B \chi_{S_2} = \nu(S_1) + \nu(S_2)$$

and

$$\int_B (\chi_{S_1 \cup S_2} + \chi_{S_1 \cap S_2}) = \int_B \chi_{S_1 \cup S_2} + \int_B \chi_{S_1 \cap S_2} = \nu(S_1 \cup S_2) + \nu(S_1 \cap S_2).$$

We conclude from (12.7) and these two integral formulas that

$$\nu(S_1) + \nu(S_2) = \nu(S_1 \cup S_2) + \nu(S_1 \cap S_2),$$

which is the desired result.

2. Since S_1 and S_2 have no interior points in common, if $\mathbf{x} \in S_1 \cap S_2$, then either $\mathbf{x} \in \partial S_1$ or $\mathbf{x} \in \partial S_2$, so $S_1 \cap S_2 \subseteq \partial S_1 \cup \partial S_2$. Since both boundaries have volume zero, so does $S_1 \cap S_2$. The formula for $\nu(S_1 \cup S_2)$ then follows from statement 1.

3. If $S_1 \subseteq S_2$, then $S_2 - S_1 = S_2 \cap S_1^c$ is bounded, and $\partial(S_2 \cap S_1^c) \subseteq \partial S_1 \cup \partial S_2$. Hence, $\partial(S_2 \cap S_1^c)$ has volume zero, so $S_2 \cap S_1^c$ has volume. Since $S_2 = S_2 \cup S_1 = (S_2 \cap S_1^c) \cup S_1$ is a disjoint union, statements 1 and 2 imply that $\nu(S_2) = \nu(S_2 \cap S_1^c) + \nu(S_1)$, which is the desired result.

4. This is immediate from statement 3 and the fact that $\nu(S_2 \cap S_1^c) \geq 0$. □

Given the result of Theorem 12.6.2 (statement 1), an induction argument shows that if each set in a finite collection $\{S_1, \dots, S_N\}$ has volume (these are bounded sets, by definition), then the union $\bigcup_{i=1}^N S_i$ is also bounded and has volume. However, in general, countable unions of sets having volume need not have volume, as we have seen.

The next theorem summarizes some basic properties of the integral.

Theorem 12.6.3. *Let A have volume in $\mathbf{R}^n$, and let $f, g : A \to \mathbf{R}$ be integrable on A. Then the following statements are true:*

1. *If $f \geq 0$ on A, then $\int_A f \geq 0$.*
2. *If $f(\mathbf{x}) \leq g(\mathbf{x})$ for all $\mathbf{x} \in A$, then $\int_A f \leq \int_A g$.*
3. *The function $|f|$ is integrable on A, and $\left|\int_A f\right| \leq \int_A |f|$.*
4. *If $m \leq |f(\mathbf{x})| \leq M$ for all $\mathbf{x} \in A$, then*

$$m\,\nu(A) \leq \int_A |f| \leq M\,\nu(A).$$

Proof. 1. If $f \geq 0$ on A, then $U(f, P) \geq L(f, P) \geq 0$ for any partition P of a closed interval B containing A. Hence, $\int_A f = \inf U(f, P) = \sup L(f, P) \geq 0$.

2. Since $g - f \geq 0$, this follows from statement 1 by the linearity of the integral.

3. Since f is integrable on A, f is continuous a.e. in $\operatorname{Int} A$; therefore $|f|$ is continuous a.e. in $\operatorname{Int} A$, and it follows that $|f|$ is integrable on A (Corollary 12.5.6). Since $-f, f \leq |f|$ on A, we have $-\int_A f \leq \int_A |f|$ and $\int_A f \leq \int_A |f|$ by statement 2. Thus, $\left|\int_A f\right| \leq \int_A |f|$.

4. By statement 2, together with $m\,\nu(A) = m \int_A \chi_A = \int_A m$, and similarly for $M\,\nu(A)$, we have

$$m\,\nu(A) = \int_A m \leq \int_A |f| \leq \int_A M = M\,\nu(A).$$

This completes the proof. □

Finally, we again consider an open interval $S = (a_1, b_1) \times \cdots \times (a_n, b_n)$, and its closure, $\overline{S} = [a_1, b_1] \times \cdots \times [a_n, b_n]$, whose volumes equal $\prod_{i=1}^n (b_i - a_i)$, by axiom; thus, we defined

$$\int_S \chi_S = \int_{\overline{S}} \chi_{\overline{S}} = \prod_{i=1}^n (b_i - a_i)$$

as part of Definition 12.3.1 so that these two types of sets have consistently defined volumes. Now, if $\hat{S}$ is any variation of S such that some factors are open and some closed, and/or some factors are half-open, then $S \subset \hat{S} \subset \overline{S}$. We could define

$\nu(\hat{S}) = \int_{\hat{S}} \chi_{\hat{S}} = \prod_{i=1}^{n}(b_i - a_i)$, or simply notice that on any closed interval B containing $\overline{S}$ (including $\overline{S}$ itself), Theorem 12.6.3 (statement 2) shows that

$$\chi_S \leq \chi_{\hat{S}} \leq \chi_{\overline{S}} \quad \Longrightarrow \quad \int_B \chi_S \leq \int_B \chi_{\hat{S}} \leq \int_B \chi_{\overline{S}},$$

and hence $\nu(S) = \nu(\hat{S}) = \nu(\overline{S}) = \prod_{i=1}^{n}(b_i - a_i)$.

The final theorem of the section characterizes integrability in terms of Riemann sums defined by selections of points from the intervals of a partition. First, some definitions.

Definition 12.6.4. *Let Q be a closed interval in $\mathbf{R}^n$ and suppose $f : Q \to \mathbf{R}$ is bounded. Suppose P is a partition of Q and $\{S_1, \ldots, S_m\}$ is the collection of intervals of P. Then the* **mesh** *of P is the maximum length of any edge (factor) of the intervals $S_1, \ldots, S_m$, and a* **selection** *for P is a set $\mathcal{S} = \{\mathbf{x}_1, \ldots, \mathbf{x}_m\}$ such that $\mathbf{x}_i \in S_i$ for $1 \leq i \leq m$. The* **Riemann sum** *for f corresponding to the partition P and selection $\mathcal{S}$ is*

$$R(f, P, \mathcal{S}) = \sum_{i=1}^{m} f(\mathbf{x}_i)\, \nu(S_i).$$

Theorem 12.6.5. *Let Q be an interval in $\mathbf{R}^n$ and suppose $f : Q \to \mathbf{R}$ is bounded and is zero outside Q. Then f is integrable on Q and $\int_Q f = \mathcal{I}$ if and only if, given $\epsilon > 0$, there exists $\delta = \delta(\epsilon) > 0$ such that*

$$|R(f, P, \mathcal{S}) - \mathcal{I}| < \epsilon$$

whenever P partitions Q with mesh less than δ and $\mathcal{S}$ is a selection for P.

Proof. Suppose that f is integrable on Q and $\int_Q f = \mathcal{I}$. Then f is bounded so there is an M such that $|f(\mathbf{x})| \leq M$ for all $\mathbf{x}$ in Q. Given $\epsilon > 0$, there is a partition P_0 of Q with intervals $Q_1, \ldots, Q_m$ such that

$$U(f, P_0) - L(f, P_0) = \sum_{i=1}^{m} (M_i - m_i)\, \nu(Q_i) < \frac{\epsilon}{2},$$

where $M_i = \sup_{Q_i} f(\mathbf{x})$ and $m_i = \inf_{Q_i} f(\mathbf{x})$. If $A = Q - \bigcup_{i=1}^{m} \operatorname{Int} Q_i$, then A has volume zero. Then there exists $\delta > 0$ such that, if P refines P_0 and has mesh less than δ, then the sum of the volumes of the intervals $P_1, \ldots, P_k$ of P that intersect A is less than $\epsilon/4M$. Let $P_{k+1}, \ldots, P_l$ be the remaining intervals of P, which all lie interior to the Q_i.

Letting $\mathcal{S} = \{\mathbf{x}_1, \ldots, \mathbf{x}_l\}$ be a selection for the partition P, then

$$\inf_{P_i} f \leq f(\mathbf{x}_i) \leq \sup_{P_i} f$$

for $i = k+1, \ldots, l$. Thus each of the sums

$$\sum_{i=k+1}^{l} f(\mathbf{x}_i)\, \nu(P_i) \quad \text{and} \quad \sum_{i=k+1}^{l} \int_{P_i} f$$

lies between $\sum_{i=k+1}^{l}(\inf_{P_i} f)\,\nu(P_i)$ and $\sum_{i=k+1}^{l}(\sup_{P_i} f)\,\nu(P_i)$, and it follows that

$$\Big|\sum_{i=k+1}^{l}\int_{P_i} f - \sum_{i=k+1}^{l} f(\mathbf{x}_i)\,\nu(P_i)\Big| < \frac{\epsilon}{2}, \tag{12.8}$$

since the difference between the upper and lower sums for P is less than $\epsilon/2$.

Since $-M \le f(\mathbf{x}) \le M$ for all $\mathbf{x}$, each of the sums

$$\sum_{i=1}^{k}\int_{P_i} f \quad \text{and} \quad \sum_{i=1}^{k} f(\mathbf{x}_i)\,\nu(P_i)$$

lies between $-M\sum_{i=1}^{k}\nu(P_i) > -\epsilon/4$ and $M\sum_{i=1}^{k}\nu(P_i) < \epsilon/4$. It follows that

$$\Big|\sum_{i=1}^{k}\int_{P_i} f - \sum_{i=1}^{k} f(\mathbf{x}_i)\,\nu(P_i)\Big| < \frac{\epsilon}{2}. \tag{12.9}$$

Since $\mathcal{I} = \int_Q f = \sum_{k=1}^{l}\int_{P_i} f$, inequalities (12.8) and (12.9) and the triangle inequality imply that

$$|\mathcal{I} - R(f,P,\mathcal{S})| < \epsilon,$$

as we wished to show.

For the converse implication, suppose now that $\epsilon > 0$ is given, and $\delta = \delta(\epsilon) > 0$ is such that whenever P is a partition of Q with mesh less than δ and $\mathcal{S}$ is a selection for P, we have

$$|R(f,P,\mathcal{S}) - \mathcal{I}| < \frac{\epsilon}{4}.$$

Let P be such a partition of Q with intervals $P_1, \ldots, P_p$. Let

$$a_i = \inf_{P_i} f(\mathbf{x}) \quad \text{and} \quad b_i = \sup_{P_i} f(\mathbf{x}).$$

Then $U(f,P) = \sum_{i=1}^{p} b_i\,\nu(P_i)$ and $L(f,P) = \sum_{i=1}^{p} a_i\,\nu(P_i)$. We may choose selections $\mathcal{S}' = \{\mathbf{x}'_1, \ldots, \mathbf{x}'_p\}$ and $\mathcal{S}'' = \{\mathbf{x}''_1, \ldots, \mathbf{x}''_p\}$ for P such that

$$|a_i - f(\mathbf{x}'_i)| < \frac{\epsilon}{4\,\nu(Q)} \quad \text{and} \quad |b_i - f(\mathbf{x}''_i)| < \frac{\epsilon}{4\,\nu(Q)}$$

for $1 \le i \le p$. Then, by the triangle inequality,

$$\begin{aligned}\Big|R(f,P,\mathcal{S}') - L(f,P)\Big| &= \Big|\sum_{i=1}^{p}\big(f(\mathbf{x}'_i) - a_i\big)\,\nu(P_i)\Big| \\ &\le \frac{\epsilon}{4\,\nu(Q)}\sum_{i=1}^{p}\nu(P_i) \\ &= \frac{\epsilon}{4}.\end{aligned}$$

Similarly, for the selection $\mathcal{S}''$, we have

$$\begin{aligned}\left|R(f,P,\mathcal{S}'')-U(f,P)\right| &= \left|\sum_{i=1}^{p}\left(f(\mathbf{x}_i'')-b_i\right)\nu(P_i)\right| \\ &\leq \frac{\epsilon}{4\,\nu(Q)}\sum_{i=1}^{p}\nu(P_i) \\ &= \frac{\epsilon}{4}.\end{aligned}$$

Combining these results yields, again by the triangle inequality,

$$\begin{aligned}U(f,P)-L(f,P) &\leq \left|U(f,P)-R(f,P,\mathcal{S}'')\right| + \left|R(f,P,\mathcal{S}'')-\mathcal{I}\right| \\ &\quad + \left|\mathcal{I}-R(f,P,\mathcal{S}')\right| + \left|R(f,P,\mathcal{S}')-L(f,P)\right| \\ &< \frac{\epsilon}{4}+\frac{\epsilon}{4}+\frac{\epsilon}{4}+\frac{\epsilon}{4}=\epsilon\end{aligned}$$

by the hypothesis on P. Therefore f satisfies the Riemann criterion for integrability, and hence f is integrable on Q. Since $\mathcal{I}$ is trapped between $U(f,P)$ and $L(f,P)$, $\mathcal{I}=\int_Q f$. □

Exercises.

Exercise 12.6.1. Let $f(x,y)=y^2+\cos(1/(2x-y))$ for $2x-y\neq 0$ and $f(x,y)=y^2$ for $2x-y=0$. Show that f is integrable on the open ball $B=\{(x,y)\in\mathbf{R}^2 : x^2+y^2<1\}$.

Exercise 12.6.2. Prove: For sets S_1 and S_2, $\chi_{S_1}+\chi_{S_2}=\chi_{S_1\cup S_2}+\chi_{S_1\cap S_2}$.

Exercise 12.6.3. Let $S=[0,1]\times[0,1]$. Give an example of functions $f,g:S\to\mathbf{R}$, neither of them integrable on S, such that $f+g$ is integrable on S.

Exercise 12.6.4. Show that if A is bounded and has volume zero, and $g:A\to\mathbf{R}$ is integrable on A, then $\int_A g=0$.

Exercise 12.6.5. Let A and B be sets that have volume. Show that if $A\cap B$ is empty, then $\int_{A\cup B} f=\int_A f+\int_B f$.

Exercise 12.6.6. Compute $\int_B f$ where f and B are as in Example 12.5.7.

Exercise 12.6.7. Let A and B be sets that have volume. Show that if $A\cap B$ has volume zero, then $\int_{A\cup B} f=\int_A f+\int_B f$.

Exercise 12.6.8. Prove the following statements:

1. If f and g are continuous on A, $g\geq 0$ on A, $\int_A g>0$, and A is compact and connected, then there is a point $\mathbf{c}\in A$ such that
$$\int_A fg=f(\mathbf{c})\int_A g.$$
2. If A has positive volume, is compact and connected, and $f:A\to\mathbf{R}$ is continuous, then $\int_A f=f(\mathbf{c})\,\nu(A)$ for some $\mathbf{c}\in A$.

12.7. Multiple Integrals

Readers will recall earlier experience with the computation of integrals over certain planar regions or domains in 3-space by means of multiple integrals. Theorems that establish the value of an integral as being equal to certain multiple integrals are most often associated with the work of G. Fubini (1879-1943) and referred to as Fubini's theorem(s). We begin with the following version of Fubini's theorem for planar integrals.

Theorem 12.7.1. *Let $A = [a,b] \times [c,d]$ in $\mathbf{R}^2$ and write (x,y) for points of the plane.*

1. *If $f : A \to \mathbf{R}$ is continuous, then*
$$\int_A f = \int_a^b \Big(\int_c^d f(x,y)\,dy \Big)\,dx = \int_c^d \Big(\int_a^b f(x,y)\,dx \Big)\,dy.$$
2. *If $f : A \to \mathbf{R}$ is integrable on A, and for each fixed $x \in [a,b]$, the function $f_x(y) = f(x,y)$ is integrable on $[c,d]$, then*
$$\int_A f = \int_a^b \Big(\int_c^d f(x,y)\,dy \Big)\,dx.$$
3. *If $f : A \to \mathbf{R}$ is integrable on A, and for each fixed $y \in [c,d]$, the function $f_y(x) = f(x,y)$ is integrable on $[a,b]$, then*
$$\int_A f = \int_c^d \Big(\int_a^b f(x,y)\,dx \Big)\,dy.$$

Proof. If f is continuous on A, then f is integrable on A. Moreover, each section $f_x(y)$, for fixed $x \in [a,b]$, is continuous and hence integrable on $[c,d]$, and each section $f_y(x)$, for fixed $y \in [c,d]$, is continuous and hence integrable on $[a,b]$. Thus parts 2 and 3 together cover statement 1, so we only need to prove 2 and 3. (See also Exercise 12.7.7 for a different argument for statement 1.)

2. Assume $f : A \to \mathbf{R}$ is integrable on A, and for each fixed $x \in [a,b]$ the function $f_x : [c,d] \to \mathbf{R}$ defined by $f_x(y) = f(x,y)$ is integrable on $[c,d]$. Let us write
$$g(x) = \int_c^d f_x(y)\,dy.$$

We want to show that g is integrable on $[a,b]$ and that $\int_A f = \int_a^b g(x)\,dx$. Suppose $[a,b]$ is partitioned by $a = x_0 < x_1 < \cdots < x_n = b$ with $V_i = [x_{i-1}, x_i]$, and $[c,d]$ is partitioned by $c = y_0 < y_1 < \cdots < y_m = d$ with $W_j = [y_{j-1}, y_j]$. Let $P_{[a,b]}$ denote the partition of $[a,b]$ and $P_{[c,d]}$ the partition of $[c,d]$. Let P be the partition of A given by the rectangles
$$S_{ij} = [x_{i-1}, x_i] \times [y_{j-1}, y_j], \quad 1 \le i \le n,\ 1 \le j \le m.$$

The upper sum for P is
$$U(f,P) = \sum_{i,j} M_{S_{ij}}(f)\nu(S_{ij}) = \sum_i \sum_j M_{S_{ij}}(f)\nu(V_i)\nu(W_j),$$

where $M_S(f)$ is the supremum of f on a set S. If $x \in V_i$, then for $f_x(y) = f(x, y)$ we have

$$M_{S_{ij}}(f) \geq M_{W_j}(f_x).$$

Therefore

$$\sum_j M_{S_{ij}}(f)\nu(W_j) \geq \sum_j M_{W_j}(f_x)\nu(W_j) \geq \int_c^d f_x(y)\,dy = g(x).$$

Since this inequality holds for any $x \in V_i$, we have

$$\sum_j M_{S_{ij}}(f)\nu(W_j) \geq M_{V_i}(g).$$

Consequently,

$$U(f, P) \geq \sum_i M_{V_i}(g)\nu(V_i) = U(g, P_{[a,b]}).$$

A similar argument shows that for greatest lower bounds on S_{ij} and lower sums, we have

$$L(f, P) \leq L(g, P_{[a,b]}).$$

The integrability of f on A and the last two inequalities imply that g is integrable on $[a, b]$ and

$$\int_A f = \int_a^b g(x)\,dx = \int_a^b \Big(\int_c^d f(x, y)\,dy\Big)\,dx.$$

3. Now observe that if f is integrable on A and if for each fixed $y \in [c, d]$, the function $f_y(x) = f(x, y)$ is integrable on $[a, b]$, then a similar argument using

$$h(y) = \int_a^b f_y(x)\,dx = \int_a^b f(x, y)\,dx$$

shows that

$$\int_A f = \int_c^d h(y)\,dy = \int_c^d \int_a^b f_y(x)\,dx\,dy = \int_c^d \Big(\int_a^b f(x, y)\,dx\Big)\,dy.$$

This completes the proof. □

As the reader is aware from introductory calculus, many complicated regions may be decomposed into a union of regions of the form

$$A = \{(x, y) : a \leq x \leq b,\ \phi_1(x) \leq y \leq \phi_2(x)\} \tag{12.10}$$

or

$$B = \{(x, y) : c \leq y \leq d,\ \psi_1(y) \leq x \leq \psi_2(y)\}, \tag{12.11}$$

where $\phi_1, \phi_2, \psi_1, \psi_2$ are continuous functions. Assuming that f is continuous on a set such as A or B, the integration of f can proceed by multiple integrals, as indicated in the following result.

Theorem 12.7.2. *Let f be a continuous real valued function defined on a domain of the form (12.10) or (12.11), where $\phi_1, \phi_2, \psi_1, \psi_2$ are continuous functions.*

1. *If $f : A \to \mathbf{R}$, then*

$$\int_A f = \int_a^b \Big(\int_{\phi_1(x)}^{\phi_2(x)} f(x, y)\,dy\Big)\,dx.$$

2. *If* $f : B \to \mathbf{R}$, *then*

$$\int_B f = \int_c^d \Big(\int_{\psi_1(y)}^{\psi_2(y)} f(x,y)\,dx \Big)\,dy.$$

Proof. 1. Let $R = [a,b] \times [\bar{c},\bar{d}]$ be an interval containing A and consider the extension of f to R by zero. The graph of ϕ_1 and the graph of ϕ_2 both have measure zero, by Example 12.4.6. The set of discontinuities of f on R is contained in the union of these graphs, and therefore f is integrable on R. Similarly, for each fixed $x \in [a,b]$, $f_x(y) = f(x,y)$ is continuous for $y \in [\bar{c},\bar{d}]$ except possibly at $y = \phi_1(x)$ and $y = \phi_2(x)$. So $f_x(y)$ is integrable on $[\bar{c},\bar{d}]$. By Theorem 12.7.1,

$$\int_A f = \int_R f = \int_a^b \Big(\int_{\bar{c}}^{\bar{d}} f_x(y)\,dy \Big)\,dx = \int_a^b \Big(\int_{\phi_1(x)}^{\phi_2(x)} f_x(y)\,dy \Big)\,dx,$$

which proves statement 1.

A similar argument establishes statement 2 and is left as an exercise. □

The reasoning behind Theorem 12.7.2 can be extended to obtain results like the following one on triple integrals over a region that is a *cylinder* over a plane region of the types considered in the theorem.

Theorem 12.7.3. *Let* $\alpha(x)$ *and* $\beta(x)$ *be continuous functions for* $a \le x \le b$, *with* $\alpha(x) \le \beta(x)$, *and let* $\gamma(x,y)$ *and* $\delta(x,y)$ *be continuous functions for* $a \le x \le b$, $\alpha(x) \le y \le \beta(x)$, *with* $\gamma(x,y) \le \delta(x,y)$. *Let*

$$D = \{(x,y,z) \in \mathbf{R}^3 : a \le x \le b,\ \alpha(x) \le y \le \beta(x),\ \gamma(x,y) \le z \le \delta(x,y)\}.$$

If $f : D \to \mathbf{R}$ *is continuous, then* $\int_D f$ *exists and*

$$\int_D f = \int_a^b \Big(\int_{\alpha(x)}^{\beta(x)} \Big(\int_{\gamma(x,y)}^{\delta(x,y)} f(x,y,z)\,dz \Big)\,dy \Big)\,dx.$$

Variations on Theorem 12.7.3 are possible if one integrates over a solid region D which is a cylinder over (that is, projects onto) a plane region in either the xz-plane or the yz-plane of the types considered in Theorem 12.7.2. Rather than formulate such variations, we give an illustration of the idea in an example.

Example 12.7.4. Consider the solid tetrahedron D in the first octant of $\mathbf{R}^3$, bounded by the coordinate planes and the plane $4x + 6y + 6z = 12$. This plane passes through three of the four vertex points of D, namely $(0,0,2)$, $(3,0,0)$ and $(0,2,0)$; the fourth vertex is $(0,0,0)$. We will set up a triple integral for the volume of D using the order of integration indicated by

$$\text{Volume of D} = \int\int\int (1)\,dz\,dy\,dx.$$

The base of the solid in the xy-plane is a triangle with sides on the coordinate axes and on the line $4x + 6y = 12$, or $y = 2 - \frac{2}{3}x$. For each (x,y) in that base triangle, we have $z = 2 - \frac{2}{3}x - y$. Thus,

$$\text{Volume of D} = \int_0^3 \int_0^{2-\frac{2}{3}x} \int_0^{2-\frac{2}{3}x-y} (1)\,dz\,dy\,dx.$$

See Exercise 12.7.4 for other orderings for the volume integral. △

Some regions of integration that are not cylindrical in the global sense of Theorem 12.7.3 might be expressed as unions of cylindrical subregions, extending the applicability of Theorem 12.7.3 to such unions. Integrals over regions that are not cylindrical can sometimes be transformed by coordinate change so as to yield an integral over one of the simpler types of domain considered above. The general result for such a transformation of integrals is the change of variable formula in Theorem 13.4.4.

Thus far the results on integration by multiple integrals have dealt only with the case of a function of two or three real variables. However, the results can be extended to integrals of functions of finitely many real variables. If $A \subset \mathbf{R}^n$ and $B \subset \mathbf{R}^m$ are closed intervals and $f : A \times B \subset \mathbf{R}^n \times \mathbf{R}^m \to \mathbf{R}$, then we define $\mathbf{x}$-sections and $\mathbf{y}$-sections of f as follows. For each $\mathbf{x} \in A$, define

$$f_{\mathbf{x}} : B \subset \mathbf{R}^m \to \mathbf{R} \quad \text{by} \quad f_{\mathbf{x}}(\mathbf{y}) = f(\mathbf{x}, \mathbf{y}).$$

And for each $\mathbf{y} \in B$, define

$$f_{\mathbf{y}} : A \subset \mathbf{R}^n \to \mathbf{R} \quad \text{by} \quad f_{\mathbf{y}}(\mathbf{x}) = f(\mathbf{x}, \mathbf{y}).$$

Theorem 12.7.5. *Let $A \subset \mathbf{R}^n$ and $B \subset \mathbf{R}^m$ be intervals and suppose $f : A \times B \subset \mathbf{R}^n \times \mathbf{R}^m \to \mathbf{R}$.*

1. *If f is continuous on $A \times B$, then*

$$\int_{A\times B} f = \int_A \Big(\int_B f_{\mathbf{x}}(\mathbf{y})\, d\mathbf{y} \Big)\, d\mathbf{x} = \int_B \Big(\int_A f_{\mathbf{y}}(\mathbf{x})\, d\mathbf{x} \Big)\, d\mathbf{y}.$$

2. *If f is integrable on $A \times B$ and $f_{\mathbf{x}}$ is integrable on B for each fixed $\mathbf{x} \in A$, then*

$$\int_{A\times B} f = \int_A \Big(\int_B f_{\mathbf{x}}(\mathbf{y})\, d\mathbf{y} \Big)\, d\mathbf{x} = \int_A \Big(\int_B f(\mathbf{x}, \mathbf{y})\, d\mathbf{y} \Big)\, d\mathbf{x}.$$

3. *If f is integrable on $A \times B$ and $f_{\mathbf{y}}$ is integrable on A for each fixed $\mathbf{y} \in B$, then*

$$\int_{A\times B} f = \int_B \Big(\int_A f_{\mathbf{y}}(\mathbf{x})\, d\mathbf{x} \Big)\, d\mathbf{y} = \int_B \Big(\int_A f(\mathbf{x}, \mathbf{y})\, d\mathbf{x} \Big)\, d\mathbf{y}.$$

The proof can be patterned on the arguments required to prove Theorem 12.7.1 and is left to the interested reader. For certain domains Theorem 12.7.5 (part 1) can be applied repeatedly to reduce a multiple integral of a continuous function f to a sequence of integrals, as in the next result.

Theorem 12.7.6. *If $A = [a_1, b_1] \times \cdots \times [a_n, b_n]$ is a closed interval in $\mathbf{R}^n$ and $f : A \to \mathbf{R}$ is continuous, then*

$$\int_A f = \int_{a_n}^{b_n} \Big(\cdots \Big(\int_{a_1}^{b_1} f(x_1, \ldots, x_n)\, dx_1 \Big) \cdots \Big)\, dx_n.$$

Moreover, the same value, $\int_A f$, is obtained if the integrals on the right-hand side are reordered by any other possible permutation of the n integrations.

An application to the computation of the volume of n-dimensional balls appears in Exercise 13.4.6.

Exercises.

Exercise 12.7.1. Prove part 2 of Theorem 12.7.2.

Exercise 12.7.2. Write the area of the disk $D = \{(x, y) \in \mathbf{R}^2 : x^2 + y^2 \leq 1\}$ as an iterated integral in two different ways.

Exercise 12.7.3. Write the two-dimensional volume of the region $D = \{(x, y) \in \mathbf{R}^2 : 0 \leq x \leq 1,\ x^4 \leq y \leq x^{1/3}\}$ as an iterated integral in two different ways.

Exercise 12.7.4. Set up volume integrals for the volume of the tetrahedron D in Example 12.7.4 using the orderings $\int\int\int dx\, dy\, dz$ and $\int\int\int dy\, dz\, dx$. (Observe that, beyond the example and these two orderings, there are an additional three orderings for iterated integrals for this volume.)

Exercise 12.7.5. Give a proof of Theorem 12.7.5.

Exercise 12.7.6. Sketch the parallelogram

$$P = \{(x, y) \in \mathbf{R}^2 : a \leq x \leq b,\ mx + c \leq y \leq mx + d\},$$

where $m > 0$ and $c < d$, and find its area using an iterated integral.

Exercise 12.7.7. *Equality of iterated integrals of continuous functions*
It can be interesting to see how the equality of iterated integrals of a continuous function f follows from the theorem on differentiation under the integral (Theorem 8.10.24). Assume $a < b$ and $c < d$, and suppose $f : [a, b] \times [c, d] \to \mathbf{R}$ is continuous.

1. Show that the function $h(t, y) = \int_a^t f(x, y)\, dx$ is continuous on $[a, b] \times [c, d]$ and differentiable with respect to t, and $\partial h/\partial t(t, y) = f(t, y)$.
2. Show that the function
$$g(x) = \int_c^d f(x, y)\, dy$$
is a continuous function of $x \in [a, b]$.
3. Let
$$H(t) = \int_a^t \Big(\int_c^d f(x, y)\, dy\Big)\, dx - \int_c^d \Big(\int_a^t f(x, y)\, dx\Big)\, dy.$$
Verify that $H(a) = 0$, and then use the functions $g(x)$ and $h(t, y)$ to show that $H(b) = 0$. *Hint*: Show that $H'(t) = 0$ for all $t \in [a, b]$.

Chapter 13

Transformation of Integrals

The previous chapter has provided us with an integral and a technique for its computation by iterated real integrals in Fubini's theorem. However, experience with integration by substitution in introductory calculus shows that integrals are sometimes easier to handle when transformed by a change of variable. In this chapter we develop the general change of variables formula for the integral. This result has many important practical and theoretical consequences. In this development we must consider what types of mappings are suitable as coordinate transformations and also how Jordan measurable sets transform under the mappings. In particular, we know that a bounded set S has volume if and only if ∂S has Lebesgue measure zero (Corollary 12.5.5), and we need to understand how such sets, and other Jordan measurable sets, transform under suitable coordinate transformations.

The major issues addressed in this chapter may be stated in the form of the following questions:

If a set S has volume and $\mathbf{g} : S \to \mathbf{R}^n$ is a mapping, does $\mathbf{g}(S)$ have volume, and under what conditions on $\mathbf{g}$? If a function f is integrable on a set D, and we express the function f in new coordinates (for the purpose of simplifying an integral computation) by means of an invertible mapping $\mathbf{g} : S \subset \mathbf{R}^n \to D \subset \mathbf{R}^n$ with $\mathbf{g}(S) = D$, is $f \circ \mathbf{g}$ integrable on S, and under what conditions on $\mathbf{g}$? These questions are dealt with in Section 2.

If $f \circ \mathbf{g}$ integrable on S, how is the integral of f on D transformed by means of the transformation $\mathbf{g}$? The answer is given in the change of variables formula, developed for linear mappings in Section 3 and for more general mappings in Section 4.

In Section 5, we apply the change of variable formula to examine the definition of surface integrals from multivariable calculus.

In order to generate interest in the issues involved, and to show that smoothness beyond continuity is required for our transformations, we begin in Section 1 with an example which shows that a continuous mapping need not map a set of volume zero to an image set of volume zero. This example, of interest in its own right, is a *space-filling curve.*

13.1. A Space-Filling Curve

In 1890, the mathematician G. Peano (1858-1932) gave the first example of a continuous function that maps an interval onto the unit square $[0, 1] \times [0, 1]$ in the plane. Such space-filling curves are often called Peano curves. The example given here is due to Schoenberg [**56**].

Let f be the real valued function such that $f(t) = f(-t)$ (f is even), f has period two, and f is defined for $t \in [0, 1]$ by

$$f(t) = \begin{cases} 0 & 0 \leq t \leq 1/3, \\ 3t - 1 & 1/3 \leq t \leq 2/3, \\ 1 & 2/3 \leq t \leq 1. \end{cases}$$

Let D be the subset of the plane given by

$$D = \{(t, 0) \in \mathbf{R}^2 : 0 \leq t \leq 1\}.$$

Clearly D has two-dimensional volume zero. Let $\Gamma : D \to \mathbf{R}^2$ be the curve given by the parametric equations

$$\begin{aligned} (13.1)\quad x(t) &= \frac{1}{2}f(t) + \frac{1}{2^2}f(3^2t) + \frac{1}{2^3}f(3^4t) + \cdots + \frac{1}{2^k}f(3^{2k-2}t) + \cdots, \\ (13.2)\quad y(t) &= \frac{1}{2}f(3t) + \frac{1}{2^2}f(3^3t) + \frac{1}{2^3}f(3^5t) + \cdots + \frac{1}{2^k}f(3^{2k-1}t) + \cdots \end{aligned}$$

for $0 \leq t \leq 1$. Since $0 \leq f(t) \leq 1$ and $\sum_{k=1}^{\infty} 1/2^k = 1$, we have $0 \leq x(t) \leq 1$ and $0 \leq y(t) \leq 1$ for $0 \leq t \leq 1$. Also, by the Weierstrass test, the series for $x(t)$ and $y(t)$ converge uniformly on $[0, 1]$, and therefore $x(t)$ and $y(t)$ are continuous on $[0, 1]$. We now show that the image of Γ is the unit square $[0, 1] \times [0, 1]$. Thus, although D has two-dimensional volume zero, its continuous image $\Gamma(D)$ has two-dimensional volume one.

Let (x_0, y_0) be a point of the square $[0, 1] \times [0, 1]$, and write

$$(13.3)\qquad x_0 = \frac{a_0}{2} + \frac{a_2}{2^2} + \frac{a_4}{2^3} + \cdots, \qquad y_0 = \frac{a_1}{2} + \frac{a_3}{2^2} + \frac{a_5}{2^3} + \cdots$$

for the nonterminating binary expansions of x_0 and y_0, where each a_k equals 0 or 1. Let

$$(13.4)\qquad t_0 = \frac{2a_0}{3} + \frac{2a_1}{3^2} + \frac{2a_2}{3^3} + \cdots + \frac{2a_{k-1}}{3^k} + \frac{2a_k}{3^{k+1}} + \cdots.$$

Then $t_0 \in [0, 1]$, and $\Gamma(t_0) = (x_0, y_0)$. Indeed, if $a_0 = 0$, then

$$0 \leq t_0 \leq \frac{2}{3^2} + \frac{2}{3^3} + \cdots = \frac{2}{3^2}\frac{3}{2} = \frac{1}{3},$$

and hence $f(t_0) = 0$; if $a_0 = 1$, then

$$\frac{2}{3} \leq t_0 \leq 1,$$

and hence $f(t_0) = 1$. In either case, we have $f(t_0) = a_0$ where t_0 is given by (13.4). Now observe that for each positive integer k,

$$3^k t_0 = (\text{an even integer}) + \frac{2a_k}{3} + \frac{2a_{k+1}}{3^2} + \cdots$$

and hence

$$f(3^k t_0) = f\Big(\frac{2a_k}{3} + \frac{2a_{k+1}}{3^2} + \cdots\Big) = a_k,$$

by the same calculation used above for a_0. Then, by (13.1)-(13.2) and the definitions of x_0, y_0 in (13.3), we have

$$\Gamma(t_0) = (x(t_0), y(t_0)) = (x_0, y_0).$$

Since (x_0, y_0) is an arbitrary point of $[0,1] \times [0,1]$, this shows that Γ is onto the square.

This space-filling curve shows that a continuous image of a set with volume zero need not have volume zero. Motivated by this situation, we strengthen the hypotheses of the mappings we consider in the next section.

13.2. Volume and Integrability under C^1 Maps

Both the conceptual understanding and computation of certain integrals often involve a coordinate change, or substitution, that is, an invertible mapping. These invertible mappings must meet certain requirements. In particular, recall that the volume measure of a bounded set is based on the Riemann integral, and we can integrate certain functions over a bounded domain provided the domain has volume (Corollary 12.5.6). The domain has volume if and only if the boundary of the domain has volume zero. Thus we need invertible mappings for which the image of a set having volume zero is another set having volume zero, and the image of a set having volume is another set having volume.

We now examine the images of volume zero sets.

Proposition 13.2.1. *Suppose $S \subset \mathbf{R}^n$ has volume zero and $\mathbf{g} : S \to \mathbf{R}^n$ satisfies a Lipschitz condition. Then $\mathbf{g}(S)$ has volume zero.*

Proof. We use the norm $|\mathbf{x}|_\infty = \max_{1 \le i \le n} |x_i|$ and recall that the Lipschitz condition means that there is an M such that for any $\mathbf{x}$ and $\mathbf{y}$ in S,

$$|\mathbf{g}(\mathbf{x}) - \mathbf{g}(\mathbf{y})|_\infty \le M|\mathbf{x} - \mathbf{y}|_\infty.$$

Since S is bounded, it can be enclosed in a cube R. Since S has volume zero, we can invoke the integrability of the characteristic function χ_A of A, and Theorem 12.6.5 implies that S can be covered by a finite collection of cubes $S_1, \dots, S_m$, intervals of a partition of R with mesh δ, such that each S_i intersects S, and

$$\nu(S_1) + \cdots + \nu(S_m) < \epsilon.$$

Since each S_i has side length δ, each image set $\mathbf{g}(A \cap S_i)$ is contained in a cube S_i' with side length $M\delta$. Thus,

$$\nu(S_i') = M^n \delta^n = M^n \nu(S_i).$$

Therefore $\mathbf{g}(S)$ is covered by the cubes $S'_1, \ldots, S'_m$ such that

$$\nu(S'_1) + \cdots + \nu(S'_m) = M^n \sum_{i=1}^{m} \nu(S_i) \leq M^n \epsilon.$$

This is true for every $\epsilon > 0$, so $\nu(\mathbf{g}(S)) = 0$. □

The typical setup to evaluate multiple integrals as iterated integrals involves a description of the boundary of the domain of integration. This description requires a decomposition of the boundary into a finite number of pieces, each of which is parameterized by a C^1 mapping defined on a set of lower dimension. This is why the next proposition is important.

Proposition 13.2.2. *Let S be a bounded subset of $\mathbf{R}^k$, where $k < n$. If $\mathbf{g} : S \to \mathbf{R}^n$ satisfies a Lipschitz condition, then $\mathbf{g}(S)$ has volume zero.*

Proof. We may view $\mathbf{R}^k$ as embedded in $\mathbf{R}^n$ by viewing the first k components of $\mathbf{x} \in \mathbf{R}^n$ as describing a point in $\mathbf{R}^k$. Thus, we can view S as embedded in $\mathbf{R}^n$, by writing

$$\tilde{S} = \{(x_1, \ldots, x_k, 0, \ldots, 0) \in \mathbf{R}^n : (x_1, \ldots, x_k) \in S\}.$$

Since S is bounded, it is contained in some closed cube C. Then for arbitrarily small $\delta > 0$, $\tilde{S}$ is covered by the single n-dimensional closed interval $C \times [0, \delta] \times \cdots \times [0, \delta]$, which has volume $\nu(C)\delta^{n-k}$. Therefore $\tilde{S}$ has volume zero. The mapping $\mathbf{g} : S \to \mathbf{R}^n$ yields a mapping $\tilde{\mathbf{g}} : \tilde{S} \to \mathbf{R}^n$ defined by

$$\tilde{\mathbf{g}}(x_1, \ldots, x_k, 0, \ldots, 0) = \mathbf{g}(x_1, \ldots, x_k),$$

and $\tilde{\mathbf{g}}$ is Lipschitz on $\tilde{S}$ since $\mathbf{g}$ is Lipschitz on S. Consequently, by Proposition 13.2.1, we conclude that $\mathbf{g}(S)$ has volume zero in $\mathbf{R}^n$. □

In each of the previous two propositions, the Lipschitz condition on the mapping $\mathbf{g}$ can be replaced by the stronger condition that $\mathbf{g}$ is C^1 on an open set containing the closure $\overline{S}$ of S, since this guarantees a uniform Lipschitz condition on the compact $\overline{S}$, and thus on S.

Proposition 13.2.3. *Let S be a set in $\mathbf{R}^n$ that has volume, and let U be an open set that contains $\overline{S}$. Let $\mathbf{g} : U \to \mathbf{R}^n$ be C^1 and C^1-invertible on $\operatorname{Int} S$. Then $\mathbf{g}(S)$ has volume and*

$$\partial\mathbf{g}(S) \subseteq \mathbf{g}(\partial S).$$

Proof. The interior of S, $\operatorname{Int} S$, is open, and by the continuous inverse, $\mathbf{g}(\operatorname{Int} S)$ is also open. The mapping $\mathbf{g}$ is a C^1-invertible mapping of $\operatorname{Int} S$ onto $\mathbf{g}(\operatorname{Int} S)$. Since $\bar{S} = \operatorname{Int} S \cup \partial S$ and $\partial\overline{S} = \partial S$, we have

$$\mathbf{g}(\operatorname{Int} S) \subseteq \mathbf{g}(S) \subseteq \mathbf{g}(\overline{S}) = \mathbf{g}(\operatorname{Int} S) \cup \mathbf{g}(\partial S).$$

This shows that $\partial\mathbf{g}(S) \subseteq \mathbf{g}(\partial S)$ and that $\partial\mathbf{g}(S)$ has volume zero since $\mathbf{g}(\partial S)$ has volume zero by Proposition 13.2.1. Therefore $\mathbf{g}(S)$ has volume. □

The polar coordinate mapping provides a good illustration of this proposition.

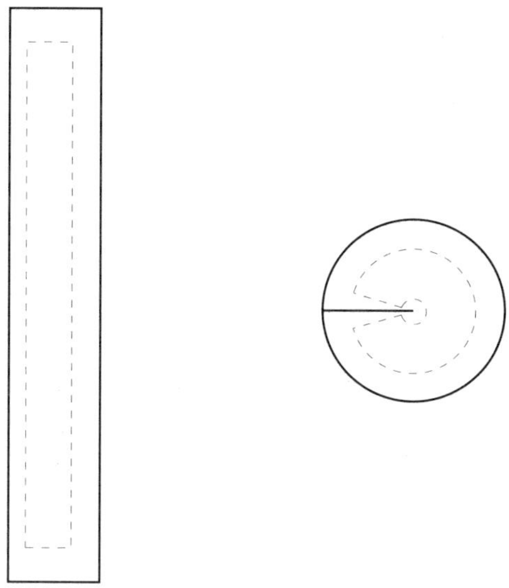

Figure 13.1. The polar coordinate mapping, $\mathbf{g}(x, y) = (x \cos y, x \sin y)$, maps the rectangle defined by $0 < x \leq 1$, $-\pi < y \leq \pi$ onto the unit disk minus the origin.

Example 13.2.4 (Polar Coordinates)**.** Define $S = \{(x, y) \in \mathbf{R}^2 : 0 < x \leq 1,\ -\pi < y \leq \pi\}$. Define $\mathbf{g} : S \to \mathbf{R}^2$ by

$$\mathbf{g}(x, y) = (x \cos y, x \sin y).$$

Then $\mathbf{g}(S) = \{(z_1, z_2) \in \mathbf{R}^2 : 0 < z_1^2 + z_2^2 \leq 1\}$, the unit disk minus the origin. The mapping $\mathbf{g}$ is C^1. Observe that

$$J_{\mathbf{g}}(x, y) = \det \begin{bmatrix} \cos y & -x \sin y \\ \sin y & x \cos y \end{bmatrix} = x,$$

so $J_{\mathbf{g}}(x, y) \neq 0$ at every point of $\operatorname{Int} S$. In fact, $\mathbf{g}$ is C^1-invertible on $\operatorname{Int} S$, as we show below, but $\mathbf{g}$ fails to be one-to-one on portions of the boundary. In detail, if (x_1, y_1) and (x_2, y_2) are in $\operatorname{Int} S$ and $\mathbf{g}(x_1, y_1) = \mathbf{g}(x_2, y_2)$, then $(x_1 \cos y_1, x_1 \sin y_1) = (x_2 \cos y_2, x_2 \sin y_2)$ implies that $x_1^2 = x_2^2$, hence $x_1 = x_2$. Then

$$\cos y_1 = \cos y_2 \qquad \text{and} \qquad \sin y_1 = \sin y_2,$$

but this cannot happen for $y_1, y_2 \in (-\pi, \pi]$, unless $y_1 = y_2$. Therefore $\mathbf{g}$ is one-to-one on $\operatorname{Int} S$, hence C^1-invertible on $\operatorname{Int} S$. We conclude that Proposition 13.2.3 applies and that $\mathbf{g}(S)$, the disk minus the origin, has volume. Of course, adding the origin in, the complete unit disk has volume. Finally, consider the boundary of S and how it is mapped. Note that $\mathbf{g}$ maps the vertical left side boundary segment of S, where $x = 0$, onto the point $(0, 0)$, and the top and bottom boundary segments of S, where $y = \pi$ and $y = -\pi$, respectively, onto the line segment $\{(-x, 0) \in \mathbf{R}^2 : 0 \leq x \leq 1\}$. The vertical right side boundary segment where $x = 1$ is mapped onto the circumference of the unit circle. Consequently, we do have $\partial \mathbf{g}(S) \subseteq \mathbf{g}(\partial S)$. But observe that $\partial \mathbf{g}(S) \neq \mathbf{g}(\partial S)$. (See Figure 13.1.) $\triangle$

If a function f is integrable on a set A that has volume and we express f in new coordinates by considering $f \circ \mathbf{g}$, where $\mathbf{g} : U \to \mathbf{R}^n$ is a C^1 mapping on an open set U, with a C^1 inverse, such that $A = \mathbf{g}(S)$ where S has volume and $\overline{S} \subset U$, then we want to know whether $f \circ \mathbf{g}$ is integrable on S. The next theorem gives the affirmative answer.

Theorem 13.2.5. *Let S be a set in $\mathbf{R}^n$ that has volume, and let U be a bounded open set in $\mathbf{R}^n$ such that $\overline{S} \subset U$. Let $\mathbf{g} : U \to \mathbf{R}^n$ be C^1 and C^1-invertible on U. If f is integrable on $\mathbf{g}(S)$, then $f \circ \mathbf{g}$ is integrable on S.*

Proof. By Proposition 13.2.3, $\mathbf{g}(S)$ has volume, as does $\mathbf{g}(\overline{S})$, since $\mathbf{g}(\overline{S}) = \mathbf{g}(\text{Int } S) \cup \mathbf{g}(\partial S)$ and $\mathbf{g}(\partial S)$ has volume zero and contains the boundary of $\mathbf{g}(\overline{S})$. We can extend f to $\mathbf{g}(\overline{S})$ by zero to points where it was not originally defined, and we continue to denote the extension by f. This extended f is still integrable on $\mathbf{g}(\overline{S})$, since $\partial \mathbf{g}(S)$ has volume zero.

It remains to show that $f \circ \mathbf{g}$ is integrable on S. We want to show that $f \circ \mathbf{g}$ is continuous almost everywhere on S. Let D be a closed set, with volume zero, contained in $\mathbf{g}(\overline{S})$ and containing $\partial \mathbf{g}(S)$ and all points where f is not continuous. Then D is compact and contained in $\mathbf{g}(U)$. Since D is compact, the C^1 inverse mapping $\mathbf{g}^{-1}$ satisfies a Lipschitz condition on D; hence, $\mathbf{g}^{-1}(D)$ has volume zero by Proposition 13.2.1. □

Exercises.

Exercise 13.2.1. Prove the following statements:

1. If $\phi_1, \phi_2 : [a, b] \to \mathbf{R}$ are continuous functions with $\phi_1(x) \leq \phi_2(x)$ for all $x \in [a, b]$, then the region
$$Q_1 = \{(x, y) \in \mathbf{R}^2 : a \leq x \leq b,\ \phi_1(x) \leq y \leq \phi_2(x)\}$$
between the graphs of ϕ_1 and ϕ_2 is Jordan measurable.
2. If $\psi_1, \psi_2 : [c, d] \to \mathbf{R}$ are continuous functions with $\psi_1(y) \leq \psi_2(y)$ for all $y \in [c, d]$, then the region
$$Q_2 = \{(x, y) \in \mathbf{R}^2 : c \leq y \leq d,\ \psi_1(y) \leq x \leq \psi_2(y)\}$$
between the graphs of ψ_1 and ψ_2 is Jordan measurable.

Exercise 13.2.2. Let $D = \{(\rho, \phi, \theta) \in \mathbf{R}^3 : 0 < \rho \leq 1, 0 \leq \phi \leq \pi, 0 \leq \theta \leq 2\pi\}$. Let $\mathbf{G} : D \subset \mathbf{R}^3 \to \mathbf{R}^3$, $\mathbf{G} = (x, y, z)$, be defined by

$$\begin{aligned} x &= \rho \sin\phi \cos\theta, \\ y &= \rho \sin\phi \sin\theta, \\ z &= \rho \cos\phi. \end{aligned}$$

Show that $\mathbf{G}(D)$ has volume, that is, $\mathbf{G}(D)$ is Jordan measurable. *Hint*: Does Proposition 13.2.3 apply?

13.3. Linear Images of Sets with Volume

Let $\mathcal{L} : \mathbf{R}^n \to \mathbf{R}^n$ be a linear transformation, and let $S \subset \mathbf{R}^n$ be a set that has volume. The main result of this section describes the volume of the image, $\mathcal{L}(S)$, in terms of the volume of S. If $\mathcal{L}$ is represented by the $n \times n$ matrix L, then we write

$$\det \mathcal{L} = \det L,$$

the determinant of the matrix L. We show in this section that

$$\nu(\mathcal{L}(S)) = |\det \mathcal{L}|\, \nu(S) \tag{13.5}$$

for any set S with volume and any linear transformation $\mathcal{L}$ of $\mathbf{R}^n$. Thus the determinant of a linear mapping acts as a magnification factor for volume under the transformation. This result, which is of most interest for invertible transformations, is an important step in the proof of the general change of variables formula in the next section.

Consider first a nonsingular linear mapping. Proposition 13.2.3 applies in the case of a nonsingular linear transformation, and thus if S has volume, then so does the image $\mathcal{L}(S)$. It remains to establish the volume formula (13.5) for the image.

We shall need two important facts about determinants:

(i) $\det T_1 T_2 = \det T_1 \det T_2$ for square matrices; similarly, $\det(\mathcal{T}_1\mathcal{T}_2) = \det \mathcal{T}_1 \det \mathcal{T}_2$ for linear transformations of $\mathbf{R}^n$;

(ii) any nonsingular matrix (linear transformation) can be written as a product of elementary matrices (a composition of linear transformations) of the three types described below:

(1) Transformations of type 1 take the form

$$\mathcal{L}_1(x_1, \ldots, x_n) = (x_1, \ldots, x_{k-1}, \lambda x_k, x_{k+1}, \ldots, x_n),$$

where λ is a *nonzero* real number. Equivalently, $\mathcal{L}_1 \mathbf{e}_j = \mathbf{e}_j$ for $j \neq k$, and $\mathcal{L}_1(\mathbf{e}_k) = \lambda \mathbf{e}_k$. The matrix representation of $\mathcal{L}_1$ in the standard basis is

$$L_1 = \operatorname{diag}(1, \ldots, 1, \lambda, 1, \ldots, 1).$$

The determinant is $\det \mathcal{L}_1 = \det L_1 = \lambda \neq 0$. Premultiplication of a matrix by L_1 corresponds to multiplying row k of the matrix by λ. The determinant of a type 1 transformation is λ. The inverse of the transformation $\mathcal{L}_1$ shown above is the mapping

$$\mathcal{L}_1^{-1}(x_1, \ldots, x_n) = (x_1, \ldots, x_{k-1}, \frac{1}{\lambda} x_k, x_{k+1}, \ldots, x_n),$$

which is also a Type 1 transformation.

(2) Type 2 transformations take the form

$$\mathcal{L}_2(x_1, \ldots, x_n) = (x_{k_1}, x_{k_2}, \ldots, x_{k_n}),$$

where $k : \{1, 2, \ldots, n\} \to \{1, 2, \ldots, n\}$ is a permutation map with images denoted $k(i) = k_i$, $1 \leq i \leq n$. Premultiplication of a given matrix by a type 2 transformation interchanges (permutes) the rows of the matrix according to the permutation k.

The matrix representation of $\mathcal{L}_2$ given above can be expressed, by displaying the columns, as

$$L_2 = \begin{bmatrix} \mathbf{e}_{k_1} & \mathbf{e}_{k_2} & \cdots & \mathbf{e}_{k_n} \end{bmatrix}.$$

Observe that premultiplication of a matrix by L_2 places the first row of the matrix into row k_1, the second row of the matrix into row k_2, and so on. The inverse of the matrix L_2 shown above is the transpose of L_2, as a direct calculation shows:

$$L_2^T = \begin{bmatrix} \mathbf{e}_{k_1}^T \\ \mathbf{e}_{k_2}^T \\ \cdots \\ \mathbf{e}_{k_n}^T \end{bmatrix} \qquad \Longrightarrow \qquad L_2^T L_2 = I.$$

Since a matrix and its transpose have the same determinant,

$$\det(L_2^T L_2) = (\det L_2)^2 = \det(I) = 1,$$

hence $\det L_2 = \pm 1$. Thus the determinant of any type 2 transformation equals ± 1.

(3) The type 3 transformation takes the form

$$\mathcal{L}_3(x_1, \ldots, x_n) = (x_1 + x_2, x_2, x_3, \ldots, x_n) = (x_1, \ldots, x_n) + x_2\mathbf{e}_1.$$

The matrix representation can be expressed, by displaying the columns, as

$$L_3 = \begin{bmatrix} \mathbf{e}_1 & \mathbf{e}_1 + \mathbf{e}_2 & \mathbf{e}_3 & \cdots & \mathbf{e}_n \end{bmatrix}.$$

Matrix L_3 has ones in the first two entries of row 1; otherwise, the i-th row, for $i \geq 2$, has a one as the (i,i) entry and zeros elsewhere. Premultiplication of a given matrix by this transformation matrix corresponds to the elementary row operation of adding row 2 to row 1 of the matrix. The determinant of the type 3 transformation is $\det \mathcal{L}_3 = \det L_3 = 1$. The inverse of the matrix L_3 above is the product of three elementary matrices, each of which has determinant equal to ± 1. We illustrate this with 2×2 matrices, since in the general case, the first two rows are extended by zero entries, and rows 3 through n of these matrices have ones on the main diagonal and zeros elsewhere. Thus, we note that

$$\begin{bmatrix} 1 & 1 \\ 0 & 1 \end{bmatrix}^{-1} = \begin{bmatrix} 1 & 0 \\ 0 & -1 \end{bmatrix}\begin{bmatrix} 1 & 1 \\ 0 & 1 \end{bmatrix}\begin{bmatrix} 1 & 0 \\ 0 & -1 \end{bmatrix} = \begin{bmatrix} 1 & -1 \\ 0 & 1 \end{bmatrix}.$$

We note that it is possible to add any row to any other row of a matrix by premultiplying by type 2 permutations to place the desired rows in row positions 1-2, and then applying the type 3 transformation. Also observe that the inverse of one of these three types of transformation matrices is another transformation matrix of the same type or a product of these types. Recall from linear algebra that if T is a nonsingular matrix, then premultiplication by transformations of types 1, 2 and 3 suffice to reduce the augmented matrix $[T \quad I]$ to the form $[I \quad T^{-1}]$. It follows that *any nonsingular matrix may be expressed as a product of matrices of types 1,2 and 3.*

We consider a useful geometric example in two dimensions.

Example 13.3.1. We consider the rectangle $B = [a_1, b_1] \times [a_2, b_2]$ in the plane and examine how it is mapped by the linear transformation $\mathcal{L} : \mathbf{R}^2 \to \mathbf{R}^2$ given by $\mathcal{L}(x_1, x_2) = (x_1 + x_2, x_2)$. The matrix representation of $\mathcal{L}$ in the standard basis is

$$L = \begin{bmatrix} 1 & 1 \\ 0 & 1 \end{bmatrix}.$$

$\mathcal{L}$ maps B onto the parallelogram $P_2 = \mathcal{L}(B)$. (See Figure 13.2.) The vertices of P_2 are the images of the four vertices of B, given by

(1) $\mathcal{L}(a_1, a_2) = (a_1 + a_2, a_2)$,

(2) $\mathcal{L}(b_1, a_2) = (a_2 + b_1, a_2)$,

(3) $\mathcal{L}(a_1, b_2) = (a_1 + b_2, b_2)$,

(4) $\mathcal{L}(b_1, b_2) = (b_1 + b_2, b_2)$.

It is straightforward to verify that the horizontal boundary segments of B map to horizontal boundary segments of $\mathcal{L}(B)$, and the vertical boundary segments of B map to the sheared boundary segments of $\mathcal{L}(B)$. (This is in accordance with Proposition 13.2.3.) The interior of B maps to the interior of $\mathcal{L}(B)$. At this point, we have not established the volume of triangles, and we have not yet proved that congruent triangles have the same volume, so we will not say that we know the volume of P_2 by elementary geometric considerations. However, we have Fubini's Theorem 12.7.2. So we express P_2 as the area between the graphs of

$$x = \psi_1(y) = y + a_1 \qquad \text{and} \qquad x = \psi_2(y) = y + b_1,$$

and apply Theorem 12.7.2 to write

$$\begin{aligned} \nu(P_2) &= \int_{a_2}^{b_2} \int_{\psi_2(y)}^{\psi_1(y)} (1)\, dx\, dy \\ &= \int_{a_2}^{b_2} [\psi_2(y) - \psi_1(y)]\, dy \\ &= \int_{a_2}^{b_2} (b_1 - a_1)\, dy \\ &= (b_1 - a_1)(b_2 - a_2). \end{aligned}$$

Since $\det \mathcal{L} = 1$, indeed we have $\nu(\mathcal{L}(B)) = |\det \mathcal{L}|\, \nu(B)$. △

As in this example, the type 3 transformation $\mathcal{L}_3$ leaves all coordinates of a point unchanged except for the first coordinate, and the first and second coordinates are mapped in the same way as the two coordinates of the planar mapping in Example 13.3.1.

We first establish the formula for the volume of the image of a *closed interval* under the elementary linear transformations.

Lemma 13.3.2. *If B is a closed interval in $\mathbf{R}^n$, then for $j = 1, 2, 3$, $\mathcal{L}_j(B)$ has volume, and*

$$\nu(\mathcal{L}_j(B)) = |\det \mathcal{L}_j|\, \nu(B), \qquad \textit{for } j = 1, 2, 3. \tag{13.6}$$

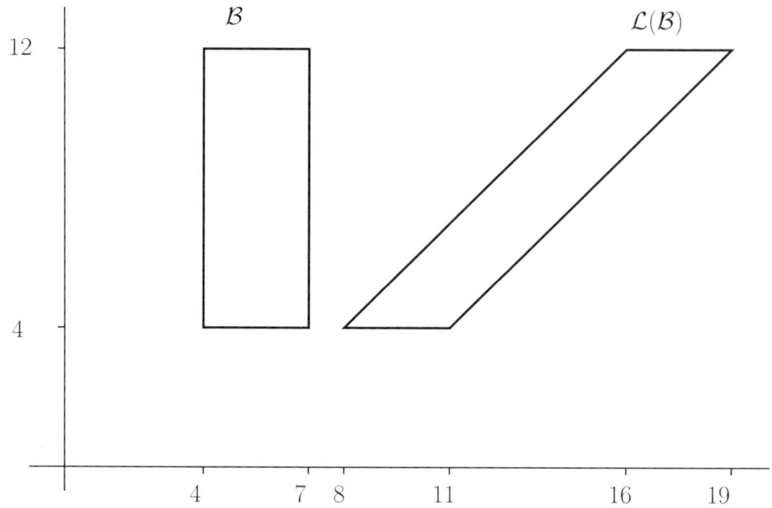

Figure 13.2. A two-dimensional interval $\mathcal{B}$ and its image $\mathcal{L}(\mathcal{B})$ under the linear shear mapping, $\mathcal{L}(x_1, x_2) = (x_1 + x_2, x_2)$, a type (3) elementary transformation.

Proof. Recall that $\lambda \neq 0$ for $\mathcal{L}_1$. We know that $\mathcal{L}_j(B)$ has volume for $j = 1, 2, 3$ by Proposition 13.2.3. Now we establish formula (13.6) for these invertible $\mathcal{L}_j$. Now $\mathcal{L}_1$ maps $B = [a_1, b_1] \times \cdots \times [a_n, b_n]$ into either

$$\mathcal{L}_1(B) = [a_1, b_1] \times \cdots \times [\lambda a_k, \lambda b_k] \times \cdots \times [a_n, b_n] \quad (\text{if } \lambda > 0)$$

or

$$\mathcal{L}_1(B) = [a_1, b_1] \times \cdots \times [\lambda b_k, \lambda a_k] \times \cdots \times [a_n, b_n] \quad (\text{if } \lambda < 0).$$

Hence, in either case, by direct calculation we have

$$\nu(\mathcal{L}_1(B)) = |\lambda| \nu(B) = |\det \mathcal{L}_1|\, \nu(B).$$

Now consider the transformation

$$\mathcal{L}_2(x_1, \dots, x_n) = (x_{k_1}, x_{k_2}, \dots, x_{k_n}),$$

defined by a permutation map $k : \{1, 2, \dots, n\} \to \{1, 2, \dots, n\}$ with $k(i) = k_i$, $1 \leq i \leq n$. Then $\mathcal{L}_2$ maps $B = [a_1, b_1] \times \cdots \times [a_n, b_n]$ onto the interval

$$[a_{k_1}, b_{k_1}] \times \cdots \times [a_{k_n}, b_{k_n}],$$

from which it follows that

$$\nu(\mathcal{L}_2(B)) = \prod_{i=1}^{n} (b_{k_i} - a_{k_i}) = \nu(B) = |\det \mathcal{L}_2|\, \nu(B),$$

since $\det \mathcal{L}_2 = \pm 1$.

The transformation $\mathcal{L}_3$ leaves the third through last coordinates of every point unchanged, and maps the first two coordinates in the same way as the planar mapping in Example 13.3.1. Thus, we have

$$\mathcal{L}_3(B) = P_2 \times D,$$

where $D = [a_3, b_3] \times \cdots \times [a_n, b_n]$ and P_2 is the parallelogram described in Example 13.3.1. The set $\mathcal{L}_3(B)$ is described by the inequalities

$$y + a_1 \le x \le y + b_1, \quad a_2 \le y \le b_2, \quad \text{and} \quad a_i \le x_i \le b_i, \text{ for } i = 3, \ldots, n.$$

(See Example 13.3.1.) Then by Example 13.3.1 and Theorem 12.7.5, we have

$$\begin{aligned} \nu(\mathcal{L}_3(B)) &= \int_D \Big(\int_{P_2} (1)\, dx_1\, dx_2 \Big)\, dx_3 \cdots dx_n \\ &= \int_D (b_1 - a_1)(b_2 - a_2)\, dx_3 \cdots dx_n \\ &= (b_1 - a_1)(b_2 - a_2)(b_3 - a_3) \cdots (b_n - a_n) \\ &= \nu(B) \\ &= |\det \mathcal{L}_3|\, \nu(B), \end{aligned}$$

since $|\det \mathcal{L}_3| = 1$, and this completes the proof. □

Using Lemma 13.3.2, we can describe the volume of the image $\mathcal{L}_j(A)$, for any set A that has volume, for the elementary linear transformations $\mathcal{L}_j$, $j = 1, 2, 3$.

Lemma 13.3.3. *If $A \subset \mathbf{R}^n$ has volume, then for each $j = 1, 2, 3$, $\mathcal{L}_j(A)$ has volume, and*

$$\nu(\mathcal{L}_j(A)) = |\det \mathcal{L}_j|\, \nu(A), \qquad \text{for } j = 1, 2, 3. \tag{13.7}$$

Proof. Again by Proposition 13.2.3, $\mathcal{L}_j(A)$ has volume for $j = 1, 2, 3$. It remains to show equality (13.7). We first get an estimate of the volume of A using intervals, which will lead to an estimate of the volume of $\mathcal{L}_j(A)$ for the elementary linear transformations. Let λ be a nonzero real number. Since A has volume, if B is any closed interval containing A, then given any $\epsilon > 0$ there is a partition P_ϵ of B such that

$$\nu(A) = \int_B \chi_A = \inf_P U(\chi_A, P) > U(\chi_A, P_\epsilon) - \frac{\epsilon}{|\lambda|}, \tag{13.8}$$

the infimum being taken over all possible partitions P of B. Let $S_1, \ldots, S_N$ be a listing of all the intervals of P_ϵ that contain points of A. Then $A \subseteq \bigcup_{i=1}^N S_i$ and

$$U(\chi_A, P_\epsilon) = \sum_{i=1}^N \nu(S_i).$$

Hence, by a rearrangement of (13.8),

$$\sum_{i=1}^N \nu(S_i) < \nu(A) + \frac{\epsilon}{|\lambda|}. \tag{13.9}$$

We now argue for the equality (13.7) specifically for $\mathcal{L}_1$; afterwards, we discuss how the same argument also applies to $\mathcal{L}_2$ and $\mathcal{L}_3$. Since $A \subseteq \bigcup_{i=1}^N S_i$, we have

$$\mathcal{L}_1(A) \subseteq \bigcup_{i=1}^N \mathcal{L}_1(S_i),$$

and each set $\mathcal{L}_1(S_i)$ has volume, so the union does as well. Consequently, by Theorem 12.6.2 (4),

$$\nu(\mathcal{L}_1(A)) \le \nu\Big(\bigcup_{i=1}^{N} \mathcal{L}_1(S_i)\Big). \tag{13.10}$$

Since $\mathcal{L}_1$ is one-to-one and the intervals S_i have no interior points in common, the sets $\mathcal{L}_1(S_i)$ have no interior points in common. Hence, by Theorem 12.6.2 (statement 2),

$$\nu\Big(\bigcup_{i=1}^{N} \mathcal{L}_1(S_i)\Big) = \sum_{i=1}^{N} \nu(\mathcal{L}_1(S_i)). \tag{13.11}$$

By Lemma 13.3.2, we have

$$\nu(\mathcal{L}_1(S_i)) = |\det \mathcal{L}_1|\, \nu(S_i) \quad \text{for } 1 \le i \le N. \tag{13.12}$$

Hence, from (13.9), (13.10), (13.11) and (13.12), we have

$$\begin{aligned} \nu(\mathcal{L}_1(A)) &\le \sum_{i=1}^{N} \nu(\mathcal{L}_1(S_i)) \\ &= \sum_{i=1}^{N} |\det \mathcal{L}_1|\, \nu(S_i) \\ &< |\det \mathcal{L}_1| \Big(\nu(A) + \frac{\epsilon}{|\lambda|}\Big) \\ &= |\det \mathcal{L}_1|\, \nu(A) + \epsilon, \end{aligned}$$

if indeed we choose $\lambda = \det \mathcal{L}_1$. Since this estimate holds for any $\epsilon > 0$, we conclude that

$$\nu(\mathcal{L}_1(A)) \le |\det \mathcal{L}_1|\, \nu(A).$$

However, the inverse, $\mathcal{L}_1^{-1}$, of $\mathcal{L}_1$ is an elementary linear transformation of the same type, with $1/\lambda$ in place of λ, so the argument just given shows that

$$\nu(A) = \nu\Big(\mathcal{L}_1^{-1}(\mathcal{L}_1(A))\Big) \le \frac{1}{|\lambda|} \nu(\mathcal{L}_1(A))$$

and hence

$$\nu(\mathcal{L}_1(A)) \ge |\lambda|\, \nu(A) = |\det \mathcal{L}_1|\, \nu(A).$$

We conclude that (13.7) holds when $j = 1$.

To complete the proof, we observe that all the steps of the argument from (13.8) on are also valid for the elementary linear transformations $\mathcal{L}_2$ and $\mathcal{L}_3$, except that we replace "$|\lambda|$" by "1", and "$|\det \mathcal{L}_1| = |\lambda|$" by "$|\det \mathcal{L}_j| = 1$", as appropriate for $j = 2, 3$. (See also Exercise 13.3.2.) This completes the proof of Lemma 13.3.3. □

We are now ready to establish the main result of the section, which is the change of variables formula for linear mappings.

Proposition 13.3.4. *If $\mathcal{L} : \mathbf{R}^n \to \mathbf{R}^n$ is a nonsingular linear mapping and A is a set that has volume, that is, $\int_A \chi_A$ exists, then $\mathcal{L}(A)$ has volume and*

$$\nu(\mathcal{L}(A)) = |\det \mathcal{L}|\, \nu(A) = \int_A |\det \mathcal{L}|. \tag{13.13}$$

Proof. Since A has volume and $\mathcal{L}$ is nonsingular, Proposition 13.2.3 guarantees that $\mathcal{L}(A)$ has volume. Observe that the equality on the right holds since the definition of volume of a set implies that

$$|\det \mathcal{L}|\,\nu(A) = |\det \mathcal{L}| \int_A \chi_A = \int_A |\det \mathcal{L}|.$$

We proceed to show that the equality on the left holds, that is, $\nu(\mathcal{L}(A)) = |\det \mathcal{L}|\,\nu(A)$. Since $\mathcal{L}$ is nonsingular, it may be written as a composition of elementary linear transformations, each of which is nonsingular; thus,

$$\mathcal{L} = \mathcal{L}_{k_1} \circ \mathcal{L}_{k_2} \circ \cdots \circ \mathcal{L}_{k_m},$$

where $k_i \in \{1, 2, 3\}$ for $1 \leq i \leq m$. The matrix representation is given by the m-fold matrix product

$$L = L_{k_1} L_{k_2} \cdots L_{k_m}.$$

We have

$$\det \mathcal{L} = (\det \mathcal{L}_{k_1})(\det \mathcal{L}_{k_2}) \cdots (\det \mathcal{L}_{k_m}).$$

Repeated application of Lemma 13.3.3 yields

$$\begin{aligned}
\nu(\mathcal{L}(A)) &= |\det \mathcal{L}_{k_1}|\,\nu(\mathcal{L}_{k_2}\mathcal{L}_{k_3} \cdots \mathcal{L}_{k_m}(A)) \\
&= |\det \mathcal{L}_{k_1}|\,|\det \mathcal{L}_{k_2}|\,\nu(\mathcal{L}_{k_3} \cdots \mathcal{L}_{k_m}(A)) \\
\cdots &\quad \cdots \\
&= |\det \mathcal{L}_{k_1}|\,|\det \mathcal{L}_{k_2}| \cdots |\det \mathcal{L}_{k_m}|\,\nu(A) \\
&= |\det \mathcal{L}|\,\nu(A),
\end{aligned}$$

which proves (13.13) and completes the proof. □

Proposition 13.3.4 is a preliminary version of the change of variables formula of the following section, since it states that if the linear transformation $\mathcal{L}$ is nonsingular, then

$$\int_{\mathcal{L}(A)} 1 = \int_A (1 \circ \mathcal{L})\,|\det \mathcal{L}|.$$

The integral of the characteristic function (constant 1) over the image set $\mathcal{L}(A)$ is the same as the integral of the composition $1 \circ \mathcal{L}$ (which is constant 1) multiplied by $|\det \mathcal{L}|$, over the domain set A. In the next section, the formula will be generalized to cover the case of C^1-invertible coordinate transformations $\mathbf{g}$ defined on an open neighborhood of A, and any integrable function f on the image $\mathbf{g}(A)$, so that

$$\int_{\mathbf{g}(A)} f = \int_A f \circ \mathbf{g}\,|\det D\mathbf{g}|.$$

Given Proposition 13.3.4, we can now describe an important property possessed by volume measure. Exercise 12.4.5 shows that the volume measure of a set is invariant under translations. Exercise 13.3.3 shows that volume measure is invariant under orthogonal transformations. Any orthogonal transformation can be written as a composition of a finite number of rotations. Thus, these two exercises show that volume measure is invariant under any **Euclidean rigid motion**, that is, under any finite sequence of translations and rotations. Two sets are **congruent** if one of them can be obtained from the other by a rigid motion. Therefore congruent sets with volume have the same volume.

Finally, we comment on the case of a *singular* linear transformation $\mathcal{L}$, for which $\det \mathcal{L} = 0$. In this case (13.5) is equivalent to the statement that for any set $S \subset \mathbf{R}^n$ that has volume, the volume of the image $\mathcal{L}(S)$ is zero. Intuitively, this is reasonable since $\operatorname{rank} \mathcal{L} = k < n$ so that $\mathcal{L}$ maps S onto a bounded subset of a lower-dimensional subspace $W \subset \mathbf{R}^n$. Since W has dimension $k < n$, we may map it by a Euclidean rigid motion so as to identify it with the first k coordinates of $\mathbf{R}^n$, and thus we equate the volume of $\mathcal{L}(S)$ with the volume of its image in $\mathbf{R}^n$ after the rigid motion. In Proposition 13.2.2 we may use $\mathbf{g}$ as the inclusion map of $\mathbf{R}^k$ into $\mathbf{R}^n$ to conclude that $\mathcal{L}(S)$ has volume zero. This shows that (13.13) also holds when $\mathcal{L}$ is a singular linear transformation.

Exercises.

Exercise 13.3.1. Consider the transformation matrix

$$L = \begin{bmatrix} 0 & 0 & 1 \\ 1 & 0 & 0 \\ 0 & 1 & 0 \end{bmatrix}.$$

Show that its inverse is its transpose.

Exercise 13.3.2. Trace through the proof of Lemma 13.3.3 to verify that replacing "$|\lambda|$" by "1" and "$|\det \mathcal{L}_1| = |\lambda|$" by "$|\det \mathcal{L}_j| = 1$" as appropriate for $j = 2, 3$, establishes that (13.7) holds for $j = 2, 3$.

Exercise 13.3.3. *Volume is invariant under orthogonal transformations*
A real $n \times n$ matrix L is called **orthogonal** if its columns form an orthonormal basis of $\mathbf{R}^n$, or, as is equivalent, L satisfies the condition $L^T L = I$, where I is the $n \times n$ identity matrix. Prove: If L is an orthogonal matrix, then $\det L = \pm 1$ and $\nu(L(A)) = \nu(A)$ for any set $A \subset \mathbf{R}^n$ that has volume.

13.4. The Change of Variables Formula

Recall that for a C^1 mapping $\mathbf{g}$, we write $D\mathbf{g}(\mathbf{x})$ for the derivative of $\mathbf{g}$ at $\mathbf{x}$, and $J_{\mathbf{g}}(\mathbf{x})$ for the Jacobian matrix representation of $D\mathbf{g}(\mathbf{x})$ (the matrix representation of $D\mathbf{g}(\mathbf{x})$ with respect to the standard basis). We shall write $\Delta_{\mathbf{g}}(\mathbf{x})$ to denote the determinant of the Jacobian matrix $J_{\mathbf{g}}(\mathbf{x})$, which is also the determinant of the linear transformation $D\mathbf{g}(\mathbf{x})$. Thus,

$$\Delta_{\mathbf{g}}(\mathbf{x}) = \det J_{\mathbf{g}}(\mathbf{x}) = \det D\mathbf{g}(\mathbf{x}). \tag{13.14}$$

We shall prove the following *Change of Variables Formula*:

Theorem.[1] *Let Q be an interval in $\mathbf{R}^n$, and $\mathbf{g} : U \to \mathbf{R}^n$ a C^1 mapping which is C^1-invertible on an open set U containing Q. If $f : \mathbf{g}(Q) \to \mathbf{R}$ is an integrable function such that $f \circ \mathbf{g}$ is also integrable, then*

$$\int_{\mathbf{g}(Q)} f = \int_Q (f \circ \mathbf{g}) \, |\Delta_{\mathbf{g}}|.$$

The intuition behind this formula is not difficult to understand. If $\mathbf{g}$ were linear, then, by Proposition 13.3.4, we would have $\nu(\mathbf{g}(S)) = |\det \mathbf{g}| \, \nu(S)$ for any set S

[1] We will eventually establish this change of variables formula under a slightly less restrictive assumption on $\mathbf{g}$: that $\mathbf{g}$ is C^1 on the interior of the rectangle Q.

that has volume. We expect that when $\mathbf{g}$ is a nonlinear C^1 mapping, we should have

$$\nu(\mathbf{g}(S)) \approx |\Delta_{\mathbf{g}}|\, \nu(S)$$

for sets S with small volume. Hence, for an interval S of a partition P of Q, we expect to have

$$M_{\mathbf{g}(S)}(f)\, \nu(\mathbf{g}(S)) \approx M_S\Big((f \circ \mathbf{g})\, |\Delta_{\mathbf{g}}|\Big)\, \nu(S),$$

where as usual the M-notation denotes supremum of the indicated function over the indicated set. There should be a similar approximation with M replaced by m, denoting the infimum of a function over a set. Of course, $\mathbf{g}(S)$ need not be an interval of a partition of a closed interval containing $\mathbf{g}(S)$; however, since the intervals S are disjoint except possibly for boundary points in common, the images $\mathbf{g}(S)$ are also disjoint except possibly for boundary points in common, by Proposition 13.2.3. Thus we expect that

$$\begin{aligned}\int_{\mathbf{g}(Q)} f &\approx \sum_S M_{\mathbf{g}(S)}(f)\, \nu(\mathbf{g}(S)) \\ &\approx \sum_S M_S\Big((f \circ \mathbf{g})\, |\Delta_{\mathbf{g}}|\Big)\, \nu(S) \\ &\approx \int_Q (f \circ \mathbf{g})\, |\Delta_{\mathbf{g}}|.\end{aligned}$$

We recall here the useful result of Theorem 11.2.3, part 1, which we repeat for convenience as a lemma.

Lemma 13.4.1. *Let U be an open set in $\mathbf{R}^n$ containing the origin and $\mathbf{F} : U \to \mathbf{R}^n$ a C^1 mapping such that $\mathbf{F}(\mathbf{0}) = \mathbf{0}$ and $D\mathbf{F}(\mathbf{0}) = I$. If $0 < \epsilon < 1$, then there exists $r > 0$ such that*

$$\|D\mathbf{F}(\mathbf{x}) - I\|_\infty < \epsilon \tag{13.15}$$

for all $\mathbf{x}$ in the cube C_r, and consequently

$$C_{(1-\epsilon)r} \subseteq \mathbf{F}(C_r) \subseteq C_{(1+\epsilon)r}.$$

In the next two results we make more precise the idea that for small mesh intervals Q, we should have

$$\nu(\mathbf{g}(Q)) \approx |\Delta_{\mathbf{g}}|\, \nu(Q).$$

Let Q be a closed interval centered at the point $\mathbf{a}$. (The coordinates of $\mathbf{a}$ are the center points of each edge factor of Q.) Suppose $\mathbf{g} : U \to \mathbf{R}^n$ is a C^1-invertible mapping on an open set U containing Q. Suppose the derivative of $\mathbf{g}$ is near constant on Q, say $D\mathbf{g}(\mathbf{x}) \approx D\mathbf{g}(\mathbf{a})$. Then, in view of the inequality

$$\begin{aligned}\|[D\mathbf{g}(\mathbf{a})]^{-1} D\mathbf{g}(\mathbf{x}) - I\|_\infty &= \|[D\mathbf{g}(\mathbf{a})]^{-1}(D\mathbf{g}(\mathbf{x}) - D\mathbf{g}(\mathbf{a}))\|_\infty \\ &\le \|[D\mathbf{g}(\mathbf{a})]^{-1}\|_\infty\, \|D\mathbf{g}(\mathbf{x}) - D\mathbf{g}(\mathbf{a})\|_\infty,\end{aligned}$$

suppose that

$$\|[D\mathbf{g}(\mathbf{a})]^{-1} D\mathbf{g}(\mathbf{x}) - I\|_\infty < \epsilon. \tag{13.16}$$

We use the lemma to show that (13.16) implies that the image $\mathbf{g}(Q)$ closely approximates the parallelepiped defined by the linear image $D\mathbf{g}(\mathbf{a})(Q)$. The linear mapping Proposition 13.3.4 will then provide a better estimate for the approximation

$$\nu(\mathbf{g}(Q)) \approx |\det D\mathbf{g}(\mathbf{a})|\,\nu(Q).$$

We will need to move from cubes to general intervals. For example, think of the polar or spherical coordinate mappings, which are not defined on cubes. Therefore we establish our estimate in two stages, Lemma 13.4.2 (for cubes Q) followed by Lemma 13.4.3 (for noncubes Q).

Lemma 13.4.2. *Let $Q = C_r(\mathbf{a})$ be a cube centered at the point $\mathbf{a} \in \mathbf{R}^n$, and $\mathbf{g} : U \to \mathbf{R}^n$ a C^1-invertible mapping on an open set U containing Q. If there exists $\epsilon \in (0,1)$ such that*

$$\|[D\mathbf{g}(\mathbf{a})]^{-1} \circ D\mathbf{g}(\mathbf{x}) - I\|_\infty \le \epsilon \tag{13.17}$$

for all $\mathbf{x} \in Q$, then $\mathbf{g}(Q)$ has volume and

$$(1-\epsilon)^n\,|\det D\mathbf{g}(\mathbf{a})|\,\nu(Q) \le \nu(\mathbf{g}(Q)) \le (1+\epsilon)^n\,|\det D\mathbf{g}(\mathbf{a})|\,\nu(Q). \tag{13.18}$$

Proof. That $\mathbf{g}(Q)$ has volume follows from Proposition 13.2.3. We now consider the proof of (13.18). Recall that C_r is the cube of radius r centered at the origin. Also recall that $\tau_{\mathbf{a}}(\mathbf{x}) = \mathbf{a} + \mathbf{x}$ is the translation by $\mathbf{a}$; then $\tau_{\mathbf{a}}(C_r) = C_r(\mathbf{a}) = Q$. Let $b = \mathbf{g}(\mathbf{a})$, and define

$$\mathbf{F} = [D\mathbf{g}(\mathbf{a})]^{-1} \circ \tau_{\mathbf{b}}^{-1} \circ \mathbf{g} \circ \tau_{\mathbf{a}};$$

then $\mathbf{F}$ maps an open neighborhood of the origin onto another open neighborhood of the origin. By the chain rule,

$$D\mathbf{F} = [D\mathbf{g}(\mathbf{a})]^{-1} \circ D\tau_{\mathbf{b}}^{-1} \circ D\mathbf{g} \circ D\tau_{\mathbf{a}},$$

and since the derivative of a translation is the identity, it follows that

$$D\mathbf{F}(\mathbf{x}) = [D\mathbf{g}(\mathbf{a})]^{-1} \circ D\mathbf{g}(\tau_{\mathbf{a}}(\mathbf{x})),$$

for all $\mathbf{x}$ in C_r, and thus for all $\tau_{\mathbf{a}}(\mathbf{x})$ in Q. Since $\mathbf{F}(\mathbf{0}) = \mathbf{0}$ and $D\mathbf{F}(\mathbf{0}) = I$, by (13.17) we can apply Lemma 13.4.1 to conclude that

$$C_{(1-\epsilon)r} \subseteq \mathbf{F}(C_r) \subseteq C_{(1+\epsilon)r}.$$

(See Figure 13.3.) It follows that

$$(1-\epsilon)^n\nu(C_r) \le \nu(\mathbf{F}(C_r)) \le (1+\epsilon)^n\nu(C_r). \tag{13.19}$$

From the definition of $\mathbf{F}$, we have $D\mathbf{g}(\mathbf{a})(\mathbf{F}(C_r)) = \tau_{\mathbf{b}}^{-1}(\mathbf{g}(Q))$, and by the linear mapping Proposition 13.3.4, we obtain

$$\nu(\mathbf{g}(Q)) = |\det D\mathbf{g}(\mathbf{a})|\,\nu(\mathbf{F}(C_r)).$$

Since $\nu(Q) = \nu(C_r)$, it follows that multiplication of (13.19) by $|\det D\mathbf{g}(\mathbf{a})|$ yields exactly (13.18), as we wished to show. □

Given a noncube interval Q, we could enclose Q in a large cube and extend the function to be integrated over Q by zero to the large cube, then proceed from there to subdivide the cube by smaller cubes. We would have to show that in the limit as the partition mesh goes to zero, the portion of Q that is covered by fractions of

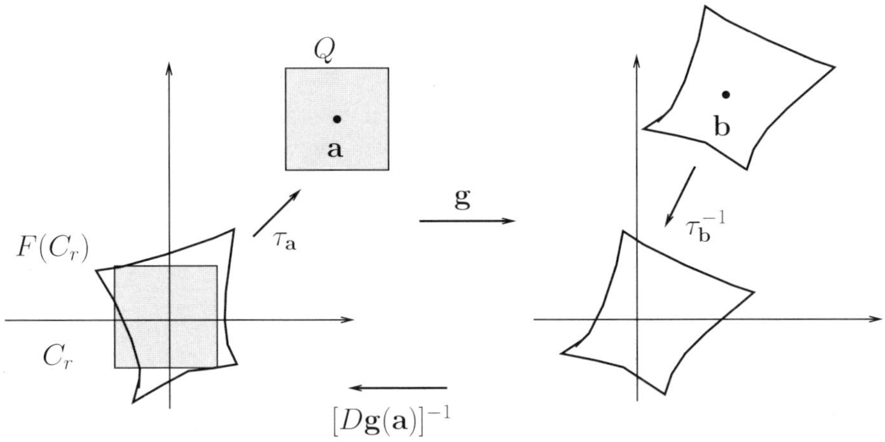

Figure 13.3. Mapping cubes to provide a volume estimate: The mapping $F = [D\mathbf{g}(\mathbf{a})]^{-1} \circ \tau_{\mathbf{b}}^{-1} \circ \mathbf{g} \circ \tau_{\mathbf{a}}$ takes the cube C_r centered at the origin onto the image $F(C_r)$, and $C_{(1-\epsilon)r} \subseteq \mathbf{F}(C_r) \subseteq C_{(1+\epsilon)r}$. This mapping yields bounds on the volume of the image $\mathbf{g}(Q)$ of the cube Q centered at $\mathbf{a}$. See Lemma 13.4.2.

small cubes has a negligible contribution to the integral. Instead of doing that, we now deal directly with the noncube interval Q.

If Q is a noncube closed interval in $\mathbf{R}^n$, let $Q = \mathcal{I}_1 \times \mathcal{I}_2 \times \cdots \times \mathcal{I}_n$. Let $\mathbf{a}$ be the center point of Q, that is, the coordinates of $\mathbf{a}$ are the midpoints of each factor of Q. Denote the factor lengths by $s_1, s_2, \ldots, s_n$, and the longest factor length by $s_{\max}$ and the shortest factor length by $s_{\min}$. We can map the unit cube in the max norm, $C_1 = \{\mathbf{x} \in \mathbf{R}^n : |\mathbf{x}|_\infty \leq 1\}$ one-to-one and onto an interval congruent to Q but centered at the origin, by the mapping

$$\rho(\mathbf{x}) = \Big(\frac{1}{2}s_1 x_1,\ \frac{1}{2}s_2 x_2,\ \ldots,\ \frac{1}{2}s_n x_n\Big), \tag{13.20}$$

since boundary points of C_1 have some coordinate $x_j = \pm 1$, and the corresponding image coordinate equals $\pm\frac{1}{2}s_j$. Then map $\rho(C_1)$ to Q by the translation $\tau(\mathbf{x}) = \mathbf{x} + \mathbf{a}$. The matrix representation of ρ is diagonal with main diagonal entries

$$\frac{1}{2}s_1,\ \frac{1}{2}s_1,\ \ldots,\ \frac{1}{2}s_n.$$

Therefore

$$\det \rho = \frac{1}{2^n}\prod_{i=1}^{n} s_i = \frac{1}{2^n}\,\nu(Q).$$

The inverse mapping is

$$\rho^{-1}(\mathbf{x}) = \Big(\frac{2}{s_1}x_1, \frac{2}{s_2}x_2, \ldots, \frac{2}{s_n}x_n\Big),$$

and the norms of ρ and ρ^{-1} induced by the vector norm $|\cdot|_\infty$ are easily seen to be

$$\|\rho\|_\infty = \frac{1}{2}s_{\max} \quad \text{and} \quad \|\rho^{-1}\|_\infty = \frac{2}{s_{\min}}.$$

Consequently, we have

$$R := \|\rho^{-1}\|_\infty \, \|\rho\|_\infty = \frac{s_{\max}}{s_{\min}},$$

a quantity useful to us below. We observe that if we start with this Q and partition it by similar intervals, that is, we subdivide each factor with the same number N of real subintervals, then the intervals of the resulting partition are all similar to Q and thus maintain the same ratio of $s_{\max}/s_{\min}$ as Q has. Thus,

$$R = \frac{s_{\max}}{s_{\min}}$$

remains the same for every interval of the partition.

Lemma 13.4.3. *Let Q be a closed interval centered at the point $\mathbf{a} \in \mathbf{R}^n$, and $\mathbf{g} : U \to \mathbf{R}^n$ a C^1-invertible mapping on an open set U containing Q. Let ρ map the unit cube C_1 to Q as defined in (13.20). Let $R = |\rho^{-1}|_\infty \, |\rho|_\infty$. If $\epsilon > 0$ such that*

$$\|[D\mathbf{g}(\mathbf{a})]^{-1} \circ D\mathbf{g}(\mathbf{x}) - I\|_\infty \leq \epsilon \tag{13.21}$$

for all $\mathbf{x} \in Q$, and $\epsilon R < 1$, then $\mathbf{g}(Q)$ has volume, and

$$(1 - \epsilon R)^n \, |\det D\mathbf{g}(\mathbf{a})| \, \nu(Q) \;\leq\; \nu(\mathbf{g}(Q)) \;\leq\; (1 + \epsilon R)^n \, |\det D\mathbf{g}(\mathbf{a})| \, \nu(Q)\,. \tag{13.22}$$

Proof. That $\mathbf{g}(Q)$ has volume follows from Proposition 13.2.3. We proceed to the proof of (13.22). We map the unit cube C_1 in the maximum norm, where $|\mathbf{x}|_\infty \leq 1$, to the interval Q, taking the origin to $\mathbf{a}$, by the mapping $\tau_{\mathbf{a}} \circ \rho$ as defined above. Let $\mathbf{b} = \mathbf{g}(\mathbf{a})$ and let

$$\mathbf{S} = \mathbf{g} \circ \tau_{\mathbf{a}} \circ \rho.$$

Then for $\mathbf{x} \in C_1$ and $\mathbf{y} = \tau_{\mathbf{a}}(\mathbf{x}) \in Q$, we have $D\mathbf{S}(\mathbf{x}) = D\mathbf{g}(\mathbf{y}) \circ \rho$, since ρ is linear and the derivative of the translation is the identity. By the chain rule, for $\mathbf{x} \in C_1$ and $\mathbf{y} = \tau_{\mathbf{a}}(\mathbf{x}) \in Q$,

$$\begin{aligned} [D\mathbf{S}(\mathbf{0})]^{-1} D\mathbf{S}(\mathbf{x}) &= \left[D\mathbf{g}(\mathbf{a}) \circ \rho\right]^{-1} D\mathbf{g}(\mathbf{y}) \circ \rho \\ &= \rho^{-1} \, [D\mathbf{g}(\mathbf{a})]^{-1} D\mathbf{g}(\mathbf{y}) \, \rho. \end{aligned}$$

Hence,

$$\begin{aligned} \left\|[D\mathbf{S}(\mathbf{0})]^{-1} D\mathbf{S}(\mathbf{x}) - I\right\|_\infty &= \left\|\rho^{-1} \, [D\mathbf{g}(\mathbf{a})]^{-1} D\mathbf{g}(\mathbf{y}) \, \rho - I\right\|_\infty \\ &= \left\|\rho^{-1} \left([D\mathbf{g}(\mathbf{a})]^{-1} D\mathbf{g}(\mathbf{y}) - I\right) \rho\right\|_\infty \\ &\leq \left\|\rho^{-1}\right\|_\infty \left\|[D\mathbf{g}(\mathbf{a})]^{-1} D\mathbf{g}(\mathbf{y}) - I\right\|_\infty \left\|\rho\right\|_\infty \\ &= \left\|\rho^{-1}\right\|_\infty \left\|\rho\right\|_\infty \left\|[D\mathbf{g}(\mathbf{a})]^{-1} D\mathbf{g}(\mathbf{y}) - I\right\|_\infty \\ &\leq R\,\epsilon, \end{aligned}$$

where we have used the hypothesis that $\|[D\mathbf{g}(\mathbf{a})]^{-1} D\mathbf{g}(\mathbf{y}) - I\|_\infty \leq \epsilon$ for all $\mathbf{y}$ in Q. If we also have $R\epsilon < 1$, then $\mathbf{S}$ satisfies hypothesis (13.17) in Lemma 13.4.2 with ϵ replaced by $\epsilon R < 1$ and Q being the cube C_1 (for which $r = 1$). By invoking Lemma 13.4.2, we conclude that

$$(1 - \epsilon R)^n |\det D\mathbf{S}(\mathbf{0})| \nu(C_1) \;\leq\; \nu(\mathbf{S}(C_1)) \;\leq\; (1 + \epsilon R)^n |\det D\mathbf{S}(\mathbf{0})| \nu(C_1).$$

Since $\nu(C_1) = 2^n$ and $\det \rho = \nu(Q)/2^n$, it follows that

$$|\det D\mathbf{S}(\mathbf{0})|\nu(C_1) = |\det D\mathbf{g}(\mathbf{a})|\,|\det \rho|\,\nu(C_1) = |\det D\mathbf{g}(\mathbf{a})|\,\nu(Q).$$

Since $\mathbf{S}(C_1) = \mathbf{g}(Q)$, we conclude that $\nu(\mathbf{g}(Q))$ is bounded according to

$$(1-\epsilon R)^n\,|\det D\mathbf{g}(\mathbf{a})|\,\nu(Q) \leq \nu(\mathbf{g}(Q)) \leq (1+\epsilon R)^n\,|\det D\mathbf{g}(\mathbf{a})|\,\nu(Q),$$

which is exactly (13.22), as desired. □

In Theorem 13.4.4 below, we invoke Theorem 12.6.5, which allows us to choose partitions of the noncube interval Q for which the intervals are similar to Q; that is, we can always subdivide the factors of Q by the same number N of real subintervals, so that we get N^n intervals of the partition, with the partition having mesh equal to $s_{\max}/N$, where $s_{\max}$ is the length of the longest factor of Q. Then the ratio of longest to shortest side of Q, $R = s_{\max}/s_{\min}$, is maintained for every interval Q_i of our partitions. Thus we may always map the cube C_1 to such intervals Q_i having center point $\mathbf{a}_i$ by a mapping $\tau_{\mathbf{a}_i} \circ \rho$ satisfying $R = |\rho^{-1}|_\infty\,|\rho|_\infty = s_{\max}/s_{\min}$, since this ratio is the same for all intervals Q_i of our partitions.

Theorem 13.4.4. *Let Q be an interval in $\mathbf{R}^n$, and $\mathbf{g}: U \to \mathbf{R}^n$ a C^1 mapping which is C^1-invertible on an open set U containing Q. If $f : \mathbf{g}(Q) \to \mathbf{R}$ is an integrable function such that $f \circ \mathbf{g}$ is also integrable, then*

$$\int_{\mathbf{g}(Q)} f = \int_Q (f \circ \mathbf{g})\,|\Delta_\mathbf{g}|, \tag{13.23}$$

where $\Delta_\mathbf{g}\mathbf{x} = \det D\mathbf{g}(\mathbf{x})$ for $\mathbf{x} \in Q$.

Proof. Let $\eta > 0$, let P be a partition of Q by similar intervals $Q_1, \ldots, Q_k$, and let $\mathcal{S} = \{\mathbf{a}_1, \ldots, \mathbf{a}_k\}$ be the selection such that $\mathbf{a}_i$ is the center point of Q_i, for each i. By the integrability of $f \circ \mathbf{g}$ and Theorem 12.6.5, there exists a $\delta_1 = \delta_1(\eta) > 0$ such that if the mesh of P is less than δ_1, then

$$\left| R\Big((f \circ \mathbf{g})\,|\Delta_\mathbf{g}|,\, P,\, \mathcal{S}\Big) - \int_Q f \circ \mathbf{g}\,|\Delta_\mathbf{g}| \right| < \frac{\eta}{2} \tag{13.24}$$

for the Riemann sum

$$R\Big((f \circ \mathbf{g})\,|\Delta_\mathbf{g}|,\, P,\, \mathcal{S}\Big) = \sum_{i=1}^{k} f(\mathbf{g}(\mathbf{a}_i))\,|\Delta_\mathbf{g}(\mathbf{a}_i)|\,\nu(Q_i).$$

Since $\Delta_\mathbf{g}$ is continuous and $f \circ \mathbf{g}$ is integrable on Q, there are bounds A and B such that

$$|\Delta_\mathbf{g}(\mathbf{x})| \leq A \qquad \text{and} \qquad |f \circ \mathbf{g}(\mathbf{x})| \leq M$$

for all $\mathbf{x}$ in Q. As usual, write

$$\begin{aligned} m_i &= \inf_{Q_i}(f \circ \mathbf{g})(\mathbf{x}), \\ M_i &= \sup_{Q_i}(f \circ \mathbf{g})(\mathbf{x}). \end{aligned}$$

Then our Riemann sum is bracketed by the lower and upper sums for the partition P:

(13.25)

$$\alpha := \sum_{i=1}^{k} m_i |\Delta_{\mathbf{g}}(\mathbf{a}_i)| \nu(Q_i) \leq R\Big((f \circ \mathbf{g})\,|\Delta_{\mathbf{g}}|,\, P,\, \mathcal{S}\Big) \leq \sum_{i=1}^{k} M_i |\Delta_{\mathbf{g}}(\mathbf{a}_i)| \nu(Q_i) =: \beta.$$

By Theorem 12.6.5, there is a $\delta_2 = \delta_2(\eta) > 0$ such that if the mesh of a partition is less than δ_2, then any two Riemann sums for $f \circ \mathbf{g}$ differ by less than $\eta/(12A)$. For such partitions, we continue to write Q_i for the intervals and k for their total number; hence, the upper and lower sums differ by at most $\eta/(6A)$, and thus

$$\sum_{i=1}^{k} (M_i - m_i)\nu(Q_i) \leq \frac{\eta}{6A}. \tag{13.26}$$

Consequently for a partition with mesh less than $\min\{\delta_1, \delta_2\}$, (13.25) and (13.26) imply that

$$\beta - \alpha = \sum_{i=1}^{k} (M_i - m_i)\, |\Delta_{\mathbf{g}}(\mathbf{a}_i)|\, \nu(Q_i) < \frac{\eta}{6}, \tag{13.27}$$

since $|\Delta_{\mathbf{g}}(\mathbf{a}_i)| \leq A$. To this point in the argument, we have a Riemann sum that differs from $\int_Q (f \circ \mathbf{g})\,|\Delta_{\mathbf{g}}|$ by less than $\eta/2$, and lies between numbers α and β which differ by less than $\eta/6$.

The goal now is to bracket the value $\int_{\mathbf{g}(Q)} f$ between two numbers α' and β' that are close to α and β. The intervals Q_i of our partition intersect only along their boundaries, if at all, and therefore

$$\int_{\mathbf{g}(Q)} f = \sum_{i=1}^{k} \int_{\mathbf{g}(Q_i)} f.$$

Therefore we define α' and β', a lower and upper approximation for $\int_{\mathbf{g}(Q)} f$, by

$$\alpha' := \sum_{i=1}^{k} m_i\, \nu(\mathbf{g}(Q_i)) \leq \int_{\mathbf{g}(Q)} f \leq \sum_{i=1}^{k} M_i\, \nu(\mathbf{g}(Q_i)) =: \beta', \tag{13.28}$$

with m_i and M_i as defined previously. We are now able to estimate the differences $\alpha' - \alpha$ and $\beta' - \beta$ using Lemma 13.4.3.

Let ϵ satisfy $0 < \epsilon R < 1$, where $R = s_{\max}/s_{\min}$, the ratio of maximum to minimum side length for Q and all intervals Q_i of our partitions. Suppose that ϵ also satisfies

$$(1 + \epsilon R)^n - (1 - \epsilon R)^n < \frac{\eta}{6AM\nu(Q)}. \tag{13.29}$$

Since $\mathbf{g}$ is C^1-invertible on the open set U containing Q, there is an upper bound B for the norm $\|[D\mathbf{g}(\mathbf{x})]^{-1}\|_\infty$ for $\mathbf{x}$ in Q, thus

$$\|[D\mathbf{g}(\mathbf{x})]^{-1}\|_\infty \leq B \quad \text{for } \mathbf{x} \in Q.$$

By the uniform continuity of $D\mathbf{g}$ on the compact set Q, there exists a $\delta_3 = \delta_3(\epsilon)$ such that, if the mesh of our partition is less than δ_3, then

$$\|D\mathbf{g}(\mathbf{x}) - D\mathbf{g}(\mathbf{a}_i)\|_\infty < \frac{\epsilon}{B}$$

for all $\mathbf{x}$ in Q_i. And this implies that

$$\begin{aligned} \| [D\mathbf{g}(\mathbf{a}_i)]^{-1} \circ D\mathbf{g}(\mathbf{x}) - I \|_\infty &= \left\| [D\mathbf{g}(\mathbf{a}_i)]^{-1} \Big(D\mathbf{g}(\mathbf{x}) - D\mathbf{g}(\mathbf{a}_i) \Big) \right\|_\infty \\ &\le \|[D\mathbf{g}(\mathbf{a}_i)]^{-1}\|_\infty \, \|D\mathbf{g}(\mathbf{x}) - D\mathbf{g}(\mathbf{a}_i)\|_\infty \\ &\le B\frac{\epsilon}{B} = \epsilon \end{aligned}$$

for all $\mathbf{x}$ in Q_i. Hence, Lemma 13.4.3 applies and shows that the volume $\nu(\mathbf{g}(Q_i))$ satisfies the bounds

$$(1-\epsilon R)^n \, |\det D\mathbf{g}(\mathbf{a}_i)| \, \nu(Q_i) \le \nu(\mathbf{g}(Q_i)) \le (1+\epsilon R)^n \, |\det D\mathbf{g}(\mathbf{a}_i)| \, \nu(Q_i),$$

and clearly $|\det D\mathbf{g}(\mathbf{a}_i)| \, \nu(Q_i)$ has these same lower and upper bounds as well. Therefore if the mesh of the partition is less than δ_3, then, again writing $\Delta_\mathbf{g}(\mathbf{a}_i) = \det D\mathbf{g}(\mathbf{a}_i)$, we have

$$\Big| \nu(\mathbf{g}(Q_i)) - |\Delta_\mathbf{g}(\mathbf{a}_i)| \, \nu(Q_i) \Big| \le [(1+\epsilon R)^n - (1-\epsilon R)^n] \, |\Delta_\mathbf{g}(\mathbf{a}_i)| \, \nu(Q_i). \tag{13.30}$$

Now assume that the partition P has mesh less than $\min\{\delta_1, \delta_2, \delta_3\}$. We continue to write k for the total number of intervals of P. Now we estimate the difference $\beta' - \beta$ between upper sum approximations of the two integrals of interest, and the difference $\alpha' - \alpha$ between lower sum approximations of those integrals. By (13.25) and (13.28), (13.29) and (13.30), we have

$$\begin{aligned} |\beta' - \beta| &\le \sum_{i=1}^k |M_i| \Big| \nu(\mathbf{g}(Q_i)) - |\Delta_\mathbf{g}(\mathbf{a}_i)| \nu(Q_i) \Big| \\ &\le M \sum_{i=1}^k [(1+\epsilon R)^n - (1-\epsilon R)^n] \, |\Delta_\mathbf{g}(\mathbf{a}_i)| \, \nu(Q_i) \\ &\le M A \, \nu(Q) \, [(1+\epsilon R)^n - (1-\epsilon R)^n] < \frac{\eta}{6}. \end{aligned} \tag{13.31}$$

Also by (13.25) and (13.28), (13.29) and (13.30),

$$\begin{aligned} |\alpha' - \alpha| &\le \sum_{i=1}^k |m_i| \Big| \nu(\mathbf{g}(Q_i)) - |\Delta_\mathbf{g}(\mathbf{a}_i)| \nu(Q_i) \Big| \\ &\le M \sum_{i=1}^k [(1+\epsilon R)^n - (1-\epsilon R)^n] \, |\Delta_\mathbf{g}(\mathbf{a}_i)| \, \nu(Q_i) \\ &\le M A \, \nu(Q) \, [(1+\epsilon R)^n - (1-\epsilon R)^n] < \frac{\eta}{6}. \end{aligned} \tag{13.32}$$

If we let $\alpha^* = \min\{\alpha, \alpha'\}$ and $\beta^* = \max\{\beta, \beta'\}$, then (13.25) and (13.28) imply that $\int_{\mathbf{g}(Q)} f$ and the Riemann sum $R\Big((f \circ \mathbf{g}) \, |\Delta_\mathbf{g}|, \, P, \, \mathcal{S}\Big)$ must lie between α^* and β^*. Now (13.27), (13.31) and (13.32) imply that $\beta^* - \alpha^* < \eta/2$, and therefore

$$\left| R\Big((f \circ \mathbf{g}) \, |\Delta_\mathbf{g}|, \, P, \, \mathcal{S}\Big) - \int_{\mathbf{g}(Q)} f \right| < \frac{\eta}{2}. \tag{13.33}$$

Thus, in view of (13.24), both $\int_{\mathbf{g}(Q)} f$ and $\int_Q f \circ \mathbf{g}\,|\Delta_{\mathbf{g}}|$ differ from $R\Big((f \circ \mathbf{g})\,|\Delta_{\mathbf{g}}|, P, \mathcal{S}\Big)$ by less than $\eta/2$, and hence

$$\Big| \int_{\mathbf{g}(Q)} f - \int_Q f \circ \mathbf{g}\,|\Delta_{\mathbf{g}}| \Big| < \eta.$$

Since this holds for every $\eta > 0$, the desired result (13.23) follows. □

Theorem 13.4.4 is an important result, and yet it does not in itself cover examples likely to appear in applications. Consider specifically Example 13.2.4, the polar coordinate map which is not C^1-invertible on an open neighborhood of the closed interval $[0,1] \times [-\pi, \pi]$. A similar situation occurs in the spherical coordinate map from (ρ, ϕ, θ) coordinates to (x, y, z) coordinates.

In fact, the hypothesis that the transformation $\mathbf{g}$ is C^1-invertible on U and $Q \subset U$ is stronger than necessary. The next statement confirms that our mapping $\mathbf{g}$ only needs to be C^1-invertible on the *interior* of Q, and this can be important in applications.

Theorem 13.4.5. *Let Q be a closed interval in $\mathbf{R}^n$. Suppose $\mathbf{g}$ is C^1 on an open set U containing Q and C^1-invertible on the interior of Q. If f is an integrable function such that $f \circ \mathbf{g}$ is also integrable, then*

$$\int_{\mathbf{g}(Q)} f = \int_Q f \circ \mathbf{g}\,|\Delta_{\mathbf{g}}|.$$

Proof. Recall that $Q = \operatorname{Int} Q \cup \partial Q$ and $\nu(\partial Q) = 0$. We may partition Q by closed intervals Q_k such that $Q_1, \ldots, Q_m$, $m < k$, lie within $\operatorname{Int} Q$; thus

$$Q^* := \bigcup_{k=1}^{m} Q_k \subset \operatorname{Int} Q.$$

Write $K = Q - \operatorname{Int} Q^*$. Given any $\epsilon > 0$, we can choose the partition so that $\nu(K) < \epsilon$, since ∂Q has volume zero. Then the Theorem 13.4.4 as already proved gives

$$\int_{\mathbf{g}(Q^*)} f = \int_{Q^*} f \circ \mathbf{g}\,|\Delta_{\mathbf{g}}|.$$

But

$$\int_{\mathbf{g}(Q)} f = \int_{\mathbf{g}(Q^*)} f + \int_{\mathbf{g}(K)} f \tag{13.34}$$

since $\operatorname{Int} K$ and $\operatorname{Int} Q^*$ are disjoint and their common boundary has volume zero, and hence $\operatorname{Int} \mathbf{g}(K)$ and $\operatorname{Int} \mathbf{g}(Q^*)$ are disjoint and their common boundary has volume zero. We also have

$$\int_Q f \circ \mathbf{g}\,|\Delta_{\mathbf{g}}| = \int_{Q^*} f \circ \mathbf{g}\,|\Delta_{\mathbf{g}}| + \int_K f \circ \mathbf{g}\,|\Delta_{\mathbf{g}}| \tag{13.35}$$

by the same facts about K and Q^*. Since f is integrable on $\mathbf{g}(Q)$ it is bounded on $\mathbf{g}(K)$, and since $\mathbf{g}$ is C^1 on U, and hence $|\Delta_{\mathbf{g}}|$ is bounded on Q, its bound provides a Lipschitz condition for $\mathbf{g}$ on all of Q. Thus the integral of f over $\mathbf{g}(K)$ in (13.34), and the integral of $f \circ \mathbf{g}\,|\Delta_{\mathbf{g}}|$ over K in (13.35), can be made as small as desired by making $\nu(K)$, and hence $\nu(\mathbf{g}(K))$, sufficiently small. Then by (13.34) and (13.35),

$$\begin{aligned}\left|\int_{\mathbf{g}(Q)} f - \int_Q f \circ \mathbf{g}\,|\Delta_{\mathbf{g}}|\right| &\le \left|\int_{\mathbf{g}(Q^*)} f - \int_{Q^*} f \circ \mathbf{g}\,|\Delta_{\mathbf{g}}|\right| \\ &\quad + \left|\int_{\mathbf{g}(K)} f - \int_K f \circ \mathbf{g}\,|\Delta_{\mathbf{g}}|\right| \\ &\le 0 + \left|\int_{\mathbf{g}(K)} f\right| + \left|\int_K f \circ \mathbf{g}\,|\Delta_{\mathbf{g}}|\right|,\end{aligned}$$

and the last two terms on the right may be made arbitrarily small, by choice of K through our choice of partition of Q. This completes the proof. □

It is worth remarking that we called on the Riemann sum result in Theorem 12.6.5 in the proof of the change of variables formula. Theorem 12.6.5 also covers the single variable case, as written, but we did not need it to establish the change of variables formula in that case (Theorem 6.7.7), because the earlier result followed directly from the fundamental theorem of calculus. Of course, estimating the volume of a C^1 image in the multivariable case is more involved than knowing the interval image of a substitution in Theorem 6.7.7.

The most frequent applications of the change of variables formula in introductory multivariable calculus occur with the use of polar coordinates in the plane and cylindrical coordinates or spherical coordinates in space. The change of variables formula is often described simply by stating the change in the area element dA or the volume element dV.

Example 13.4.6 (Cylindrical Coordinates)**.** The cylindrical coordinate transformation $(x, y, z) = \mathbf{g}(r, \theta, z)$ given by $x = r\cos\theta$, $y = r\sin\theta$ and $z = z$ is defined on an open neighborhood of the closed interval $Q = [a_1, a_2] \times [0, 2\pi] \times [c, d]$ and is C^1-invertible on the interior of Q. It has Jacobian

$$|\Delta_{\mathbf{g}}| = \begin{vmatrix} \cos\theta & -r\sin\theta & 0 \\ \sin\theta & r\cos\theta & 0 \\ 0 & 0 & 1 \end{vmatrix} = r,$$

and hence

$$\int_{\mathbf{g}(Q)} f = \int_Q (f \circ \mathbf{g})\,|\Delta_{\mathbf{g}}| = \int_Q f(\mathbf{g}(r, \theta, z))\,r$$

for an integrable function f such that $f \circ \mathbf{g}$ is integrable. A version of Fubini's theorem can then be used to organize the calculation of the integral. △

Example 13.4.7 (Spherical Coordinates)**.** The spherical coordinate transformation $(x, y, z) = \mathbf{g}(\rho, \phi, \theta)$ given by

$$\begin{aligned} x &= \rho\sin\phi\cos\theta, \\ y &= \rho\sin\phi\sin\theta, \\ z &= \rho\cos\phi \end{aligned}$$

is defined on an open neighborhood of the closed interval $Q = [0, a_2] \times [0, \pi] \times [0, 2\pi]$ and is C^1-invertible on the interior of Q. It has Jacobian determinant

$$|\Delta_{\mathbf{g}}| = \begin{vmatrix} \sin\phi\cos\theta & \rho\cos\phi\cos\theta & -\rho\sin\phi\sin\theta \\ \sin\phi\sin\theta & \rho\cos\phi\sin\theta & \rho\sin\phi\cos\theta \\ \cos\phi & -\rho\sin\phi & 0 \end{vmatrix} = \rho^2\sin\phi,$$

and hence

$$\int_{\mathbf{g}(Q)} f = \int_Q (f\circ\mathbf{g})\,|\Delta_{\mathbf{g}}| = \int_Q f(\mathbf{g}(\rho,\phi,\theta))\,\rho^2\sin\phi$$

for an integrable function f such that $f\circ\mathbf{g}$ is integrable. A version of Fubini's theorem can then organize the calculation. △

Some applications of these formulas appear in the exercises.

Exercises.

Exercise 13.4.1. Show that if A has volume and S is an interval, then exactly one of the following is true: (i) $S \subset \operatorname{Int} A$, (ii) $S \subset (\overline{A})^c$, or (iii) $S \cap \partial A$ is nonempty.

Exercise 13.4.2. The rectangle $S = \{(x_1, x_2) \in \mathbf{R}^2 : 0 \le r \le 1,\ 0 \le \theta \le 2\pi\}$ is mapped by $(x_1, x_2) = \mathbf{g}(r, \theta) = (r\cos\theta, r\sin\theta)$ onto the unit disk, $\mathbf{g}(S) = \{(x_1, x_2) \in \mathbf{R}^2 : x_1^2 + x_2^2 \le 1\}$. Write the change of variables formula for the integration of a function f over this disk. Then integrate the constant function $f \equiv 1$ and thus compute the area of the disk.

Exercise 13.4.3. *Spherical and Cylindrical Coordinates*

1. Write the change of variables formula for the integration of a function $f(x, y, z)$ over the upper half sphere $z \ge 0$ of radius 3, centered at the origin, using spherical coordinates.
2. Write the change of variables formula for the integration of a function $f(x, y, z)$ over the upper half sphere $z \ge 0$ of radius 3, centered at the origin, using cylindrical coordinates.

Exercise 13.4.4. Let S be the solid sphere of diameter 4 centimeters, centered at the origin in $\mathbf{R}^3$. A solid cylindrical core of diameter 2 centimeters is machined out of the center of this solid sphere.

1. Write an integral using cylindrical coordinates in $\mathbf{R}^3$ to represent the volume of this cylindrical core.
2. Write an integral over a disk in the plane that represents the volume of this cylindrical core.
3. Find the volume of the portion of the sphere remaining after this cylindrical core is removed.

Exercise 13.4.5. Interpret the integral

$$\int_0^{2\pi}\int_0^{\pi}\int_0^{(1-\cos\phi)/2} \rho^2\sin\phi\,d\rho\,d\phi\,d\theta$$

geometrically. Describe or sketch the solid region of integration.

Exercise 13.4.6. *Volume of balls*
The closed unit ball about the origin in n dimensions, with respect to the Euclidean norm, we denote here by B_1^n, and its volume by $\nu_n(1)$. The closed ball about the origin of radius $r > 0$ we denote by B_r^n, and its volume by $\nu_n(r)$.

1. Show that $\nu_n(r) = r^n \nu_n(1)$. *Hint*: Define $\mathbf{g} : \mathbf{R}^n \to \mathbf{R}^n$ by $\mathbf{g}(\mathbf{x}) = r\mathbf{x}$.
2. Verify that $\nu_1(1) = 2$, $\nu_2(1) = \pi$, and $\nu_3(1) = \frac{4}{3}\pi$.
3. Show that

$$\nu_n(1) = \nu_{n-2}(1)\frac{2\pi}{n}.$$

Hint for part 3: Write points in $\mathbf{R}^n$ with coordinates $(x, y, x_3, \dots, x_n)$ and note that

$$B_1^n = \{(x, y, x_3, \dots, x_n) : x^2 + y^2 + x_3^2 + \cdots + x_n^2 \le 1\}.$$

Enclose B_1^n in the n-dimensional interval $[-1,1] \times [-1,1] \times \cdots \times [-1,1]$. If $\chi_1^n(x, y, x_3, \dots, x_n)$ denotes the characteristic function of B_1^n, then we may write

$$\nu_n(1) = \int_{-1}^{1}\int_{-1}^{1}\left[\int_{R_{n-2}} \chi_1^n(x, y, x_3, \dots, x_n)\, dx_3 \cdots dx_n\right] dx\, dy \tag{13.36}$$

where R_{n-2} is the $(n-2)$-dimensional interval with all factors equal to $[-1,1]$. If $x^2 + y^2 > 1$, then $\chi_1^n(x, y, x_3, \dots, x_n) = 0$. Let $D = \{(x, y) : x^2 + y^2 \le 1\}$. If $(x, y) \in D$, then $\chi_1^n(x, y, x_3, \dots, x_n)$, viewed as a function of $(x_3, \dots, x_n)$, is the characteristic function of the $(n-2)$-dimensional Euclidean ball of radius $\sqrt{1 - x^2 - y^2}$ centered at the origin. Thus the inner integral in (13.36) equals

$$\int_{R_{n-2}} \chi_1^n(x, y, x_3, \dots, x_n)\, dx_3 \cdots dx_n = (1 - x^2 - y^2)^{(n-2)/2}\, \nu_{n-2}(1).$$

Hence,

$$\nu_n(1) = \nu_{n-2}(1) \int_D (1 - x^2 - y^2)^{(n-2)/2}\, dx\, dy.$$

Now complete the argument by showing that

$$\int_D (1 - x^2 - y^2)^{(n-2)/2}\, dx\, dy = \frac{2\pi}{n}.$$

4. Conclude by induction that

$$\nu_{2k}(1) = \frac{\pi^k}{k!} \qquad \text{and} \qquad \nu_{2k-1}(1) = \frac{2^k \pi^{k-1}}{1 \cdot 3 \cdot 5 \cdots (2k-1)}.$$

Exercise 13.4.7. *Volume of balls and the gamma function*
Recall the gamma function of Exercise 7.5.9,

$$\Gamma(\alpha) := \int_0^\infty x^{\alpha-1} e^{-x}\, dx,$$

defined for $\alpha > 0$. It satisfies $\Gamma(\alpha + 1) = \alpha\Gamma(\alpha)$ for $\alpha > 0$ and, in particular, $\Gamma(n + 1) = n!$ for $n = 0, 1, 2, 3, \dots$.

1. Verify that $\Gamma(\frac{1}{2}) = \sqrt{\pi}$. *Hint*: Use the substitution $x = u^2$ and the fact that

$$\int_0^\infty e^{-u^2}\, du = \frac{\sqrt{\pi}}{2}.$$

2. Find the values $\Gamma(\frac{3}{2})$, $\Gamma(\frac{5}{2})$ and $\Gamma(\frac{7}{2})$.
3. Show that
$$\Gamma\left(n+\frac{1}{2}\right) = \sqrt{\pi}\,\frac{1\cdot 3\cdot 5\cdots(2n-1)}{2^n}, \qquad \text{for } n = 1,2,3,\ldots.$$
4. Refer to Exercise 13.4.6, part 4. Use the properties $\Gamma(\alpha+1) = \alpha\Gamma(\alpha)$, $\Gamma(1) = 1$ and $\Gamma(\frac{1}{2}) = \sqrt{\pi}$ to show that the volume $\nu_n(1)$ of the unit ball in n dimensions is given by
$$\nu_n(1) = \frac{\pi^{n/2}}{\Gamma(\frac{n}{2}+1)}, \qquad \text{for } n = 1,2,3,\ldots.$$
Conclude that the volume $\nu_n(r)$ of the ball of radius $r > 0$ is given by
$$\nu_n(r) = r^n\frac{\pi^{n/2}}{\Gamma(\frac{n}{2}+1)}, \qquad \text{for } n = 1,2,3,\ldots.$$

13.5. The Definition of Surface Integrals

The change of variables formula is useful not only as a computational tool but also as an essential conceptual tool. We demonstrate this in the present section by illustrating the role of the change of variables formula in the definition of surface integrals from multivariable calculus. We then briefly recall the divergence theorem and discuss a coordinate-free interpretation of the divergence of a vector field at a point, as well as a coordinate-free interpretation of the Laplacian of a real valued function.

Let us call $\Omega \subset \mathbf{R}^2$ a **region** if $\Omega = \operatorname{Int} Q$ where Q is a bounded and Jordan measurable set, that is, $\partial\Omega = \partial Q$ has volume zero. Then any continuous, bounded real valued function defined on Ω is integrable there.

Definition 13.5.1. *Let $\Omega \subset \mathbf{R}^2$ be a region with $\Omega = \operatorname{Int} Q$. Suppose $\Phi : Q \to \mathbf{R}^3$ is a C^1 mapping that satisfies the following conditions:*

1. *Φ is one-to-one on Ω.*

2. *The component functions of $\Phi = (\phi_1, \phi_2, \phi_3)$ have bounded first-order partial derivatives on Ω.*

3. *For each point $(u, v) \in \Omega$,*
$$\Phi_u(u,v) \times \Phi_v(u,v) \neq \mathbf{0},$$
where
$$\Phi_u(u,v) = \frac{\partial\Phi}{\partial u}(u,v) = \left(\frac{\partial\phi_1}{\partial u}(u,v), \frac{\partial\phi_2}{\partial u}(u,v), \frac{\partial\phi_3}{\partial u}(u,v)\right)$$
and
$$\Phi_v(u,v) = \frac{\partial\Phi}{\partial v}(u,v) = \left(\frac{\partial\phi_1}{\partial v}(u,v), \frac{\partial\phi_2}{\partial v}(u,v), \frac{\partial\phi_3}{\partial v}(u,v)\right).$$

Then the image $S = \Phi(\Omega)$ is called a **surface** *(or* **surface patch***) in $\mathbf{R}^3$, and Φ is called a* **parametrization** *of the surface $S = \Phi(\Omega)$.*

It is convenient to view $\frac{\partial\Phi}{\partial u}(u,v)$ and $\frac{\partial\Phi}{\partial v}(u,v)$ as row vectors here; when viewed as columns, they are the columns of the Jacobian matrix $J_\Phi(u,v)$.

Condition 3 of Definition 13.5.1 guarantees that a surface $S = \Phi(\Omega)$ has a well-defined normal vector at each point $\Phi(u,v) \in S$, defined by the cross product indicated there, and this normal vector varies smoothly with the parameters (u,v) in Ω. Thus, the surface has a well-defined tangent plane at each of its points, the tangent plane being orthogonal to the normal vector and spanned by $\frac{\partial\Phi}{\partial u}(u,v)$ and $\frac{\partial\Phi}{\partial v}(u,v)$. At each (u,v), the normal vector shown then completes a basis that forms a right-handed coordinate system. Condition 2 of Definition 13.5.1 ensures that the Euclidean norm of the cross product remains bounded, and hence integrable, since it is continuous.

Example 13.5.2. Consider an open spherical cap S about the north pole on the sphere in $\mathbf{R}^3$ of radius $\rho = 2$. Let $0 < a < 2$, and suppose this cap S lies above the disk region in the xy-plane described by $x^2 + y^2 < a^2$. Then S is the graph of the function $z = g(x,y) = (4 - x^2 - y^2)^{1/2}$ over that disk. And S is parametrized by $\Phi(u,v) = (u, v, g(u,v)) = (u, v, (4 - u^2 - v^2)^{1/2}$ according to Definition 13.5.1, since the first order partial derivatives g_u and g_v are bounded on the open disk $u^2 + v^2 < a^2 < 4$:

$$g_u(u,v) = -\frac{u}{(4 - u^2 - v^2)^{1/2}} \qquad \text{and} \qquad g_v(u,v) = -\frac{v}{(4 - u^2 - v^2)^{1/2}}.$$

However, if we let S be the entire open upper hemisphere, then the same function g defined on the open disk $u^2 + v^2 < 4$ (when $a = 2$) yields S as its graph, but the same Φ mapping now has unbounded partial derivatives g_u and g_v on the disk $u^2 + v^2 < 4$. Thus, Φ does not parametrize the open upper hemisphere in accord with Definition 13.5.1. For that purpose, however, consider the spherical coordinates mapping $\Psi : Q \to \mathbf{R}^3$,

$$\Psi(\phi, \theta) = \big(2\sin\phi\cos\theta, 2\sin\phi\sin\theta, 2\cos\phi\big),$$

where $Q = \{(\phi,\theta) : 0 \le \phi \le \pi/2,\ 0 \le \theta \le 2\pi\}$. Then Ψ is one-to-one on $\Omega = \operatorname{Int} Q$, is C^1 on Q, and has bounded first order partial derivatives, in accordance with Definition 13.5.1. The image surface $\Psi(\Omega)$ is the open upper hemisphere, minus the north pole and the meridian where $\theta = 0$ (consisting of all points of the form $\Psi(\phi, 0) = (2\sin\phi, 0, 2\cos\phi)$). However, this excluded set has volume zero and therefore cannot affect the value of a surface integral over the hemisphere using the parametrization Ψ. (Once we have defined surface integrals, this can be confirmed for this example in Exercise 13.5.3.) $\triangle$

Definition 13.5.3. *Let Ω be a region in $\mathbf{R}^2$ and S a surface parametrized by $\Phi : \Omega \to \mathbf{R}^3$. If $f : S \to \mathbf{R}$ is a continuous and bounded function on S, the* **surface integral** *of f over S is defined by*

$$\int_S f\,d\sigma = \int_\Omega (f \circ \Phi)\,|\Phi_u \times \Phi_v|_2\,du\,dv$$

where $|\cdot|_2$ is the Euclidean norm. In particular, the **area** *of S is defined to be*

$$\operatorname{area} S = \int_\Omega |\Phi_u \times \Phi_v|_2\,du\,dv.$$

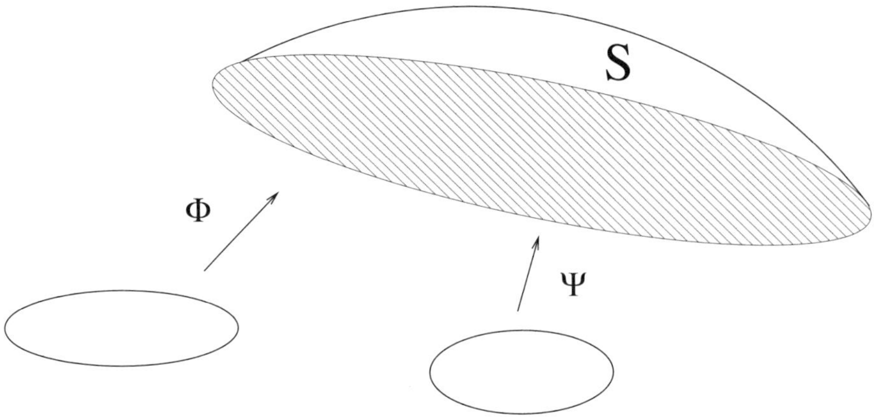

Figure 13.4. Coordinate independence of the surface integral definition: Two different parametrizations, Φ and Ψ, of a surface patch S are related by $\Psi \circ \mathbf{g} = \Phi$, where $\mathbf{g}$ is C^1-invertible; that is, $\Psi^{-1} \circ \Phi$ is C^1-invertible, with C^1 inverse $\Phi^{-1} \circ \Psi$. Theorem 13.5.4 uses this fact to establish that the value of a surface integral over S does not depend on the parametrization.

We have included the **element of surface area** $d\sigma = du\,dv$ in this definition since there is a need to express coordinates in the domain. Observe that the integral representing the area of a surface is the special surface integral involving the function $f \equiv 1$.

If S is a surface which is the graph of a continuously differentiable function g of u and v, and if S is parametrized by $\Phi(u, v) = (u, v, g(u, v))$ as in Definition 13.5.1, then it is straightforward to check that

$$\Phi_u(u, v) \times \Phi_v(u, v) = (-g_u, -g_v, 1) \tag{13.37}$$

is the normal to the surface at $\Phi(u, v)$. By taking the Euclidean norm in (13.37), the surface integral becomes

$$\int_S f\, d\sigma = \int_\Omega (f \circ \Phi) \sqrt{1 + [g_u]^2 + [g_v]^2}\, du\, dv.$$

This yields a multiple integral to compute $\int_S f\, d\sigma$ indicated by

$$\begin{aligned} \int_S f\, d\sigma &= \int\int_\Omega (f \circ \Phi)(u, v)\, |\Phi_u(u, v) \times \Phi_v(u, v)|_2\, du\, dv \\ &= \int\int_\Omega (f \circ \Phi)(u, v) \sqrt{1 + [g_u(u, v)]^2 + [g_v(u, v)]^2}\, du\, dv, \end{aligned}$$

with appropriate limits inserted for the double integrals on the right side depending on Ω, and the possibility of reversing the order of integration to $dv\, du$ if necessary.

Our main goal now is to show that Definition 13.5.3 unambiguously defines the value of the surface integral $\int_S f\, d\sigma$; in other words, the integral value does not depend on the parametrization used for S. (See Figure 13.4.)

Theorem 13.5.4. *Let Ω and Ω' be regions in $\mathbf{R}^2$, and let the surface S in $\mathbf{R}^3$ be parametrized by $\Phi : \Omega \to \mathbf{R}^3$ and $\Psi : \Omega' \to \mathbf{R}^3$, so that $S = \Phi(\Omega) = \Psi(\Omega')$. If $f : S \to \mathbf{R}$ is continuous and bounded, then*

$$\int_\Omega (f \circ \Phi)\, |\Phi_u \times \Phi_v|_2 \, du\, dv = \int_{\Omega'} (f \circ \Psi)\, |\Psi_{u'} \times \Psi_{v'}|_2 \, du'\, dv'. \tag{13.38}$$

Proof. By assumption, the mappings $\Phi : \Omega \to \mathbf{R}^3$ and $\Psi : \Omega' \to \mathbf{R}^3$ are one-to-one and onto S. If (u, v) is in Ω, then we define $\mathbf{g}(u, v) = (u', v')$ to be the unique point in Ω' at which

$$\Psi(u', v') = \Phi(u, v). \tag{13.39}$$

This defines $\mathbf{g} : \Omega \to \mathbf{R}^2$, and $\mathbf{g}$ is one-to-one with image equal to Ω', since both Φ and Ψ are onto S.

Denote the component form of these mappings by $\Phi = (\phi_1, \phi_2, \phi_3)$, $\Psi = (\psi_1, \psi_2, \psi_3)$, and $\mathbf{g} = (g_1, g_2)$. Then equation (13.39) is equivalent to the system

$$\begin{aligned} \psi_1(g_1(u,v), g_2(u,v)) &= \phi_1(u,v), \\ \psi_2(g_1(u,v), g_2(u,v)) &= \phi_2(u,v), \\ \psi_3(g_1(u,v), g_2(u,v)) &= \phi_3(u,v). \end{aligned} \tag{13.40}$$

We want to show that $\mathbf{g} : \Omega \to \mathbf{R}^2$ is a C^1 mapping that is C^1-invertible on Ω. This will show that it is a legitimate change of variables.

Let (u_0, v_0) be a point in Ω. By hypothesis,

$$\frac{\partial \Psi}{\partial u'}(\mathbf{g}(u_0, v_0)) \times \frac{\partial \Psi}{\partial v'}(\mathbf{g}(u_0, v_0)) \neq \mathbf{0},$$

hence there is at least one component of this cross product that is nonzero. With no loss in generality we may assume that the last component of this cross product is nonzero. This assumption is equivalent to the condition that the determinant

$$\det \begin{bmatrix} \frac{\partial \psi_1}{\partial u'}(\mathbf{g}(u_0, v_0)) & \frac{\partial \psi_2}{\partial u'}(\mathbf{g}(u_0, v_0)) \\ \frac{\partial \psi_1}{\partial v'}(\mathbf{g}(u_0, v_0)) & \frac{\partial \psi_2}{\partial v'}(\mathbf{g}(u_0, v_0)) \end{bmatrix} \neq 0.$$

By the inverse function theorem, the mapping $(\psi_1, \psi_2) : \Omega' \to \mathbf{R}^2$ is locally invertible at the point $\mathbf{g}(u_0, v_0)$ with a C^1 inverse. From the first two equations of the system (13.40), we conclude that there is an open neighborhood O of the point (u_0, v_0) on which

$$(g_1, g_2) = (\psi_1, \psi_2)^{-1}(\phi_1, \phi_2).$$

Thus, on the neighborhood O, (g_1, g_2) is a composition of C^1 mappings and therefore $\mathbf{g}$ is C^1 there. Since (u_0, v_0) was an arbitrary point in Ω, we conclude that $\mathbf{g}$ is a C^1 mapping taking Ω one-to-one and onto Ω'. It remains to show that $D\mathbf{g}(u, v)$ is invertible for each (u, v) in Ω.

System (13.40) can be written as $\Phi(u, v) = \Psi \circ \mathbf{g}(u, v)$. Since we now know that $\mathbf{g}$ is C^1, the chain rule applies and we have

$$D\Phi(u, v) = D\Psi(\mathbf{g}(u, v)) D\mathbf{g}(u, v), \quad (u, v) \in \Omega,$$

and this implies the Jacobian matrix identity

$$J_\Phi(u, v) = J_\Psi(\mathbf{g}(u, v))\, J_{\mathbf{g}}(u, v), \quad (u, v) \in \Omega.$$

This equality of 3×2 matrices implies the equality of all corresponding 2×2 submatrices on the left and right sides. The determinants of these 2×2 submatrices determine the components of the cross products that appear in (13.38). To be precise, suppose C and A are 3×2 matrices and B is a 2×2 matrix such that

$$C = AB.$$

Denote the columns, left to right, of C and A by C_1, C_2 and A_1, A_2. Then

$$C_1 \times C_2 = (\det B)\, A_1 \times A_2.$$

(See Exercise 13.5.1.) It follows that

$$\Phi_u(u,v) \times \Phi_v(u,v) = [\det D\mathbf{g}(u,v)]\, \Psi_{u'}(\mathbf{g}(u,v)) \times \Psi_{v'}(\mathbf{g}(u,v)). \tag{13.41}$$

Each component of the cross product on the left equals the corresponding component of the cross product on the right multiplied by $\det D\mathbf{g}(u,v)$; that is, for each standard basis vector $\mathbf{e}_i$, $1 \le i \le 3$, we have

$$\Big(\Phi_u(u,v) \times \Phi_v(u,v)\Big) \cdot \mathbf{e}_i = [\det D\mathbf{g}(u,v)]\Big(\Psi_{u'}(\mathbf{g}(u,v)) \times \Psi_{v'}(\mathbf{g}(u,v))\Big) \cdot \mathbf{e}_i.$$

Taking Euclidean norms in (13.41) gives

$$|\Phi_u(u,v) \times \Phi_v(u,v)|_2 = |\det D\mathbf{g}(u,v)|\, |\Psi_{u'}(\mathbf{g}(u,v)) \times \Psi_{v'}(\mathbf{g}(u,v))|_2. \tag{13.42}$$

We know that $\det D\mathbf{g}(u,v) \neq 0$ for each $(u,v) \in \Omega$, since $\mathbf{g}$ is C^1-invertible on Ω. Recalling that system (13.40) says that $\Phi(u,v) = \Psi \circ \mathbf{g}(u,v)$, the change of variables formula and (13.42) allow us to express the right-hand side of (13.38) by

$$\begin{aligned}
&\int_{\Omega'} (f \circ \Psi)\, |\Psi_{u'} \times \Psi_{v'}|_2\, du'\, dv' \\
&\quad = \int_{\Omega} (f \circ \Psi \circ \mathbf{g})(u,v)\, |\Psi_{u'}(\mathbf{g}(u,v)) \times \Psi_{v'}(\mathbf{g}(u,v))|_2\, |\det D\mathbf{g}(u,v)|\, du\, dv \\
&\quad = \int_{\Omega} (f \circ \Phi)(u,v))\, |\Phi_u(u,v) \times \Phi_v(u,v)|_2\, du\, dv,
\end{aligned}$$

which is the left-hand side of (13.38), as we wished to show. □

Theorem 13.5.4 confirms, via the change of variables formula, that surface area and surface integrals as defined in Definition 13.5.3 do not depend on the particular parametrization used for a surface S.

Readers may recall the divergence theorem from a first multivariable calculus course. Let $B_r(\mathbf{a})$ be the closed ball of radius $r > 0$ centered at the point $\mathbf{a}$ in $\mathbf{R}^3$, and let $S_r(\mathbf{a}) = \partial B_r(\mathbf{a})$ be the boundary spherical surface. If $\mathbf{F}(\mathbf{x}) = (f_1(\mathbf{x}), f_2(\mathbf{x}), f_3(\mathbf{x}))$ is a continuously differentiable vector field on an open set containing $\mathbf{a}$, then the divergence theorem applied to the region $B_r(\mathbf{a})$ says that

$$\int_{S_r(\mathbf{a})} \mathbf{F} \cdot \mathbf{n}\, d\sigma = \int_{B_r(\mathbf{a})} \operatorname{div} \mathbf{F}\, dV \tag{13.43}$$

where $\mathbf{n}$ denotes the outward unit normal to $S_r(\mathbf{a})$ at each point, $d\sigma$ is the element of surface area and dV is the element of volume. The integral on the left side measures the net flux (flow) of the vector field outward across the surface of the sphere. The integral on the right side is the integral of the divergence of the field $\mathbf{F}$ over the ball, where

$$\operatorname{div} \mathbf{F} = \frac{\partial f_1}{\partial x_1} + \frac{\partial f_2}{\partial x_2} + \frac{\partial f_3}{\partial x_3}.$$

We will not prove the result of (13.43), but we wish to use it to illustrate a coordinate-free interpretation of the divergence of $\mathbf{F}$. There is also a coordinate-free interpretation of the Laplacian of a real function f. Recall the Laplacian of f is the differential operator

$$\Delta f(\mathbf{x}) = \operatorname{div} \nabla f(\mathbf{x}) = \frac{\partial^2 f}{\partial x_1^2} + \frac{\partial^2 f}{\partial x_2^2} + \frac{\partial^2 f}{\partial x_3^2}.$$

This discussion will add some insight for readers who will go on to the chapter on the Dirichlet problem and Fourier series.

Suppose the vector field $\mathbf{F}$ is C^1 on an open set U containing the point $\mathbf{a}$ in $\mathbf{R}^3$. Let r be such that the ball $B_r(\mathbf{a}) \subset U$, and let (r_k) be a decreasing sequence of numbers less than r, such that $r_k \to 0$ as $k \to \infty$. Using the continuity of $\operatorname{div} \mathbf{F}$, let us write

$$M_k = \max |\operatorname{div} \mathbf{F}(\mathbf{x})| \qquad \text{and} \qquad m_k = \min |\operatorname{div} \mathbf{F}(\mathbf{x})|$$

for $\mathbf{x}$ in $B_{r_k}(\mathbf{a})$. By the divergence theorem applied to each of these balls about $\mathbf{a}$,

$$m_k\, \nu(B_{r_k}(\mathbf{a})) \le \int_{S_{r_k}(\mathbf{a})} \mathbf{F} \cdot \mathbf{n}\, d\sigma = \int_{B_{r_k}(\mathbf{a})} \operatorname{div} \mathbf{F}\, dV \le M_k\, \nu(B_{r_k}(\mathbf{a})).$$

Division by $\nu(B_{r_k}(\mathbf{a}))$ yields

$$m_k \le \frac{1}{\nu(B_{r_k}(\mathbf{a}))} \int_{S_{r_k}(\mathbf{a})} \mathbf{F} \cdot \mathbf{n}\, d\sigma = \frac{1}{\nu(B_{r_k}(\mathbf{a}))} \int_{B_{r_k}(\mathbf{a})} \operatorname{div} \mathbf{F}\, dV \le M_k.$$

As $k \to \infty$, we have $M_k \to \operatorname{div} F(\mathbf{a})$ and $m_k \to \operatorname{div} F(\mathbf{a})$ by continuity of the divergence. Hence, we conclude that

$$\operatorname{div} \mathbf{F}(\mathbf{a}) = \lim_{\rho \to 0} \frac{1}{\nu(B_\rho(\mathbf{a}))} \int_{S_\rho(\mathbf{a})} \mathbf{F} \cdot \mathbf{n}\, d\sigma.$$

This limit provides a coordinate-free interpretation of the divergence of $\mathbf{F}$ at $\mathbf{a}$ in terms of the flux of $\mathbf{F}$ across spheres centered at $\mathbf{a}$ as the spherical radius approaches zero.

We also see from this result that if $\mathbf{F}$ is a gradient vector field, $\mathbf{F} = \nabla f$, where f is a C^2 real valued function, then

$$\Delta f(\mathbf{a}) = \lim_{\rho \to 0} \frac{1}{\nu(B_\rho(\mathbf{a}))} \int_{S_\rho(\mathbf{a})} \nabla f \cdot \mathbf{n}\, d\sigma.$$

The limit gives a coordinate-free interpretation of the Laplacian of f at $\mathbf{a}$ in terms of the flux of ∇f across spheres centered at $\mathbf{a}$ as the spherical radius approaches zero.

Exercises.

Exercise 13.5.1. Suppose C and A are 3×2 matrices and B is a 2×2 matrix such that $C = AB$. Denoting the columns, left to right, of C and A by C_1, C_2 and A_1, A_2, show that

$$C_1 \times C_2 = (\det B)\, A_1 \times A_2.$$

Exercise 13.5.2. Let $h : \mathbf{R}^3 \to \mathbf{R}$ be a continuously differentiable function and $\mathbf{p}$ a point at which $\nabla h(\mathbf{p}) \neq \mathbf{0}$. Let $y = h(\mathbf{p})$. Show that there is an open ball B about $\mathbf{p}$ such that the set

$$S = \{\mathbf{x} \in B : h(\mathbf{x}) = y\} = h^{-1}(y)$$

is a surface in $\mathbf{R}^3$. *Hint*: Apply the implicit function theorem.

Exercise 13.5.3. Consider the upper hemisphere of radius $\rho = 2$ discussed in Example 13.5.2, along with the parametrization $\Psi : Q \to \mathbf{R}^3$ by spherical coordinates given there, where $Q = \{(\phi, \theta) : 0 \le \phi \le \pi/2,\ 0 \le \theta \le 2\pi\}$, and

$$\Psi(\phi, \theta) = \big(2 \sin\phi \cos\theta, 2 \sin\phi \sin\theta, 2\cos\phi\big).$$

Recall that $\Omega = \operatorname{Int} Q$. Find the surface area of the complete spherical surface of radius 2 by computing the upper hemispherical surface area as the multiple integral

$$\int_{\Omega}\!\!\int |\Psi_\phi(\phi, \theta) \times \Psi_\theta(\phi, \theta)|_2 \, d\phi \, d\theta$$

and then using symmetry to add in the area of the lower hemisphere. Verify the relation $V = \frac{\rho}{3} A$ between volume V and surface area A for this sphere of radius $\rho = 2$. Then verify the relation for the general sphere of radius ρ. *Hint*: Carry general ρ throughout the calculation.

13.6. Notes and References

Sagan [**55**] was the source for the Schoenberg space-filling curve in the first section of this chapter. The development of the integral in this and the preceding chapter drew mainly from Edwards [**10**] and Sagan [**54**], and to some extent from Lang [**42**]. The development of the change of variables formula drew mainly from Edwards [**10**], which also has a discussion of the classical notation and terminology associated with the change of variables formula that some readers may find useful. The material on the definition of surface integrals was influenced by Guillemin and Pollack [**22**] and Fitzpatrick [**12**].

The theory of the Riemann integral is adequate for many applications and purposes, including the study of finite-dimensional smooth manifolds and the integration of smooth functions and smooth differential forms defined on them. Readers who have successfully completed the section on surface integrals should have no trouble with the definition of a manifold, whether 2-dimensional or k-dimensional, modeled locally on Euclidean space of the same dimension. See Boothby [**6**], Lee [**44**], or Munkres [**48**] for introductions to finite-dimensional manifolds.

Chapter 14

Ordinary Differential Equations

The goal of this chapter is to present some fundamental facts about systems of ordinary differential equations. The chapter involves applications of the fundamental theorem of calculus, the completeness of the spaces $C[a,b]$ and $C_n[a,b]$, and the contraction mapping theorem. We begin with scalar differential equations and then extend the discussion to systems, presenting results on local existence and uniqueness for initial value problems, extension of solutions and behavior at time boundaries, and continuous dependence of solutions on initial conditions, parameters, and vector fields.

14.1. Scalar Differential Equations

Let I be an open interval and $x : I \to \mathbf{R}$ a differentiable function. We will use the notation $\dot{x}$ for the derivative function dx/dt. This is a standard notation in the study of differential equations.

Let $f : I \to \mathbf{R}$ be a continuous function. Then the differential equation $\dot{x}(t) = f(t)$, with initial condition at time t_0 given by $x(t_0) = x_0$, has a unique solution $x(t)$ for any $(t_0, x_0) \in I \times \mathbf{R}$. By the fundamental theorem of calculus (differentiation of integrals) we have

$$\frac{d}{dt} \int_{t_0}^{t} f(s)\, ds = f(t) \quad \text{for all } t \in I.$$

Thus the function $x(t)$ defined by the integral formula

$$x(t) = x_0 + \int_{t_0}^{t} f(s)\, ds, \qquad t \in I,$$

is a solution of the stated initial value problem. To see that the solution is uniquely determined by x_0, suppose that both $u(t)$ and $v(t)$ satisfy $\dot{x}(t) = f(t)$, $x(t_0) = x_0$;

that is, for all $t \in I$,

$$\dot{u}(t) = f(t), \quad u(t_0) = x_0 \qquad \text{and} \qquad \dot{v}(t) = f(t), \quad v(t_0) = x_0.$$

Thus, by the fundamental theorem of calculus (integration of derivatives), we have

$$u(t) - x_0 = \int_{t_0}^{t} f(s)\, ds$$

and

$$v(t) - x_0 = \int_{t_0}^{t} f(s)\, ds$$

for all $t \in I$ and therefore $u(t) = v(t)$ on I.

The main goal of this section is to extend this statement about existence and uniqueness of a solution to cover the more general **initial value problem (IVP)**

$$\dot{x} = f(t, x), \qquad x(t_0) = x_0, \tag{14.1}$$

where $f : \mathcal{D} \subseteq \mathbf{R}^2 \to \mathbf{R}$, $\mathcal{D}$ is an open set, and $(t_0, x_0) \in \mathcal{D}$. The function f now depends on both x and t, and appropriate assumptions on f will be stated below. A **solution** of the initial value problem (14.1) is a differentiable function $x : I \to \mathbf{R}$, defined on an open interval I containing t_0, such that $x(t_0) = x_0$, $(t, x(t)) \in \mathcal{D}$ for all $t \in I$, and $\dot{x}(t) = f(t, x(t))$.

The next result says that if f is continuous on $\mathcal{D}$, then (14.1) is equivalent to an integral equation.

Theorem 14.1.1. *Suppose that $f : \mathcal{D} \subseteq \mathbf{R}^2 \to \mathbf{R}$ is continuous, and let I be an open interval. A continuous function $x : I \to \mathbf{R}$ such that $(s, x(s)) \in \mathcal{D}$ for all $s \in I$ satisfies the integral equation*

$$x(t) = x_0 + \int_{t_0}^{t} f(s, x(s))\, ds, \qquad t \in I, \tag{14.2}$$

*if and only if x is a solution on I of the initial value problem (*14.1*).*

Proof. Suppose $x : I \to \mathbf{R}$ is continuous on I and satisfies the integral equation (14.2). Then $x(t_0) = x_0$. Since f is continuous on $\mathcal{D}$, the integrand $f(s, x(s))$ is continuous on I, and by the fundamental theorem, $x(t)$ is differentiable for each $t \in I$ and

$$\dot{x}(t) = f(t, x(t)), \quad t \in I. \tag{14.3}$$

Therefore $x : I \to \mathbf{R}$ is a solution of (14.1) on I. Conversely, if $x : I \to \mathbf{R}$ satisfies the IVP (14.1), then x is differentiable on I and hence continuous on I, so $f(t, x(t))$ is continuous on I. The evaluation half of the fundamental theorem then implies that (14.2) holds. □

The advantage of the integral equation is that we can view the initial value problem as a fixed point problem. Indeed, the proof of the next theorem is an application of the contraction mapping theorem.

Theorem 14.1.2. *Let $f : \mathcal{D} \subseteq \mathbf{R}^2 \to \mathbf{R}$ be continuous on $\mathcal{D}$, and suppose that for some $a > 0$, $b > 0$, the function f satisfies the Lipschitz condition*

$$|f(t, x_1) - f(t, x_2)| \leq K|x_1 - x_2|$$

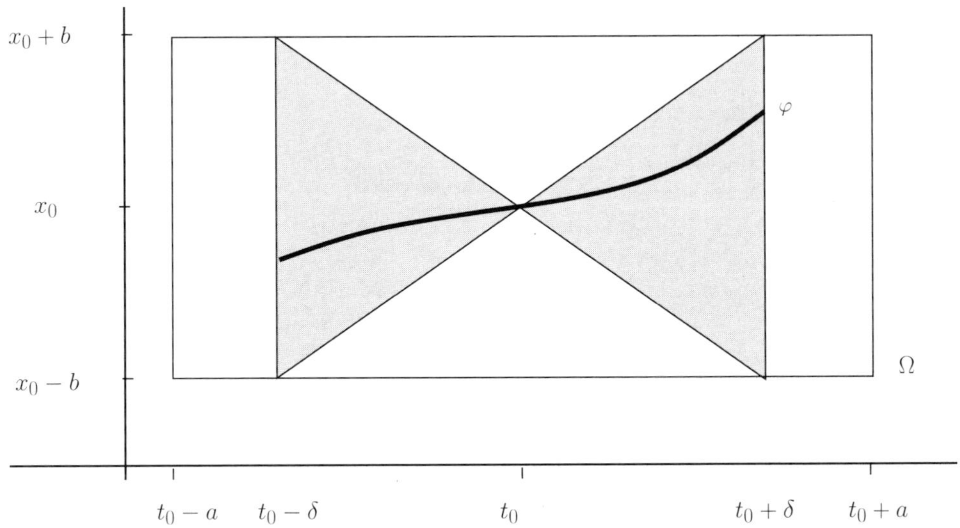

Figure 14.1. A function ϕ satisfying $\phi(t_0) = x_0$ as a candidate for local solution of the initial value problem $\dot{x} = f(t, x)$, $x(t_0) = x_0$. The set Ω is restricted by choice of δ.

for all pairs (t, x_1), (t, x_2) *in the set*

$$\Omega = \{(t, x) : |t - t_0| \le a, |x - x_0| \le b\} \subset \mathcal{D}.$$

Let M *be such that* $|f(t, x)| \le M$ *for all* (t, x) *in* Ω. *If* δ *satisfies*

$$0 < \delta < \min\{a,\, b/M,\, 1/K\},$$

there is a unique solution $x(t)$ *of the initial value problem* (14.1) *defined on the interval* $[t_0 - \delta, t_0 + \delta]$.

Proof. Choose any $a > 0$ and $b > 0$ such that the set Ω is contained in $\mathcal{D}$. Since f is continuous on $\mathcal{D}$ and Ω is compact, there is a bound M such that $|f(t, x)| \le M$ for all $(t, x) \in \Omega$. Let δ satisfy $0 < \delta < a$, and consider the space of continuous functions

$$B = \big\{\phi : \phi \in C[t_0 - \delta, t_0 + \delta] \text{ and } |\phi(t) - x_0| \le b \text{ for all } t \in [t_0 - \delta, t_0 + \delta]\big\}.$$

We think of B as our space of candidates for a solution. (See Figure 14.1.)

Note that B contains all continuous functions on $[t_0 - \delta, t_0 + \delta]$ that satisfy $\phi(t_0) = x_0$ and $|\phi(t) - x_0| \le b$ for all $t \in [t_0 - \delta, t_0 + \delta]$. Also, if $\phi \in B$, then $(t, \phi(t)) \in \Omega$ for all $t \in [t_0 - \delta, t_0 + \delta]$. Thus, if $\phi \in B$, then the Lipschitz estimate and bound on f will certainly apply. The space B depends on the choice of $\delta > 0$, but for any $\delta > 0$, B is a closed subset of the complete space $C[t_0 - \delta, t_0 + \delta]$, so B is a complete metric space in the metric determined by the sup norm, by Theorem 9.1.11. (See also Exercise 9.1.5.) Define a mapping T on $C[t_0 - \delta, t_0 + \delta]$ by the right-hand side of the integral equation (14.2), that is,

$$T[\phi](t) = x_0 + \int_{t_0}^{t} f(s, \phi(s))\, ds.$$

Our goal is to choose δ such that

(1) $T : B \to B$;

(2) T is a contraction mapping on B.

Suppose $\phi \in B$. We want to be sure that the image $T[\phi]$ is in B. Since $\phi \in B$, ϕ is continuous on $[t_0 - \delta, t_0 + \delta]$ and $(t, \phi(t)) \in \Omega$ for all $t \in [t_0 - \delta, t_0 + \delta]$. Since f is continuous on Ω, $f(t, \phi(t))$ is defined and continuous for $t \in [t_0 - \delta, t_0 + \delta]$. It follows that $T[\phi]$ is defined for $t \in [t_0 - \delta, t_0 + \delta]$, and $T[\phi]$ is a continuous function on $[t_0 - \delta, t_0 + \delta]$. In fact, by the fundamental theorem (differentiation of integrals), $T[\phi]$ is differentiable on $(t_0 - \delta, t_0 + \delta)$ and has one-sided derivatives at $t_0 \pm \delta$. We need the image $T[\phi] \in B$, so we must show that $|T[\phi](t) - x_0| \le b$. Recalling that we may have $t < t_0$ as well as $t > t_0$, we can estimate as follows:

$$
\begin{aligned}
|T[\phi](t) - x_0| &= \left| \int_{t_0}^{t} f(s, \phi(s))\, ds \right| \\
&\le \left| \int_{t_0}^{t} |f(s, \phi(s))|\, ds \right| \\
&\le M|t - t_0| \;\le\; M\delta.
\end{aligned}
$$

If we choose $\delta < b/M$, then $T[\phi]$ is in B for each ϕ in B, and thus $T : B \to B$. Assume we have now chosen $0 < \delta < \min\{a, b/M\}$.

We now want to show that T is a contraction on B. We may do so by maintaining the previous restrictions on δ or by imposing a tighter restriction (and thus accepting a shorter interval of definition for our local solution). What is required to make T a contraction on B? We estimate as follows:

$$
\begin{aligned}
|(T[\phi] - T[\psi])(t)| &\le \left| \int_{t_0}^{t} f(s, \phi(s))\, ds - \int_{t_0}^{t} f(s, \psi(s))\, ds \right| \\
&= \left| \int_{t_0}^{t} [f(s, \phi(s)) - f(s, \psi(s))]\, ds \right| \\
&\le \left| \int_{t_0}^{t} |f(s, \phi(s)) - f(s, \psi(s))|\, ds \right| \\
&\le \left| K \int_{t_0}^{t} |\phi(s) - \psi(s)|\, ds \right| \\
&\le K|t - t_0|\, \|\phi - \psi\| \;\le\; K\delta\, \|\phi - \psi\|,
\end{aligned}
$$

since $|\phi(s) - \psi(s)| \le \|\phi - \psi\|$, the sup norm of $\phi - \psi$. It follows that if $0 < \delta < \min\{a, b/M\}$, then $T : B \to B$ and

$$
\| T[\phi] - T[\psi] \| = \max_{|t - t_0| \le \delta} |(T[\phi] - T[\psi])(t)| \le K\delta\, \|\phi - \psi\|.
$$

Therefore by choosing $0 < \delta < \min\{a, b/M, 1/K\}$, we have $T : B \to B$ and T is a contraction mapping on B with contraction constant $K\delta < 1$. (See Figure 14.1.1.) By the contraction mapping theorem, T has a unique fixed point in B, and the fixed point ϕ satisfies the integral equation (14.2) on $[t_0 - \delta, t_0 + \delta]$. Hence, ϕ must be a solution of the initial value problem (14.1) on $[t_0 - \delta, t_0 + \delta]$, by Theorem 14.1.1. This proves the *existence* of a solution of (14.1) on $[t_0 - \delta, t_0 + \delta]$.

On the other hand, if ψ is any solution of the initial value problem (14.1) defined on the interval $[t_0 - \delta, t_0 + \delta]$, then ψ is a solution of the integral equation (14.2) on $[t_0 - \delta, t_0 + \delta]$, and by our choice of $\delta > 0$, the image point $\psi(t)$ lies in Ω for all t. Thus, the function ψ must be an element of the space B, which shows that ψ must be the fixed point of $T : B \to B$. This proves the *uniqueness* statement. □

We remark here that if $f : \mathcal{D} \subseteq \mathbf{R}^2 \to \mathbf{R}$ in Theorem 14.1.2 is continuously differentiable in (t, x) throughout $\mathcal{D}$, then the mean value theorem implies that f satisfies a local Lipschitz condition with respect to x in a neighborhood of each point (t_0, x_0) in $\mathcal{D}$. We examine this in more detail later for systems.

Consider an autonomous (time-independent) equation, $\dot{x} = f(x)$, $x(t_0) = x_0$, where f has no explicit dependence on time t. We observe that both Theorem 14.1.1 and Theorem 14.1.2 apply to such equations with no changes in statements or proofs except to replace "$f(t,x)$" throughout by "$f(x)$" and domain "$\mathcal{D}$" in $\mathbf{R}^{n+1}$ by an open domain "D" in $\mathbf{R}^n$ for $f : D \subseteq \mathbf{R}^n \to \mathbf{R}^n$. The definition of the set Ω in Theorem 14.1.2 remains the same. A careful review of the statements and proofs will verify this. However, for an autonomous equation, as for the general nonautonomous case, there are generally restrictions on the time domain for individual solutions. As an example, consider the equation $\dot{x} = x^2$, $x(0) = x_0$. The solution with $x(0) = 1$ is $x(t) = 1/(1-t)$, which is defined for $-\infty < t < 1$. The solution with $x(0) = -1$ is $-1/(1+t)$, defined for $-1 < t < \infty$. The solution with $x(0) = 0$ is $x(t) \equiv 0$ on $-\infty < t < \infty$. The general solution of this IVP can be written $\phi(t, x_0) = x_0/(1 - t\,x_0)$, where $x(0) = x_0$, obtained by direct integration.

Exercises.

Exercise 14.1.1. Let $t_0 = 0$ and $x_0 = 0$. The initial value problem

$$\dot{x} = f(t,x) = 1 + x^2, \qquad x(0) = 0,$$

has solution $x(t) = \tan x$ defined on $(-\pi/2, \pi/2)$. Using estimates for $|f(t,x)|$ and $|f(t,x_1) - f(t,x_2)|$ determined by $\Omega = \{(t,x) : |t| \leq \pi/4, |x| \leq 2\}$, what is the maximum local interval of existence described by the statement of Theorem 14.1.2? *Hint*: Find the bound M and Lipschitz constant K determined by Ω.

Exercise 14.1.2. For the initial value problem $\dot{x} = 1/(1+t^2)$, $x(0) = 0$, show that, in defining Ω, b can be taken as large, and K as small positive, as we wish, and a can be taken less than $\min\{b/M, 1/K\}$. What can you conclude? What is the unique solution of this IVP?

14.2. Systems of Ordinary Differential Equations

Our purpose is to study the existence and uniqueness of solutions to initial value problems for systems of ordinary differential equations in $\mathbf{R}^n$. We begin with some standard notation for representing the problem and its solution. An **initial value problem** has the form

$$\dot{\mathbf{x}} = \mathbf{f}(t, \mathbf{x}), \qquad \mathbf{x}(t_0) = \mathbf{x}_0, \tag{14.4}$$

where $\mathbf{x} \in \mathbf{R}^n$, $\dot{\mathbf{x}} = d\mathbf{x}/dt$, $\mathbf{f}(t, \mathbf{x})$ is an $\mathbf{R}^n$-valued function of $(t, \mathbf{x})$ in some open set $\mathcal{D} \subseteq \mathbf{R}^{n+1}$, and $(t_0, \mathbf{x}_0) \in \mathcal{D}$. The function $\mathbf{f}$ is called the **vector field** of (14.4).

The most fundamental case to consider is the case of an **autonomous system**, defined by a time-independent vector field and described by equations of the form

$$\dot{\mathbf{x}} = \mathbf{f}(\mathbf{x}), \qquad \mathbf{x}(t_0) = \mathbf{x}_0, \tag{14.5}$$

where $\mathbf{f} : D \subseteq \mathbf{R}^n \to \mathbf{R}^n$, D is an open set, and $\mathbf{x}_0$ a given point, in $\mathbf{R}^n$. Thus (14.5) represents the system of n ordinary differential equations

$$\begin{aligned} \dot{x}_1 &= f_1(x_1, \ldots, x_n), \\ \dot{x}_2 &= f_2(x_1, \ldots, x_n) \\ \vdots &= \vdots \\ \dot{x}_n &= f_n(x_1, \ldots, x_n) \end{aligned}$$

for the unknown functions $x_i = x_i(t)$, $1 \leq i \leq n$, where f_i is the i-th component function of $\mathbf{f}$, along with the initial condition

$$\mathbf{x}(t_0) = (x_1(t_0), x_2(t_0), \ldots, x_n(t_0)) = \mathbf{x}_0.$$

The nonautonomous initial value problem (14.4) is described by similar equations except that the functions f_i may depend explicitly on t, so that the right-hand sides are functions of $(t, \mathbf{x}) \in \mathcal{D} \subseteq \mathbf{R}^{n+1}$.

Remark. *Observe that we denote the domain of an autonomous system by $D \subseteq \mathbf{R}^n$ and the domain of a nonautonomous system by a script letter, $\mathcal{D} \subseteq \mathbf{R}^{n+1}$.*

14.2.1. Definition of Solution and the Integral Equation.

We begin with the definition of solution for an initial value problem of the general form (14.4), which includes the autonomous case (14.5).

Definition 14.2.1. *Let $\mathcal{D}$ be an open set in $\mathbf{R}^{n+1}$, let $\mathbf{f} : \mathcal{D} \subseteq \mathbf{R}^{n+1} \to \mathbf{R}^n$ be a C^1 mapping, and let $(t_0, \mathbf{x}_0) \in \mathcal{D}$. A* **solution** *of the initial value problem (14.4) is a differentiable function $\mathbf{x} : I \to \mathbf{R}$, defined on an open interval I containing t_0, such that $\mathbf{x}(t_0) = \mathbf{x}_0$ and for all $t \in I$, $(t, \mathbf{x}(t)) \in \mathcal{D}$ and $\dot{\mathbf{x}}(t) = f(t, \mathbf{x}(t))$. A solution of the autonomous problem (14.5) is a differentiable function $\mathbf{x} : I \to \mathbf{R}$, defined on an open interval I containing t_0, such that $\mathbf{x}(t_0) = \mathbf{x}_0$ and for all $t \in I$, $\mathbf{x}(t) \in D \subseteq \mathbf{R}^n$ and $\dot{\mathbf{x}}(t) = f(\mathbf{x}(t))$.*

If the vector field $\mathbf{f}$ is continuous on $\mathcal{D}$ for (14.4) (or D, for (14.5)), then the initial value problem is equivalent to an integral equation. We state and prove this for the nonautonomous case, and note, as for the scalar case, that only minor notational changes are needed for the autonomous case.

Theorem 14.2.2. *Suppose that $\mathbf{f} : \mathcal{D} \subseteq \mathbf{R}^{n+1} \to \mathbf{R}^n$ is continuous, and let I be an open interval containing t_0. A continuous function $\mathbf{x} : I \to \mathbf{R}^n$, such that $(t, \mathbf{x}(t)) \in \mathcal{D}$ for all $t \in I$, satisfies the integral equation*

$$\mathbf{x}(t) = \mathbf{x}_0 + \int_{t_0}^{t} \mathbf{f}(s, \mathbf{x}(s))\, ds, \quad t \in I,$$

if and only if $\mathbf{x}$ is a solution on I of the initial value problem (14.4).

Proof. Suppose $\mathbf{x} : I \to \mathbf{R}$ is continuous on I, $(t, \mathbf{x}(t)) \in \mathcal{D}$ for all $t \in I$, and $\mathbf{x}$ satisfies the integral equation

$$\mathbf{x}(t) = \mathbf{x}_0 + \int_{t_0}^{t} \mathbf{f}(s, \mathbf{x}(s))\, ds, \quad t \in I. \tag{14.6}$$

Since $\mathbf{f}$ is continuous and the composition of continuous functions is continuous, the integrand $\mathbf{f}(s, \mathbf{x}(s))$ is continuous on I, and $\mathbf{x}(t_0) = x_0$ by the integral equation with $t = t_0$. By the fundamental theorem of calculus, the right side of (14.6) is differentiable for $t \in I$, so $\mathbf{x}(t)$ is differentiable for $t \in I$ and

$$\dot{\mathbf{x}}(t) = \mathbf{f}(t, \mathbf{x}(t)), \quad t \in I. \tag{14.7}$$

Therefore $\mathbf{x}$ is a solution of the initial value problem on I.

Conversely, suppose that $\mathbf{x} : I \to \mathbf{R}^n$ satisfies (14.7) and $\mathbf{x}(t_0) = \mathbf{x}_0$. Then $\mathbf{x}$ is differentiable on I and hence continuous on I, so $\mathbf{f}(t, \mathbf{x}(t))$ is continuous on I. The evaluation half of the fundamental theorem of calculus then implies that

$$\mathbf{x}(t) = \mathbf{x}_0 + \int_{t_0}^{t} \dot{\mathbf{x}}(s)\, ds = \mathbf{x}_0 + \int_{t_0}^{t} \mathbf{f}(s, \mathbf{x}(s))\, ds \quad t \in I,$$

and the proof is completed. □

Exercise.

Exercise 14.2.1. Consider the autonomous system (14.5).

1. Show that if $\phi(t)$ is a solution on the interval $(-\delta, \delta)$ of $\dot{\mathbf{x}} = \mathbf{f}(\mathbf{x})$ with $\phi(0) = \mathbf{x}_0$, then $\psi(t) := \phi(t - \tau)$ is a solution on the interval $(\tau - \delta, \tau + \delta)$ of $\dot{\mathbf{x}} = \mathbf{f}(\mathbf{x})$ with $\psi(\tau) = \mathbf{x}_0$.
2. Show that if the IVP for $\dot{\mathbf{x}} = \mathbf{f}(\mathbf{x})$ with initial condition $\mathbf{x}(0) = \mathbf{x}_0$ has a unique solution in some interval about 0, then the initial value problem with initial condition $\mathbf{x}(\tau) = \mathbf{x}_0$ has a unique solution on some interval about τ.

14.2.2. Completeness of $C_n[a, b]$. For systems of ordinary differential equations we must consider continuous functions taking values in $\mathbf{R}^n$. We need to know that the relevant space of continuous functions is complete with respect to an appropriate norm. For the local existence and uniqueness theorem for systems, it is convenient to work with the space of continuous functions defined on a closed interval $[a, b]$ and taking values in $\mathbf{R}^n$. We denote this space by $C_n[a, b]$. We also find it convenient to work with the norm $|\cdot|_1$ on $\mathbf{R}^n$ defined by

$$|\mathbf{v}|_1 = \sum_{i=1}^{n} |v_i|$$

for $\mathbf{v} = (v_1, \ldots, v_n) \in \mathbf{R}^n$. We may norm the vector space $C_n[a, b]$ by defining

$$\|\mathbf{u}\| = \max_{t \in [a,b]} |\mathbf{u}(t)|_1, \quad \mathbf{u} \in C_n[a, b]. \tag{14.8}$$

We may refer to this norm as the **max norm** on $C_n[a,b]$. The proof that (14.8) defines a norm on $C_n[a,b]$ is left to Exercise 14.2.2, but we observe that the maximum in (14.8) exists since the composition $t \mapsto |\mathbf{u}(t)|_1$ is continuous on $[a,b]$ and therefore achieves its maximum at some point $t_0 \in [a,b]$.[1]

In accordance with the norm (14.8) on $C_n[a,b]$, we see that a sequence $(\mathbf{x}_k)$ in $C_n[a,b]$ is a Cauchy sequence if for every $\epsilon > 0$ there is an $N = N(\epsilon)$ such that if $k, j \geq N(\epsilon)$, then

$$|\mathbf{x}_k(t) - \mathbf{x}_j(t)|_1 < \epsilon \quad \text{for all } t \in [a,b].$$[2]

Convergence of a sequence $(\mathbf{x}_k)$ in the norm on $C_n[a,b]$ means exactly uniform convergence of $(\mathbf{x}_k)$ on $[a,b]$; see Exercises 14.2.3 and 14.2.4 for a reminder of the definition of uniform convergence.

Theorem 14.2.3. *The space $C_n[a,b]$ is complete in the norm given by (14.8).*

Proof. Let $(\mathbf{u}_k)$ be a Cauchy sequence in $C_n[a,b]$. We must show that there is an element $\mathbf{u}$ in $C_n[a,b]$ such that

$$\lim_{k\to\infty} \|\mathbf{u}_k - \mathbf{u}\| = 0.$$

Since $\mathbf{u}_k$ is a Cauchy sequence, for every $\epsilon > 0$ there is an $N(\epsilon)$ such that if $k, m \geq N(\epsilon)$, then for all $t \in [a,b]$,

$$|\mathbf{u}_k(t) - \mathbf{u}_m(t)|_1 \leq \|\mathbf{u}_k - \mathbf{u}_m\| < \epsilon.$$

Thus for each $t \in [a,b]$, the sequence $\mathbf{u}_k(t)$ is a Cauchy sequence in $\mathbf{R}^n$. Since $\mathbf{R}^n$ is complete, $\lim_{k\to\infty} \mathbf{u}_k(t)$ exists for each $t \in [a,b]$, and we define

$$\mathbf{u}(t) = \lim_{k\to\infty} \mathbf{u}_k(t), \quad t \in [a,b].$$

Now, using only the knowledge that we have a pointwise limit function $\mathbf{u}$, we must show two things: (i) $\lim_{k\to\infty} \|\mathbf{u}_k - \mathbf{u}\| = 0$, and (ii) $\mathbf{u} \in C_n[a,b]$.

(i) Let $\epsilon > 0$. There is an N such that for all $k, m \geq N$ and all $t \in [a,b]$,

$$|\mathbf{u}_k(t) - \mathbf{u}_m(t)|_1 < \epsilon.$$

By the continuity of the norm on $\mathbf{R}^n$, if we let $m \to \infty$, then for fixed $k \geq N$ and all $t \in [a,b]$,

$$\lim_{m\to\infty} |\mathbf{u}_k(t) - \mathbf{u}_m(t)|_1 = |\mathbf{u}_k(t) - \lim_{m\to\infty} \mathbf{u}_m(t)|_1 = |\mathbf{u}_k(t) - \mathbf{u}(t)|_1 \leq \epsilon.$$

Thus for every $\epsilon > 0$ there is an N such that if $k \geq N$, then

$$\|\mathbf{u}_k - \mathbf{u}\| \leq \epsilon.$$

Hence $\lim_{k\to\infty} \|\mathbf{u}_k - \mathbf{u}\| = 0$.

(ii) Now we show that the limit function $\mathbf{u}$ is continuous, hence in $C_n[a,b]$. Let $t_0 \in I$. For any fixed k, the triangle inequality implies that

$$|\mathbf{u}(t) - \mathbf{u}(t_0)|_1 \leq |\mathbf{u}(t) - \mathbf{u}_k(t)|_1 + |\mathbf{u}_k(t) - \mathbf{u}_k(t_0)|_1 + |\mathbf{u}_k(t_0) - \mathbf{u}(t_0)|_1.$$

By the convergence of $\mathbf{u}_k$ to $\mathbf{u}$ in norm, for any $\epsilon > 0$ there is an N such that if $k \geq N$, then the first and third terms on the right-hand side are each less than $\epsilon/3$.

---

[1] Note that elsewhere we have been writing, and will continue to write, $C[a,b]$ instead of $C_1[a,b]$ for the *real-valued* continuous functions on $[a,b]$.

[2] Such a sequence is sometimes called a *uniformly Cauchy* sequence of functions, the uniformity being with respect to t in the interval $[a,b]$.

For fixed $k \geq N$, continuity of $\mathbf{u}_k$ at t_0 implies that there is a $\delta > 0$ such that if $|t - t_0| < \delta$, then

$$|\mathbf{u}_k(t) - \mathbf{u}_k(t_0)|_1 < \epsilon/3.$$

Thus, for any $\epsilon > 0$ there is a $\delta > 0$ such that if $|t - t_0| < \delta$, then

$$|\mathbf{u}(t) - \mathbf{u}(t_0)|_1 < \epsilon/3 + \epsilon/3 + \epsilon/3 = \epsilon.$$

Therefore $\mathbf{u}$ is continuous at t_0. Since t_0 was an arbitrary point of $[a, b]$, $\mathbf{u}$ is continuous on $[a, b]$. This completes the proof. □

Exercises.

Exercise 14.2.2. Verify that $C_n[a, b]$ is a normed vector space with the max norm defined by (14.8).

Exercise 14.2.3. A sequence of functions $\mathbf{x}_k : [a, b] \to \mathbf{R}^n$ **converges uniformly** on $[a, b]$ to the function $\mathbf{x} : [a, b] \to \mathbf{R}^n$ if for every $\epsilon > 0$ there is an $N = N(\epsilon)$ such that if $k \geq N(\epsilon)$, then

$$|\mathbf{x}_k(t) - \mathbf{x}(t)|_1 < \epsilon \quad \text{for all } t \in [a, b].$$

Convergence of a sequence in the normed space $C_n[a, b]$ is equivalent to uniform convergence of the sequence to its pointwise limit. Show this in two steps:

1. Suppose $\mathbf{x}_k \in C_n[a, b]$ for each k and $(\mathbf{x}_k)$ converges to $\mathbf{x} \in C_n[a, b]$ in the max norm. Show that the sequence of functions $\mathbf{x}_k$ converges uniformly on $[a, b]$ to $\mathbf{x}$.
2. Show that if $(\mathbf{x}_k)$, $\mathbf{x}_k \in C_n[a, b]$, converges uniformly on $[a, b]$ to $\mathbf{x}$, then $\mathbf{x}$ is continuous on $[a, b]$, and $\mathbf{x}_k$ converges to $\mathbf{x}$ in the max norm on $C_n[a, b]$. *Hint*: Examine the proof of Theorem 14.2.3.

Exercise 14.2.4. Refer to the definition of uniform convergence in Exercise 14.2.3. Let $\mathbf{u}_k = (u_{k1}, \ldots, u_{kn}) : [a, b] \to \mathbf{R}^n$ and $\mathbf{u} = (u_1, \ldots, u_n) : [a, b] \to \mathbf{R}^n$. Prove the following:

1. If $\mathbf{u}_k$ converges uniformly on $[a, b]$ to $\mathbf{u}$, then each component sequence u_{ki}, $1 \leq i \leq n$, converges uniformly on $[a, b]$ to the corresponding component $u_i : I \to \mathbf{R}$ of $\mathbf{u}$.
2. If each component sequence u_{ki}, $1 \leq i \leq n$, converges uniformly on $[a, b]$ to a function u_i, then the sequence $\mathbf{u}_k = (u_{k1}, \ldots, u_{kn})$ converges uniformly on $[a, b]$ to $\mathbf{u} = (u_1, \ldots, u_n)$.

14.2.3. The Local Lipchitz Condition. In order to guarantee that initial value problems have a unique solution, we must impose some smoothness condition on the vector field. Continuity of the vector field at (t_0, x_0) (or at x_0 for an autonomous field) does not guarantee a unique solution for the initial value problem. For example, the initial value problem $\dot{x} = x^{1/3}$, $x(0) = 0$, has more than one solution. One solution is $\phi(t) \equiv 0$, by inspection. But another solution is defined by $\psi(t) = 0$ for $t \leq 0$ and $\psi(t) = (2/3)^{3/2} t^{3/2}$ for $t \geq 0$. This function is differentiable, including at the origin, satisfies $\psi(0) = 0$, and satisfies the differential equation.

We impose the following stronger form of continuity on the vector field.

Definition 14.2.4. *Let U be an open set in $\mathbf{R}^n$ and let $\mathbf{f} : U \to \mathbf{R}^n$. The mapping* $\mathbf{f}$ *is said to be* **locally Lipschitz on** U *if for every $\mathbf{a} \in U$ there is an $r > 0$, and a number L such that*

$$|\mathbf{f}(\mathbf{x}_1) - \mathbf{f}(\mathbf{x}_2)|_1 \le L|\mathbf{x}_1 - \mathbf{x}_2|_1 \quad \text{for all } \mathbf{x}_1, \mathbf{x}_2 \in B_r(\mathbf{a}).$$

The number L, which depends on $\mathbf{a}$ *and r, is called a* **local Lipschitz constant**, *and we say that* $\mathbf{f}$ *satisfies a* **local Lipschitz condition** *on $B_r(\mathbf{a})$. (Thus* $\mathbf{f}$ *is locally Lipschitz on U if and only if* $\mathbf{f}$ *satisfies a local Lipschitz condition on a neighborhood of each point* $\mathbf{a}$ *in U.)*

The condition in Definition 14.2.4 is also known as **local Lipschitz continuity** of $\mathbf{f}$ on U. It should be clear that if $\mathbf{f}$ satisfies a local Lipschitz condition on $B_r(\mathbf{a})$, then $\mathbf{f}$ is continuous at $\mathbf{a}$. The converse is not true, as Exercise 14.2.5 illustrates with the example $f(x) = x^{1/3}$.

By the equivalence of norms on $\mathbf{R}^n$, a mapping that is locally Lipschitz on U with respect to one norm, say the norm $|\cdot|_1$ as in Definition 14.2.4, is also locally Lipschitz on U with respect to any other norm, with, generally speaking, a different local Lipschitz constant for each norm.

Our goal is to show that continuous differentiability of $\mathbf{f}$ implies a local Lipschitz condition in a neighborhood of each point in the domain.

Theorem 14.2.5. *Let U be an open set in $\mathbf{R}^n$ and let $\mathbf{x} \in U$. Suppose $\mathbf{f} : U \to \mathbf{R}^n$ is C^1, and let $\mathbf{h} \in \mathbf{R}^n$. Suppose the line segment consisting of the points $\mathbf{x} + t\mathbf{h}$, $0 \le t \le 1$, is contained in U. Then*

$$\mathbf{f}(\mathbf{x}+\mathbf{h}) - \mathbf{f}(\mathbf{x}) = \int_0^1 D\mathbf{f}(\mathbf{x}+t\mathbf{h})\mathbf{h}\,dt = \int_0^1 D\mathbf{f}(\mathbf{x}+t\mathbf{h})\,dt \cdot \mathbf{h}.$$

Proof. Consider the function of t defined by $\mathbf{g}(t) = \mathbf{f}(\mathbf{x}+t\mathbf{h})$. By the chain rule, $D\mathbf{g}(t) = D\mathbf{f}(\mathbf{x}+t\mathbf{h})\mathbf{h}$ for all t. By the Fundamental Theorem of Calculus, we have

$$\mathbf{g}(1) - \mathbf{g}(0) = \int_0^1 D\mathbf{g}(t)\,dt.$$

Since $\mathbf{g}(1) = \mathbf{f}(\mathbf{x}+\mathbf{h})$ and $\mathbf{g}(0) = \mathbf{f}(\mathbf{x})$, and since $\mathbf{h}$ can be taken outside the integral, the result follows. □

Now assume that $\mathbf{f} : U \subseteq \mathbf{R}^n \to \mathbf{R}^n$ is C^1. Let $\mathbf{x} \in U$ and take any closed ball B about $\mathbf{x}$ which is contained within U. Each of the first order partial derivatives $\partial f_i/\partial x_j(\mathbf{x})$ of each component f_i of $\mathbf{f}$ is a continuous real valued function on the compact set B, as is $|\partial f_i/\partial x_j(\mathbf{x})|$. Thus there exist numbers $m_{ij} > 0$ such that

$$\left|\frac{\partial f_i}{\partial x_j}(\mathbf{x})\right| \le m_{ij} \quad \text{for all } \mathbf{x} \in B.$$

We then have

$$\|D\mathbf{f}(\mathbf{x})\|_{\text{as}} \le \sum_{i=1}^n \sum_{j=1}^n m_{ij} =: L \quad \text{for all } \mathbf{x} \in B,$$

using the absolute sum norm on the left-hand side. By Theorem 14.2.5, if $\mathbf{x}_1$ and $\mathbf{x}_2$ are any two points in B, then, writing $\mathbf{h} = \mathbf{x}_2 - \mathbf{x}_1$, we have

$$|\mathbf{f}(\mathbf{x}_2) - \mathbf{f}(\mathbf{x}_1)|_1 = \left| \int_0^1 D\mathbf{f}(\mathbf{x}_1 + t\mathbf{h})\mathbf{h}\, dt \right|_1 \leq \int_0^1 |D\mathbf{f}(\mathbf{x}_1 + t\mathbf{h})\mathbf{h}|_1\, dt.$$

We have used the fact that for any vector function $\mathbf{v}(t)$,

$$\left| \int_0^1 \mathbf{v}(t)\, dt \right|_1 \leq \int_0^1 |\mathbf{v}(t)|_1\, dt.$$

Since the norm $|\cdot|_1$ is compatible with the absolute sum matrix norm,

$$|D\mathbf{f}(\mathbf{x}_1 + t\mathbf{h})\mathbf{h}|_1 \leq \|D\mathbf{f}(\mathbf{x}_1 + t\mathbf{h})\|_{\text{as}}\, |\mathbf{h}|_1 \leq L\, |\mathbf{h}|_1.$$

It follows that

$$|\mathbf{f}(\mathbf{x}_2) - \mathbf{f}(\mathbf{x}_1)|_1 \leq L|\mathbf{x}_2 - \mathbf{x}_1|_1$$

for all $\mathbf{x}_1$, $\mathbf{x}_2$ in B. The Lipschitz constant L depends on the neighborhood B. We record the result in the next theorem.

Theorem 14.2.6. *Let $U \subseteq \mathbf{R}^n$ be an open set. If $\mathbf{f}$ is C^1 on U, then $\mathbf{f}$ is locally Lipschitz on U.*

We remark that this theorem can also be established by invoking the mean value theorem on a neighborhood $B_r(\mathbf{a})$.

Continuous differentiability of a vector field $\mathbf{f}$ implies a local Lipschitz condition in a neighborhood of each point of the domain, but not necessarily a global Lipschitz constant and global Lipschitz condition over the entire domain. As a simple example, $f(x) = 1/x$ is locally Lipschitz on $(0, 1)$ since it is C^1 on this interval, but there is no global Lipschitz constant L such that $|f(x) - f(y)| \leq L|x - y|$ for all $x, y \in (0, 1)$.

The vector field in the next example is globally Lipschitz on the entire plane.

Example 14.2.7. The equations of motion for a pendulum with friction are

$$\begin{aligned} \dot{x}_1 &= x_2, \\ \dot{x}_2 &= -\sin x_1 - x_2. \end{aligned}$$

We may consider this system to be defined on the entire plane. The Jacobian matrix of the vector field $\mathbf{f}$ at $\mathbf{x}$ is

$$J_{\mathbf{f}}(\mathbf{x}) = \begin{bmatrix} 0 & 1 \\ -\cos x_1 & -1 \end{bmatrix}.$$

Using the the sum of absolute entries matrix norm, we have a global bound for the norm of the Jacobian, given by

$$\|J_{\mathbf{f}}(\mathbf{x})\|_{\text{as}} \leq 3 \quad \text{for all } \mathbf{x} \in \mathbf{R}^2,$$

which gives a global Lipschitz constant for $\mathbf{f}$. △

Exercises.

Exercise 14.2.5. 1. Show that the function $f(x) = x^{1/3}$ does not satisfy a Lipschitz condition in any interval about the origin.
2. Show that the function $f(x) = 1/x$ does not satisfy a global Lipschitz condition on the interval $(0, 1)$.

Exercise 14.2.6. 1. Show that $f(t, x) = t\,x^{1/3}$ does not satisfy a Lipschitz condition with respect to x in any open neighborhood of $(t_0, x_0) = (0, 0)$.
2. Find two different solutions of the initial value problem $\dot{x} = t\,x^{1/3}$, $x(0) = 0$.

Exercise 14.2.7. Show that for any vector function $\mathbf{v}(t)$ defined for $a < t < b$,

$$\left| \int_a^b \mathbf{v}(t)\,dt \right|_1 \leq \int_a^b |\mathbf{v}(t)|_1\,dt.$$

14.2.4. Existence and Uniqueness of Solutions. For the proof of the existence and uniqueness theorem for initial value problems, we will focus on the case of **autonomous systems**,

$$\dot{\mathbf{x}} = \mathbf{f}(\mathbf{x}), \quad \mathbf{x}(t_0) = \mathbf{x}_0, \tag{14.9}$$

where $\mathbf{f} : U \to \mathbf{R}^n$ is a locally Lipschitz mapping on the open set $U \subseteq \mathbf{R}^n$. The proof is an application of the contraction mapping theorem in Banach spaces. In order to have a fixed point problem, we formulate the system of differential equations as an integral equation for an unknown function in the complete space $C_n(I)$ for a suitable interval I.

We shall use the vector norm $|\cdot|_1$ for vectors in $\mathbf{R}^n$, and the norm

$$\|\phi\| = \sup_{t \in I} |\phi(t)|_1$$

for elements of the function space $C_n(I)$. For the theorem on local solutions, we can take I to be a closed and bounded interval, so that the supremum in the definition of the norm can be taken to be a maximum. Recall that convergence in this norm means uniform convergence on the interval I.

We observe that for a vector function $\phi(t)$, we have

$$\left| \int_I \phi(t)\,dt \right|_1 \leq \int_I |\phi(t)|_1\,dt.$$

A **solution** of (14.9) on an interval I containing t_0 is a function $\mathbf{x} : I \to \mathbf{R}^n$ such that $\mathbf{x}(t_0) = \mathbf{x}_0$ and $\dot{\mathbf{x}}(t) = \mathbf{f}(\mathbf{x}(t))$ for each $t \in I$. We may speak about a solution on a finite closed interval by considering the appropriate one-sided derivatives at the endpoints.

Associated with (14.9) we have the integral equation

$$\mathbf{x}(t) = \mathbf{x}_0 + \int_{t_0}^t \mathbf{f}(\mathbf{x}(s))\,ds, \qquad t \in I,$$

where $\mathbf{x}_0$ is given and I is an interval containing t_0. Theorem 14.2.2, applied to an autonomous vector field $\mathbf{f}$, shows that if $\mathbf{f}$ is continuous, then a function $\mathbf{x} : I \to \mathbf{R}$ is a solution of the initial value problem (14.9) on I if and only if $\mathbf{x}$ is a solution on I of the integral equation. (See also Exercise 14.2.8.) However, as we have seen, mere continuity of $\mathbf{f}$ does not guarantee a *unique* solution of (14.9). We now apply the contraction mapping theorem (Theorem 9.2.3) to prove existence and uniqueness of a solution of (14.9) for *locally Lipschitz* vector fields $\mathbf{f}$.

Theorem 14.2.8 (Existence and Uniqueness for Initial Value Problems). *Let U be an open set in $\mathbf{R}^n$, and suppose that $f : U \to \mathbf{R}^n$ is locally Lipschitz on U. For any $\mathbf{x}_0 \in U$ and any initial time $t_0 \in \mathbf{R}$, there exists a $\delta > 0$ such that the initial value problem (14.9) has a unique solution $\mathbf{x} : [t_0 - \delta, t_0 + \delta] \to \mathbf{R}^n$.*

Proof. A function $\mathbf{x}(t)$ is a solution of the initial value problem on an interval $[t_0 - \delta, t_0 + \delta]$ if and only if

$$\mathbf{x}(t) = \mathbf{x}_0 + \int_{t_0}^{t} \mathbf{f}(\mathbf{x}(s))\, ds \quad \text{for } t \in [t_0 - \delta, t_0 + \delta]. \tag{14.10}$$

Motivated by (14.10), we define a mapping T of continuous functions by using the right-hand side of (14.10) to write

$$(T\phi)(t) = \mathbf{x}_0 + \int_{t_0}^{t} \mathbf{f}(\phi(s))\, ds. \tag{14.11}$$

We need an appropriate $\delta > 0$ and an appropriate subset Y of $C_n[t_0 - \delta, t_0 + \delta]$ such that $T : Y \to Y$ and T is a contraction mapping on Y. Our desired unique local solution will then be the unique fixed point of T in Y.

Since $\mathbf{f}$ is C^1 on U, we may choose an $r > 0$ sufficiently small such that the closed ball $\overline{B}_r(\mathbf{x}_0)$ about $\mathbf{x}_0$ is contained in U,

$$|\mathbf{f}(\mathbf{x})|_1 \le M \quad \text{for all } \mathbf{x} \in \overline{B}_r(\mathbf{x}_0),$$

and $\mathbf{f}$ satisfies the local Lipschitz condition

$$|\mathbf{f}(\mathbf{x}_1) - \mathbf{f}(\mathbf{x}_2)|_1 \le L|\mathbf{x}_1 - \mathbf{x}_2|_1, \quad \text{for } \mathbf{x}_1, \mathbf{x}_2 \in \overline{B}_r(\mathbf{x}_0).$$

Choose $\delta > 0$ such that $\delta M < r$ and $\delta L < 1$. With δ so chosen, the space $C_n[t_0 - \delta, t_0 + \delta]$ is complete in the sup norm.

Let $\mathbf{x}_0$ denote also the constant function in $C_n[t_0 - \delta, t_0 + \delta]$ taking the value $\mathbf{x}_0$. Consider the subset Y of $C_n[t_0 - \delta, t_0 + \delta]$ defined by

$$Y = \{\phi \in C_n[t_0 - \delta, t_0 + \delta] : \phi(t_0) = \mathbf{x}_0 \text{ and } \phi(t) \in \overline{B}_r(\mathbf{x}_0) \text{ for all } |t - t_0| \le \delta\}.$$

Equivalently, Y consists of those continuous functions ϕ on $[t_0 - \delta, t_0 + \delta]$ for which $\phi(t_0) = \mathbf{x}_0$ and $\|\phi - \mathbf{x}_0\| \le r$. Then Y is a closed subset of $C_n[t_0 - \delta, t_0 + \delta]$ (Exercise 14.2.9). Note that if ϕ is continuous on $[t_0 - \delta, t_0 + \delta]$, then the image $T(\phi)$ defined by (14.11) is also continuous on $[t_0 - \delta, t_0 + \delta]$, so T maps Y into $C_n[t_0 - \delta, t_0 + \delta]$.

We now show that $T : Y \to Y$ and that T is a contraction mapping. Let $\phi \in Y$. We have $(T\phi)(t_0) = \mathbf{x}_0$, and, for $|t - t_0| \le \delta$,

$$\begin{aligned} |(T\phi)(t) - \mathbf{x}_0|_1 &= \left| \int_{t_0}^{t} \mathbf{f}(\phi(s))\, ds \right|_1 \\ &\le \int_{t_0}^{t} |\mathbf{f}(\phi(s))|_1\, ds \\ &\le \delta M, \end{aligned}$$

where we have used the bound for $\mathbf{f}$ on $\overline{B}_r(\mathbf{x}_0)$. Since $\delta M < r$, if follows that $T\phi \in Y$. We now show that T is a contraction mapping on Y. For $\phi_1, \phi_2 \in Y$ and $|t - t_0| \le \delta$, we have

$$\begin{aligned}
|(T\phi_1)(t) - (T\phi_2)(t)|_1 &= \left| \int_{t_0}^t \mathbf{f}(\phi_1(s))\, ds - \int_{t_0}^t \mathbf{f}(\phi_2(s))\, ds \right|_1 \\
&\le \int_{t_0}^t |\mathbf{f}(\phi_1(s)) - \mathbf{f}(\phi_2(s))|_1\, ds \\
&\le L\delta \max_{|s-t_0|\le\delta} |\phi_1(s) - \phi_2(s)|_1 \\
&= L\delta\, \|\phi_1 - \phi_2\|,
\end{aligned}$$

where we have used the Lipschitz condition for $\mathbf{f}$ on $\overline{B}_r(\mathbf{x}_0)$. Taking the supremum for $|t - t_0| \le \delta$ on the left-hand side, we have

$$\|T(\phi_1) - T(\phi_2)\| \le L\delta \|\phi_1 - \phi_2\|.$$

Since ϕ_1 and ϕ_2 were arbitrary in Y and $L\delta < 1$, T is a contraction mapping on Y. By the contraction mapping Theorem 9.2.3, T has a unique fixed point in the closed set Y, and this fixed point is the desired unique local solution of (14.10), and hence (14.9), on $[t_0 - \delta, t_0 + \delta]$. □

As we have seen, Theorem 14.2.8 applies to any initial value problem when the vector field $\mathbf{f}$ is C^1 on U, since $\mathbf{f}$ is then locally Lipschitz on U. A similar approach can be applied to the nonautonomous initial value problem

$$\dot{\mathbf{x}} = \mathbf{f}(t, \mathbf{x}), \quad \mathbf{x}(t_0) = \mathbf{x}_0,$$

under the assumptions that $\mathbf{f}$ is continuous on a closed rectangle

$$\Omega = \{(t, \mathbf{x}) \in \mathbf{R}^{n+1} : |t - t_0| \le a,\ |\mathbf{x} - \mathbf{x}_0|_1 \le b\}$$

about the initial point $(t_0, \mathbf{x}_0)$, and $\mathbf{f}$ satisfies a Lipschitz condition with respect to $\mathbf{x}$ on Ω, as in the scalar Theorem 14.1.2. (See Exercises 14.2.11 and 14.2.12.)

The properties of a contraction mapping have been used to prove the existence and uniqueness of a locally defined solution for initial value problems. There is no claim that this iterative technique is an especially efficient technique for approximating solutions.[3] However, we do know that the iterates of the T mapping must converge uniformly to the local solution on the interval $[t_0 - \delta, t_0 + \delta]$.

In the next section we consider the extension of a local solution to a maximal interval of existence.

Exercises.

Exercise 14.2.8. Verify that if the autonomous vector field $\mathbf{f}$ is continuous, then a function $\mathbf{x} : I \to \mathbf{R}$ is a solution on I of the initial value problem $\dot{\mathbf{x}} = \mathbf{f}(\mathbf{x})$, $\mathbf{x}(t_0) = \mathbf{x}_0$, if and only if $\mathbf{x}$ is a solution on I of the integral equation

$$\mathbf{x}(t) = \mathbf{x}_0 + \int_{t_0}^t \mathbf{f}(\mathbf{x}(s))\, ds, \qquad t \in I.$$

[3] Contraction mappings do provide efficient practical approximations in many situations; see for example the basic iteration schemes in Exercises 9.5.8-9.5.9 for linear algebraic systems.

Exercise 14.2.9. Show that the set $Y = \{\phi \in C_n[t_0-\delta, t_0+\delta] : \phi(t_0) = \mathbf{x}_0 \text{ and } \|\phi - \mathbf{x}_0\| \leq r\}$ is a closed set in the normed space $C_n[t_0 - \delta, t_0 + \delta]$.

Exercise 14.2.10. Carry out the iteration of the contraction mapping T in Theorem 14.2.8 for the following initial value problems, using the stated initial condition:

1. $\dot{x} = x$, $x(0) = 1$;
2. $\dot{x} = x^2$, $x(0) = 1$;
3. $\dot{x} = 1 + x^2$, $x(0) = 0$.

In each case, compute the iterates $\phi_k = T(\phi_{k-1})$ for $k = 1, 2, 3$, and compare ϕ_3 with the exact solution.

Exercise 14.2.11. Suppose that $\mathbf{f}(t, \mathbf{x})$ is continuous on a closed rectangle $\Omega = \{(t, \mathbf{x}) \in \mathbf{R}^{n+1} : |t - t_0| \leq a, \ |\mathbf{x} - \mathbf{x}_0| \leq b\}$ about the initial point $(t_0, \mathbf{x}_0)$, and assume that $\mathbf{f}$ satisfies a Lipschitz condition with respect to $\mathbf{x}$ on Ω, that is,

$$|\mathbf{f}(t, \mathbf{x}) - \mathbf{f}(t, \mathbf{y})|_1 \leq L\, |\mathbf{x} - \mathbf{y}|_1 \quad \text{for } (t, \mathbf{x}), (t, \mathbf{y}) \in \Omega.$$

Use a contraction mapping argument to show that the initial value problem $\dot{\mathbf{x}} = \mathbf{f}(t, \mathbf{x})$, $\mathbf{x}(t_0) = \mathbf{x}_0$, has a unique solution on an interval $|t - t_0| < \delta$ for some $\delta > 0$. *Hint*: Make the minor modifications to the Theorem 14.2.8 proof, or see the scalar Theorem 14.1.2 if needed.

Exercise 14.2.12. Suppose that $\mathbf{f}(t, \mathbf{x})$ is continuously differentiable on a closed rectangle $\Omega = \{(t, \mathbf{x}) \in \mathbf{R}^{n+1} : |t - t_0| \leq a, \ |\mathbf{x} - \mathbf{x}_0| \leq b\}$ about the initial point $(t_0, \mathbf{x}_0)$. Show that $\mathbf{f}$ satisfies a Lipschitz condition with respect to $\mathbf{x}$ on Ω, that is,

$$|\mathbf{f}(t, \mathbf{x}) - \mathbf{f}(t, \mathbf{y})|_1 \leq L\, |\mathbf{x} - \mathbf{y}|_1 \quad \text{for } (t, \mathbf{x}), (t, \mathbf{y}) \in \Omega.$$

Hint: Employ the mean value theorem.

14.3. Extension of Solutions

The basic existence and uniqueness theorem for initial value problems provides us with a unique local solution to the problem

$$\dot{\mathbf{x}} = \mathbf{f}(\mathbf{x}), \quad \mathbf{x}(0) = \mathbf{x}_0, \tag{14.12}$$

where $\mathbf{f} : D \to \mathbf{R}^n$ and D is an open subset of $\mathbf{R}^n$. With no loss in generality we take the initial time to be $t_0 = 0$ in (14.12). We now address the extension of solutions to their maximal interval of definition.

14.3.1. The Maximal Interval of Definition. We assume that the vector field $\mathbf{f}$ is C^1, or at least continuous and locally Lipschitz in $\mathbf{x}$ on an open domain D. A local solution of the initial value problem with $\mathbf{x}(0) = \mathbf{x}_0$ exists on an interval $[-\delta, \delta]$ for some $\delta > 0$. With $\mathbf{x}(\delta) \in D$, we may extend this local solution to an interval $[-\delta, \delta + \delta_1]$ for some $\delta_1 > 0$, using the local existence and uniqueness theorem, given the initial condition $\mathbf{x}(\delta)$ at time δ. We may continue this extension process in forward time so long as the solution has a well-defined finite value in D at the new right-hand endpoint. Similarly we may extend the solution by steps to the left. Of course, the culmination of this extension process may be that the endpoints of these local extensions accumulate at some finite time value, beyond which the solution can be extended no further. More on this in a moment.

We first establish that the initial value problem has a maximal open interval of existence J. Let J be the union of all open intervals on which there is a solution of the initial value problem. Clearly J is an open interval, and the uniqueness result may be used to show that the solution of our initial value problem is uniquely determined on J, since local solutions on any two open subintervals of J must agree on their intersection.

Why can the solution not be extended beyond J? Suppose $J = (\alpha, \beta)$ and $\beta < \infty$. If the solution has a finite limit as $t \to \beta^-$, and this finite limit point is in the domain of the vector field, then the solution can be uniquely extended through this limit point on some time interval $(\alpha, \beta + \delta)$ for some $\delta > 0$, by the existence and uniqueness theorem. But such an extension contradicts our definition of J. Thus, either the solution has no finite limit as $t \to \beta^-$, or, if a finite limit exists, then it is not a point of the domain of the vector field. In either case, clearly the solution cannot be extended to time $t = \beta$. (Unless, of course, there is the possibility of extending the definition of the vector field to be C^1, or continuous and locally Lipschitz, on an extended domain; however, we are assuming that no such extension is made.) Similar considerations apply at the left-hand time boundary α. Thus, the maximal interval of definition is open, $J = (\alpha, \beta)$. Theorem 14.3.3 below gives a precise statement about the behavior of solutions as a time boundary is approached.

Let $\mathbf{f}$ be locally Lipschitz on the open set $D \subseteq \mathbf{R}^n$ and let $\phi(t, \mathbf{x}_0)$ be the solution of the system $\dot{\mathbf{x}} = \mathbf{f}(\mathbf{x})$ with $\phi(0, \mathbf{x}_0) = \mathbf{x}_0 \in D$. We also write $\phi_t(\mathbf{x}_0) = \phi(t, \mathbf{x}_0)$. Given $\mathbf{x}_0$, let $(\alpha(\mathbf{x}_0), \beta(\mathbf{x}_0))$ be the maximal interval of definition of the solution $\phi_t(\mathbf{x}_0)$. The set

$$\{\phi_t(\mathbf{x}_0) \, : \, \alpha(\mathbf{x}_0) < t < \beta(\mathbf{x}_0)\}$$

is called the **orbit of** $\mathbf{x}_0$, and the set

$$\{\phi_t(\mathbf{x}_0) \, : \, 0 \le t < \beta(\mathbf{x}_0)\}$$

is the **forward orbit of** $\mathbf{x}_0$. The set

$$\{\phi_t(\mathbf{x}_0) \, : \, \alpha(\mathbf{x}_0) < t \le 0\}$$

is the **backward orbit of** $\mathbf{x}_0$. We speak of the solution mapping $\phi(t, \mathbf{x}_0)$, which is a function defined on some subset of the product space $\mathbf{R} \times \mathbf{R}^n$, as the **flow** of the system $\dot{\mathbf{x}} = \mathbf{f}(\mathbf{x})$, or the flow of the vector field $\mathbf{f}$. The next result describes a basic composition property of this solution mapping.

Theorem 14.3.1. *Let $\{\phi_t\}$ be the flow for the system $\dot{\mathbf{x}} = \mathbf{f}(\mathbf{x})$ on D. Let $\mathbf{x}_0 \in D$. For all t, s for which these solution values are defined,*

$$\phi_{t+s}(\mathbf{x}_0) = \phi_t \circ \phi_s(\mathbf{x}_0) = \phi_t(\phi_s(\mathbf{x}_0)). \tag{14.13}$$

This property can also be written $\phi(t + s, \mathbf{x}_0) = \phi(t, \phi(s, \mathbf{x}_0))$.

Proof. Let s be fixed but arbitrary such that $\phi_s(\mathbf{x}_0)$ is defined. Let $\mathbf{x}(t) = \phi_{t+s}(\mathbf{x}_0)$ and $\mathbf{y}(t) = \phi_t(\phi_s(\mathbf{x}_0))$. By definition, $\mathbf{x}(t)$ and $\mathbf{y}(t)$ are solutions of $\dot{\mathbf{x}} = \mathbf{f}(\mathbf{x})$ defined at least for t near zero. The solution $\mathbf{x}(t)$ has initial condition $\mathbf{x}(0) = \phi_s(\mathbf{x}_0)$, as does $\mathbf{y}(t)$. By uniqueness, $\mathbf{x}(t) = \mathbf{y}(t)$ for all t for which these solutions are defined. This argument applies for any choice of the point $\phi_s(\mathbf{x}_0)$ on the orbit through $\mathbf{x}_0$. This completes the proof. □

Example 14.3.2. The flow of the scalar differential equation $\dot{x} = x^2$ is given by

$$\phi_t(x) = \phi(t, x) = \frac{x}{1 - xt}.$$

The domain for the flow is described as follows. For $x > 0$, the maximal interval of definition is the time interval $(-\infty, 1/x)$. For $x < 0$, the maximal interval of definition is $(1/x, \infty)$, and for $x = 0$, the constant solution $x = 0$ is defined for $-\infty < t < \infty$. For this flow, we may verify directly that we have

$$\phi_t(\phi_s(x_0)) = \frac{\phi_s(x_0)}{1 - \phi_s(x_0)t} = \frac{\frac{x_0}{1-x_0 s}}{1 - \frac{x_0}{1-x_0 s}t} = \frac{x_0}{1 - x_0(t+s)} = \phi_{t+s}(x_0),$$

which illustrates the property in Theorem 14.3.1. △

We now address the behavior of solutions at the time boundaries of the maximal interval of definition. The next theorem says that if a solution $\phi_t(\mathbf{x}_0)$ is not defined for all time, then it must leave any compact subset of the domain D as $t \to \alpha(\mathbf{x}_0)^+$ or as $t \to \beta(\mathbf{x}_0)^-$.

Theorem 14.3.3 (Behavior of Solutions at Time Boundaries)**.** *Let D be an open set in $\mathbf{R}^n$, $\mathbf{x}_0 \in D$, and let $\mathbf{f} : D \to \mathbf{R}^n$ be continuous and locally Lipschitz.*

1. *Let $(\alpha(\mathbf{x}_0), \beta(\mathbf{x}_0))$ be the maximal interval of definition of the solution $\phi_t(\mathbf{x}_0)$. Let K be an arbitrary compact subset of D. If $\mathbf{x}_0 \in K$ and $\beta(\mathbf{x}_0) < \infty$, then there is a time t_K with $0 < t_K < \beta(\mathbf{x}_0)$ such that $\phi_{t_K}(\mathbf{x}_0) \notin K$. Similarly, if $\mathbf{x}_0 \in K$ and $\alpha(\mathbf{x}_0) > -\infty$, then there is a time t_K with $\alpha(\mathbf{x}_0) < t_K < 0$ such that $\phi_{t_K}(\mathbf{x}_0) \notin K$.*
2. *Let K be a compact subset of D. If K contains the entire forward orbit of $\mathbf{x}_0$, then $\phi_t(\mathbf{x}_0)$ exists for all $t > 0$, that is, $\beta(\mathbf{x}_0) = \infty$. If K contains the entire backward orbit of $\mathbf{x}_0$, then $\phi_t(\mathbf{x}_0)$ exists for all $t < 0$, that is, $\alpha(\mathbf{x}_0) = -\infty$.*

Proof. 1. Let K be a compact subset of D. Since $\mathbf{f}$ is C^1, there are constants $M > 0$ and $L > 0$ such that $|\mathbf{f}(\mathbf{x})| \le M$ and $|\mathbf{f}(\mathbf{x}) - \mathbf{f}(\mathbf{z})| \le L|\mathbf{x} - \mathbf{z}|$ for $\mathbf{x}, \mathbf{z} \in K$. Thus for as long as the solution $\phi_t(\mathbf{x}_0)$ remains in K, the mean value theorem implies that

$$|\phi_t(\mathbf{x}_0) - \phi_s(\mathbf{x}_0)| \le M|t - s|.$$

Suppose that $\phi_t(\mathbf{x}_0) \in K$ for all t such that $0 \le t < \beta(\mathbf{x}_0)$. Then the limit

$$\lim_{t \to \beta(\mathbf{x}_0)^-} \phi_t(\mathbf{x}_0) =: \phi_{\beta(\mathbf{x}_0)}(\mathbf{x}_0)$$

exists as a finite limit in K. (See Exercise 14.3.1.) By the existence and uniqueness theorem, there is a $\delta > 0$ and a solution defined on the interval $(\beta(\mathbf{x}_0) - \delta, \beta(\mathbf{x}_0) + \delta)$ which equals $\phi_{\beta(\mathbf{x}_0)}(\mathbf{x}_0)$ at time $t = \beta(\mathbf{x}_0)$. But this contradicts the definition of $(\alpha(\mathbf{x}_0), \beta(\mathbf{x}_0))$ as the maximal interval of definition of $\phi_t(\mathbf{x}_0)$. Therefore $\phi_t(\mathbf{x}_0)$ must leave the set K before time $t = \beta(\mathbf{x}_0)$.

The argument for the behavior as $t \to \alpha(\mathbf{x}_0)^+$ is similar.

2. If K is compact and contains the entire forward orbit of $\mathbf{x}_0 \in K$, then the contrapositive of the implication in part 1 immediately implies that $\beta(\mathbf{x}_0) = \infty$. Similarly, if a compact K contains the entire backward orbit of $\mathbf{x}_0 \in K$, then by part 1, $\alpha(\mathbf{x}_0) = -\infty$. □

Exercise.

Exercise 14.3.1. Suppose that K is a compact set, and for $0 \le t < \beta(\mathbf{x}_0)$, $\phi_t(\mathbf{x}_0) \in K$ and $|\phi_t(\mathbf{x}_0) - \phi_s(\mathbf{x}_0)| \le M|t-s|$ for $t, s > 0$. Show that

$$\lim_{t\to\beta(\mathbf{x}_0)^-} \phi_t(\mathbf{x}_0)$$

exists as a finite limit in K.

14.3.2. An Example of a Newtonian System. We consider an application involving a second-order ordinary differential equation of the form $\ddot{y} + f(y) = 0$, referred to as a Newtonian equation. By setting $x_1 = y$ and $x_2 = \dot{x}_1 = \ddot{y}$, a Newtonian equation may be written as a first-order system

$$\begin{aligned}\dot{x}_1 &= x_2,\\ \dot{x}_2 &= -f(x_1),\end{aligned}$$

a standard formulation in the study of higher order equations. The dot notation for derivatives with respect to the independent time variable t is also a standard notation; thus $\dot{x}_1 = dx_1/dt$ and $\dot{x}_2 = dx_2/dt$. We define the total mechanical energy of the system at the point (x_1, x_2) by

$$\begin{aligned}E(x_1, x_2) &= \frac{1}{2}x_2^2 + \int_0^{x_1} f(s)\,ds\\ &= \text{(kinetic energy)} + \text{(potential energy)}.\end{aligned}$$

In the next example, we take $f(y) = f(x_1) = \sin(x_1)$, which defines a harmonic oscillator system studied in introductory differential equations courses, also called a nonlinear pendulum without friction.

Example 14.3.4. Let $f(x_1) = \sin(x_1)$ and consider the system

$$\begin{aligned}\dot{x}_1 &= x_2,\\ \dot{x}_2 &= -\sin(x_1).\end{aligned}$$

The sine function has a Taylor expansion about $x_1 = 0$ beginning with

$$\sin(x_1) = x_1 - \frac{1}{3!}x_1^3 + \cdots,$$

since $f'(0) = 1$, $f''(0) = 0$ and $f'''(0) = -1$. For any initial condition, the system has a unique solution defined on an interval about $t = 0$. Since $\dot{x}_1 = \dot{x}_2 = 0$ when $x_1 = x_2 = 0$, the origin $(0, 0)$ is an equilibrium (constant) solution of the system. We now examine the behavior of solutions that start at initial conditions $(x_1(0), x_2(0))$ close to the origin. The total mechanical energy is given by

$$E(x_1, x_2) = \frac{1}{2}x_2^2 + \int_0^{x_1} \sin(s)\,ds = \frac{1}{2}x_2^2 + 1 - \cos(x_1).$$

The rate of change of energy E along a solution $(x_1(t), x_2(t))$ is

$$\begin{aligned}\frac{d}{dt}E(x_1(t), x_2(t)) &= f(x_1)\dot{x}_1 + x_2\dot{x}_2\\ &= f(x_1)x_2 + x_2(-f(x_1)) = 0.\end{aligned}$$

This is the statement of *conservation of energy* for such a system. It implies that any solution must remain within a level set of the total energy function $E(x_1, x_2)$; that is,

$$\frac{1}{2}x_2^2(t) + 1 - \cos(x_1(t)) = E(x_1(0), x_2(0)) = C \quad \text{constant}$$

for all t for which the solution is defined. Observe that E has a local minimum value at the origin, where $E = 0$. By a simple algebraic step, we can express the level curves of E by solving for x_2 in terms of x_1, yielding closed curves of constant energy described by

$$\frac{1}{2}x_2^2 + (1 - \cos x_1) = C,$$

that is, $x_2 = \pm[2(C - (1 - \cos x_1))]^{1/2}$, for C positive and sufficiently small. For initial conditions sufficiently close to the equilibrium at the origin, the corresponding solutions are trapped within a constant energy curve, and that curve is a closed curve surrounding the origin. By the results of the current section, the solutions are therefore defined for all forward time, and solutions must orbit repeatedly around the origin as $t \to \infty$ (Exercise 14.3.2). In other words, these closed curves represent periodic solutions of the system. △

Exercise.

Exercise 14.3.2. This exercise shows that the constant energy curves for small values of E in Example 14.3.4 must be traversed completely by a solution and thus constitute periodic solutions.

1. Recall that the origin is an isolated equilibrium of the system, and observe that the closed curves are compact subsets of the plane. Conclude that the solution must exist for all forward time.
2. Let (x_{10}, x_{20}) be sufficiently close to the origin that its level curve of energy E is a closed curve about the origin. Show that the velocity vectors $(\dot{x}_1(t), \dot{x}_2(t))$ have Euclidean norm $(x_2^2 + [\sin(x_1)]^2)^{1/2}$ bounded below for all t by a positive constant. Conclude that the solution $(x_1(t), x_2(t))$ must trace out the entire closed curve (repeatedly) since the closed curve has finite total length.

14.4. Continuous Dependence

Systems of ordinary differential equations provide valuable models for many deterministic dynamical processes where the solution behavior should exhibit a continuous dependence on initial conditions, physical or other estimated parameters, and right-hand sides. This section includes precise statements of such properties for systems of ordinary differential equations.

14.4.1. Continuous Dependence on Initial Conditions, Parameters, and Vector Fields. We begin with a simple inequality known as *Gronwall's inequality*.

Lemma 14.4.1 (Gronwall's Inequality). *Suppose that $u(t)$ and $v(t)$ are nonnegative continuous functions on $[0, T)$ that satisfy*

$$u(t) \leq M + \int_0^t v(s)u(s)\,ds, \quad \textit{for } t \in [0, T),$$

where $M \geq 0$ is constant. Then

$$u(t) \leq Me^{\int_0^t v(s)\,ds}, \quad \text{for } t \in [0, T).$$

Proof. Let $U(t) := M + \int_0^t v(s)u(s)\,ds$ for $0 \leq t < T$. By assumption, $u(t) \leq U(t)$ for $t \in [0, T)$. By the fundamental theorem of calculus,

$$\dot{U}(t) = v(t)u(t) \leq v(t)U(t),$$

where we have used the nonnegativity. Multiply both sides by $e^{-\int_0^t v(s)\,ds}$. Recall the product rule, to obtain

$$\frac{d}{dt}\left[U(t)e^{-\int_0^t v(s)\,ds}\right] \leq 0.$$

Now integrate both sides from 0 to t, to yield

$$U(t)e^{-\int_0^t v(s)\,ds} - U(0) \leq 0.$$

Since $U(0) = M$ and $u(t) \leq U(t)$, we have

$$u(t) \leq U(t) \leq Me^{\int_0^t v(s)\,ds} \quad \text{for } t \in [0, T).$$

This completes the proof of Gronwall's inequality. □

In Gronwall's inequality we may have T finite or infinite. Note that all statements in the proof hold for $0 \leq t < T$.

Gronwall's inequality leads to a basic result on the continuous dependence of solutions on initial conditions.

Theorem 14.4.2 (Continuous Dependence on Initial Conditions). *Suppose that in some open set U, the vector field $\mathbf{f}$ satisfies an estimate*

$$|\mathbf{f}(\mathbf{x}) - \mathbf{f}(\mathbf{y})|_1 \leq K|\mathbf{x} - \mathbf{y}|_1,$$

where $K > 0$ is constant. Suppose $\phi_1(t)$ and $\phi_2(t)$ are solutions of $\dot{\mathbf{x}} = \mathbf{f}(\mathbf{x})$ with initial conditions $\phi_1(0)$, $\phi_2(0)$ in U. Then, for all $t \geq 0$ for which these solutions exist and the estimate

$$|\mathbf{f}(\phi_1(t)) - \mathbf{f}(\phi_2(t))|_1 \leq K|\phi_1(t) - \phi_2(t)|_1 \tag{14.14}$$

holds, we have

$$|\phi_1(t) - \phi_2(t)|_1 \leq |\phi_1(0) - \phi_2(0)|_1\, e^{Kt}.$$

Proof. Write $\mathbf{a} = \phi_1(0)$ and $\mathbf{b} = \phi_2(0)$. For the solutions in question, we have

$$\phi_1(t)) = \mathbf{a} + \int_0^t \mathbf{f}(\phi_1(s))\,ds$$

and

$$\phi_2(t) = \mathbf{b} + \int_0^t \mathbf{f}(\phi_2(s))\,ds.$$

Now we estimate using the absolute sum vector norm, $|\cdot|_1$, obtaining

$$|\phi_1(t) - \phi_2(t)|_1 \leq |\mathbf{a} - \mathbf{b}|_1 + \int_0^t |\mathbf{f}(\phi_1(s)) - \mathbf{f}(\phi_2(s))|_1\,ds.$$

As long as the Lipschitz estimate (14.14) holds, we have

$$|\phi_1(t) - \phi_2(t)|_1 \leq |\mathbf{a} - \mathbf{b}|_1 + \int_0^t K|\phi_1(s) - \phi_2(s)|_1 \, ds.$$

Now we may apply Gronwall's inequality with $u(t) := |\phi_1(t) - \phi_2(t)|_1$, $v(t) := K$ and $M := |\mathbf{a} - \mathbf{b}|_1$ to conclude that

$$|\phi_1(t) - \phi_2(t)|_1 \leq |\mathbf{a} - \mathbf{b}|_1 \, e^{Kt},$$

for as long as the solutions exist and the estimate (14.14) holds. □

We observe that the required estimates of Theorem 14.4.2 can be shown to hold for a C^1 vector field by choosing U sufficiently small, and then for any two initial conditions $\phi_1(0)$ and $\phi_2(0)$ in U, the corresponding solutions will be defined at least over some common finite time interval $[0, \beta]$. On any such interval, we have

$$|\phi_1(t) - \phi_2(t)|_1 \leq |\phi_1(0) - \phi_2(0)|_1 \, e^{K\beta} \quad \text{for all } t \in [0, \beta].$$

Thus, as $\phi_2(0)$ approaches $\phi_1(0)$ within U, the solution $\phi_2(t)$ approaches $\phi_1(t)$ uniformly in t over the interval $[0, \beta]$.

Suppose the system depends on several constant parameters which might be physical constants or other estimated constant parameters. If there are m such parameters involved, then we can model them by a parameter vector $\mathbf{p} \in \mathbf{R}^m$. Then the system takes the form

$$\dot{\mathbf{x}} = \mathbf{f}(\mathbf{x}, \mathbf{p}),$$

where $\mathbf{x} \in \mathbf{R}^n$ and $\mathbf{p} \in \mathbf{R}^m$. We can make this into a first order system in the vector variable $(\mathbf{x}, \mathbf{p}) \in \mathbf{R}^{n+m}$ by writing

$$\begin{aligned} \dot{\mathbf{x}} &= \mathbf{f}(\mathbf{x}, \mathbf{p}), \\ \dot{\mathbf{p}} &= \mathbf{0}. \end{aligned}$$

We may use the norm $|\cdot|_1$ on $\mathbf{R}^{n+m}$ and then apply Theorem 14.4.2 to this augmented system. We conclude that if the initial condition vectors $(\mathbf{x}_{10}, \mathbf{p}_1)$ and $(\mathbf{x}_{20}, \mathbf{p}_2)$ are sufficiently close in the 1-norm, then the corresponding solutions $(\mathbf{x}_1(t, \mathbf{x}_{10}, \mathbf{p}_1), \mathbf{p}_1)$ and $(\mathbf{x}_2(t, \mathbf{x}_{20}, \mathbf{p}_2), \mathbf{p}_2)$ satisfy

$$\big|(\mathbf{x}_1(t, \mathbf{x}_{10}, \mathbf{p}_1), \mathbf{p}_1) - (\mathbf{x}_2(t, \mathbf{x}_{20}, \mathbf{p}_2), \mathbf{p}_2)\big|_1 \leq (|\mathbf{x}_{10} - \mathbf{x}_{20}|_1 + |\mathbf{p}_1 - \mathbf{p}_2|_1)e^{K\beta}$$

for all t in a common finite interval of definition $[0, \beta]$. Thus, as $\mathbf{x}_{20}$ approaches $\mathbf{x}_{10}$ and $\mathbf{p}_2$ approaches $\mathbf{p}_1$, the solution $\mathbf{x}_2(t, \mathbf{x}_{20}, \mathbf{p}_2)$ approaches $\mathbf{x}_1(t, \mathbf{x}_{10}, \mathbf{p}_1)$ uniformly in t over the interval $[0, \beta]$.

The next result is a statement about the continuous dependence of solutions on the right-hand side of the differential equation.

Theorem 14.4.3 (Continuous Dependence on Right-Hand Sides). *Suppose $W \subset \mathbf{R}^n$ is an open set and the vector fields $\mathbf{f} : W \to \mathbf{R}^n$ and $\mathbf{g} : W \to \mathbf{R}^n$ are C^1 on W. Suppose $\mathbf{f}$ satisfies the Lipschitz condition*

$$|\mathbf{f}(\mathbf{x}) - \mathbf{f}(\mathbf{y})|_1 \leq K|\mathbf{x} - \mathbf{y}|_1$$

for $\mathbf{x}, \mathbf{y}$ in W. Suppose in addition that for all $\mathbf{x} \in W$,

$$|\mathbf{f}(\mathbf{x}) - \mathbf{g}(\mathbf{x})|_1 < \epsilon.$$

If $\mathbf{x}(t)$ *and* $\mathbf{y}(t)$ *are solutions of* $\dot{\mathbf{x}} = \mathbf{f}(\mathbf{x})$, $\dot{\mathbf{y}} = \mathbf{g}(\mathbf{y})$, *respectively, on an interval* $J = [0, T)$, *and* $\mathbf{x}(0) = \mathbf{y}(0)$, *then*

$$|\mathbf{x}(t) - \mathbf{y}(t)|_1 \leq \frac{\epsilon}{K}\left(e^{Kt} - 1\right)$$

for $t \in [0, T)$.

Proof. For $t > 0$ we have, using $\mathbf{x}(0) = \mathbf{y}(0)$,

$$\mathbf{x}(t) - \mathbf{y}(t) = \int_0^t [\dot{\mathbf{x}}(s) - \dot{\mathbf{y}}(s)]\, ds = \int_0^t [\mathbf{f}(\mathbf{x}(s)) - \mathbf{g}(\mathbf{y}(s))]\, ds.$$

Hence,

$$\begin{aligned} |\mathbf{x}(t) - \mathbf{y}(t)|_1 &\leq \int_0^t |\mathbf{f}(\mathbf{x}(s)) - \mathbf{g}(\mathbf{y}(s))|_1\, ds \\ &\leq \int_0^t |\mathbf{f}(\mathbf{x}(s)) - \mathbf{f}(\mathbf{y}(s))|_1\, ds + \int_0^t |\mathbf{f}(\mathbf{y}(s)) - \mathbf{g}(\mathbf{y}(s))|_1\, ds \\ &\leq K \int_0^t \left(|\mathbf{x}(s)) - \mathbf{y}(s)|_1 + \frac{\epsilon}{K}\right) ds. \end{aligned}$$

Letting $u(t) = |\mathbf{x}(t)) - \mathbf{y}(t)|_1$, we have

$$u(t) + \frac{\epsilon}{K} \leq \frac{\epsilon}{K} + \int_0^t K\left[u(s) + \frac{\epsilon}{K}\right] ds.$$

With $v(s) := K$, an application of Gronwall's inequality implies that

$$u(t) + \frac{\epsilon}{K} \leq \frac{\epsilon}{K} e^{Kt} \qquad \text{for } t \in [0, T),$$

from which the result follows. □

Exercises.

Exercise 14.4.1. Let $h(t)$ be real valued and continuous on an interval J containing $t_0 = 0$. Let a be a real constant. Show that the unique solution of the initial value problem $\dot{x} = ax + h(t)$, $x(0) = x_0$, is given by

$$x(t) = e^{at}\left[x_0 + \int_0^t e^{-as} h(s)\, ds\right] = e^{at} x_0 + e^{at} \int_0^t e^{-as} h(s)\, ds$$

for t in J. This is the **variation of parameters** formula for a scalar nonhomogeneous equation.

Exercise 14.4.2. A study of the planar system

$$\begin{aligned} \dot{x} &= -x + x^2 y, \\ \dot{y} &= -y \end{aligned}$$

can help in understanding continuous dependence on initial conditions.

1. Explicit solutions can be obtained. Since $y(t) = y_0 e^{-t}$, the equation for x is a Bernoulli equation, which can be integrated using the change of variable $u = 1/x$ and variation of parameters.

2. Use the exact solutions to show that each branch of the hyperbola $xy = 2$ is a solution trajectory. This hyperbola forms the boundary of the basin of attraction of the equilibrium (constant) solution $(x, y) = (0, 0)$, as you will see in part 3.

3. Using the exact solutions, study the forward time behavior of solutions with initial condition in a small ball centered on the point $(1, 2)$ on the hyperbola $x_1x_2 = 2$. You should find that solutions starting at points (x_0, y_0) with $x_0y_0 > 2$ become unbounded in finite forward time. Determine this finite escape time in terms of the initial condition. In contrast, solutions starting at points (x_0, y_0) with $x_0y_0 < 2$ will satisfy $\lim_{t\to\infty}(x(t), y(t)) = (0, 0)$. Observe that this does not violate Theorem 14.4.2, because continuous dependence of solutions on initial conditions is a statement about solution behavior over some finite time interval, common to the interval of definition for the nearby solutions being considered.

Exercise 14.4.3. Show that if $f(t)$ is continuous and nonnegative for $t \in [0, \beta)$ and satisfies $f(t) \leq \int_0^t f(s)\,ds$ for $t \in [0, \beta)$, then $f(t) \equiv 0$.

14.4.2. Newtonian Equations and Examples of Stability.

This subsection provides an initial exploration of the concept of stability. We consider the stability of equilibrium (constant) solutions in some examples of Newtonian systems.

The stability definition captures the idea that sufficiently small deviations of the initial condition from equilibrium will result in solutions that remain close to the equilibrium for all forward time.

Definition 14.4.4. *Suppose* $\mathbf{x}_0$ *is an equilibrium solution of the system* $\dot{\mathbf{x}} = \mathbf{F}(\mathbf{x})$. *We say that this equilibrium is* **stable** *if for every* $\epsilon > 0$ *there is a* $\delta = \delta(\epsilon) > 0$ *such that if* $|\mathbf{x} - \mathbf{x}_0|_2 < \delta$, *then the solution* $\phi(t, \mathbf{x})$ *with* $\phi(0, \mathbf{x}) = \mathbf{x}$ *satisfies* $|\phi(t, \mathbf{x}) - \mathbf{x}_0|_2 < \epsilon$ *for all* $t \geq 0$. *(See Figure* 14.2.*)*

The stability concept might be viewed in the following light. We know that solutions exhibit continuous dependence on *finite* intervals (Theorem 14.4.2). The stability condition for an equilibrium says that the flow of the differential equation exhibits a continuous dependence on initial conditions on the *infinite* time interval $[0, \infty)$ in a neighborhood of the equilibrium solution.

In many applications, we want to know that solutions starting near equilibrium not only remain close to equilibrium but approach the equilibrium asymptotically as $t \to \infty$. The appropriate definition is stated next.

Definition 14.4.5. *Suppose* $\mathbf{x}_0$ *is an equilibrium solution of the system* $\dot{\mathbf{x}} = \mathbf{F}(\mathbf{x})$. *We say that* $\mathbf{x}_0$ *is* **asymptotically stable** *if it is stable and there exists a number* $r > 0$ *such that if* $|\mathbf{x} - \mathbf{x}_0|_2 < r$, *then the solution* $\phi(t, \mathbf{x})$ *with* $\phi(0, \mathbf{x}) = \mathbf{x}$ *satisfies*

$$\lim_{t\to\infty} \phi(t, \mathbf{x}) = \mathbf{x}_0.$$

(See Figure 14.3.*)*

Interested readers can consider a Newtonian system with friction in Exercises 14.4.4-14.4.5.

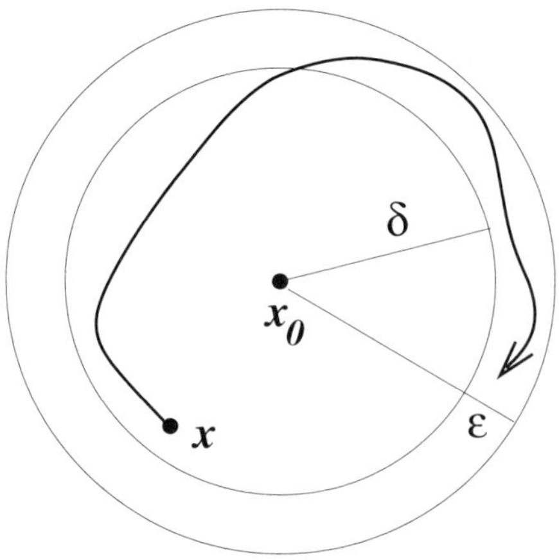

Figure 14.2. Definition 14.4.4: The stability of an equilibrium solution $\mathbf{x}_0$. The solution $\phi(t, \mathbf{x})$ with initial condition $\phi(0, \mathbf{x}) = \mathbf{x}$ remains within ϵ of $\mathbf{x}_0$ in forward time provided $\mathbf{x}$ is chosen within a distance $\delta = \delta(\epsilon)$.

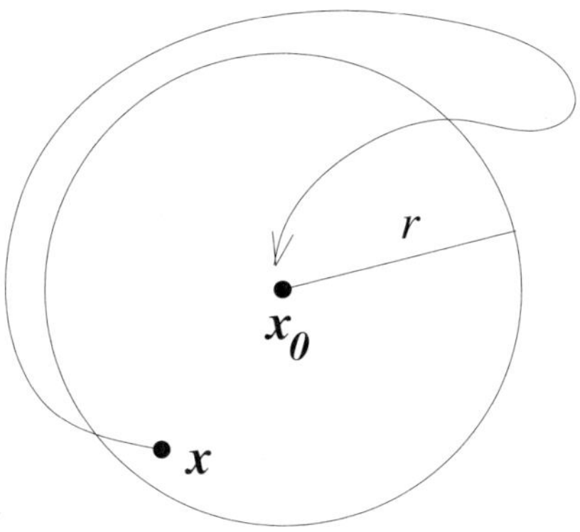

Figure 14.3. Definition 14.4.5: Asymptotic stability of an equilibrium solution $\mathbf{x}_0$ equals stability plus the attractivity property that $\lim_{t\to\infty} \phi(t, \mathbf{x}) = \mathbf{x}_0$ for solutions $\phi(t, \mathbf{x})$ with initial condition $\phi(0, \mathbf{x}) = \mathbf{x} \in B_r(\mathbf{x}_0)$ for some $r > 0$.

Exercises.

Exercise 14.4.4. *Stability in a Newtonian system*
This exercise continues the discussion of a Newtonian system begun in Example 14.3.4. Consider the Newtonian equation with a friction term proportional to the velocity, $\ddot{y} + b\dot{y} + f(y) = 0$, $b > 0$, where $f(y) = \sin y$. Write this as the system

$$\begin{aligned} \dot{x}_1 &= x_2, \\ \dot{x}_2 &= -bx_2 - f(x_1), \qquad b > 0, \end{aligned}$$

by setting $x_1 = y$ and $x_2 = \dot{x}_1 = \ddot{y}$.

1. Verify that the origin is an isolated equilibrium (constant) solution.
2. Use the total mechanical energy function E from Example 14.3.4 to conclude that the origin $(0, 0)$ is stable in the sense of Definition 14.4.4. Also verify that

the origin is stable even without the friction term. *Reminder*: Verify that for $|\mathbf{x}_0|_2$ sufficiently small, solutions are defined for all forward time.

Exercise 14.4.5. *Asymptotic stability in a Newtonian system*
We further consider a Newtonian system $\dot{\mathbf{x}} = \mathbf{F}(\mathbf{x})$ with a friction term,

$$\begin{aligned} \dot{x}_1 &= x_2, \\ \dot{x}_2 &= -bx_2 - f(x_1), \qquad b > 0. \end{aligned}$$

Assume that f is any C^1 function such that $f(0) = 0$ and $f'(0) = a > 0$. (For example, let $f(x_1) = \sin x_1$ if you wish.) Then the origin is an isolated equilibrium solution. Given the presence of friction in the system, we know from the analysis in Exercise 14.4.4 that the origin is stable. This exercise will establish the asymptotic stability of the origin when f satisfies the stated conditions.

1. We define a new energy function $V(\mathbf{x}) = \mathbf{x}^T P\mathbf{x}$ which is a quadratic form with P symmetric and positive definite, to be determined below. Then $V(\mathbf{x})$ will serve as a norm squared in a precise sense, because if P is symmetric positive definite, then the bilinear form $(\mathbf{x}, \mathbf{y}) = \mathbf{x}^T P\mathbf{y}$ is an inner product for $\mathbf{R}^2$; hence, $(V(\mathbf{x}))^{1/2} = (\mathbf{x}^T P\mathbf{x})^{1/2}$ is a norm on $\mathbf{R}^2$. Verify these facts about the bilinear form.

2. Suppose the system $\dot{\mathbf{x}} = \mathbf{F}(\mathbf{x})$ is written as $\dot{\mathbf{x}} = A\mathbf{x}+\mathbf{F}_2(\mathbf{x})$, where $A = D\mathbf{F}(\mathbf{0})$ and $\mathbf{F}_2(\mathbf{x})$ involves terms of second or higher order in the Taylor expansion of the components of $\mathbf{F}$. The rate of change of the energy V along a solution $\mathbf{x}(t)$ is given by

$$\begin{aligned} \frac{d}{dt}V(\mathbf{x}(t)) &= \frac{d}{dt}\mathbf{x}^T(t)P\mathbf{x}(t) = \dot{\mathbf{x}}(t)^T P\mathbf{x}(t) + \mathbf{x}(t)^T P\dot{\mathbf{x}}(t) \\ &= \mathbf{x}^T(t)A^T P\mathbf{x}(t) + \mathbf{F}_2^T(\mathbf{x}(t))P\mathbf{x}(t) + \mathbf{x}^T(t)PA\mathbf{x}(t) + \mathbf{x}^T(t)P\mathbf{F}_2(\mathbf{x}(t)) \\ &= \mathbf{x}(t)^T(A^T P + PA)\mathbf{x}(t) + \mathbf{F}_2^T(\mathbf{x}(t))P\mathbf{x}(t) + \mathbf{x}^T(t)P\mathbf{F}_2(\mathbf{x}(t)) \\ &= \mathbf{x}^T(t)(A^T P + PA)\mathbf{x}(t) + 2\,\mathbf{F}_2^T(\mathbf{x})(t)P\mathbf{x}(t). \end{aligned} \tag{14.15}$$

Thus we define P to be the unique solution of the linear matrix equation

$$A^T P + PA = -k\,I,$$

where I is the 2×2 identity matrix.[4] The positive constant k is for convenience. With P thus defined, we have

$$\frac{d}{dt}V(\mathbf{x}(t)) = -k\,\mathbf{x}^T(t)\mathbf{x}(t) + 2\mathbf{F}_2^T(\mathbf{x})P\mathbf{x}(t) = -k\,|\mathbf{x}(t)|_2^2 + 2\mathbf{F}_2^T(\mathbf{x})P\mathbf{x}(t). \tag{14.16}$$

Since the last term on the right is of at least third order, it follows that $\frac{d}{dt}V(\mathbf{x}(t)) < 0$ for solutions with initial condition sufficiently close to the origin. For later reference, we define the function $\dot{V}(\mathbf{x})$ by

$$\dot{V}(\mathbf{x}) = \nabla V(\mathbf{x}) \cdot \mathbf{F}(\mathbf{x}) = \nabla V(\mathbf{x}) \cdot (A\mathbf{x} + \mathbf{F}_2(\mathbf{x})).$$

Substitution of a solution $\mathbf{x}(t)$ for $\mathbf{x}$ in the formula for $\dot{V}$ gives exactly the rate of change of V along that solution at any time t. To complete this part,

[4] The unique solution for P is ensured by the fact that all eigenvalues of $A = D\mathbf{F}(\mathbf{0})$ have negative real part; see [**66**]. For our system, the eigenvalues are $(-b \pm \sqrt{b^2 - 4a})/2$. These eigenvalues are real and negative if $b^2 - 4a \geq 0$ (overdamping), and complex conjugates with negative real part $-b/2$ if $b^2 - 4a < 0$ (underdamping).

compute the unique solution of $A^T P + PA = -k\,I$, where A is the Jacobian matrix of our planar system above at the equilibrium at the origin. *Hint*: Let $k = 2$ to eliminate fractions in P. Note that $\nabla V(\mathbf{x}) = 2P\mathbf{x}$.

3. With P in hand from part 2, write $\dot{V}(\mathbf{x})$ explicitly as indicated in part 2. Identify $\mathbf{F}_2(\mathbf{x})$ and show explicitly at least two terms in its Taylor expansion about the origin. Conclude that $\dot{V}(\mathbf{x}) < 0$ in some ball about the origin.
4. Show that the origin is asymptotically stable (Definition 14.4.5). *Hint and Outline*: If the positive valued $V(x_1(t), x_2(t))$ strictly decreases on $[0, \infty)$, then it is bounded below by zero and therefore has a limit, say
$$\lim_{t\to\infty} V(x_1(t), x_2(t)) = \xi.$$
We want to show that $\xi = 0$, because then a solution $\mathbf{x}(t)$ with initial condition sufficiently close to the origin must satisfy $\mathbf{x}(t) \to \mathbf{0}$ as $t \to \infty$, since V is positive-definite in a neighborhood of the origin. Suppose to the contrary that $\xi > 0$, and argue for a contradiction: Use continuity of $V(\mathbf{x})$ to argue that the solution must remain outside some ball $B_r(\mathbf{0})$, $r > 0$. For fixed initial condition $\mathbf{x}_0$, use the fact that $\dot{V}(\mathbf{x})$ is continuous on the compact set
$$K = \{\mathbf{x} \in \mathbf{R}^2 : r \le |\mathbf{x}|_2 \le |\mathbf{x}_0|_2\}$$
to conclude that the number
$$-\nu = \max_{r\le|\mathbf{x}|_2\le|\mathbf{x}_0|_2} \dot{V}(\mathbf{x})$$
exists and $-\nu < 0$. (The number $-\nu$ is the slowest rate of change of V along any solution in the set K.) Then use the fundamental theorem of calculus in the form
$$V(\mathbf{x}(t)) = V(\mathbf{x}_0) + \int_0^t \dot{V}(\mathbf{x}(s))\,ds$$
to examine the behavior of $V(\mathbf{x}(t))$ for large t, and reach a contradiction.

14.5. Matrix Exponentials and Linear Autonomous Systems

The matrix exponential is a basic tool in the advanced theory of ordinary differential equations. This section defines the matrix exponential and explains its role in the fundamental existence and uniqueness theorem for linear systems of ordinary differential equations with constant coefficients.

Recall that if a is a fixed real number, then the real exponential function e^{at}, $t \in \mathbf{R}$, is defined by the real infinite series of functions
$$e^{at} := 1 + at + \frac{1}{2!}(at)^2 + \cdots + \frac{1}{k!}(at)^k + \cdots = \sum_{k=0}^{\infty} \frac{t^k}{k!}a^k.$$
Since $\frac{d}{dt}e^{ta} = ae^{ta}$, we see that the function $x(t) = e^{at}x_0$, $-\infty < t < \infty$, is the unique solution of the initial value problem $\dot{x} = ax$, $x(0) = x_0$.

There is a similar matrix construction that yields the unique solution of the initial value problem

$$\dot{\mathbf{x}} = A\mathbf{x}, \qquad \mathbf{x}(0) = \mathbf{x}_0 \tag{14.17}$$

where $\mathbf{x} \in \mathbf{R}^n$ and A is a real $n \times n$ matrix. Written out in component detail, such a system takes the form

$$\begin{aligned}\dot{x}_1 &= a_{11}x_1 + a_{12}x_2 + \cdots + a_{1n}x_n,\\ \dot{x}_2 &= a_{21}x_1 + a_{22}x_2 + \cdots + a_{2n}x_n\\ \cdots &= \cdots\\ \dot{x}_n &= a_{n1}x_1 + a_{n2}x_2 + \cdots + a_{nn}x_n.\end{aligned}$$

If $A \in \mathbf{R}^{n\times n}$, we may consider the **matrix exponential series** defined by

$$e^{tA} := I + tA + \frac{t^2}{2!}A^2 + \cdots + \frac{t^k}{k!}A^k + \cdots = \sum_{k=0}^{\infty} \frac{t^k}{k!}A^k. \tag{14.18}$$

See Exercise 14.5.1. We want to show that for any matrix A, and for any real number t, the series (14.18) converges. To accomplish this, we wish to apply Theorem 9.5.11 to the matrix exponential series (14.18) in the normed space $\mathbf{R}^{n\times n}$, which is complete by Theorem 9.5.9. Suppose that $0 < \beta < \infty$. For each real t with $|t| \le \beta$, we can estimate the general term of this series by

$$\left\| \frac{1}{k!}t^k A^k \right\| = \frac{|t|^k}{k!}\|A^k\| \le \frac{\beta^k}{k!}\|A\|^k =: M_k.$$

The series $\sum M_k$ converges by the ratio test, its limit being $e^{\beta\|A\|}$. Therefore by Theorem 9.5.11 we may conclude that the matrix series (14.18) converges for each real t. Observe that each component entry of the matrix series is a power series in t, which can be differentiated term-by-term with respect to t. Thus the series for e^{tA} can be differentiated term-by-term with respect to t to obtain

$$\begin{aligned}\frac{d}{dt}e^{tA} &= A + tA^2 + \frac{t^2}{2!}A^3 + \cdots + \frac{t^k}{k!}A^{k+1} + \cdots\\ &= Ae^{tA} = e^{tA}A.\end{aligned}$$

Thus, e^{tA} is a matrix solution of the differential equation $\dot{\mathbf{x}} = A\mathbf{x}$. Equivalently, each column of e^{tA} is a vector function that solves $\dot{\mathbf{x}} = A\mathbf{x}$. Thus $\mathbf{x}(t) = e^{tA}\mathbf{x}_0$ is the unique solution of the initial value problem for $\dot{\mathbf{x}} = A\mathbf{x}$ with $\mathbf{x}(0) = \mathbf{x}_0$.

The next theorem gives some useful properties of e^{tA}, all of which are corollaries of the uniqueness of solutions of initial value problems (Theorem 14.2.8).

Theorem 14.5.1. *If $A \in \mathbf{R}^{n\times n}$, then the following properties hold:*

1. *If $AB = BA$, then $e^{t(A+B)} = e^{tA}e^{tB}$ for all t.*
2. *e^{tA} is nonsingular for each t, and $(e^{tA})^{-1} = e^{-tA}$.*
3. *If S is nonsingular, then $S^{-1}e^{tA}S = e^{t(S^{-1}AS)}$ for all t.*

Proof. 1. We show that the function $X(t) := e^{tA}e^{tB}$ is a solution of $\dot{X} = (A+B)X$ satisfying $X(0) = I$; the result then follows by the uniqueness of solutions. Differentiate $X(t)$, using the product rule to get

$$\dot{X}(t) = Ae^{tA}e^{tB} + e^{tA}Be^{tB}.$$

Note that $AB = BA$ implies that $e^{tA}B = Be^{tA}$ (consider the defining series). It follows easily that $\dot{X}(t) = (A+B)X(t)$, $X(0) = I$, hence $X(t) = e^{t(A+B)}$.

2. Let $B = -A$ in statement 1. Then $I = e^{t(A-A)} = e^{tA}e^{-tA}$, and statement 2 follows immediately.

3. This property can be deduced from a power series argument, but again we will use uniqueness of solutions. Let $\bar{A} = S^{-1}AS$ and let $X(t) = S^{-1}e^{tA}S$. Clearly, $X(0) = I$. We now show that $X(t)$ satisfies $\dot{X}(t) = \bar{A}X(t)$. Differentiate $X(t)$ to get

$$\dot{X}(t) = S^{-1}Ae^{tA}S = S^{-1}ASS^{-1}e^{tA}S = \bar{A}S^{-1}e^{tA}S = \bar{A}X(t).$$

By uniqueness, $X(t) = e^{t\bar{A}} = e^{t(S^{-1}AS)}$, as we wished to show. □

We summarize the situation for the initial value problem (14.17) as follows.

Theorem 14.5.2 (Fundamental Theorem for Linear Autonomous Systems). *For any* $\mathbf{x}_0 \in \mathbf{R}^n$ *there exists a unique solution of the initial value problem* (14.17), *given by* $\mathbf{x}(t) = e^{tA}\mathbf{x}_0$, *and this solution is defined for all real* t.

Consider the following simple harmonic oscillator system.

Example 14.5.3. We compute e^{tA} for the matrix

$$A = \begin{bmatrix} 0 & 1 \\ -1 & 0 \end{bmatrix}.$$

This is the coefficient matrix for the simple linear harmonic oscillator system

$$\begin{aligned} \dot{x}_1 &= x_2 \\ \dot{x}_2 &= -x_1, \end{aligned}$$

which comes from the scalar second-order equation $\ddot{y} + y = 0$ by setting $x_1 = y$ and $x_2 = \dot{y}$. The powers of A required for the exponential series follow a periodic cycle of length four:

$$A^1 = A, \qquad A^2 = -I, \qquad A^3 = -A, \qquad A^4 = I, \qquad A^5 = A, \dots .$$

From this fact it is straightforward to find that (14.18) implies

$$e^{tA} = \begin{bmatrix} (1 - \frac{t^2}{2!} + \frac{t^4}{4!} - \cdots) & (t - \frac{t^3}{3!} + \frac{t^5}{5!} - \cdots) \\ (-t + \frac{t^3}{3!} - \frac{t^5}{5!} + \cdots) & (1 - \frac{t^2}{2!} + \frac{t^4}{4!} - \cdots) \end{bmatrix} = \begin{bmatrix} \cos t & \sin t \\ -\sin t & \cos t \end{bmatrix}.$$

The first column of e^{tA} solves the system with initial condition $(x_1(0), x_2(0)) = (1, 0)$, and the second column is the solution with $(x_1(0), x_2(0)) = (0, 1)$. △

Let A be a real $n \times n$ matrix. A matrix solution $X(t)$ of $\dot{\mathbf{x}} = A\mathbf{x}$ that is nonsingular for all t is called a **fundamental matrix solution**. When A has complex eigenvalues, a fundamental matrix can have complex entries. Theorem 14.5.1 (statement 2) shows that e^{tA} is a fundamental matrix solution, and it can always be used to produce the general real valued solution of the system. It is straightforward to check that if $X(t)$ is any fundamental matrix solution and C is any invertible matrix of the same size, then $X(t)C$ is also a fundamental matrix solution (Exercise 14.5.4). From this fact and the uniqueness of solutions of initial value problems, we have

$$e^{tA} = X(t)X(0)^{-1},$$

as each side is a matrix solution of $\dot{\mathbf{x}} = A\mathbf{x}$, and each side equals the identity matrix when $t = 0$.

There are some special solutions of the system $\dot{\mathbf{x}} = A\mathbf{x}$ that are easily identified. In particular, let λ be an eigenvalue of A and let $\mathbf{v}$ be a corresponding eigenvector (possibly complex). Thus $A\mathbf{v} = \lambda\mathbf{v}$. Then the function $\phi(t) = e^{\lambda t}\mathbf{v}$ is a (possibly complex valued) solution, since

$$\dot{\phi}(t) = \frac{d}{dt}[e^{\lambda t}\mathbf{v}] = \left(\frac{d}{dt}e^{\lambda t}\right)\mathbf{v} = \lambda e^{\lambda t}\mathbf{v} = e^{\lambda t}A\mathbf{v} = A\phi(t).$$

Suppose now that A is diagonalizable with eigenvalues $\lambda_1, \ldots, \lambda_n$. If the vectors $\mathbf{v}_1, \ldots, \mathbf{v}_n$ form a basis of n linearly independent eigenvectors of A (which exist since A is diagonalizable) with $A\mathbf{v}_j = \lambda_j\mathbf{v}_j$, then we have the n special solutions

$$\phi_1(t) = e^{\lambda_1 t}\mathbf{v}_1, \quad \ldots \quad, \phi_n(t) = e^{\lambda_n t}\mathbf{v}_n.$$

Place these solutions as columns in the square matrix

$$\Phi(t) = [\phi_1(t) \quad \cdots \quad \phi_n(t)].$$

Then $\Phi(t)$ is a fundamental matrix solution of $\dot{\mathbf{x}} = A\mathbf{x}$, with $\Phi(0) = [\mathbf{v}_1 \quad \cdots \quad \mathbf{v}_n]$. It follows that

$$e^{tA} = \Phi(t)\Phi(0)^{-1} = \Phi(t)[\mathbf{v}_1 \quad \cdots \quad \mathbf{v}_n]^{-1}.$$

Example 14.5.4. Consider the system

$$\dot{\mathbf{x}} = \begin{bmatrix} 3 & -2 \\ 1 & 1 \end{bmatrix}\mathbf{x}.$$

We will construct the general solution using the matrix solution e^{tA}. The characteristic equation is $\det(\lambda I - A) = (\lambda - 3)(\lambda - 1) + 2 = \lambda^2 - 4\lambda + 5 = 0$, so the eigenvalues are $\lambda = 2 \pm i$. An eigenvector for $2 + i$ is

$$\mathbf{v}_1 = \begin{bmatrix} 1+i \\ 1 \end{bmatrix}.$$

An eigenvector for $\lambda = 2 - i$ is

$$\mathbf{v}_2 = \begin{bmatrix} 1-i \\ 1 \end{bmatrix}.$$

These two eigenvectors are linearly independent since

$$\det\begin{bmatrix} 1+i & 1-i \\ 1 & 1 \end{bmatrix} = 2i \neq 0.$$

Therefore two solutions are given by

$$\phi_1(t) = e^{(2+i)t}\begin{bmatrix} 1+i \\ 1 \end{bmatrix}, \qquad \phi_2(t) = e^{(2-i)t}\begin{bmatrix} 1-i \\ 1 \end{bmatrix}.$$

Thus,

$$\Phi(t) := \begin{bmatrix} e^{(2+i)t}(1+i) & e^{(2-i)t}(1-i) \\ e^{(2+i)t} & e^{(2-i)t} \end{bmatrix}$$

is a matrix solution. One can check that the columns $\phi_1(t)$, $\phi_2(t)$ are linearly independent for each t, and thus $\Phi(t)$ is a fundamental matrix solution. The general real valued solution using e^{tA} is given by

$$\mathbf{x}(t) = e^{tA}\mathbf{x}_0 = \Phi(t)\Phi(0)^{-1}\mathbf{x}_0,$$

and a straightforward calculation gives

$$e^{tA} = \begin{bmatrix} e^{2t}(\cos t + \sin t) & -2e^{2t} \sin t \\ e^{2t} \sin t & e^{2t}(\cos t - \sin t) \end{bmatrix}.$$

A detailed component description of the general solution can now be given (Exercise 14.5.3). △

The general solution of nonhomogenous linear autonomous systems can be expressed using the variation of parameters formula for systems; see Exercise 14.5.5.

Exercises.

Exercise 14.5.1. Show that if we carry out the iteration of the contraction mapping T in the existence and uniqueness Theorem 14.2.8 for the system $\dot{\mathbf{x}} = A\mathbf{x}$, using the initial approximation $\mathbf{x}_0$, we generate the partial sums of $e^{tA}\mathbf{x}_0$.

Exercise 14.5.2. Show that for each real number t, $\|e^{tA}\| \le e^{|t|\,\|A\|}$.

Exercise 14.5.3. Verify the computation of e^{tA} in Example 14.5.4, and write the detailed component description of the general solution of the system given there.

Exercise 14.5.4. Show that if $X(t)$ is any fundamental matrix solution of $\dot{\mathbf{x}} = A\mathbf{x}$ and C is any invertible matrix of the same size, then $X(t)C$ is also a fundamental matrix solution. Deduce that $e^{tA} = X(t)X(0)^{-1}$.

Exercise 14.5.5. Let $\mathbf{h}(t)$ be a real vector function taking values in $\mathbf{R}^n$ and continuous on an interval J containing $t_0 = 0$. Let A be a real $n \times n$ matrix. Show that the unique solution of the initial value problem $\dot{\mathbf{x}} = A\mathbf{x} + \mathbf{h}(t)$, $\mathbf{x}(0) = \mathbf{x}_0$, is given by

$$\mathbf{x}(t) = e^{tA}\Big[\mathbf{x}_0 + \int_0^t e^{-sA}\mathbf{h}(s)\,ds\Big] = e^{tA}\mathbf{x}_0 + e^{tA}\int_0^t e^{-sA}\mathbf{h}(s)\,ds$$

for t in J. This is the **variation of parameters** formula for a nonhomogeneuous linear system with constant coefficient matrix A.

14.6. Notes and References

The text by Hirsch and Smale [**28**] is an interesting and rigorous introduction to systems of ordinary differential equations at the advanced undergraduate level, with many of the later chapters also appropriate for beginning graduate students. See also the second edition by Hirsch, Smale and Devaney [**29**] as well as Brauer and Nohel [**7**].

We concentrated on linear autonomous systems in Section 14.5. The text by Hale [**24**] includes material on linear nonautonomous systems and their variation of parameters formula.

Ordinary differential equations play an important role in mathematical control theory; Terrell [**66**] is an introduction to some core ideas for advanced undergraduates and beginning graduate students, and has many additional references.

Chapter 15

The Dirichlet Problem and Fourier Series

Fourier series representations of functions are important in many applications, especially problems that involve partial differential equations. Fourier series use integration to define the series coefficients, in contrast to Taylor series representations, which use differentiation. Fourier series can be useful in problems where a function can be discontinuous at a limited number of points, or might not have a continuous derivative. We introduce Fourier series by way of an important problem in the area of partial differential equations involving Laplace's equation.

Section 1 provides a brief introduction to Laplace's equation, indicating why it appears in many of the equations of mathematical physics.

Section 2 introduces the important orthogonality relations satisfied by the basic sine and cosine functions on $[-\pi, \pi]$. These functions are the building blocks for Fourier series. We then define the Fourier coefficients for a Riemann integrable function on $[-\pi, \pi]$.

Section 3 introduces the Dirichlet problem for the unit disk and describes how to construct special product solutions by means of the separation of variables method. An infinite linear combination of the special product solutions (an infinite series) represents the unique solution to the Dirichlet problem.

Section 4 offers guided exercises to show how the ideas from Section 3 can be used to construct series solutions to some basic problems for the one-dimensional heat equation (temperature in a thin metal rod) and the one-dimensional wave equation (displacement of a vibrating string).

Section 5 shows that the Fourier coefficients of a function provide the best mean square approximation of the function using the basic sine and cosine functions on $[-\pi, \pi]$. (This is approximation in the L^2 norm for Riemann integrable functions). The result on mean square approximation leads to Bessel's inequality and also to

Parseval's equality for continuous functions on $[-\pi, \pi]$. Parseval's equality is an analogue of the Pythagorean theorem in Euclidean space $\mathbf{R}^n$.

Section 6 presents a useful pointwise convergence result for Fourier series. We note the general difficulty of the pointwise convergence problem, and then prove a theorem on the uniform convergence of the Fourier series of a continuous function with a piecewise continuous derivative.

Finally, Section 7 presents Fejér's theorem, which says that the Cesàro means of the Fourier series of a continuous function converge uniformly to the function.

15.1. Introduction to Laplace's Equation

We will introduce the use of Fourier series by way of an important problem in the area of partial differential equations involving Laplace's equation

$$\Delta u(x, y, z) = 0, \quad (x, y, z) \in U.$$

Recall that $\Delta u = \operatorname{div} \operatorname{grad} u = u_{xx} + u_{yy} + u_{zz}$ is called the **Laplacian** of u, defined in an open set U in $\mathbf{R}^3$. The Laplacian operator Δ, and in particular, Laplace's equation, appears throughout mathematical physics in a variety of contexts. The dependent variable u should be thought of as a concentration (or density) of some quantity in equilibrium. Consider a vector field $\mathbf{F}$ giving the flux density in U, so that by the equilibrium hypothesis, about any point $\mathbf{a}$ in U the flux of u across a small sphere $S_r(\mathbf{a})$ centered at $\mathbf{a}$ is zero, that is,

$$\int_{S_r(\mathbf{a})} \mathbf{F} \cdot \mathbf{n} \, d\sigma = 0,$$

where $\mathbf{n}$ is the outward unit normal to $S_r(\mathbf{a})$. By the divergence theorem (13.43),

$$\int_{B_r(\mathbf{a})} \operatorname{div} \mathbf{F} \, dV = \int_{S_r(\mathbf{a})} \mathbf{F} \cdot \mathbf{n} \, d\sigma = 0$$

where $B_r(\mathbf{a})$ is the ball of radius r centered at $\mathbf{a}$. If $\mathbf{F}$ is C^1, this implies

$$\operatorname{div} \mathbf{F} = 0 \quad \text{throughout } U.$$

In many physical contexts, it is reasonable to assume that the flux density $\mathbf{F}$ is proportional to the gradient of u, $\operatorname{grad} u$. If the flow is from higher density regions to lower density regions, then

$$\mathbf{F} = -k \operatorname{grad} u, \qquad k > 0.$$

In these situations, we have

$$\operatorname{div} \mathbf{F} = 0 \quad \Longrightarrow \quad \operatorname{div} \operatorname{grad} u = \Delta u = 0,$$

which is Laplace's equation.

Laplace's equation in the plane is

$$u_{xx}(x, y) + u_{yy}(x, y) = 0, \qquad (x, y) \in U.$$

It describes steady-state temperature (heat) distributions over a given domain U. The heat equation in planar regions is

$$u_t(x, y, t) = u_{xx}(x, y, t) + u_{yy}(x, y, t), \qquad (x, y, t) \in U \times [0, \infty).$$

In the steady state or equilibrium state, a solution $u = u(x, y, t)$ of the heat equation is independent of time t, thus $u_t(x, y, t) = 0$ for all (x, y, t), and hence $u = u(x, y)$ is a function of position alone and u satisfies Laplace's equation in U. The problem of solving Laplace's equation over a region U when the values of u are specified on the boundary of U is called the **Dirichlet problem** for U. We shall consider the Dirichlet problem for the unit disk in the plane, and its solution by means of Fourier series.

First, we need the building blocks of Fourier series.

15.2. Orthogonality of the Trigonometric Set

The building blocks of Fourier series are the elements of the **trigonometric set** given by

$$\{1,\ \cos x,\ \sin x,\ \cos 2x,\ \sin 2x,\ \ldots,\ \cos nx,\ \sin nx,\ \ldots\}. \tag{15.1}$$

If f and g are two Riemann integrable functions defined on an interval $[a, b]$, their inner product is defined by

$$(f, g) = \int_a^b f(x)g(x)\, dx,$$

and we say that f and g are **orthogonal** if $(f, g) = 0$, sometimes denoted by $f \perp g$. A development of Fourier series based on the trigonometric set could be based on any interval of length 2π. We also employ the extensions of functions defined on that interval to periodic functions of period 2π on the whole real line. We choose to work with the interval $[-\pi, \pi]$ as the basic interval.

From the basic trigonometric identities,

$$\begin{aligned} 2\cos nx\ \cos mx &= \cos(n+m)x + \cos(n-m)x; \\ 2\cos nx\ \sin mx &= \sin(n+m)x - \sin(n-m)x; \\ 2\sin nx\ \sin mx &= \cos(n-m)x - \cos(n+m)x, \end{aligned}$$

we have the following orthogonality relations satisfied by the elements of the trigonometric set:

$$\int_{-\pi}^{\pi} \cos nx\, dx = \int_{-\pi}^{\pi} \sin nx\, dx = 0 \qquad (n \geq 1), \tag{15.2}$$

that is, $\cos nx \perp 1$ and $\sin nx \perp 1$ for all $n \geq 1$;

$$\int_{-\pi}^{\pi} \cos nx\ \sin mx\, dx = 0 \qquad (n, m \geq 1,\ n \neq m), \tag{15.3}$$

$$\int_{-\pi}^{\pi} \cos nx\ \cos mx\, dx = \int_{-\pi}^{\pi} \sin nx\ \sin mx\, dx = 0 \qquad (n, m \geq 1,\ n \neq m), \tag{15.4}$$

and finally, $\int_{-\pi}^{\pi} (1)(1)\, ds = 2\pi$, and

$$\int_{-\pi}^{\pi} \cos^2 nx\, dx = \int_{-\pi}^{\pi} \sin^2 nx\, dx = \pi \qquad (n \geq 1). \tag{15.5}$$

We may summarize (15.2), (15.3) and (15.4) by the statement that the elements of the trigonometric set are pairwise orthogonal. Such a set is called simply an **orthogonal set**.

Now suppose that a series in the elements of the orthogonal set converges *uniformly* to a function $f(x)$ on $[-\pi, \pi]$, that is, for certain constants a_k and b_k, we have

$$f(x) = \frac{1}{2}a_0 + \sum_{k=1}^{\infty}(a_k \cos kx + b_k \sin kx), \qquad x \in [-\pi, \pi],$$

where the partial sums converge uniformly to f. The explanation for the choice of the constant term in the form $\frac{1}{2}a_0$ will be given in a moment. Being the sum of a uniformly convergent series of continuous functions, the function f is continuous on $[-\pi, \pi]$. The uniform convergence also implies that term-by-term integration of the series is valid, and this, together with the orthogonality relations, allows us to determine the coefficients. For example, integration of the series for f implies

$$\int_{-\pi}^{\pi} f(x)\,dx = \int_{-\pi}^{\pi} \frac{1}{2}a_0\,dx = 2\pi\frac{1}{2}a_0 = \pi a_0.$$

For fixed $n \geq 1$, multiplication of the series by $\cos nx$ and term-by-term integration yields

$$\int_{-\pi}^{\pi} f(x)\cos nx\,dx = \int_{-\pi}^{\pi} a_n \cos^2 nx\,dx = \pi a_n.$$

Thus the coefficients a_n are given by the formulas

$$a_n = \frac{1}{\pi}\int_{-\pi}^{\pi} f(x)\cos nx\,dx, \qquad n = 0, 1, 2, \ldots.$$

(The case $n = 0$ here explains our choice of the form $\frac{1}{2}a_0$ for the constant term in the series.) Similarly, for fixed $n \geq 1$, multiplication of the series by $\sin nx$ yields

$$\int_{-\pi}^{\pi} f(x)\sin nx\,dx = \int_{-\pi}^{\pi} b_n \sin^2 nx\,dx = \pi b_n.$$

Thus the coefficients b_n are given by the formulas

$$b_n = \frac{1}{\pi}\int_{-\pi}^{\pi} f(x)\sin nx\,dx, \qquad n = 1, 2, \ldots.$$

With the preceding construction as *motivation*, we now consider a Riemann integrable function f, which need not be continuous at every point of $[-\pi, \pi]$, and we *define* the numbers

$$a_n = \frac{1}{\pi}\int_{-\pi}^{\pi} f(x)\cos nx\,dx \qquad (n = 0, 1, 2, \ldots) \tag{15.6}$$

and

$$b_n = \frac{1}{\pi}\int_{-\pi}^{\pi} f(x)\sin nx\,dx \qquad (n = 1, 2, \ldots) \tag{15.7}$$

to be the **Fourier coefficients** of f with respect to the trigonometric set. The series

$$\frac{1}{2}a_0 + \sum_{k=1}^{\infty}(a_k \cos kx + b_k \sin kx) \tag{15.8}$$

with the coefficients defined by (15.6), (15.7) is then called the **Fourier series** of f. If f is actually the sum of a series of the form (15.8) on $[-\pi, \pi]$, then the argument in the preceding paragraph shows that the Fourier series of f is the only possible such series that can converge *uniformly* to f. However, given a Riemann integrable function f, the question of whether the Fourier series of f actually converges to f uniformly, or even pointwise at every point, remains to be addressed. We present a basic result on pointwise convergence in Section 15.6. The question of the validity of term-by-term integration of the Fourier series of a general integrable function, or of the series multiplied by one of the elements of the trigonometric series, as carried out above, was answered fully only after the introduction of the Lebesgue integral.

In this book we shall work with Fourier series in the form (15.8) for real valued functions f. However, (15.8) may also be expressed in complex number form, based on the Euler identity $e^{ix} = \cos x + i \sin x$ and the inner product

$$(f, g) = \int_{-\pi}^{\pi} f(x)\overline{g(x)}\, dx$$

for functions that may take on complex values on $[-\pi, \pi]$. The complex conjugate of $g(x)$ is $\overline{g(x)}$. The functions e^{inx}, $n \in \mathbf{Z}$, are pairwise orthogonal on the interval $[-\pi, \pi]$:

$$(e^{inx}, e^{imx}) = \int_{-\pi}^{\pi} e^{inx}e^{-imx}\, dx = \begin{cases} 0 & \text{for } n \neq m, \\ 2\pi & \text{for } n = m, \end{cases}$$

as is easily verified (Exercise 15.2.2). The complex form of (15.8) allows for extensions of the theory to complex valued functions. For the interested reader, Exercise 15.2.3 shows the complex coefficients.

Exercises.

Exercise 15.2.1. Verify the relations (15.2), (15.3), (15.4) and (15.5) for the trigonometric set in (15.1) on the interval $[-\pi, \pi]$.

Exercise 15.2.2. Verify the pairwise orthogonality of the functions e^{inx}, $n \in \mathbf{Z}$, on $[-\pi, \pi]$, and the fact that $(e^{inx}, e^{inx}) = 2\pi$ for all n.

Exercise 15.2.3. *Complex form of Fourier series*
Let f be a Riemann integrable function. Then the Fourier coefficients are defined. Show that the real Fourier series (15.8) with coefficients a_n, b_n given by (15.6), (15.7) can be expressed in the form $\sum_{-\infty}^{\infty} c_k e^{ikx}$, where the complex Fourier coefficients c_n, $n \in \mathbf{Z}$, are given by

$$c_n = \frac{1}{2\pi}(f, e^{inx}) = \frac{1}{2\pi}\int_{-\pi}^{\pi} f(x)e^{-inx}\, dx, \qquad n = 0, \pm 1, \pm 2, \ldots. \tag{15.9}$$

More precisely, show that this definition of c_n implies that $c_n = \frac{1}{2}(a_n - ib_n)$ for $n \neq 0$, and that $c_0 = \frac{1}{2}a_0$. Show that $c_{-n} = \overline{c_n}$, and hence show that for each integer n,

$$\sum_{k=-n}^{n} c_k e^{ikx} = \frac{1}{2}a_0 + \sum_{k=1}^{n}(a_k \cos kx + b_k \sin kx),$$

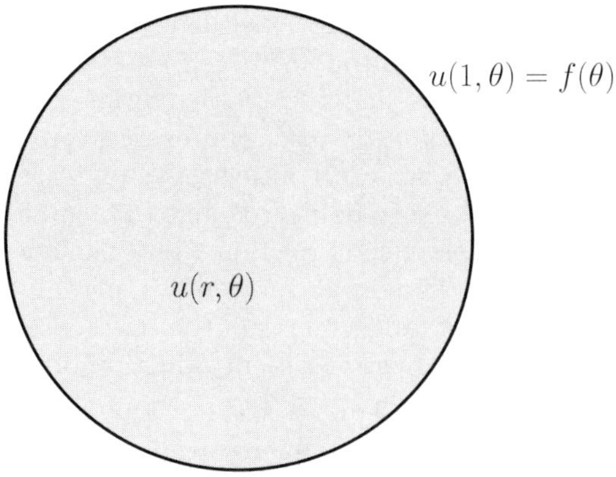

Figure 15.1. The Dirichlet problem for the unit disk: Determine $u(r,\theta)$ in the interior of the disk if $u(1,\theta) = f(\theta)$ is known on the boundary circle.

the n-th partial sum of the Fourier series, expressed in complex form. If the limit exists for a particular x, then, by definition,

$$\sum_{-\infty}^{\infty} c_k e^{ikx} = \lim_{n\to\infty} \sum_{k=-n}^{n} c_k e^{ikx} = \lim_{n\to\infty} \left[\frac{1}{2}a_0 + \sum_{k=1}^{n}(a_k \cos kx + b_k \sin kx)\right].$$

Verify that if term-by-term integration of the series $\sum_{-\infty}^{\infty} c_k e^{ikx}$ is valid, then the orthogonality relations of the set $\{e^{inx} : n \in \mathbf{Z}\}$ imply that the coefficients c_n must be given by (15.9).

15.3. The Dirichlet Problem for the Disk

The Dirichlet problem for the closed unit disk is the boundary value problem for Laplace's equation

$$u_{xx}(x,y) + u_{yy}(x,y) = 0, \qquad \text{for } x^2 + y^2 < 1,$$

where

$$u(x,y) = f(x,y), \qquad \text{for } x^2 + y^2 = 1.$$

The function f is the given boundary condition. The problem is expressed in polar coordinates for the unit disk as follows: The function $u(r,\theta)$ is to satisfy the differential equation

$$u_{rr}(r,\theta) + \frac{1}{r}u_r(r,\theta) + \frac{1}{r^2}u_{\theta\theta}(r,\theta) = 0 \tag{15.10}$$

for (r,θ) in the interior of the disk, that is, for $r < 1$, and the boundary condition

$$u(1,\theta) = f(\theta), \quad -\pi \le \theta \le \pi. \tag{15.11}$$

See Exercise 15.3.1. For the boundary condition $f(\theta)$ in (15.11), we require that $f(\theta + 2\pi) = f(\theta)$ for all θ. (See Figure 15.1.)

Following the approach of Fourier, we look for special solutions of Laplace's equation (15.10) in product form $u(r,\theta) = R(r)\Theta(\theta)$. Substituting this product into (15.10) yields

$$R''(r)\Theta(\theta) + \frac{1}{r}R'(r)\Theta(\theta) + \frac{1}{r^2}R(r)\Theta''(\theta) = 0.$$

Multiplication by r^2 and division by $u = R\Theta$ (we seek nonzero solutions) yields

$$\frac{r^2R''(r) + rR'(r)}{R(r)} = -\frac{\Theta''(\theta)}{\Theta(\theta)}.$$

Since this is to be an identity in (r, θ), and the left-hand side depends only on r while the right-hand side depends only on θ, both sides equal a common constant λ,

$$\frac{r^2R''(r) + rR'(r)}{R(r)} = \lambda \quad \text{and} \quad -\frac{\Theta''(\theta)}{\Theta(\theta)} = \lambda.$$

The variables have been separated, hence the name *separation of variables method* for this approach. Since $u(r, \theta)$ is to be 2π-periodic in θ, we require that $\Theta(\theta)$ be 2π-periodic in θ, so $\Theta(\theta)$ must also satisfy the periodic boundary condition $\Theta(-\pi) = \Theta(\pi)$. We now construct these product solutions.

The ordinary differential equations to be solved may be written as the boundary value problem,

$$\Theta''(\theta) + \lambda\Theta(\theta) = 0, \qquad \Theta(-\pi) = \Theta(\pi), \tag{15.12}$$

together with

$$r^2R''(r) + rR'(r) - \lambda R(r) = 0. \tag{15.13}$$

The idea is to first solve the boundary value problem (15.12) for those values of λ yielding nonzero solutions for $\Theta(\theta)$. Then, for each such value of λ, solve (15.13), yielding a nonzero solution for $R(r)$.

The reader is invited to show (Exercise 15.3.2) that the boundary value problem (15.12) requires that $\lambda \geq 0$, and, in fact, that $\lambda = n^2$ for nonnegative integer n. For these λ, we obtain for $n > 0$ the solutions

$$\Theta_n(\theta) = a_n \cos n\theta + b_n \sin n\theta,$$

and for $n = 0$,the solution

$$\Theta_0(\theta) = \frac{1}{2}a_0,$$

where the a_n and b_n are real constants.

Then, in (15.13), set $\lambda = \lambda_n = n^2$ to obtain

$$r^2R''(r) + rR'(r) - n^2R(r) = 0,$$

which is an Euler equation. With $\lambda_n = n^2$ we find the solutions

$$R(r) = A_0 + B_0 \ln r \quad \text{if } n = 0$$

and

$$R(r) = A_n r^n + B_n r^{-n} \quad \text{if } n = 1, 2, \ldots.$$

Since our solutions should be continuous at the origin $r = 0$ we require that $B_n = 0$ for all n. By choosing unit constants for the A_n, we have $R_n(r) = r^n$. (See Exercise 15.3.3.)

We now have a sequence of basic solutions $R_n(r)\Theta_n(\theta)$, indexed by the nonnegative integers n, which we may write in the form

$$\begin{aligned}
u_0(r,\theta) &= \frac{1}{2}a_0 \quad \text{(constant)},\\
u_1(r,\theta) &= r(a_1\cos\theta + b_1\sin\theta),\\
u_2(r,\theta) &= r^2(a_2\cos 2\theta + b_2\sin 2\theta)\\
&\cdots\\
u_n(r,\theta) &= r^n(a_n\cos n\theta + b_n\sin n\theta)\\
&\cdots
\end{aligned}$$

Since Laplace's equation is linear in the dependent variable, any finite sum of these product solutions is also a solution of Laplace's equation. We now consider the possibility of a solution $u(r,\theta)$ in the form of an infinite series,

$$u(r,\theta) = \frac{1}{2}a_0 + \sum_{n=1}^{\infty} r^n(a_n\cos n\theta + b_n\sin n\theta), \tag{15.14}$$

and ask whether the series sum, assuming it exists, also satisfies the boundary condition, $u(r,\theta) = f(\theta)$, $-\pi \le \theta < \pi$. On the one hand, by setting $r = 1$ in (15.14), it appears we must have

$$u(1,\theta) = f(\theta) = \frac{1}{2}a_0 + \sum_{n=1}^{\infty}(a_n\cos n\theta + b_n\sin n\theta), \tag{15.15}$$

which means we desire the convergence of the series to the sum $f(\theta)$ when $r = 1$. On the other hand, if, in a purely formal way, we let $r \to 1^-$ in (15.14), we hope to have the same result, namely $\lim_{r\to 1^-} u(r,\theta) = f(\theta)$. These wishes deal with two different processes. The first deals with the convergence of the Fourier series of f to f itself, since we will choose the usual Fourier coefficients below for the series determined by f. The second wish concerns the question of recovering the solution value at a boundary point from knowledge of the solution $u(r,\theta)$ at points in the interior of the disk.

Fourier used the pairwise orthogonality property of the trigonometric set

$$\{1, \cos\theta, \sin\theta, \cos 2\theta, \sin 2\theta, \dots\}$$

on $[-\pi,\pi]$ to *define* the appropriate coefficients a_n and b_n for the expansion (15.15) of $f(\theta)$ by the formulas

$$a_n = \frac{1}{\pi}\int_{-\pi}^{\pi} f(t)\cos nt\,dt \qquad \text{and} \qquad b_n = \frac{1}{\pi}\int_{-\pi}^{\pi} f(t)\sin nt\,dt.$$

As we have seen in Section 15.2, these are necessarily the values of the coefficients if the series on the right-hand side of (15.15) converges and can be integrated term-by-term. Given these Fourier coefficients, we assume for now that the series (15.14) converges and the sum $u(r,\theta)$ is a solution of the Dirichlet problem.

If $f(\theta)$ is *continuous and piecewise smooth,*[1] its Fourier series converges absolutely and uniformly to f, as we will see later in Theorem 15.6.7. For any (r,θ) in

[1]This means that $f'(x)$ exists at all but finitely many points in any bounded interval, and the left-hand and right-hand limits of f' exist at those points where f' fails to exist. See Definition 15.6.3 for the general definition of **piecewise smooth** function.

the interior of the disk, the series for $u(r,\theta)$ is dominated termwise by the series for f, so the series for $u(r,\theta)$ also converges absolutely and uniformly in the disk, under those conditions on f. The differentiated series (with respect to either r or θ) have continuous terms, and converge uniformly in the interior of the disk, so $u(r,\theta)$ can be differentiated term-by-term to verify that it really is a solution of Laplace's equation in the interior of the disk, under the stated assumptions on f. Thus there is a large class of functions f that serve as boundary data in solvable Dirichlet problems.

It turns out that not every continuous function $f(\theta)$ is the limit of the partial sums of its Fourier series at every point of its domain. In fact, there are continuous functions for which the Fourier partial sums diverge at certain points. However, even if the Fourier partial sums defined by f do not converge to f at certain points, there are other modes of convergence for Fourier series, for example Cesàro summability in Fejér's theorem on continuous functions (Theorem 15.7.3), and convergence in the L^2 norm (Theorem 18.4.7), that provide useful information about large classes of functions.

We now consider a useful alternative representation for the solution $u(r,\theta)$, which will also allow a reconstruction of the boundary data for continuous f. The resulting integral formula for $u(r,\theta)$ is called the *Poisson integral formula*. To derive this integral formula for the solution, we begin with the Fourier coefficients of f as defined above. Substitution of these coefficient formulas into the series for $u(r,\theta)$ gives

$$u(r,\theta)=\frac{a_0}{2}+\sum_{n=1}^{\infty}r^n\Big(\frac{1}{\pi}\int_{-\pi}^{\pi}f(t)\cos nt\,dt\,\cos n\theta+\frac{1}{\pi}\int_{-\pi}^{\pi}f(t)\sin nt\,dt\,\sin n\theta\Big).$$

An equivalent expression is

$$u(r,\theta)=\frac{a_0}{2}+\frac{1}{\pi}\sum_{n=1}^{\infty}r^n\int_{-\pi}^{\pi}f(t)(\cos nt\cos n\theta+\sin nt\sin n\theta)\,dt.$$

Since $\cos n(\theta-t)=\cos nt\cos n\theta+\sin nt\sin n\theta$, we have

$$u(r,\theta)=\frac{a_0}{2}+\frac{1}{\pi}\sum_{n=1}^{\infty}r^n\int_{-\pi}^{\pi}f(t)\cos n(\theta-t)\,dt.$$

Remembering the formula for a_0, we can express $u(r,\theta)$ by

$$u(r,\theta)=\frac{1}{\pi}\int_{-\pi}^{\pi}f(t)\Big(\frac{1}{2}+\sum_{n=1}^{\infty}r^n\cos n(\theta-t)\Big)\,dt.$$

The sum in parentheses in the integrand can be simplified further. Let $\alpha=\theta-t$, and define the complex number $z=re^{i\alpha}=r(\cos\alpha+i\sin\alpha)$. Then $z^n=r^ne^{in\alpha}=r^n(\cos n\alpha+i\sin n\alpha)$. Let $\operatorname{Re}w$ denote the real part of a complex number w. Since

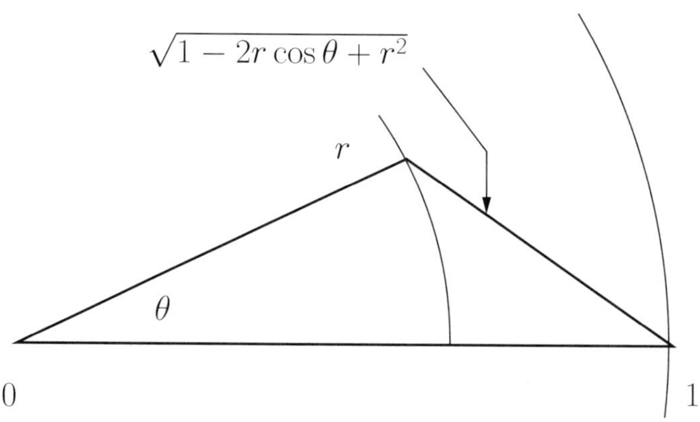

Figure 15.2. Some geometry of the Poisson kernel $P(r,\theta)$: The side of the triangle that joins the polar point $(1,0)$ to (r,θ) has length determined by the law of cosines, and the Poisson kernel $P(r,\theta)$ equals $1/2\pi$ times the ratio of $1-r^2$ to the square of this length.

the series converges for $|z| = r < 1$ we may write

$$\begin{aligned}
\frac{1}{2} + \sum_{n=1}^{\infty} r^n \cos n(\theta - t) &= \operatorname{Re}\Big(\frac{1}{2} + \sum_{n=1}^{\infty} z^n\Big) \\
&= \operatorname{Re}\Big(-\frac{1}{2} + \frac{1}{1-z}\Big) \qquad \text{(for } |z|<1\text{)} \\
&= \operatorname{Re}\frac{1+z}{2(1-z)} \\
&= \operatorname{Re}\frac{(1+z)(1-\bar{z})}{2|1-z|^2} \\
&= \frac{1-|z|^2}{2|1-z|^2} \\
&= \frac{1-r^2}{2(1-2r\cos\alpha + r^2)}. \qquad (15.16)
\end{aligned}$$

Substituting this into the previous integral formula for $u(r,\theta)$, and recalling that $\alpha = \theta - t$, we have the **Poisson integral formula**,

$$u(r,\theta) = \frac{1}{2\pi}\int_{-\pi}^{\pi} f(t)\frac{1-r^2}{1-2r\cos(\theta-t)+r^2}\,dt. \tag{15.17}$$

This formula expresses the solution of the Dirichlet problem with boundary data f as an integral of f times a shifted version of the **Poisson kernel** $P(r,\theta)$ defined by

$$P(r,\theta) := \frac{1}{2\pi}\frac{1-r^2}{1-2r\cos\theta+r^2}, \tag{15.18}$$

which is independent of f. (See Figure 15.2.) Thus the Poisson integral formula takes the form

$$u(r,\theta) = \int_{-\pi}^{\pi} f(t)P(r,\theta-t)\,dt.$$

The integral formula is valid for points (r, θ) in the *interior* of the disc, due to our summation of the geometric series in its derivation. If we set $r = 0$ in the integral formula, we find that

$$u(0, \theta) = \frac{1}{2\pi} \int_{-\pi}^{\pi} f(t)\, dt,$$

which says that the value of the steady state temperature distribution at the center of the disc is the average value of f over the boundary circle. This also follows from the series representation of $u(r, \theta)$ in view of the definition of the coefficient a_0.

Note that if we set $r = 1$ in the Poisson integral formula, we cannot get the correct result. However, we have *Poisson's theorem.*

Theorem 15.3.1 (Poisson). *If f is continuous and 2π-periodic, and $u(r, \theta)$ is given by (15.17), then*

$$\lim_{r \to 1^-} u(r, \theta) = f(\theta),$$

uniformly in θ.

Poisson's theorem follows from key properties of the Poisson kernel $P(r, \theta)$ as a function of θ. These properties are not difficult to establish. In particular:

(1) $P(r, \theta) = P(r, -\theta)$, that is, $P(r, \theta)$ is an even function of θ. This is clear from (15.18) since $\cos\theta$ is an even function.

(2) For each $r < 1$, the maximum of $P(r, \theta)$ occurs at $\theta = 0$, and the maximum value is $(1 + r)/2\pi(1 - r)$.

(3) For each $r < 1$, $P(r, \theta)$ is decreasing on $0 \leq \theta \leq \pi$ with a minimum value of $(1 - r)/2\pi(1 + r)$ at $\theta = \pi$. Thus, $P(r, \theta) > 0$.

(4) For each $r < 1$, we have $\int_{-\pi}^{\pi} P(r, \theta)\, d\theta = 1$.

(See Figure 15.3.) We can view the Poisson integral formula (15.17) as a weighted sum of the values of the boundary data $f(\theta)$.

We now turn to the proof of Poisson's theorem.

Proof of Poisson's Theorem 15.3.1. We must show that given any $\epsilon > 0$, there is an $r_0 < 1$ such that for all r with $r_0 < r < 1$, we have

$$|u(r, \theta) - f(\theta)| < \epsilon.$$

When we integrate a function 2π-periodic in θ over the interval $[-\pi, \pi]$, the result is the same as integrating that function over the interval $[a + \pi, a - \pi]$ for any number a. Since $P(r, \theta)$ is positive and 2π-periodic in θ, it follows that for fixed $r < 1$,

$$\begin{aligned} |u(r, \theta) - f(\theta)| &= \left| \int_{\theta-\pi}^{\theta+\pi} P(r, \theta - t)[f(t) - f(\theta)]\, dt \right| \\ &\leq \int_{\theta-\pi}^{\theta+\pi} P(r, \theta - t)\, |f(t) - f(\theta)|\, dt. \end{aligned}$$

For t near θ, we can make $|f(t) - f(\theta)|$ small by the continuity of f. For larger differences $\theta - t$, we can still make P small by taking r close to 1. So we write the

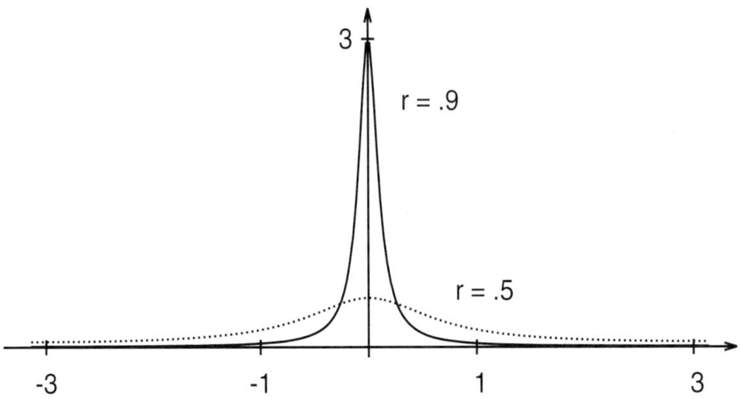

Figure 15.3. Graphs of the Poisson kernel $P(r,\theta)$, for $r = .5$ (dotted curve) and $r = .9$ (solid curve).

last integral above as the sum of two integrals, as follows. For any δ with $0 < \delta < \pi$, we have

$$\begin{aligned} |u(r,\theta) - f(\theta)| &\leq \int_{\theta-\pi}^{\theta+\pi} P(r,\theta - t)\,|f(t) - f(\theta)|\,dt \\ &= \int_{|\theta-t|\leq\delta} P(r,\theta - t)\,|f(t) - f(\theta)|\,dt \\ &\quad + \int_{\delta<|\theta-t|<\pi} P(r,\theta - t)\,|f(t) - f(\theta)|\,dt. \end{aligned}$$

Denote the sum of integrals on the right by $I_1 + I_2$.

To estimate I_1, first note that $\theta - \pi, \theta + \pi \in [-2\pi, 2\pi]$, and f is uniformly continuous on $[-2\pi, 2\pi]$. Given $\epsilon > 0$, there is a δ with $0 < \delta < \pi$ such that if $|\theta - t| < \delta$, then

$$|f(t) - f(\theta)| < \frac{\epsilon}{2}.$$

For this δ,

$$I_1 \leq \frac{\epsilon}{2}\int_{|\theta-t|\leq\delta} P(r,\theta - t)\,dt < \frac{\epsilon}{2},$$

where we have used the integral property (4) of the Poisson kernel.

For the integral I_2 over $\delta < |\theta - t| < \pi$, observe that

$$\max_{\delta<|\theta-t|<\pi} P(r,\theta - t) = \frac{1}{2\pi}\frac{1 - r^2}{1 - 2r\cos\delta + r^2}$$

by property (3) of the Poisson kernel. Thus,

$$I_2 \le \frac{1}{2\pi}\frac{1-r^2}{1-2r\cos\delta+r^2}\int_{\delta<|\theta-t|<\pi}|f(t)-f(\theta)|\,dt.$$

With δ fixed such that $0<\delta<\pi$, we have

$$\lim_{r\to 1^-}\frac{1}{2\pi}\frac{1-r^2}{1-2r\cos\delta+r^2}\int_{\delta<|\theta-t|<\pi}|f(t)-f(\theta)|\,dt = 0,$$

since $(1-2r\cos\delta+r^2)\to 2-2\cos\delta\neq 0$ and $(1-r^2)\to 0$. Hence there is an $r_0<1$ such that if $r_0<r<1$, then the bound on I_2 is less than $\epsilon/2$.

Combining the estimates for I_1 and I_2, we have that for $r_0<r<1$,

$$|u(r,\theta)-f(\theta)| \le I_1+I_2 < \frac{\epsilon}{2}+\frac{\epsilon}{2} = \epsilon,$$

as was to be shown. □

Poisson's theorem has the following interesting consequence.

Theorem 15.3.2. *Let $f,g:\mathbf{R}\to\mathbf{R}$ be continuous functions of period 2π, or, equivalently, $f,g\in CP[-\pi,\pi]$. If f and g have the same Fourier series, that is, the same Fourier coefficients, then $f(x)=g(x)$ for all $x\in\mathbf{R}$.*

Proof. By assumption, f and g have the same Fourier coefficients a_0, a_n and b_n. By Theorem 15.3.1, for each fixed θ, the function

$$u(r,\theta) = \frac{1}{2}a_0 + \sum_{n=1}^{\infty} r^n(a_n\cos n\theta + b_n\sin n\theta)$$

converges to $f(\theta)$ and to $g(\theta)$ as $r\to 1-$. Hence $f(\theta)=g(\theta)$ for all θ by uniqueness of limits. □

We note that Theorem 15.3.2 implies that the mapping from $CP[-\pi,\pi]$ to the sequence space l^2 defined by $f\mapsto(a_0,a_1,b_1,a_2,b_2,\ldots)$, where a_0,a_n,b_n are the Fourier coefficients of f, is a one-to-one mapping.

We have found a solution of the Dirichlet problem for the closed unit disk, at least for boundary data f that is continuous and piecewise smooth, and we have given two representations for that solution, by series and by the Poisson integral formula. But we have not proved that the Dirichlet problem has a *unique* solution for a given f. We now fill that gap.

The uniqueness of the solution to the Dirichlet problem follows from an important property of any solution of Laplace's equation $\Delta u = 0$ that is continuous on the boundary of the disk D.

Theorem 15.3.3. *Let u be a harmonic function in the open unit disk D that is continuous on $\overline{D} = D\cup\partial D$. Then u achieves its maximum and minimum values at points on the boundary,*

$$\max_{\overline{D}} u = \max_{\partial D} u \qquad \textit{and} \qquad \min_{\overline{D}} u = \min_{\partial D} u.$$

Proof. We prove the statement about the maximum value. The argument for the minimum is similar. The proof is by contradiction. We assume that $\Delta u = u_{xx} + u_{yy} = 0$ in D and that the maximum of u does not occur on ∂D. Then there is a point $(x_0, y_0) \in D$ such that

$$d := u(x_0, y_0) - \max_{\partial D} u > 0.$$

Consider the function $v : \overline{D} \to \mathbf{R}$ defined by

$$v(x, y) = u(x, y) + \epsilon[(x - x_0)^2 + (y - y_0)^2],$$

where $\epsilon > 0$. Then v is C^2 on D and continuous on ∂D. We have

$$|v(x, y) - u(x, y)| \leq \epsilon, \qquad (x, y) \in \overline{D},$$

and $v(x_0, y_0) = u(x_0, y_0)$. Hence, if we take $\epsilon < d/2$, then

$$v(x_0, y_0) = u(x_0, y_0) > \max_{\partial D} v.$$

This implies that the maximum value of v is achieved at some point (x_1, y_1) in the open disk D. Necessary conditions for this maximum of v are that $v_x(x_1, y_1) = v_y(x_1, y_1) = 0$, and $v_{xx}(x_1, y_1) \leq 0$ and $v_{yy}(x_1, y_1) \leq 0$. Consequently, $\Delta v(x_1, y_1) \leq 0$. However,

$$\begin{aligned} \Delta v(x_1, y_1) &= \Delta u(x_1, y_1) + \epsilon \Delta[(x - x_0)^2 + (y - y_0)^2] \\ &= 0 + 4\epsilon > 0. \end{aligned}$$

This is the contradiction we were seeking. Therefore u achieves its maximum value on the boundary of D. The argument for the minimum value is similar and is left to Exercise 15.3.9. □

The uniqueness of a solution to the Dirichlet problem now follows easily.

Theorem 15.3.4. *There is at most one solution of the Dirichlet problem, $\Delta u = 0$ on D with $u = f$ on ∂D, where u is C^2 on D and f is a continuous function on ∂D.*

Proof. Let f be continuous on ∂D, and suppose that u_1 and u_2 both solve the Dirichlet problem with boundary data f. Let $w = u_1 - u_2$. Then

$$\Delta w = \Delta u_1 - \Delta u_2 = 0 \quad \text{on } D,$$

and $w = 0$ on ∂D. By Theorem 15.3.3, necessarily $w \leq 0$ on D. The same argument applies to $-w$, so $-w \leq 0$. (Or, by the minimum property of Theorem 15.3.3, we have $w \geq 0$.) Hence, $w = 0$ on D, so $u_1 = u_2$ on D. Since $u_1 = u_2 = f$ on ∂D, this completes the proof. □

We end the section with an example to provide encouragement for readers to study complex variable techniques in the Dirichlet problem and in other problems in partial differential equations. The example shows that complex variables and complex analysis can be helpful. The example also shows the importance of knowing about the uniqueness of solutions.

Example 15.3.5. If $f(\theta) = \cos 2\theta$, then in our expression for the Poisson integral formula, the solution of the Dirichlet problem is

$$u(r,\theta) = \frac{1}{2\pi}\int_{-\pi}^{\pi} \cos 2t \frac{1-r^2}{1-2r\cos(\theta-\rho)+r^2}\,dt.$$

This is difficult to evaluate exactly by integration techniques or tables. But we show here that $f(\theta) = \cos 2\theta = \operatorname{Re} z^2$, where $z = x + iy = re^{i\theta} = e^{i\theta}$ when $r = 1$, on the boundary circle. For any point within the disk,

$$\operatorname{Re} z^2 = \operatorname{Re}\left[(x+iy)(x+iy)\right] = \operatorname{Re}(x^2 - y^2 + i2xy) = x^2 - y^2,$$

and $x^2 - y^2$ is a solution to Laplace's equation $u_{xx} + u_{yy} = 0$. Moreover, on the boundary circle, we have $x^2 - y^2 = \cos^2\theta - \sin^2\theta = \cos 2\theta$, as we wished to show. Therefore $\operatorname{Re} z^2 = x^2 - y^2 = r^2\cos^2\theta - r^2\sin^2\theta = r^2\cos 2\theta$ is the unique solution of the Dirichlet problem with boundary data $f(\theta) = \cos 2\theta$. More generally, from complex analysis, both the real part $u(x,y)$ and the imaginary part $v(x,y)$ of a complex analytic function $f(z) = u(x,y) + iv(x,y)$ are solutions of Laplace's equation, by virtue of the Cauchy-Riemann equations,

$$u_x(x,y) = v_y(x,y) \qquad \text{and} \qquad u_y(x,y) = -v_x(x,y),$$

which must be satisfied by the real and imaginary parts of any complex analytic function. △

One can consider the Dirichlet problem for any open planar domain Ω that is simply connected (informally, the domain has no holes) with piecewise smooth boundary. The famous *Riemann mapping theorem* of complex analysis asserts that such a domain can be mapped conformally (that is, by a complex differentiable mapping) onto the open unit disk with the boundary of Ω being mapped to the unit circle. Thus, in principle, the Dirichlet problem for Ω can be transformed to the Dirichlet problem for the unit disk, whose solution we have obtained in this section.

Exercises.

Exercise 15.3.1. Show that the polar coordinate transformation $x = r\cos\theta$, $y = r\sin\theta$ transforms the equation $u_{xx}(x,y) + u_{yy}(x,y) = 0$ into the polar form

$$u_{rr}(r,\theta) + \frac{1}{r}u_r(r,\theta) + \frac{1}{r^2}u_{\theta\theta}(r,\theta) = 0.$$

Hint: Think of $u = u(r,\theta) = u(r(x,y),\theta(x,y))$. Start with $u_x = u_r r_x + u_\theta \theta_x$ and $u_y = u_r r_y + u_\theta \theta_y$. Recall that $r = (x^2+y^2)^{1/2}$ and $\theta = \arctan(y/x)$.

Exercise 15.3.2. Show that the boundary value problem $\Theta''(\theta) + \lambda\Theta(\theta) = 0$, $\Theta(-\pi) = \Theta(\pi)$, has nonzero solutions only for $\lambda \geq 0$, and that the allowable values of λ are given by $\lambda_n = n^2$ for nonnegative integer n. Thus verify the solutions for $\Theta(\theta)$ stated in the text. *Hint*: Consider the cases $\lambda < 0$, $\lambda = 0$ and $\lambda > 0$ separately.

Exercise 15.3.3. Verify that the equation $r^2R''(r) + rR'(r) - n^2R(r) = 0$, for nonnegative integer n, has the solutions stated in the text. *Note*: To *construct* these solutions, instead of simply verifying them, we note that an Euler equation with independent variable r may be transformed to a constant coefficient linear equation in the independent variable t by means of the change of variable $r = e^t$.

Exercise 15.3.4. Verify statements (2) and (3) concerning the maximum and minimum values of the Poisson kernel $P(r, \theta)$ for fixed $r < 1$.

Exercise 15.3.5. Verify property (4) concerning the integral of the Poisson kernel $P(r, \theta)$ for $r < 1$. *Hint*: From (15.18) and the left-hand side of (15.16), we have

$$P(r, \theta) = \frac{1}{\pi}\left[\frac{1}{2} + \sum_{n=1}^{\infty} r^n \cos n\theta\right].$$

Justify the term-by-term integration of the series.

Exercise 15.3.6. Argue that if f is 2π-periodic and piecewise continuous, that is, f has at most finitely many discontinuities in any finite interval, all jump discontinuities, then

$$\lim_{r \to 1^-} \int_{-\pi}^{\pi} f(t)P(r, \theta_0 - t)\, dt = f(\theta_0)$$

at any point θ_0 where f is continuous.

Exercise 15.3.7. Establish the following results:

1. If f is continuous of period 2π, or equivalently, $f \in CP[-\pi, \pi]$, and if the Fourier series of f converges uniformly to a function g, then $g = f$.
2. The Fourier series of $f(\theta) = |\theta|$, $-\pi \leq \theta \leq \pi$, converges uniformly. *Hint*: Looking forward a bit, you may use the statement of Theorem 15.6.7 to confirm that the series converges to $f(\theta)$ for all θ.

Exercise 15.3.8. Suppose f is continuous and 2π-periodic, with

$$M = \max_{-\pi \leq \theta \leq \pi} f(\theta) \qquad \text{and} \qquad m = \min_{-\pi \leq \theta \leq \pi} f(\theta).$$

Show that the solution $u(r, \theta)$ of the Dirichlet problem with boundary data f satisfies $m \leq u(r, \theta) \leq M$.

Exercise 15.3.9. Complete the proof of Theorem 15.3.3 by showing that a function u harmonic on the unit disk D and continuous on $\overline{D}$ achieves its minimum value at a point on ∂D.

Exercise 15.3.10. Let $D_n = B_1(\mathbf{0}) = \{\mathbf{x} \in \mathbf{R}^n : |\mathbf{x}| < 1\}$, the open unit ball in $\mathbf{R}^n$, and let $\Delta u := u_{x_1 x_1} + \cdots + u_{x_n x_n}$ be the Laplacian operator on C^2 functions of n variables. Prove that there is at most one solution of the Dirichlet problem, $\Delta u = 0$ on D_n with $u = f$ on ∂D_n, where u is C^2 on D and f is a continuous function on ∂D.

Exercise 15.3.11. Suppose f is a function whose Fourier series converges to f on $[-\pi, \pi]$. Show that the Dirichlet problem for the circle $x^2 + y^2 = R^2$, with boundary data $f(\theta)$, has the solution

$$u(r, \theta) = \frac{1}{2}a_0 + \sum_{n=1}^{\infty} \left(\frac{r}{R}\right)^n (a_n \cos n\theta + b_n \sin n\theta),$$

where a_n and b_n are the Fourier coefficients of $f(\theta)$, and that the Poisson integral formula for this solution is

$$u(r,\theta) = \frac{1}{2\pi}\int_{-\pi}^{\pi} \frac{R^2 - r^2}{R^2 - 2Rr\cos(\theta - t) + r^2} f(t)\, dt.$$

15.4. More Separation of Variables

This section is offered as a brief exploration, via guided exercises, to apply the separation of variables technique to some basic problems for the heat equation and the wave equation.

15.4.1. The Heat Equation: Two Basic Problems.

The one-dimensional heat equation is

$$u_t(x,t) = u_{xx}(x,t).$$

It models the distribution of heat in a thin metallic rod of uniform density, assuming that the rod is insulated along its lateral surface so that there is no exchange of heat with the surrounding medium through this surface. More precisely, $u(x,t)$ is the temperature of the rod at position x at time t. The time rate of change of the temperature, u_t, is proportional to the one-dimensional Laplacian of u, u_{xx}. (Take the proportionality constant to be 1.) We consider two problems for the heat equation that combine elements discussed previously for Fourier series.

Exercises.

Exercise 15.4.1. *The rod with fixed temperature at the ends*
For convenience we consider a metal rod of length π. The temperature of the rod at position x at time t is $u(x,t)$. The temperature at the ends $x = 0$ and $x = \pi$ is fixed at zero, and the initial temperature distribution is given. Thus we assume that

$$u_t(x,t) = u_{xx}(x,t), \quad 0 < x < \pi, \quad t > 0,$$

$u(0,t) = 0$, $u(\pi,t) = 0$, and $u(x,0) = f(x)$ for $0 < x < \pi$, where f is assumed to be piecewise smooth.

1. Using the boundary conditions at the endpoints, show that the search for product solutions $\phi(x)T(t)$ leads to the boundary value problem (BVP)

   $$\phi''(x) + \lambda\phi(x) = 0, \quad \phi(0) = 0, \quad \phi(\pi) = 0$$

   together with $T'(t) + \lambda T(t) = 0$, where λ is a constant.
2. Show that the BVP of part 1 has nonzero solutions *only for positive* λ. (Consider the cases $\lambda < 0$, $\lambda = 0$ and $\lambda > 0$ separately.) Show that, in fact, the eigenvalues are $\lambda_n = n^2$, $n = 1, 2, 3, \ldots$, with corresponding eigenfunctions $\phi_n(x) = \sin nx$, $n = 1, 2, 3, \ldots$. (The λ_n are the eigenvalues of the linear differential operator $-d^2/dt^2$.)
3. Show that for each $\lambda_n = n^2$, $n = 1, 2, 3, \ldots$, we may take $T_n(t) = e^{-n^2 t}$, giving the product solutions $u_n(x,t) = \phi_n(x)T_n(t) = \sin(nx)e^{-n^2 t}$, $n = 1, 2, 3, \ldots$.

4. To match the initial condition defined by f, use a series of the form

$$u(x,t) = \sum_{n=1}^{\infty} b_n \sin(nx) e^{-n^2 t}$$

and require that

$$u(x,0) = f(x) = \sum_{n=1}^{\infty} b_n \sin(nx).$$

This requires a *sine series* for f valid on $0 < x < \pi$. To achieve it, extend f to an *odd* function f_{odd} on the interval $-\pi < x < \pi$, and use the Fourier series of f_{odd} to get the sine series required, with coefficients

$$b_n := \frac{1}{\pi} \int_{-\pi}^{\pi} f_{odd}(x) \sin(nx)\, dx.$$

In fact, since the integrand is an even function (odd times odd = even), we can actually write

$$b_n = \frac{2}{\pi} \int_{0}^{\pi} f(x) \sin(nx)\, dx,$$

since $f_{odd}(x) = f(x)$ for $0 < x < \pi$.

5. Construct the solution in the case where $f(x) = x$ for $0 < x < \pi$. First, show that f_{odd} has Fourier sine series

$$f_{odd}(x) = \sum_{n=1}^{\infty} (-1)^{n+1} \frac{2}{n} \sin(nx).$$

Justify the term-by-term differentiations required to show that the solution of the heat equation problem with fixed end temperatures is then given by

$$u(x,t) = \sum_{n=1}^{\infty} (-1)^{n+1} \frac{2}{n} \sin(nx) e^{-n^2 t}.$$

Note that by construction, the sum of this series satisfies the boundary conditions and the initial condition.

Exercise 15.4.2. *The rod with insulated ends*
Again we consider a metal rod of length π. The temperature at position x at time t is $u(x,t)$, and the problem of insulated ends is described by

$$u_t(x,t) = u_{xx}(x,t), \qquad 0 < x < \pi, \quad t > 0,$$

with boundary conditions

$$u_x(0,t) = 0, \qquad u_x(\pi,t) = 0,$$

and initial condition

$$u(x,0) = f(x), \quad 0 < x < \pi,$$

where f is assumed to be piecewise smooth. The boundary conditions say that there is no heat flow into or out of the rod ends.

1. Using the boundary conditions at the endpoints, show that the search for product solutions leads to the boundary value problem (BVP)

$$\phi''(x) + \lambda\phi(x) = 0, \quad \phi'(0) = 0, \quad \phi'(\pi) = 0$$

together with $T'(t) + \lambda T(t) = 0$, where λ is a constant.

2. Show that in this case the boundary value problem for ϕ has eigenvalues $\lambda_0 = 0$, $\lambda_n = n^2$, for $n = 1, 2, 3, \ldots$, with corresponding eigenfunctions $\phi_0(x) = 1$, $\phi_n(x) = \cos nx$, for $n = 1, 2, 3, \ldots$.

3. Show that we may take $T_0(t) = 1$ for $\lambda_0 = 0$, and $T_n(t) = e^{-n^2 t}$ for $n = 1, 2, 3, \ldots$, giving the product solutions $u_n(x,t) = \phi_n(x)T_n(t) = \cos(nx)e^{-n^2 t}$, $n = 1, 2, 3, \ldots$, and $u_0(x,t) = 1$.

4. To match the initial condition defined by f, use a series of the form

$$u(x,t) = \frac{1}{2}a_0 + \sum_{n=1}^{\infty} a_n \cos(nx)e^{-n^2 t}$$

and require that

$$u(x,0) = f(x) = \frac{1}{2}a_0 + \sum_{n=1}^{\infty} a_n \cos(nx).$$

This requires a *cosine series* for f valid on $0 < x < \pi$. To achieve it, extend f to an *even* function f_{even} on the interval $-\pi < x < \pi$, and use the Fourier series of f_{even} to get the cosine series required, with coefficients

$$a_n := \frac{1}{\pi}\int_{-\pi}^{\pi} f_{even}(x)\cos(nx)\,dx.$$

In fact, since the integrand is an even function (even times even = even), we can actually write

$$a_n = \frac{2}{\pi}\int_0^{\pi} f(x)\sin(nx)\,dx,$$

since $f_{even}(x) = f(x)$ for $0 < x < \pi$.

5. Construct the solution in the case where $f(x) = x$ for $0 < x < \pi$. First, show that f_{even} has Fourier cosine series

$$f_{even}(x) = \frac{\pi}{2} - \frac{4}{\pi}\sum_{n=1}^{\infty} \frac{1}{(2n-1)^2}\cos(2n-1)x.$$

Justify the term-by-term differentiations required to show that the solution of the heat equation problem with insulated ends is given by

$$u(x,t) = \frac{\pi}{2} - \frac{4}{\pi}\sum_{n=1}^{\infty} \frac{1}{(2n-1)^2}\cos[(2n-1)x]\,e^{-(2n-1)^2 t}.$$

Note that by construction, the sum of this series satisfies the boundary conditions and the initial condition.

Exercise 15.4.3. Consider the function

$$f(x) = 100, \quad 0 < x < \pi.$$

Construct three different Fourier series for f, as follows:

1. Extend f to an even function on $[-\pi, \pi]$. Find the Fourier series for this extension. This Fourier series is called the Fourier cosine series for f on $[0, \pi]$.
2. Extend f to an odd function on $[-\pi, \pi]$. Find the Fourier series for this extension. This Fourier series is called the Fourier sine series for f on $[0, \pi]$.
3. Extend f to $[-\pi, \pi]$ by setting $f(x) = 0$ for $-\pi < x < 0$. Find the Fourier series for this extension.
4. Solve the heat equation problem with fixed temperature at the ends, if $f(x) = 100$ for $0 < x < \pi$.
5. Solve the heat equation problem with insulated ends, if $f(x) = 100$ for $0 < x < \pi$.

15.4.2. The Wave Equation with Fixed Ends. The one-dimensional wave equation is

$$u_{tt}(x, t) = u_{xx}(x, t).$$

It models the vibrations of a stretched string under additional boundary conditions on the ends of the string. Think of a guitar string, for example. The real function $u(x, t)$ is the displacement of the string, at position x at time t, from its equilibrium position, which is taken to be $u = 0$. We consider a basic problem involving the wave equation in which the string ends are fixed in position with displacement zero. Notice that initial conditions for both position and velocity are needed since the equation is second order in the time t.

Exercise.

Exercise 15.4.4. *The vibrating string with fixed ends*
For convenience we consider a string of length π. The fixed end problem is described by

$$u_{tt}(x, t) = u_{xx}(x, t), \qquad 0 < x < \pi, \quad t > 0,$$

with boundary conditions

$$u(0, t) = 0, \qquad u(\pi, t) = 0,$$

and the initial displacement and initial velocity of the string modeled by

$$u(x, 0) = f(x), \qquad u_t(x, 0) = g(x), \qquad 0 < x < \pi,$$

where f and g are assumed to be piecewise smooth.

1. Using the separation of variables method, construct a series solution for the wave equation with fixed ends.
2. Carry out the solution for a "plucked string" where the initial displacement is given by
$$f(x) = \begin{cases} \frac{1}{50}x, & 0 < x \le \pi/2, \\ \frac{1}{50}(\pi - x), & \pi/2 \le x < \pi \end{cases}$$
and the initial velocity is $g(x) = 0$ for $0 < x < \pi$. (See Figure 15.4.)

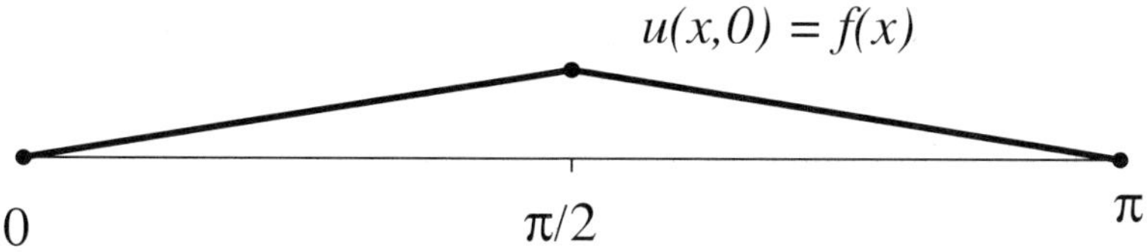

Figure 15.4. The initial profile, $u(x,0) = f(x)$, of the plucked string. The initial velocity is $u_t(x,0) = g(x) = 0$.

15.5. The Best Mean Square Approximation

If f is Riemann integrable over $[-\pi, \pi]$, then also its square f^2 is Riemann integrable over $[-\pi, \pi]$, so

$$\int_{-\pi}^{\pi} f^2(x)\, dx < \infty.$$

Suppose we approximate f by a trigonometric polynomial of the form

$$T(x) = \frac{1}{2}c_0 + \sum_{k=1}^{n}(c_k \cos kx + d_k \sin kx), \tag{15.19}$$

that is, a **trigonometric polynomial of degree** n, as we shall call it. The next theorem states that the Fourier partial sums of a Riemann integrable function f on $[-\pi, \pi]$ provide the best mean square approximation of f among all trigonometric polynomials.

Theorem 15.5.1. *If f is Riemann integrable over $[-\pi, \pi]$ and $T(x)$ is a trigonometric polynomial of degree at most n as in (15.19), then the integral*

$$\int_{-\pi}^{\pi} [f(x) - T(x)]^2\, dx$$

is minimized when the coefficients c_k and d_k are equal to the Fourier coefficients of f, that is, $c_k = a_k$ in (15.6) for $0 \le k \le n$, and $d_k = b_k$ in (15.7) for $1 \le k \le n$.

Proof. If $T(x)$ has the form (15.19), then expansion of the squared integrand yields

$$\begin{aligned}\int_{-\pi}^{\pi} [f(x) - T(x)]^2\, dx &= \int_{-\pi}^{\pi} f^2(x)\, dx - 2\int_{-\pi}^{\pi} f(x)T(x)\, dx + \int_{-\pi}^{\pi} T^2(x)\, dx \\ &= \int_{-\pi}^{\pi} f^2(x)\, dx - \pi a_0 c_0 - 2\pi \sum_{k=1}^{n}(a_k c_k + b_k d_k) \\ &\quad + \frac{1}{2}\pi c_0^2 + \pi \sum_{k=1}^{n}(c_k^2 + d_k^2). \end{aligned} \tag{15.20}$$

Write

$$s_n(x) = \frac{1}{2}a_0 + \sum_{k=1}^{n}(a_k \cos kx + b_k \sin kx) \tag{15.21}$$

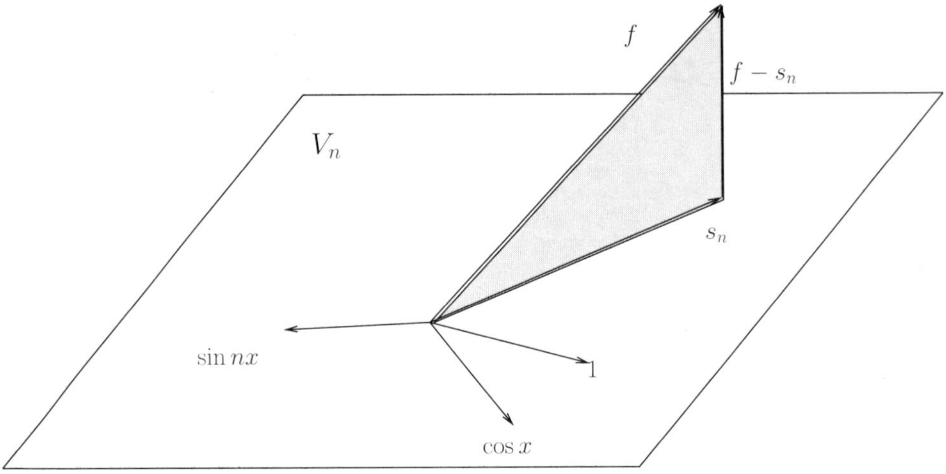

Figure 15.5. The Fourier partial sum s_n is the best mean square approximation to f from $V_n = \text{span}\,\{1, \cos x, \sin x, \ldots, \cos nx, \sin nx\}$. The error $f - s_n$ is orthogonal to V_n.

for the n-th partial sum of the Fourier series of f. If we let $T(x) = s_n(x)$, then (15.20) immediately implies

$$\int_{-\pi}^{\pi} [f(x) - s_n(x)]^2\,dx = \int_{-\pi}^{\pi} f^2(x)\,dx - \frac{1}{2}\pi a_0^2 - \pi \sum_{k=1}^{n} (a_k^2 + b_k^2). \tag{15.22}$$

Then a direct comparison of (15.20) and (15.22) shows that

$$\begin{aligned}\int_{-\pi}^{\pi} [f(x) - T(x)]^2\,dx &= \int_{-\pi}^{\pi} [f(x) - s_n(x)]^2\,dx \\ &\quad + \frac{1}{2}\pi(a_0 - c_0)^2 + \pi \sum_{k=1}^{n} [(a_k - c_k)^2 + (b_k - d_k)^2].\end{aligned}$$

It follows that

$$\int_{-\pi}^{\pi} [f(x) - s_n(x)]^2\,dx \le \int_{-\pi}^{\pi} [f(x) - T(x)]^2\,dx,$$

and equality holds if and only if $c_0 = a_0$ and $c_k = a_k$, $d_k = b_k$ for $k = 1, \ldots, n$. This completes the proof. □

Theorem 15.5.1 says that the best mean square approximation to f from the subspace spanned by the functions

$$1, \cos x, \sin x, \ldots, \cos nx, \sin nx,$$

is the Fourier partial sum $s_n = \frac{1}{2}a_0 + \sum_{k=1}^{n}(a_k \cos kx + b_k \sin kx)$. (See Figure 15.5.)

We can obtain more from the identity (15.22) in the proof. A simple rearrangement yields the inequality

$$\frac{1}{2}\pi a_0^2 + \pi \sum_{k=1}^{n} (a_k^2 + b_k^2) \leq \int_{-\pi}^{\pi} f^2(x)\, dx,$$

which holds for each positive integer n. Letting $n \to \infty$, we conclude that

$$\frac{1}{2}\pi a_0^2 + \pi \sum_{k=1}^{\infty} (a_k^2 + b_k^2) \leq \int_{-\pi}^{\pi} f^2(x)\, dx, \tag{15.23}$$

where the series on the left converges. This inequality is known as **Bessel's inequality**. The convergence of the series on the left implies the convergence of both $\sum a_k^2$ and $\sum b_k^2$. This yields the important fact known as the *Riemann-Lebesgue theorem.*

Theorem 15.5.2 (Riemann-Lebesgue). *If f is Riemann integrable over $[-\pi, \pi]$, then the Fourier coefficients of f satisfy $\lim_{k\to\infty} a_k = 0$ and $\lim_{k\to\infty} b_k = 0$.*

For an arbitrary Riemann integrable function f, the Fourier series is defined, but it need not be the case that the two series $\sum a_k$ and $\sum b_k$ converge. If these series converge *absolutely*, then the Fourier series converges uniformly, by the Weierstrass test. However, there are Riemann integrable functions for which the Fourier series does not converge uniformly, for example any function with finitely many jump discontinuities in $[-\pi, \pi]$.

We now show that equality holds in (15.23) for *continuous* functions of period 2π, yielding the identity

$$\frac{1}{2}\pi a_0^2 + \pi \sum_{k=1}^{\infty} (a_k^2 + b_k^2) = \int_{-\pi}^{\pi} f^2(x)\, dx, \tag{15.24}$$

called **Parseval's equality**. Observe that by (15.22), Parseval's equality (15.24) is equivalent to the statement that

$$\lim_{n\to\infty} \int_{-\pi}^{\pi} [f(x) - s_n(x)]^2\, dx = 0.$$

We will see that Parseval's equality for continuous functions follows from a trigonometric version of the Weierstrass approximation theorem, stated next.

Theorem 15.5.3 (Trigonometric Weierstrass Approximation Theorem). *Suppose f is a continuous function of period 2π, or equivalently, f is continuous on $[-\pi, \pi]$, $f(-\pi) = f(\pi)$, and f is extended to the real line by the definition $f(x + 2\pi) = f(x)$ for all x. Then for every $\epsilon > 0$ there is a trigonometric polynomial $T(x)$ of the form (15.19), such that*

$$|f(x) - T(x)| < \epsilon$$

for all real x.

Proof. Consider f as the boundary data in the Dirichlet problem for the unit disk. By Poisson's theorem, given $\epsilon > 0$ there is an $r_0 = r_0(\epsilon)$ such that for $r_0 < r < 1$,

$$|u(r, x) - f(x)| < \frac{\epsilon}{2}$$

for $-\pi \le x \le \pi$, where

$$u(r,x) = \frac{1}{2}a_0 + \sum_{k=1}^{\infty} r^n(a_n \cos nx + b_n \sin nx)$$

solves the Dirichlet problem with boundary data f, and

$$\begin{aligned} a_n &= \frac{1}{\pi}\int_{-\pi}^{\pi} f(x)\cos nx\,dx, \qquad n = 0,1,2,\ldots, \\ b_n &= \frac{1}{\pi}\int_{-\pi}^{\pi} f(x)\sin nx\,dx, \qquad n = 1,2,\ldots . \end{aligned}$$

Let $M = \max_{-\pi\le x\le\pi} |f(x)|$. Then $|a_n| \le 2M$ and $|b_n| \le 2M$. Hence, for any positive integer N,

$$\Big|\sum_{n\ge N} r^n(a_n \cos nx + b_n \sin nx)\Big| \le 4M \sum_{n\ge N} r^n = 4M\frac{r^N}{1-r},$$

where we have used the geometric series sum with $0 < r < 1$. We may choose $N = N(\epsilon)$ sufficiently large that the bound on the right-hand side is less than $\epsilon/2$. Then, in particular for

$$T_N(x) = \frac{1}{2}a_0 + \sum_{k=1}^{N} r^n(a_n \cos nx + b_n \sin nx),$$

the N-th partial sum of $u(r,x)$, and any fixed r with $r_0 < r < 1$, we have

$$\begin{aligned} |f(x) - T_N(x)| &= |f(x) - u(r,x)| + |u(r,x) - T_N(x)| \\ &< \frac{\epsilon}{2} + \frac{\epsilon}{2} = \epsilon \end{aligned}$$

for all $x \in [-\pi,\pi]$, and by periodicity for all real x. Thus the trigonometric polynomial $T_N(x)$ fulfills the stated requirement. □

It is worth remarking here that Theorem 15.5.3 does not contradict the fact that the Fourier series of a continuous function need not converge pointwise everywhere, much less uniformly, to the function, since the trigonometric polynomial T in this theorem is not a partial sum of the Fourier series of f.

Parseval's equality for continuous functions of period 2π is now a direct consequence of Theorem 15.5.3.

Theorem 15.5.4 (Parseval's Theorem). *If f is continuous on $[-\pi,\pi]$, $f(-\pi) = f(\pi)$, and a_n, b_n are the Fourier coefficients of f, then*

$$\frac{1}{2}\pi a_0^2 + \pi\sum_{k=1}^{\infty}(a_k^2 + b_k^2) = \int_{-\pi}^{\pi} f^2(x)\,dx. \tag{15.25}$$

Proof. Let $s_n(x)$ be the n-th partial sum of the Fourier series of f, as in (15.21). Theorem 15.5.3 applies to the extension of f to the real line, and for any $n \ge N = N(\epsilon)$ as in Theorem 15.5.3, Theorem 15.5.1 implies that

$$\int_{-\pi}^{\pi} [f(x) - s_n(x)]^2\,dx \le \int_{-\pi}^{\pi} [f(x) - T_N(x)]^2\,dx < 2\pi\epsilon^2.$$

This shows that

$$\lim_{n\to\infty} \int_{-\pi}^{\pi} [f(x) - s_n(x)]^2\, dx = 0,$$

and by (15.22) the proof is complete. □

The results of this section are reminiscent of the results on orthogonal expansion in $\mathbf{R}^n$. In particular, Parseval's equality (15.25) is the analogue of the Pythagorean theorem for continuous functions on $[-\pi, \pi]$. This may not be obvious from (15.25), because our elements $u = 1$, $\cos nx$, $\sin nx$, $n \in \mathbf{N}$, were not normalized so that $(u, u) = \int_{-\pi}^{\pi} u^2(x)\, dx = 1$. See Exercise 15.5.2.

The development of Fourier series can be carried out for Riemann integrable functions defined over any finite interval. For a brief start on such a program, see Exercise 15.5.4.

Exercises.

Exercise 15.5.1. Find the best mean square approximation of $f(x) = \sin^4(x)$ by a trigonometric polynomial of degree four or less, and display the approximating trigonometric polynomial. *Hint*: No integrations are required.

Exercise 15.5.2. We write $(h, g) := \int_{-\pi}^{\pi} h(x)g(x)\, dx$. We have seen in Example 8.2.5 and Exercise 8.2.4 that this product (h, g) has all the properties of an inner product except that $(h, h) = 0$ does not imply $h \equiv 0$, only that $h(x) = 0$ almost everywhere in $[-\pi, \pi]$. Thus, if we define $\|f\|_2 = (f, f)^{1/2}$, then we obtain a norm on the vector space of equivalence classes of Riemann integrable functions determined by the equivalence relation that $h \approx g$ if and only if $h(x) = g(x)$ almost everywhere in $[-\pi, \pi]$. Show that if we list the normalized trigonometric set

$$\left\{ \frac{1}{\sqrt{2\pi}}, \frac{\cos(x)}{\sqrt{\pi}}, \frac{\sin(x)}{\sqrt{\pi}}, \frac{\cos(2x)}{\sqrt{\pi}}, \frac{\sin(2x)}{\sqrt{\pi}}, \dots, \right\}$$

as the sequence $u_0, u_1, u_2, u_3, u_4, \dots$, then Parseval's identity (15.25), repeated here for convenience,

$$\int_{-\pi}^{\pi} f^2(x)\, dx = \frac{1}{2}\pi a_0^2 + \pi \sum_{k=1}^{\infty} (a_k^2 + b_k^2),$$

where a_0, a_k, b_k are the Fourier coefficients of f, takes the form

$$\|f\|_2^2 = (f, f) = \int_{-\pi}^{\pi} f^2(x)\, dx = \sum_{k=0}^{\infty} (f, u_k)^2.$$

Exercise 15.5.3. *Orthogonal sequences on* $[0, \pi]$
Note: The result of part 2 of this exercise is used later in the proof of Theorem 15.6.2 on pointwise convergence of Fourier series.

1. Show that each of the sequences $(\cos(nt))$, $n \geq 0$, and $(\sin(nt))$, $n \geq 1$, are pairwise orthogonal on the interval $[0, \pi]$, that is, for $n \neq m$,

$$\int_0^{\pi} \cos(nt)\cos(mt)\, dt = 0 \qquad \text{and} \qquad \int_0^{\pi} \sin(nt)\sin(mt)\, dt = 0.$$

2. Use the ideas of this section to deduce that for any Riemann integrable function f on $[0, \pi]$,

$$\lim_{n\to\infty} \int_0^{\pi} f(t)\cos(nt)\,dt = 0 \qquad \text{and} \qquad \lim_{n\to\infty} \int_0^{\pi} f(t)\sin(nt)\,dt = 0.$$

Hint: We have the Fourier cosine series for f_{even} and the Fourier sine series for f_{odd}. Establish Bessel's inequality for the coefficients of these two series.

Exercise 15.5.4. *Orthogonal sequences on* $[-L, L]$

1. Show that the functions

$$1,\ \cos\Big(\frac{\pi x}{L}\Big),\ \sin\Big(\frac{\pi x}{L}\Big),\ \ldots,\ \cos\Big(\frac{k\pi x}{L}\Big),\ \sin\Big(\frac{k\pi x}{L}\Big),\ \ldots$$

are pairwise orthogonal on $[-L, L]$.

2. Define the Fourier coefficients of a function f defined on $[-L, L]$ with respect to this orthogonal set.

15.6. Convergence of Fourier Series

As we mentioned earlier, there are functions whose Fourier series diverges at a point of continuity. Thus, to assure the convergence of the Fourier partial sums $s_n(x)$ to $f(x)$ for a particular x, we need a stronger hypothesis than continuity of f at the point x. The goal of this section is to establish a sufficient condition for the convergence of the Fourier series of f at x to the value $f(x)$, a condition general enough to cover many applications.

Let f be Riemann integrable over $[-\pi, \pi]$ and extend f to a periodic function on the whole real line with period 2π. (The actual values of f at $x = 2n\pi$, $n \in \mathbf{Z}$, do not matter, since a change in these values cannot affect the definition of the Fourier coefficients.) We begin with an alternative expression for the Fourier partial sums $s_n(x)$. Using the integral definitions for the Fourier coefficients a_n and b_n, in (15.6), (15.7), respectively, we have

$$\begin{aligned} s_n(x) &= \frac{1}{2}a_0 + \sum_{k=1}^{n}(a_k\cos(kx) + b_k\sin(kx)) \\ &= \frac{1}{\pi}\int_{-\pi}^{\pi}\Big[\frac{1}{2} + \sum_{k=1}^{n}(\cos(kx)\cos(kt) + \sin(kx)\sin(kt))\Big] f(t)\,dt \\ &= \frac{1}{\pi}\int_{-\pi}^{\pi}\Big[\frac{1}{2} + \sum_{k=1}^{n}\cos k(x-t)\Big] f(t)\,dt \end{aligned}$$

by the formula for the cosine of the difference of angles. Thus, we define the function

$$D_n(x) = \frac{1}{\pi}\Big[\frac{1}{2} + \sum_{k=1}^{n}\cos(kx)\Big] = \frac{1}{2\pi} + \frac{1}{\pi}\sum_{k=1}^{n}\cos(kx), \tag{15.26}$$

called the **Dirichlet kernel**. For each n, $D_n(x)$ is an even function of period 2π, since each of the functions $\cos(kx)$ is even and has period 2π. Then the Fourier partial sum $s_n(x)$ may be expressed as

$$s_n(x) = \int_{-\pi}^{\pi} D_n(x-t)\, f(t)\, dt. \tag{15.27}$$

For fixed x, if we let $u = x - t$, $du = -dt$, then the integral in (15.27) is

$$\int_{-\pi}^{\pi} D_n(u)\, f(x-u)\, (-1) du.$$

Now let $u = -t$, $du = -dt$, and use the fact that D_n is an even function, to obtain the same integral in the form

$$\int_{-\pi}^{\pi} D_n(t)\, f(x+t)\, dt.$$

Finally, since D_n and f have periodic 2π and D_n is even, we may replace t by $-t$ in the integrand, and write

$$s_n(x) = \int_{-\pi}^{\pi} D_n(t)\, f(x-t)\, dt. \tag{15.28}$$

We summarize the most important properties of the Dirichlet kernel $D_n(x)$.

Lemma 15.6.1. *The Dirichlet kernel $D_n(x)$ satisfies*

$$D_n(x) = \frac{1}{2\pi} + \frac{1}{\pi}\sum_{k=1}^{n} \cos(kx) = \frac{\sin(n+\frac{1}{2})x}{2\pi \sin\left(\frac{1}{2}x\right)} \qquad \text{for } n = 1, 2, 3, \dots,$$

with the value determined by the limit of the right-hand side at points where the denominator is zero. Moreover,

$$\int_{-\pi}^{\pi} D_n(x)\, dx = 1, \qquad \text{for } n = 1, 2, 3, \dots.$$

Proof. By the result on the finite geometric sum $\sum_{k=1}^{n} e^{ikx}$ in Example 3.11.10, the real part of that sum is given by

$$\sum_{k=1}^{n} \cos(kx) = \frac{\cos\frac{1}{2}(n+1)x\, \sin\frac{1}{2}nx}{\sin\frac{1}{2}x},$$

and hence,

$$\frac{1}{2} + \sum_{k=1}^{n} \cos(kx) = \frac{1}{2} + \frac{\cos\frac{1}{2}(n+1)x\, \sin\frac{1}{2}nx}{\sin\frac{1}{2}x}.$$

Multiplying both sides by $2\sin\frac{1}{2}x$, we have

$$2\sin\frac{1}{2}x \left[\frac{1}{2} + \sum_{k=1}^{n} \cos(kx)\right] = \sin\frac{1}{2}x + 2\cos\frac{1}{2}(n+1)x\, \sin\frac{1}{2}nx.$$

The final product term on the right can be written as the difference of the identities

$$\sin(A+B) = \sin A \cos B + \cos A \sin B$$

and

$$\sin(A-B) = \sin A \cos B - \cos A \sin B,$$

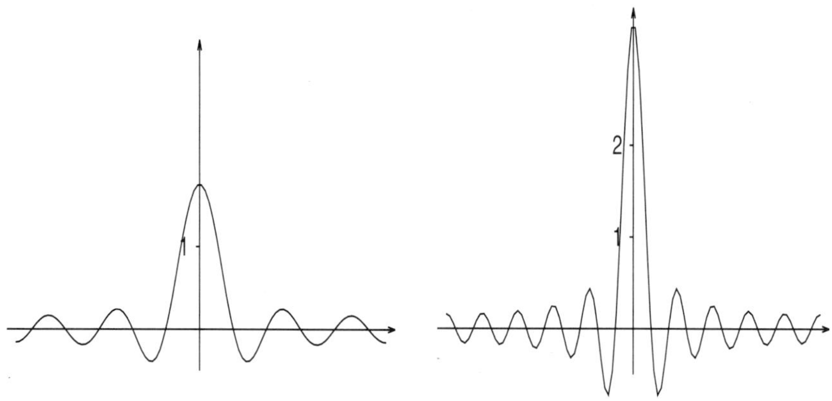

Figure 15.6. Graphs of the Dirichlet kernel $D_n(x)$ over $[-\pi, \pi]$: Left, $D_5(x)$; Right, $D_{10}(x)$.

with $A = \frac{1}{2}(n+1)x$ and $B = \frac{1}{2}nx$. This yields

$$2\cos\frac{1}{2}(n+1)x\ \sin\frac{1}{2}nx = \sin\left(n+\frac{1}{2}\right)x - \sin\frac{1}{2}x.$$

Thus,

$$2\sin\frac{1}{2}x\left[\frac{1}{2}+\sum_{k=1}^{n}\cos(kx)\right] = \sin\left(n+\frac{1}{2}\right)x,$$

and hence

$$\frac{1}{2}+\sum_{k=1}^{n}\cos(kx) = \frac{\sin(n+\frac{1}{2})x}{2\sin\left(\frac{1}{2}x\right)}.$$

A division by π yields the stated formula for $D_n(x)$.

Within the interval $[-\pi, \pi]$, $\sin\left(\frac{1}{2}x\right)$ is zero only when $x = 0$, and each $D_n(x)$ has a finite limit there, since an application of l'Hôpital's rule gives

$$\lim_{x\to 0}\frac{\sin(n+\frac{1}{2})x}{2\sin\left(\frac{1}{2}x\right)} = \lim_{x\to 0}\frac{(n+\frac{1}{2})\cos(n+\frac{1}{2})x}{\cos\left(\frac{1}{2}x\right)} = n+\frac{1}{2}.$$

Thus, each $D_n(x)$ is Riemann integrable over $[-\pi, \pi]$. In fact,

$$\int_{-\pi}^{\pi} D_n(x)\,dx = \int_{-\pi}^{\pi}\frac{1}{2\pi}\,dx = 1,$$

since $\int_{-\pi}^{\pi}\cos(kx)\,dx = 0$ for each k. □

The properties of the Dirichlet kernel help to motivate the pointwise convergence theorem given below. See Figure 15.6 for the graphs of $D_5(x)$ and $D_{10}(x)$ over $[-\pi, \pi]$.

Now consider $D_n(x-t)$ as a function of t for fixed x. By Lemma 15.6.1, the roots of $D_n(x-t)$ are at the points $t = x + k\delta$, where k is a nonzero integer and

$$\delta = \frac{2\pi}{2n+1}.$$

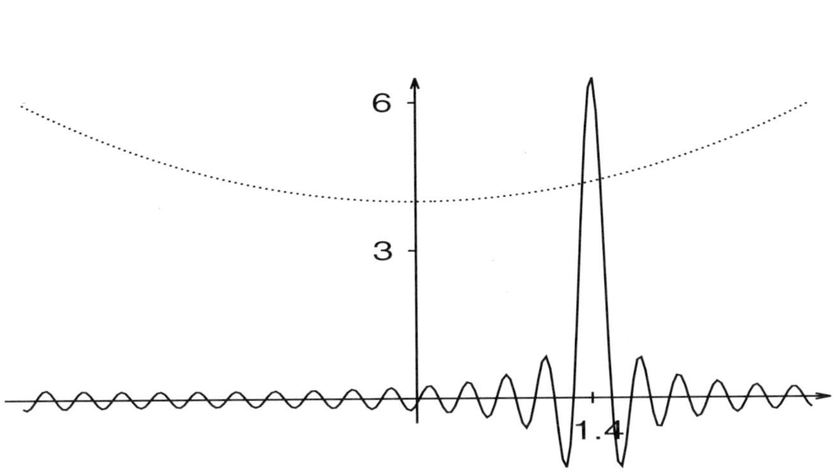

Figure 15.7. The sifting property of the Dirichlet kernel: The dotted graph of $f(t) = 4 + \frac{1}{5}t^2$ and the graph of $D_{20}(1.4 - t)$ as a function of t. The integral of $D_n(1.4-t)f(t)$ over the interval $(x-\delta, x+\delta)$, for small $\delta > 0$, approximates $f(1.4)$. In general, $\int_{x-\delta}^{x+\delta} D_n(x-t)\, f(t)\, dt \approx f(x)$.

The graph of $D_n(x-t)$ has a main arch of almost triangular shape over the interval $(x-\delta, x+\delta)$, and most of the weighting by $D_n(x-t)$ is concentrated near x. The integration from $-\pi$ to π used in computing $s_n(x)$ in (15.27) produces a near negligible contribution from outside the interval $(x-\delta, x+\delta)$, but inside that interval, $f(t) \approx f(x)$. The height of the main arch is $(2n+1)/2\pi$. Thus the area under the main arch is approximately

$$\frac{1}{2}(\text{base})\,(\text{height}) = \delta\frac{(2n+1)}{2\pi} = \frac{2\pi}{2n+1}\,\frac{(2n+1)}{2\pi} = 1.$$

(See Figure 15.7, which shows the graph of a continuous function f along with the graph of $D_n(x-t)$ as a function of t.) Thus the formula (15.27) for $s_n(x)$ implies that for large n,

$$s_n(x) \approx \int_{x-\delta}^{x+\delta} D_n(x-t)\, f(t)\, dt \approx f(x) \int_{x-\delta}^{x+\delta} D_n(x-t)\, dt \approx f(x).$$

With the considerations above as motivation, the next theorem gives a verifiable sufficient condition for the Fourier series of f to converge. Before proving it, we note that the value of f at a single point x may be reassigned in any way we want without changing the values of the Fourier coefficients of f and thus without changing the Fourier series of f. When the indicated one-sided limits exist, we write

$$f(x-) = \lim_{t \to x-} f(t)$$

for the left-hand limit of f at x, and

$$f(x+) = \lim_{t \to x+} f(t)$$

for the right-hand limit of f at x. Similarly, when $f(x-)$ and $f(x+)$ exist, we write

$$f'(x-) = \lim_{t\to 0+} \frac{f(x-t) - f(x-)}{t}$$

for the left-hand derivative of f at x, and

$$f'(x+) = \lim_{t\to 0+} \frac{f(x+t) - f(x+)}{t}$$

for the right-hand derivative of f at x.

Theorem 15.6.2. *Let f be Riemann integrable over $[-\pi, \pi]$, and extended to the entire real line by the condition $f(x + 2\pi) = f(x)$. Let $s_n(x)$ be the n-th partial sum of the Fourier series of f. Then $s_n(x) \to [f(x+) + f(x-)]/2$ at every point x at which the limiting values $f(x-)$, $f(x+)$, $f'(x-)$ and $f'(x+)$ exist.*

Proof. The integrability hypothesis on f ensures that the Fourier series of f is defined. By the integral statement in Lemma 15.6.1, for any x we may write

$$f(x) = f(x)\int_{-\pi}^{\pi} D_n(t)\,dt = \int_{-\pi}^{\pi} f(x)D_n(t)\,dt,$$

since x is treated as a constant for the integration with respect to t. Then, using expression (15.28) for $s_n(x)$, we have

$$s_n(x) - f(x) = \int_{-\pi}^{\pi} [f(x-t) - f(x)]D_n(t)\,dt. \tag{15.29}$$

At a given value x, the Fourier series of f converges to $f(x)$ if and only if this difference has limit zero as $n \to \infty$. We fix a value x for the discussion to follow. In the integral on the right side, we may replace t by $-t$, using the fact that the integrand is periodic with period 2π and D_n is even, and also write

$$s_n(x) - f(x) = \int_{-\pi}^{\pi} [f(x+t) - f(x)]D_n(t)\,dt. \tag{15.30}$$

On adding (15.29) and (15.30), we obtain

$$2[s_n(x) - f(x)] = \int_{-\pi}^{\pi} [f(x-t) + f(x+t) - 2f(x)]D_n(t)\,dt,$$

and hence

$$s_n(x) - f(x) = \frac{1}{2}\int_{-\pi}^{\pi} [f(x-t) + f(x+t) - 2f(x)]D_n(t)\,dt. \tag{15.31}$$

In (15.31), the integrand is an even function, so we may write

$$s_n(x) - f(x) = \int_{0}^{\pi} [f(x-t) + f(x+t) - 2f(x)]D_n(t)\,dt. \tag{15.32}$$

For the fixed x in this discussion, we now impose the hypothesis that the limiting values $f(x-)$, $f(x+)$, $f'(x-)$ and $f'(x+)$ all exist. As noted before the theorem statement, we may replace the value $f(x)$ by the expression $[f(x+) + f(x-)]/2$. This means that if f is continuous at this x, then $f(x) = [f(x+) + f(x-)]/2$ is unchanged, and if f is discontinuous at this x, then we are reassigning the value of f at this single point x to be the average of the left-hand and right-hand limits there. Thus, in (15.32) we set $f(x) = [f(x+) + f(x-)]/2$ on both sides and split the resulting integral into two parts, to obtain

$$\begin{aligned} s_n(x) - [f(x+) + f(x-)]/2 &= \int_0^\pi [f(x-t) - f(x-)]D_n(t)\,dt \\ &\quad + \int_0^\pi [f(x+t) - f(x+)]D_n(t)\,dt. \end{aligned} \tag{15.33}$$

Our goal is to show that each of these integrals approaches zero as $n \to \infty$. Consider the second integral on the right in (15.33). Using the compact formula for $D_n(t)$ from Lemma 15.6.1, we may write it as

$$\frac{1}{\pi}\int_0^\pi \frac{f(x+t) - f(x+)}{t}\,\frac{t}{2\sin\left(\frac{1}{2}t\right)}\,\sin\left(n + \frac{1}{2}\right)t\,dt. \tag{15.34}$$

If $f'(x+)$ is finite (our hypothesis), then the difference quotient for f in this integrand has the finite limit $f'(x+)$ as $t \to 0+$. Also,

$$\lim_{t\to 0} \frac{t}{2\sin\left(\frac{1}{2}t\right)} = \lim_{t\to 0} \frac{1}{\cos\left(\frac{1}{2}t\right)} = 1,$$

by l'Hôpital's rule. We may take these limiting values as the values of the corresponding factors in the integrand when $t = 0$, and thus define the function

$$\phi(t) = \frac{f(x+t) - f(x+)}{t}\,\frac{t}{2\sin\left(\frac{1}{2}t\right)} \qquad \text{for } 0 \le t \le \pi.$$

(Recall that x is fixed.) Then ϕ is Riemann integrable on $[0, \pi]$. Moreover, (15.34) is now

$$\frac{1}{\pi}\int_0^\pi \phi(t)\,\sin\left(n + \frac{1}{2}\right)t\,dt. \tag{15.35}$$

Since $\sin(n + \frac{1}{2})t = \sin\left(\frac{t}{2}\right)\cos(nt) + \cos\left(\frac{t}{2}\right)\sin(nt)$, (15.35) equals the sum

$$\frac{1}{\pi}\int_0^\pi \phi(t)\,\sin\left(\frac{t}{2}\right)\cos(nt)\,dt + \frac{1}{\pi}\int_0^\pi \phi(t)\,\cos\left(\frac{t}{2}\right)\sin(nt)\,dt.$$

The functions $\phi(t)\sin(\frac{t}{2})$ and $\phi(t)\cos(\frac{t}{2})$ are both Riemann integrable over $[0, \pi]$. Thus by Exercise 15.5.3, both integrals approach zero as $n \to \infty$. This completes the proof that the second integral on the right in (15.33) approaches zero as $n \to \infty$. A similar argument shows that the first integral on the right in (15.33) approaches zero as $n \to \infty$. Consequently, (15.33) implies that $s_n(x) \to [f(x+) - f(x-)]/2$. This is true for each point x at which the limiting values $f(x-)$, $f(x+)$, $f'(x-)$ and $f'(x+)$ exist, and the theorem is proved. □

Definition 15.6.3. *A function f is* **piecewise smooth** *if, on any bounded interval, f is C^1 except possibly at finitely many points, at each of which the one-sided limits $f(x-)$, $f(x+)$, $f'(x-)$ and $f'(x+)$ exist as finite values.*

(See Figure 15.8.) From this definition it follows that Theorem 15.6.2 applies to piecewise smooth functions f of period 2π, and we see that the Fourier series of such a function converges at every point to the value $[f(x+) + f(x-)]/2$. We record two corollaries of Theorem 15.6.2 as theorems in their own right.

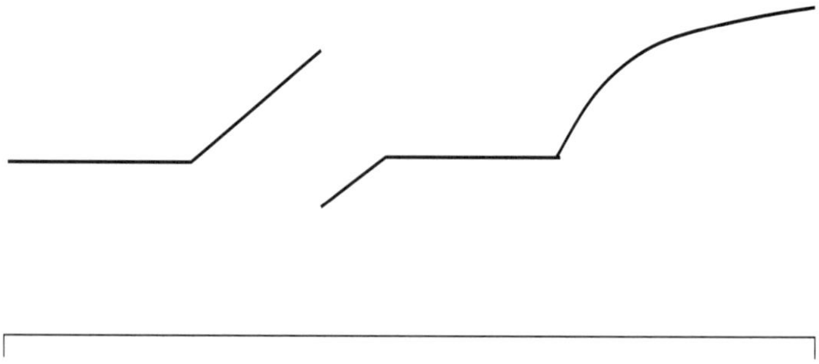

Figure 15.8. A piecewise smooth function graph over its domain interval.

Theorem 15.6.4. *If f is piecewise smooth and has period 2π, then the Fourier series of f converges to $f(x)$ at any point x where f is continuous.*

Proof. At any point x where f is continuous, we have $f(x-) = f(x+) = f(x)$. By Theorem 15.6.2, the Fourier series of f converges at x to the value $f(x)$. □

Theorem 15.6.5. *If f is Riemann integrable and of period 2π, then the Fourier series of f converges to $f(x)$ at any point x where f is differentiable.*

Proof. At any point x where f is differentiable, f is continuous and we have $f(x-) = f(x+) = f(x)$ and $f'(x-) = f'(x+) = f'(x)$. By Theorem 15.6.2, the Fourier series of f converges at x to the value $f(x)$. □

The sums for many specific numerical series follow from a knowledge of the convergence of specific Fourier series or from Parseval's equality. We offer one example here.

Example 15.6.6. Consider the continuous, piecewise smooth function $g(x) = |x|$ for $-\pi \le x < \pi$. The Fourier series of g is

$$\frac{\pi}{2} - \frac{4}{\pi}\sum_{k=1}^{\infty} \frac{\cos(2k-1)x}{(2k-1)^2},$$

and it converges uniformly to g, as we will see in Theorem 15.6.7 below. See Figure 15.9 for the graph of g and the first three terms of the Fourier series of g. Since $g(0) = 0$, we have

$$0 = \frac{\pi}{2} - \frac{4}{\pi}\sum_{k=1}^{\infty} \frac{1}{(2k-1)^2},$$

from which we find that

$$\frac{\pi^2}{8} = \sum_{k=1}^{\infty} \frac{1}{(2k-1)^2} = 1 + \frac{1}{3^2} + \frac{1}{5^2} + \cdots .$$

On the other hand, an application of Parseval's equality for this continuous function gives

$$\frac{1}{\pi}\int_{-\pi}^{\pi} |x|^2\, dx = \frac{1}{2}\pi^2 + \frac{16}{\pi^2}\Big(1 + \frac{1}{3^4} + \frac{1}{5^4} + \cdots\Big).$$

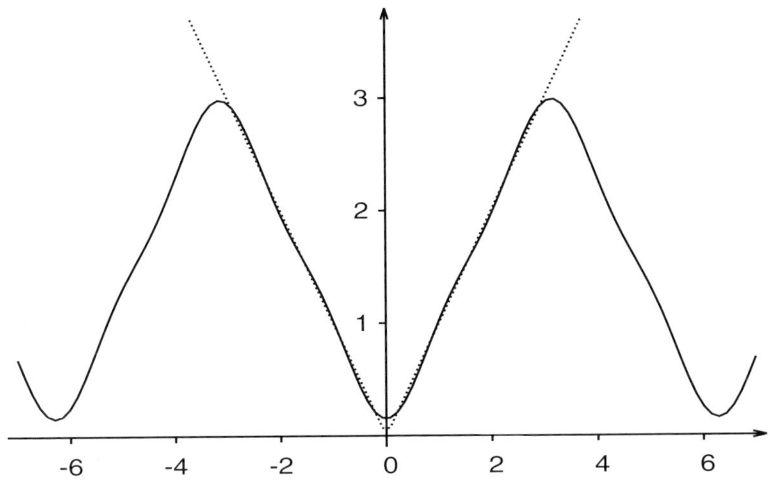

Figure 15.9. Three terms of the Fourier series for $g(x) = |x|$, whose dotted graph is nearly indistinguishable from the partial sum over $-\pi \le x < \pi$.

The left-hand side here equals $2\pi^2/3$, and after rearrangement, we have

$$\frac{\pi^4}{96} = 1 + \frac{1}{3^4} + \frac{1}{5^4} + \cdots .$$

Since we have the fourth powers of the *odd* positive integers only, this is not the full p-series with $p = 4$. But suppose we want $S = \sum_{n=1}^{\infty} 1/n^4$, the full p-series with $p = 4$. Since the p-series is known to converge, we may write

$$\begin{aligned} S &= \left(1 + \frac{1}{3^4} + \frac{1}{5^4} + \cdots\right) + \left(\frac{1}{2^4} + \frac{1}{4^4} + \frac{1}{6^4} + \cdots\right) \\ &= \frac{\pi^4}{96} + \frac{1}{2^4}\left(1 + \frac{1}{2^4} + \frac{1}{3^4} + \frac{1}{4^4} + \frac{1}{5^4} + \cdots\right) \\ &= \frac{\pi^4}{96} + \frac{1}{16}S. \end{aligned}$$

Now solve for S to find that $S = \sum_{n=1}^{\infty} 1/n^4 = \pi^4/90$. △

For those who have read or worked out Exercise 15.5.2, the following comment is of interest. For a function that is square integrable, that is,

$$\int_{-\pi}^{\pi} [f(x)]^2 \, dx < \infty,$$

the square of the L^2 norm of f is defined by

$$\|f\|_2^2 = \int_{-\pi}^{\pi} [f(x)]^2 \, dx.$$

(This notation was first introduced in (8.2) of Section 8.3.) The norm $\|f\|_2$ serves as the norm in the space of square integrable functions on $[-\pi, \pi]$, or, more precisely, the space of equivalence classes of Riemann integrable functions on $[-\pi, \pi]$, as described in Exercise 15.5.2. Parseval's equality for a continuous function f gives this L^2 norm in terms of the Fourier coefficients of f. After the Lebesgue integral is defined, and we consider the larger space of functions that are square integrable in the sense of Lebesgue, it can be shown that Parseval's identity holds for all functions in that space.

The final result of the section is the result alluded to earlier that the Fourier series of a continuous, piecewise smooth function f converges absolutely and uniformly to f.

Theorem 15.6.7. *If $f : [-\pi, \pi] \to \mathbf{R}$ is continuous and piecewise smooth (that is, f' is piecewise continuous) and $f(-\pi) = f(\pi)$, then the Fourier series of f converges absolutely and uniformly to f.*

Proof. The key ideas in the proof are Bessel's inequality for the derivative f' and the Weierstrass test for uniform convergence. Let

$$\frac{1}{2}a_0 + \sum_{k=1}^{\infty} \big(a_k \cos(kx) + b_k \sin(kx)\big) \tag{15.36}$$

be the Fourier series of f. If we show that for each $k \geq 1$ there is a number M_k such that

$$\big|a_k \cos(kx) + b_k \sin(kx)\big| \leq M_k$$

and the series $\sum_{k=1}^{\infty} M_k$ converges, then the uniform convergence of (15.36) follows from the Weierstrass test.

From the inequality $0 \leq (r-s)^2 = r^2 - 2rs + s^2$ for real numbers r and s, we have

$$2rs \leq r^2 + s^2, \tag{15.37}$$

and hence

$$(r+s)^2 = r^2 + 2rs + s^2 \leq 2(r^2 + s^2).$$

It follows that for all $k \geq 1$,

$$\big(a_k \cos(kx) + b_k \sin(kx)\big)^2 \leq 2\big(a_k^2 \cos^2(kx) + b_k^2 \sin^2(kx)\big) \leq 2(a_k^2 + b_k^2).$$

Consequently, for $k \geq 1$,

$$\begin{aligned}\big|a_k \cos(kx) + b_k \sin(kx)\big| &\leq \sqrt{2}\sqrt{a_k^2 + b_k^2} \\ &< 2\frac{1}{k}\sqrt{k^2(a_k^2 + b_k^2)} \\ &\leq \frac{1}{k^2} + k^2(a_k^2 + b_k^2) =: M_k,\end{aligned}$$

where the final line follows from (15.37). Since $\sum_{k=1}^{\infty} 1/k^2$ converges, it only remains to show that $\sum_{k=1}^{\infty} k^2(a_k^2 + b_k^2)$ converges.

The derivative f' is piecewise continuous and hence integrable on $[-\pi, \pi]$, so the Fourier series of f' is defined. In fact, kb_k and $-ka_k$ are the Fourier coefficients of f' with respect to $\cos kx$ and $\sin kx$, respectively. To see this, we integrate by parts in the definition of these coefficients and use $f(-\pi) = f(\pi)$ to find, for $k \geq 0$,

$$\begin{aligned}
\frac{1}{\pi}\int_{-\pi}^{\pi} f'(x)\cos(kx)\,dx &= \frac{1}{\pi}\left[f(x)\cos(kx)\Big|_{-\pi}^{\pi} + \int_{-\pi}^{\pi} kf(x)\sin(kx)\,dx\right] \\
&= \frac{1}{\pi}\left[f(\pi)(\cos(k\pi) - \cos(-k\pi)) + k\int_{-\pi}^{\pi} f(x)\sin(kx)\,dx\right] \\
&= \frac{k}{\pi}\int_{-\pi}^{\pi} f(x)\sin(kx)\,dx \\
&= kb_k.
\end{aligned}$$

A similar argument shows that, for $k \geq 1$,

$$\frac{1}{\pi}\int_{-\pi}^{\pi} f'(x)\sin(kx)\,dx = -ka_k.$$

Then Bessel's inequality for f' implies that

$$\sum_{k=1}^{\infty} k^2(a_k^2 + b_k^2) \leq \frac{1}{\pi}\int_{-\pi}^{\pi} [f'(x)]^2\,dx < \infty,$$

and hence the series on the left-hand side converges, as desired. Therefore the Fourier series of f converges absolutely and uniformly by the Weierstrass test, and by Theorem 15.6.4 it must converge to f itself. □

Exercises.

Exercise 15.6.1. In Example 15.6.6, we found the Fourier series for the periodic extension of $g(x) = |x|$, $-\pi \leq x < \pi$ to all of $\mathbf{R}$. Show that the Fourier series converges to $g(x)$ for every real x, and that the convergence is uniform.

Exercise 15.6.2. The 2π-periodic extension of the function

$$f(x) = \begin{cases} -1, & \text{if } -\pi < x < 0, \\ 1, & \text{if } 0 < x < \pi \end{cases}$$

is called a square wave function.

1. Find the Fourier series of f, noting that f is an odd function. Graph f and several partial sums of the Fourier series, for example, the first 4 nonzero terms, then the first 8 nonzero terms.
2. Show that the Fourier series converges for every real x, and graph the sum of the series. Notice the overshoot (or undershoot) in the plot of the partial sums near the discontinuity at $x = 0$. Then see Exercise 15.6.3.

Exercise 15.6.3. Estimate by visual examination the size of the maximum overshoot (or undershoot) of the partial sums approximating the square wave function in Exercise 15.6.2 on either side of the discontinuity at $x = 0$. *Hint*: Use a mathematical software package or the trace feature on a graphing calculator. The overshoot (or undershoot) is an unavoidable characteristic of the Fourier partial sums near a jump discontinuity. It is generally designated as the *Gibbs phenomenon* after the mathematical physicist J. W. Gibbs who studied it.

Exercise 15.6.4. Consider the 2π-periodic extension of the function

$$f(x) = \frac{1}{2}(\pi - x), \quad 0 \le x < 2\pi.$$

We call the extension the sawtooth function, denoted also by f. Notice that f has discontinuities at $x = 2n\pi$, $n \in \mathbf{Z}$.

1. Find the Fourier series of f, noting that f is an odd function. Graph f and several partial sums of the Fourier series, for example, the first 5 nonzero terms, then the first 10 nonzero terms.
2. Show that the Fourier series converges for every real x, and graph the sum of the series. Then see Exercise 15.6.5.

Exercise 15.6.5. Estimate by visual examination the size of the maximum overshoot (or undershoot) of the partial sums approximating the sawtooth function in Exercise 15.6.4 on either side of the discontinuity at $x = 0$.

Exercise 15.6.6. *Smoothness and the decay of Fourier coefficients*
Let f be a periodic function of period 2π.

1. Show that if f is C^1, then there is a number $M_1 > 0$ such that the Fourier coefficients a_k, b_k of f satisfy
$$|a_k| \le \frac{M_1}{k} \quad \text{and} \quad |b_k| \le \frac{M_1}{k} \quad \text{for all } k.$$
Hint: Integrate by parts.
2. Show that if f is C^n, then there is a number $M_n > 0$ such that the Fourier coefficients a_k, b_k of f satisfy
$$|a_k| \le \frac{M_n}{k^n} \quad \text{and} \quad |b_k| \le \frac{M_n}{k^n} \quad \text{for all } k.$$
3. Use the Riemann-Lebesgue theorem (Theorem 15.5.2) to improve the result of part 2 to read that if f is C^n, then the Fourier coefficients a_k and b_k of f satisfy
$$\lim_{k\to\infty} k^n |a_k| = 0 \quad \text{and} \quad \lim_{k\to\infty} k^n |b_k| = 0.$$

15.7. Fejér's Theorem

We noted earlier that there exist functions for which the Fourier partial sums diverge at some points of continuity. There is a different notion of summability of a Fourier series that provides a more reliable representation of function values for continuous functions. In this section we explore this alternative summation process, known as *Cesàro summability*, which means we consider not the Fourier partial sums themselves, but rather their *arithmetic means*.

Let $s_n(x)$ be the n-th partial sum of the Fourier series of f, given by

$$s_n(x) = \frac{1}{2}a_0 + \sum_{k=1}^{\infty} \big(a_k \cos(kx) + b_k \sin(kx)\big).$$

Define a new sequence of sums, $\sigma_n(x)$, by

$$\sigma_n(x) = \frac{1}{n+1}\sum_{k=0}^{n} s_n(x), \qquad n = 0, 1, 2, \dots . \tag{15.38}$$

Then $\sigma_n(x)$ is the arithmetic mean of the first $n+1$ Fourier partial sums of f, also called the n-th **Cesàro sum** of f, or the n-th **Fejér mean** of f. (See Exercise 15.7.1.) The importance of Fejér's theorem, to be proved below, is that for any continuous function f, the Cesàro sums σ_n converge uniformly to f.

We shall write the n-th Fejér mean $\sigma_n(x)$ in a form similar to the form (15.28) for the n-th Fourier partial sum $s_n(x)$, that is,

$$\sigma_n(x) = \frac{1}{\pi}\int_{-\pi}^{\pi} K_n(t)\, f(x-t)\, dt, \tag{15.39}$$

but with a kernel function $K_n(t)$ different from the Dirichlet kernel $D_n(t)$.

First we need a compact formula for the n-th Fejér mean, $\sigma_n(x)$.

Lemma 15.7.1. *If $x \neq 2m\pi$, $m \in \mathbf{Z}$, then*

$$\sum_{k=0}^{n} \sin\left(k+\frac{1}{2}\right)x = \frac{\sin^2 \frac{1}{2}(n+1)x}{\sin(\frac{1}{2}x)}, \qquad n = 1, 2, 3, \dots .$$

Proof. The statement to be proved is equivalent to

$$\sin(\frac{1}{2}x)\sum_{k=0}^{n} \sin\left(k+\frac{1}{2}\right)x = \sin^2\frac{1}{2}(n+1)x. \tag{15.40}$$

If we let $A = \frac{1}{2}x$ and $B = (k+\frac{1}{2})x$ in the formulas

$$\cos(A \pm B) = \cos A \cos B \mp \sin A \sin B,$$

then the difference $\cos(A-B) - \cos(A+B)$ equals

$$2\sin(\frac{1}{2}x)\, \sin\left(k+\frac{1}{2}\right)x = \cos(kx) - \cos(k+1)x.$$

If we sum both sides from $k=0$ to $k=n$, the terms on the right-hand side telescope to yield

$$2\sin(\frac{1}{2}x)\sum_{k=0}^{n} \sin\left(k+\frac{1}{2}\right)x = 1 - \cos(n+1)x.$$

From the identity $\sin^2\theta = (1-\cos 2\theta)/2$, we have

$$\sin^2\frac{1}{2}(n+1)x = \frac{(1-\cos(n+1)x)}{2},$$

and in view of (15.40), the lemma is proved. □

Let f be continuous and periodic with period 2π. From (15.28), we have

$$s_k(x) = \int_{-\pi}^{\pi} D_k(t)\, f(x-t)\, dt,$$

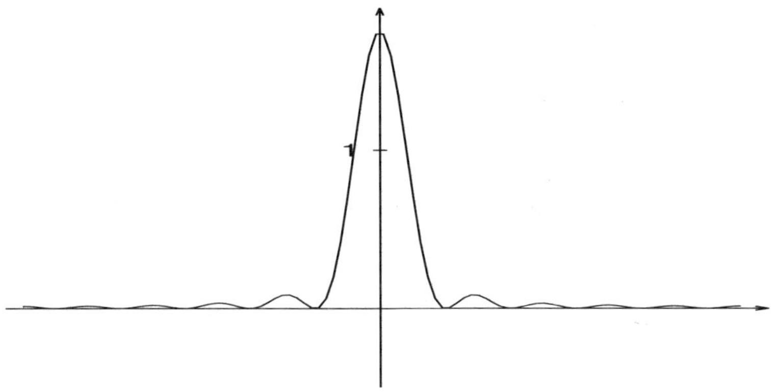

Figure 15.10. The graph of the Fejér kernel $K_{10}(x)$ over $[-\pi, \pi]$.

where $D_k(t)$ is the Dirichlet kernel. Hence,

$$\sigma_n(x) = \frac{1}{n+1}\sum_{k=0}^{n} s_k(x) = \frac{1}{n+1}\int_{-\pi}^{\pi}\Big[\sum_{k=0}^{n} D_k(t)\Big] f(x+t)\,dt,$$

where again we have used the 2π periodicity of D_k and f and the fact that D_k is an even function, to replace t by $-t$ in the integrand. By Lemma 15.6.1,

$$D_k(t) = \frac{\sin(k+\frac{1}{2})x}{2\pi\sin(\frac{1}{2}x)}, \qquad k = 0, 1, 2, \ldots,$$

and thus it follows from Lemma 15.7.1 that

$$\sum_{k=0}^{n} D_k(t) = \frac{\sin^2\frac{1}{2}(n+1)x}{2\pi\sin^2(\frac{1}{2}x)}.$$

Consequently, if we define the **Fejér kernels** $K_n(x)$ by

$$K_n(x) := \frac{1}{n+1}\Big[\sum_{k=0}^{n} D_k(x)\Big] = \frac{1}{n+1}\,\frac{\sin^2\frac{1}{2}(n+1)x}{2\pi\sin^2(\frac{1}{2}x)}, \tag{15.41}$$

then

$$\sigma_n(x) = \int_{-\pi}^{\pi} K_n(t)f(x+t)\,dt. \tag{15.42}$$

Figure 15.10 shows the graph of $K_{10}(x)$ over $[-\pi, \pi]$.

From the definition, $K_n(x) \geq 0$ for all x, and also $\int_{-\pi}^{\pi} K_n(x)\,dx = 1$ since $\int_{-\pi}^{\pi} D_k(x)\,dx = 1$ for $0 \leq k \leq n$. We summarize these facts, and one additional estimate, in the following lemma.

Lemma 15.7.2. *The Fejér kernels $K_n(x)$ satisfy, for each integer $n \geq 0$,*

$$K_n(x) = \frac{1}{n+1}\Big[\sum_{k=0}^{n} D_k(x)\Big] = \frac{1}{n+1}\,\frac{\sin^2\frac{1}{2}(n+1)x}{2\pi\sin^2(\frac{1}{2}x)}, \tag{15.43}$$

with the value determined by the limit of the right-hand side at points where the denominator is zero, and hence $K_n(x) \ge 0$ *for all* x*. Moreover, we have*

$$\int_{-\pi}^{\pi} K_n(x)\,dx = 1, \qquad n \ge 0. \tag{15.44}$$

In addition, for any δ *such that* $0 < \delta < \pi$,

$$0 \le K_n(x) \le \frac{1}{2(n+1)}\frac{\pi}{\delta^2}, \qquad \delta \le |x| \le \pi. \tag{15.45}$$

Proof. It only remains to prove (15.45). To see it, note that the graph of the line $y = x/\pi$ lies below the graph of $\sin(\frac{1}{2}x)$ for $0 \le x \le \pi$, and since both of these functions are odd, we have $|\sin(\frac{1}{2}x)| \ge |x|/\pi$ for $-\pi \le x \le \pi$. Hence, $\sin^2(\frac{1}{2}x) \ge x^2/\pi^2$ for $-\pi \le x \le \pi$. By (15.43), if $0 < \delta < \pi$, then

$$0 \le K_n(x) \le \frac{1}{2\pi(n+1)}\frac{\pi^2}{x^2} \le \frac{1}{2(n+1)}\frac{\pi}{\delta^2}$$

for $\delta \le |x| \le \pi$. □

We can now prove Fejér's theorem for continuous 2π-periodic functions.

Theorem 15.7.3 (Fejér)**.** *Let* f *be continuous on* $[-\pi, \pi]$ *with* $f(-\pi) = f(\pi)$*, and extend* f *to the entire real line by the condition* $f(x + 2\pi) = f(x)$*. Let* σ_n *be the* n*-th Cesàro sum of* f*. Then* σ_n *converges uniformly to* f *on* $\mathbf{R}$*.*

Proof. Since f is continuous and periodic, there is a number M such that $|f(x)| \le M$ for all x, and f is uniformly continuous on $\mathbf{R}$. By (15.42) and (15.44), we may write

$$\sigma_n(x) - f(x) = \int_{-\pi}^{\pi} K_n(t)[f(x+t) - f(x)]\,dt.$$

The idea is that for t small, we can make the difference in function values small by uniform continuity of f, and for t larger, we can bound K_n for large n and use the uniform bound for f. Thus the integral on the right-hand side will be split into three parts, over the intervals $[-\pi, -\delta]$, $[-\delta, \delta]$ and $[\delta, \pi]$ for appropriate $0 < \delta < \pi$. By uniform continuity of f, given $\epsilon > 0$, there is a $\delta = \delta(\epsilon) > 0$ with $0 < \delta < \pi$

such that $|f(x+t) - f(x)| < \epsilon/2$ if $|t| < \delta$. By (15.45) and (15.44), we then have

$$\begin{aligned} \big|\sigma_n(x) - f(x)\big| &\le \int_{-\pi}^{\pi} K_n(t)\big|f(x+t) - f(x)\big|\,dt \\ &\le \int_{-\pi}^{-\delta} \frac{\pi}{2(n+1)\delta^2}\, 2M\,dt \\ &\quad + \int_{-\delta}^{\delta} K_n(t)\frac{\epsilon}{2}\,dt \\ &\quad + \int_{\delta}^{\pi} \frac{\pi}{2(n+1)\delta^2}\, 2M\,dt \\ &\le 2(\pi-\delta)\frac{\pi M}{(n+1)\delta^2} + \frac{\epsilon}{2}\int_{-\delta}^{\delta} K_n(t)\,dt \\ &\le \frac{2M\pi^2}{(n+1)\delta^2} + \frac{\epsilon}{2}, \end{aligned}$$

where we have used $K_n(t) \ge 0$ and (15.44) in the last line. This estimate holds for all $x \in \mathbf{R}$ and all n. Finally, for n sufficiently large, we will have

$$\big|\sigma_n(x) - f(x)\big| \le \frac{2M\pi^2}{(n+1)\delta^2} + \frac{\epsilon}{2} < \frac{\epsilon}{2} + \frac{\epsilon}{2} = \epsilon.$$

Therefore σ_n converges uniformly to f on $\mathbf{R}$. □

Fejér's theorem states that every continuous function of period 2π can be approximated arbitrarily closely in the uniform norm by linear combinations of elements in the trigonometric set

$$\left\{\frac{1}{\sqrt{2\pi}}, \frac{\cos(x)}{\sqrt{\pi}}, \frac{\sin(x)}{\sqrt{\pi}}, \frac{\cos(2x)}{\sqrt{\pi}}, \frac{\sin(2x)}{\sqrt{\pi}}, \dots\right\}.$$

Later in Theorem 18.3.4, we will examine mean square convergence (that is, L^2 norm convergence), in the space $CP[-\pi,\pi]$ of continuous 2π-periodic functions, as an application of Fejér's theorem.

Exercises.

Exercise 15.7.1. Given a series $\sum_{k=0}^{\infty} a_k$ and its partial sums $s_n = \sum_{k=0}^{n} a_k$, define the arithmetic means

$$\sigma_n = \frac{s_0 + s_1 + \cdots + s_n}{n+1}, \qquad n = 0, 1, 2, \dots .$$

(If the series indexing starts with $k = 1$, then $\sigma_n = (s_1 + \cdots + s_n)/n$.) Then σ_n is called the n-th **Cesàro sum** of the series $\sum_{k=0}^{\infty} a_k$. If $\lim_{n\to\infty} \sigma_n = L$ exists, we say the series $\sum_{k=0}^{\infty} a_k$ is **Cesàro summable** to L.

1. A series that diverges may be Cesàro summable. As an example, show that the divergent series $\sum_{k=0}^{\infty}(-1)^k = 1 - 1 + 1 - 1 + \cdots$ is Cesàro summable to $1/2$.
2. Prove: If $\sum_{k=0}^{\infty} a_k = L$ exists, that is, $\lim_{n\to\infty} s_n = L$, then $\sum_{k=0}^{\infty} a_k$ is Cesàro summable to L, $\lim_{n\to\infty} \sigma_n = L$. *Hint*: By considering the shifted sequences $s_n - L$ and $\sigma_n - L$, we may assume that $L = 0$. Then, for $n_0 < n$, write
$$\sigma_n = \frac{s_0 + \cdots + s_{n_0}}{n+1} + \frac{s_{n_0+1} + \cdots + s_n}{n+1}.$$
Use boundedness of $(s_n)_{n=0}^{\infty}$ to bound the first term on the right, and use convergence of $(s_n)_{n=0}^{\infty}$ to make the second term less than a given $\epsilon > 0$.

Exercise 15.7.2. Deduce the trigonometric Weierstrass Theorem 15.5.3 from Fejér's Theorem.

Exercise 15.7.3. Deduce Theorem 15.3.2 from Fejér's Theorem.

Exercise 15.7.4. The Cesàro sums σ_n exhibit no overshoot (no Gibbs phenomenon) when used to approximate the sawtooth function of Exercise 15.6.4. Illustrate this by computing and graphing σ_5 and σ_{10} for that function.

15.8. Notes and References

The presentation of the Dirichlet problem and some of its consequences follows Seeley [**58**] with additional help from Friedman [**17**]. For much more on Fourier analysis and partial differential equations, see the detailed introductions by Folland [**14**], Gonzalez-Velasco [**20**], Haberman [**23**] or Strauss [**63**]. Gonzalez-Velasco [**20**] integrates some historical material on Fourier analysis with the mathematics of partial differential equations, and provides an interesting portrait of the development of analysis concepts from the eighteenth to the twentieth century.

The reader has probably observed the simplicity in form of the series solution to the Dirichlet problem for the disk, given the Fourier series for the boundary data f. The construction of the product solutions, in effect, introduced the factor r^n in front of each nonconstant term of the Fourier series for f:
$$u(r,\theta) = \frac{1}{2}a_0 + \sum_{n=1}^{\infty} r^n \big(a_n \cos n\theta + b_n \sin n\theta\big).$$
In effect, this construction pulls the boundary data into the interior of the disk. This result is related to the concept of Abel summation of a given series of real or complex numbers. See Stein and Shakarchi [**61**] for this, as well as an extensive introduction to Fourier analysis.

Another side of Fourier analysis is the theory of *Fourier transforms*, which are useful in representing functions on the real line that are not periodic and in solving certain problems in partial differential equations with unbounded domains. For substantial introductions, see Folland [**14**], Körner [**37**], or Stein and Shakarchi [**61**].

The subject of pointwise convergence of Fourier series is difficult. Dirichlet proved in 1829 a version of Theorem 15.6.2 for piecewise continuous and piecewise monotone functions, and Fejér's theorem appeared in 1904. These theorems are sufficient for most problems a reader might encounter in an undergraduate introduction to partial differential equations. One indication of the difficulty in studying pointwise convergence is that it is possible for the Fourier series of a continuous function to diverge at some (in fact, infinitely many) points. For specific examples and discussion, see Duren [**9**] or Stein and Shakarchi [**61**]. In 1966, L. Carleson proved that the Fourier series of every function whose square is integrable in the sense of Lebesgue converges almost everywhere, so that the points of divergence constitute a set of Lebesgue measure zero. See Folland [**15**] for more information and references.

For more on the Gibbs phenomenon, mentioned in the exercises for the section on pointwise convergence of Fourier series, see Gonzalez-Velasco [**20**] or Strauss [**63**].

Chapter 16

Measure Theory and Lebesgue Measure

The theory of measure is an extension of the concepts of length of an interval, area of a planar region, and volume of a solid region. Both single variable and multivariable calculus deal with these notions by means of the Riemann integral of real functions. Chapters 12 and 13 have extended the scope of that work with the theory of Jordan measure (the theory of volume) based on the Riemann integral over bounded intervals in $\mathbf{R}^n$.

During the latter part of the nineteenth century and the early twentieth century, the deficiencies of the Riemann integral with respect to limit processes led to a reexamination of the notions of integral and measure. Fundamentally, there are two approaches to the topics of integral and measure: (i) Develop measure theory first, and then define integration based on the theory of measure; and (ii) first develop integration theory over certain sets and then define the measure of other sets in terms of the integral. Chapter 12 followed approach (ii) and developed Jordan measure (volume measure) based on the Riemann integral over bounded intervals. In this and the following chapter, we follow approach (i). The present chapter includes a unified development of Lebesgue measure for $\mathbf{R}$ and $\mathbf{R}^n$, and the following chapter develops the Lebesgue integral for functions defined on subsets of $\mathbf{R}$ and $\mathbf{R}^n$. This allows us to measure a larger class of sets and to integrate a larger class of functions over those sets. As we will see, the Lebesgue integral behaves well under limit processes without the assumption of uniform convergence.

After some introductory motivational comments, Section 1 discusses the properties of σ-algebras, the set-theoretic foundation for measure theory. Section 2 briefly summarizes the extended real number system. Section 3 presents the general properties of measures. Section 4 details the construction of a measure from a simpler structure known as an outer measure. Section 5 applies the construction from Section 4 to define Lebesgue measure on $\mathbf{R}$ and $\mathbf{R}^n$.

Motivations. The essential concepts needed for the development of Lebesgue measure are few. Based on ideas from the study of area and volume, there are some simple properties that any measure should have, and these requirements also say something about the class of sets we want to measure.

Let us write μ for our measure function on sets. If we can measure sets A and B, then we know $\mu(A)$ and $\mu(B)$, the measure of A and the measure of B, and we should also be able to measure $A \cap B$ and $A \cup B$. If $A \cap B$ is empty, then we should have $\mu(A \cup B) = \mu(A) + \mu(B)$. For example, by considering the disjoint union $(a, b] = (a, b) \cup \{b\}$, and naturally requiring $\mu((a, b)) = b - a$, we necessarily have $\mu(\{x\}) = 0$ for any point x, and hence the measure of any finite set of real numbers should be zero. (We have already seen other sets in $\mathbf{R}$ or $\mathbf{R}^n$ which have Lebesgue measure zero.) In general, we should have $\mu(A \cup B) = \mu(A) + \mu(B) - \mu(A \cap B)$, since the measure of the intersection is counted twice already in the sum $\mu(A) + \mu(B)$. This also shows that the measure of the empty set must be zero. If $A \subset B$, then we want $\mu(A) \leq \mu(B)$, but this follows from the measure of disjoint sets just stated, since $\mu(B) = \mu(A) + \mu(B - A)$. However, for *that* statement to be meaningful, we need to be able to measure the complement $B - A$ for any measurable sets A and B.

We must allow that certain sets A have $\mu(A) = \infty$, for example, any interval of the form $[n, \infty) \subset \mathbf{R}$. Similar considerations apply to certain unbounded sets in $\mathbf{R}^n$. So the measure function μ can take values in the set $\mathbf{R} \cup \{\infty\}$, which we denote by $[0, \infty]$. It will be essential to measure any open set. In particular, by the structure theorem for open subsets of $\mathbf{R}$, this means we must be able to measure countable disjoint unions. If countable unions and complements can be measured, then countable intersections can also be measured.

The theory discussed here develops a satisfactory theory of measure having these, and other, important properties. Most importantly, this theory of measure leads to a satisfactory concept of the integral for a sufficiently large class of functions to meet the needs of modern mathematical analysis. In particular, *the resulting new integral will have more satisfactory behavior under the limit processes of analysis without requiring uniform convergence.*

Additional motivation for measure theory comes from the theory of probability, in which events are modeled by subsets of a measure space. As noted above, events (sets) of interest often arise as countable unions or countable intersections of other events. Whether we study a probability measure or a more general measure, we need the set-theoretic structure known as a σ-algebra.

16.1. Algebras and σ-Algebras

Let X be a set. When $A \subset X$, the notation A^c means the complement of the set A in X.

Definition 16.1.1. *An* **algebra** *is a collection $\mathcal{A}$ of subsets of X such that*

1. *$\mathcal{A}$ contains X and the empty set.*
2. *If $A \in \mathcal{A}$, then $A^c \in \mathcal{A}$.*

3. *If $A, B \in \mathcal{A}$, then $A \cup B \in \mathcal{A}$. An algebra $\mathcal{A}$ is a* **σ-algebra** *if, in addition, the following property holds.*

4. *If $A_j \in \mathcal{A}$ for $j \in \mathbf{N}$, then $\bigcup_{j=1}^{\infty} A_j$ is in $\mathcal{A}$.*

When a reminder of the ambient set X is needed, we may say that an algebra (or σ-algebra) $\mathcal{A}$ is an algebra (or σ-algebra) of X.

If A, B belong to an algebra $\mathcal{A}$, then, since $A \cap B = (A^c \cup B^c)^c$, 2 and 3 imply that $A \cap B$ also belongs to $\mathcal{A}$. By an induction argument, an algebra contains the union of any finite collection of its elements and the intersection of any finite collection of its elements.

If sets A_j, $j \in \mathbf{N}$, belong to a σ-algebra $\mathcal{A}$, then, since $\bigcap_{j=1}^{\infty} A_j = (\bigcup_{j=1}^{\infty} A_j^c)^c$, properties 2 and 4 of the definition imply that $\bigcap_{j=1}^{\infty} A_j$ also belongs to $\mathcal{A}$. Therefore a σ-algebra contains the union of any countable collection of its elements and the intersection of any countable collection of its elements.

For any set X, the collection of all subsets of X is a σ-algebra.

Example 16.1.2. Consider the set $X = \{1, 2, \ldots, N\}$. There are 2^N subsets of X. In fact, there is a one-to-one correspondence between the collection of these subsets and the collection of strings of zeros and ones of length N, since each subset of $\{1, 2, \ldots, N\}$ corresponds uniquely to a yes or no answer about whether each integer k, $1 \leq k \leq N$, belongs to that subset. These 2^N subsets of X form a σ-algebra on X. But they do not form a σ-algebra on $\mathbf{N}$, the full set of positive integers. For example, the collection is not closed under complements in $\mathbf{N}$. A counting argument similar to the one just given shows that the collection of all subsets of $\mathbf{N}$ is a σ-algebra with uncountably many elements. △

A pair $(X, \mathcal{A})$ consisting of a set and a σ-algebra (of X) is called a **measurable space**. A set $A \in \mathcal{A}$ is said to be **measurable**, or **$\mathcal{A}$-measurable**. The terminology implies an intention to assign some measure to the sets in $\mathcal{A}$. We define measures on σ-algebras in the next section.

Proposition 16.1.3. *If $\mathcal{A}$ is an algebra of subsets of X and $(A_i)_{i=1}^{\infty}$ is a sequence of sets in $\mathcal{A}$, then there is a sequence $(B_i)_{i=1}^{\infty}$ of sets in $\mathcal{A}$ which are pairwise disjoint ($B_i \cap B_j$ is empty for $i \neq j$) and such that*

$$\bigcup_{i=1}^{\infty} B_i = \bigcup_{i=1}^{\infty} A_i .$$

Moreover, $\bigcup_{k=1}^{n} B_k = \bigcup_{k=1}^{n} A_k$ for each n.

Proof. Let $B_1 = A_1$, and define inductively the sets B_n for each integer $n \geq 2$, by

$$B_n = A_n - \bigcup_{i=1}^{n-1} A_i.$$

Since $\mathcal{A}$ is an algebra of subsets of X, each set B_n is in $\mathcal{A}$. By definition, $B_n \subseteq A_n$ for each n. (Thus $\bigcup_{k=1}^{n} B_k \subseteq \bigcup_{k=1}^{n} A_k$ for each n.) If $m \neq n$, say $m > n$, then $B_m = A_m - \bigcup_{i=1}^{m-1} A_i$, and since $B_n \subseteq A_n \subseteq B_m^c$ for $m > n$, we have that $B_m \bigcap B_n$

is empty. Therefore the sets B_i are pairwise disjoint. Moreover, $\bigcup_{i=1}^{\infty} B_i \subseteq \bigcup_{i=1}^{\infty} A_i$ since each $B_i \subseteq A_i$.

If $x \in \bigcup_{i=1}^{\infty} A_i$, then $x \in A_i$ for some i. Let n be the smallest value of i for which $x \in A_i$. Then it must be that $x \in B_n = A_n - \bigcup_{i=1}^{n-1} A_i$, and therefore $x \in \bigcup_{i=1}^{\infty} B_i$. Thus $\bigcup_{i=1}^{\infty} A_i \subseteq \bigcup_{i=1}^{\infty} B_i$.

Finally, given n, if $x \in \bigcup_{k=1}^{n} A_k$, then $x \in A_j$ for some $1 \le j \le n$. Let m be the least such index in this range. Then $x \in A_m - \bigcup_{k=1}^{m-1} A_k = B_m$, and therefore $x \in \bigcup_{k=1}^{n} B_k$. Given the reverse containment noted above, the proof is now complete. □

Proposition 16.1.3 does not assume that $\mathcal{A}$ is a σ-algebra, as there is no assertion that the countable union is an element of $\mathcal{A}$; however, the proposition automatically holds for any σ-algebra $\mathcal{A}$.

The next result shows that the intersection of any collection of σ-algebras of X is also a σ-algebra of X.

Proposition 16.1.4. *Let X be a set.*

1. *If $\mathcal{A}_\alpha$ is a σ-algebra of X for each α in a nonempty index set I, then $\bigcap_{\alpha \in I} \mathcal{A}_\alpha$ is also a σ-algebra.*
2. *If C is a collection of subsets of X, then the intersection of all σ-algebras of X that contain C is a σ-algebra of X. This is called the* **σ-algebra of X generated by** *C, denoted $\sigma(C)$. If $\mathcal{B}$ is any σ-algebra of X that contains C, then $\sigma(C) \subseteq \mathcal{B}$.*

Proof. 1. If each $\mathcal{A}_\alpha$ is a σ-algebra of X, then the intersection $\bigcap_{\alpha \in I} \mathcal{A}_\alpha$ satisfies the defining properties 1-4 for a σ-algebra.

2. Note that $\sigma(C)$, the intersection of all σ-algebras that contain C, exists, since the collection of all subsets of X is a σ-algebra containing C. Then clearly $\sigma(C) \subseteq \mathcal{B}$ for any σ-algebra $\mathcal{B}$ that contains C. □

With a view toward our main interest, we make a few comments on $\mathbf{R}$ and $\mathbf{R}^n$ before continuing the general development. A topological space, such as $\mathbf{R}$ or $\mathbf{R}^n$, has a collection of open sets that define the topology of the space. We want all the open sets to be measurable. Our task is to define an appropriate σ-algebra $\mathcal{M}$ of subsets of $\mathbf{R}^n$ and a positive measure μ on $\mathcal{M}$, with all the open sets being elements of $\mathcal{M}$. The next definition provides a start to that task.

Definition 16.1.5. *The σ-algebra $\mathcal{B}_n$ generated by the collection of open subsets of $\mathbf{R}^n$ is called the* **Borel σ-algebra** *of $\mathbf{R}^n$. The elements of $\mathcal{B}_n$ are called* **Borel sets**, *and they are also said to be* **Borel measurable**. *More generally, if X is a metric space, the Borel σ-algebra of X is the smallest σ-algebra that contains all the open sets of X, and its elements are called the Borel sets of X.*

The Borel σ-algebra $\mathcal{B}_n$ is the intersection of all σ-algebras that contain all open sets in $\mathbf{R}^n$. (See Proposition 16.1.4.) The next proposition shows several equivalent ways to generate the Borel σ-algebra $\mathcal{B}_1$ of the real line.

Proposition 16.1.6. *The Borel σ-algebra $\mathcal{B}_1$ of the real numbers is generated by each of these collections:*

1. $C_1 = \{(a, b) : a, b \in \mathbf{R}, a < b\}$;
2. $C_2 = \{[a, b] : a, b \in \mathbf{R}, a < b\}$;
3. $C_3 = \{[a, b) : a, b \in \mathbf{R}, a < b\}$;
4. $C_4 = \{(a, b] : a, b \in \mathbf{R}, a < b\}$;
5. $C_5 = \{(a, \infty) : a \in \mathbf{R}\}$.

Proof. We want to show that $\mathcal{B}_1 = \sigma(C_j)$ for $1 \le j \le 5$. We will prove that $\mathcal{B}_1 = \sigma(C_1)$ and that $\sigma(C_2) = \sigma(C_1)$. The remaining equalities are left as an exercise.

1. $\mathcal{B}_1 = \sigma(C_1)$: First, observe that, the Borel σ-algebra $\mathcal{B}_1$ contains $\sigma(C_1)$ because $\mathcal{B}_1$ contains every bounded open interval. On the other hand, $\sigma(C_1)$ contains not only all bounded open intervals, but unbounded ones as well, since

$$(-\infty, b) = \bigcup_{k=1}^{\infty} (b - k, b) \in \sigma(C_1) \quad \text{and} \quad (a, \infty) = \bigcup_{k=1}^{\infty} (a, a + k) \in \sigma(C_1).$$

Thus, by the structure theorem for open sets of the real line, $\sigma(C_1)$ contains all the open sets, and hence $\sigma(C_1)$ contains $\mathcal{B}_1$. Therefore $\mathcal{B}_1 = \sigma(C_1)$.

2. $\sigma(C_2) = \sigma(C_1)$: Given $[a, b]$, with $a < b$, we can write

$$[a, b] = \bigcap_{k=1}^{\infty} \left(a - \frac{1}{k}, b + \frac{1}{k}\right),$$

which shows that $[a, b]$ is in $\sigma(C_1)$, and it follows that $\sigma(C_2) \subseteq \sigma(C_1)$. On the other hand, given (a, b), with $a < b$,

$$(a, b) = \bigcup_{k=k_0}^{\infty} \left[a + \frac{1}{k}, b - \frac{1}{k}\right], \quad \text{where } \frac{2}{k_0} < b - a.$$

This shows that (a, b) is in $\sigma(C_2)$, hence $\sigma(C_1) \subseteq \sigma(C_2)$. Therefore $\sigma(C_2) = \sigma(C_1) = \mathcal{B}_1$.

The remaining equalities $\sigma(C_5) = \sigma(C_4) = \sigma(C_3) = \mathcal{B}_1$ are left to Exercise 16.1.1. □

The Lebesgue σ-algebra $\mathcal{M}$ to be defined later will contain the Borel σ-algebra,[1] but there will be Lebesgue measurable sets that are not Borel sets. Since $\mathcal{M}$ is to be a σ-algebra, all closed sets will be in $\mathcal{M}$ and consequently all compact sets will be in $\mathcal{M}$. The development of the Lebesgue σ-algebra $\mathcal{M}$ and the Lebesgue measure μ on $\mathcal{M}$ can be handled in a unified way for any $\mathbf{R}^n$, $n \ge 1$. This is done in Section 16.5. We cannot measure every subset of $\mathbf{R}$ (or $\mathbf{R}^n$) and maintain the desirable properties we require of a measure. There are subsets of $\mathbf{R}$ that are not Lebesgue measurable. One such set will be constructed in Exercise 16.5.13.

[1] This is proved in Theorem 16.5.8 after the construction of Lebesgue measure.

Our goals for the next three sections are to discuss arithmetic in the extended real numbers, to set out the fundamental properties of measures, and then to construct a measure from an outer measure. These results prepare the way for the definition of Lebesgue measure.

Exercise.

Exercise 16.1.1. Prove the remaining equalities $\sigma(C_5) = \sigma(C_4) = \sigma(C_3) = \mathcal{B}_1$ asserted in Proposition 16.1.6.

16.2. Arithmetic in the Extended Real Numbers

This section briefly summarizes the necessary arithmetic for the extended real number system, that is, the real number field with the two elements $+\infty$ and $-\infty$ appended. The ordering is given by $-\infty < x < +\infty$ for any real number x. We often write ∞ for $+\infty$. The set of extended real numbers may be denoted by $[-\infty, \infty]$, and the nonnegative extended reals may be denoted by $[0, \infty]$.

In $[-\infty, \infty]$, the usual field operations hold for addition and multiplication of real numbers. The following laws hold for arithmetic involving real numbers x and $\pm\infty$.

For addition and subtraction:

1. $x + (\pm\infty) = (\pm\infty) + x = \pm\infty$.
2. $x - (\pm\infty) = -(\pm\infty) + x = \mp\infty$.
3. $\infty + \infty = \infty$ and $-\infty - \infty = -\infty$.

The operations $(\pm\infty) + (\mp\infty)$ and $(\pm\infty) - (\pm\infty)$ are **undefined**.

For multiplication:

4. For each $x \in \mathbf{R}$,

$$x(\pm\infty) = (\pm\infty)x = \begin{cases} \pm\infty & \text{if } x > 0, \\ 0 & \text{if } x = 0, \\ \mp\infty & \text{if } x < 0. \end{cases}$$

5. $(\pm\infty)(\pm\infty) = +\infty$ and $(\pm\infty)(\mp\infty) = -\infty$

As we see, in measure theory it is useful to define $0(\infty) = (\infty)0 = 0$. For example, we want the integral of the constant zero function over the infinite interval $[0, \infty)$ to be zero.

The use of the elements $\pm\infty$ and the laws for the extended real numbers provide a language for speaking about cases that are likely to arise often enough in measure theory to justify some terminology.

What about upper and lower bounds for sets of extended real numbers? If a set S has no real number upper bound, we may write $\sup S = \infty$. Similarly, if a set S has no real number lower bound, we may write $\inf S = -\infty$. Thus $\sup S$ and $\inf S$ are defined in $[-\infty, \infty]$ for any nonempty set S.

Given the definitions above and the operations in the field **R** itself, we say that $\mathbf{R} \cup \{+\infty, -\infty\} = [-\infty, \infty]$ is the system of **extended real numbers**. A function f that takes values in $[-\infty, \infty]$ is an **extended real valued function**.

16.3. Measures

In this section we present the general properties of measures. We shall use the symbol Σ for a σ-algebra on an arbitrary set X. We start with a general definition of a measure on an arbitrary set, so that we can consider some very simple examples of measures. With some basic properties of measures in place, we will be in a better position to think about how a measure might be constructed.

Definition 16.3.1. *Let (X, Σ) be a measurable space, that is, Σ is a σ-algebra of X. A* **positive measure** *(or simply, a* **measure***) on (X, Σ) is a function $\mu : \Sigma \to [0, \infty]$ such that $\mu(A) < \infty$ for at least one set $A \in \Sigma$, and μ is* **countably additive**, *which means that if $\{A_j\}$ is a countable collection of pairwise disjoint members of Σ, then*

$$\mu\Big(\bigcup_{j=1}^{\infty} A_j\Big) = \sum_{j=1}^{\infty} \mu(A_j).$$

The triple (X, Σ, μ) is then called a **positive measure space** *(or simply, a* **measure space***).*

We assume there is some set that has finite measure in order to avoid a measure of no interest to us. This assumption also has the desirable consequence that the measure of the empty set is zero, for if $\mu(A) < \infty$, then we can define the countable, pairwise disjoint collection consisting of $A_1 = A$ and A_j equal to the empty set for $j \geq 2$. By countable additivity, $\mu(A) = \mu(\bigcup_{j=1}^{\infty} A_j) = \mu(A) + \mu(\emptyset) + 0$. By subtracting the finite $\mu(A)$, we find the empty set has measure zero. By a similar argument, we see that countable additivity implies that a measure is **finitely additive**, which means that if $A_j \in \Sigma$, for $1 \leq j \leq n$, and the sets A_j are pairwise disjoint, then

$$\mu\Big(\bigcup_{j=1}^{n} A_j\Big) = \sum_{j=1}^{n} \mu(A_j).$$

A pairwise disjoint collection of sets may be called a disjoint collection or a collection of disjoint sets. Any countable union of disjoint sets is also called a disjoint union.

More advanced treatments of measure theory include other types of measures. In this book we consider only positive measures, and we will say *measure* and *measure space* without the qualifier *positive*. When the σ-algebra on X is understood, we say that μ is a measure on X.

Example 16.3.2. Let $X = \mathbf{N}$, the set of positive integers, and let Σ be the σ-algebra of all subsets of $\mathbf{N}$. Define $\mu : \Sigma \to [0, \infty]$ by letting $\mu(A)$ be the cardinality of A, for each subset A of X. Then μ is a measure on X, called the **counting measure** on X. △

If (X, Σ, μ) is a measure space and $\mu(X) = 1$, then (X, Σ, μ) is a **probability measure space**.

An interesting probability measure space is the interval $I = (0, 1]$, which we now briefly discuss. When we have defined Lebesgue measure on a σ-algebra of subsets of the real line, we will have a measure on the unit interval I by simply intersecting the sets of that σ-algebra with I itself. Since Lebesgue measure assigns every interval a measure equal to the interval length, the Lebesgue measure of I is 1, and hence Lebesgue measure is a probability measure on I.

There is a reason for omitting 0 from our unit interval. Every $\omega \in I$ can be written in the form

$$\omega = \sum_{i=1}^{\infty} \frac{a_i}{2^i}, \qquad a_i \in \{0, 1\}.$$

Since the a_i completely determine ω, we may write

$$\omega = .a_1\, a_2\, a_3 \, \cdots ,$$

which is the binary expansion of ω. With a given ω, we associate a sequence β of coin tosses, known as a Bernoulli trial (or Bernoulli sequence) by placing an H as the n-th term of the sequence if $a_n = 1$ and a T as the n-th term if $a_n = 0$. Let B be the set of all sequences β of the symbols H and T. To make our mapping $\omega \mapsto \beta \in B$ well-defined, we agree to use only nonterminating binary expansions for the elements of I. This means that we toss out all terminating binary expansions, the ones that have an infinite tail of zeros. For example $\omega = .01\bar{0} = .00\bar{1}$, and we choose to use the expansion $\omega = .00\bar{1}$ for the number $1/4$. By this choice, we also toss out all Bernoulli sequences that end in all tails. (We should not be concerned about their loss. And this is why we omit 0 from I.) Then the mapping from I to B given by $\omega \mapsto \beta$ is well-defined and one-to-one, but not onto. The set B_T of Bernoulli sequences that end in all tails is countable, because the set of sequences that end in tails after the n-th term in the sequence is finite, having 2^n elements, for each n. (See Exercise 1.4.3.) Thus B_T is a countable union of finite sets, hence countable. Therefore we have the following correspondence.

Proposition 16.3.3. *Let B_T denote the set of Bernoulli sequences that end in all tails. Then $B - B_T$ is indexed by the points in $I = (0, 1]$ using nonterminating binary expansions for elements $\omega \in I$.*

We can make use of the probability space I even though, at this point, we know only the Lebesgue measure of finite unions of disjoint intervals. We use the following *Borel principle.*

Borel Principle. *Suppose E is a probabilistic event occurring in certain Bernoulli sequences. Let B_E denote the subset of B for which the event occurs, and let S_E be the corresponding subset of I. Then the probability that E occurs is equal to $m(S_E)$.*

Here is a sampling of events for which we can determine the probability.

1. Let E be the event that a head is thrown on the first toss. We expect the probability of this event is $1/2$ for a fair coin. Then E corresponds to the subset S_E of I consisting of all those binary representations that begin with a 1 in the first digit. A number ω is in S_E if and only if $\omega = .1a_2a_3\ldots$. Thus, ω is in S_E if and only if $\omega \geq .1000\ldots$ and $\omega \leq .1111\ldots$. Therefore $S_E = [1/2, 1]$, and the measure of S_E is $1/2$, as expected.

2. Let E be the event that the first 10 tosses are some prescribed sequence and everything else is arbitrary. The corresponding subset of I is

$$S_E = \{\omega \in I : \omega = .a_1\, a_2\, \ldots\, a_{10}\, \ldots\}$$

where only $a_1, \ldots, a_{10}$ are prescribed. If $s = .a_1\, a_2 \ldots a_{10}\bar{0}$, then $S_E = [s, s + (1/2^{10}))$. Since the measure of the interval S_E is 2^{10}, this is the probability of event E. This is expected, as the first ten tosses can occur in 2^{10} different ways, equally likely for a fair coin. By a similar analysis, if E is the event that the first n tosses are some prescribed sequence, then the probability of E is the measure of the interval $S_E = [s, s + (1/2^n))$, where $s = .a_1\, \ldots\, a_n\bar{0}$, which is $1/2^n$. Again, this meets our expectations.

3. Suppose E is the event that in the first 10 tosses, exactly 3 heads appear. The associated subset of I is

$$S_E = \{\omega \in I : \omega = .a_1\, a_2\, \ldots\, a_{10}\, \ldots, \text{ where exactly 3 of the first } n\ a_i \text{ are } 1\}.$$

Now fix the first n digits, exactly three of which equal 1. If $s = .a_1\, a_2\, \ldots\, a_{10}\bar{0}$, then S_E contains the interval $[s, s + (1/2^{10}))$. With other choices of the first 10 digits (each choice having exactly three digits equal to 1) there are a total of

$$\binom{10}{3} = \frac{10!}{7!\,3!} = \frac{10 \cdot 9 \cdot 8}{3 \cdot 2} = 120$$

intervals (the number of combinations of three things chosen from ten), each having measure $1/2^{10}$, all of which belong to S_E. Moreover, these intervals are pairwise disjoint, since the distance between any two of these s numbers represented by the first ten digits must be greater than $1/2^{10}$. Thus the total measure of these intervals is

$$\frac{1}{2^{10}}120 = \frac{120}{1024} = .1171875,$$

yielding a probability of E equal to .1171875.

4. (Gambler's Ruin) This example is more involved, but provides important insight and motivation. Suppose a gambler starts with $\$X$ and bets on a sequence of coin tosses. At each toss he wins \$1 if a head appears and loses \$1 if a tail appears. What is the probability of event E, *gambler's ruin*, that he will lose his entire original stake? We need an appropriate notation to describe the amount won or lost at each toss, the payoff or loss after k tosses, and the ways in which the gambler can lose the entire stake. For $\omega \in I$, define the k-th Rademacher function $R_k : I \to \mathbf{R}$ by

$$R_k(\omega) = 2a_k - 1,$$

where $\omega = .a_1\, a_2\, a_3\, \ldots$ in the binary expansion we are using. Thus

$$R_k(\omega) = \begin{cases} +1 & \text{if } a_k = 1, \\ -1 & \text{if } a_k = 0, \end{cases}$$

so $R_k(\omega)$ is the amount won or lost on the k-th toss.

See Figure 16.1 for the graphs of the first three Rademacher functions. The functions R_k allow a description of event E, the gambler's ruin. The

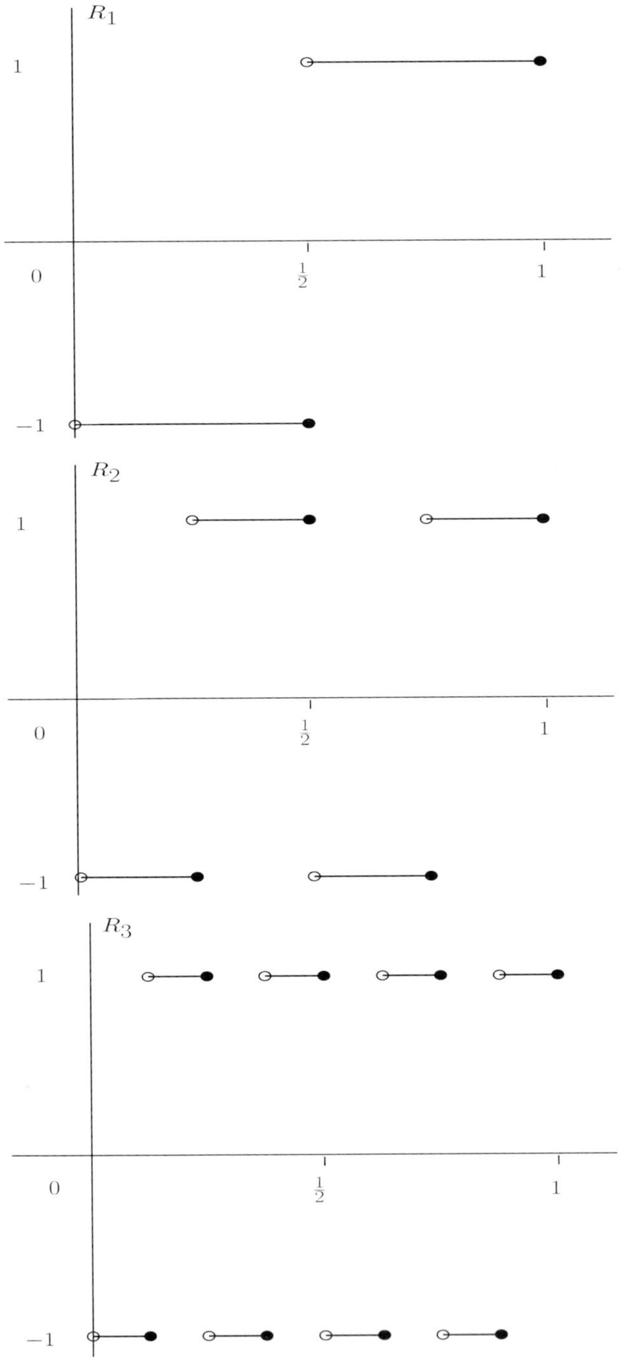

Figure 16.1. The first three Rademacher functions: R_1, R_2 and R_3.

total amount won or lost after k tosses is given by

$$f_k(\omega) := \sum_{j \le k} R_j(\omega).$$

Then the event E_k, that the gambler loses his stake at the k-th toss, corresponds to the subset S_{E_k} of I given by

$$S_{E_k} = \{\omega \in I : f_j(\omega) > -X \text{ for } j < k, \text{ and } f_k(\omega) = -X\}.$$

Any one of these events E_k signals the gambler's ruin. Hence the complete loss of the original stake, event E, is represented by the union

$$S_E = \bigcup_{k=1} S_{E_k} = \bigcup_{k=1} \{\omega \in I : f_j(\omega) > -X \text{ for } j < k, \text{ and } f_k(\omega) = -X\}.$$

At least two observations are important here. The first is that event E is described by a countably infinite union of sets, which needs to be measurable. The second is the realization that it will be important to be able to measure sets of the form

$$\{\omega \in I : f(\omega) > \alpha\} \qquad \text{and} \qquad \{\omega \in I : f(\omega) \le \alpha\}$$

for suitable functions f. The first observation emphasizes the need for a σ-algebra of measurable sets, and the second observation points toward the later definition of measurable function.

So as not to leave the reader hanging about the gambler's fate, it can be shown that the Lebesgue measure, $m(S_E)$, of the set S_E equals

$$m(S_E) = \sum_{k=1}^{\infty} m(S_{E_k}) = 1.$$

Thus, if the gambler bets long enough, he eventually loses his initial stake, no matter how large that stake may be. Exercise 16.3.4 may develop some initial intuition towards this result.

We cannot delve deeply into probability measures and problems in this book. However, probability can provide accessible motivation and concrete problems when learning about measure and integration.

We can now establish a few properties of measures. Suppose a set is the union of a nested sequence of sets or the intersection of a nested sequence of sets. Measures have a nice continuity property with respect to such monotone sequences, as seen in the next proposition.

Proposition 16.3.4. *Let (X, Σ, μ) be a measure space. Then the following properties hold:*

1. *If $A, B \in \Sigma$ and $A \subseteq B$, then $\mu(A) \le \mu(B)$. If $\mu(B) < \infty$, then $\mu(B - A) = \mu(B) - \mu(A)$.*

2. *If $A_k \in \Sigma$ for $k = 1, 2, \ldots$, then $\mu(\bigcup_{k=1}^{\infty} A_k) \le \sum_{k=1}^{\infty} \mu(A_k)$.*

3. *If* $A_k \in \Sigma$ *for* $k = 1, 2, \ldots,$ *and* $A_k \subseteq A_{k+1}$ *for all* k, *then*

$$\mu\Big(\bigcup_{k=1}^{\infty} A_k\Big) = \lim_{k\to\infty} \mu(A_k).$$

4. *If* $A_k \in \Sigma$ *for* $k = 1, 2, \ldots,$ *with* $A_{k+1} \subseteq A_k$ *for all* k, *and* $\mu(A_{k_0}) < \infty$ *for some positive integer* k_0, *then*

$$\mu\Big(\bigcap_{k=1}^{\infty} A_k\Big) = \lim_{k\to\infty} \mu(A_k).$$

Proof. 1. Write $B = A \cup (B - A)$, a disjoint union, and thus

$$\mu(B) = \mu(A) + \mu(B - A).$$

Since $\mu(B - A) \geq 0$, $\mu(B) - \mu(A) \geq 0$, hence $\mu(A) \leq \mu(B)$. If $\mu(B) < \infty$, then $\mu(A) < \infty$ as well, so $\mu(B - A) = \mu(B) - \mu(A)$ is well defined.

2. By Proposition 16.1.3, we may write $\bigcup_{k=1}^{\infty} A_k = \bigcup_{k=1}^{\infty} B_k$ for a sequence of pairwise disjoint sets B_k with $B_k \subseteq A_k$ for each k. Then

$$\mu\Big(\bigcup_{k=1}^{\infty} A_k\Big) = \mu\Big(\bigcup_{k=1}^{\infty} B_k\Big) = \sum_{k=1}^{\infty} \mu(B_k) \leq \sum_{k=1}^{\infty} \mu(A_k).$$

3. Letting $A_0 = \emptyset$, we may write $\bigcup_{k=1}^{\infty} A_k = \bigcup_{k=1}^{\infty}(A_k - A_{k-1})$, which is now a disjoint union. Hence,

$$\begin{aligned}
\mu\Big(\bigcup_{k=1}^{\infty} A_k\Big) &= \mu\Big(\bigcup_{k=1}^{\infty}(A_k - A_{k-1})\Big) = \sum_{k=1}^{\infty} \mu(A_k - A_{k-1}) \\
&= \lim_{m\to\infty} \sum_{k=1}^{m} \mu(A_k - A_{k-1}) = \lim_{m\to\infty} \mu\Big(\bigcup_{k=1}^{m}(A_k - A_{k-1})\Big) \\
&= \lim_{m\to\infty} \mu(A_m).
\end{aligned}$$

4. By hypothesis, the sequence $A_{k_0} - A_k$ is an increasing sequence of sets, and

$$\bigcup_{k=1}^{\infty}(A_{k_0} - A_k) = A_{k_0} - \bigcap_{k=1}^{\infty} A_k.$$

Thus, by item 1, we have

$$\mu\Big(\bigcup_{k=1}^{\infty}(A_{k_0} - A_k)\Big) = \mu(A_{k_0}) - \mu\Big(\bigcap_{k=1}^{\infty} A_k\Big).$$

The sequence $\mu(A_k)$ is decreasing and bounded below by zero. By items 3 and 1, we have

$$\begin{aligned}
\mu\Big(\bigcup_{k=1}^{\infty}(A_{k_0} - A_k)\Big) &= \lim_{k\to\infty} \mu(A_{k_0} - A_k) \\
&= \lim_{k\to\infty} [\mu(A_{k_0}) - \mu(A_k)] \\
&= \mu(A_{k_0}) - \lim_{k\to\infty} \mu(A_k).
\end{aligned}$$

Comparing the two results, we have $\mu\big(\bigcap_{k=1}^{\infty} A_k\big) = \lim_{k\to\infty} \mu(A_k)$. □

Definition 16.3.5. *Let (X, Σ, μ) be a measure space. A measure μ is said to be* **complete** *if every subset of a set having measure zero is measurable (and hence has measure zero). That is, if $A \subset B$ and $\mu(B) = 0$, then $A \in \Sigma$ and $\mu(A) = 0$.*

Working with a complete measure avoids some annoying technical issues, so it is worth knowing that any measure μ on X which is not complete may be extended to a larger σ-algebra of X on which it is complete. However, this extension result is not needed in this book. The main measures of interest to us, Lebesgue measure on the real line and Lebesgue measure on $\mathbf{R}^n$, are complete measures. Nevertheless, the notion of complete measure should be kept in mind in a few of the statements later on. For example, Lebesgue measure on the real line will be defined on a σ-algebra containing, but strictly larger than, the Borel σ-algebra $\mathcal{B}_1$, and Lebesgue measure restricted to $\mathcal{B}_1$ is not a complete measure.

Exercises.

Exercise 16.3.1. Use the counting measure of Example 16.3.2 to give an example of $A \subset B$ where $\mu(B) = \infty$ and the expression $\mu(B) - \mu(A)$ is undefined. Also give an example of a sequence $A_{k+1} \subset A_k$ for each k, where the limit statement in Proposition 16.3.4 (item 4) does not hold.

Exercise 16.3.2. Is the counting measure of Example 16.3.2 a complete measure?

Exercise 16.3.3. Determine the probabilities of these events using measure arguments:

1. Find the probability of the coin toss event E, that the first 3 coin tosses are all heads. That is, identify the corresponding subset S_E of $X = (0, 1]$ and determine its Lebesgue measure.
2. Find the probability of the event E that the first 3 coin tosses are all tails. Identify the relevant subset S_E of $X = (0, 1]$ and determine its Lebesgue measure.

Exercise 16.3.4. Suppose the gambler has an initial stake of one dollar. Determine the probability of his ruin occuring at the first, third, and fifth toss.

16.4. Measure from Outer Measure

A standard way to construct a measure is to first construct a preliminary rougher measurement of sets called an outer measure. A measure may then be constructed from an outer measure. In particular, the Lebesgue measures on $\mathbf{R}$ and $\mathbf{R}^n$ will be constructed from suitable outer measures on those spaces. Thus we begin with the definition of outer measure for an arbitrary set X.

Definition 16.4.1. *Let X be a set. An* **outer measure** *on X is a function μ^* on the σ-algebra of all subsets of X such that*

1. *$\mu^*(A) \geq 0$ for every subset A of X, and $\mu^*(\emptyset) = 0$;*
2. *if $A \subseteq B$, then $\mu^*(A) \leq \mu^*(B)$;*
3. *if $A_k \subseteq X$ for each k, then $\mu^*(\bigcup_{k=1}^{\infty} A_k) \leq \sum_{k=1}^{\infty} \mu^*(A_k)$.*

Remark. *Property 2 of this definition is called* **monotonicity** *of the outer measure, and property 3 is called* **countable subadditivity** *of the outer measure. By taking A_k empty for $k \geq n$, property 3 implies that an outer measure is finitely subadditive:* $\mu^*(\bigcup_{k=1}^n A_k) \leq \sum_{k=1}^n \mu^*(A_k)$ *for any subsets A_k, $1 \leq k \leq n$, of X.*

If a measure μ is defined on the σ-algebra of all subsets of X, then μ is an outer measure on X by Proposition 16.3.4, part 2. For example, the counting measure on $\mathbf{N}$ is an outer measure.

An outer measure on X is easy to construct. Let $\mathcal{K}$ be any collection of subsets of X that includes the empty set, and that covers X in the following sense: for any subset A of X, there is a sequence E_k in $\mathcal{K}$ such that $A \subseteq \bigcup_{k=1}^\infty E_k$. Such a collection $\mathcal{K}$ is called a **sequential covering class** of X. The idea is that for a given X, the covering class should be chosen to consist of sets that generate a desired σ-algebra of measurable sets. For example, the collection of intervals of the form $[a, b)$, $a < b$, is a sequential covering class of $\mathbf{R}$, and it generates the Borel σ-algebra.

Let $\mathcal{K}$ be a sequential covering class of X, and let $\lambda : \mathcal{K} \to [0, \infty]$ be any extended real valued function that satisfies $\lambda(\emptyset) = 0$. For each subset A of X, define

$$\mu^*(A) := \inf\Big\{\sum_{k=1}^\infty \lambda(E_k) : E_k \in \mathcal{K} \text{ and } A \subseteq \bigcup_{k=1}^\infty E_k\Big\}. \tag{16.1}$$

The function λ provides the mechanism for ensuring that the covering sets are assigned their appropriate measure. For $\mathbf{R}$ and $\mathbf{R}^n$, λ is the volume function on certain intervals. The role of λ becomes clearer when we define Lebesgue measure in the next section.

Theorem 16.4.2. *For any sequential covering class $\mathcal{K}$ on X and for any extended real valued function $\lambda : \mathcal{K} \to [0, \infty]$ with $\lambda(\emptyset) = 0$, the function μ^* defined by (16.1) is an outer measure on X.*

Proof. Properties 1 and 2 of outer measure are immediate from (16.1). It only remains to show subadditivity of μ^*. Let A_k be subsets of X. Let $\epsilon > 0$. By (16.1), for each k, there is a sequence of sets $E_{kj} \in \mathcal{K}$ such that $A_k \subseteq \bigcup_{j=1}^\infty E_{kj}$ and

$$\sum_{j=1}^\infty \lambda(E_{kj}) \leq \mu^*(A_k) + \frac{\epsilon}{2^k}.$$

This is valid even when $\mu^*(A_k) = \infty$, since the left-hand side must then be ∞, by (16.1). We have $\bigcup_{k=1}^\infty A_k \subseteq \bigcup_{k,j=1}^\infty E_{kj}$, which is a countable union of sets from $\mathcal{K}$, and hence

$$\begin{aligned}
\mu^*\Big(\bigcup_{k=1}^{\infty} A_k\Big) &\le \sum_{k,j=1}^{\infty} \lambda(E_{kj}) \quad \text{(by (16.1))} \\
&\le \sum_{k=1}^{\infty} \big(\mu^*(A_k) + \frac{\epsilon}{2^k}\big) \\
&\le \Big(\sum_{k=1}^{\infty} \mu^*(A_k)\Big) + \epsilon.
\end{aligned}$$

Observe that the last line is valid if $\mu^*(A_k) = \infty$ for some k, or if $\mu^*(A_k)$ is finite for all k and $\sum_{k=1}^{\infty} \mu^*(A_k) = \infty$. Since ϵ was arbitrary, this proves subadditivity of μ^*. □

Now the main task is to construct a measure from a given outer measure. An outer measure is a way of assigning a number to every subset of a set X by approximating each set using sets from a relevant sequential covering class. A relevant covering class should consist of sets we want to be measurable, and which can be used to generate, or approximate well, the sets of most interest to us. For example, in the construction of Lebesgue measure on $\mathbf{R}$ it will be convenient to use the sequential covering class consisting of the intervals $[a, b)$ for $a, b \in \mathbf{R}$. The closed intervals in $\mathbf{R}^n$ will provide an appropriate sequential covering class for $n \ge 2$. The development below shows that the outer measure generates a σ-algebra of measurable sets that includes, but is larger than, the Borel σ-algebra.

To help motivate the definition of measurability, we make a couple of observations. Recall that a measure is necessarily finitely additive on disjoint sets. If a set A is to be measurable, and we split an arbitrary set E into the sets $E \cap A$ and $E \cap A^c$, then we could say that the set A behaves well with respect to the outer measure of E if $\mu^*(E) = \mu^*(E \cap A) + \mu^*(E \cap A^c)$. As our measurable sets, we want only those sets A that behave well with respect to the outer measure of *every* set E. This observation motivates the next definition.

Definition 16.4.3. *Let μ^* be an outer measure on X. A subset A of X is μ^*-* **measurable** *if for each set $E \subseteq X$,*

$$\mu^*(E) = \mu^*(E \cap A) + \mu^*(E \cap A^c). \tag{16.2}$$

It follows from this definition that the empty set and X are both μ^*-measurable (Exercise 16.4.1). The definition is symmetric in A and its complement A^c, so a set is μ^*-measurable if and only if its complement is μ^*-measurable. Since $E = (E \cap A) \cup (E \cap A^c)$, and an outer measure is subadditive, we always have

$$\mu^*(E) \le \mu^*(E \cap A) + \mu^*(E \cap A^c).$$

Hence to establish (16.2) for a set A, it suffices to show that

$$\mu^*(E) \ge \mu^*(E \cap A) + \mu^*(E \cap A^c). \tag{16.3}$$

Moreover, since (16.3) clearly holds in case $\mu^*(E) = \infty$, it is only necessary to consider sets E with $\mu^*(E) < \infty$.

The next result says that the restriction of an outer measure μ^* to its μ^*-measurable sets is a measure on a σ-algebra.

Theorem 16.4.4. *If μ^* is an outer measure on X, then the collection $\mathcal{M}$ of μ^*-measurable sets is a σ-algebra of X, and $\mu := \mu^*|_{\mathcal{M}}$ is a complete measure on X.*

Proof. We show first that $\mathcal{M}$ is an algebra. As noted above, the empty set and X are both μ^*-measurable, and $A \in \mathcal{M}$ if and only if $A^c \in \mathcal{M}$, by the symmetry of the definition. Let $A, B \in \mathcal{M}$. We want to show that $A \cup B \in \mathcal{M}$. Let E be any subset of X. Since $A \in \mathcal{M}$ and $B \in \mathcal{M}$,

$$\begin{aligned} \mu^*(E) &= \mu^*(E \cap A) + \mu^*(E \cap A^c) && \text{(since } A \in \mathcal{M}) \\ &= \mu^*(E \cap A \cap B) + \mu^*(E \cap A \cap B^c) \\ &\quad + \mu^*(E \cap A^c \cap B) + \mu^*(E \cap A^c \cap B^c). \end{aligned} \tag{16.4}$$

Since $B \in \mathcal{M}$, (16.4) follows from (16.2) with E replaced first by $E \cap A$, then by $E \cap A^c$. We have $(A \cup B)^c = A^c \cap B^c$ and $A \cup B = (A \cap B) \cup (A \cap B^c) \cup (A^c \cap B)$. After intersecting both $A \cup B$ and $(A \cup B)^c$ with E, and using subadditivity of μ^*, we have

$$\begin{aligned} \mu^*(E \cap (A \cup B)^c)) &= \mu^*(E \cap (A^c \cap B^c)) \\ \mu^*(E \cap (A \cup B)) &\leq \mu^*(E \cap A \cap B) + \mu^*(E \cap A \cap B^c) + \mu^*(E \cap A^c \cap B). \end{aligned}$$

By adding these expressions, left and right, and using (16.4), we obtain

$$\mu^*(E) \geq \mu^*(E \cap (A \cup B)) + \mu^*(E \cap (A \cup B)^c),$$

and conclude that $A \cup B \in \mathcal{M}$. This shows that $\mathcal{M}$ is an algebra.

The next step is to show that $\mathcal{M}$ is a σ-algebra. Since $\mathcal{M}$ is an algebra, Proposition 16.1.3 implies that any countable union of sets of $\mathcal{M}$ is equal to a countable disjoint union of sets of $\mathcal{M}$. Thus, to show that $\mathcal{M}$ is a σ-algebra it suffices to show that any countable disjoint union of sets of $\mathcal{M}$ is again in $\mathcal{M}$. Let A_j be a sequence of disjoint sets, with $A_j \in \mathcal{M}$. For each positive integer n, let $B_n = \bigcup_{j=1}^n A_j$, and let $B = \bigcup_{j=1}^\infty A_j$. If E is any subset of X, then, by measurability of A_n (and replacing E by $E \cap B_n$ in (16.2)),

$$\begin{aligned} \mu^*(E \cap B_n) &= \mu^*(E \cap B_n \cap A_n) + \mu^*(E \cap B_n \cap A_n^c) \\ &= \mu^*(E \cap A_n) + \mu^*(E \cap B_{n-1}), \end{aligned}$$

since $B_n \cap A_n = A_n$ and $B_n \cap A_n^c = B_{n-1}$. We can apply the same argument to the last term on the right, to get

$$\mu^*(E \cap B_{n-1}) = \mu^*(E \cap A_{n-1}) + \mu^*(E \cap B_{n-2}).$$

After applying this argument $n-1$ times, we obtain

$$\mu^*(E \cap B_n) = \sum_{j=2}^n \mu^*(E \cap A_j) + \mu^*(E \cap B_1) = \sum_{j=1}^n \mu^*(E \cap A_j), \tag{16.5}$$

since $B_1 = A_1$. Since $\mathcal{M}$ is an algebra, $B_n \in \mathcal{M}$. Since $B_n \subseteq B$, we have $B^c \subseteq B_n^c$. For every set E, and each n, (16.5) implies that

$$\begin{aligned}\mu^*(E) &= \mu^*(E \cap B_n) + \mu^*(E \cap B_n^c) \\ &= \Big(\sum_{j=1}^{n} \mu^*(E \cap A_j)\Big) + \mu^*(E \cap B_n^c) \\ &\geq \Big(\sum_{j=1}^{n} \mu^*(E \cap A_j)\Big) + \mu^*(E \cap B^c),\end{aligned}$$

since $B^c \subseteq B_n^c$ for each n. We may let $n \to \infty$, and conclude that

$$\begin{aligned}\mu^*(E) &\geq \Big(\sum_{j=1}^{\infty} \mu^*(E \cap A_j)\Big) + \mu^*(E \cap B^c) \\ &\geq \mu^*\Big(\bigcup_{j=1}^{\infty} E \cap A_j\Big) + \mu^*(E \cap B^c) \\ &= \mu^*(E \cap B) + \mu^*(E \cap B^c) \;\geq\; \mu^*(E),\end{aligned} \tag{16.6}$$

by subadditivity of μ^* in each of the last two lines. This shows that the countable disjoint union $B = \bigcup_{j=1}^{\infty} A_j$ is in $\mathcal{M}$. As we noted above, this result implies, by Proposition 16.1.3, that $\mathcal{M}$ is a σ-algebra. The other important fact about (16.6) is that all three of the *greater than or equal to* signs must be equalities.

It remains to show that μ^* restricted to $\mathcal{M}$ is a measure, and that if $\mu^*(E) = 0$, then $E \in \mathcal{M}$. First, μ^* restricted to $\mathcal{M}$ is a measure if it is countably additive, since we know already that $\mu^*(\emptyset) = 0 < \infty$. The equality (16.6), developed for countable disjoint unions, and expressed here in its true form, as

$$\mu^*(E) = \Big(\sum_{j=1}^{\infty} \mu^*(E \cap A_j)\Big) + \mu^*(E \cap B^c),$$

holds for any set E and any disjoint union $B = \bigcup_{j=1}^{\infty} A_j$, where each A_j is in $\mathcal{M}$. Now let $E = B$ in this equation, to obtain

$$\mu^*(B) = \sum_{j=1}^{\infty} \mu^*(A_j).$$

This proves the countable additivity of $\mu^*|_{\mathcal{M}}$. Therefore $\mu := \mu^*|_{\mathcal{M}}$ is a measure on X.

Now suppose that $A \subset B$, $B \in \mathcal{M}$, and $\mu^*(B) = 0$. We have $\mu^*(A) \leq \mu^*(B)$ since an outer measure has this monotonic property by definition. Hence $\mu^*(A) = 0$. It remains to show that A is μ^*-measurable. If E is any subset of X, then two applications of monotonicity yield

$$\mu^*(E \cap A) + \mu^*(E \cap A^c) \leq 0 + \mu^*(E \cap A^c) \leq \mu^*(E).$$

This inequality shows that A is μ^*-measurable. Therefore $\mu = \mu^*|_{\mathcal{M}}$ is a complete measure. $\square$

Exercises.

Exercise 16.4.1. Let μ^* be an outer measure on X. Verify that the empty set and X are both μ^*-measurable.

Exercise 16.4.2. Using the structure theorem for open sets in $\mathbf{R}$, we could define the total length of an open set O, $\lambda(O)$, to be the sum of the lengths of the (countably many) disjoint open intervals whose union equals O. Then, if E is a subset of $\mathbf{R}$, define

$$\mu^*(E) = \inf\{\lambda(O) : O \text{ is open and } E \subset O\}.$$

1. Verify that the open intervals constitute a sequential covering class of $\mathbf{R}$.
2. Verify that Theorem 16.4.2 applies, and thus μ^* is an outer measure on $\mathbf{R}$.

16.5. Lebesgue Measure in Euclidean Space

This section includes the definitions of Lebesgue measure on $\mathbf{R}$, and on $\mathbf{R}^n$, by applying the results of the preceding section. These measures will be denoted by m (for the case of the real line) and m_n, for $n \geq 2$, if a reference to dimension is needed. There is some simplification possible for the case of the real line. Some readers may have a primary interest in the real line, so we consider that case separately. For each space $\mathbf{R}^n$, we show that the Lebesgue σ-algebra $\mathcal{M}_n$ defined below contains the Borel σ-algebra $\mathcal{B}_n$, and hence $\mathcal{M}_n$ contains all open sets and all closed sets.

16.5.1. Lebesgue Measure on the Real Line.

Let $\mathcal{K}$ be the collection of real intervals $J = [a, b)$, with $a < b$. Then $\mathcal{K}$ is a sequential covering class of $\mathbf{R}$. Let $\lambda(\emptyset) = 0$, and define $\lambda(J) = \lambda([a, b)) = b - a$ for each element $J = [a, b) \in \mathcal{K}$. Then we define m^* on the σ-algebra of all subsets of $\mathbf{R}$ in accordance with (16.1). Therefore define

$$\mu^*(A) := \inf\Big\{\sum_{k=1}^{\infty} \lambda(J_k) : J_k \in \mathcal{K} \text{ and } A \subseteq \bigcup_{k=1}^{\infty} J_k\Big\} \tag{16.7}$$

for $A \subseteq \mathbf{R}$. Then m^* is an outer measure, by Theorem 16.4.2, called the **Lebesgue outer measure** on $\mathbf{R}$. Let $\mathcal{M} = \mathcal{M}_1$ be the collection of m^*-measurable sets in accordance with Definition 16.4.3. Then the restriction $m = m^*|_{\mathcal{M}}$ is a complete measure on $\mathbf{R}$ by Theorem 16.4.4. The collection $\mathcal{M}$ is the **Lebesgue σ-algebra** of $\mathbf{R}$, and m is **Lebesgue measure** on $\mathbf{R}$. The elements of $\mathcal{M}$ are known as the **Lebesgue measurable sets** in $\mathbf{R}$. The triple $(\mathbf{R}, \mathcal{M}, m)$ is the one-dimensional **Lebesgue measure space**.

It might seem that much is left mysterious by this definition of $(\mathbf{R}, \mathcal{M}, m)$ by way of Theorem 16.4.4 from the previous section, since we have not determined the Lebesgue measure of many sets. However, the usefulness of Definition 16.4.3 and Theorem 16.4.4 lies in their generality. As we noted when defining outer measure, the function λ on the sequential covering class $\mathcal{K}$ is the mechanism that ensures that the covering sets are assigned their appropriate (outer) measure. In the present case, our interest is in showing that $m^*([a, b)) = \lambda([a, b))$, where $\lambda([a, b)) = b - a$ by definition. This result is not obvious from the definition of $m^*([a, b))$, though it is clear that it should be true, and we need it in the proof of Theorem 16.5.3 below. Lemma 16.5.1 is a nontrivial result that depends on the completeness of $\mathbf{R}$.

Lemma 16.5.1. *For every bounded interval $[a, b)$ with $a < b$,*

$$m^*([a, b)) = \lambda([a, b)) = b - a.$$

Proof. By definition, $\lambda([a, b)) = b - a$ and $[a, b)$ is in the covering class $\mathcal{K}$, so $m^*([a, b)) \le \lambda([a, b))$. Suppose $[a, b)$ is covered by a union $\bigcup_{k=1}^{\infty} J_k$, where the J_k belong to $\mathcal{K}$. Then $[a, b) = [a, b) \bigcap_{k=1}^{\infty} \left(\bigcup_{k=1}^{\infty} J_k\right) = \bigcup_{k=1}^{\infty} \left([a, b) \bigcap J_k\right)$. Each intersection $[a, b) \bigcap J_k$ is either empty or another interval of the same type; we assume without loss of generality that each intersection is nonempty. Thus the union $\bigcup_{k=1}^{\infty} \left([a, b) \bigcap J_k\right)$ of intervals closed at the left end and open at the right end covers $[a, b)$. (Recall we have not established that intervals of this type are m^*-measurable, nor do we know the value of $m^*([a, b) \bigcap J_k)$ yet, since it is an interval of the same type as $[a, b)$.)

The proof of the lemma is by contradiction.

We suppose that the lengths of the covering intervals $[a, b) \bigcap J_k$ sum to a number less than $b - a$. That is, we assume that $\lambda([a, b))$ is not the infimum that defines $m^*([a, b))$. Then there are numbers $a' > a$, $b' < b$, and $\delta > 0$, such that

$$\left[\sum_{k=1}^{\infty} \lambda\Big([a, b) \bigcap J_k\Big)\right] + \delta \;<\; b' - a' \;<\; b - a.$$

Let $0 < \epsilon < \delta$. The interval $[a', b']$ is covered by these half-open intervals. For each k, we can extend $[a, b) \bigcap J_k$ past its left endpoint a distance $\epsilon/2^k$, creating an open interval by the extension, and thus we conclude that the interval $[a', b']$ is covered by countably many open intervals whose lengths sum to less than $b' - a'$. But $[a', b']$ is compact.[2] So there are finitely many of these open intervals that cover $[a', b']$. Let $\{a_{k_1}, a_{k_2}, \ldots, a_{k_N}\}$ be an ordered enumeration of the left endpoints of these finitely many open intervals, and write the intervals as $\{(a_{k_i}, b_{k_i}) : 1 \le i \le N\}$. Then we have $\sum_{i=1}^{N} (b_{k_i} - a_{k_i}) < b' - a'$. This is the contradiction we were seeking. Therefore $m^*([a, b)) = \lambda([a, b)) = b - a$. □

Lemma 16.5.2. *The outer measure m^* is finitely additive on disjoint intervals of the form $[a, b)$.*

Proof. It suffices to prove the result for two disjoint intervals of this form, which we shall label as $[a, b)$ and $[c, d)$. Subadditivity of m^* implies that

$$m^*\big([a, b) \cup [c, d)\big) \le m^*([a, b)) + m^*([c, d)) = (b - a) + (d - c),$$

where the last equality is from Lemma 16.5.1. On the other hand, the definition of Lebesgue outer measure implies that for any $\eta > 0$, there is a covering of $[a, b) \cup [c, d)$ by intervals E_k such that

$$\sum_{k=1}^{\infty} \lambda(E_k) \le m^*\big([a, b) \cup [c, d)\big) + \eta.$$

We may assume without loss of generality that each E_k intersects only one of the intervals $[a, b)$ and $[c, d)$. (Otherwise, we may subdivide each E_k, if necessary, so

[2] The fact that $[a', b']$ is compact was proved in Theorem 4.2.7, which depended on the completeness of $\mathbf{R}$ by way of the nested interval theorem.

that each of the (new) intervals E_k intersects only one of the intervals $[a,b)$ and $[c,d)$.) Thus we split the covering $\{E_k\}$ into two coverings, $\{E'_k\}$ covering $[a,b)$ and $\{E''_k\}$ covering $[c,d)$. Then

$$\begin{aligned} m^*([a,b)) + m^*([c,d)) &\le \sum_k \lambda(E'_k) + \sum_k \lambda(E''_k) \\ &= \sum_{k=1}^{\infty} \lambda(E_k) \\ &\le m^*\big([a,b) \cup [c,d)\big) + \eta. \end{aligned}$$

Since $\eta > 0$ is arbitrary, this shows that $m^*\big([a,b) \cup [c,d)\big) \ge (b-a) + (d-c)$. $\square$

Observe now that it remains to show that each interval $[a,b)$ is m^*-measurable. This is done in the next result, which shows that all open sets and all closed sets of the real line are Lebesgue measurable.

Theorem 16.5.3. *The Lebesgue σ-algebra $\mathcal{M}$ contains the Borel σ-algebra $\mathcal{B}$.*

Proof. Since $\mathcal{M}$ is a σ-algebra and $\mathcal{B}$ is the σ-algebra generated by the collection of bounded intervals of the form $J = [a,b)$, $a < b$, it suffices to show that each interval $J = [a,b)$ is m^*-measurable. To do so, we must show that

$$m^*(E) \ge m^*(E \cap J) + m^*(E \cap J^c)$$

for every set E with $m^*(E) < \infty$. (This inequality is satisfied automatically if $m^*(E) = \infty$.) Let E be an arbitrary set with finite outer measure and let $J = [a,b)$. By definition of $m^*(E)$, given $\epsilon > 0$, there exists a sequence of intervals $I_j = [a_j, b_j)$ such that $E \subseteq \bigcup_j I_j$ and

$$\sum_j \lambda([a_j,b_j)) = \sum_j (b_j - a_j) \le m^*(E) + \epsilon. \tag{16.8}$$

Since $E \subseteq \bigcup_j I_j$, intersecting E with J and with J^c yields

$$m^*(E \cap J) \le \sum_j m^*(I_j \cap J)$$

and

$$m^*(E \cap J^c) \le \sum_j m^*(I_j \cap J^c),$$

by subadditivity of m^*. Summing these results gives

$$m^*(E \cap J) + m^*(E \cap J^c) \le \sum_j \Big[m^*(I_j \cap J) + m^*(I_j \cap J^c)\Big]. \tag{16.9}$$

Consider a term in this sum, with j fixed. First, $I_j \cap J$ is an interval that is either empty or is closed on the left and open on the right. Second, depending on the relative locations of I_j and J, the set $I_j \cap J^c$ is a union of exactly zero, one, or two disjoint intervals of the same type. (See Exercise 16.5.1.) It follows that I_j is a disjoint union of intervals of the same type, and hence, by Lemma 16.5.1 and Lemma 16.5.2,

$$m^*(I_j) = b_j - a_j = m^*(I_j \cap J) + m^*(I_j \cap J^c). \tag{16.10}$$

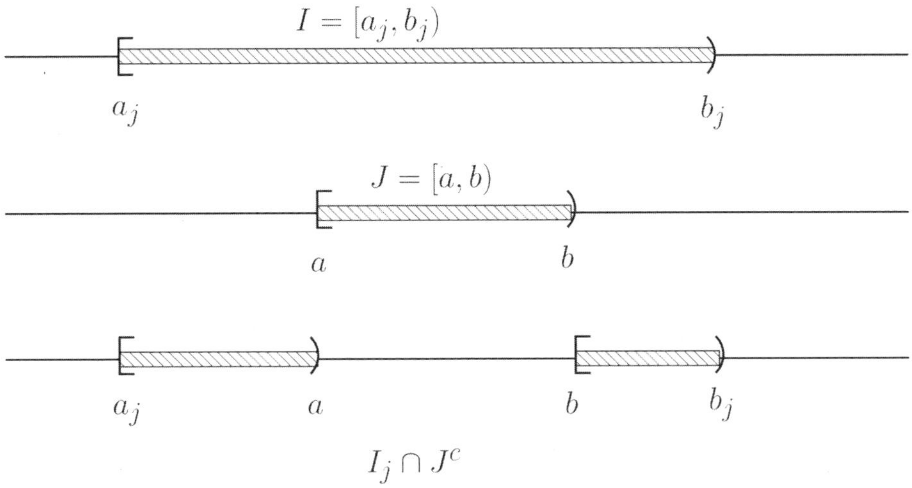

Figure 16.2. The illustration shows the case where $I_j = [a_j, b_j)$, $J = [a, b)$, and $a_j < a < b < b_j$, so that the set $I_j \cap J^c$ consists of the two disjoint intervals shown on the bottom number line. Lemma 16.5.1 and Lemma 16.5.2 establish equality in (16.10) for this case as well as the other possible cases of relative positioning of I_j and J.

Note that each term on the right is an interval length or the sum of two disjoint interval lengths. (See Figure 16.2, which illustrates $I_j \cap J^c$ in the case where $J \subset I_j$ as a proper subset with $a_j < a < b < b_j$, and then $I_j \cap J^c$ is the union of the two intervals, $[a_j, a)$ and $[b, b_j)$.)

Now back to our arbitrary set E with $m^*(E) < \infty$. It follows from (16.8), (16.9) and (16.10), that

$$m^*(E \cap J) + m^*(E \cap J^c) \le \sum_j m^*(I_j) = \sum_j (b_j - a_j) \le m^*(E) + \epsilon.$$

Since $\epsilon > 0$ was arbitrary, this shows that J is m^*-measurable. $\square$

The outer measure of any point on the line is zero, and since every singleton set belongs to $\mathcal{B}$, we have $m(\{\text{point}\}) = 0$. By countable additivity, any countable set has Lebesgue measure zero. But there are uncountable sets having Lebesgue measure zero, for example, the Cantor set (Example 6.4.3). We also have

$$m([a, b]) = m(\{b\}) + m([a, b)) = 0 + m([a, b)) = b - a,$$

and

$$m((a, b)) = \lim_{n\to\infty} m([a + 1/n, b)) = \lim_{n\to\infty} (b - a - 1/n) = b - a,$$

by Proposition 16.3.4. Similarly, it now follows that

$$m((a, b]) = \lim_{n\to\infty} m((a, b + 1/n)) = \lim_{n\to\infty} (b + 1/n - a) = b - a.$$

If an open set O is the countable disjoint union of open intervals (a_j, b_j), then countable additivity of m implies that

$$m(O) = \sum_j m((a_j, b_j)) = \sum_j (b_j - a_j),$$

as we expected. For example, the open set $\bigcup_{k=1}^{\infty}(k - 1/2^k, k + 1/2^k)$ has Lebesgue measure

$$\sum_{k=1}^{\infty} \frac{2}{2^k} = \sum_{k=1}^{\infty} \frac{1}{2^{k-1}} = 2.$$

It is again worth remarking that every closed set is measurable, since $\mathcal{M}$ is a σ-algebra so that the complement of every open set is measurable. Hence every compact set, being closed, is measurable. We remark that m^* is not a measure on the σ-algebra of *all* subsets of $\mathbf{R}$, because there exist nonmeasurable sets. An example of a nonmeasurable set appears at the end of the chapter.

Though not the last word on Lebesgue measure on the real line, this section equips us with the Lebesgue σ-algebra $\mathcal{M}$, a sufficiently rich supply of measurable sets to help develop the Lebesgue integral of real valued functions on subsets of $\mathbf{R}$ in the following chapter.

Exercises.

Exercise 16.5.1. Verify the statement in the proof of Theorem 16.5.3 that the set $I_j \cap J^c$ is a union of either zero, one, or two intervals closed on the left and open on the right, and hence, for each j, $m^*(I_j \cap J) + m^*(I_j \cap J^c) = m^*(I_j)$.

Exercise 16.5.2. Refer to Exercise 16.4.2 and the outer measure determined there by the covering class consisting of the empty set and the open intervals (a, b), with $a < b$, and the definition that $\lambda(\emptyset) = 0$ and $\lambda((a, b)) = b - a$. Prove that $\mu^*((a, b)) = \lambda((a, b)) = b - a$. *Hint*: Follow Lemma 16.5.1, as needed.

16.5.2. Metric Outer Measure; Lebesgue Measure on Euclidean Space. Now let $X = \mathbf{R}^n$ and let $\mathcal{K}$ be the collection consisting of the empty set and all intervals in $\mathbf{R}^n$ of the form

$$E = \{\mathbf{x} = (x_1, \dots, x_n) : a_i \le x_i \le b_i \text{ for } i = 1, \dots, n\} \tag{16.11}$$

where $\mathbf{a} = (a_1, \dots, a_n)$ and $\mathbf{b} = (b_1, \dots, b_n)$. In what follows we often use the symbol E, sometimes subscripted, for closed intervals. Then $\mathcal{K}$ is a sequential covering class of $\mathbf{R}^n$. Define λ on $\mathcal{K}$ by $\lambda(\emptyset) = 0$ and

$$\lambda(E) = \prod_{i=1}^{n} (b_i - a_i),$$

which is of course the n-dimensional volume of (16.11). Then define m_n^* on the σ-algebra of all subsets of $\mathbf{R}^n$ in accordance with (16.1). Therefore define

$$\mu^*(A) := \inf\Big\{\sum_{k=1}^{\infty} \lambda(E_k) : E_k \in \mathcal{K} \text{ and } A \subseteq \bigcup_{k=1}^{\infty} E_k\Big\} \tag{16.12}$$

for $A \subseteq \mathbf{R}^n$. Then m_n^* is an outer measure, by Theorem 16.4.2, called the **Lebesgue outer measure** on $\mathbf{R}^n$. Let $\mathcal{M}_n$ be the collection of m_n^*-measurable sets in accordance with Definition 16.4.3. By Theorem 16.4.4, the restriction $m_n = m_n^*|_{\mathcal{M}_n}$ is a complete measure on $\mathbf{R}^n$, called **Lebesgue measure** on $\mathbf{R}^n$, and $\mathcal{M}_n$ is the **Lebesgue σ-algebra** of $\mathbf{R}^n$. The elements of $\mathcal{M}_n$ are known as the **Lebesgue measurable sets** in $\mathbf{R}^n$. The triple $(\mathbf{R}^n, \mathcal{M}_n, m_n)$ is the n-dimensional **Lebesgue measure space**.

We could verify now that the Lebesgue outer measure of any interval in $\mathbf{R}^n$ is given by its n-dimensional volume. Let us focus on an open interval E in $\mathbf{R}^n$. An argument similar to the one in Lemma 16.5.1, using compactness of a closed interval interior to E and proof by contradiction, shows that we cannot have $m_n^*(E) < \nu_n(E) = \lambda(E)$, and hence $m_n^*(E) = \nu_n(E) = \lambda(E)$. Actually, the same argument works to show that for any interval E, whether open, closed, or otherwise, $m_n^*(E) = \nu_n(E) = \lambda(E)$. Since every interval is a Borel set, we will revisit the intervals after we show that every Borel set is Lebesgue measurable.

There is an important property of Lebesgue outer measure for $\mathbf{R}^n$ which will help in showing that the Borel σ-algebra, $\mathcal{B}_n$, is included in $\mathcal{M}_n$ for $n \geq 2$. First, some definitions. Let X be a metric space with metric ρ. If B is any subset of X and $x \in X$, then the **distance from x to B** is

$$\rho(x, B) := \inf\{\rho(x, y) : y \in B\}.$$

If A is also a subset of X, then

$$\rho(A, B) := \inf\{\rho(x, y) : x \in A,\ y \in B\}$$

defines the distance between A and B. Clearly, $\rho(\{x\}, B) = \rho(x, B)$. The **diameter** of a set A is

$$d(A) := \sup\{\rho(x, y) : x, y \in A\},$$

and A is **bounded** if $d(A) < \infty$.

Definition 16.5.4. *Let μ^* be an outer measure defined on the σ-algebra of all subsets of a metric space X with metric ρ. Then μ^* is a* **metric outer measure** *if, whenever A and B are subsets of X such that $\rho(A, B) > 0$, then*

$$\mu^*(A \cup B) = \mu^*(A) + \mu^*(B).$$

If μ^* is a metric outer measure, this definition implies that if, in a finite union $\bigcup_{k=1}^n A_k$, the distance between any two of the sets is positive, then $\mu^*\left(\bigcup_{k=1}^n A_k\right) = \sum_{k=1}^n \mu^*(A_k)$.

The next result shows that Lebesgue outer measure is a metric outer measure.

Theorem 16.5.5. *For each positive integer n, the Lebesgue outer measure m_n^* is a metric outer measure on the σ-algebra of all subsets of $\mathbf{R}^n$.*

Proof. We shall write ρ for the metric on $\mathbf{R}^n$. Let A and B be subsets of $\mathbf{R}^n$ such that $\rho(A, B) > 0$. By subadditivity of outer measure,

$$m_n^*(A \cup B) \leq m_n^*(A) + m_n^*(B).$$

We must show the reverse inequality. Let $\epsilon > 0$. By definition of the outer measure, there is a collection $\{E_k\}_{k=1}^{\infty}$ of closed intervals such that

$$\sum_{k=1}^{\infty} \lambda(E_k) \leq m_n^*(A \cup B) + \epsilon.$$

We may assume that the diameter of each E_k is less than $\rho(A, B)$, because otherwise we can subdivide each E_k into a finite number of subintervals that have this property. Therefore we can split the collection $\{E_k\}$ into two collections $\{E_k'\}$ and $\{E_k''\}$ such that $\{E_k'\}$ covers A and $\{E_k''\}$ covers B. Then

$$\begin{aligned} m_n^*(A) + m_n^*(B) &\leq \sum_k \lambda(E_k') + \sum_k \lambda(E_k'') \\ &= \sum_{k=1}^{\infty} \lambda(E_k) \\ &\leq m_n^*(A \cup B) + \epsilon. \end{aligned}$$

Since $\epsilon > 0$ is arbitrary, this shows that $m_n^*(A \cup B) \geq m_n^*(A) + m_n^*(B)$. □

We remark here that Lebesgue outer measure for the real line is also a metric outer measure, since this theorem applies to sets A and B in $\mathbf{R}$ with $\rho(A, B) > 0$ and covering intervals $E_k = [a_k, b_k)$. A special case of this property was proved in establishing (16.10), in the proof of Theorem 16.5.3.

We return now to general metric outer measures on a metric space X. The next two theorems will establish that if μ^* is a metric outer measure, then every Borel set is indeed μ^*-measurable.

First, if μ^* is a metric outer measure and $\mu^*(A) < \infty$, then $\mu^*(A)$ is the limit of the outer measures of an increasing sequence of closed subsets of A.

Theorem 16.5.6. *Let X be a metric space with metric ρ, and let μ^* be a metric outer measure on X. Let A and B be any subsets of X such that $A \subset B$, B is an open set, and $\mu^*(A) < \infty$. For each positive integer n, let*

$$A_n = \{x \in A : \rho(x, B^c) \geq 1/n\}.$$

Then A_n is closed, $A_n \subseteq A_{n+1}$, and

$$\mu^*(A) = \lim_{n\to\infty} \mu^*(A_n).$$

Proof. Each A_n is closed: Let $x_k \in A_n$ and suppose $x_k \to x$ as $k \to \infty$. If $y \in B^c$, then $\rho(x_k, y) \geq 1/n$ for all k, and therefore

$$\rho(x, y) = \lim_{k\to\infty} \rho(x_k, y) \geq \frac{1}{n}.$$

Since this is true for every $y \in B^c$, by taking the infimum over y in B^c, it follows that $\rho(x, B^c) \geq 1/n$. Hence, $x \in A_n$.

By definition of the sets A_n, we have $A_n \subseteq A_{n+1} \subseteq A$ for each n, and therefore the sequence $\mu^*(A_n)$ is increasing and $\lim_{n\to\infty} \mu^*(A_n) \leq \mu^*(A) < \infty$. (See Figure 16.3.) In order to prove the theorem, it suffices to show that $\lim_{n\to\infty} \mu^*(A_{2n}) = \mu^*(A)$, by Theorem 2.4.17.

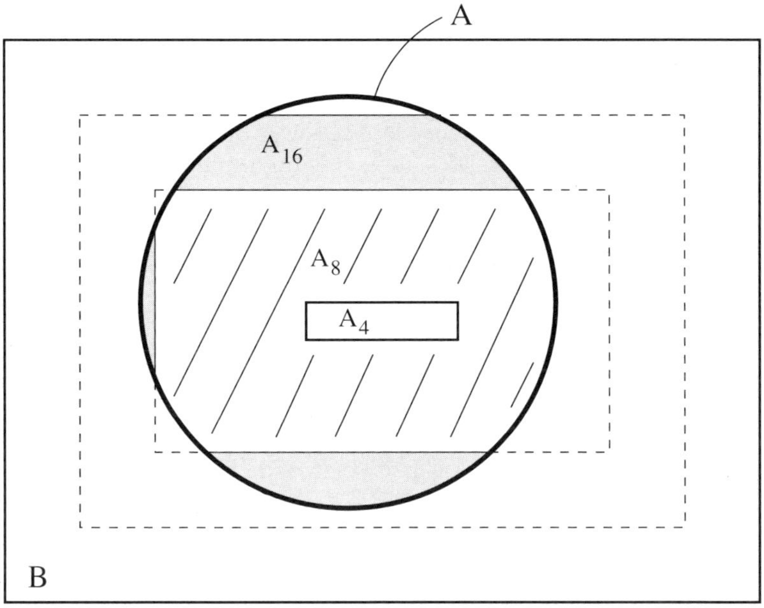

Figure 16.3. To accompany Theorem 16.5.6: The closed sets A_n approximate a set A having finite outer measure.

Let us show that $A = \bigcup_{n=1}^{\infty} A_n$. If $x \in A$, then $x \in B$, and since B is open, there is an $\epsilon > 0$ such that the open ϵ-ball about x is contained in B. Hence, $\rho(x, B^c) \geq \epsilon$. This shows that $x \in A_n$ if $\epsilon \geq 1/n$. Hence, $A \subseteq \bigcup_{n=1}^{\infty} A_n$. Since each $A_n \subseteq A$ by definition, it follows that $A = \bigcup_{n=1}^{\infty} A_n$.

Let $G_n = A_{n+1} - A_n$ for each $n \geq 1$. Then $A = \bigcup_{n=1}^{\infty} A_n = \bigcup_{n=1}^{\infty} G_n$, and we have

$$A = A_{2n} \cup \Big[\bigcup_{k=2n}^{\infty} G_k \Big] = A_{2n} \cup \Big[\bigcup_{k=n}^{\infty} G_{2k} \Big] \cup \Big[\bigcup_{k=n}^{\infty} G_{2k+1} \Big].$$

We will prove that as $n \to \infty$, the outer measure of the two unions in brackets on the right-hand side approaches zero. We note that the sets G_n are pairwise disjoint. By the subadditivity of outer measure,

$$\mu^*(A) \leq \mu^*(A_{2n}) + \sum_{k=n}^{\infty} \mu^*(G_{2k}) + \sum_{k=n}^{\infty} \mu^*(G_{2k+1}). \tag{16.13}$$

By definition of the G_n, if $x \in G_{2l}$ and $y \in G_{2k+2}$, with $l \leq k$, then $x \in A_{2l+1}$ and $y \in A_{2k+3} - A_{2k+2}$, and therefore

$$\rho(x, B^c) > \frac{1}{2l+1} \qquad \text{and} \qquad \rho(y, B^c) < \frac{1}{2k+2}.$$

Thus,

$$\rho(x, y) > \frac{1}{2l+1} - \frac{1}{2k+2} > \frac{1}{2k+1} - \frac{1}{2k+2} > 0.$$

(See Exercise 16.5.6.) Hence,

$$\rho(G_{2l}, G_{2k+2}) = \inf\left\{\rho(x,y) : x \in G_{2l},\, y \in G_{2k+2}\right\} \geq \frac{1}{2k+1} - \frac{1}{2k+2} > 0.$$

We observe that $A_{2n} \supseteq \bigcup_{k=1}^{n-1} G_{2k}$, and by the lower bound for $\rho(G_{2l}, G_{2k+2})$ when $l \leq k$, the distance between any two sets in this finite union is positive. Since μ^* is a metric outer measure,

$$\mu^*(A) \geq \mu^*(A_{2n}) \geq \mu^*\Big(\bigcup_{k=1}^{n-1} G_{2k}\Big) = \sum_{k=1}^{n-1} \mu^*(G_{2k}).$$

Since this inequality holds for each n, and $\mu^*(A) < \infty$, it follows that the series

$$\sum_{k=1}^{\infty} \mu^*(G_{2k})$$

converges. Similarly, since $A_{2n} \supseteq \bigcup_{k=1}^{n-1} G_{2k+1}$, and the distance between any two sets in this union is positive,

$$\mu^*(A) \geq \mu^*(A_{2n}) \geq \mu^*\Big(\bigcup_{k=1}^{n-1} G_{2k+1}\Big) = \sum_{k=1}^{n-1} \mu^*(G_{2k+1}).$$

Therefore the series

$$\sum_{k=1}^{\infty} \mu^*(G_{2k+1})$$

converges. Now, letting $n \to \infty$ in (16.13), we conclude that

$$\mu^*(A) \leq \lim_{n\to\infty} \mu^*(A_{2n}) + 0 + 0 = \lim_{n\to\infty} \mu^*(A_{2n}).$$

Hence, we have

$$\mu^*(A) \leq \lim_{n\to\infty} \mu^*(A_{2n}) = \lim_{n\to\infty} \mu^*(A_n) \leq \mu^*(A),$$

and the theorem follows. □

The next theorem is the main result about metric outer measures. Recall that the Borel sets are the elements of the σ-algebra generated by the open sets (Definition 16.1.5). Of course this is the same as the σ-algebra generated by the closed sets.

Theorem 16.5.7. *If μ^* is a metric outer measure on a metric space X, then every closed set is μ^*-measurable. Consequently, every Borel set in X is μ^*-measurable.*

Proof. Let F be a closed set in X. For any set A with $\mu^*(A) < \infty$, we have $\mu^*(A - F) < \infty$ and $A - F \subset F^c$, and F^c is an open set. By Theorem 16.5.6, there is an increasing sequence of closed sets A_n contained in $A - F$ such that

$$\rho(A_n, F) \geq 1/n \tag{16.14}$$

and

$$\lim_{n\to\infty} \mu^*(A_n) = \mu^*(A - F). \tag{16.15}$$

By (16.14), we have $\rho(A_n, A \cap F) > 0$ for each n, and hence

$$\mu^*(A) \geq \mu^*[(A \cap F) \cup A_n] = \mu^*(A \cap F) + \mu^*(A_n)$$

for each n. By letting $n \to \infty$ and using (16.15), we obtain

$$\mu^*(A) \geq \mu^*(A \cap F) + \lim_{n\to\infty} \mu^*(A_n) = \mu^*(A \cap F) + \mu^*(A - F).$$

Since A is an arbitrary set with finite outer measure, this shows that F is μ^*-measurable. Since F was an arbitrary closed subset of X, every closed set is μ^*-measurable. We know that the μ^*-measurable sets form a σ-algebra, and since this σ-algebra contains every closed set, it contains every open set, and hence contains the Borel σ-algebra of X. □

The desired result for the Lebesgue measure space $(\mathbf{R}^n, \mathcal{M}_n, m_n)$ now follows immediately from Theorem 16.5.5 and Theorem 16.5.7.

Theorem 16.5.8. *The Lebesgue σ-algebra $\mathcal{M}_n$ contains the Borel σ-algebra $\mathcal{B}_n$.*

Theorem 16.5.8 assures us that we can measure all open sets and all closed sets, along with many other sets, as the σ-algebras $\mathcal{B}_n$ and $\mathcal{M}_n$ do not have the same cardinality. We note also that, as in the case of the real line, there are nonmeasurable subsets of $\mathbf{R}^n$.

Every interval, whether open, closed, or otherwise, is an element of $\mathcal{B}_n$, and hence is measurable. As mentioned earlier, if E is an open interval in $\mathbf{R}^n$, then $m_n(E) = \nu_n(E) = \lambda(E)$. If E is a closed interval, then $E = \operatorname{Int} E \cup \partial E$, and we know that $\nu_n(\partial E) = 0$, hence $m_n(\partial E) = 0$, and therefore $m_n(E) = m_n(\operatorname{Int} E) + m_n(\partial E) = \nu_n(E)$. Finally, if E is any interval properly contained in $\overline{E}$ and properly containing $\operatorname{Int} E$, then monotonicity of measure implies

$$\nu_n(E) = m_n(\operatorname{Int} E) \leq m_n(E) \leq m_n(\overline{E}) = \nu_n(E).$$

This is all as expected for intervals.

There is an interesting structure theorem for the open sets in $\mathbf{R}^n$ that allows us to make some further observations about Lebesgue measure. Two intervals in $\mathbf{R}^n$ (whether open, closed, or otherwise) are **nonoverlapping** if their interiors are disjoint, that is, they intersect only in some boundary points, if at all. Thus the intersection of the two intervals equals the intersection of their boundaries. Similarly, the intervals in an arbitrary collection of intervals are called nonoverlapping if any two of them are nonoverlapping.

Theorem 16.5.9 (Structure of Open Sets in $\mathbf{R}^n$). *Every open set in $\mathbf{R}^n$, $n \geq 1$, can be expressed as a countable union of nonoverlapping closed cubes.*

Proof. Consider the collection of all points in $\mathbf{R}^n$ with integer coordinates and the corresponding collection C_0 of closed cubes with edge length 1 having vertices at these points. From bisecting each edge of a given cube in C_0, we obtain 2^n new closed cubes with edge length $1/2$. Denote the total collection of such cubes from all the cubes of C_0 by C_1. On continued bisection of these cubes, we generate, for each positive integer k, a collection C_k of cubes having edge length $1/2^k$, and each cube in C_k is the union of 2^n nonoverlapping cubes from C_{k+1}. By induction, each collection C_k is countable.

Let O be any open set in $\mathbf{R}^n$. Let S_0 be the collection of all cubes in C_0 which lie entirely within O. Then let S_1 be the collection of those cubes in C_1 which lie entirely within O but are not subcubes of any cube in S_0. Then define inductively for $k \geq 1$ the collection S_k of cubes in C_k which lie entirely within O but are not subcubes of any cube in $S_0, S_1, \ldots, S_{k-1}$. Write $S = \bigcup_{k=1}^{\infty} S_k$, the collection of all cubes that belong to one of the collections S_k, $k \in \mathbf{N}$. Then S is countable, since each S_k is countable, and by construction, the cubes in S are nonoverlapping and contained entirely within O. Hence $S \subseteq O$.

We now show that $O \subseteq S = \bigcup_{k=1}^{\infty} S_k$. Let us denote an arbitrary cube in S by Q. Since O is open, for any point in O there is an open cube B_δ, with edge length $\delta > 0$, about the point, such that $B_\delta \subset O$. Since the edge length of the cubes in C_k approaches zero as $k \to \infty$, the point is eventually enclosed by a cube in some S_k. Hence, $O = \bigcup_{Q \in S} Q$. □

If O is open in $\mathbf{R}^n$, then we can write $O = \bigcup_{k=1}^{\infty} Q_k$, where the Q_k are nonoverlapping closed cubes as in Theorem 16.5.9, and it is true that $m_n(O) = \sum_{k=1}^{\infty} m_n(Q_k)$, since each Q_k intersects other cubes in the union only within ∂Q_k, and therefore its intersection with them has measure zero. (See Exercise 16.5.8.)

The next result shows that every Lebesgue measurable set in $\mathbf{R}^n$ can be approximated arbitrarily closely in measure by open sets and closed sets.

Theorem 16.5.10. *Let A be a subset of $\mathbf{R}^n$. The following two statements are true and equivalent to each other:*

1. *If A is measurable, then for every $\epsilon > 0$ there is a closed set $F \subseteq A$ such that $m_n(A - F) < \epsilon$.*
2. *If A is measurable, then for every $\epsilon > 0$ there is an open set $O \supseteq A$ such that $m_n(O - A) < \epsilon$.*

Proof. 1. Suppose $m_n^*(A) < \infty$. Let $\epsilon > 0$. By Theorem 16.5.6, there is a closed set $F \subset A$ such that $m_n^*(A) - m_n^*(F) < \epsilon$. If A is measurable, then so is $A - F$. Additivity of measure gives $m_n(A) = m_n(F) + m_n(A - F)$, hence $m_n(A - F) = m_n(A) - m_n(F)$, so $m_n(A - F) < \epsilon$. Thus, statement 1 holds for sets A with finite outer measure.

Now suppose A has infinite outer measure. Then A is measurable. Let B_j denote the open Euclidean ball of radius j centered at the origin, for each positive integer j. Set $A_1 = A \cap B_1$, and for each j, let $A_j = A \cap (B_j - B_{j-1})$. Then $A = \bigcup_{j=1}^{\infty} A_j$, and each A_j is measurable with finite measure. By Theorem 16.5.6, for each j there is a closed set $F_j \subseteq A_j$ such that $m_n(A_j - F_j) = m_n(A_j) - m_n(F_j) < \epsilon/2^j$. Let $F = \bigcup_j F_j$. Then $F \subseteq A$. We observe that F is a closed set: Any cluster point of F must belong to either B_1 or to $B_j - B_{j-1}$ for some $j \geq 2$, and therefore it must be a cluster point of either F_1 or of some F_j with $j \geq 2$. Since the sets F_j are closed, the cluster point in question belongs either to F_1 or to some F_j with $j \geq 2$, and hence belongs to F. Therefore F is closed. Since $A - F = \bigcup_{j=1}^{\infty}(A_j - F_j)$ is a disjoint union of measurable sets, it follows that

$$m_n(A - F) = \sum_{j=1}^{\infty} m_n(A_j - F_j) < \sum_{j=1}^{\infty} \epsilon/2^j = \epsilon,$$

as we wished to show.

2. If A is measurable, then so is A^c. By statement 1, given any $\epsilon > 0$ there is a closed set $F \subseteq A^c$ such that $m_n(A^c - F) < \epsilon$. Then $A \subseteq F^c$, F^c is open, and $m_n(F^c - A) = m_n(F^c \cap A^c) = m_n(A^c - F) < \epsilon$. This proves the desired statement with $O := F^c$.

We have shown that statement 1 implies 2. Now assume that statement 2 holds. If A is measurable, then so is A^c. Given $\epsilon > 0$, statement 2 implies that there exists an open set $O \supseteq A^c$ such that $m_n(O - A^c) < \epsilon$. Then $O^c \subseteq A$ and O^c is closed. Moreover, $m_n(A - O^c) = m_n(A \cap O) = m_n(O - A^c) < \epsilon$. Thus, statement 1 holds. □

If we replace m_n by the outer measure m_n^* in each of the necessary conditions for measurability in Theorem 16.5.10, we obtain *sufficient* conditions for statement (16.2) of Definition 16.4.3. Thus, each of the modified conditions provides a characterization of Lebesgue measurability. To be precise, one can prove the following statements about subsets A of $\mathbf{R}^n$:

1^*. If for every $\epsilon > 0$ there is a closed set $F \subseteq A$ such that $m_n^*(A - F) < \epsilon$, then A satisfies $m_n^*(E) = m_n^*(E \cap A) + m_n^*(E \cap A^c)$ for every set E in $\mathbf{R}^n$. Therefore A is Lebesgue measurable.

2^*. If for every $\epsilon > 0$ there is an open set $O \supseteq A$ such that $m_n^*(O - A) < \epsilon$, then A satisfies $m_n^*(E) = m_n^*(E \cap A) + m_n^*(E \cap A^c)$ for every set E in $\mathbf{R}^n$. Therefore A is Lebesgue measurable.

We will not prove statements 1^* and 2^* here, but see the Notes and References for this chapter.

The definition of Lebesgue outer measure used intervals with edges parallel to the standard coordinate axes. But the outer measure of a set should not depend on the position in space of the coordinate axes. We prove below that the outer measure does not depend on the position of the orthogonal coordinate axes. To do this, we consider a fixed rotation of the standard coordinate axes.

A rotation of $\mathbf{R}^n$ is determined by a mapping of the standard ordered basis, $\{\mathbf{e}_1, \ldots, \mathbf{e}_n\}$, to a new orthonormal ordered basis, $\{\mathbf{u}_1, \ldots, \mathbf{u}_n\}$, with $\mathbf{e}_j \mapsto \mathbf{u}_j$, $1 \leq j \leq n$, to be definite. Let E be an interval with edges parallel to the $\mathbf{e}_k$ and let $\tilde{E}$ denote the corresponding rotated interval with edges parallel to the $\mathbf{u}_k$. The inverse orthogonal transformation such that $\mathbf{u}_j \mapsto \mathbf{e}_j$, $1 \leq j \leq n$, maps $\tilde{E}$ back to E. The volume $\lambda(E) = \nu_n(E)$ is unchanged by rotation: $\lambda(\tilde{E}) = \lambda(E)$.

Let $\tilde{m}_n^*(A)$ denote the outer measure of a set A relative to these rotated intervals; thus,

$$\tilde{m}_n^*(A) = \inf\Big\{\sum_k \lambda(\tilde{E}_k) : A \subseteq \bigcup_k \tilde{E}_k\Big\} \tag{16.16}$$

where the $\tilde{E}_k$ are closed intervals with edges parallel to the $\mathbf{u}_k$.

Theorem 16.5.11. *For every $A \subseteq \mathbf{R}^n$, $\tilde{m}_n^*(A) = m_n^*(A)$.*

Proof. All intervals in the proof are closed intervals.

Given an interval $\tilde{E}$ with edges parallel to the $\mathbf{u}_k$, and any $\epsilon > 0$, there exists a countable collection $\{E_k\}$ of standard basis intervals such that $\tilde{E} \subseteq \bigcup_{k=1}^{\infty} E_k$ and

$\sum_{k=1}^{\infty} \lambda(E_k) \le \lambda(\tilde{E}) + \epsilon$. To see this, let $\tilde{E}_1$ be an interval containing $\tilde{E}$ in its interior, such that $\lambda(\tilde{E}_1) \le \lambda(\tilde{E}) + \epsilon$. By Theorem 16.5.9, the interior of $\tilde{E}_1$ can be expressed as the union of nonoverlapping closed intervals E_k. Thus,

$$\tilde{E} \subset \bigcup_{k=1}^{\infty} E_k = \operatorname{Int} \tilde{E}_1 .$$

Since the E_k are nonoverlapping and $\bigcup_{k=1}^{N} E_k \subset \tilde{E}_1$,

$$\sum_{k=1}^{N} \lambda(E_k) \le \lambda(\tilde{E}_1)$$

for each N. Letting $N \to \infty$ proves the claim that $\sum_{k=1}^{\infty} \lambda(E_k) \le \lambda(\tilde{E}) + \epsilon$.

Given a standard interval E with edges parallel to the $\mathbf{u}_k$, and $\epsilon > 0$, similar reasoning shows that there exists a countable collection $\{\tilde{E}_k\}$ of the rotated intervals such that $E \subseteq \bigcup_{k=1}^{\infty} \tilde{E}_k$ and $\sum_{k=1}^{\infty} \lambda(\tilde{E}_k) \le \lambda(E) + \epsilon$.

Let A be any subset of $\mathbf{R}^n$. Given $\epsilon > 0$, let $\{E_k\}_{k=1}^{\infty}$ be such that $A \subseteq \bigcup_{k=1}^{\infty} E_k$ and $\sum_{k=1}^{\infty} \lambda(E_k) \le m_n^*(A) + \epsilon/2$. For each k, choose $\{\tilde{E}_{k,j}\}_{j=1}^{\infty}$ such that

$$(16.17) \qquad E_k \subseteq \bigcup_{j=1}^{\infty} \tilde{E}_{k,j} \qquad \text{and} \qquad \sum_{j=1}^{\infty} \lambda(\tilde{E}_{k,j}) \le \lambda(E_k) + \frac{\epsilon}{2^{k+1}} .$$

From (16.17), we have

$$\begin{aligned} \sum_{k,j=1}^{\infty} \lambda(\tilde{E}_{k,j}) &\le \sum_{k=1}^{\infty} \lambda(E_k) + \sum_{k=1}^{\infty} \frac{\epsilon}{2^{k+1}} \\ &= \Big[\sum_{k=1}^{\infty} \lambda(E_k) \Big] + \frac{\epsilon}{2} \le \Big[m_n^*(A) + \frac{\epsilon}{2} \Big] + \frac{\epsilon}{2} = m_n^*(A) + \epsilon . \end{aligned}$$

Since $A \subseteq \bigcup_{k,j=1}^{\infty} \tilde{E}_{k,j}$, it follows that $\tilde{m}_n^*(A) \le m_n^*(A) + \epsilon$. Since $\epsilon > 0$ is arbitrary, $\tilde{m}_n^*(A) \le m_n^*(A)$.

Now observe that, given A and any $\epsilon > 0$, a symmetric argument can be made with the covering of A by the E_k replaced by a covering by the rotated intervals $\tilde{E}_k$. One can choose $\{E_{k,j}\}_{j=1}^{\infty}$ such that $A \subseteq \bigcup_{k,j=1}^{\infty} E_{k,j}$ and $m_n^*(A) \le \tilde{m}_n^*(A) + \epsilon$. Hence, $m_n^*(A) \le \tilde{m}_n^*(A)$. We conclude that $m_n^*(A) = \tilde{m}_n^*(A)$. Since A is an arbitrary subset of $\mathbf{R}^n$, this completes the proof. □

The outer measure could also be defined by using the sequential covering class consisting of the empty set and all parallelepipeds with edges parallel to the elements of any fixed ordered basis of $\mathbf{R}^n$. The proof follows the pattern in the argument for Theorem 16.5.11.

Looking toward the next chapter for a moment, we note that subsets of a measure space are naturally given the structure of a measure space in their own right; see Exercise 16.5.10. This allows us to define spaces of integrable functions defined on subintervals of $\mathbf{R}$ or subsets of $\mathbf{R}^n$. In this connection, we note that every Jordan measurable set is Lebesgue measurable with the same volume measure: for the intervals, this is known to us now, and for general sets with volume, it follows from the fact that if χ_A is Riemann integrable, then it is integrable in the sense of Lebesgue, with the same value.

The final item of this section is an example of a nonmeasurable set of real numbers.

Example 16.5.12 (G. Vitali's Nonmeasurable Set)**.** Let us divide $\mathbf{R}$ into equivalence classes based on the relation that $x \sim y$ if and only if $x - y$ is a rational number. It is easy to verify that this is an equivalence relation on $\mathbf{R}$. One of the equivalence classes consists of all the rational numbers. A different equivalence class contains the Euler number e, and if e is equivalent to w, then $e - w = q \in \mathbf{Q}$, so $w = e - q$ for rational q. Thus the equivalence class containing e consists of all numbers of the form $e - q$, where q is rational. Similarly, the equivalence class of $\sqrt{5}$ consists of all numbers of the form $\sqrt{5} - r$, with r rational. From each equivalence class, we choose one number which lies in the interval $(0, 1)$, and denote the collection of these numbers by $\mathcal{N}$. Such a choice in $(0, 1)$ from each equivalence class is possible since $\mathbf{Q}$ is dense in $\mathbf{R}$. △

Theorem 16.5.13 (G. Vitali's Nonmeasurable Set)**.** *The set $\mathcal{N}$ defined in Example 16.5.12 is not Lebesgue measurable.*

Proof. For any rational number q, define the translated set

$$\mathcal{N} + q = \{w + q : w \in \mathcal{N}\}.$$

By adding a rational to each element of $\mathcal{N}$, we obtain another set that has exactly one element from each equivalence class: If w_1 and w_2 are in $\mathcal{N}$ and $w_1 \neq w_2$, then $w_1 + q - (w_2 + q) = w_1 - w_2$, which is not rational, so w_1, w_2 belong to distinct classes. If q_1 and q_2 are distinct rational numbers, then the translations $\mathcal{N} + q_1$ and $\mathcal{N} + q_2$ are disjoint (Exercise 16.5.9).

For every real number α in the interval $(0, 1)$, there is exactly one rational number q such that x is contained in $\mathcal{N} + q$, and this q lies in the interval $(-1, 1)$: Given $\alpha \in (0, 1)$, let $\beta \in \mathcal{N}$ be the one number in $\mathcal{N}$ equivalent to α. By definition of the equivalence, $\alpha - \beta \in \mathbf{Q}$, and since both α and β are in $(0, 1)$, we have $-1 < \alpha - \beta < 1$. Thus, $\alpha = \beta + (\alpha - \beta) \in \mathcal{N} + (\alpha - \beta)$ as claimed.

For rational $q \in (-1, 1)$, the sets $\mathcal{N} + q$ are pairwise disjoint, and since $\mathcal{N} \subset (0, 1)$, each of these sets $\mathcal{N} + q$ is contained in $(-1, 2)$. By the previous paragraph, $(0, 1)$ is contained in the union of these sets. Hence,

$$(0, 1) \subseteq \bigcup_{q \in \mathbf{Q} \cap (-1,1)} (\mathcal{N} + q) \subseteq (-1, 2).$$

Lebesgue outer measure is translation invariant, since the measure of intervals is translation invariant (Exercise 16.5.4). Hence, $m^*(\mathcal{N} + q) = m^*(\mathcal{N})$. Monotonicity and subadditivity of the outer measure gives

$$1 \leq m^*\Big(\bigcup_{q \in \mathbf{Q} \cap (-1,1)} (\mathcal{N} + q)\Big) \leq \sum_{q \in \mathbf{Q} \cap (-1,1)} m^*(\mathcal{N} + q) = \sum_{q \in \mathbf{Q} \cap (-1,1)} m^*(\mathcal{N}).$$

This implies $m^*(\mathcal{N}) > 0$.

If $\mathcal{N}$ is measurable, then so are all the translations, $\mathcal{N} + q$. Therefore the measure of the countable disjoint union $\bigcup_{q\in\mathbf{Q}\cap(-1,1)}(\mathcal{N}+q)$ satisfies

$$1 \le m\Big(\bigcup_{q\in\mathbf{Q}\cap(-1,1)} (\mathcal{N}+q)\Big) = \sum_{q\in\mathbf{Q}\cap(-1,1)} m(\mathcal{N}) \le m\big((-1,2)\big) = 3.$$

This is not possible, since $m(\mathcal{N}) > 0$ implies $\sum_{q\in\mathbf{Q}\cap(-1,1)} m(\mathcal{N}) = \infty$. We conclude that the set $\mathcal{N}$ is not Lebesgue measurable. □

Exercises.

Exercise 16.5.3. Let E be an open interval in $\mathbf{R}^n$. Show, from the definition of outer measure, that $\mu_n^*(E) = \nu_n(E) = \lambda(E)$. *Hint*: Lemma 16.5.1.

Exercise 16.5.4. *Translation invariance of Lebesgue measure*

1. Show that Lebesgue outer measure for $\mathbf{R}$ is invariant under translations: For every set $E \subset \mathbf{R}$ and any real number y,
$$m^*(E) = m^*(E+y), \quad \text{where} \quad E + y := \{x + y : x \in E\}.$$
2. Show that if A is a Lebesgue measurable subset of $\mathbf{R}$, and $B = \{a+b : a \in A\}$ for some fixed number b, then B is measurable and $m(B) = m(A)$.
3. Show that n-dimensional Lebesgue measure m_n on $\mathbf{R}^n$ is translation invariant. *Hint*: Show it for Lebesgue outer measure.

Exercise 16.5.5. Let L be any line, or line segment, in $\mathbf{R}^n$, and let H be any hyperplane of dimension $k < n$, or a subset thereof, in $\mathbf{R}^n$.

1. Show that L has Lebesgue measure zero. *Hint*: Lebesgue measure is rotation invariant, by Theorem 16.5.11, and translation invariant, by Exercise 16.5.4.
2. Show that H has Lebesgue measure zero.

Exercise 16.5.6. Let X be a metric space with metric ρ and let $A \subset X$.

1. Show that $|\rho(x, A) - \rho(y, A)| \le \rho(x, y)$ for any two points $x, y \in X$. *Hint*: For any $z \in A$, $\rho(x, z) \le \rho(x, y) + \rho(y, z)$. Take the infimum over z in A.
2. Prove: If $\rho(x, A) > \alpha$ and $\rho(y, A) < \beta$ and $\alpha > \beta$, then $\rho(x, y) > \alpha - \beta$.

Exercise 16.5.7. Let ρ be a metric on $\mathbf{R}^n$. Show that if A and B are nonempty compact sets in $\mathbf{R}^n$ and $A \cap B$ is empty, then $\rho(A, B) > 0$. *Hint*: Use Exercise 16.5.6.

Exercise 16.5.8. If O is open in $\mathbf{R}^n$, then write $O = \bigcup_{k=1}^\infty Q_k$, where $\{Q_k\}$ is a countable collection of nonoverlapping closed cubes, as in Theorem 16.5.9. Show that $m_n(O) = \sum_{k=1}^\infty m_n(Q_k)$. *Hint*: Use the fact that each Q_k intersects other cubes in the collection only within ∂Q_k, and therefore its intersection with them has measure zero, but write O as a disjoint union.

Exercise 16.5.9. Consider the nonmeasurable set $\mathcal{N}$ at the end of this section. Show that if q_1 and q_2 are distinct rational numbers, then the translations $\mathcal{N} + q_1$ and $\mathcal{N} + q_2$ are disjoint.

Exercise 16.5.10. Assume that Σ is a σ-algebra of the set X.

1. Let $E \subset X$. Show that $\Sigma_E := \{A \cap E : A \in \Sigma\}$ is a σ-algebra of E.
2. Suppose that (X, Σ, μ) is a measure space, and suppose $E \in \Sigma$. Show that (E, Σ_E, μ_E) is a measure space, where $\mu_E(A \cap E) := \mu(A \cap E)$ for each set $A \cap E \in \Sigma_E$.

16.6. Notes and References

This chapter was influenced by Friedman [**17**], Royden [**51**], Wheeden and Zygmund [**67**], and Bass [**3**]. Other useful resources that develop Lebesgue measure and the Lebesgue integral with a variety of different approaches include Fleming [**13**], Hoffman [**30**], Jones [**33**] and Rudin [**52**].

The definition of measurability (Definition 16.4.3) based on Lebesgue outer measure is due to C. Carathéodory. Given the Lebesgue outer measure, an alternative definition of measurability is that a set E in $\mathbf{R}^n$ is measurable if for every $\epsilon > 0$ there is an open set O such that $E \subset O$ and $m_n^*(O - E) < \epsilon$. This is the definition of measurability in Wheeden and Zygmund [**67**] for Lebesgue measure on $\mathbf{R}^n$. The statement (16.2) of measurability is then proved as a theorem that characterizes measurable sets.

A metric outer measure is also called a *Carathéodory outer measure.*

For the theorem that any incomplete measure can be extended to a larger σ-algebra on which it is complete, see Rudin [**53**].

The material on Bernoulli trials and the event of *gambler's ruin* is from Adams and Guillemin [**1**].

The definition of the nonmeasurable set $\mathcal{N}$ in Theorem 16.5.13 uses the Axiom of Choice. Readers intrigued by nonmeasurable sets, or those who want to learn more about views on the Axiom of Choice at the time of G. Vitali's publication of his nonmeasurable set in 1905, might want to read Bressoud [**8**] for an appreciation of the issue, including historical perspective and indications of surprising consequences of the axiom. In addition, Bressoud [**8**] outlines (by means of exercises on page 158) the construction of a Jordan measurable set which is not a Borel set. Every Jordan measurable set is Lebesgue measurable, so the example gives a set belonging to $\mathcal{M}$ but not to $\mathcal{B}$. In fact, the Borel σ-algebra $\mathcal{B}$ and Lebesgue σ-algebra $\mathcal{M}$ do not have the same cardinality.

Chapter 17

The Lebesgue Integral

In this chapter we develop the integral for real valued, and extended real valued, functions defined on a measure space (X, Σ, μ). The results include the Lebesgue integral of functions defined on the Lebesgue measure space $(\mathbf{R}^n, \mathcal{M}_n, m_n)$, for any fixed positive integer dimension n. A reader who wishes to focus only on the real line or on $\mathbf{R}^n$ may assume that (X, Σ, μ) is the Lebesgue measure space of their choice throughout the chapter, but there is no advantage or simplification in the development by such a specification.

Let us first consider *why* we are interested in the Lebesgue integral. Then we offer some brief comments to motivate *how* the integral will be developed.

Although the Riemann integral is useful for many purposes, it has two deficiencies that make it unsatisfactory as a general theory of integration for analysis. Stated simply, these are the deficiencies: (1) Not enough functions are Riemann integrable to constitute complete spaces of integrable functions, and (2) limiting operations require uniform convergence for satisfactory results. Analysis generally requires complete spaces to obtain the most definitive and satisfactory results, and uniform convergence is too restrictive as a general requirement for limiting operations involving integrals.

Consider for a moment the geometric idea of integrating a nonnegative function. The strategy of the Riemann integral over $[a, b]$ is to partition $[a, b]$ into subintervals, multiply the supremum and infimum of the function over each subinterval by the length of the subinterval, and sum the results over all the subintervals. Thus we obtain two sums, an upper sum and lower sum, that approximate the area between function graph and horizontal axis. We imagine letting the mesh size of the partition approach zero, and label as integrable those functions for which the infimum of all the upper sums equals the supremum of all the lower sums. We observe that this process involves the approximation of f by **step functions**. (The Riemann integral implicitly uses step functions, and we formally define a **step function** in this chapter as a function that has a finite range such that each function value is assumed on an *interval* in the domain.)

Lebesgue's idea was to subdivide the range of the nonnegative function rather than the domain. His definition reads similarly to a Riemann sum, except that, for a subinterval $[y_{i-1}, y_i]$ of the range of f, the contribution to the estimate of area now takes the form of the real number product

$$[y_{i-1}]\, \mu(f^{-1}([y_{i-1}, y_i])),$$

where μ is our measure in the domain space. Clearly, we must require of our functions f that the inverse images $f^{-1}([y_{i-1}, y_i])$ are measurable. The choice of the lower value y_{i-1} to approximate f over the set $f^{-1}([y_{i-1}, y_i])$ means an approximation of f from below. To be specific, if for each positive integer n, we divide the interval $[0, n]$ in the range space into $n2^n$ subintervals of equal length $1/2^n$, then we estimate the contribution to area from each range subinterval by

$$\left(\frac{k-1}{2^n}\right) \mu\left(\left\{x \in \mathbf{R} : \frac{k-1}{2^n} \le f(x) \le \frac{k}{2^n}\right\}\right),$$

where $1 \le k \le n2^n$. By summing these contributions from $k = 1$ to $k = n2^n$ we arrive at the area estimate

$$\sum_{k=1}^{n2^n} \left(\frac{k-1}{2^n}\right) \mu\left(\left\{x \in \mathbf{R} : \frac{k-1}{2^n} \le f(x) \le \frac{k}{2^n}\right\}\right).$$

This estimates the area under the graph of f as the area under a **simple function** with $n2^n$ distinct function values. (A **simple function** is defined in this chapter as a function that has a finite range such that each function value is assumed on a *measurable set* in the domain.) By continuing to expand the coverage of the range space using the interval $[0, n]$ for increasing n, and continuing the dyadic subdivision of the range intervals, Lebesgue arrived at a definition of the integral of a nonnegative function on the real line, which reads

$$\int_{\mathbf{R}} f\, dm = \lim_{n\to\infty} \sum_{k=1}^{n2^n} \left(\frac{k-1}{2^n}\right) m\left(\left\{x \in \mathbf{R} : \frac{k-1}{2^n} \le f(x) \le \frac{k}{2^n}\right\}\right),$$

where the measure m indicated here is Lebesgue measure on the real line. Such an approach requires that all these inverse images for f are measurable sets in the domain. This leads us to the first order of business, the definition of measurable functions.

17.1. Measurable Functions

Let (X, Σ) be a measurable space. We define measurable real functions using only the set-theoretic properties involving the σ-algebra Σ, independently of any particular measure that might be defined on Σ. This is motivated by the idea noted above that we must be able to measure the inverse image of any interval. We observe that if E is any subset of X, then E inherits the structure of a measurable space from X and Σ: (E, Σ_E) is a measurable space as in Exercise 16.5.10.

Definition 17.1.1. *Let (X, Σ) be a measurable space. If $f : X \to \mathbf{R}$, then f is* **measurable** *if for every real number a, the set $\{x \in X : f(x) < a\} = f^{-1}((-\infty, a))$ is measurable. Suppose $E \subseteq X$ and $f : E \to \mathbf{R}$. Then f is measurable if for every real number a, $\{x \in E : f(x) < a\} = E \cap f^{-1}((-\infty, a))$ is measurable.*

Proposition 17.1.2. *If (X, Σ) is a measurable space, $E \subseteq X$, and $f : E \to \mathbf{R}$ is measurable, then E is measurable, $E \in \Sigma$.*

Proof. We have $E = \bigcup_{n=1}^{\infty} f^{-1}\big((-\infty, n)\big)$, so E is a countable union of measurable sets, hence E is measurable. □

The next proposition gives several equivalent conditions for measurability of a real function.

Proposition 17.1.3. *Suppose (X, Σ) is a measurable space, $E \subseteq X$ and $f : E \to \mathbf{R}$. The following statements are equivalent:*

1. *For each real number a, the set $\{x \in E : f(x) < a\}$ is measurable.*
2. *For each real number a, the set $\{x \in E : f(x) \leq a\}$ is measurable.*
3. *For each real number a, the set $\{x \in E : f(x) > a\}$ is measurable.*
4. *For each real number a, the set $\{x \in E : f(x) \geq a\}$ is measurable.*

Each of statements 1-4 implies the following statement:

5. *For each real number a, the set $\{x \in E : f(x) = a\}$ is measurable.*

Proof. The inverse images in statements 1 and 4 are complements of each other in E. So statement 1 is equivalent to 4. Similarly, the inverse images in statements 2 and 3 are complementary in E, so statements 2 and 3 are equivalent. We can write

$$\{x : f(x) < a\} = \bigcup_{n=1}^{\infty} \Big\{x : f(x) \leq a - \frac{1}{n}\Big\},$$

so if statement 2 holds, then so does 1, since a countable union of measurable sets is measurable. On the other hand, we have

$$\{x : f(x) \leq a\} = \bigcap_{n=1}^{\infty} \Big\{x : f(x) < a + \frac{1}{n}\Big\},$$

so if statement 1 holds, then so does 2, since a countable intersection of measurable sets is measurable. Thus statements 1 and 2 are equivalent. Similar arguments show the equivalence of statements 3 and 4. Hence, statements 1-4 are equivalent.

Finally, suppose that one, and hence all, of 1-4 hold. If a is a real number, then

$$\{x : f(x) = a\} = \{x : f(x) \leq a\} \cap \{x : f(x) \geq a\},$$

which is an intersection of measurable sets, and hence measurable. □

Example 17.1.4. Any constant function $f : X \to \mathbf{R}$, $f(x) \equiv C \in \mathbf{R}$, is measurable, since the inverse image $\{x \in X : f(x) < a\}$ is either the empty set (if $C \geq a$) or all of X (if $C < a$). △

Measurable functions link the topology of $\mathbf{R}$ and the σ-algebra of the measurable space, according to the next result.

Proposition 17.1.5. *Let (X, Σ) be a measurable space. A function $f : X \to \mathbf{R}$ is measurable if and only if for any open set O in $\mathbf{R}$, the set*

$$f^{-1}(O) = \{x \in X : f(x) \in O\}$$

is measurable, that is, $f^{-1}(O) \in \Sigma$. *If* $E \subseteq X$ *and* $f : E \to \mathbf{R}$, *then* f *is measurable if and only if for any open set* O *in* $\mathbf{R}$, *the set*

$$E \cap f^{-1}(O) = \{x \in E : f(x) \in O\}$$

is measurable.

Proof. It suffices to consider the domain X. Suppose f is measurable. For any $a < b$, we have $(a, b) = (-\infty, b) - (-\infty, a]$, using the difference notation for set complement. Thus,

$$f^{-1}((a, b)) = f^{-1}((-\infty, b)) - f^{-1}((-\infty, a]),$$

the difference of two measurable sets, so $f^{-1}((a, b))$ is measurable for any open interval (a, b). Since any open set O of $\mathbf{R}$ is a countable union $\bigcup_{k=1}^{\infty}(a_k, b_k)$ of open intervals, it follows that $f^{-1}(O) = \bigcup_{k=1}^{\infty} f^{-1}((a_k, b_k))$ is measurable, for each open set O of $\mathbf{R}$.

Suppose that $f^{-1}(O)$ is measurable for each open set O. Then $f^{-1}((-\infty, a))$ is measurable for each real number a. Hence, by definition, f is measurable. $\square$

Note the similarity of Proposition 17.1.5 with the statement that a function $f : \mathbf{R} \to \mathbf{R}$ is continuous if and only if the inverse image of any open set in $\mathbf{R}$ is open. In each context, the relevant structure is preserved under inverse images.

If $(X, \Sigma) = (\mathbf{R}^n, \mathcal{M})$, the Lebesgue measurable space, the measurable functions are called the **Lebesgue measurable functions**, or simply the **measurable** functions on $\mathbf{R}^n$.

Example 17.1.6. Let n be a fixed positive integer. The characteristic function of a set $E \subseteq \mathbf{R}^n$, $\chi_E : \mathbf{R}^n \to \mathbf{R}$, is defined by $\chi_E(\mathbf{x}) = 1$ if $\mathbf{x} \in E$ and $\chi_E(\mathbf{x}) = 0$ otherwise. Then χ_E is measurable if and only if the set E is measurable. $\triangle$

The real valued continuous functions on $\mathbf{R}^n$ constitute an important class of measurable functions (Exercise 17.1.1).

We now consider the measurability of various combinations of measurable functions.

Proposition 17.1.7. *Let* (X, Σ) *be a measurable space and* $E \subseteq X$. *If* $f, g : E \to \mathbf{R}$ *are measurable and* c *is a real constant, then* $f+g$, $f-g$, fg *and* cf *are measurable.*

Proof. Constant functions are measurable (Example 17.1.4), so we only need to show that sums and products of measurable functions are measurable.

Assume f and g are measurable. We now show that condition 1 of Proposition 17.1.3 holds for the sum $f + g$. The inequality $f(x) + g(x) < a$ holds if and only if $f(x) < -g(x) + a$, and by the density of the rationals in $\mathbf{R}$, this holds if and only if there is a rational number r such that $f(x) < r < -g(x) + a$, which is equivalent to the statement that there is a rational number r such that $f(x) < r$ *and* $g(x) < a - r$. Thus,

$$\{x \in E : f(x) + g(x) < a\} = \bigcup_{r \in \mathbf{Q}} \{x \in E : f(x) < r\} \cap \{x \in E : g(x) < a - r\}.$$

Since this is a countable union of measurable sets, it is measurable. Since a was arbitrary, it follows that $f + g$ is measurable.

For products, observe first that f measurable implies f^2 is measurable (Exercise 17.1.4). Since we can write

$$fg = \frac{1}{2}[(f+g)^2 - f^2 - g^2],$$

measurability of f and g implies that fg is measurable. □

We recall the definitions of sup, inf, lim sup and lim inf of a sequence of real numbers. (See Definition 3.10.2 for lim sup and lim inf.) We apply the definitions to a sequence of functions f_n having a common domain E, and thus define the functions $\sup f_n$, $\inf f_n$, $\limsup f_n$ and $\liminf f_n$, as follows:

- ⋆ $(\sup f_n)(x) := \sup\{f_n(x) : n \in \mathbf{N}\}$;
- ⋆ $(\inf f_n)(x) := \inf\{f_n(x) : n \in \mathbf{N}\}$;
- ⋆ $(\limsup f_n)(x) := \inf_m \sup\{f_n(x) : n \geq m\}$;
- ⋆ $(\liminf f_n)(x) := \sup_m \inf\{f_n(x) : n \geq m\}$.

Remember that the sequence $(\sup\{f_n(x) : n \geq m\})_{m=1}^\infty$ is decreasing with limit being its infimum, and the sequence $(\inf\{f_n(x) : n \geq m\})_{m=1}^\infty$ is increasing with limit being its supremum. There is little risk of ambiguity in dropping the parentheses around the function labels on the left side, and we do so from here on.

In integration theory we sometimes deal with functions that can take on the value ∞ or $-\infty$. As usual, we write $[-\infty, \infty] = \mathbf{R} \cup \{-\infty\} \cup \{\infty\}$. We say that $f : E \subseteq X \to [-\infty, \infty]$ is an **extended real valued measurable function** if f satisfies one (and hence all) of statements 1-4 of Proposition 17.1.3, and, in addition, the sets

$$\{x \in E : f(x) = -\infty\} \qquad \text{and} \qquad \{x \in E : f(x) = \infty\}$$

are measurable.

Proposition 17.1.8. *Let (X, Σ) be a measurable space. Let f be an extended real valued function defined on a set $E \subseteq X$. Then f is measurable if and only if the sets $f^{-1}(\{-\infty\})$ and $f^{-1}(\{\infty\})$ are measurable and $f^{-1}(O) = \{x : f(x) \in O\}$ is measurable for every open set O in $\mathbf{R}$.*

Proof. This is immediate from Proposition 17.1.5 and the definition of extended real valued measurable function stated above before this proposition. □

Proposition 17.1.9. *Let (X, Σ) be a measurable space. Suppose that for each positive integer n, f_n is a measurable real or extended real valued function defined on $E \subseteq X$. Then the functions*

$$\sup f_n, \qquad \inf f_n, \qquad \limsup f_n, \qquad \liminf f_n$$

are defined on E and measurable.

Proof. Let $f(x) = \sup f_n(x)$, $x \in E$. If $a \in \mathbf{R}$, then $f(x) > a$ if and only if $f_n(x) > a$ for some n, hence

$$\{x \in E : f(x) > a\} = \bigcup_{n=1}^{\infty} \{x \in E : f_n(x) > a\},$$

and since each of the sets in this countable union is measurable, their union is measurable. This is true for each real number a, so $f = \sup f_n$ is measurable. If $f(x) = \infty$ for some x, we have

$$\{x \in E : f(x) = \infty\} = \bigcap_{m=1}^{\infty} \{x \in E : f(x) > m\},$$

which is a countable intersection of measurable sets, and hence measurable.

Since $\inf f_n = -\sup(-f_n)$, it follows that $\inf f_n$ is measurable since $-f_n$, $\sup(-f_n)$ and $-\sup(-f_n)$ are measurable. (Alternatively, letting $g = \inf f_n$, we see that $g(x) < a$ if and only if $f_n(x) < a$ for some n, hence

$$\{x \in E : g(x) < a\} = \bigcup_{n=1}^{\infty} \{x \in E : f_n(x) < a\},$$

and thus $\{x \in E : g(x) < a\}$ is measurable, for each real number a. Also, $\{x \in E : g(x) = -\infty\} = \bigcap_{m=1}^{\infty}\{x \in E : g(x) < -m\}$, a countable intersection of measurable sets.)

Now $\limsup f_n(x) = \inf_m \sup\{f_n(x) : n \geq m\}$, and for each m, $\sup_{n \geq m} f_n$ is measurable by the argument of the first paragraph above. By the argument of the second paragraph above, $\inf_m \sup_{n \geq m} f_n$ is measurable, so $\limsup f_n$ is measurable. Similarly, $\liminf f_n(x) = \sup_m \inf\{f_n(x) : n \geq m\}$ is the supremum of a sequence of measurable functions, and therefore $\liminf f_n$ is measurable. $\square$

Let us review Proposition 17.1.7, which was proved for real valued f and g. If f and g are extended real valued, then in order to define $f + g$ and $f - g$ we must rule out the possibility of the undefined expressions $-\infty + \infty$ and $\infty - \infty$. If these expressions never occur for the pair f, g, and if f and g are extended real valued measurable functions, then so are $f + g$ and $f - g$. (See also Exercise 17.1.8.)

Most of the terminology in the next definition is familiar from our experience with the Riemann integral.

Definition 17.1.10. *Let (X, Σ, μ) be a measure space. A property holds* **almost everywhere (a.e.)** *if the set on which it fails to hold is a set of measure zero. In particular:*

1. *Two functions f and g are* **equal almost everywhere**, *written $f = g$ a.e., if they have a common domain $E \subseteq X$ and $\mu(\{x \in E : f(x) \neq g(x)\}) = 0$. (We also say that $f = g$ a.e. in E.)*
2. *A sequence of functions f_n having a common domain $E \subseteq X$* **converges almost everywhere** *to a function g, written $f_n \to g$ a.e., if there is a set N with $\mu(N) = 0$ such that $f_n(x) \to g(x)$ for each $x \in E - N$. (We also say that $f_n \to g$ a.e in E.)*

The next results assume a measure μ is defined on the σ-algebra Σ of X. We can also work with the measure space (E, Σ_E, μ_E) for any measurable set $E \in \Sigma$.

Proposition 17.1.11. *Let (X, Σ, μ) be a measure space. Let f and g be functions defined on a subset E of X. Suppose μ is a complete measure. If f is real valued or extended real valued, and measurable, and if $f = g$ a.e. in E, then g is measurable.*

Proof. Since f is measurable, the domain E is measurable. Let $N = \{x \in E : f(x) \neq g(x)\}$. Then $\mu(N) = 0$ by hypothesis. Let a be any real number. We can write

$$\{x : g(x) < a\} = \{x \in E : f(x) < a\} \cup \{x \in N : g(x) < a\} - \{x \in N : g(x) \geq a\}.$$

The first set on the right is measurable since f is measurable. The last two sets on the right are measurable since they are subsets of N, which has measure zero, and the measure μ is a complete measure. Since Σ contains the union and difference of measurable sets, and a was arbitrary, $\{x : g(x) < a\}$ is measurable for every real number a. If f takes on real values only, it now follows that g is measurable.

If f takes on the value $-\infty$, then we can write

$$\begin{aligned}\{x : g(x) = -\infty\} &= \Big(\bigcap_{m=1}^{\infty} \{x : f(x) < -m\}\Big) \cup \Big(\bigcap_{m=1}^{\infty} \{x \in N : g(x) < -m\}\Big) \\ &\quad - \{x \in N : g(x) > -\infty\},\end{aligned}$$

which is a measurable set. If f takes on the value ∞, then we can write

$$\begin{aligned}\{x : g(x) = \infty\} &= \Big(\bigcap_{m=1}^{\infty} \{x : f(x) > m\}\Big) \cup \Big(\bigcap_{m=1}^{\infty} \{x \in N : g(x) > m\}\Big) \\ &\quad - \{x \in N : g(x) < \infty\},\end{aligned}$$

which is a measurable set. This completes the proof that g is measurable. □

We have seen by examples that the space of continuous functions on $[a, b]$ and the space of Riemann integrable functions on $[a, b]$ are not closed under pointwise limits. A great advantage of the class of measurable functions is that it is closed under pointwise limits.

Proposition 17.1.12. *Let (X, Σ, μ) be a measure space. Suppose the functions f_n are measurable and have a common domain $E \subseteq X$. The following statements are true:*

1. *If the sequence f_n converges everywhere in E to a function g, then g is measurable.*
2. *If μ is a complete measure and $f_n \to g$ a.e. in E, then g is measurable.*

Proof. By Proposition 17.1.9, $\limsup f_n$ is a measurable function.

1. If $f_n \to g$ everywhere in E, then $g = \lim_{n\to\infty} f_n = \limsup f_n$, so g is measurable.

2. If there is a set $N \subseteq E$ with $\mu(N) = 0$ such that $f_n(x) \to g(x)$ at all points $x \in E - N$, then $g(x) = \limsup f_n(x)$ for all $x \in E - N$, so g is a measurable function defined on $E \cap N^c$. Thus, g is equal to the measurable function $\limsup f_n$

almost everywhere in E. Since the measure μ is complete, Proposition 17.1.11 applies, and we conclude that g is measurable. □

In Proposition 17.1.12, the limit function g can be defined, or redefined, however we might wish on N and still remain measurable, because any subset of N is measurable (with measure zero) since μ is a complete measure.

Exercises.

Exercise 17.1.1. Show that if $f : \mathbf{R}^n \to \mathbf{R}$ is continuous, then f is measurable.

Exercise 17.1.2. Show that if f and g are measurable functions defined on $E \subseteq X$, then so are $\max(f, g)$ and $\min(f, g)$, defined on E by

$$\max\{f, g\}(x) = \max\{f(x), g(x)\} \qquad \text{and} \qquad \min\{f, g\}(x) = \min\{f(x), g(x)\}.$$

Then show that the maximum, $\max\{f_1, \ldots, f_n\}$, and minimum, $\min\{f_1, \ldots, f_n\}$, of any finite collection of measurable functions on E is measurable.

Exercise 17.1.3. Prove: If (X, Σ) is a measurable space and $f : X \to \mathbf{R}$ is a measurable function, then for any Borel set B of $\mathbf{R}$, the set $f^{-1}(B)$ is measurable. *Hint*: Consider the collection of sets E such that $f^{-1}(E)$ is measurable, and show that it is a σ-algebra.

Exercise 17.1.4. Let (X, Σ) be a measurable space and $E \in \Sigma$. If f is defined on E and measurable, then $|f|$ and $|f|^2 = f^2$ are measurable.

Exercise 17.1.5. Show that if $f : \mathbf{R} \to \mathbf{R}$ is measurable and $g : \mathbf{R} \to \mathbf{R}$ is continuous, then $g \circ f$ is measurable.

Exercise 17.1.6. Prove: A monotone function $f : \mathbf{R} \to \mathbf{R}$ is measurable.

Exercise 17.1.7. Let (X, Σ) be a measurable space and $E \in \Sigma$. The **positive part** of a function f defined on E is the function f^+ defined on E by

$$f^+(x) = \begin{cases} f(x), & \text{if } f(x) > 0, \\ 0, & \text{if } f(x) \leq 0. \end{cases}$$

The **negative part** of f is the function f^- defined on E by

$$f^-(x) = \begin{cases} -f(x), & \text{if } f(x) < 0, \\ 0, & \text{if } f(x) \geq 0. \end{cases}$$

1. Verify that $f^+(x) = \max\{f, 0\}$, $f^-(x) = \min\{f, 0\}$, and $f = f^+ - f^-$.
2. Let $f : X \to \mathbf{R}$. Show that f is measurable if and only if f^+ and f^- are measurable.

Exercise 17.1.8. Suppose f and g are measurable extended real valued functions having the same domain in $\mathbf{R}$. Prove: If f and g are finite almost everywhere, then $f + g$ is measurable no matter how the sum is redefined at points where it takes the form $\infty - \infty$ or $-\infty + \infty$.

Exercise 17.1.9. Suppose f and g are measurable extended real valued functions having the same domain in $\mathbf{R}$, and f and g are finite almost everywhere. Show that their product fg is measurable.

17.2. Simple Functions and the Integral

The building blocks of integration theory are the *simple functions*. These functions have finite range, but constitute a much larger class of functions than the *step functions* used to approximate Riemann integrals.

Definition 17.2.1. *Let (X, Σ) be a measurable space. A function $s : X \to \mathbf{R}$ is a* **simple function** *if it has a finite range $\{a_1, \dots, a_m\}$ and the sets*

$$E_i = \{x \in X : s(x) = a_i\}, \qquad 1 \le i \le m,$$

are measurable. Thus,

$$s(x) = \sum_{i=1}^{m} a_i \chi_{E_i}. \tag{17.1}$$

We note that in the canonical expression (17.1) in this definition, the sets E_i are disjoint due to the listing of the finite range set. It is possible that one of the distinct values a_i listed for s is 0. The class of simple functions includes the characteristic functions of measurable sets. Every simple function is measurable; this follows readily from the definition, or, with reference to the representation (17.1), measurability follows from Proposition 17.1.7 and the measurability of the characteristic functions χ_{E_i}.

Recalling the definition of the Riemann integral of a function, we observe that it involves special simple functions, the *step functions*, where each of the sets E_i is an interval and the domain $[a, b]$ is partitioned by the intervals E_i.

Definition 17.2.2. *A real valued function ϕ defined on $[a, b]$ is a* **step function** *if there is a partition $\{x_0 = a, x_1, \dots, x_{n-1}, x_n = b\}$ of $[a, b]$ such that for each $k = 1, \dots, n$, the function ϕ assumes only one value on the interval $[x_{k-1}, x_k]$.*

With the introduction of simple functions in Definition 17.2.1, we considerably enlarge the basic building blocks to be used in developing the theory of measurable functions and, ultimately, the Lebesgue integral. It is especially important that nonnegative simple functions can be used to approximate any nonnegative measurable function, whether bounded or unbounded.

Theorem 17.2.3. *Let (X, Σ) be a measurable space. If f is a nonnegative measurable function, then there exists an increasing sequence s_n of nonnegative simple functions such that $f(x) = \lim_{n\to\infty} s_n(x)$ for all x.*

Proof. The function f may be unbounded. In any case, for each positive integer n, we partition the interval $[0, n)$ in the range space into $n2^n$ subintervals of equal length $1/2^n$, and define

$$s_n(x) = \begin{cases} (k-1)/2^n & \text{if } (k-1)/2^n \le f(x) < k/2^n \quad (1 \le k \le n2^n); \\ n & \text{if } f(x) \ge n. \end{cases}$$

Since f is nonnegative and measurable, each function s_n is a nonnegative simple function. At each x, $s_n(x) \le s_{n+1}(x) \le f(x)$, because in the passage from n to $n + 1$, we halve each of the previous subintervals of the range interval $[0, n)$ and add 2^{n+1} subintervals of the range interval $[n, n+1)$. Suppose that at the point x, $f(x) < \infty$. For any n such that $f(x) < n$,

$$0 \leq f(x) - s_n(x) \leq \frac{1}{2^n},$$

and thus $\lim_{n\to\infty} s_n(x) = f(x)$. At any point x where $f(x) = \infty$, $s_n(x) = n$ for each n, so for every x, $\lim_{n\to\infty} s_n(x) = f(x)$. □

Given a measure μ on X, we can define the Lebesgue integral of simple functions.

Definition 17.2.4. *Let (X, Σ, μ) be a measure space. If s is a simple function on X with range $\{a_1, \ldots, a_m\}$, and $E_i = \{x : s(x) = a_i\}$, then the* **Lebesgue integral** *of s over X is defined to be*

$$\int_X s\, d\mu = \sum_{i=1}^{m} a_i\, \mu(E_i).$$

(If $a_i = 0$ for some i and $\mu(E_i) = \infty$, then $a_i\mu(E_i) = 0 \cdot \infty = 0$.)

A simple function might be expressed in more than one way as a linear combination of characteristic functions. For example, we have $\chi_{A\cup B} = \chi_A + \chi_B$ if A and B are disjoint, and $\chi_{A\cup B} = \chi_A + \chi_B - \chi_{A\cap B}$ if A and B have nonempty intersection. So we should check that the definition of $\int_X s\, d\mu$ does not depend on how s is written. If s has the canonical expression (17.1) in Definition 17.2.1 and we also have $s = \sum_{j=1}^{n} b_j \chi_{B_j}$, then

$$\sum_{i=1}^{m} a_i\mu(E_i) = \sum_{j=1}^{n} b_j\mu(B_j).$$

The verification of this is left to the reader.

We note that it is possible to have $\int_X s\, d\mu = \pm\infty$.

Definition 17.2.5. *If s is a simple function on X with range $\{a_1, \ldots, a_m\}$, we say that s is* **integrable** *(on X) if $\int_X |s|\, d\mu < \infty$.*

The proof of the next result is left to Exercise 17.2.4.

Theorem 17.2.6. *If s is a simple function on X with range $\{a_1, \ldots, a_m\}$, then s is integrable if and only if $\mu(E_i) < \infty$ for each set $E_i = \{x : s(x) = a_i\}$ for which $a_i \neq 0$.*

The next theorem lists some basic properties of the integral for simple functions. The reader is invited to supply the proofs in Exercise 17.2.5.

Theorem 17.2.7. *Let ϕ and ψ be integrable simple functions on a measure space (X, Σ, μ). Then the following properties hold:*

1. *If $\phi \geq 0$ a.e., then $\int_X \phi\, d\mu \geq 0$.*
2. *If $\phi \leq \psi$ a.e., then $\int_X \phi\, d\mu \leq \int_X \psi\, d\mu$.*
3. *$|\phi|$ is an integrable simple function, and*

$$\left| \int_X \phi\, d\mu \right| \leq \int_X |\phi|\, d\mu.$$

4. *$\int_X |\phi + \psi|\, d\mu \leq \int_X |\phi|\, d\mu + \int_X |\psi|\, d\mu$.*

Exercises.

Exercise 17.2.1. Let ϕ and ψ be simple functions on X. Show that

1. $|\phi|$ is a simple function.
2. $\phi + \psi$ and $\phi\psi$ are simple functions.
3. If E is any measurable set, then the product $\chi_E \phi$ is a simple function.

Exercise 17.2.2. In Theorem 17.2.3, there is no guarantee that the approximating simple functions are zero outside some set of finite measure. Suppose f is nonnegative and measurable on $\mathbf{R}$. Show that there exists an increasing sequence s_n of simple functions with pointwise limit f such that $m(\{x : s_n(x) \neq 0\}) < \infty$ for each n. *Hint*: $\mathbf{R}$ is a countable union of sets having finite Lebesgue measure.

Exercise 17.2.3. Suppose a simple function $s : X \to \mathbf{R}$ can be expressed in two ways as

$$s(x) = \sum_{i=1}^{m} a_i \chi_{A_i}(x) \quad \text{and} \quad s(x) = \sum_{j=1}^{n} b_j \chi_{B_j}(x).$$

Show that

$$\sum_{i=1}^{m} a_i \mu(A_i) = \sum_{j=1}^{n} b_j \mu(B_j),$$

and hence the integral of a simple function is well defined by Definition 17.2.4.

Exercise 17.2.4. Prove Theorem 17.2.6.

Exercise 17.2.5. Prove Theorem 17.2.7.

17.3. Definition of the Lebesgue Integral

We now deal with a measure space (X, Σ, μ). The definition of the Lebesgue integral proceeds from nonnegative simple functions to nonnegative measurable functions, and finally to general measurable functions. Extended real valued functions may also be handled. Such functions can appear as pointwise limits of sequences of measurable functions.

The integral of simple functions is considered at the end of the previous section, but the definition for nonnegative simple functions is included here. Recall that every simple function is measurable, by Definition 17.2.1.

Definition 17.3.1. *Let (X, Σ, μ) be a measure space.*

1. *If s is a nonnegative simple function on X with range $\{a_1, \ldots, a_m\}$, and $E_i = \{x \in X : s(x) = a_i\}$, then the* **Lebesgue integral** *of s is*

$$\int_X s\, d\mu = \sum_{i=1}^{m} a_i\, \mu(E_i). \tag{17.2}$$

   (If $a_i = 0$ for some i and $\mu(E_i) = \infty$, then $a_i \mu(E_i) = 0 \cdot \infty = 0$.)

2. *If f is a nonnegative measurable function, we define the* **Lebesgue integral** *of f to be*

$$\int_X f\, d\mu = \sup\left\{ \int_X s\, d\mu : 0 \leq s \leq f \text{ and } s \text{ is simple} \right\}. \tag{17.3}$$

3. *If f is measurable, then the* **Lebesgue integral** *of f is*

$$\int_X f\,d\mu = \int_X f^+\,d\mu - \int_X f^-\,d\mu, \tag{17.4}$$

where $f^+(x) = \max(f(x), 0)$ and $f^-(x) = \max(-f(x), 0)$ are the positive and negative parts of f, respectively, provided at least one of the integrals $\int_X f^+\,d\mu$ and $\int_X f^-\,d\mu$ is finite.

Remarks on the Definition. Consider definition (17.3) for nonnegative measurable functions. If $f = s \geq 0$ is a simple function in (17.3), then we obtain the same value for the integral of $f = s$ there as we do in (17.2). Also, (17.3) is a natural definition, given the approximation of f from below by simple functions in Theorem 17.2.3. This is analogous to the approximation of a nonnegative Riemann integrable function by lower sums based on a partition of the domain interval. However, we can handle a considerably larger class of functions with (17.3) because we employ a σ-algebra of sets in the domain space X rather than only the closed and bounded intervals in $\mathbf{R}$ or $\mathbf{R}^n$. Much of the power and effectiveness of the Lebesgue integral in the limit theorems of Section 17.4 is due to (17.3). Now consider (17.4). If $f = s$ is a simple function that takes on both positive and negative values, then (17.4) is consistent with Definition 17.2.4 in the previous section. Moreover, a simple function is integrable if and only if $\mu(E_i) < \infty$ for each set $E_i = \{x : s(x) = a_i\}$ for which $a_i \neq 0$, in accord with Theorem 17.2.6.

Definition 17.3.2. *If $f : X \to \mathbf{R}$ is measurable and $\int_X |f|\,d\mu < \infty$, then we say that f is (Lebesgue)* **integrable** *on X.*

If f is integrable on X, then for any measurable subset E of X,

$$|\chi_E(x) f(x)| \leq |f(x)|$$

for all x, and we define the Lebesgue **integral of f over E** by

$$\int_E f\,d\mu = \int_X \chi_E f\,d\mu. \tag{17.5}$$

Then f is integrable on E, since $\int_E |f|\,d\mu = \int_X |\chi_E f|\,d\mu \leq \int_X |f|\,d\mu < \infty$.

We can also relax the assumption that our functions are defined on all of X to begin with, and define the integral over measurable sets $E \subset X$ as follows: If f is defined on a measurable set E, we can extend f by zero to all of X. If $\chi_E f$ is measurable, then equation (17.5) defines the integral of f over E. If $\chi_E f$ is integrable, that is, $\int_E |f|\,d\mu = \int_X |\chi_E f|\,d\mu < \infty$, then we say that f is **integrable on** E. This is the same as working directly with the measure space (E, Σ_E, μ_E), by restriction of (X, Σ, μ) as in Exercise 16.5.10, and the measurable functions and integrable functions on E.

We have noted that one difference between the Lebesgue integral and the Riemann integral is that the Riemann approach subdivides the domain of the function and the Lebesgue approach subdivides the range of the function. But we emphasize that subdividing the range requires that we be able to measure a much larger class of subsets of the domain than just the intervals. See also Exercise 17.3.3.

There are a few more properties of the integral that can be established now; see Exercises 17.3.1-17.3.2. These properties are sufficient to allow us to prove the monotone convergence theorem in the next section (Theorem 17.4.1). Using the monotone convergence theorem, we will establish the additivity of the integral in Theorem 17.4.2, and combined with Exercise 17.3.1, this will establish the linearity of the integral.

Exercises.

Exercise 17.3.1. Show that if f is a real valued measurable function on X and α a real constant, then $\int_X \alpha f\, d\mu = \alpha \int_X f\, d\mu$. *Hint*: Start with the simple functions, then consider the nonnegative measurable functions, and then the general measurable functions.

Exercise 17.3.2. Prove the following, where all functions involved are assumed to be nonnegative and measurable on X, and all sets involved are assumed to be measurable subsets of X:

1. If $0 \le f \le g$ everywhere in E, then $0 \le \int_E f\, d\mu \le \int_E g\, d\mu$.
2. If $A \subseteq B$, then $\int_A f\, d\mu \le \int_B f\, d\mu$.
3. If $f \equiv 0$ on E, then $\int_E f\, d\mu = 0$, even when $\mu(E) = \infty$.
4. If $\mu(E) = 0$, then $\int_E f\, d\mu = 0$, even if $f = \infty$ everywhere on E.

Exercise 17.3.3. Let $f : \mathbf{R} \to \mathbf{R}$ be a bounded nonnegative measurable function on the real line and let m denote Lebesgue measure. Let s_n be the increasing sequence of nonnegative simple functions in Theorem 17.2.3 with $\lim_{n\to\infty} s_n(x) = f(x)$ for all $x \in \mathbf{R}$. Verify that

$$\lim_{n\to\infty} \int_{\mathbf{R}} s_n\, dm = \lim_{n\to\infty} \sum_{k=1}^{n2^n} \Big(\frac{k-1}{2^n}\Big) m\Big(\Big\{x \in \mathbf{R} : \frac{k-1}{2^n} \le f(x) \le \frac{k}{2^n}\Big\}\Big).$$

This formula is Lebesgue's original definition of the integral of f. The proof that this limit equals $\int_{\mathbf{R}} f\, dm$ in agreement with (17.3) in Definition 17.3.1 is part of the proof of the monotone convergence theorem in the next section.

17.4. The Limit Theorems

The limit theorems of this section are a primary reason the Lebesgue integral is a major improvement on the Riemann integral and the preferred choice for work in advanced analysis. We are given a measure space (X, Σ, μ), and, as usual in this book, the measure μ is a complete measure. In particular, the results hold for the Lebesgue measure spaces $\mathbf{R}$ and $\mathbf{R}^n$.

Theorem 17.4.1 (Monotone Convergence). *Suppose the functions f_n are nonnegative and measurable on X, and*

$$f_n(x) \le f_{n+1}(x)$$

for all x and all n. If $f(x) = \lim_{n\to\infty} f_n(x)$ for all $x \in X$, then

$$\int_X f\, d\mu = \lim_{n\to\infty} \int_X f_n\, d\mu.$$

Proof. Since $f_n \le f_{n+1}$, the sequence $\int_X f_n \, d\mu$ is a monotone increasing sequence of real numbers. Thus, the limit of the integrals exists as an extended real number, either finite or infinite. By Proposition 17.1.9, the limit function f is measurable. Since $f_n \le f$, we have $\int_X f_n \, d\mu \le \int_X f \, d\mu$ for each n (Exercise 17.3.2 (statement 1)). Hence,

$$\lim_{n\to\infty} \int_X f_n \, d\mu \le \int_X f \, d\mu.$$

Let $L = \lim_{n\to\infty} \int_X f_n \, d\mu$. We want to show that $L \ge \int_X f \, d\mu$. This will follow from Definition 17.3.1 (statement 2) if we show that $L \ge \int_X s \, d\mu$ for every nonnegative simple function s with $s \le f$. Let $s = \sum_{k=1}^m a_k \chi_{E_k}$ be a nonnegative simple function with $s \le f$, and let β be a number with $0 < \beta < 1$. Let $A_n = \{x \in X : f_n(x) \ge \beta \, s(x)\}$. Then A_n is measurable for each n. Since the sequence f_n increases with limit f everywhere, $A_n \subset A_{n+1}$ for each n, and $\bigcup_{n=1}^\infty A_n = X$. For each n,

$$\begin{aligned} \int_X f_n \, d\mu &\ge \int_{A_n} f_n \, d\mu \ge \beta \int_{A_n} s \, d\mu \\ &= \beta \int_{A_n} \sum_{k=1}^m a_k \chi_{E_k} \, d\mu = \beta \int_X \sum_{k=1}^m a_k \chi_{E_k \cap A_n} \, d\mu \\ &= \beta \sum_{k=1}^m a_k \mu(E_k \cap A_n). \end{aligned} \tag{17.6}$$

We have $E_k \cap A_n \subseteq E_k \cap A_{n+1}$ for each n, and $\bigcup_{n=1}^\infty E_k \cap A_n = E_k$. We have $\lim_{n\to\infty} \mu(E_k \cap A_n) = \mu(E_k)$ by Proposition 16.3.4 (property 3). Thus, letting $n \to \infty$ in (17.6) yields

$$L \ge \beta \sum_{k=1}^m a_k \mu(E_k) = \beta \int_X s \, d\mu.$$

Since β was arbitrary in $(0, 1)$, $L \ge \int_X s \, d\mu$. Since s was arbitrary with $s \le f$, $L \ge \int_X f \, d\mu$, and the proof is complete. □

Neither of the hypotheses, nonnegative f_n or monotone increasing sequence (f_n), can be dropped from Theorem 17.4.1. If the functions f_n are *not* nonnegative, and f_n increases with limit f, then the conclusion of Theorem 17.4.1 need not hold (Exercise 17.4.1). If the f_n are nonnegative, but are *not increasing* to a limit f, then the conclusion of Theorem 17.4.1 need not hold (Exercise 17.4.2).

We can now prove the additivity of the integral for a measure space (X, Σ, μ). This, combined with Exercise 17.3.1, will complete the proof of the linearity of the integral.

Theorem 17.4.2. *If f and g are nonnegative and measurable on X, or if f and g are real valued and integrable, then*

$$\int_X (f + g) \, d\mu = \int_X f \, d\mu + \int_X g \, d\mu.$$

Proof. The result will be proved first for nonnegative simple functions, then for nonnegative measurable functions, and then for real valued integrable functions.

Suppose f and g are nonnegative simple functions. Then we may write $f = \sum_{j=1}^{m_1} a_j \chi_{A_j}$ where the A_j are pairwise disjoint and have union equal to X, and $g = \sum_{j=1}^{m_2} b_j \chi_{B_j}$ where the B_j are pairwise disjoint and have union equal to X. (Recall that $a_j = 0$ and $b_j = 0$ are possible values of f and g.) Then

$$f + g = \sum_{i=1}^{m_1} \sum_{j=1}^{m_2} (a_i + b_j) \chi_{A_i \cap B_j},$$

and we compute that

$$\begin{aligned} \int_X f + g \, d\mu &= \sum_{i=1}^{m_1} \sum_{j=1}^{m_2} (a_i + b_j) \, \mu(A_i \cap B_j) \\ &= \sum_{i=1}^{m_1} \sum_{j=1}^{m_2} a_i \, \mu(A_i \cap B_j) + \sum_{i=1}^{m_1} \sum_{j=1}^{m_2} b_j \, \mu(A_i \cap B_j) \\ &= \sum_{i=1}^{m_1} a_i \, \mu(A_i) + \sum_{j=1}^{m_2} b_j \, \mu(B_j) \\ &= \int_X f \, d\mu + \int_X g \, d\mu. \end{aligned}$$

Therefore the integral is additive for nonnegative simple functions.

Suppose now that f and g are nonnegative and measurable. There exist sequences s_n and γ_n of nonnegative simple functions such that s_n is increasing with pointwise limit f and γ_n is increasing with pointwise limit g. Then $s_n + \gamma_n$ is a sequence of nonnegative simple functions with limit $f + g$, and by the monotone convergence theorem,

$$\int_X (f + g) \, d\mu = \lim_{n \to \infty} \int_X (s_n + \gamma_n) \, d\mu.$$

By the linearity of the integral for simple functions, the existence of the limits, and the monotone convergence theorem, we have

$$\begin{aligned} \lim_{n \to \infty} \int_X (s_n + \gamma_n) \, d\mu &= \lim_{n \to \infty} \int_X s_n \, d\mu + \lim_{n \to \infty} \int_X \gamma_n \, d\mu \\ &= \int_X f \, d\mu + \int_X g \, d\mu. \end{aligned}$$

Thus the integral is additive for nonnegative measurable functions.

Now suppose that f and g are real valued and integrable. The sum $f + g$ is integrable, since $|f|$ and $|g|$ are nonnegative measurable functions, and additivity for them implies that

$$\int_X |f + g| \, d\mu \leq \int_X (|f| + |g|) \, d\mu = \int_X |f| \, d\mu + \int_X |g| \, d\mu < \infty.$$

Now we use the positive and negative parts of f, g and $f + g$. Write

$$f + g = (f + g)^+ - (f + g)^- \qquad \text{and} \qquad f + g = f^+ - f^- + g^+ - g^-.$$

Then we have

$$(f + g)^+ + f^- + g^- = f^+ + g^+ + (f + g)^-.$$

By the additivity established above for nonnegative measurable functions,

$$\int_X (f+g)^+\,d\mu + \int_X f^-\,d\mu + \int_X g^-\,d\mu = \int_X f^+\,d\mu + \int_X g^+\,d\mu + \int_X (f+g)^-\,d\mu.$$

By Definition 17.3.1 (item 3) and a rearrangement of the previous equation,

$$\begin{aligned} \int_X (f+g)\,d\mu &= \int_X (f+g)^+\,d\mu - \int_X (f+g)^-\,d\mu \\ &= \int_X f^+\,d\mu - \int_X f^-\,d\mu + \int_X g^+\,d\mu - \int_X g^-\,d\mu \\ &= \int_X f\,d\mu + \int_X g\,d\mu, \end{aligned}$$

which proves additivity for real valued integrable functions. □

Corollary 17.4.3. *If the functions f_k, $k \in \mathbf{N}$, are nonnegative and measurable on X, then*

$$\int_X \Big(\sum_{k=1}^{\infty} f_k\Big)\,d\mu = \sum_{k=1}^{\infty} \int_X f_k\,d\mu.$$

Proof. Consider the partial sums, $s_n = \sum_{k=1}^n f_k$. The sequence (s_n) is increasing and $\lim_{n\to\infty} s_n = \sum_{k=1}^{\infty} f_k$. We have

$$\begin{aligned} \int_X \Big(\sum_{k=1}^{\infty} f_k\Big)\,d\mu &= \int_X \Big(\lim_{n\to\infty} \sum_{k=1}^{n} f_k\Big)\,d\mu \\ &= \int_X \Big(\lim_{n\to\infty} s_n\Big)\,d\mu \\ &= \lim_{n\to\infty} \int_X s_n\,d\mu, \end{aligned}$$

by the monotone convergence theorem. We also have

$$\begin{aligned} \lim_{n\to\infty} \int_X s_n\,d\mu &= \lim_{n\to\infty} \int_X \Big(\sum_{k=1}^{n} f_k\Big)\,d\mu \\ &= \lim_{n\to\infty} \sum_{k=1}^{n} \int_X f_k\,d\mu \\ &= \sum_{k=1}^{\infty} \int_X f_k\,d\mu, \end{aligned}$$

by the linearity of the integral and the definition of series sum. □

Theorem 17.4.4 (Fatou's Lemma). *If the functions f_n are nonnegative and measurable on X, then*

$$\int_X \liminf f_n\,d\mu \le \liminf \int_X f_n\,d\mu.$$

Proof. The proof uses only the definition of the lim inf of a sequence of real numbers and the monotone convergence theorem. Let $g_n(x) = \inf_{k\ge n} f_k(x)$ for all x. Then the functions g_n are nonnegative, they satisfy $g_n \le g_{n+1}$ for all n, and

$\lim_{n\to\infty} g_n(x) = \liminf f_n$ for all x. By definition, $g_n \leq f_k$ for all $k \geq n$, so $\int_X g_n\,d\mu \leq \int_X f_k\,d\mu$ for all $k \geq n$. It follows that

$$\int_X g_n\,d\mu \leq \inf_{k\geq n} \int_X f_k\,d\mu. \tag{17.7}$$

Let $n \to \infty$ in (17.7). The monotone convergence theorem gives the limit of the left-hand side of (17.7), and the definition of lim inf gives the limit of the right-hand side. Thus,

$$\int_X \liminf f_n\,d\mu \leq \liminf \int_X f_n\,d\mu,$$

which proves Fatou's lemma. □

As an immediate consequence of Fatou's lemma, if $\liminf \int_X f_n\,d\mu < \infty$, then $f := \liminf f_n$ is an integrable function on X. See Exercise 17.4.3 for an application.

Theorem 17.4.5 (Dominated Convergence). *Suppose the functions f_n are measurable on X and $f_n(x) \to f(x)$ for all $x \in X$. If there exists a nonnegative integrable function g on X such that $|f_n(x)| \leq g(x)$ for all $x \in X$, then f is integrable and*

$$\lim_{n\to\infty} \int_X f_n\,d\mu = \int_X f\,d\mu.$$

Proof. The function $g - f_n$ is measurable and nonnegative, and we have

$$\lim_{n\to\infty} [g(x) - f_n(x)] = g(x) - f(x)$$

for all x. By Fatou's lemma (Theorem 17.4.4),

$$\int_X \liminf (g - f_n)\,d\mu = \int_X (g - f)\,d\mu \leq \liminf \int_X (g - f_n)\,d\mu.$$

Since $f_n \to f$ pointwise, we have $|f_n(x)| \to |f(x)|$, and hence $|f(x)| \leq g(x)$ for all x. This proves that $|f|$ is integrable, hence f is integrable. Then

$$\begin{aligned}
\int_X g\,d\mu - \int_X f\,d\mu &\leq \liminf \int_X (g - f_n)\,d\mu \\
&= \int_X g\,d\mu + \liminf \Big(- \int_X f_n\,d\mu \Big) \\
&= \int_X g\,d\mu - \limsup \int_X f_n\,d\mu.
\end{aligned}$$

Hence, by a simple rearrangement,

$$\int_X f\,d\mu \geq \limsup \int_X f_n\,d\mu.$$

By considering $g + f_n$, Fatou's lemma yields

$$\int_X g\,d\mu + \int_X f\,d\mu \leq \liminf \int_X (g + f_n)\,d\mu = \int_X g\,d\mu + \liminf \int_X f_n\,d\mu,$$

and therefore

$$\int_X f\,d\mu \leq \liminf \int_X f_n\,d\mu.$$

It follows that $\liminf \int_X f_n\,d\mu = \limsup \int_X f_n\,d\mu$, and thus, $\lim_{n\to\infty} \int_X f_n\,d\mu$ exists and equals $\int_X f\,d\mu$. □

If we think for a moment in terms of functions defined on the real line, the hypothesis in Theorem 17.4.5 that $g \geq 0$ and integrable ensures that the area under the graph of g has finite area, and the hypothesis that $|f_n(x)| \leq g(x)$ for all x and all n ensures that the graphs of $|f_n|$ are trapped inside the region under the graph of g. Simple examples show that such hypotheses are needed to ensure the conclusion that the integral and limit operations can be interchanged. For example, consider the functions f_n on $\mathbf{R}$, for integers $n \geq 0$, given by

$$f_n(x) = \begin{cases} 1 & \text{for } n < x < n+1, \\ 0 & \text{otherwise.} \end{cases}$$

Then f_n converges pointwise to the zero function on $\mathbf{R}$. However, for all n, $\int_{\mathbf{R}} f_n \, dm = 1$, so that

$$\lim_{n\to\infty} \int_{\mathbf{R}} f_n \, dm \neq \int_{\mathbf{R}} (0) \, dm.$$

For this sequence f_n there is no integrable function g such that $|f_n(x)| \leq g(x)$ for all x and all n.

The limit property of the dominated convergence theorem is a considerable improvement over the Riemann integral, as it does not require uniform convergence of the sequence. Consider for example the sequence of functions f_n from Exercise 7.1.3, repeated here for convenience.

Example 17.4.6. Let $\{q_1, q_2, q_3, \ldots\} = \mathbf{Q} \cap [0, 1]$ be an enumeration of the rational numbers in the interval $[0, 1]$. Define $f_n : [0, 1] \to \mathbf{R}$ by

$$f_n(x) = \begin{cases} 0 & \text{if } x \in \{q_1, q_2, q_3, \ldots, q_n\}, \\ 1 & \text{if } x \in [0, 1] - \{q_1, q_2, q_3, \ldots, q_n\}. \end{cases}$$

Then each f_n is measurable on $[0, 1]$, and f_n converges pointwise to the Dirichlet function f, where $f(x) = 0$ if $x \in \mathbf{Q} \cap [0, 1]$ and $f(x) = 1$ if $x \in [0, 1] - \mathbf{Q}$, which is not Riemann integrable. However, $|f_n(x)| \leq g(x) \equiv 1$ for all $x \in [0, 1]$, so f is integrable in the Lebesgue sense, and

$$\lim_{n\to\infty} \int_{[0,1]} f_n \, dm = \int_{[0,1]} f \, dm = 1,$$

by Theorem 17.4.5. △

The fact that the Dirichlet function is Lebesgue integrable is not so important in and of itself; however, the ease with which the Lebesgue integral handles it is impressive, and this function is just one of the many holes in the space of Riemann integrable functions that are now filled due to the Lebesgue integral. We return to this issue in the final section of the chapter when we discuss the completeness of the space of functions that are integrable in the sense of Lebesgue.

Corollary 17.4.7. *Suppose the functions f_k, $k \in \mathbf{N}$, are integrable on X and such that*

$$\sum_{k=1}^{\infty} \int_X |f_k| \, d\mu < \infty.$$

Then the series $\sum_{k=1}^{\infty} f_k$ *converges absolutely a.e. in* X*, the sum is integrable on* X*, and*

$$\int_X \Big(\sum_{k=1}^{\infty} f_k\Big)\, d\mu = \sum_{k=1}^{\infty} \int_X f_k \, d\mu.$$

Proof. Let $g = \sum_{k=1}^{\infty} |f_k| = \lim_{n\to\infty} \sum_{k=1}^{n} |f_k|$. By Corollary 17.4.3 of the monotone convergence theorem, we have

$$\int_X g \, d\mu = \lim_{n\to\infty} \sum_{k=1}^{n} \int_X |f_k| \, d\mu = \sum_{k=1}^{\infty} \int_X |f_k| \, d\mu < \infty,$$

using the present hypothesis. Therefore g is integrable. This implies, in particular, that g must be finite almost everywhere on X. Hence, the series $\sum_{k=1}^{\infty} f_k$ converges absolutely almost everywhere on X.

Let $s_n = \sum_{k=1}^{n} f_k$. Then we have

$$|s_n| \le \sum_{k=1}^{n} |f_k| \le \sum_{k=1}^{\infty} |f_k| = g.$$

The dominated convergence theorem applies to the sequence s_n, and we conclude that $\lim_{n\to\infty} s_n$ is integrable and

$$\int_X \Big(\sum_{k=1}^{\infty} f_k\Big)\, d\mu = \int_X \Big(\lim_{n\to\infty} s_n\Big)\, d\mu = \lim_{n\to\infty} \int_X s_n \, d\mu = \lim_{n\to\infty} \sum_{k=1}^{n} \int_X f_k \, d\mu,$$

as we wished to show. □

We expect the integral of a function f to be countably additive on disjoint sets. Indeed, for any measure space (X, Σ, μ), the integral of a nonnegative measurable function f on X defines another measure on (X, Σ).

Theorem 17.4.8. *Let* (X, Σ, μ) *be a measure space and let* f *be a nonnegative measurable function on* X*. Suppose* $\int_A f \, d\mu < \infty$ *for some set* $A \in \Sigma$*. Then the function* $\lambda : \Sigma \to [0, \infty]$ *given by*

$$\lambda(E) = \int_E f \, d\mu$$

defines a measure on (X, Σ)*.*

Proof. We only need to show countable additivity of λ. Let $E = \bigcup_{n=1}^{\infty} E_n$ where the sets E_n are measurable and pairwise disjoint. Let

$$f_n(x) = \sum_{k=1}^{n} f(x) \chi_{E_k}(x), \qquad x \in X.$$

Then the f_n are nonnegative and measurable, $f_n \le f_{n+1}$ for each n, and $\lim_{n\to\infty} f_n(x) = f(x)\, \chi_E(x)$. By Theorem 17.4.1,

$$\lim_{n\to\infty} \int_X f_n \, d\mu = \int_X f\, \chi_E \, d\mu = \int_E f \, d\mu.$$

By the linearity of the integral and the definition of λ,

$$\int_X f_n \, d\mu = \int_X \Big(\sum_{k=1}^{n} f \chi_{E_k}\Big) \, d\mu = \sum_{k=1}^{n} \int_X f \chi_{E_k} \, d\mu = \sum_{k=1}^{n} \lambda(E_k).$$

Hence,

$$\lambda(E) = \int_X f \chi_E \, d\mu = \lim_{n\to\infty} \sum_{k=1}^{n} \lambda(E_k) = \sum_{k=1}^{\infty} \lambda(E_k),$$

as we wished to show. □

The assumption in Theorem 17.4.8 that $\int_A f \, d\mu < \infty$ for some set $A \in \Sigma$ implies that $\lambda(\emptyset) = 0$.

Theorem 17.4.8 allows us to break up a measurable set E into a countable disjoint union of measurable sets, and then integrate f over the separate pieces, summing the results to get $\int_E f \, d\mu$. In particular, if E has measure zero, then

$$\int_X f \, d\mu = \int_E f \, d\mu + \int_{E^c} f \, d\mu = \int_{E^c} f \, d\mu. \tag{17.8}$$

Thus it is clear that the integrability of a function, and the value of the integral of a function, are not affected by altering the function on a set of measure zero. We may even ignore a set of measure zero in many statements about integrals, if convenient. For example, in the monotone convergence theorem we may replace the phrase "$f(x) = \lim_{n\to\infty} f_n(x)$ for all x" with the phrase "$f(x) = \lim_{n\to\infty} f_n(x)$ a.e.". In the dominated convergence theorem, we may replace the phrase "$f_n(x) \to f(x)$ for all x" by "$f_n(x) \to f(x)$ a.e. in X", and the phrase "$|f_n(x)| \le g(x)$ for all x" by "$|f_n(x)| \le g(x)$ a.e. in X".

The next result is another application of dominated convergence.

Theorem 17.4.9. *Suppose f and g are measurable functions on X. If g is integrable and $|f(x)| \le g(x)$ a.e. in X, then f is integrable.*

Proof. It suffices to show that $|f|$ is integrable. Since $|f|$ is measurable, there is an increasing sequence s_n of nonnegative simple functions such that $\lim_{n\to\infty} s_n(x) = |f(x)|$ for all x. Then $0 \le s_n \le g(x)$ a.e. in X. Let $E = \{x \in X : |f(x)| > g(x)\}$. By Exercise 17.3.2,

$$\int_X s_n \, d\mu = \int_{X-E} s_n \, d\mu \le \int_{X-E} g \, d\mu = \int_X g \, d\mu < \infty.$$

Thus, s_n is integrable for each n. By Theorem 17.4.5, $|f|$ is integrable. □

In the following theorem, we use (17.8) implicitly when $\mu(E) = 0$.

Theorem 17.4.10. *Let f and g be real valued measurable functions on a measure space (X, Σ, μ). Then the following properties hold:*

1. *If $f \ge 0$ a.e. in X, then $\int_X f \, d\mu \ge 0$.*
2. *If $f \le g$ a.e. in X, then $\int_X f \, d\mu \le \int_X g \, d\mu$.*

3. *If b_1 and b_2 are real constants such that $b_1 \leq f(x) \leq b_2$ a.e. in a measurable set E with $\mu(E) < \infty$, then*

$$b_1\, \mu(E) \leq \int_E f\, d\mu \leq b_2\, \mu(E).$$

4. $\int_X |f+g|\, d\mu \leq \int_X |f|\, d\mu + \int_X |g|\, d\mu$.
5. *If f is integrable, then $|f|$ is integrable, and $|\int_X f\, d\mu| \leq \int_X |f|\, d\mu$.*

Proof. 1. If $f \geq 0$ a.e. in X, then for any simple function s with $0 \leq s \leq f$, we have $\int_X f\, d\mu \geq \int_X s\, d\mu \geq 0$ by (17.3).

Statement 2 follows from statement 1 by considering $g - f$ and using additivity.

Statement 3 follows from statement 2 and the integral of constant simple functions.

Statement 4 follows from the triangle inequality, $|f(x)+g(x)| \leq |f(x)|+|g(x)|$, the estimate in statement 3, and additivity of the integral.

5. If f is integrable, the integrability of $|f|$ is immediate from Definition 17.3.2. Since $-f, f \leq |f|$, we have $-\int_X f\, d\mu \leq \int_X |f|\, d\mu$ and $\int_X f\, d\mu \leq \int_X |f|\, d\mu$ by statement 2, and thus $|\int_X f\, d\mu| \leq \int_X |f|\, d\mu$. $\square$

The theory of the integral developed here can be extended to complex valued functions on a measure space. See Exercise 17.4.12 for a quick look at complex valued functions on a subset of the real line with Lebesgue measure.

Exercises.

Exercise 17.4.1. Let $X = (0, \infty)$ with Lebesgue measure. Let $f_n(x) = -1/n$ for all $x \in (0, \infty)$. Show that the conclusion of Theorem 17.4.1 does not hold.

Exercise 17.4.2. Let $X = (0, \infty)$ with Lebesgue measure. Let $f_n = n\chi_{(0,1/n)}$. Show that the conclusion of Theorem 17.4.1 does not hold.

Exercise 17.4.3. Suppose E is a measurable set and the functions f_n are measurable on E with $\lim_{n\to\infty} f_n(x) = f(x)$, $x \in E$. Show that if $\sup_n \int_E |f_n|\, d\mu \leq M < \infty$, then $\int_E |f|\, d\mu \leq M$. *Hint*: Use Fatou's lemma.

Exercise 17.4.4. Show that Fatou's lemma implies the monotone convergence theorem.

Exercise 17.4.5. Suppose $f : \mathbf{R} \to \mathbf{R}$ is an integrable function. Show that

$$\int_{\mathbf{R}} f\, dm = \lim_{n\to\infty} \int_{\mathbf{R}} f(x) e^{-|x|^2/n}\, dx.$$

Exercise 17.4.6. Let $f_n = \frac{1}{n}\chi_{(0,n)}$.

1. Show that $f_n \to 0$ uniformly on $\mathbf{R}$, although $\int_{\mathbf{R}} f_n\, dm = 1$ for each n. Why does this example not contradict Theorem 17.4.5?
2. Compare $\liminf_{n\to\infty} \int_{\mathbf{R}} f_n\, dm$ and $\int_{\mathbf{R}} \liminf_{n\to\infty} f_n\, dm$.

Exercise 17.4.7. Prove: If f is a measurable function on $E \subseteq X$ and $f = 0$ a.e. on E, then $\int_E f\, d\mu = 0$. If f and g are measurable on E and $f = g$ a.e. on E, then $\int_E f\, d\mu = \int_E g\, d\mu$.

Exercise 17.4.8. Prove: If f is a nonnegative measurable function on $E \subseteq X$ and $\int_E f\,d\mu = 0$, then $f = 0$ a.e. on E. *Hint*: Write $P = \{x \in E : f(x) > 0\} = \bigcup_{n=1}^{\infty} P_n$, where $P_n = \{x \in E : f(x) > 1/n\}$.

Exercise 17.4.9. Suppose $f : X \to \mathbf{R}$ is measurable. Show that f is integrable if and only if $|f|$ is integrable. Show that f is integrable if and only if f^+ and f^- are integrable.

Exercise 17.4.10. Suppose f and g are real valued functions on X. If f is integrable and g is measurable, and $g = f$ a.e. in X, then g is integrable and $\int_X g\,d\mu = \int_X f\,d\mu$.

Exercise 17.4.11. Let $(\mathbf{R}^n, \mathcal{M}_n, m_n)$ be the Lebesgue measure space and suppose $f : \mathbf{R}^n \to \mathbf{R}$ is an integrable function. Prove: If E is a measurable set and $f \geq \delta > 0$ on E, then $m(E) < \infty$.

Exercise 17.4.12. Let $f = u + iv$ be a complex valued function defined on the measure space $(\mathbf{R}, \mathcal{M}, m)$, where u and v are the real and imaginary parts of f. We say that f is a **complex valued measurable function** if the real part u and imaginary part v are measurable functions. If $f = u + iv$ is measurable, we say that f is **integrable** over $\mathbf{R}$ if u and v are integrable over $\mathbf{R}$, and then the **Lebesgue integral** of f is defined by

$$\int_{\mathbf{R}} f\,dm = \int_{\mathbf{R}} u\,dm + i \int_{\mathbf{R}} v\,dm.$$

1. Suppose u and v are real valued measurable functions on $\mathbf{R}$. Show that $f = u + iv$ is integrable if and only if $|f|$ is integrable.
2. Prove the linearity of the integral for complex valued integrable functions on $\mathbf{R}$, using complex scalars.

Exercise 17.4.13. Suppose $f : \mathbf{R} \to \mathbf{R}$ is integrable and bounded. Show that for every $\epsilon > 0$ there is a $\delta = \delta(\epsilon) > 0$ such that if A is any set with $m(A) < \delta$, then $\int_A |f|\,dm < \epsilon$ and hence $|\int_A f\,dm| < \epsilon$. (This property is called the **absolute continuity** of the integral.)

Exercise 17.4.14. 1. Suppose $f : \mathbf{R} \to \mathbf{R}$ is nonnegative and measurable. Let

$$f_n(x) = \begin{cases} f(x) & \text{if } f(x) \leq n, \\ n & \text{if } f(x) > n. \end{cases}$$

Show that $\lim_{n\to\infty} \int_{\mathbf{R}} f_n\,dm = \int_{\mathbf{R}} f\,dm$. *Hint*: Note that $f_n = \min(f, n)$.

2. Show that a similar result holds if f is a general integrable function, and we define functions f_n by

$$f_n(x) = \begin{cases} f(x) & \text{if } |f(x)| \leq n \text{ and } |x| \leq n, \\ 0 & \text{otherwise.} \end{cases}$$

Exercise 17.4.15. Suppose (X, Σ, μ) is a measure space and $E \subseteq X$ is a measurable set. If $(f_n)_{k=1}^{\infty}$ is a decreasing sequence of nonnegative integrable functions on E with pointwise limit f on E, then $\lim_{n\to\infty} \int_E f_n\,dm = \int_E f\,dm$.

17.5. Comparison with the Riemann Integral

The goal of this section is to show that a Riemann integrable function on $[a, b]$ is measurable and Lebesgue integrable on $[a, b]$, and the two integrals have the same value. Thus the Lebesgue integral is a true extension of the Riemann integral.

We consider bounded functions $f : [a, b] \to \mathbf{R}$. If f is Riemann integrable over $[a, b]$, we denote its Riemann integral as usual by $\int_a^b f(x)\, dx$. If f is measurable and Lebesgue integrable over $[a, b]$, we indicate its Lebesgue integral by $\int_{[a,b]} f\, dm$.

Recall that a step function $s : \mathbf{R} \to \mathbf{R}$ is a simple function $s = \sum_{k=1}^{m} a_k \chi_{E_k}$ where each set E_k is an interval. Thus, every step function is a Lebesgue measurable simple function. With this in view, it is immediate from the definitions that

$$\int_a^b s(x)\, dx = \int_{[a,b]} s\, dm$$

for every step function s on $[a, b]$.

Theorem 17.5.1. *If $f : [a, b] \to \mathbf{R}$ is Riemann integrable over $[a, b]$, then f is measurable and Lebesgue integrable, and*

$$\int_a^b f(x)\, dx = \int_{[a,b]} f\, dm.$$

Proof. Suppose $f : [a, b] \to \mathbf{R}$ is Riemann integrable. Then f is bounded on $[a, b]$ and continuous on $[a, b] - E$, where the set E of discontinuities of f has measure zero, $m(E) = 0$. Write $f = f^+ - f^-$, where

$$f^+(x) = \max(f(x), 0) \quad \text{and} \quad f^-(x) = \max(-f(x), 0)$$

are the positive and negative parts of f, respectively. Let E^+ be the set of discontinuities of f^+ and let E^- be the set of discontinuities of f^-. Then $E^+ \subseteq E$ and $E^- \subseteq E$, so $m(E^+) = 0$ and $m(E^-) = 0$. Thus f^+ and f^- are nonnegative Riemann integrable functions.

Now, the definition of the Lebesgue integral of f requires that f be measurable and that

$$\int_{[a,b]} f\, dm = \int_{[a,b]} f^+\, dm - \int_{[a,b]} f^-\, dm,$$

where at least one of the integrals on the right-hand side is finite. Thus our task is to establish the theorem statement for nonnegative Riemann integrable functions.

Consider the positive part f^+. By definition of the Riemann integral,

$$\int_a^b f^+(x)\, dx = \sup_{s \le f^+} \int_{[a,b]} s\, dm = \inf_{t \ge f^+} \int_{[a,b]} t\, dm,$$

where s and t are step functions. If we know that f^+ is measurable, then, since every step function is a simple function, we conclude that

$$\sup_{s \le f^+} \int_{[a,b]} s\, dm \le \sup_{\phi \le f^+} \int_{[a,b]} \phi\, dm \le \inf_{\psi \ge f^+} \int_{[a,b]} \phi\, dm \le \inf_{t \ge f^+} \int_{[a,b]} t\, dm, \tag{17.9}$$

where the inner supremum and infimum are over simple functions ϕ and ψ, and it is the supremum over simple $\phi \le f^+$ that defines $\int_{[a,b]} f^+\, dm$, *provided* f^+ is

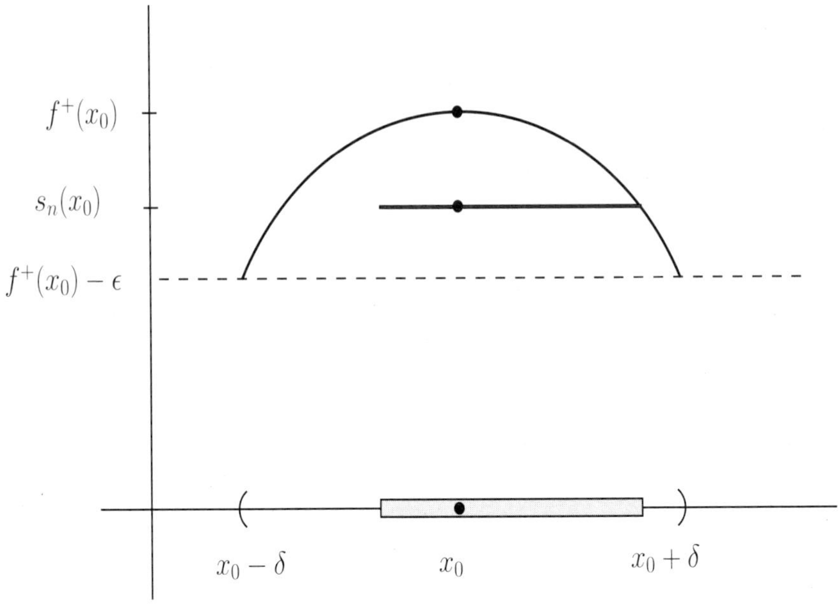

Figure 17.1. A shaded subinterval of a partition P_n of $[a, b]$, on which the step function s_n equals the infimum of f^+ over the subinterval. As $n \to \infty$, the subinterval length for the partitions P_n approaches zero, and $\lim_{n\to\infty} s_n(x_0) = f^+(x_0)$, if f^+ is continuous at x_0.

measurable. Thus, to conclude that f^+ is integrable in the Lebesgue sense and the theorem statement is true, we only need to establish that f^+ is measurable.

Let P_n be a sequence of partitions of $[a, b]$ using equal length subintervals, such that the mesh of the partition (the length of the subintervals) goes to zero as $n \to \infty$. For each n, let s_n be the step function defined by this partition such that $s_n(x) = \inf_{x\in J} f^+(x)$, where J is any subinterval of the partition P_n. We claim that s_n converges pointwise to f^+ at every point of continuity of f^+. In order to prove this claim, suppose that x_0 is a point of continuity of f^+. Then given $\epsilon > 0$, there exists $\delta(\epsilon) > 0$ such that if $|x - x_0| < \delta(\epsilon)$, then $0 \le f^+(x_0) - s_n(x_0) < \epsilon$. As $n \to \infty$, the sequence $s_n(x_0)$ increases, and if x_0 lies within interval J of partition P_n, then

$$s_n(x_0) = \inf_{x\in J} f^+(x) \le f^+(x_0)$$

for all n. (See Figure 17.1.) Hence, $\lim_{n\to\infty} s_n(x_0) = f^+(x_0)$ when f^+ is continuous at x_0. Therefore s_n converges to f^+ pointwise outside its set E^+ of discontinuities, hence almost everywhere in $[a, b]$, since f^+ is Riemann integrable.

Now the step functions s_n are measurable, and by Proposition 17.1.12, their pointwise limit a.e., f^+, is measurable. Hence, Riemann integrability and measurability of f^+, combined with (17.9), imply that

$$\int_a^b f^+(x)\,dx = \sup_{\phi\le f^+} \int_{[a,b]} \phi\,dm = \int_{[a,b]} f^+\,dm.$$

The argument is the same for f^-, and thus

$$\int_a^b f^-(x)\,dx = \sup_{\phi \le f^-} \int_{[a,b]} \phi\,dm = \int_{[a,b]} f^-\,dm.$$

Hence, we have

$$\begin{aligned} \int_a^b f(x)\,dx &= \int_a^b (f^+(x) - f^-(x))\,dx \\ &= \int_a^b f^+(x)\,dx - \int_a^b f^-(x)\,dx \\ &= \int_{[a,b]} f^+\,dm - \int_{[a,b]} f^-\,dm \\ &= \int_{[a,b]} f\,dm, \end{aligned}$$

as desired. □

See Exercise 17.5.1 for a different proof that a Riemann integrable function is Lebesgue measurable.

For unbounded functions, or for unbounded intervals of integration, the Riemann integral is extended by means of improper integrals, when they are convergent. For nonnegative functions on finite intervals, a convergent improper integral agrees with the Lebesgue integral; see Exercise 17.5.2 for a typical case.

The improper integral of a function on $\mathbf{R}$ which is not nonnegative may exist without the function being Lebesgue integrable; an example is given in Exercise 17.5.3. On the other hand, functions which are absolutely integrable on $\mathbf{R}$ in either the Riemann or the Lebesgue sense are assigned the same integrals by these theories; for precise statements, see Exercises 17.5.4, 17.5.5.

Finally, if J is an interval in $\mathbf{R}^n$, then we define a function $\mathbf{F} : J \to \mathbf{R}^m$ to be Lebesgue measurable if and only if each real component function is measurable with respect to the n-dimensional Lebesgue σ-algebra $\mathcal{M}_n$ (Definition 17.1.1). We define $\mathbf{F}$ to be Lebesgue integrable if and only if each real component function is Lebesgue integrable with respect to n-dimensional Lebesgue measure m_n (Definition 17.3.2). We recall that a function $\mathbf{F} : J \to \mathbf{R}^m$ is Riemann integrable if and only if each real component function is Riemann integrable (Definition 12.1.2). Using the techniques we have been working with, it is possible to show that every Riemann integrable $\mathbf{F} : J \to \mathbf{R}^m$ is Lebesgue measurable and integrable, and the two integrals agree.

In the final section of this chapter, we define the normed vector space of Lebesgue integrable functions on a measure space (X, Σ, μ) and show that it is a complete normed space, a Banach space. By the result of the present section, in the case of $X = [a, b]$ with Lebesgue measure, this complete space contains $\mathcal{R}[a, b]$.

Exercises.

Exercise 17.5.1. Here is a different proof that a Riemann integrable function f on $[a, b]$ is Lebesgue measurable. Let V be any open subset of $\mathbf{R}$. Show that $f^{-1}(V)$ is measurable as follows:

1. Let $G = \{x \in [a, b] : f \text{ is continuous at } x\}$. Show that for each $x \in G$, there is an open interval U_x such that $x \in U_x \subset f^{-1}(V)$.
2. Let $U = \bigcup_{x \in G} U_x$. Show that $f^{-1}(V) - U \subseteq [a, b] - G$.
3. Conclude that $f^{-1}(V)$ is measurable.

Exercise 17.5.2. 1. Suppose f is nonnegative on $[a, b]$ and Riemann integrable on every subinterval $[a + \epsilon, b]$, $\epsilon > 0$. Suppose that the improper integral

$$I = \lim_{\epsilon \to 0} \int_{a+\epsilon}^{b} f(x)\, dx$$

exists. Show that f is Lebesgue integrable on $[a, b]$ and $\int_{[a,b]} f\, dm = I$.

2. Compute $\int_{[0,1]} f\, dm$ if $f(x) = 1/\sqrt{x}$ for $0 < x \le 1$, and $f(0) = 1$.

Exercise 17.5.3. Show that the function $f(x) = \sin x/x$ is not Lebesgue integrable on $(1, \infty)$, but the improper Riemann integral exists. *Hint*: Let $u = 1/x$ and $dv = \sin x\, dx$.

Exercise 17.5.4. Show that if f is Lebesgue integrable on $(-\infty, \infty)$ and if the improper Riemann integral $\lim_{n\to\infty} \int_{-n}^{n} f(x)\, dx$ exists, then

$$\int_{\mathbf{R}} f\, dm = \lim_{n\to\infty} \int_{-n}^{n} f(x)\, dx.$$

Hint: Consider $f_n^+ = f^+ \chi_{[-n,n]}$, and similarly for f^-.

Exercise 17.5.5. Let f be a bounded Riemann integrable function on bounded intervals of $\mathbf{R}$, and suppose that $|f|$ is Riemann integrable in the sense that

$$\int_{-\infty}^{\infty} |f(x)|\, dx := \lim_{n\to\infty} \int_{-n}^{n} |f(x)|\, dx < \infty.$$

Show that f is Lebesgue integrable on $(-\infty, \infty)$, and

$$\int_{\mathbf{R}} f\, dm = \int_{-\infty}^{\infty} f(x)\, dx.$$

17.6. Banach Spaces of Integrable Functions

In this section we define the normed vector space of integrable functions on a measure space (X, Σ, μ) and show that it is a complete normed space, a Banach space.

Definition 17.6.1. *Let (X, Σ, μ) be a measure space. We denote by $L^1(X, \Sigma, \mu)$ the set of functions $f : X \to \mathbf{R}$ such that f is measurable and $\int_X |f|\, d\mu < \infty$, with the identification that $f = g$ if and only if the set of points x where $f(x) \neq g(x)$ is a set of measure zero.*

Thus the elements of $L^1((X, \Sigma, \mu))$ are actually equivalence classes of functions; however, we operate in practice with individual functions, always keeping in mind the equivalence involved in the definition. Recalling that a function $f : X \to \mathbf{R}$ is Lebesgue integrable on X if f is measurable and $\int_X |f|\, d\mu < \infty$ (Definition 17.3.2), we see that $L^1(X, \Sigma, \mu)$ is the set of Lebesgue integrable functions on the measure space.

Theorem 17.6.2. *The vector space $L^1(X, \Sigma, \mu)$ of Lebesgue integrable functions is a normed space with norm*

$$\|f\|_1 = \int_X |f|\, d\mu$$

for $f \in L^1(X, \Sigma, \mu)$.

Proof. If f and g are in $L^1(X, \Sigma, \mu)$ and α, β are real numbers, then

$$\begin{aligned}
\int_X |\alpha f + \beta g|\, d\mu &\leq \int_X (|\alpha|\,|f| + |\beta|\,|g|)\, d\mu \\
&= \int_X |\alpha|\,|f|\, d\mu + \int_X |\beta|\,|g|\, d\mu \\
&= |\alpha| \int_X |f|\, d\mu + |\beta| \int_X |g|\, d\mu < \infty,
\end{aligned}$$

by the linearity of the integral. Thus, $\alpha f + \beta g \in L^1(X, \Sigma, \mu)$. It follows easily that $L^1(X, \Sigma, \mu)$ is a vector space.

Clearly $\|f\|_1 \geq 0$ for every $f \in L^1(X, \Sigma, \mu)$, and we have seen that $\|f\|_1 = 0$ if and only if $f = 0$ almost everywhere on X, hence, if and only if f is a representative of the zero equivalence class in $L^1(X, \Sigma, \mu)$. We also see easily that $\|\alpha f\|_1 = |\alpha|\,\|f\|_1$. The proof of the triangle inequality appears in the estimate above by taking $\alpha = 1$ and $\beta = 1$. □

A sequence (f_k) in $L^1(X, \Sigma, \mu)$ converges to f in the L^1 norm if

$$\lim_{k\to\infty} \|f_k - f\|_1 = 0.$$

Theorem 17.6.3. *$L^1(X, \Sigma, \mu)$ is a Banach space, complete in the norm $\|\cdot\|_1$.*

Proof. For simplicity in the proof, we denote by L^1 the space $L^1(X, \Sigma, \mu)$.

Let (f_k) be a Cauchy sequence in L^1. There is an n_1 such that if $m, n \geq n_1$, then $\|f_m - f_n\|_1 < 1/4$. There is an $n_2 > n_1$ such that if $m, n \geq n_2$, then $\|f_m - f_n\|_1 < 1/8$. Continuing in this way, we inductively define $n_k > n_{k-1}$ such that if $m, n \geq n_k$, then $\|f_m - f_n\|_1 < 1/2^{k+1}$. The indices so chosen thus define a subsequence (f_{n_k}) of our original Cauchy sequence (f_k).

We now show that the subsequence (f_{n_k}) converges pointwise almost everywhere in X and the limit is an integrable function. By construction of the subsequence, we have

$$\|f_{n_{k+1}} - f_{n_k}\|_1 < \frac{1}{2^{k+1}}.$$

Let $g_1 = f_{n_1}$, and for $k \geq 2$, let $g_k = f_{n_k} - f_{n_{k-1}}$. Then for each k, $g_k \in L^1$, $\|g_k\|_1 < 1/2^k$ and

$$f_{n_k} = \sum_{j=1}^{k} g_j. \tag{17.10}$$

Now consider the series $\sum_{k=1}^{\infty} g_k$. For the integrals of the g_k, we have

$$\sum_{k=1}^{\infty} \int_X |g_k|\, d\mu = \sum_{k=1}^{\infty} \|g_k\|_1 < \sum_{k=1}^{\infty} \frac{1}{2^k} < \infty.$$

By Corollary 17.4.7 of the dominated convergence theorem, the series $\sum_{k=1}^{\infty} g_k$ converges absolutely almost everywhere in X and the sum is integrable on X. By (17.10), we may write this sum as

$$\lim_{k\to\infty} f_{n_k} = \sum_{j=1}^{\infty} g_j$$

almost everywhere in X. Let f be the pointwise limit of (f_{n_k}) where the sequence converges, and define f to be zero on the complementary set of measure zero. Then f is integrable, as noted, by Corollary 17.4.7.

We want to show that our Cauchy sequence f_k converges to f in norm in L^1. It suffices to show that the subsequence (f_{n_k}) converges to f in norm (Exercise 17.6.1). Given $\epsilon > 0$, there is an n_0 such that if $m, k \geq n_0$, then $\|f_m - f_k\|_1 < \epsilon$, and since the subsequence indexing has $n_m \geq m$ and $n_k \geq k$, it follows that if $m, k \geq n_0$, then

$$\|f_{n_m} - f_{n_k}\|_1 = \int_X |f_{n_m} - f_{n_k}|\, d\mu < \epsilon.$$

Hold k fixed, and let $m \to \infty$, to obtain by Fatou's lemma,

$$\begin{aligned}
\|f - f_{n_k}\|_1 &= \int_X |f - f_{n_k}|\, d\mu \\
&= \int_X \liminf_m |f_{n_m} - f_{n_k}|\, d\mu \\
&\leq \liminf_m \int_X |f_{n_m} - f_{n_k}|\, d\mu \ \leq\ \epsilon.
\end{aligned}$$

Thus, given $\epsilon > 0$, there is an n_0 such that if $k \geq n_0$, then $\|f - f_{n_k}\|_1 \leq \epsilon$. Therefore the subsequence (f_{n_k}) converges to f in norm. Hence the Cauchy sequence (f_k) itself converges to f in norm. This proves the completeness of $L^1 = L^1(X, \Sigma, \mu)$. $\square$

Theorem 17.6.3 covers the space of most interest to us, the Lebesgue measure space of the interval $[a, b]$ of the real number line, which we denote simply by $L^1[a, b]$. It also covers the spaces $L^1(\mathbf{R}, \mathcal{M}, m)$ and $L^1(\mathbf{R}^n, \mathcal{M}_n, m_n)$, as well as the Lebesgue measure spaces defined for any fixed measurable subset of $\mathbf{R}$ or $\mathbf{R}^n$. All of these spaces are Banach spaces, by Theorem 17.6.3.

An important part of the proof of Theorem 17.6.3 is the analysis of the pointwise convergent subsequence (f_{n_k}). However, we should be careful not to read too much into the existence of such a subsequence. The original Cauchy sequence (f_k) need not converge pointwise at any point. Consider the following example.

Example 17.6.4. Let $X = [0,1]$. For each n, we divide $[0,1]$ into n subintervals of equal length $1/n$. For fixed n, the subintervals are

$$[k/n, (k+1)/n], \quad 0 \le k \le n-1.$$

Now define a sequence of functions $f_j \in L^1[0,1]$ as follows: Let f_1 be the characteristic function of $[0,1]$; let

$$f_2 = \chi_{[0,1/2]} \quad \text{and} \quad f_3 = \chi_{[1/2,1]};$$

then let

$$f_4 = \chi_{[0,1/3]}, \quad f_5 = \chi_{[1/3,2/3]}, \quad f_6 = \chi_{[2/3,1]},$$

and so on. Then $f_j \to 0$ in norm in $L^1[0,1]$, since $\|f_j\|_1 = 1/n$ if f_j is the characteristic function of an interval of length $1/n$. However, the sequence f_j does not converge pointwise anywhere in $[0,1]$. To see this, note that any given x in $[0,1]$ will lie outside infinitely many of the subintervals $[k/n, (k+1)/n]$ as $n \to \infty$, hence $f_j(x) = 0$ for infinitely many j. However, x will also lie within infinitely many of the subintervals $[k/n, (k+1)/n]$ as $n \to \infty$, hence $f_j(x) = 1$ for infinitely many j. Therefore $\lim_{j\to\infty} f_j(x)$ does not exist for any $x \in [0,1]$. Nevertheless, as in the proof of Theorem 17.6.3, there exists a subsequence of (f_j) that converges pointwise almost everywhere in $[0,1]$ (Exercise 17.6.2). △

Exercise 17.6.3 provides an example to show that pointwise convergence does not imply convergence in the L^1 norm.

The final chapter of the book returns to some ideas of inner product spaces and Fourier series and applies the ideas of this chapter to the study of Hilbert space and orthonormal sets.

Exercises.

Exercise 17.6.1. Prove: If (f_k) is a Cauchy sequence in a normed space V, and a subsequence (f_{n_k}) converges in norm to an element f of V, then (f_k) converges to f in norm. *Hint*: Recall the proof of Theorem 2.7.2.

Exercise 17.6.2. Identify a subsequence of the sequence (f_j) of Example 17.6.4 that converges pointwise almost everywhere (or everywhere) in $[0,1]$.

Exercise 17.6.3. Let $X = (0,1)$ with Lebesgue measure. Consider the functions $f_k(x) = k^2\chi_{(0,1/k)}$, $k \in \mathbf{N}$. Show that $f_k \in L^1\big((0,1)\big)$ for each k and $f_k \to 0$ pointwise on $(0,1)$, but f_k does not converge to zero in the L^1 norm.

17.7. Notes and References

The presentation of the Lebesgue integral in this chapter was influenced by Friedman [**17**], Royden [**51**], Wheeden and Zygmund [**67**], and Bass [**3**].

Fubini's theorem for Lebesgue integrable functions, not mentioned in the text, requires product measures and can be found in Folland [**15**], Kolmogorov and Fomin [**36**], Rudin [**53**], or Wheeden and Zygmund [**67**].

For a comprehensive introduction to Lebesgue measure and integration motivated by the historical questions that led to their development, see the excellent text by Bressoud [**8**]. The work of H. Lebesgue (1875-1941) on the integral began to appear in 1900, and quickly began to influence work in mathematical analysis. See Lebesgue [**43**] for his own informal account of his ideas. Lebesgue measure and integration theory became the widely recognized foundation of modern probability theory with the systematic presentation by A. N. Kolmogorov in 1933; the original German presentation was followed by a Russian translation in 1936, and the first English translation in 1950. See Kolmogorov [**35**].

What about the fundamental theorem of calculus in the context of the Lebesgue integral? Definitive results on this topic involve the concepts of *functions of bounded variation* and *absolutely continuous functions*, and this is left for the reader to explore; see Bressoud [**8**], Folland [**15**], Gariepy and Ziemer [**19**], Kolmogorov and Fomin [**36**], or Royden [**51**].

Chapter 18

Inner Product Spaces and Fourier Series

The basic concepts of inner product spaces appear in Section 8.2 in the discussion of the Euclidean metric space $\mathbf{R}^n$. Several important inner product spaces were introduced there, including $C[a,b]$, the space of continuous real valued functions on $[a,b]$, and $\mathcal{R}[a,b]$, the space of (equivalence classes of) Riemann integrable functions on $[a,b]$. The Cauchy-Schwarz inequality (Theorem 8.2.3) for an inner product space is proved there. A formal definition of norm for vector spaces and an introduction to the norm induced by an inner product appear in Section 8.3. Section 8.4 introduced orthogonal sets and orthogonal expansion of vectors in $\mathbf{R}^n$.

This chapter begins with examples of orthonormal sets and then presents the basic theory of orthonormal (Fourier) expansions. We examine the role of orthonormal sets in important spaces introduced earlier and highlight the importance of complete inner product spaces. In the last section, which is based on the Lebesgue integral, we show, in particular, that the Lebesgue space of square integrable functions on an interval is a complete inner product space, a Hilbert space. In particular, the sequence space l^2 and the Lebesgue function space $L^2([-\pi,\pi],\mathcal{M},m)$ are isometrically isomorphic Hilbert spaces.

18.1. Examples of Orthonormal Sets

Since our main interest now is in function spaces, we use the common function symbols, such as f, g for general elements in the space. In an inner product space V, we have the norm $\|f\| = \sqrt{(f,f)}$ for f in V. We begin by recalling the definition of an orthonormal set.

Definition 18.1.1. *Let V be an inner product space. A set X of vectors in V is an* **orthogonal set** *if the elements of X are pairwise orthogonal, that is, $(f,g)=0$ for any two distinct elements f and g in X. A set X is an* **orthonormal set** *if X is an orthogonal set and $\|f\|=1$ for every f in X.*

We begin with important examples of orthonormal sets in $\mathbf{R}^n$ and l^2.

Example 18.1.2. The standard basis vectors $\mathbf{e}_1, \ldots, \mathbf{e}_n$ in $\mathbf{R}^n$ form an orthonormal set in $\mathbf{R}^n$. The full set of these n vectors is a basis for $\mathbf{R}^n$, hence an orthonormal basis for $\mathbf{R}^n$. Any proper subset of this basis, say $\{\mathbf{e}_1, \ldots, \mathbf{e}_k\}$ for $1 \leq k < n$, is still an orthonormal set, though not a basis, in $\mathbf{R}^n$. △

Example 18.1.3. In the sequence space l^2, the set $\{e_k : k \in \mathbf{N}\}$, with the ordered components of e_k denoted e_{kj}, $j \in \mathbf{N}$, and defined by

$$e_{kj} = \begin{cases} 1 & \text{if } j = k, \\ 0 & \text{if } j \neq k, \end{cases}$$

is an orthonormal set. △

We have the following examples of orthonormal sets in the spaces $\mathcal{R}[-\pi, \pi]$ and $\mathcal{R}[0, \pi]$.

Example 18.1.4. In the space $\mathcal{R}[-\pi, \pi]$, consider the set $\{\phi_k : k = 0, 1, 2, \ldots\}$ given by $\phi_0(x) = 1/\sqrt{2\pi}$ and, for each integer $k \geq 1$,

$$\phi_{2k-1}(x) = \frac{1}{\sqrt{\pi}} \cos kx, \quad \phi_{2k}(x) = \frac{1}{\sqrt{\pi}} \sin kx.$$

Displayed in sequence, the elements ϕ_k are given by

$$\frac{1}{\sqrt{2\pi}}, \quad \frac{1}{\sqrt{\pi}} \cos x, \quad \frac{1}{\sqrt{\pi}} \sin x, \quad \frac{1}{\sqrt{\pi}} \cos 2x, \quad \frac{1}{\sqrt{\pi}} \sin 2x, \quad \ldots.$$

These elements constitute an orthonormal set in $\mathcal{R}[-\pi, \pi]$, called the standard trigonometric set. △

Example 18.1.5. In each of the spaces $C[0, \pi]$ and $\mathcal{R}[0, \pi]$, the sets

$$\{\sin x,\ \sin 2x,\ \sin 3x,\ \ldots\} \quad \text{and} \quad \{1,\ \cos x,\ \cos 2x,\ \cos 3x,\ \ldots\}$$

are orthogonal sets. (See Exercise 15.5.3.) From a somewhat deeper point of view, these orthogonal sets can be recognized as being analogous to the tip of an iceberg: each is the set of eigenfunctions of a boundary value problem for a linear Sturm-Liouville differential operator. There are many important Sturm-Liouville boundary value problems that give rise to other orthonormal sets of functions on an interval. We saw examples of such problems involving the heat equation and the wave equation when the space variable belongs to $[0, \pi]$. For more on other Sturm-Liouville boundary value problems, see the Notes and References for this chapter. △

Exercises.

Exercise 18.1.1. Let X be an orthogonal set in an inner product space V. Prove that X is a linearly independent set (Definition 8.1.11).

Exercise 18.1.2. *Gram-Schmidt orthogonalization procedure*
Let V be an inner product space. Suppose $\{f_1, f_2, f_3, \ldots\}$ is a linearly independent set in V. Define $v_1 = f_1$, and then define

$$v_n = f_n - \sum_{k=1}^{n-1} \frac{(f_n, v_k)}{\|v_k\|^2} v_k, \qquad \text{for } n \geq 2.$$

Prove that $\{v_1, v_2, v_3, \ldots\}$ is an orthogonal set in V, and that for each n,

$$\operatorname{span}\{v_1, \ldots, v_n\} = \operatorname{span}\{f_1, \ldots, f_n\}.$$

Exercise 18.1.3. Consider the inner product space $C[-1,1]$.

1. Using the Gram-Schmidt procedure, find an orthogonal basis for the four-dimensional subspace of $C[-1,1]$ spanned by the functions 1, t, t^2, and t^3.
2. The Legendre polynomials $P_0, P_1, P_2, \ldots$ are obtained from the Gram-Schmidt procedure applied to the functions $1, t, t^2, \ldots$ except that the resulting functions are normalized to satisfy $P_k(1) = 1$ for each k (rather than requiring $(P_k, P_k) = 1$). The first four *Legendre polynomials* are $P_0(t) = 1$, $P_1(t) = t$, $P_2(t) = \frac{1}{2}(3t^2 - 1)$ and $P_3(t) = \frac{1}{2}(5t^3 - 3t)$. Show that these functions are pairwise orthogonal in $C[-1,1]$. Do they span the same subspace of $C[-1,1]$ described in part 1?
3. Find the Legendre polynomial $P_4(t)$.

18.2. Orthonormal Expansions

The main goal of this section is to establish useful generalizations for real inner product spaces of the results on orthonormal expansion of vectors in $\mathbf{R}^3$ given in Section 8.4. Along the way, certain facts discussed earlier for Fourier series with respect to the orthonormal trigonometric set on $[-\pi, \pi]$ now can be viewed in a more general setting.

In what follows, V is always an infinite-dimensional real inner product space and we denote the norm of an element f in V by $\|f\|$. Thus, $\|f\| = (f,f)^{1/2}$. Elements of orthogonal sets are generally denoted by u_k or v_k. The results in subsection 18.2.1 do not require completeness of the inner product space V. However, in subsection 18.2.2, we assume that V is a Hilbert space (Definition 9.3.3), a complete inner product space.

18.2.1. Basic Results for Inner Product Spaces.

The Pythagorean theorem for finite sums of orthogonal elements in V is a direct generalization of the situation for an orthogonal set in $\mathbf{R}^n$ (Exercise 8.4.3).

Theorem 18.2.1 (Pythagorean theorem). *Let V be a real inner product space. The following statements are true:*

1. *If $\{v_k : 1 \le k \le m\}$ is a finite orthogonal set in V, then*

$$\left\| \sum_{k=1}^{m} v_k \right\|^2 = \sum_{k=1}^{m} \|v_k\|^2.$$

2. *If $\{u_k : 1 \le k \le m\}$ is a finite orthonormal set in V, then for any real numbers a_k,*

$$\left\| \sum_{k=1}^{m} a_k u_k \right\|^2 = \sum_{k=1}^{m} a_k^2.$$

Proof. Let $F = \{1, 2, \ldots, m\}$. Statement 1 follows by expanding the inner product to obtain

$$\Big\| \sum_{k \in F} v_k \Big\|^2 = (\sum_{k \in F} v_k, \sum_{k \in F} v_k) = \sum_{i,j \in F} (v_i, v_j),$$

and then using orthogonality to eliminate the terms that involve (v_i, v_j) for $i \neq j$. The terms that remain form the sum

$$\sum_{k \in F} (v_k, v_k) = \sum_{k \in F} \|v_k\|^2.$$

Statement 2 follows from 1 by setting $v_k = a_k u_k$, since $\|a_k u_k\|^2 = a_k^2$. □

Suppose $\{u_k : k \in \mathbf{N}\}$ is an orthonormal set in V. We address the following questions, motivated by the results in $\mathbf{R}^3$ (and $\mathbf{R}^n$) in Section 8.4, and by the best mean square approximation result for Fourier series in Section 15.5:

(1) If f is represented by a convergent expansion, $f = \sum_{k=1}^{\infty} c_k u_k$, relative to the u_k, for some real numbers c_k, is it true that $c_k = (f, u_k)$?

(2) If we use only the first n elements u_k in approximating f by a linear combination, $f \approx \sum_{k=1}^{n} c_k u_k$, do the coefficients $c_k = (f, u_k)$ give the best result as measured by the norm on V?

(3) Under what conditions can an arbitrary element f in V be represented by a convergent expansion, $f = \sum_{k=1}^{\infty} c_k u_k$, relative to the u_k?

We proceed to answer the first two questions.

If we take the inner product of elements of V with a fixed element z, we get a continuous mapping from V into the real numbers. This is the content of the following theorem.

Theorem 18.2.2. *Let V be a real inner product space and let z be any fixed element of V. The mapping defined by*

$$f \mapsto (f, z)$$

is continuous. Consequently, if $\sum_{k=1}^{\infty} f_k$ converges in norm to f in V, then

$$(f, z) = \Big(\sum_{k=1}^{\infty} f_k, z \Big) = \sum_{k=1}^{\infty} (f_k, z).$$

Proof. First note that if $z = 0$, then the mapping $f \mapsto (f, 0)$ is the constant zero function, which is continuous. Now assume $z \neq 0$. Since the inner product is linear in the first argument, we have $|(f, z) - (g, z)| = |(f - g, z)|$. Thus, by the Cauchy-Schwarz inequality,

$$|(f, z) - (g, z)| = |(f - g, z)| \leq \|f - g\| \, \|z\|.$$

Given $\epsilon > 0$, we have $|(f, z) - (g, z)| < \epsilon$ provided we choose $\|f - g\| < \delta(\epsilon) = \epsilon / \|z\|$. This shows that the inner product with a fixed z is uniformly continuous on V.

Suppose $\sum_{k=1}^{\infty} f_k$ converges in norm to f in V. Letting $s_n = \sum_{k=1}^{n} f_k$, we have

$$(f, z) = \Big(\sum_{k=1}^{\infty} f_k, z \Big) = \Big(\lim_{n \to \infty} s_n, z \Big) = \lim_{n \to \infty} (s_n, z).$$

By the definition of s_n, the linearity of the inner product in the first argument, and the definition of the sum of an infinite numerical series, we have

$$(f, z) = \lim_{n\to\infty} (s_n, z) = \lim_{n\to\infty} \sum_{k=1}^{n} (f_k, z) = \sum_{k=1}^{\infty} (f_k, z),$$

as was to be shown. □

Given an orthonormal set $\{u_k : k \in \mathbf{N}\}$ in V and $f \in V$, we define the **Fourier coefficients** of f with respect to this set (or, with respect to the u_k) to be the numbers $c_k = (f, u_k)$. The next result says that if an element f in V equals a convergent series expansion, $f = \sum_{k=1}^{\infty} c_k u_k$, with respect to the orthonormal set, then the coefficients c_k must be the Fourier coefficients of f with respect to the u_k.

Theorem 18.2.3. *Let V be a real inner product space and let $\{u_k : k \in \mathbf{N}\}$ be an orthonormal set in V. If $f \in V$ and $f = \sum_{k=1}^{\infty} c_k u_k$ for some real numbers c_k, then for all k,*

$$c_k = (f, u_k).$$

Proof. If $f = \sum_{j=1}^{\infty} c_j u_j$, then by Theorem 18.2.2, we have

$$(f, u_k) = \Big(\sum_{j=1}^{\infty} c_j u_j, u_k \Big) = \sum_{j=1}^{\infty} c_j (u_j, u_k) = c_k$$

for each positive integer k. □

This answers question (1) posed at the beginning of the section. The only way to represent $f \in V$ exactly by a convergent expansion with respect to an orthonormal set $\{u_k : k \in \mathbf{N}\}$, if possible at all, is that the coefficients are the Fourier coefficients $c_k = (f, u_k)$.

Now suppose we approximate an element f in V using only the first n elements u_k of an orthonormal set, and we use a linear combination, $\sum_{k=1}^{n} c_k u_k$. What choice of coefficients gives the closest approximation to f in the norm on V? As in the case of the trigonometric set in Section 15.5, the best choice is the Fourier coefficients. (Think of the normalized trigonometric set in Exercise 15.5.2.)

Lemma 18.2.4. *Let V be a real inner product space and $\{u_k : k \in \mathbf{N}\}$ an orthonormal set in V. If $y_n = \sum_{k=1}^{n} c_k u_k$, then*

$$\|f - y_n\|^2 = \|f\|^2 - \sum_{k=1}^{n} (f, u_k)^2 + \sum_{k=1}^{n} [c_k - (f, u_k)]^2.$$

Proof. We calculate, starting from $\|f - y_n\|^2 = (f - y_n, f - y_n)$ and using orthonormality of the u_k, as follows: If $y_n = \sum_{k=1}^{n} c_k u_k$, then the bilinearity of the inner product implies that

$$\begin{aligned} \|f - y_n\|^2 &= (f - y_n, f - y_n) \\ &= (f, f) - 2(f, y_n) + (y_n, y_n) \\ &= \|f\|^2 - 2\sum_{k=1}^{n} c_k (f, u_k) + \sum_{k=1}^{n} c_k^2. \end{aligned}$$

For each k, completion of the square gives

$$c_k^2 - 2c_k(f, u_k) = [c_k - (f, u_k)]^2 - (f, u_k)^2,$$

and the desired result follows by summing these terms from $k = 1$ to $k = n$. □

Lemma 18.2.4 leads directly to the approximation of f by a finite linear combination $y_n = \sum_{k=1}^n c_k u_k$ that minimizes $\|f - y_n\|$ over all such linear combinations.

Theorem 18.2.5. *Let V be a real inner product space and $\{u_k : k \in \mathbf{N}\}$ an orthonormal set in V. For any $f \in V$ and positive integer n, and for any choice of real numbers $c_1, \ldots, c_n$,*

$$\Big\|f - \sum_{k=1}^{n}(f, u_k)u_k\Big\| \le \Big\|f - \sum_{k=1}^{n} c_k u_k\Big\|. \tag{18.1}$$

Proof. For any choice of real numbers $c_1, \ldots, c_n$ in the sum $y_n = \sum_{k=1}^n c_k u_k$, Lemma 18.2.4 implies that

$$\|f - y_n\|^2 = \|f\|^2 - \sum_{k=1}^{n}(f, u_k)^2 + \sum_{k=1}^{n}[c_k - (f, u_k)]^2. \tag{18.2}$$

Only the last sum on the right-hand side, $\sum_{k=1}^n [c_k - (f, u_k)]^2$, which is nonnegative, depends on the choice of the c_k, and clearly $\|f - y_n\|^2$ is minimized when $c_k = (f, u_k)$. This proves (18.1). □

Thus far we have seen that for a given f, the Fourier coefficients $c_k = (f, u_k)$ yield the best approximation to f, $\sum_{k=1}^n (f, u_k)u_k$, that uses only the first n elements u_k in a linear combination. In addition, we know that if $f = \sum_{k=1}^\infty c_k u_k$, then necessarily $c_k = (f, u_k)$.

There is another important consequence of Lemma 18.2.4: It implies the convergence of the series $\sum_{k=1}^\infty (f, u_k)^2$ for any f in V.

Theorem 18.2.6 (Bessel's Inequality). *Let V be a real inner product space and let $\{u_k : k \in \mathbf{N}\}$ be an orthonormal set in V. For any $f \in V$ the series $\sum_{k=1}^\infty (f, u_k)^2$ converges, and*

$$\sum_{k=1}^{\infty}(f, u_k)^2 \le \|f\|^2.$$

Proof. By Lemma 18.2.4, if $s_n = \sum_{k=1}^n (f, u_k)u_k$, then

$$\|f - s_n\|^2 = \|f\|^2 - \sum_{k=1}^{n}(f, u_k)^2 \ge 0.$$

Hence, for each n,

$$\sum_{k=1}^{n}(f, u_k)^2 \le \|f\|^2,$$

which bounds the n-th partial sum of the series $\sum_{k=1}^\infty (f, u_k)^2$ by $\|f\|^2$. Since the sequence of partial sums is increasing and bounded above by $\|f\|^2$, the series converges. Letting $n \to \infty$ yields

$$\sum_{k=1}^{\infty}(f, u_k)^2 = \lim_{n\to\infty}\sum_{k=1}^{n}(f, u_k)^2 \leq \|f\|^2,$$

as we wished to show. □

Another way to state the conclusion of Bessel's inequality is to say that for any f in V, the sequence of Fourier coefficients (f, u_k) is an element of the sequence space l^2.

The convergence of the series $\sum_{k=1}^{\infty}(f, u_k)^2$ immediately implies the following corollary of Bessel's inequality, a general *Riemann-Lebesgue theorem.*

Theorem 18.2.7 (Riemann-Lebesgue). *If $\{u_k : k \in \mathbf{N}\}$ is an orthonormal set in V, then for any $f \in V$,*

$$\lim_{k\to\infty}(f, u_k) = 0.$$

We encountered the special case of this result involving the space $\mathcal{R}[-\pi, \pi]$ and the (normalized) standard trigonometric set in Theorem 15.5.2 (also known as the *Riemann-Lebesgue theorem*), which took the following form: If $f \in \mathcal{R}[-\pi, \pi]$, then

$$\lim_{k\to\infty}\int_{-\pi}^{\pi} f(x)\cos kx\,dx \;=\; 0 \;=\; \lim_{k\to\infty}\int_{-\pi}^{\pi} f(x)\sin kx\,dx.$$

We have omitted the normalization constants $1/\pi$ in these limit statements.

18.2.2. Complete Spaces and Complete Orthonormal Sets. Thus far we have not assumed that our inner product space V is a Hilbert space (Definition 9.3.3). The previous results did not require that V be complete in the norm induced by the inner product. Our previous examples for these results in an infinite-dimensional setting are: the space of continuous 2π-periodic functions, denoted $CP[-\pi, \pi]$; the space of Riemann integrable functions, $\mathcal{R}[-\pi, \pi]$; and l^2. We have seen that $CP[-\pi, \pi]$ and $\mathcal{R}[-\pi, \pi]$ are not complete in the norm given by $\|f\|^2 = \int_{-\pi}^{\pi} f^2(x)\,dx$. We know from Theorem 9.3.4 that l^2 is a Hilbert space.

In the labeled results for the remainder of the section we assume that V is a Hilbert space.

Lemma 18.2.8. *Suppose V is a Hilbert space, and $\{u_k : k \in \mathbf{N}\}$ is an orthonormal set in V. Then for every $f \in V$, the series $\sum_{k=1}^{\infty}(f, u_k)u_k$ converges in norm to an element of V, and*

$$\Big\|\sum_{k=1}^{\infty}(f, u_k)u_k\Big\| \leq \|f\|.$$

Proof. Let $f \in V$. By Bessel's inequality, the series $\sum_{k=1}^{\infty}(f, u_k)^2$ converges. Write $s_n = \sum_{k=1}^{n}(f, u_k)u_k$ for the n-th partial sum of the series $\sum_{k=1}^{\infty}(f, u_k)u_k$.

For $n > m$, the partial sums satisfy

$$\begin{aligned}\Big\| \sum_{k=1}^{n} (f, u_k) u_k - \sum_{k=1}^{m} (f, u_k) u_k \Big\|^2 &= \Big\| \sum_{k=m+1}^{n} (f, u_k) u_k \Big\|^2 \\ &= \sum_{k=m+1}^{n} (f, u_k)^2,\end{aligned}$$

by orthonormality of the u_k. By the convergence of $\sum_{k=1}^{\infty}(f, u_k)^2$, the partial sums s_n form a Cauchy sequence in V. Since V is complete, $\sum_{k=1}^{\infty}(f, u_k)u_k$ converges in norm to an element of V. Now the Pythagorean theorem and Bessel's inequality yields

$$\begin{aligned}\Big\| \sum_{k=1}^{\infty} (f, u_k) u_k \Big\|^2 &= \lim_{n\to\infty} \Big\| \sum_{k=1}^{n} (f, u_k) u_k \Big\|^2 \\ &= \lim_{n\to\infty} \sum_{k=1}^{n} (f, u_k)^2 = \sum_{k=1}^{\infty} (f, u_k)^2 \le \|f\|^2.\end{aligned}$$

Taking the square root of both sides completes the proof. □

Without the assumption that V is complete in Lemma 18.2.8, the conclusion of this lemma can fail.

Example 18.2.9. Let V be the subspace of l^2 consisting of finite linear combinations of the unit vectors e_k, for $k \ge 2$, and the vector

$$\xi := e_1 + \sum_{k\ge 2} \frac{1}{k^2} e_k,$$

where e_k has a 1 in the k-th entry and zeros elsewhere. The series in this definition of ξ converges in l^2 since the partial sums form a Cauchy sequence and l^2 is complete. Elements in V have the form

$$c_1\xi + \sum_{k\in F} c_k e_k,$$

where F is any finite subset of $\{2, 3, 4, \dots\}$, and c_1 and the c_k are real numbers. Then V has the inner product and norm of l^2, and the set $\{e_k : k \ge 2\}$ is an orthonormal set in V. Note that e_1 is not an element of V. However, $\xi \in V$, and the sequence of partial sums of the Fourier series of ξ with respect to the set $\{e_k : k \ge 2\}$ is Cauchy, but has no limit in V: We can see this from the calculation, valid in l^2, that

$$\sum_{m=2}^{\infty} (\xi, e_m) e_m = \sum_{m\ge 2} \Big(e_1 + \sum_{k\ge 2} \frac{1}{k^2} e_k \,,\, e_m \Big) e_m = \sum_{m\ge 2} \frac{1}{m^2} e_m = \xi - e_1,$$

which is not an element of V since e_1 is not in V. This shows that V is not complete and that the conclusion of Lemma 18.2.8 fails in V. △

Lemma 18.2.8 shows that the Fourier series of every f in a Hilbert space converges, but the lemma provides only a norm inequality relating the series sum and f. For example, in l^2, using the orthonormal set $E = \{e_2, e_4, e_6, \dots\}$ of evenly indexed standard unit vectors, we cannot exactly represent every element by its

Fourier series with respect to E, and many elements will be poorly approximated by that Fourier series. This clearly suggests that some condition on the orthonormal set $\{u_k : k \in \mathbf{N}\}$, in addition to completeness of V, is needed to ensure that for every f in V,

$$f = \sum_{k=1}^{\infty} (f, u_k) u_k. \tag{18.3}$$

Let us first consider some necessary conditions for (18.3), and to do that we need not assume completeness of V:

If (18.3) holds for *every* f in V, then there cannot be an orthonormal set in V that properly contains $\{u_k : k \in \mathbf{N}\}$, for if there is a vector h in V such that $(h, u_k) = 0$ for all $k \in \mathbf{N}$ and $h = \sum_{k=1}^{\infty}(h, u_k)u_k$, then $h = 0$.

Also, if the series $\sum_{k=1}^{\infty}(f, u_k)u_k$ converges and equals f, the norm of the sum must equal the norm of f, hence $\| \sum_{k=1}^{\infty}(f, u_k)u_k\| = \|f\|$. We know that for *finite* sums, the Pythagorean theorem implies that for every n,

$$\Big\| \sum_{k=1}^{n} (f, u_k) u_k \Big\|^2 = \sum_{k=1}^{n} (f, u_k)^2.$$

Thus, if (18.3) holds, then by taking the limit as $n \to \infty$, we have

$$\|f\|^2 = \Big\| \sum_{k=1}^{\infty} (f, u_k) u_k \Big\|^2 = \sum_{k=1}^{\infty} (f, u_k)^2.$$

The equality $\|f\|^2 = \sum_{k=1}^{\infty}(f, u_k)^2$ is called **Parseval's equation**. (We have seen it in Theorem 15.5.4 for the non-normalized trigonometric set in the space $CP[-\pi, \pi]$.) Conversely, if Parseval's equation holds for every f in V, then every f in V is the sum of its Fourier series with respect to the set $\{u_k : k \in \mathbf{N}\}$: This follows from a direct calculation of $\|f - s_n\|^2 = (f - s_n, f - s_n)$ by expanding the inner product and using orthonormality. (See Exercise 18.2.8.) We note as well that Parseval's equation, if holding for all f in V, implies that if $h \in V$ and $(h, u_k) = 0$ for all $k \in \mathbf{N}$, then $\|h\|^2 = 0$ and hence $h = 0$.

In view of Lemma 18.2.8, it is natural to consider the three conditions just noted in a Hilbert space, and to try to close the loop of implications. This leads to the following theorem.

Theorem 18.2.10. *Let V be a Hilbert space and $\{u_k : k \in \mathbf{N}\}$ an orthonormal set in V. The following conditions are equivalent:*

(1) *If $h \in V$ and $(h, u_k) = 0$ for all $k \in \mathbf{N}$, then $h = 0$.*

(2) *For every $f \in V$, we have $f = \sum_{k=1}^{\infty}(f, u_k)u_k$, where the series converges in norm.*

(3) *Every $f \in V$ satisfies* **Parseval's equation**, $\|f\|^2 = \sum_{k=1}^{\infty}(f, u_k)^2$.

Proof. In our comments before the theorem we noted that conditions (2) and (3) are equivalent, and that each implies condition (1). (We remark here that these implications did not require the completeness of the space V.)

Thus it remains to show that (1) implies (2), and this is where we use the assumption that V is a Hilbert space.

Assume that (1) holds. For any $f \in V$, the series $\sum_{k=1}^{\infty}(f, u_k)u_k$ converges to an element of V, by Lemma 18.2.8. To see that the limit must be f, we consider the difference $f - \sum_{k=1}^{\infty}(f, u_k)u_k$, and note that for every $m \in \mathbf{N}$, orthogonality of the u_k and continuity of the inner product imply that

$$\Big(f - \sum_{k=1}^{\infty}(f, u_k)u_k, u_m\Big) = (f, u_m) - \Big(\sum_{k=1}^{\infty}(f, u_k)u_k, u_m\Big) = (f, u_m) - (f, u_m) = 0.$$

Hence, by (1), we have $f - \sum_{k=1}^{\infty}(f, u_k)u_k = 0$. This is true for every f in V and thus (2) holds. □

Theorem 18.2.3 and Theorem 18.2.10 together provide the answer to the third question posed at the beginning of this section, when V is a Hilbert space: Under what conditions is an arbitrary element f in V represented by a convergent expansion, $f = \sum_{k=1}^{\infty} c_k u_k$, relative to the orthonormal set $\{u_k : k \in \mathbf{N}\}$?

With every element f in a Hilbert space V with orthonormal set $\{u_k : k \in \mathbf{N}\}$, we can associate the following three objects:

1. the sequence of Fourier coefficients, (f, u_k);
2. the series sum, $g = \sum_{k=1}^{\infty}(f, u_k)u_k$, which is an element of V;
3. the difference $h := f - g$.

Condition (2) of Theorem 18.2.10 states that for every $f \in V$, the corresponding g equals f; condition (1) states that for every $f \in V$, the difference $h = f - g$ is zero; and (3) says that for every $f \in V$, $\|f\| = \|g\|$. These are equivalent properties of an orthonormal set $\{u_k : k \in \mathbf{N}\}$ in a Hilbert space V.

To provide some terminology, we consider the conditions (1), (2) and (3) for a moment outside of their connection with Theorem 18.2.10.

Definition 18.2.11. *Let V be a real inner product space. An orthonormal set $\{u_k : k \in \mathbf{N}\}$ in V is called* **maximal in** V *if condition (1) of Theorem* 18.2.10 *holds: If $h \in V$ and $(h, u_k) = 0$ for all $k \in \mathbf{N}$, then $h = 0$. If condition (2) holds, then $\{u_k : k \in \mathbf{N}\}$ is called an* **orthonormal basis** *for V.*

Remark. *The term* **orthonormal basis** *in Definition* 18.2.11 *can be understood in the sense that under condition (2), every element of the space V can be approximated arbitrarily closely in norm by a finite linear combination of the elements in the set $\{u_k : k \in \mathbf{N}\}$, in particular by the Fourier partial sums. We note that some resources refer to an orthonormal set that is* **maximal in** V *as being* **complete in** V*; this is another standard term, but it is of course a different use of the term* complete *than the one meaning convergence of Cauchy sequences. We use the terminology* **maximal in** V *in Definition* 18.2.11 *for condition (1) stated there, to avoid any potential ambiguity.*

It can be convenient to work with an orthogonal set that is not normalized, as we have seen with the trigonometric set on $[-\pi, \pi]$, and the term *maximal* (respectively, *basis*) can be used for an orthogonal set $\{\phi_k : k \in \mathbf{N}\}$ in a space V, if the normalized set with elements $\phi_k/\|\phi_k\|$ is a maximal orthonormal set (respectively, an orthonormal basis) in V.

If V is a Hilbert space, then an orthonormal set in V is maximal if and only if it is an orthonormal basis for V. It is not difficult to see that the orthonormal set $\{e_k\}_{k=1}^{\infty}$ in the Hilbert space l^2 is a maximal orthonormal set in l^2 and an orthonormal basis for l^2. (Exercise 18.2.5.)

In general, without V being a complete space, condition (1) does not imply condition (2), as the next example shows.

Example 18.2.12. Consider again the space V of Example 18.2.9 with the orthonormal set $U = \{e_k : k \geq 2\}$. We claim that U is maximal in V: To see this, suppose $h \in V$ and $(h, e_k) = 0$ for all $k \geq 2$. Since $h \in l^2$ and

$$\{e_1\} \cup U = \{e_k : k \in \mathbf{N}\}$$

is a maximal orthonormal set in l^2, we have $h = \sum_{k=1}^{\infty}(h, e_k)e_k$, and since $(h, e_k) = 0$ for $k \geq 2$, it follows that $h = (h, e_1)e_1$. However, since e_1 is not in V, the only possibility that h is an element of V is that $(h, e_1) = 0$ and hence $h = 0$. Therefore U is maximal in V. However, condition (2) fails for the orthonormal set U in V: As we saw in Example 18.2.9, the vector

$$\xi := e_1 + \sum_{k \geq 2} \frac{1}{k^2} e_k,$$

an element of V, is not the sum of its Fourier series $\sum_{m \geq 2}(\xi, e_m)e_m$ with respect to U. Of course, condition (3) also fails for U, since $\|\xi\|^2 \neq \sum_{k=2}^{\infty}(\xi, e_k)^2$. △

We remarked in the proof of Theorem 18.2.10 that completeness of V was not needed for the implications (2) implies (3) and (3) implies (1), but only to close the loop by proving (1) implies (2) and (1) implies (3). There are important examples of function spaces V where conditions (2), (3) and (1) hold without V being complete: The proof of Theorem 15.5.4 shows that (2) holds for the trigonometric set in $CP[-\pi, \pi]$. Theorem 18.3.8 in the next section shows that (2) holds for the trigonometric set in the space $\mathcal{R}[-\pi, \pi]$. These are useful results concerning the trigonometric set, though each of these spaces fails to be a Hilbert space with the L^2 norm. This success of the trigonometric set in $CP[-\pi, \pi]$ and $\mathcal{R}[-\pi, \pi]$ is explained within a larger framework in the final section of the book. There we show that the space $L^2[-\pi, \pi]$ of functions square integrable in the Lebesgue sense, which includes all Riemann integrable functions on $[-\pi, \pi]$, is a Hilbert space (Theorem 18.4.3), and that the trigonometric set is an orthogonal basis for that space (Theorem 18.4.7) and maximal in $L^2[-\pi, \pi]$.

Exercises.

Exercise 18.2.1. Find the closest approximation to the point $(1, 2, 3)$, as measured in the standard Euclidean norm, using an approximation that lies within the plane given by the equation

$$x + y + z = 0.$$

Here x, y, z are the coordinates of a point in $\mathbf{R}^3$. *Hint*: You will need orthogonal basis vectors in order to use the simple inner product coefficient formulas.

Exercise 18.2.2. Let A be $m \times n$ and $b \in \mathbf{R}^m$. When $A\mathbf{x} = \mathbf{b}$ is not consistent, that is, when $\mathbf{b}$ is not in the range space (column space) of A, then there is no exact solution of this linear algebraic system. However, if A has full column rank, that is, if A is one-to-one, then there is a unique vector $\mathbf{c}$ that *minimizes* the squared error over all vectors $\mathbf{x}$; thus,

$$||A\mathbf{c} - \mathbf{b}||_2^2 = \min_{\mathbf{x}\in\mathbf{R}^n} ||A\mathbf{x} - \mathbf{b}||_2^2.$$

Find this *least squares solution* $\mathbf{x} = \mathbf{c}$ for the system

$$\begin{bmatrix} -4 & 2 \\ 0 & 5 \\ 5 & 0 \end{bmatrix} \begin{bmatrix} x_1 \\ x_2 \end{bmatrix} = \begin{bmatrix} 1 \\ 2 \\ 3 \end{bmatrix}.$$

Exercise 18.2.3. Find the best approximation in the L^2 norm to the function

$$f(x) = x, \quad 0 \le x \le \pi:$$

(i) by an element of the space $U_1 = \text{span}\,\{\cos x, \cos 2x\}$;

(ii) by an element of the space $U_2 = \text{span}\,\{\sin x, \sin 2x\}$;

(iii) by an element of the space $U_3 = \text{span}\,\{\sin x, \cos x\}$.

Exercise 18.2.4. Apply Theorem 18.2.5 in the inner product space $C[-1,1]$ to find the polynomial function of degree two, $ax^2 + bx + c$, which provides the best approximation to e^x over the interval $[-1,1]$, in the least squares sense of the L^2 norm in this space. *Hint*: See Exercise 18.1.3.

Exercise 18.2.5. Recall that the sequence space l^2 is a Hilbert space, by Theorem 9.3.4.

1. Show that the orthonormal set $\{e_k : k \in \mathbf{N}\}$ in l^2 is a maximal orthonormal set in l^2, where e_k has k-th entry 1 and all other entries zero.
2. Identify an infinite orthonormal set in l^2 which is not maximal in l^2.

Exercise 18.2.6. Let V be a real Hilbert space and suppose that $\{u_k : k \in \mathbf{N}\}$ is a maximal orthonormal set in V. Show that for all f and g in V,

$$(f, g) = \sum_{k=1}^{\infty} (f, u_k)(g, u_k).$$

(Note that the case $f = g$ yields Parseval's equality.)

Exercise 18.2.7. Suppose $\{u_k : k \in \mathbf{N}\}$ is a maximal orthonormal set in a Hilbert space V. Let $\Phi : V \to l^2$ be defined by $\Phi(f) = \hat{f}$ where $\hat{f}_k = (f, u_k)$ for each k. Prove the following statements.

1. Φ is linear, that is, $\Phi(\alpha f) = \alpha\Phi(f)$ and $\Phi(f + g) = \Phi(f) + \Phi(g)$ for all $f, g \in V$ and $\alpha \in \mathbf{R}$. Thus, if $\hat{f}_k$ and $\hat{g}_k$ are the Fourier coefficients of f and g, respectively, then $\alpha\hat{f}_k$ are the Fourier coefficients of αf, and $\hat{f}_k + \hat{g}_k$ are the Fourier coefficients of $f + g$.
2. Φ is one-to-one on V: If f and g have the same sequence of Fourier coefficients with respect to $\{u_k : k \in \mathbf{N}\}$, then $f = g$.

Exercise 18.2.8. Let V be a real inner product space (not necessarily a Hilbert space) and $\{u_k : k \in \mathbf{N}\}$ an orthonormal set in V. Show that if $\|f\|^2 = \sum_{k=1}^{\infty}(f, u_k)^2$ for every f in V, then the Fourier series of every f in V converges to f. *Hint*: Let $s_n = \sum_{k=1}^{n}(f, u_k)u_k$ and expand $\|f - s_n\|^2 = (f - s_n, f - s_n)$.

18.3. Mean Square Convergence

Let V be an inner product space. In Theorem 18.2.5 we considered the best mean square approximation to an element f in V using a finite linear combination of orthogonal elements, as we did earlier for the Fourier partial sums of Riemann integrable functions in Theorem 15.5.1.

The previous section emphasized the importance of working with a complete space. Suppose V is a Hilbert space and $U = \{u_k : k \in \mathbf{N}\}$ is an orthonormal set in V. If U is maximal in V, then we have $f = \sum_{k=1}^{\infty}(f, u_k)u_k$ for all f in V. If, on the other hand, U is not maximal in V, then what is the object to which the series $\sum_{k=1}^{\infty}(f, u_k)u_k$ converges? By Lemma 18.2.8, it converges, but the limit will not be f, in general, if U is not maximal in V. The limit of the series is the best approximation to f in norm from among all functions that have the form $\sum_{k=1}^{\infty} c_k u_k$, where the coefficient sequence (c_k) satisfies $\sum_{k=1}^{\infty} |c_k|^2 < \infty$. (We recall that the vector series $\sum_{k=1}^{\infty} c_k u_k$ converges if and only if the real series $\sum_{k=1}^{\infty} |c_k|^2$ converges, by the proof of Lemma 18.2.8.)

We record this result on infinite linear combinations as a theorem.

Theorem 18.3.1. *If $\{u_k : k \in \mathbf{N}\}$ is an orthonormal set in an infinite-dimensional Hilbert space V, then*

$$\left\| f - \sum_{k=1}^{\infty}(f, u_k)u_k \right\| \le \left\| f - \sum_{k=1}^{\infty} c_k u_k \right\|$$

for all choices of real numbers c_k with $\sum_{k=1}^{\infty} |c_k|^2 < \infty$. Equality holds only when $c_k = (f, u_k)$ for all k.

Proof. Write

$$f - \sum_{k=1}^{\infty} c_k u_k = \left[f - \sum_{k=1}^{\infty}(f, u_k)u_k \right] + \sum_{k=1}^{\infty} \big((f, u_k) - c_k\big)u_k.$$

The difference $f - \sum_{k=1}^{\infty}(f, u_k)u_k$ is orthogonal to all the u_k, as seen in the argument for Theorem 18.2.10, and thus it is orthogonal to every term in the second series on the right-hand side. A simple limit argument using the continuity of the inner product then shows that the difference $f - \sum_{k=1}^{\infty}(f, u_k)u_k$ is orthogonal to the sum of the second series on the right. Thus, by the Pythagorean theorem (for two orthogonal vectors), we have

$$\left\| f - \sum_{k=1}^{\infty} c_k u_k \right\|^2 = \left\| f - \sum_{k=1}^{\infty}(f, u_k)u_k \right\|^2 + \left\| \sum_{k=1}^{\infty} \big((f, u_k) - c_k\big)u_k \right\|^2.$$

However,

$$\left\| \sum_{k=1}^{\infty} \big((f, u_k) - c_k\big)u_k \right\|^2 = \sum_{k=1}^{\infty} |(f, u_k) - c_k|^2,$$

as the proof of Lemma 18.2.8 shows. Hence,

$$\left\| f - \sum_{k=1}^{\infty} c_k u_k \right\|^2 = \left\| f - \sum_{k=1}^{\infty}(f, u_k)u_k \right\|^2 + \sum_{k=1}^{\infty} |(f, u_k) - c_k|^2.$$

Since the last sum on the right is nonnegative, this equation proves the inequality in the theorem statement, as well as the necessary condition for equality asserted there. □

Theorem 18.3.1 extends the results given earlier for approximation of f by finite sums $\sum_{k=1}^{n} c_k u_k$. This theorem also highlights the geometric significance of the Fourier series of an element f in V.

18.3.1. Comparison of Pointwise, Uniform, and L^2 Norm Convergence. Our main interest in this chapter is the inner product space of functions on a closed interval $[a, b]$ that are square integrable in the sense of Lebesgue, denoted by $L^2[a, b]$. We formally define this space and study it in the next section. One goal is to establish that $L^2[a, b]$ is a complete inner product space, a Hilbert space. We will also show that the normalized standard trigonometric set is a maximal orthonormal set in $L^2[a, b]$ if $b - a = 2\pi$. Before doing that, however, we consider the relations among the three modes of convergence of main interest to us in this book: pointwise, uniform, and L^2 norm convergence. This seems an appropriate place to do this, since mean square convergence is relatively newer to us at this point, and yet it is our primary interest in this chapter.

We first considered the L^2 norm in Section 8.3 for Riemann integrable functions. Suppose for the moment that f_n and f are Riemann integrable. Convergence of f_n to f in the L^2 norm means convergence in mean square, that is,

$$\|f_n - f\|_2 = \Big(\int_a^b |f_n(x) - f(x)|^2 \, dx \Big)^{1/2} \to 0$$

as $n \to \infty$.

We now consider the relation between any two of these modes of convergence of a sequence f_n to f: pointwise, uniform, and mean square convergence.

The functions given by $f_n(x) = x^n$, $x \in [0, 1]$, define a sequence in $C[0, 1]$ with discontinuous pointwise limit f, since f is zero everywhere in $[0, 1]$ except at 1, where $f(1) = 1$. This example reminds us that pointwise convergence does not imply uniform convergence. (The convergence cannot be uniform since the limit is not continuous on $[0, 1]$.) It also shows that mean square convergence does not imply uniform convergence, since we have

$$\int_0^1 |f_n(x) - f(x)|^2 \, dx = \int_0^1 (x^n - 0)^2 \, dx = \int_0^1 x^{2n} \, dx = \frac{x^{2n+1}}{2n+1}\Big|_0^1 = \frac{1}{2n+1} \to 0$$

as $n \to \infty$.

We want to show that pointwise convergence of f_n to f does not imply convergence in mean square, and convergence of f_n to f in mean square does not imply pointwise convergence to f everywhere in the domain. For examples that contrast pointwise and mean square convergence, let $[a, b] = [0, 1]$, and consider the following:

Example 18.3.2. Define $f_n(x) = n$ for $0 < x < 1/n$, and $f_n(x) = 0$ elsewhere for x in $[0, 1]$. Then f_n converges pointwise to $f = 0$ on $[0, 1]$, because for $x > 0$,

$f_n(x) = 0$ for all $1/n < x$, hence for all $n > 1/x$. However, we have

$$\|f_n\|_2^2 = \|f_n - 0\|_2^2 = \int_0^1 |f_n(x)|^2\,dx = \int_0^{1/n} n^2\,dx = n,$$

and therefore f_n does not converge to the zero function in the L^2 norm. △

Example 18.3.3. Define $f_n(x) = 1$ for $0 \le x \le 1/n$, and $f_n(x) = 0$ for $1/n < x \le 1$. Then we have

$$\|f_n\|_2^2 = \|f_n - 0\|_2^2 = \int_0^1 |f_n(x)|^2\,dx = \int_0^{1/n} (1)\,dx = \frac{1}{n},$$

and therefore f_n converges to the zero function in the L^2 norm. However, since $f_n(0) = 1$ for all positive integer n, the f_n do not converge pointwise to the zero function everywhere in the domain. △

In this last example, the failure of pointwise convergence to zero occurs only at a single point in $[0, 1]$. However, Example 18.4.4 of the next section gives a sequence of functions on $[0, 1]$ that converges to the zero function in the L^2 norm, yet the sequence fails to converge pointwise at every point in $[0, 1]$.

On the other hand, there is the positive result that uniform convergence of f_n to f on a bounded interval implies convergence in the L^2 norm (Exercise 18.3.1). This result need not hold, however, for uniform convergence on unbounded intervals (Exercise 18.3.2).

Finally, with a view toward the next section, we summarize the main results in this book (proved earlier, or still to come) about the convergence of Fourier series: If f is a 2π-periodic function, then the Fourier series of f converges to f

(i) absolutely and uniformly (Theorem 15.6.7), and in L^2 norm (Theorem 18.4.7), if f is continuous and piecewise smooth;

(ii) pointwise (Theorem 15.6.2) and in L^2 norm (Theorem 18.4.7), if f is piecewise smooth;

(iii) in L^2 norm (Theorem 18.4.7), if $f \in L^2[-\pi, \pi]$.

Exercises.

Exercise 18.3.1. Suppose that f and f_n, $n \ge 1$, are square integrable on $[a, b]$. Show that if f_n converges uniformly to f on $[a, b]$, then

$$\lim_{n\to\infty} \int_a^b |f_n(x) - f(x)|^2\,dx = 0.$$

Hint: Apply the definition of uniform convergence (Definition 7.1.5).

Exercise 18.3.2. The result of Exercise 18.3.1 need not hold for uniform convergence on an unbounded interval. Give an example of square integrable functions $f_n : [0, \infty) \to \mathbf{R}$ such that $f_n \to 0$ uniformly on $[0, \infty)$, but f_n does not converge in norm to the zero function. *Hint*: Consider functions f_n with value $1/2^n$ on disjoint intervals of length 2^n.

18.3.2. Mean Square Convergence for $CP[-\pi,\pi]$. We return now to the space $CP[-\pi,\pi]$ and consider again the convergence of the Fourier partial sums in the L^2 norm. Recall that $f \in CP[-\pi,\pi]$ if f is continuous and $f(-\pi) = f(\pi)$, so that f extends to a continuous function on $\mathbf{R}$. We have established already, in the proof of Theorem 15.5.4, that the Fourier partial sums of a continuous 2π-periodic function converge to the function in mean square, that is, in the L^2 norm. Theorem 15.5.4 was an application of the trigonometric Weierstrass approximation Theorem 15.5.3 and the best mean square approximation property for Riemann integrable functions in Theorem 15.5.1.

We show here how the mean square convergence follows from Fejér's Theorem (Theorem 15.7.3).

Theorem 18.3.4. *Let $f \in CP[-\pi,\pi]$ and let*

$$s_n = \frac{1}{2}a_0 + \sum_{k=1}^{n}(a_k \cos kx + b_k \sin kx)$$

be the n-th partial sum of the Fourier series of f. Then

$$\lim_{n\to\infty} \|f - s_n\|_2^2 = \int_{-\pi}^{\pi} |f(x) - s_n(x)|^2\, dx = 0,$$

and Parseval's equality holds:

$$\frac{1}{2}\pi a_0^2 + \pi\sum_{k=1}^{\infty}(a_k^2 + b_k^2) = \int_{-\pi}^{\pi} f^2(x)\, dx.$$

Proof. Let $f \in CP[-\pi,\pi]$, and let s_n be the sequence of partial sums of the Fourier series of f, and σ_n the sequence of Cesàro sums. Given $\epsilon > 0$, by Theorem 15.7.3 there is a positive integer $N = N(\epsilon)$ such that if $n \geq N$, then

$$|\sigma_n(x) - f(x)| < \frac{\sqrt{\epsilon}}{2\sqrt{\pi}} \qquad \text{for all } x \in [-\pi,\pi].$$

Thus if $n \geq N$, then, in the norm induced by the inner product on $CP[-\pi,\pi]$,

$$\|\sigma_n - f\|_2^2 = \int_{-\pi}^{\pi} |\sigma_n(x) - f(x)|^2\, dx \leq \int_{-\pi}^{\pi} \frac{\epsilon}{4\pi}\, dx = \frac{\epsilon}{2}.$$

By definition, for each n, the n-th Cesàro sum σ_n is a linear combination of the functions $\phi_0, \phi_1, \phi_2, \ldots, \phi_{2n-1}, \phi_{2n}$, that is,

$$\frac{1}{\sqrt{2\pi}}, \frac{\cos x}{\sqrt{\pi}}, \frac{\sin x}{\sqrt{\pi}}, \ldots, \frac{\cos nx}{\sqrt{\pi}}, \frac{\sin nx}{\sqrt{\pi}},$$

and by the theorem on best mean square approximation of f by the Fourier partial sums (Theorem 15.5.1), we have

$$\|s_n - f\|_2^2 \leq \|\sigma_n - f\|_2^2 \leq \frac{\epsilon}{2} < \epsilon.$$

This proves what we wanted, as it shows that

$$\lim_{n\to\infty} \int_{-\pi}^{\pi} [f(x) - s_n(x)]^2\, dx = 0,$$

and again, by (15.22), this is equivalent to Parseval's equality. □

18.3.3. Mean Square Convergence for $\mathcal{R}[-\pi,\pi]$. We show here that every Riemann integrable function on $[-\pi,\pi]$ (or any interval of length 2π) can be approximated arbitrarily closely in the L^2 norm by its Fourier partial sums. The argument proceeds as follows: (i) every function $f \in \mathcal{R}[-\pi,\pi]$ can be approximated in norm by step functions; (ii) every step function can be approximated in norm by a continuous function on $[-\pi,\pi]$; (iii) thus, every Riemann integrable function on $[-\pi,\pi]$ can be approximated arbitrarily closely in the L^2 norm by a continuous function; and finally, (iv) an application of Theorem 18.3.4 and Theorem 15.5.1 completes the argument.

Our **step functions** $g : [-\pi,\pi] \to \mathbf{R}$ have the form

$$g = \sum_{k=1}^{m} a_k \chi_{E_k} \tag{18.4}$$

where the E_k are the disjoint open intervals corresponding to a partition $P = \{x_0 = a, x_1, \ldots, x_{m-1}, x_m = b\}$ of $[-\pi,\pi]$. For purposes of integration, it does not matter how such a g is defined at specific points x_k of the partition. We show now that any function f in $\mathcal{R}[-\pi,\pi]$ can be approximated arbitrarily closely in the L^2 norm by such a step function.

Lemma 18.3.5. *Let $f \in \mathcal{R}[-\pi,\pi]$. For every $\epsilon > 0$ there is a step function g of the form (18.4) on $[-\pi,\pi]$ such that*

$$\|f - g\|_2 < \epsilon.$$

Proof. Since $f \in \mathcal{R}[-\pi,\pi]$, there is a number M such that $|f(x)| \le M$ for all $x \in [-\pi,\pi]$. Given $\epsilon > 0$, there is a partition $P = \{a, x_1, \ldots, x_{m-1}, b\}$ of $[-\pi,\pi]$ such that

$$\int_{-\pi}^{\pi} f(x)\,dx - L(f,P) < \frac{\epsilon^2}{2M}.$$

Define $g(x) = \inf\{f(x) : x_{k-1} \le x \le x_k\} =: a_k$ if $x_{k-1} < x < x_k$, for $1 \le k \le m$. Then g is a step function taking the form (18.4), and $g(x) \le f(x) \le M$ a.e. in $[-\pi,\pi]$. Moreover, by definition of the lower sum for P, $\int_{-\pi}^{\pi} g(x)\,dx = L(f,P)$. Hence,

$$\int_{-\pi}^{\pi} [f(x) - g(x)]\,dx = \int_{-\pi}^{\pi} f(x)\,dx - L(f,P) < \frac{\epsilon^2}{2M}.$$

Since $(f-g)^2 = f^2 - 2fg + g^2 = f(f-g) + g(g-f)$, we have

$$\begin{aligned}
\int_{-\pi}^{\pi} [f(x) - g(x)]^2\,dx &\le \int_{-\pi}^{\pi} |f(x)|\,|f(x) - g(x)|\,dx + \int_{-\pi}^{\pi} |g(x)|\,|g(x) - f(x)|\,dx \\
&\le 2M \int_{-\pi}^{\pi} [f(x) - g(x)]\,dx \qquad (\text{since } g(x) \le f(x)) \\
&< 2M \frac{\epsilon^2}{2M} < \epsilon^2,
\end{aligned}$$

and therefore $\|f - g\|_2 < \epsilon$. □

The step functions (18.4) on $[-\pi,\pi]$ can be approximated arbitrarily closely in the L^2 norm by continuous functions on $[-\pi,\pi]$.

Lemma 18.3.6. *Let $g : [-\pi, \pi] \to \mathbf{R}$ be a step function of the form (18.4). Then for every $\epsilon > 0$ there is a function h in $CP[-\pi, \pi]$ such that*

$$\|g - h\|_2 < \epsilon.$$

Proof. Let $\epsilon > 0$. Suppose first that the step function $g = c\chi_{(-\pi,\pi)} \equiv c$ on $(-\pi, \pi)$. Whatever we might take g to be at $-\pi$ and π, suppose that $|g(x)| \le M$ for all $x \in [-\pi, \pi]$. Let

$$h(x) = \begin{cases} c & \text{if } -\pi + \delta \le x \le \pi - \delta, \\ 0 & \text{if } x = -\pi \text{ or } x = \pi, \end{cases}$$

where δ is to be determined, and then extend h to all of $[-\pi, \pi]$ by requiring it to be linear over the intervals $[-\pi, -\pi + \delta]$ and $[\pi - \delta, \pi]$, and continuous on $[-\pi, \pi]$.

We have $|g(x) - h(x)| \le |g(x)| + |h(x)| \le M + c \le 2M$ for all $x \in [-\pi, \pi]$. Therefore $|g(x) - h(x)|^2 \le 4M^2$ for all $x \in [-\pi, \pi]$. Hence,

$$\begin{aligned} \int_{-\pi}^{\pi} [g(x) - h(x)]^2\, dx &= \int_{-\pi}^{-\pi+\delta} [g(x) - h(x)]^2\, dx + \int_{\pi-\delta}^{\pi} [g(x) - h(x)]^2\, dx \\ &\le 4M^2\delta + 4M^2\delta = 8M^2\delta. \end{aligned}$$

We may choose δ such that $\delta < \epsilon^2/8M^2$ to complete the definition of h. Then we have $\|g - h\|_2 < \epsilon$ as desired.

Now let g be an arbitrary step function of the form (18.4) and let P be the partition associated with g. By an argument similar to the one just given, replacing $-\pi$ and π by x_{k-1} and x_k for $1 \le k \le m$, we have that for each k, there is a continuous function h_k on $[x_{k-1}, x_k]$ such that $h_k(x_{k-1}) = 0 = h_k(x_k)$ and

$$\int_{x_{k-1}}^{x_k} [g(x) - h_k(x)]^2\, dx < \frac{\epsilon^2}{m}.$$

(See Figure 18.1.) Let $h(x) = h_k(x)$ for $x_{k-1} \le x \le x_k$, $1 \le k \le m$. Then h is continuous on $[-\pi, \pi]$, because for $k = 2, \ldots, m$, $h_k(x_{k-1}) = 0 = h_k(x_k)$. Note also that $h(-\pi) = 0 = h(\pi)$. We have

$$\int_{-\pi}^{\pi} [g(x) - h(x)]^2\, dx = \sum_{k=1}^{m} \int_{x_{k-1}}^{x_k} [g(x) - h_k(x)]^2\, dx < m\frac{\epsilon^2}{m} = \epsilon^2,$$

and hence $\|g - h\|_2 < \epsilon$. □

We noted that $h(-\pi) = 0 = h(\pi)$, because it means that h can be extended to a continuous function on the real line of period 2π, and this fact is useful in Theorem 18.3.8 below.

It now follows easily that every Riemann integrable function on $[-\pi, \pi]$ can be approximated arbitrarily closely in the L^2 norm by a continuous function.

Theorem 18.3.7. *Let $f \in \mathcal{R}[-\pi, \pi]$. For every $\epsilon > 0$ there is an $h \in CP[-\pi, \pi]$ such that*

$$\|f - h\|_2 < \epsilon.$$

Proof. Let $\epsilon > 0$. By Lemma 18.3.5 there is a step function g such that $\|f - g\|_2 < \epsilon/2$. By Lemma 18.3.6 there is a continuous function $h \in CP[-\pi, \pi]$ such that $\|g - h\|_2 < \epsilon/2$. Thus, $\|f - h\|_2 \le \|f - g\|_2 + \|g - h\|_2 < \epsilon$. □

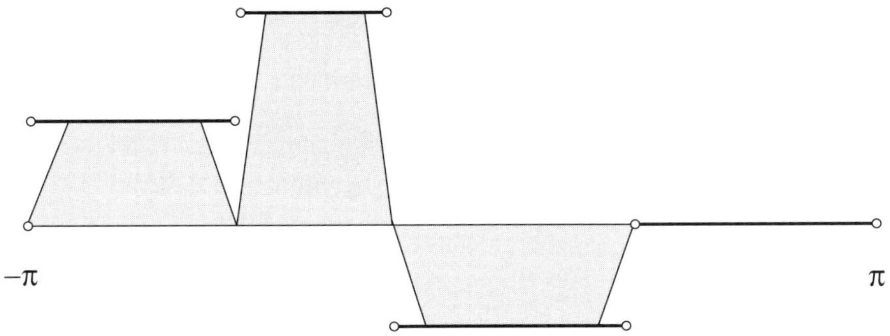

Figure 18.1. Approximating a step function by a continuous function.

The next result extends the result of Theorem 18.3.4.

Theorem 18.3.8. *Let $f \in \mathcal{R}[-\pi, \pi]$ and let*

$$s_n = \frac{1}{2}a_0 + \sum_{k=1}^{n}(a_k \cos kx + b_k \sin kx)$$

be the n-th partial sum of the Fourier series of f. Then

$$\lim_{n\to\infty} \|f - s_n\|_2^2 = \int_{-\pi}^{\pi} |f(x) - s_n(x)|^2\, dx = 0,$$

and Parseval's equality holds:

$$\frac{1}{2}\pi a_0^2 + \pi\sum_{k=1}^{\infty}(a_k^2 + b_k^2) = \int_{-\pi}^{\pi} f^2(x)\, dx.$$

Proof. Let $f \in \mathcal{R}[-\pi, \pi]$ and let s_n be the n-th partial sum of the Fourier series of f. Given $\epsilon > 0$, there is a continuous function $h \in CP[-\pi, \pi]$ such that $\|f - h\|_2 < \epsilon/2$. Let t_n be the partial sums of the Fourier series of h. By Theorem 18.3.4, $\lim_{n\to\infty} \|h - t_n\|_2 = 0$, so there is an $N = N(\epsilon)$ such that if $n \geq N$, then $\|h - t_n\|_2 < \epsilon/2$. By Theorem 15.5.1 on the best mean square approximation of f by the Fourier partial sums, if $n \geq N$, then

$$\|f - s_n\|_2 \leq \|f - t_n\|_2 \leq \|f - h\|_2 + \|h - t_n\|_2 < \epsilon/2 + \epsilon/2 = \epsilon.$$

This shows that

$$\lim_{n\to\infty} \int_{-\pi}^{\pi} [f(x) - s_n(x)]^2\, dx = 0,$$

and by (15.22), this is equivalent to Parseval's equality. □

Note that if f and g are Riemann integrable functions defined on $[-\pi, \pi]$, and f and g have the same Fourier series (the same Fourier coefficients), then Theorem 18.3.8 implies that $\|f - g\|_2 = 0$, and hence $f = g$ a.e. in $[-\pi, \pi]$.

18.4. Hilbert Spaces of Integrable Functions

The first goal of this section is to define the space of square integrable functions on a measure space, and show that it is a complete inner product space, a Hilbert space. In particular, this result applies to the square integrable functions on the Lebesgue measure space $([a,b], \mathcal{M}_{[a,b]}, m_{[a,b]})$, which we denote simply by $L^2[a,b]$.

The second goal is to show that the normalized standard trigonometric set is a maximal orthonormal set in the space $L^2[a,b]$ when $b-a=2\pi$. Thus, the trigonometric set is a maximal orthonormal set in the Hilbert space $L^2[-\pi,\pi]$ of square integrable functions on $[-\pi,\pi]$.

The section culminates in the Riesz-Fischer theorem, which establishes an isomorphism between the Hilbert spaces $L^2[-\pi,\pi]$ and l^2.

We first define the space of square integrable functions on a measure space. The definitions are stated in enough generality to cover the spaces of most interest to us. Throughout, readers can think of (X,Σ,μ) as being $(E,\mathcal{M},m)$ for a measurable set E of $\mathbf{R}$, or $(E,\mathcal{M}_n,m_n)$ for a measurable set E of $\mathbf{R}^n$.

Definition 18.4.1. *Let (X,Σ,μ) be a measure space. We denote by $L^2(X,\Sigma,\mu)$ the set of functions $f: X \to \mathbf{R}$ such that f is measurable and $\int_X |f|^2\,d\mu < \infty$, with the identification that $f = g$ if and only if the set of points where $f(x) \neq g(x)$ is a set of measure zero.*

Thus the elements of $L^2(X,\Sigma,\mu)$ are equivalence classes of functions, though we operate in practice with individual functions, always keeping in mind the equivalence involved in this definition.

Our first goal is to prove that $L^2(X,\Sigma,\mu)$ is a complete inner product space, a Hilbert space. In order to define the inner product in $L^2(X,\Sigma,\mu)$ as $(f,g) = \int_X fg\,d\mu$, we need to know that if $f,g \in L^2(X,\Sigma,\mu)$, then fg is integrable. If $f,g \in L^2(X,\Sigma,\mu)$, then f and g are measurable, hence fg is measurable, and therefore $|fg|$ is measurable. We have

$$\begin{aligned} 0 \le \int_X |fg|\,d\mu &= \int_{|f|\ge|g|} |fg|\,d\mu + \int_{|g|>|f|} |fg|\,d\mu \\ &\le \int_{|f|\ge|g|} |f|^2\,d\mu + \int_{|g|>|f|} |g|^2\,d\mu \\ &\le \int_X |f|^2\,d\mu + \int_X |g|^2\,d\mu < \infty. \end{aligned}$$

Thus, fg is integrable. We can now show that $L^2(X,\Sigma,\mu)$ is a vector space. To do so, we only need to show that it is closed under the operations of scalar multiplication and addition. Readers can easily verify closure of $L^2(X,\Sigma,\mu)$ under scalar multiplication by real numbers. Let $f,g \in L^2(X,\Sigma,\mu)$. Then

$$\begin{aligned} \int_X |f+g|^2\,d\mu &= \int_X (f^2 + 2fg + g^2)\,d\mu \\ &\le \int_X f^2\,d\mu + 2\int_X |fg|\,d\mu + \int_X g^2\,d\mu < \infty, \end{aligned}$$

and thus $f+g \in L^2(X,\Sigma,\mu)$. We summarize the facts thus far, as follows.

Theorem 18.4.2. *For any measure space (X, Σ, μ), the vector space $L^2(X, \Sigma, \mu)$ is an inner product space with inner product*

$$(f, g) = \int_X fg \, d\mu, \qquad f, g \in L^2(X, \Sigma, \mu),$$

and a normed space with norm

$$\|f\|_2 = (f, f)^{1/2} = \Big(\int_X f^2 \, d\mu\Big)^{1/2}, \qquad f \in L^2(X, \Sigma, \mu).$$

We note that $(f, f) = 0$ means $\int_X f^2 \, d\mu = 0$, hence $f = 0$ a.e. in X, by Exercise 17.4.8. The Cauchy-Schwarz inequality in $L^2(X, \Sigma, \mu)$,

$$|(f, g)| \leq \|f\|_2 \, \|g\|_2,$$

displayed in detail, is

$$\Big|\int_X fg \, d\mu\Big| \leq \Big(\int_X f^2 \, d\mu\Big)^{1/2} \Big(\int_X g^2 \, d\mu\Big)^{1/2}$$

for $f, g \in L^2(X, \Sigma, \mu)$.

Theorem 18.4.3. *The vector space $L^2(X, \Sigma, \mu)$ is a Hilbert space, complete in the norm induced by the inner product.*

Proof. Let (f_k), $k \in \mathbf{N}$, be a Cauchy sequence in $L^2(X, \Sigma, \mu)$. Then there is a subsequence (f_{n_k}), $k \in \mathbf{N}$, such that

$$\|f_{n_{k+1}} - f_{n_k}\|_2 < \frac{1}{2^k}$$

for all k. Choose any function $h \in L^2(X, \Sigma, \mu)$. By the Cauchy-Schwarz inequality,

$$\int_X \big|h(f_{n_{k+1}} - f_{n_k})\big| \, d\mu \leq \frac{1}{2^k} \|h\|_2.$$

Hence,

$$\sum_{k=1}^{\infty} \int_X \big|h(f_{n_{k+1}} - f_{n_k})\big| \, d\mu \leq \|h\|_2.$$

The monotone convergence theorem (Theorem 17.4.1) applies to the increasing sequence of nonnegative, measurable partial sums of $\sum_{k=1}^{\infty} \big|h(f_{n_{k+1}} - f_{n_k})\big|$. Thus we can interchange the summation and the integral, to obtain

$$\int_X \Big(\sum_{k=1}^{\infty} \big|h(f_{n_{k+1}} - f_{n_k})\big|\Big) \, d\mu \leq \|h\|_2.$$

Therefore we have

$$\sum_{k=1}^{\infty} \big|h(x)(f_{n_{k+1}}(x) - f_{n_k}(x))\big| = |h(x)| \sum_{k=1}^{\infty} \big|(f_{n_{k+1}}(x) - f_{n_k}(x))\big| < \infty \tag{18.5}$$

almost everywhere in X. Hence,

$$\sum_{k=1}^{\infty} \big|(f_{n_{k+1}}(x) - f_{n_k}(x))\big| < \infty$$

almost everywhere in X: To see this, note that if this series diverges on a set E of positive measure, we could choose h to be nonzero on a subset of E of positive measure, which would contradict (18.5).

The m-th partial sum of the series $\sum_{k=1}^{\infty} \left(f_{n_{k+1}}(x) - f_{n_k}(x)\right)$, which converges absolutely almost everywhere in X, is given by

$$\sum_{k=1}^{m} \left(f_{n_{k+1}}(x) - f_{n_k}(x)\right) = f_{n_{m+1}}(x) - f_{n_1}(x),$$

and hence we can define

$$f(x) = \lim_{m\to\infty} f_{n_m}(x)$$

at points where the series converges, and set $f = 0$ at points in the complementary set of measure zero.

We have a pointwise limit a.e. of the subsequence (f_{n_m}) to the function f. We now want to show that $f = \lim_{k\to\infty} f_k$ in norm. Let $\epsilon > 0$. Since (f_k) is Cauchy, there exists a positive integer $N = N(\epsilon)$ such that if $n, k \geq N$, then $\|f_n - f_k\|_2 < \epsilon$. For each $k \geq N$, Fatou's lemma (Theorem 17.4.4) implies

$$\begin{aligned} \int_X |f - f_k|^2 \, d\mu &= \int_X \liminf_{m\to\infty} |f_{n_m} - f_k|^2 \, d\mu \\ &\leq \liminf_{m\to\infty} \int_X |f_{n_m} - f_k|^2 \, d\mu \\ &= \liminf_{m\to\infty} \|f_{n_m} - f_k\|_2^2 \leq \epsilon^2, \end{aligned} \tag{18.6}$$

where we used the fact that $m \geq N$ implies $n_m \geq N$, hence $\|f_{n_m} - f_k\|_2 < \epsilon$. It follows from (18.6) that $f - f_k \in L^2(X, \Sigma, \mu)$ for $k \geq N$. Thus,

$$f = f_k + (f - f_k) \in L^2(X, \Sigma, \mu),$$

and by (18.6),

$$\|f - f_k\|_2 \leq \epsilon, \quad \text{for } k \geq N = N(\epsilon).$$

This is true for every $\epsilon > 0$, and hence $f_k \to f$ in $L^2(X, \Sigma, \mu)$. Therefore $L^2(X, \Sigma, \mu)$ is complete. □

A slightly different proof of Theorem 18.4.3 is considered in Exercise 18.4.6.

Mirroring our comments after the proof of completeness of $L^1(X, \Sigma, \mu)$ in Theorem 17.6.3, and recalling Example 17.6.4, we again should not read too much into the existence of a pointwise convergent subsequence of the Cauchy sequence (f_k) in Theorem 18.4.3.

Example 18.4.4. We again consider $X = [0, 1]$ and the sequence of functions (f_j) defined in Example 17.6.4. Since each f_j is the characteristic function of a subinterval of $[0, 1]$, each f_j is square integrable and therefore an element of $L^2[0, 1]$. Moreover, by the construction of the sequence (f_j) in Example 17.6.4, $\|f_j\|_2 = 1/\sqrt{n}$ if f_j is the characteristic function of an interval of length $1/n$. Therefore $f_j \to 0$ in norm in $L^2[0, 1]$. However, as shown in the earlier example, the sequence f_j fails to converge pointwise everywhere in $[0, 1]$, since $\limsup f_j(x) = 1$ and $\liminf f_j(x) = 0$ for every $x \in [0, 1]$. △

Now let $[a, b]$ be a real interval, and consider $(X, \Sigma, \mu) = ([a, b], \mathcal{M}, m)$, the Lebesgue measure space induced by the Lebesgue measure on $\mathbf{R}$ in accordance with the construction in Exercise 16.5.10. We shall simply write $L^2[a, b]$ for the space $L^2([a, b], \mathcal{M}, m)$. If $b - a = 2\pi$, then we know that the normalized standard trigonometric set is an orthornormal set in $L^2[a, b]$. In particular, the trigonometric set is an orthornormal set in $L^2[-\pi, \pi]$.

We wish to show that the trigonometric set is a maximal orthornormal set in $L^2[-\pi, \pi]$. This will follow in Theorem 18.4.7 below from the following facts: (a) the density of the Riemann integrable functions in $L^2[-\pi, \pi]$; and (b) the mean square convergence result for $\mathcal{R}[-\pi, \pi]$ in Theorem 18.3.8.

We consider a general closed interval $[a, b]$ for the density result.

Theorem 18.4.5. *Let $f \in L^2[a, b]$. Given any $\epsilon > 0$, there is a simple function $h \in L^2[a, b]$ such that $\|f - h\|_2 < \epsilon$. Thus, the simple functions in $L^2[a, b]$ are dense in $L^2[a, b]$.*

Proof. Suppose $f \in L^2[a, b]$ and $f \geq 0$ on $[a, b]$. By Theorem 17.2.3, there is an increasing sequence s_n of nonnegative simple functions such that $\lim_{n \to \infty} s_n = f$ pointwise on $[a, b]$. Since $0 \leq f(x) - s_n(x) \leq f(x)$ for all x,

$$|f(x) - s_n(x)|^2 \leq |f(x)|^2,$$

and $|f|^2$ is integrable. By the dominated convergence theorem (Theorem 17.4.5),

$$\lim_{n \to \infty} \int_{[a,b]} |f - s_n|^2 \, dm = \int_{[a,b]} \lim_{n \to \infty} |f - s_n|^2 \, dm = \int_{[a,b]} 0 \, dm = 0.$$

Thus, $\|f - s_n\|_2^2 \to 0$, and hence $\|f - s_n\|_2 \to 0$, as $n \to \infty$. Therefore given any $\epsilon > 0$, there is an $N = N(\epsilon)$ such that if $n \geq N$, then $\|f - s_n\|_2 < \epsilon$. This proves the result for nonnegative $f \in L^2[a, b]$.

Now let $f \in L^2[a, b]$ and suppose f takes positive and negative values. Write $f = f^+ - f^-$. Since $f^+, f^- \geq 0$, given $\epsilon > 0$, there exist simple functions h_1 and h_2 such that

$$\|f^+ - h_1\|_2 < \frac{\epsilon}{2} \qquad \text{and} \qquad \|h_2 - f^-\|_2 < \frac{\epsilon}{2}.$$

Then $h = h_1 - h_2$ is a simple function, and

$$\begin{aligned} \|f - h\|_2 &= \|f^+ - f^- - h_1 + h_2\|_2 \\ &\leq \|f^+ - h_1\|_2 + \|h_2 - f^-\|_2 < \epsilon. \end{aligned}$$

Since $\epsilon > 0$ is arbitrary, this completes the proof. □

In order to show that the Riemann integrable functions are dense in $L^2[a, b]$, it suffices to show that the step functions on $[a, b]$ are dense in $L^2[a, b]$, since the step functions are dense in $\mathcal{R}[a, b]$. (See Definition 17.2.2 for the definition of step functions.)

Theorem 18.4.6. *Let $f \in L^2[a, b]$. Given any $\epsilon > 0$, there is a step function $g \in L^2[a, b]$ such that $\|f - g\|_2 < \epsilon$. Thus, the step functions are dense in $L^2[a, b]$.*

Proof. First, suppose that $f = \chi_E$ where E is a measurable set in $[a, b]$. Let $\epsilon > 0$. Since E has finite measure, Theorem 16.5.10 (statement 2) implies that there is an open set O in $[a,b]$ such that $E \subset O$ and $m(O-E) = m(O) - m(E) < \epsilon^2/2$, hence $m(O) < m(E) + \epsilon^2/2$. We can write

$$O = \bigcup_{n=1}^{\infty} O_n,$$

where the O_n are pairwise disjoint open intervals relative to $[a,b]$. (If O is a finite union of open intervals in $[a,b]$, then we may append empty sets to form a countable union for the argument that follows, or simply note that χ_O is a step function that approximates χ_E and proceed as shown below.) Since $m(O) = \sum_{n=1}^{\infty} m(O_n)$ and $m(O) < \infty$, the series converges. Thus, there is an $N = N(\epsilon)$ such that

$$\sum_{n=N+1}^{\infty} m(O_n) < \frac{\epsilon^2}{2}.$$

Let $V = \bigcup_{n=1}^{N} O_n$. Then χ_V is a step function that approximates χ_O, and hence approximates χ_E. We now show that in the norm on $L^2[a,b]$, we have $\|\chi_V - \chi_E\|_2 < \epsilon$. To see this, note first that

$$V - E \subset O - E \quad \Longrightarrow \quad m(V-E) \le m(O-E) < \frac{\epsilon^2}{2}.$$

We also have $E - V \subset \bigcup_{n=N+1}^{\infty} O_n$, so $m(E-V) \le \epsilon^2/2$. Hence, by the definition of characteristic function, we have

$$\begin{aligned}
\int_{[a,b]} |\chi_V - \chi_E|^2\, dm &= \int_{[a,b]-(V\cup E)} |\chi_V - \chi_E|^2\, dm + \int_{V-E} |\chi_V - \chi_E|^2\, dm \\
&\quad + \int_{E-V} |\chi_V - \chi_E|^2\, dm + \int_{E\cap V} |\chi_V - \chi_E|^2\, dm \\
&= \int_{V-E} |\chi_V - \chi_E|^2\, dm + \int_{E-V} |\chi_V - \chi_E|^2\, dm \\
&= \int_{V-E} (1)^2\, dm + \int_{E-V} (1)^2\, dm \\
&= m(V-E) + m(E-V) < \epsilon^2/2 + \epsilon^2/2 = \epsilon^2.
\end{aligned}$$

It follows that $\|\chi_V - \chi_E\|_2 < \epsilon$.

Now let f be an arbitrary element of $L^2[a,b]$ and let $\epsilon > 0$. By Theorem 18.4.5, there is a simple function $h \in L^2[a,b]$ such that $\|f - h\|_2 < \epsilon/2$. We may write

$$h = \sum_{k=1}^{M} c_k \chi_{E_k},$$

where for $1 \le k \le M$, $c_k \ne 0$ and E_k is measurable in $[a,b]$. By the result proved above for characteristic functions of measurable sets, there exist step functions g_k, $1 \le k \le M$, such that

$$\|g_k - \chi_{E_k}\|_2 < \frac{\epsilon}{2M|c_k|}$$

for each k. Since the g_k are step functions, the linear combination $g := \sum_{k=1}^{M} c_k g_k$ is also a step function, and

$$\begin{aligned}
\|f-g\|_2 &\le \|f-h\|_2 + \|h-g\|_2 \\
&= \|f-h\|_2 + \left\| \sum_{k=1}^{M} c_k \chi_{E_k} - \sum_{k=1}^{M} c_k g_k \right\|_2 \\
&\le \|f-h\|_2 + \sum_{k=1}^{M} |c_k| \, \|\chi_{E_k} - g_k\|_2 \\
&< \frac{\epsilon}{2} + \sum_{k=1}^{M} |c_k| \frac{\epsilon}{2M|c_k|} \\
&= \frac{\epsilon}{2} + \frac{\epsilon}{2} = \epsilon.
\end{aligned}$$

This completes the proof that the step functions are dense in $L^2[a,b]$. □

We can now show that when $b-a = 2\pi$, the normalized trigonometric set $\{\phi_k\}$ is a maximal orthonormal set in $L^2[a,b] = L^2[a, a+2\pi]$. It suffices to prove this for the interval $[-\pi, \pi]$.

Theorem 18.4.7. *The normalized standard trigonometric set $\{\phi_k : k \ge 0\}$ is a maximal orthonormal set in $L^2[-\pi, \pi]$.*

Proof. Let $f \in L^2[-\pi,\pi]$ and let s_n be the n-th partial sum of the Fourier series of f with respect to the ϕ_k. We want to show that $\|f - s_n\|_2 \to 0$ as $n \to \infty$, as the maximality of the trigonometric set in $L^2[-\pi,\pi]$ then follows from Theorem 18.2.10. Let $\epsilon > 0$. By Theorem 18.4.6, there is a step function $g \in \mathcal{R}[-\pi,\pi]$ such that

$$\|f-g\|_2 < \frac{\epsilon}{2}.$$

If t_n is the n-th partial sum of the Fourier series of g, then by the mean square convergence result in $\mathcal{R}[-\pi,\pi]$ (Theorem 18.3.8), we have

$$\lim_{n\to\infty} \|g - t_n\|_2 = 0.$$

Thus, there is an $N = N(\epsilon)$ such that if $n \ge N$, then

$$\|g - t_n\|_2 < \frac{\epsilon}{2}.$$

By the best mean square approximation theorem (Theorem 15.5.1), we have

$$\|f - s_n\|_2 \le \|f - t_n\|_2$$

for all n, since t_n is a linear combination of $\phi_0, \phi_1, \phi_2, \ldots, \phi_{2n-1}, \phi_{2n}$. Therefore if $n \ge N$, then

$$\|f - s_n\|_2 \le \|f - t_n\|_2 \le \|f-g\|_2 + \|g - t_n\|_2 < \epsilon.$$

This shows that $\lim_{n\to\infty} \|f - s_n\|_2 = 0$ and completes the proof. □

Theorem 18.4.7 can also be expressed as the density of the trigonometric polynomials in $L^2[-\pi, \pi]$. (See also Exercise 18.4.5).

We also note that the density of $\mathcal{R}[-\pi, \pi]$ in $L^2[-\pi, \pi]$ combined with Theorem 18.3.7 shows that $C[-\pi, \pi]$ is dense in $L^2[-\pi, \pi]$.

If $f \in L^2[-\pi, \pi]$, then its Fourier coefficients with respect to the normalized trigonometric set $\{\phi_k : k \geq 0\}$ are defined. Thus we have a linear mapping $\Phi : L^2[-\pi, \pi] \to l^2$ given by

$$\Phi(f) = \big((f, \phi_k)\big)_{k=0}^{\infty},$$

and Bessel's inequality implies that $\Phi(f)$ is in l^2. Parseval's equality holds, since it is equivalent to $\lim_{n\to\infty} \|f - s_n\|_2 = 0$, and implies that Φ preserves norms,

$$\|\Phi(f)\|_2^2 = \sum_{k=0}^{\infty} (f, \phi_k)^2 = \|f\|_2^2.$$

Thus Φ is called an *isometry*. The norm-preserving property implies that Φ is one-to-one, for if $\Phi(f) = \Phi(g)$, then $\Phi(f - g) = 0$, and hence $f = g$ in $L^2[-\pi, \pi]$.

It is natural to ask whether $\Phi : L^2[-\pi, \pi] \to l^2$ is onto l^2. Does every element of l^2 correspond to a square integrable function on $[-\pi, \pi]$? Is the space $L^2[-\pi, \pi]$ large enough to contain a function whose Fourier coefficients with respect to the set $\{\phi_k\}$ equal a prescribed sequence in l^2? The Riesz-Fischer theorem tells us that Φ does map $L^2[-\pi, \pi]$ *onto* l^2. This fact was, and remains, a triumph of the Lebesgue theory of the integral.

Theorem 18.4.8 (Riesz-Fischer). *Let $\{\phi_k : k \geq 0\}$ be the normalized standard trigonometric set on $[-\pi, \pi]$. Suppose $\xi = (c_k)_{k=0}^{\infty} \in l^2$, and we define*

$$s_n = \sum_{k=0}^{n} c_k \phi_k.$$

Then there is an element $f \in L^2[-\pi, \pi]$ such that s_n converges in norm to f, and ξ is the sequence of Fourier coefficients of f with respect to the orthonormal set $\{\phi_k : k \geq 0\}$.

Proof. Since $\{\phi_k\}$ is an orthonormal set, if $n > m$, then by the Pythagorean theorem,

$$\|s_n - s_m\|_2^2 = \Big\| \sum_{k=m+1}^{n} c_k \phi_k \Big\|_2^2 = \sum_{k=m+1}^{n} c_k^2.$$

Since $(c_k) \in l^2$, (s_n) is a Cauchy sequence in $L^2[-\pi, \pi]$. Since $L^2[-\pi, \pi]$ is complete, there is an $f \in L^2[-\pi, \pi]$ such that $\lim_{n\to\infty} \|f - s_n\|_2 = 0$. Hence, $f = \sum_{k=0}^{\infty} c_k \phi_k$, the convergence being in norm.

Let $n > k$. Then

$$(f, \phi_k) - c_k = (f, \phi_k) - (s_n, \phi_k) = (f - s_n, \phi_k).$$

The Cauchy-Schwarz inequality then gives

$$\big|(f, \phi_k) - c_k\big| = \big|(f - s_n, \phi_k)\big| \leq \|f - s_n\|_2 \, \|\phi_k\|_2 = \|f - s_n\|_2,$$

since $\|\phi_k\|_2 = 1$. Letting $n \to \infty$, we conclude that $c_k = (f, \phi_k)$, and this is true for every $k \geq 0$. □

The Riesz-Fischer theorem tells us that $\Phi(f) = ((f, \phi_k))_{k=0}^{\infty}$ not only preserves norms (an isometry), but also Φ preserves inner products: Given any ξ and η in l^2, there exist f and g in $L^2[-\pi, \pi]$ such that

$$(f, g) = \sum_{k=0}^{\infty} (f, \phi_k)(g, \phi_k) = \big(\Phi(f), \Phi(g)\big) = (\xi, \eta).$$

The mapping Φ thus defines what is called a **unitary equivalence** between the Hilbert spaces $L^2[-\pi, \pi]$ and l^2. We note that the space of Riemann integrable functions $\mathcal{R}[-\pi, \pi]$ is not unitarily equivalent to l^2, as there are elements of l^2 that are not the sequence of Fourier coefficients of any Riemann integrable function.

The Riesz-Fischer theorem allows us to view $L^2[-\pi, \pi]$ as an infinite-dimensional analogue of finite-dimensional Euclidean space. In this view every function is the sum, in the sense of convergence in norm, of its Fourier series with respect to a maximal orthonormal set, with Parseval's equality playing the role of the Pythagorean theorem. The space $L^2[-\pi, \pi]$ is complete and fills the gaps where Cauchy sequences in $\mathcal{R}[-\pi, \pi]$ do not converge to an element of $\mathcal{R}[-\pi, \pi]$. Of course there are significant differences between this infinite-dimensional space and the finite-dimensional case, including the fact that the unit ball in $L^2[-\pi, \pi]$ is not compact (Exercise 18.4.1).

In this book we have concentrated on the closed interval $[a, b]$. There are many boundary value problems of classical and modern physics, known as Sturm-Liouville problems, that require for their complete solution and understanding, the space of square integrable functions on an interval D of the real numbers. We examined a few of the simplest Sturm-Liouville problems in our coverage of the heat equation and wave equation examples on $D = [0, \pi]$. In cases where the interval D is unbounded, an inner product can be defined using a weight function specific to the problem which allows for finite integrals over the unbounded domain. The space of square integrable functions on D, defined in terms of a problem-specific weighted inner product, is complete. These spaces are Hilbert spaces. Some of the best known examples of orthonormal sets arise from Sturm-Liouville boundary value problems as the eigenfunctions of second-order linear differential operators, or functions closely related to the eigenfunctions. For various intervals D, some examples are: the Legendre polynomials on $(-1, 1)$, the Hermite polynomials and associated Hermite functions on $(-\infty, \infty)$, and the Laguerre polynomials on $(0, \infty)$. The Hermite functions and the Laguerre polynomials are important in quantum mechanics. The Bessel functions are important in the study of the wave equation, in particular the study of vibrations of circular membranes such as drumheads. Folland [**14**] is an interesting text and reference for these problems and their mathematical properties.

We end with a final comment on the term-by-term integration of Fourier series. Using his new integral, Lebesgue showed that the Fourier series of an integrable function can always be integrated term-by-term (Exercise 18.4.2). Recall that it was this term-by-term integration, under the assumption of *uniform convergence* of the series, that motivated the appropriate formulas for the Fourier coefficients. Thus our definition of Fourier coefficients for any integrable function is fully justified from the standpoint of the Lebesgue integral.

Exercises.

Exercise 18.4.1. Show that the unit ball in $L^2[-\pi,\pi]$ is not compact. *Hint*: Consider l^2 and see Exercise 9.1.12.

Exercise 18.4.2. *Term-by-term integration of Fourier series*
Suppose that f is integrable and square integrable on $[-\pi,\pi]$, so that

$$\int_{-\pi}^{\pi} |f(x)|\, dx < \infty \qquad \text{and} \qquad \int_{-\pi}^{\pi} |f(x)|^2\, dx < \infty.$$

Let

$$s_n = \frac{1}{2}a_0 + \sum_{k=1}^{n} (a_k \cos kx + b_k \sin kx)$$

be the Fourier partial sums for f. Show that if $-\pi \le a \le \pi$, then the series

$$\int_a^x \frac{a_0}{2}\, d\theta + \sum_{k=1}^{\infty} \int_a^x (a_k \cos k\theta + b_k \sin k\theta)\, d\theta$$

converges uniformly to $\int_a^x f(\theta)\, d\theta$ on $[-\pi,\pi]$. *Hint*: This is equivalent to showing that the sequence $\int_a^x s_n(\theta)\, d\theta$ converges uniformly to $\int_a^x f(\theta)\, d\theta$ on $[-\pi,\pi]$, which is equivalent to showing that $\int_a^x (s_n(\theta) - f(\theta))\, d\theta$ converges uniformly to zero on $[-\pi,\pi]$. Use the Cauchy-Schwarz inequality. This result does not assume that the Fourier series of f converges uniformly or even pointwise; it says that the term-by-term integrated series is justified and yields the correct result for the integral of f.

Exercise 18.4.3. Show that the function $f(x) = 1/\sqrt{x}$ is Lebesgue integrable on $(0,1)$, but that its square $f^2(x) = 1/x$ is not Lebesgue integrable on $(0,1)$.

Exercise 18.4.4. By Exercise 8.3.17, every finite-dimensional subspace of a normed space is complete, and hence closed. On the other hand, it is clear that a closed subspace is necessarily complete. Give an example of an infinite-dimensional subspace of $L^2[-\pi,\pi]$ which is dense in $L^2[-\pi,\pi]$ and not closed. Consider the same question for l^2.

Exercise 18.4.5. Let V be a Hilbert space with an orthonormal basis $\{u_k : k \in \mathbf{N}\}$. Show that the following condition is equivalent to conditions (1)-(3) of Theorem 18.2.10:

(4) Finite linear combinations of elements of $\{u_k : k \in \mathbf{N}\}$ are dense in V.

Suggestion: Show that (4) implies (1) ($f \in V$ and $(f, u_k) = 0$ for all k implies $f = 0$), and (3) (Parseval's equality) implies (4).

Exercise 18.4.6. At the beginning of the proof that $L^2(X,\Sigma,\mu)$ is a Hilbert space, we let (f_n), $n \in \mathbf{N}$, be a Cauchy sequence in $L^2(X,\Sigma,\mu)$, such that the subsequence (f_{n_k}), $k \in \mathbf{N}$, satisfied

$$\|f_{n_{k+1}} - f_{n_k}\|_2 < \frac{1}{2^k}$$

for all k. Instead of choosing an arbitrary function $h \in L^2(X,\Sigma,\mu)$ and proceeding as in the text, show that we could obtain a pointwise limit almost everywhere in X

for this subsequence, as follows:

1. Let
$$g_m = \sum_{k=1}^{m} |f_{n_{k+1}} - f_{n_k}| \qquad \text{and} \qquad g = \sum_{k=1}^{\infty} |f_{n_{k+1}} - f_{n_k}|.$$
Verify that for each m, g_m is measurable, and by the triangle inequality for the norm in $L^2(X, \Sigma, \mu)$, $\|g_m\|_2 < 1$.
2. Use Fatou's lemma (Theorem 17.4.4) to show that
$$\|g\|_2^2 \leq \liminf_{m\to\infty} \|g_m\|_2^2 \leq 1,$$
and hence $\|g\|_2 \leq 1$.
3. Conclude that $0 \leq g(x) < \infty$ almost everywhere in X, and therefore the series defining g converges a.e. in X.
4. Then show that the series
$$f_{n_1}(x) + \sum_{k=1}^{\infty} \left(f_{n_{k+1}}(x) - f_{n_k}(x)\right)$$
converges for almost all x in X to a limit function f. (The remainder of the proof then proceeds as before.)

Exercise 18.4.7. *A case where pointwise implies mean square convergence* Suppose $f_n \in L^2[a,b]$ for all n and $f_n \to f$ pointwise on $[a,b]$. Show that if there exists $g \in L^2[a,b]$ such that $|f_n(x)| \leq |g(x)|$ for all n and all $x \in [a,b]$, then $f_n \to f$ in norm. *Hint*: Apply the dominated convergence theorem.

Exercise 18.4.8. Consider the intervals $(0,1]$ and $[1,\infty)$ as measure spaces with Lebesgue measure.

1. Show that $f(x) = x^{-3/4}$ is in $L^1((0,1])$, but not in $L^2((0,1])$.
2. Show that $f(x) = x^{-3/4}$ is in $L^2([1,\infty))$, but not in $L^1([1,\infty))$.

18.5. Notes and References

This chapter is influenced by Folland [**14**], Hoffman [**30**], Rudin [**52**], and Johnsonbaugh and Pfaffenberger [**32**]. My thanks go to an anonymous reviewer for Examples 18.2.9-18.2.12.

For more on Fourier analysis and its applications, including a comprehensive look at Sturm-Liouville problems, see Folland [**14**]. Other books that discuss these problems in detail are González-Velasco [**20**], Haberman [**23**] and Strauss [**63**].

Epstein [**11**] and Kreyszig [**41**] are introductory texts on functional analysis that cover linear operators on Hilbert space. In this connection, a background in linear algebra from Halmos [**25**] can be helpful, as it projects a point of view towards the infinite-dimensional problems of Hilbert space while covering finite-dimensional spaces.

For an interesting book on topology and modern analysis, function spaces and function approximation, see Simmons [**59**].

Appendix A

The Schroeder-Bernstein Theorem

This appendix presents a formal proof of the Schroeder-Bernstein theorem.

A.1. Proof of the Schroeder-Bernstein Theorem

Theorem A.1.1 (Schroeder-Bernstein). *Let X and Y be sets. If there exists a one-to-one mapping $f : X \to Y$ and a one-to-one mapping $g : Y \to X$, then X and Y have the same cardinality.*

Proof. Let $Y_1 = f(X)$, $\tilde{X}_1 = g(Y)$ and $X_1 = g(Y_1)$. If $Y_1 = Y$, or equivalently, $X_1 = \tilde{X}_1$, then f is one-to-one from X onto Y. Otherwise, if we construct a one-to-one mapping k from X onto $\tilde{X}_1$, then

$$h(x) := g^{-1}(k(x))$$

will be a one-to-one mapping of X onto Y, as desired. We now construct such a mapping $k : X \to \tilde{X}_1$. (See Figure A.1.)

Recall that $g \circ f$ is a one-to-one mapping and hence, by our definitions above, a one-to-one mapping from X onto its subset X_1. Let $X_0 = X$. Now define sets $\tilde{X}_n$ and X_n for $n = 1, 2, \ldots$ as follows:

$$\begin{aligned}
\tilde{X}_1 &= g(Y) \text{ (as above)} & X_1 &= g(f(X_0)) \text{ (as above)}\\
\tilde{X}_2 &= g(f(\tilde{X}_1)) & X_2 &= g(f(X_1))\\
\tilde{X}_3 &= g(f(\tilde{X}_2)) & X_3 &= g(f(X_2))\\
&\vdots & &\vdots\\
\tilde{X}_{n+1} &= g(f(\tilde{X}_n)) & X_{n+1} &= g(f(X_n)).
\end{aligned}$$

Since $X_1 \subset \tilde{X}_1$, we see by induction that $X_n \subset \tilde{X}_n$ for all $n \geq 1$. And since $\tilde{X}_1 \subset X_0$ we have $\tilde{X}_2 \subset X_1$, and by induction we conclude that $\tilde{X}_{n+1} \subset X_n$ for all

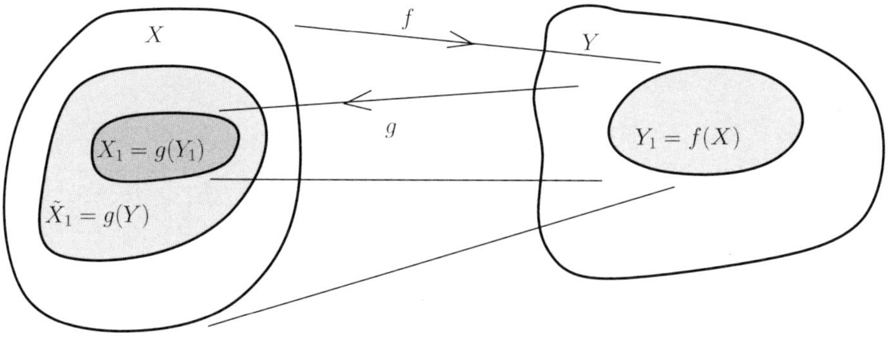

Figure A.1. Initial construction of sets in the Schroeder-Bernstein theorem.

$n \geq 1$. Thus, $X_1 \subset \tilde{X}_1 \subset X_0 = X$, and for all $n \geq 1$, we have

$$\tilde{X}_{n+1} \subset X_n \subset \tilde{X}_n.$$

Let $X_0 = X$ and define a mapping $k : X \to X$ by

$$k(x) = \begin{cases} g(f(x)) & \text{if } x \in X_n - \tilde{X}_{n+1},\ n = 0, 1, 2, \ldots, \\ x & \text{if } x \in \tilde{X}_n - X_n,\ n = 1, 2, 3, \ldots, \\ x & \text{if } x \in \bigcap_{n=1}^{\infty} X_n. \end{cases}$$

By the way the sets X_n and $\tilde{X}_n$ are constructed, k maps $(X_n - \tilde{X}_{n+1})$ onto $(X_{n+1} - \tilde{X}_{n+2})$ for $n = 0, 1, 2 \ldots$. Since the composition $g \circ f$ is one-to-one over each of these sets, we have that k is defined and one-to-one on X; moreover, the range of k must be

$$\left[\bigcup_{n=1}^{\infty}(X_n - \tilde{X}_{n+1})\right] \cup \left[\bigcup_{n=1}^{\infty}(\tilde{X}_n - X_n)\right] \cup \left[\bigcap_{n=1}^{\infty} X_n\right] = \tilde{X}_1. \tag{A.1}$$

(See Exercise A.1.1.) Therefore k is the desired one-to-one mapping from X onto $\tilde{X}_1$, and hence $h(x) = g^{-1}(k(x))$ is a one-to-one mapping of X onto Y. □

Exercise.

Exercise A.1.1. Verify that the mapping k is defined on all of X, and that the range of k equals $\tilde{X}_1$ as shown in (A.1).

Symbols and Notations

B.1. Symbols and Notations Reference List

$\triangle$	denotes end of an example
$\square$	denotes end of a proof
$\exists$	there exists
$\forall$	for every; for all
$\in$	belongs to, is an element of
$\emptyset$	the empty set
$a := b$	a is defined to be b
$a =: b$	b is defined to be a
$\equiv$	identically equal to, e.g., $\cos^2 t + \sin^2 t \equiv 1$
$S_1 \Longrightarrow S_2$	S_1 implies S_2; if S_1, then S_2;
	the **contrapositive** is: if not S_2, then not S_1
	the **converse** is: if S_2, then S_1
$S_1 \Longleftrightarrow S_2$	S_1 if and only if S_2
$\{x \in S : P\}$	the set of $x \in S$ with property P
$\subset$, $\subseteq$	proper subset of, subset of
$A \cup B$	union of A and B
$A \cap B$	intersection of A and B
B^c	complement of B, if universe understood
$A - B$	complement of B in A, that is, $A \cap B^c$
$\sup S$	supremum (least upper bound) of S
$\inf S$	infimum (greatest lower bound) of S
$\mathbf{N}$	set of natural numbers (positive integers)
$\mathbf{Q}$	rational number field
$\mathbf{R}$	real number field
$\mathbf{R}^n$	real space of dimension n

$\mathbf{R}^{n\times n}$	space of $n \times n$ matrices with real entries
$\mathbf{R}^{m\times n}$	space of $m \times n$ matrices with real entries
$\mathrm{Inv}(\mathbf{R}^n, \mathbf{R}^n)$	set of invertible $n \times n$ real matrices
$\mathbf{C}$	complex number field
$\mathbf{C}^n$	complex space of dimension n
$C[a,b]$	space of real valued continuous functions on $[a,b]$
$C_n[a,b]$	space of $\mathbf{R}^n$ valued continuous functions on $[a,b]$
$CP[-\pi,\pi]$	space of continuous functions of period 2π
$\mathcal{R}[a,b]$	space of Riemann integrable functions on $[a,b]$
$L^2[-\pi,\pi]$	space of functions square integrable on $[-\pi,\pi]$
$f(t) \to L$	means $\lim_{t\to c} f(t) = L$, if c understood
$B_r(a)$	$\{x : \|x-a\| < r\}$
$B_r(\mathbf{x}_0)$	$\{\mathbf{x} : \\|\mathbf{x}-\mathbf{x}_0\\| < r\}$, space and norm understood
$\overline{B}_r(\mathbf{x}_0)$	$\{\mathbf{x} : \\|\mathbf{x}-\mathbf{x}_0\\| \le r\}$, space and norm understood
$\|\mathbf{x}\|_2$	Euclidean norm of $\mathbf{x} \in \mathbf{R}^n$
$\|\mathbf{x}\|$	norm of vector $\mathbf{x} \in \mathbf{R}^n$, often subscripted, e.g., $\|\mathbf{x}\|_2$
$\\|A\\|$	norm of matrix A, often subscripted, e.g., $\\|A\\|_\infty$
$\\|f\\|$	norm of real valued function
$\\|\mathbf{F}\\|$	norm of vector valued function
$\dot{x}$, $\dot{\mathbf{x}}$	first derivative of x, $\mathbf{x}$ with respect to t
$\ddot{x}$, $\ddot{\mathbf{x}}$	second derivative of x, $\mathbf{x}$ with respect to t
$x^{(j)}$, $\mathbf{x}^{(j)}$	order j derivative of x, $\mathbf{x}$ with respect to t
$\mathbf{x}^T$, A^T	transpose of column vector $\mathbf{x}$, or matrix A
$P > 0$	symmetric positive definite matrix P
$P \ge 0$	symmetric positive semidefinite matrix P
$\mathrm{Re}\, z$	real part of complex number z
$\mathrm{Im}\, z$	imaginary part of complex number z
$\overline{z}$	complex conjugate of z
$N(A)$	null space of matrix A
$R(A)$	range space (columnspace) of matrix A
$\mathrm{span}\, S$	all finite linear combinations of vectors from S
$V^\perp$	orthogonal complement of V
$D_j f(\mathbf{x})$	derivative of $f(x_1,\ldots,x_n)$ with respect to x_j at $\mathbf{x}$
$\partial f/\partial x_j(\mathbf{x})$	derivative of $f(x_1,\ldots,x_n)$ with respect to x_j at $\mathbf{x}$
$D\mathbf{F}(\mathbf{a})$	derivative of $\mathbf{F} : \mathbf{R}^n \to \mathbf{R}^m$ at $\mathbf{a}$
$\partial\mathbf{F}/\partial\mathbf{x}(\mathbf{a})$	derivative $D\mathbf{F}(\mathbf{a})$
$J_{\mathbf{F}}(\mathbf{a})$	$m \times n$ Jacobian matrix of $\mathbf{F} : \mathbf{R}^n \to \mathbf{R}^m$ at $\mathbf{a}$
$\det A$	determinant of matrix A
$J_{\mathbf{g}}(\mathbf{x})$	Jacobian matrix of $\mathbf{g}$
$\Delta_{\mathbf{g}}(\mathbf{x})$	$\det J_{\mathbf{g}}(\mathbf{x})$, also, $\det D\mathbf{g}(\mathbf{x})$
$dh(\mathbf{x})$	derivative of real valued h at $\mathbf{x}$
$f\big\|_V$, $\mathbf{F}\big\|_V$	restriction of f, $\mathbf{F}$ to the domain V
$m(A)$	Lebesgue measure of set $A \subseteq \mathbf{R}$
$m_n(A)$	*n-dimensional Lebesgue measure of set $A \subseteq \mathbf{R}^n$*
$\mu(A)$	*μ-measure of set A*
$\nu(A)$	volume of set A, if dimension is understood
$\nu_n(A)$	*n-dimensional volume of set $A \subseteq \mathbf{R}^n$*

B.2. The Greek Alphabet

Twenty-one of the lowercase letters are listed, along with eleven of the uppercase letters that are sometimes useful. Names of letters appear to the right. The letters *iota*, *kappa*, and *omicron* are missing from this list, as they resemble the Roman letters i, k, and o too closely to distinguish them.

α		alpha
β		beta
γ	Γ	gamma
δ	Δ	delta
ϵ		epsilon
ζ		zeta
η		eta
θ	Θ	theta
λ	Λ	lambda
μ		mu
ν		nu
ξ	Ξ	xi
π	Π	pi
ρ		rho
σ	Σ	sigma
τ		tau
υ	Υ	upsilon
ϕ	Φ	phi
χ		chi
ψ	Ψ	psi
ω	Ω	omega

Bibliography

[1] Malcolm Adams and Victor Guillemin, *Measure theory and probability*, Birkhäuser Boston, Inc., Boston, MA, 1996. Corrected reprint of the 1986 original. MR1365744

[2] Tom M. Apostol, *Mathematical analysis*, 2nd ed., Addison-Wesley Publishing Co., Reading, Mass.-London-Don Mills, Ont., 1974. MR0344384

[3] R. F. Bass, *Real Analysis for Graduate Students*, Second Edition, Richard F. Bass, All rights reserved, 2013. First edition published 2011.

[4] Garrett Birkhoff and Saunders Mac Lane, *A survey of modern algebra*, Third edition, The Macmillan Co., New York; Collier-Macmillan Ltd., London, 1965. MR0177992

[5] Ralph P. Boas, *A primer of real functions*, 4th ed., Carus Mathematical Monographs, vol. 13, Mathematical Association of America, Washington, DC, 1996. Revised and with a preface by Harold P. Boas. MR1411907

[6] William M. Boothby, *An introduction to differentiable manifolds and Riemannian geometry*, 2nd ed., Pure and Applied Mathematics, vol. 120, Academic Press, Inc., Orlando, FL, 1986. MR861409

[7] F. Brauer and J. A. Nohel, *The Qualitative Theory of Ordinary Differential Equations: An Introduction*, W. A. Benjamin, New York, NY, 1969. Reprinted by Dover Publications, Mineola, New York, 1989.

[8] David M. Bressoud, *A radical approach to Lebesgue's theory of integration*, MAA Textbooks, Cambridge University Press, Cambridge, 2008. MR2380238

[9] Peter Duren, *Invitation to classical analysis*, Pure and Applied Undergraduate Texts, vol. 17, American Mathematical Society, Providence, RI, 2012. MR2933135

[10] C. H. Edwards Jr., *Advanced calculus of several variables*, Academic Press (a subsidiary of Harcourt Brace Jovanovich, Publishers), New York-London, 1973. MR0352341

[11] Bernard Epstein, *Linear functional analysis. Introduction to Lebesque integration and infinite-dimensional problems*, W. B. Saunders Co., Philadelphia, Pa.-London-Toronto, Ont., 1970. MR0365061

[12] P. M. Fitzpatrick, *Advanced Calculus*, Second Edition, Pure and Applied Undergraduate Texts 5, American Mathematical Society, Providence, RI, 2006.

[13] Wendell Fleming, *Functions of several variables*, 2nd ed., Undergraduate Texts in Mathematics, Springer-Verlag, New York-Heidelberg, 1977. MR0422527

[14] Gerald B. Folland, *Fourier Analysis and Its Applications*, Pure and Applied Undergraduate Texts 4, American Mathematical Society, Providence, RI, 1992.

[15] Gerald B. Folland, *Real analysis: Modern techniques and their applications; A Wiley-Interscience Publication*, 2nd ed., Pure and Applied Mathematics (New York), John Wiley & Sons, Inc., New York, 1999. MR1681462

[16] Gerald B. Folland, *Advanced Calculus*, Prentice-Hall, Upper Saddle River, NJ, 2002.

[17] Avner Friedman, *Foundations of modern analysis*, Holt, Rinehart and Winston, Inc., New York-Montreal, Que.-London, 1970. MR0275100

[18] Avner Friedman, *Advanced calculus*, Holt, Rinehart and Winston, Inc., New York-Montreal, Que.-London, 1971. MR0352342

[19] R. F. Gariepy and W. P. Ziemer, *Modern Real Analysis*, PWS Publishing Company, Boston, Massachusetts, 1995.

[20] Enrique A. González-Velasco, *Fourier analysis and boundary value problems*, Academic Press, Inc., San Diego, CA, 1996. MR1419990

[21] R. B. Guenther and J. W. Lee, *Partial Differential Equations of Mathematical Physics and Integral Equations*, Dover Publications, Mineola, NY, 1996. An unabridged, corrected and enlarged republication of the work first published by Prentice-Hall, Englewood Cliffs, NJ, 1988.

[22] Victor Guillemin and Alan Pollack, *Differential topology*, Prentice-Hall, Inc., Englewood Cliffs, N.J., 1974. MR0348781

[23] Richard Haberman, *Elementary applied partial differential equations: With Fourier series and boundary value problems*, 2nd ed., Prentice Hall, Inc., Englewood Cliffs, NJ, 1987. MR913939

[24] Jack K. Hale, *Ordinary differential equations*, 2nd ed., Robert E. Krieger Publishing Co., Inc., Huntington, N.Y., 1980. MR587488

[25] Paul R. Halmos, *Finite-dimensional vector spaces*, The University Series in Undergraduate Mathematics, D. Van Nostrand Co., Inc., Princeton-Toronto-New York-London, 1958. 2nd ed. MR0089819

[26] Paul R. Halmos, *Naive set theory*, The University Series in Undergraduate Mathematics, D. Van Nostrand Co., Princeton, N.J.-Toronto-London-New York, 1960. MR0114756

[27] G. H. Hardy, Proc. London Math. Soc. **2** (1909), no. 9, 126-144.

[28] Morris W. Hirsch and Stephen Smale, *Differential equations, dynamical systems, and linear algebra*, Pure and Applied Mathematics, Vol. 60, Academic Press [A subsidiary of Harcourt Brace Jovanovich, Publishers], New York-London, 1974. MR0486784

[29] Morris W. Hirsch, Stephen Smale, and Robert L. Devaney, *Differential equations, dynamical systems, and an introduction to chaos*, 2nd ed., Pure and Applied Mathematics (Amsterdam), vol. 60, Elsevier/Academic Press, Amsterdam, 2004. MR2144536

[30] Kenneth Hoffman, *Analysis in Euclidean space*, Prentice-Hall, Inc., Englewood Cliffs, N.J., 1975. MR0453933

[31] Kenneth Hoffman and Ray Kunze, *Linear algebra*, Second edition, Prentice-Hall, Inc., Englewood Cliffs, N.J., 1971. MR0276251

[32] Richard Johnsonbaugh and W. E. Pfaffenberger, *Foundations of mathematical analysis*, Monographs and Textbooks in Pure and Applied Math., vol. 62, Marcel Dekker, Inc., New York, 1981. MR599741

[33] Frank Jones, *Lebesgue integration on Euclidean space*, Jones and Bartlett Publishers, Boston, MA, 1993. MR1204268

[34] Nicholas D. Kazarinoff, *Analytic inequalities*, Dover Publications, Inc., Mineola, NY, 2003. Revised reprint of the 1961 original [Holt, Rinehart and Winston, New York; MR0260957 (41 #5577)]. MR2007518

[35] A. N. Kolmogorov, *Foundations of the theory of probability*, Chelsea Publishing Co., New York, 1956. Translation edited by Nathan Morrison, with an added bibliography by A. T. Bharucha-Reid. MR0079843

[36] A. N. Kolmogorov and S. V. Fomin, *Introductory real analysis*, Revised English edition. Translated from the Russian and edited by Richard A. Silverman, Prentice-Hall, Inc., Englewood Cliffs, N.Y., 1970. MR0267052

[37] T. W. Körner, *Fourier analysis*, Cambridge University Press, Cambridge, 1988. MR924154

[38] T. W. Körner, *A companion to analysis: A second first and first second course in analysis*, Graduate Studies in Mathematics, vol. 62, American Mathematical Society, Providence, RI, 2004. MR2015825

[39] Steven G. Krantz, *The elements of advanced mathematics*, 2nd ed., Studies in Advanced Mathematics, Chapman & Hall/CRC, Boca Raton, FL, 2002. MR1886441

[40] Steven G. Krantz, *Real analysis and foundations*, 2nd ed., Studies in Advanced Mathematics, Chapman & Hall/CRC, Boca Raton, FL, 2005. MR2107580

[41] Erwin Kreyszig, *Introductory functional analysis with applications*, Wiley Classics Library, John Wiley & Sons, Inc., New York, 1989. MR992618

[42] Serge Lang, *Undergraduate analysis*, 2nd ed., Undergraduate Texts in Mathematics, Springer-Verlag, New York, 1997. MR1476913

[43] Henri Lebesgue, *Measure and the integral*, Edited with a biographical essay by Kenneth O. May, Holden-Day, Inc., San Francisco, Calif.-London-Amsterdam, 1966. MR0201592

[44] John M. Lee, *Introduction to smooth manifolds*, 2nd ed., Graduate Texts in Mathematics, vol. 218, Springer, New York, 2013. MR2954043

[45] J. E. Marsden and A. J. Tromba, *Vector Calculus*, Fourth Edition, W. H. Freeman and Company, New York, NY, 1996.

[46] E. J. McShane and T. A. Botts, *Real Analysis*, Dover Publications, New York, NY, 2005. An unabridged republication of the edition published by D. Van Nostrand Company, Inc., Princeton, New Jersey, 1959.

[47] J. Milnor, *Morse theory*, Based on lecture notes by M. Spivak and R. Wells. Annals of Mathematics Studies, No. 51, Princeton University Press, Princeton, N.J., 1963. MR0163331

[48] James R. Munkres, *Analysis on manifolds*, Addison-Wesley Publishing Company, Advanced Book Program, Redwood City, CA, 1991. MR1079066

[49] Ivan Niven, *Irrational numbers*, The Carus Mathematical Monographs, No. 11, The Mathematical Association of America. Distributed by John Wiley and Sons, Inc., New York, N.Y., 1956. MR0080123

[50] M. Renardy and R. C. Rogers, *An Introduction to Partial Differential Equations*, Second Edition, Texts in Applied Mathematics 13, Springer, New York, NY, 2004.

[51] H. L. Royden, *Real analysis*, 3rd ed., Macmillan Publishing Company, New York, 1988. MR1013117

[52] Walter Rudin, *Principles of mathematical analysis: International Series in Pure and Applied Mathematics*, 3rd ed., McGraw-Hill Book Co., New York-Auckland-Düsseldorf, 1976. MR0385023

[53] Walter Rudin, *Real and complex analysis*, 3rd ed., McGraw-Hill Book Co., New York, 1987. MR924157

[54] Hans Sagan, *Advanced Calculus*, Houghton Mifflin Co., Boston, MA, 1974.

[55] Hans Sagan, *Space-filling curves*, Universitext, Springer-Verlag, New York, 1994. MR1299533

[56] I. J. Schoenberg, *On the Peano curve of Lebesgue*, Bull. Amer. Math. Soc. **44** (1938), no. 8, 519, DOI 10.1090/S0002-9904-1938-06792-4. MR1563786

[57] Michael J. Schramm, *Introduction to Real Analysis*, Dover Publications, Inc., Mineola, New York, 2008. An unabridged republication of the work originally published in 1996 by Prentice-Hall, Inc., Englewood Cliffs, NJ.

[58] Robert T. Seeley, *An introduction to Fourier series and integrals*, Dover Publications, Inc., Mineola, NY, 2006. Reprint of the 1966 original. MR2302998

[59] George F. Simmons, *Introduction to topology and modern analysis*, McGraw-Hill Book Co., Inc., New York-San Francisco, Calif.-Toronto-London, 1963. MR0146625

[60] G. D. Smith, *Numerical solution of partial differential equations: Finite difference methods*, 3rd ed., Oxford Applied Mathematics and Computing Science Series, The Clarendon Press, Oxford University Press, New York, 1985. MR827497

[61] Elias M. Stein and Rami Shakarchi, *Fourier analysis: An introduction*, Princeton Lectures in Analysis, vol. 1, Princeton University Press, Princeton, NJ, 2003. MR1970295

[62] Gilbert Strang, *Essays in linear algebra*, Wellesley-Cambridge Press, Wellesley, MA, 2012. MR3058665

[63] Walter A. Strauss, *Partial differential equations: An introduction*, John Wiley & Sons, Inc., New York, 1992. MR1159712

[64] Robert S. Strichartz, *The Way of Analysis*, Jones and Bartlett Publishers, Inc., Boston, MA, 1995.

[65] Robert S. Strichartz, *A guide to distribution theory and Fourier transforms*, Studies in Advanced Mathematics, CRC Press, Boca Raton, FL, 1994. MR1276724

[66] William J. Terrell, *Stability and stabilization: An introduction*, Princeton University Press, Princeton, NJ, 2009. MR2482799

[67] Richard L. Wheeden and Antoni Zygmund, *Measure and integral: An introduction to real analysis*, 2nd ed., Pure and Applied Mathematics (Boca Raton), CRC Press, Boca Raton, FL, 2015. MR3381284

Index

PUBLISHED TITLES IN THIS SERIES

41 **William J. Terrell,** A Passage to Modern Analysis, 2019
38 **Mark Bridger,** Real Analysis, 2019
37 **Mike Mesterton-Gibbons,** An Introduction to Game-Theoretic Modelling, Third Edition, 2019
36 **Cesar E. Silva,** Invitation to Real Analysis, 2019
35 **Álvaro Lozano-Robledo,** Number Theory and Geometry, 2019
34 **C. Herbert Clemens,** Two-Dimensional Geometries, 2019
33 **Brad G. Osgood,** Lectures on the Fourier Transform and Its Applications, 2019
32 **John M. Erdman,** A Problems Based Course in Advanced Calculus, 2018
31 **Benjamin Hutz,** An Experimental Introduction to Number Theory, 2018
30 **Steven J. Miller,** Mathematics of Optimization: How to do Things Faster, 2017
29 **Tom L. Lindstrøm,** Spaces, 2017
28 **Randall Pruim,** Foundations and Applications of Statistics: An Introduction Using R, Second Edition, 2018
27 **Shahriar Shahriari,** Algebra in Action, 2017
26 **Tamara J. Lakins,** The Tools of Mathematical Reasoning, 2016
25 **Hossein Hosseini Giv,** Mathematical Analysis and Its Inherent Nature, 2016
24 **Helene Shapiro,** Linear Algebra and Matrices, 2015
23 **Sergei Ovchinnikov,** Number Systems, 2015
22 **Hugh L. Montgomery,** Early Fourier Analysis, 2014
21 **John M. Lee,** Axiomatic Geometry, 2013
20 **Paul J. Sally, Jr.,** Fundamentals of Mathematical Analysis, 2013
19 **R. Clark Robinson,** An Introduction to Dynamical Systems: Continuous and Discrete, Second Edition, 2012
18 **Joseph L. Taylor,** Foundations of Analysis, 2012
17 **Peter Duren,** Invitation to Classical Analysis, 2012
16 **Joseph L. Taylor,** Complex Variables, 2011
15 **Mark A. Pinsky,** Partial Differential Equations and Boundary-Value Problems with Applications, Third Edition, 1998
14 **Michael E. Taylor,** Introduction to Differential Equations, 2011
13 **Randall Pruim,** Foundations and Applications of Statistics, 2011
12 **John P. D'Angelo,** An Introduction to Complex Analysis and Geometry, 2010
11 **Mark R. Sepanski,** Algebra, 2010
10 **Sue E. Goodman,** Beginning Topology, 2005
9 **Ronald Solomon,** Abstract Algebra, 2003
8 **I. Martin Isaacs,** Geometry for College Students, 2001
7 **Victor Goodman and Joseph Stampfli,** The Mathematics of Finance, 2001
6 **Michael A. Bean,** Probability: The Science of Uncertainty, 2001
5 **Patrick M. Fitzpatrick,** Advanced Calculus, Second Edition, 2006
4 **Gerald B. Folland,** Fourier Analysis and Its Applications, 1992
3 **Bettina Richmond and Thomas Richmond,** A Discrete Transition to Advanced Mathematics, 2004
2 **David Kincaid and Ward Cheney,** Numerical Analysis: Mathematics of Scientific Computing, Third Edition, 2002
1 **Edward D. Gaughan,** Introduction to Analysis, Fifth Edition, 1998